C0-AZX-353

Enzyme-Mediated Resistance to Antibiotics

Mechanisms, Dissemination, and Prospects for Inhibition

Enzyme-Mediated Resistance to Antibiotics

Mechanisms, Dissemination, and Prospects for Inhibition

Editors

Robert A. Bonomo

Section of Infectious Diseases
Louis Stokes Cleveland Department of Veterans Affairs
Medical Center
Cleveland, OH 44106

Marcelo Tolmasky

Department of Biological Science
California State University-Fullerton
Fullerton, CA 92831

Washington, D.C.

Cover image courtesy Lakshmi P. Kotra (see chapter 2)

American Society for Microbiology
1752 N Street, N.W.
Washington, DC 20036-2804

Library of Congress Cataloging-in-Publication Data

Enzyme-mediated resistance to antibiotics : mechanisms, dissemination, and prospects for inhibition / editor, Robert A. Bonomo, Marcelo E. Tolmasky.
p. ; cm.
Includes index.
ISBN-13: 978-1-55581-303-1
ISBN-10: 1-55581-303-8
1. Drug resistance in microorganisms. 2. Drug resistance in microorganisms—Genetic aspects. 3. Microbial enzymes. I. Bonomo, Robert A. II. Tolmasky, Marcelo E.
[DNLM: 1. Anti-Bacterial Agents. 2. Drug Resistance, Bacterial. 3. Enzyme Inhibitors. 4. beta-Lactamases—antagonists & inhibitors. QV 350 E615 2007]

QR177.E59 2007
616.9′041—dc22

2007002039

10 9 8 7 6 5 4 3 2 1

Address editorial correspondence to: ASM Press, 1752 N St., N.W., Washington, DC 20036-2904, U.S.A.

Send orders to: ASM Press, P.O. Box 605, Herndon, VA 20172, U.S.A.
Phone: 800-546-2416; 703-661-1593
Fax: 703-661-1501
Email: Books@asmusa.org
Online: estore.asm.org

To Rita, to whom I owe everything
—R.A.B.

To Liliana and Ryan
—M.E.T.

Contents

Contributors *ix*
Preface *xv*

A *Enzymes in Defense of the Bacterial Ribosome*

1 Overview of Aminoglycosides and Enzyme-Mediated Bacterial Resistance: Clinical Implications 3
ROBERT A. BONOMO

2 Aminoglycoside Antibiotics 7
KANCHANA MAJUMDER, LIANHU WEI, SUBHASH C. ANNEDI, AND LAKSHMI P. KOTRA

3 Structural Aspects of Aminoglycoside-Modifying Enzymes 21
GERARD D. WRIGHT AND ALBERT M. BERGHUIS

4 Aminoglycoside-Modifying Enzymes: Characteristics, Localization, and Dissemination 35
MARCELO E. TOLMASKY

5 rRNA Methylases and Resistance to Macrolide, Lincosamide, Streptogramin, Ketolide, and Oxazolidinone (MLSKO) Antibiotics 53
MARILYN C. ROBERTS

B *Enzymes in Defense of the Bacterial Cell Wall*

6 β-Lactamases: Historical Perspectives 67
KAREN BUSH AND PATRICIA A. BRADFORD

7 Resistance Mediated by Penicillin-Binding Proteins 81
MALCOLM G. P. PAGE

8 Inhibition of Class A β-Lactamases 101
SAMY O. MEROUEH, JOOYOUNG CHA, AND SHAHRIAR MOBASHERY

9 Class B β-Lactamases 115
GIAN MARIA ROSSOLINI AND JEAN-DENIS DOCQUIER

10 Crystal Structures of Class C β-Lactamases: Mechanistic Implications and Perspectives in Drug Design 145
C. BAUVOIS AND J. WOUTERS

11 Class D β-Lactamases 163
FRANCK DANEL, MALCOLM G. P. PAGE, AND DAVID M. LIVERMORE

12 Kinetics of β-Lactamases and Penicillin-Binding Proteins 195
MORENO GALLENI AND JEAN-MARIE FRÈRE

C Novel Approaches and Future Prospects

13 The Pharmaceutical Industry and Inhibitors of Bacterial Enzymes: Implications for Drug Development 217
DAVID M. SHLAES, LEFA ALKSNE, AND STEVEN J. PROJAN

14 β-Lactamase Inhibitory Proteins 227
ZHEN ZHANG AND TIMOTHY PALZKILL

15 Active Drug Efflux in Bacteria 235
JÜRG DREIER

D Dissemination of Antibiotic Resistance and Its Biological Cost

16 Overview of Dissemination Mechanisms of Genes Coding for Resistance to Antibiotics 267
MARCELO E. TOLMASKY

17 Conjugative Transposons 271
LOUIS B. RICE

18 The Dissemination of Antibiotic Resistance by Bacterial Conjugation 285
VIRGINIA L. WATERS

19 Bacterial Toxin-Antitoxin Systems as Targets for the Development of Novel Antibiotics 313
JUAN C. ALONSO, DOLORS BALSA, IZHACK CHERNY, SUSANNE K. CHRISTENSEN, MANUEL ESPINOSA, DJORDJE FRANCUSKI, EHUD GAZIT, KENN GERDES, ED HITCHIN, M. TERESA MARTÍN, CONCEPCIÓN NIETO, KARIN OVERWEG, TERESA PELLICER, WOLFRAM SAENGER, HEINZ WELFLE, KARIN WELFLE, AND JERRY WELLS

20 Integrons and Superintegrons 331
ROBERT A. BONOMO, ANDREA M. HUJER, AND KRISTINE M. HUJER

21 The Biological Cost of Antibiotic Resistance 339
DAN I. ANDERSSON, SOPHIE MAISNIER PATIN, ANNIKA I. NILSSON, AND ELISABETH KUGELBERG

Index 349

Contributors

LEFA ALKSNE
Wyeth Research, Pearl River, NY 10965

JUAN C. ALONSO
Dept. of Microbial Biotechnology, Centro Nacional de Biotecnología, CSIC,
Darwin 3, 28049 Madrid, Spain

DAN I. ANDERSSON
Dept. of Medical Biochemistry and Microbiology, Uppsala University, S-751 23
Uppsala, Sweden

SUBHASH C. ANNEDI
Dept. of Pharmaceutical Sciences, University of Toronto, Toronto, Ontario
M5G 1L7, Canada

DOLORS BALSA
Department of Pharmacological Biochemistry, Laboratorios SALVAT S.A.,
Barcelona, Spain

C. BAUVOIS
Institut de Recherches Microbiologiques Wiame, Campus Ceria, 1 Ave. E. Gryzon,
B-1070 Brussels, Belgium

ALBERT M. BERGHUIS
Dept. of Biochemistry and Dept. of Microbiology & Immunology, McGill University,
Montreal, Quebec H3A 2B4 Canada

ROBERT A. BONOMO
Section of Infectious Diseases, Louis Stokes Cleveland Department of Veterans
Affairs Medical Center, and School of Medicine, Case Western Reserve University,
Cleveland, OH 44106

PATRICIA A. BRADFORD
Wyeth Research, 401 Middletown Rd., Pearl River, NY 10965-1251

KAREN BUSH
Johnson & Johnson Pharmaceutical Research & Development, 1000 Route 202, Box 300, Raritan, NJ 08869-0602

JOOYOUNG CHA
Dept. of Chemistry and Biochemistry, University of Notre Dame, Notre Dame, IN 46556

ITZHACK CHERNY
Dept. of Molecular Microbiology and Biotechnology, George S. Wise Faculty of Life Sciences, Tel-Aviv University, Tel-Aviv 69978, Israel

SUSANNE K. CHRISTENSEN
Dept. of Biochemistry and Molecular Biology, South Denmark University, Odense M, Denmark

FRANCK DANEL
Basilea Pharmaceutica AG, Grenzacherstrasse 487, CH-4005 Basel, Switzerland

JEAN-DENIS DOCQUIER
Centre d'Ingénierie des Protéines & Laboratoire d'Enzymologie, Université de Liège, Liège, B-4000, Belgium

JÜRG DREIER
Basilea Pharmaceutica AG, Grenzacherstrasse 487, CH-4005 Basel, Switzerland

MANUEL ESPINOSA
Dept. of Protein Science, Centro de Investigaciones Biológicas, CSIC, Ramiro de Maeztu, 9, 28040-Madrid, Spain

DJORDJE FRANCUSKI
Institute for Chemistry and Biochemistry/Crystallography, Freie Universität Berlin, Takustr. 6, D-14195 Berlin, Germany

JEAN-MARIE FRÈRE
Center for Protein Engineering, University of Liège, Institut de Chimie B6, Sart Tilman, B-4000 Liège, Belgium

MORENO GALLENI
Center for Protein Engineering, University of Liège, Institut de Chimie B6, Sart Tilman, B-4000 Liège, Belgium

EHUD GAZIT
Dept. of Molecular Microbiology and Biotechnology, George S. Wise Faculty of Life Sciences, Tel-Aviv University, Tel-Aviv 69978, Israel

KENN GERDES
Dept. of Biochemistry and Molecular Biology, South Denmark University, Odense M, Denmark

ED HITCHIN
Institute of Food Research, Norwich Research Park, Norwich, United Kingdom

ANDREA M. HUJER
Research Service, Louis Stokes Cleveland Department of Veterans Affairs Medical Center, Cleveland, OH 44106

KRISTINE M. HUJER
Research Service, Louis Stokes Cleveland Department of Veterans Affairs Medical Center, Cleveland, OH 44106

LAKSHMI P. KOTRA
Center for Molecular Design and Preformulations (CMDP), Toronto General Research Institute, University Health Network, and University of Toronto, MaRS Center, TMDT 5-356, 101 College St., Toronto, Ontario M5G 1L7 Canada

ELISABETH KUGELBERG
Dept. of Bacteriology, Swedish Institute for Infectious Disease Control, S-171 82 Stockholm, and Microbiology, Tumour and Cell Biology Center, Karolinska Institute, S-171 77 Stockholm, Sweden

DAVID M. LIVERMORE
Antibiotic Resistance Monitoring & Reference Laboratory, Centre for Infections, Health Protection Agency, 61 Colindale Ave., London NW9 5EQ, United Kingdom

KANCHANA MAJUMDER
Dept. of Pharmaceutical Sciences, University of Toronto, Toronto, Ontario M5G 1L7, Canada

M. TERESA MARTÍN
Dept. of Microbial Biotechnology, Centro Nacional de Biotecnología, CSIC, Darwin 3, 28049 Madrid, Spain

SAMY O. MEROUEH
Dept. of Chemistry and Biochemistry, University of Notre Dame, Notre Dame, IN 46556

SHAHRIAR MOBASHERY
Dept. of Chemistry and Biochemistry, University of Notre Dame, Notre Dame, IN 46556

CONCEPCIÓN NIETO
Dept. of Protein Science, Centro de Investigaciones Biológicas, CSIC, Ramiro de Maeztu, 9, 28040-Madrid, Spain

ANNIKA I. NILSSON
Dept. of Medical Biochemistry and Microbiology, Uppsala University, S-751 23, Uppsala, Sweden

KARIN OVERWEG
Institute of Food Research, Norwich Research Park, Norwich, United Kingdom

MALCOLM G. P. PAGE
Basilea Pharmaceutica AG, Grenzacherstrasse 487, CH-4005 Basel, Switzerland

TIMOTHY PALZKILL
Dept. of Molecular Virology and Microbiology, Baylor College of Medicine, One Baylor Plaza, Houston, TX 77030

SOPHIE MAISNIER PATIN
Dept. of Bacteriology, Swedish Institute for Infectious Disease Control, S-171 82 Stockholm, and Microbiology, Tumour and Cell Biology Center, Karolinska Institute, S-171 77 Stockholm, Sweden

TERESA PELLICER
Dept. of Pharmacological Biochemistry, Laboratorios SALVAT S.A., Barcelona, Spain

STEVEN J. PROJAN
Biological Technologies, Wyeth Research, Cambridge, MA 02140

LOUIS B. RICE
Louis Stokes Cleveland Dept. of Veterans Affairs Medical Center and Case Western Reserve University, Cleveland, OH 44106

MARILYN C. ROBERTS
Dept. of Pathobiology and Dept. of Environmental & Occupational Health Sciences, Box 357238, School of Public Health and Community Medicine, University of Washington, Seattle, WA 98195

GIAN MARIA ROSSOLINI
Dipartimento di Biologia Molecolare, Università di Siena, Siena, I-53100, Italy

WOLFRAM SAENGER
Institute for Chemistry and Biochemistry/Crystallography, Freie Universität Berlin, Takustr. 6, D-14195 Berlin, Germany

DAVID M. SHLAES
Anti-infectives Consulting, 219 Montauk Ave., Stonington, CT 06378

MARCELO E. TOLMASKY
Dept. of Biological Science, College of Natural Sciences and Mathematics, California State University Fullerton, 800 N State College Blvd., Fullerton, CA 92831-3599

VIRGINIA L. WATERS
Dept. of Medicine, School of Medicine, University of California San Diego, La Jolla, CA 92093-0640

LIANHU WEI
Division of Cell and Molecular Biology, Toronto General Research Institute, University Health Network and Dept. of Pharmaceutical Sciences, University of Toronto, Toronto, Ontario M5G 1L7, Canada

HEINZ WELFLE
Max Delbrück Center for Molecular Medicine Berlin-Buch, Robert-Roessle-Str. 10, D-13125 Berlin, Germany

KARIN WELFLE
Max Delbrück Center for Molecular Medicine Berlin-Buch, Robert-Roessle-Str. 10, D-13125 Berlin, Germany

JERRY WELLS
Swammerdam Institute for Life Sciences, University of Amsterdam, 1018 WV, Amsterdam, The Netherlands

J. WOUTERS
Dept. of Chemistry, University of Namur, 61 Rue de Bruxelles, B-5000 Namur, Belgium

GERARD D. WRIGHT
Antimicrobial Research Centre, Dept. of Biochemistry, McMaster University, 1200 Main St. W, Hamilton, Ontario L8N 3Z5 Canada

ZHEN ZHANG
Structural & Computational Biology and Molecular Biophysics, Baylor College of Medicine, One Baylor Plaza, Houston, TX 77030

Preface

AT THE PRECIPICE

A quick look at contemporary newspapers and journals will reveal startling reports about antibiotic-resistant bacteria, "superbugs." A large number of common infectious diseases caused by bacteria were once easily treatable with antibiotics. Now many pathogens have become increasingly deadly due to antibiotic resistance. At the present time, vancomycin-resistant *Staphylococcus aureus* (VRSA), community-acquired methicillin-resistant *S. aureus* (CA-MRSA), hospital-acquired MRSA, vancomycin-resistant enterococci (VRE), penicillin-resistant *Streptococcus pneumoniae*, and multidrug-resistant (MDR) *Mycobacterium tuberculosis*, *Pseudomonas aeruginosa*, *Klebsiella pneumoniae*, and *Acinetobacter baumannii* represent a highly significant threat to children, hospitalized patients, immune-compromised individuals, and nursing home elderly. They are, in fact, a threat to us all. A recent commentary in *Nature* describes MDR *A. baumannii* as "a real danger" (1). Alarming drug resistance phenotypes in gram-negative bacteria like *P. aeruginosa* and *A. baumannii* include resistance to penicillins (piperacillin and ampicillin), extended-spectrum cephalosporins (ceftazidime and cefepime), β-lactam β-lactamase inhibitors (ampicillin/sulbactam, amoxicillin/clavulanate, and piperacillin/tazobactam), and carbapenems (meropenem and imipenem). Even resistance to colisitin, a polymyxin-class antibiotic, has emerged. It is highly disturbing to the clinician to be faced with pathogens that have become resistant to all antibiotics. It is like being at the precipice…

The history of our most trusted antibiotic, penicillin, began in 1929 when Alexander Fleming published his seminal paper in the *British Journal of Experimental Pathology* on the "mold extract" from *Penicillium* as a germ-killing compound (8). This serendipitous discovery was further developed by Howard Florey, Ernst Chain, and Norman Heatley at Oxford University. Realizing the potential of Fleming's discovery, this team developed methods for growing, extracting, and purifying enough penicillin to demonstrate its power against streptococcal and staphylococcal infections. The success of the utilization of penicillin was so spectacular that it received the appellation of "miracle drug." This early work was published in two landmark papers in *Lancet* in 1940 and 1941 (3, 5). Among the first fortunate patients to receive this "miracle" drug were the victims of the devastating Cocoanut Grove fire in Boston in November 1942 (6, 9). The high mortality of infections due to wounds sustained in battle (gangrene) and the burgeoning problem of gonorrhea and syphilis in World War II veterans also enhanced interest in the curative powers of penicillin.

Before this period of amazing discovery, E. P. Abraham and Ernst Chain reported in *Nature* (1940) the presence of an enzyme in *Bacillus* (*Escherichia*) *coli* able to inactivate penicillin (2). This significance of this report was not immediately realized. After the beginning of penicillin's use to combat infections, *S. aureus* was among the first bacteria known to become resistant to penicillin. The clinical impact of this development was staggering. This pathogen not only caused serious illness (such as pneumonia, endocarditis, osteomyelitis, and toxic shock syndrome) but was also once more untreatable. The development of semisynthetic penicillins and the discovery of other natural products stemmed this threat, but our stay was only temporary.

In 1943 Selman Waksman and his group isolated streptomycin from the soil bacterium *Streptomyces griseus* (10). Streptomycin was first tested to be effective against *M. tuberculosis*, the scourge of ancient civilizations. The use of streptomycin and other antitubercular compounds led to a significant reduction of mortality due to tuberculosis in the United States, from 39.9 deaths per 100,000 population in 1945 to 9.1 per 100,000 in 1955 (4). Waksman and his group also created the concept of systematic screening of microbial culture products, developing a technology that has provided the foundation of the early antibiotic industry. Soon, new antibiotics were discovered that provided physicians with a large number of "weapons" to combat bacterial diseases.

In spite of more than half a century of tremendous commercial and scientific investment, bacterial infectious diseases were not completely eradicated by the use of antibiotics (7). Paradoxically, several diseases re-emerged, and many of the bacterial pathogens are becoming more and more resistant to treatment with antibiotics.

The central problem is that the use of antibiotics has contributed to the inexorable rise of antibiotic-resistant bacteria. A large number of factors, including human and nonhuman use of antibiotics, have contributed to the emergence, acquisition, and spread of resistance. These factors include the use of antibiotics in food-producing animals, which leads to the development of resistance in bacteria that find their way into the human food chain; the misuse and overuse of antibiotics in humans; the demand for antibiotics by patients when they are not appropriate; noncompliance by patients who often fail to finish the antibiotic prescription; and the over-the-counter availability of antibiotics in a large number of countries. The World Health Organization has estimated that bacteria resistant to antibiotics now account for about 60% of nosocomial infections. The Centers for Disease Control and Prevention estimate that of about 60,000 deaths that occur in the United States every year due to nosocomial infections, 14,000 are the result of antibiotic-resistant bacteria. The number of deaths related to antibiotic-resistant community-acquired infections is also growing.

Years of research demonstrated that bacteria have evolved a wealth of different ways to resist the action of antibiotics as well as to transfer these capabilities. Antibiotic resistance mechanisms include (i) changes in permeability that interfere with the penetration of the antibiotic into the cell; (ii) the presence of efflux mechanisms that expel the antibiotic; (iii) modification or substitution of the target of antibiotic action; and (iv) chemical modification of the antibiotic molecule. In addition, besides vertical transmission, once the resistance trait is acquired there are several mechanisms of horizontal transfer that accelerate the dissemination of resistance.

An important component of the antibiotic resistance problem is represented by mechanisms mediated by enzymatic processes. Chemical modification of aminoglycosides and β-lactams is of great relevance in the clinical setting and is mediated by a large number of enzymes. These enzymes tend to be coded for by genes that are present in mobile elements (transposons, plasmids, etc.) that favor their quick dissemination. In this book we highlight the enzymatic capabilities of microorganisms to introduce chemical modifications that negate the biological activity of β-lactams (β-lactamases) or aminoglycoside antibiotics (aminoglycoside-modifying enzymes). These chemical modifications include destroying the β-lactam ring of β-lactamic antibiotics and introducing acetyl, nucleotidyl, or phosphate groups at different locations of the aminoglycoside molecules. We hope that the chapters describing different aspects and kinds of β-lactamases and aminoglycoside-modifying enzymes will provide the reader with a complete picture of the present state of knowledge about these important mechanisms of resistance. As an example of the existence of other enzymatic mechanisms that result in resistance, we have included a chapter on RNA methylases and resistance to erythromycin. We will illustrate the variety of scientific approaches important to their characterization and, we hope, inspire researchers to take up the still many unknowns that need to be clarified.

To complement this compilation, we will also illustrate the different ways bacteria share resistance determinants. Horizontal transfer is a big part of the problem of antibiotic resistance, and in our view, different mechanisms for dissemination of antibiotic resistance genes need to be included. In some cases, such as "transposable elements" or "plasmids," the fields have become extremely big and a chapter would not do justice to all the material that needed to be included. Recent books have been entirely devoted to these elements. Therefore, this discussion has not been included here. However, some aspects of these elements can be found through chapters in the section "Dissemination of Antibiotic Resistance and Its Biological Cost" as well as in chapters in other sections of the book. The collection included here will permit the reader to acquire information about the history and recent developments in other areas such as dissemination at the cellular level and integrons. Since the acquisition of resistance does not come free to the bacterial cell, a chapter dealing with the biological cost of resistance has been included.

Finally, one might ask, "Why catalogue these enzymatic resistance mechanisms?" Clearly, rational approaches are needed to control the dissemination of resistance genes and to combat the highly versatile inactivating enzymes. A wide variety of methods are under development, as discussed in the section "Novel Approaches and Future Prospects." It is our hope that the material included herein will inspire students of enzymology and antibiotic resistance to save us from falling over the edge.

Robert A. Bonomo
Marcelo Tolmasky

REFERENCES

1. **Abbott, A.** 2005. Medics braced for fresh superbug. *Nature* **436:**758.
2. **Abraham, E. P., and E. Chain.** 1940. An enzyme from bacteria able to destroy penicillin. *Nature* **146:**837.
3. **Abraham, E. P., E. Chain, C. M. Fletcher, A. D. Gardner, N. G. Heatley, M. A. Jennings, and H. W. Florey.** 1941. Further observations on penicillin. *Lancet* **2:**177-188.
4. **Anonymous.** 1999. Achievements in public health, 1900-1999: control of infectious diseases. *Morb. Mortal. Wkly. Rep.* **48:**621-629.

5. **Chain, E., H. Florey, A. D. Gardner, N. G. Heatley, M. Jennings, J. Orr-Ewing, and A. G. Sanders.** 1940. Penicillin as a chemotherapeutic agent. *Lancet* **ii:**226-228.
6. **Cope, O.** 1943. Care of the victims of the Cocoanut Grove fire at the Massachusetts General Hospital. *N. Engl. J. Med.* **229:**138-147.
7. **Davies, J.** 1999. In praise of antibiotics. *ASM News* **65:**304-310.
8. **Fleming, A.** 1929. On the antibacterial action of cultures of a penicillium, with special reference to their use in the isolation of *B. influenzae*. *Br. J. Exp. Pathol.* **10:**226-236.
9. **Levy, S.** 2002. *The Antibiotic Paradox: How the Misuse of Antibiotics Destroys Their Curative Powers*. Perseus Publishing, Cambridge, MA.
10. **Schatz, A., E. Bugie, and S. A. Waksman.** 1944. Streptomycin, a substance exhibiting antibiotic activity against gram-positive and gram-negative bacteria. *Proc. Soc. Exp. Biol. Med.* **55:**66-69.

Enzymes in Defense of the Bacterial Ribosome

Enzyme-Mediated Resistance to Antibiotics: Mechanisms, Dissemination, and Prospects for Inhibition
Edited by Robert A. Bonomo and Marcelo E. Tolmasky

Robert A. Bonomo

1

Overview of Aminoglycosides and Enzyme-Mediated Bacterial Resistance, Clinical Implications

Since their discovery more than 60 years ago, aminoglycosides and macrolides have been extremely valuable agents in the therapy of infectious diseases. The first aminoglycoside, streptomycin, was isolated from *Streptomyces griseus* in 1944 by Schatz, Bugie, and Waksman (1). This soil actinomycete (a bacterium) and *Micromonospora* have served as reservoirs of the isolation and purification of aminoglycosides as natural products and semisynthetic derivatives.

Aminoglycosides contain an aminocyclitol ring (streptidine or 2-deoxystreptamine) and two or more amino sugars linked by glycosidic bonds. Aminoglycosides in wide use today are amikacin, dibekacin, isepamacin, sisomicin, gentamicin, kanamycin, neomycin, netilmicin, paromomycin, streptomycin, and tobramycin (7) (Table 1.1). Alone and in combination with cell wall active agents such as β-lactams and glycopeptides, aminoglycosides have served a time-honored role in the treatment of serious infections caused by gram-negative bacilli and select gram-positive organisms. In addition, aminoglycosides have been used as part of therapy against mycobacterial infections. As antibiotics, these agents are rapidly bactericidal, demonstrate a "postantibiotic" effect (suppression of bacterial growth after short exposure), and exhibit predictable pharmacodynamics (4). In balance, these are essential antibiotics and play a central role in every pharmacy.

It is important for the mechanistic enzymologist interested in antibiotic resistance to recall the spectrum of microbiological activity and mechanism of action of aminoglycosides. Clinicians often use aminoglycosides to treat infections caused by most strains of *Escherichia coli*, *Klebsiella* spp., *Serratia*, *Enterobacter* spp., *Citrobacter* spp., *Proteus*, *Acinetobacter*, and *Pseudomonas aeruginosa*. As a class, these antibiotics find the most use against susceptible *P. aeruginosa* and other nonfermenting gram-negative bacilli (*Acinetobacter baumannii* and other *Acinetobacter* spp.). Although randomized control trials are still needed, aminoglycosides (as monotherapy or as part of combination chemotherapy) also play a critical role in the treatment of pulmonary infections in patients with cystic fibrosis.

As stated above, streptomycin has demonstrable activity against *Mycobacterium tuberculosis* and amikacin has activity against "atypical" mycobacteria (e.g., *Mycobacterium avium-M. intracellulare* complex [MAC] and *Mycobacterium kansasii*). Amikacin is the only aminoglycoside approved for use in the treatment of disease caused by MAC. Streptomycin is still listed as the drug of choice against *Yersinia pestis* (the agent of plague) and *Francisella*

Robert A. Bonomo, Section of Infectious Diseases, Louis Stokes Cleveland Veterans Affairs Medical Center, School of Medicine, Case Western Reserve University, Cleveland, OH 44106.

Table 1.1 Aminoglycosides in clinical use and their source

Source	Aminoglycoside
Streptomyces griseus	Streptomycin
Streptomyces fradiae	Neomycin, paromomycin
Streptomyces kanamyceticus	Kanamycin A, kanamycin B, amikacin, dibekacin
Micromonospora purpurea and *M. echinospora*	Gentamicin C_1, C_{1a}, C_2
Streptomyces tenebrarius	Tobramycin
Micromonospora inyonensis	Netromycin, sisomycin
Streptomyces spectabilis	Spectinomycin
Micromonospora purpurea	Isepamicin

tularensis (the agent of tularemia pneumonia). It is notable that the spectrum of activity of aminoglycosides does not extend to anaerobes (*Bacteroides* spp. and *Clostridium* spp.), *Streptococcus pneumoniae*, *Stenotrophomonas maltophilia*, *Burkholderia cepacia*, or *Mycoplasma* spp. Although *Haemophilus influenzae* and *Legionella* spp. test susceptible to aminoglycosides in vitro, these drugs are not used to treat infections caused by these pathogens. Despite the finding that aminoglycosides penetrate human cells in low concentrations, they are highly effective against tuberculosis, tularemia, and plague (caused by intracellular pathogens).

In the chapters that follow, the reader is first introduced to the varied mechanisms by which aminoglycoside-modifying enzymes (AMEs) confer resistance to aminoglycosides (Majumder et al.) (see chapter 2 in this volume). These mechanistic considerations are placed in a structural context by Wright and Berghuis (see chapter 3 in this volume) and in a genetic framework by Tolmasky (see chapter 4 in this volume). As a group, these chapters illustrate how a thorough understanding of molecular mechanisms, three-dimensional structure, and genetic variability can be leveraged in combating resistance.

The overarching message delivered in these chapters is that aminoglycosides kill bacteria by targeting the 16S rRNA of the 30S subunit of the ribosome. After penetrating the bacterial cell, these agents interfere with translation of mRNA. As a result, protein production is altered, aberrant proteins are synthesized and inserted in the cell membrane, cell permeability is increased, more aminoglycosides are taken up into the cell, and cell death ensues. The reader encounters in the chapter that follows that clinical resistance to aminoglycosides emerges by one of four mechanisms: (i) loss of cell permeability (decreased uptake); (ii) alterations in the target site (ribosome) that prevent binding; (iii) expulsion by efflux pumps; and (iv) enzymatic inactivation by AMEs. Overcoming intrinsic and acquired mechanisms is a significant challenge since the number and diversity of AMEs are great. Inhibition of AMEs by drugs such as dibekacin (a semisynthetic aminoglycoside), 5-*epi*-sisomicin, and 5-*epi*-gentamicin is discussed in the chapter that ensues. The major thrust of current efforts at AME inhibition has targeted the APH enzymes. Derivatives of neamine and kanamycin B have been tested with some success. The development of inhibitors of protein kinase has also been shown to inhibit APH (3′). Dimeric aminoglycosides and construction of aminoglycosides that undergo a cyclic process of phosphorylation and release of inorganic phosphate are exciting developments.

In understanding these mechanisms, the reader should keep in mind the microbiological difference between in vitro synergism and clinical trials showing the benefits of synergistic therapy. The activity of the β-lactam (or another cell wall active agent such as a glycopeptide) on the bacterial cell wall is thought to promote aminoglycoside uptake and result in more-rapid bacterial killing. This effect is best measured by time-kill curves.

To recall the importance of aminoglycosides as part of combination therapy, in Table 1.2 we list examples in which an aminoglycoside added to a cell wall-active agent results in synergy. Some of these observations have been based on clinical practice and animal models of infection (i.e., endocarditis due to enterococci, viridans group streptococci, and staphylococci) (2–4, 6). The clinical impact of β-lactam-aminoglycoside synergism in treating gram-positive infections other than those caused by enterococci has been more difficult to establish.

Clinicians frequently use aminoglycosides alone and in combination therapy for treating severe infections caused by gram-negative bacilli, most especially *P. aeruginosa*. It

Table 1.2 Examples of in vitro synergy of aminoglycosides and cell wall active agents

Organism	Combination
Pseudomonas aeruginosa	Gentamicin, tobramycin, amikacin + antipseudomonal β-lactams (piperacillin, aztreonam, ceftazidime, imipenem)
Staphylococcus aureus (methicillin susceptible)	Gentamicin, tobramycin + nafcillin, oxacillin, cephalothin, or vancomycin
Staphylococcus aureus (methicillin resistant)	Gentamicin, tobramycin + vancomycin, teicoplanin (± rifampin)
Streptococcus pyogenes	Gentamicin + penicillin
Enterococci	Streptomycin, gentamicin, tobramycin, amikacin + penicillin, ampicillin, vancomycin

is in these settings that the fear of drug resistance becomes paramount. Combination therapy ensures a greater chance of selection of appropriate treatment. In balance, recent studies have failed to document any advantage to using combinations to treat *P. aeruginosa* bacteremia (3, 6). Nevertheless, it is our opinion that clinicians should consider combination therapy in settings where the infected patient is severely immunocompromised and the chance of finding a resistant organism possessing an AME such as *P. aeruginosa* or *A. baumannii* is great.

Unfortunately, aminoglycosides are associated with nephrotoxicity, ototoxicity, and neuromuscular blockade. Their use in a once-a-day dosing regimen is a major advance. Once-a-day dosing is as effective as multiple daily doses and lowers the risk of drug-induced side effects. Once-a-day dosing cannot be advocated for patients with enterococcal endocarditis.

Macrolides (such as erythromycin, clarithromycin, and azithromycin) are a class of antibiotics that are used widely in many treatment regimens. Erythromycin, the parent compound, is derived from a mixture of antibiotics that includes erythromycin A, which is the active compound. Erythromycin has a 14-membered lactone ring with two sugars, L-cladinose and an amino sugar. Erythromycin was first isolated from *Streptomyces erythraeus*. Erythromycin is the parent compound for other commercially available macrolides (clarithromycin and azithromycin, a 15-membered azalide). Sixteen-membered ring macrolides are also available in Europe (spiramycin, josamycin, midecamycin, and miocamycin). Ketolides (e.g., telithromycin) are derived from clarithromycin and have two major modifications, replacement of L-cladinose by a keto function and an 11- to 12-carbamate extension with an arylalkyl modification (5). Currently, ketolides (telithromycin) are being monitored for toxicity (i.e., cases of liver failure have been reported).

Erythromycin is active against most streptococci, certain *Neisseria* spp., *Legionella*, and *H. influenzae* (only azithromycin is active enough against *H. influenzae* to be clinically reliable). Macrolides inhibit bacterial protein synthesis by binding to the 50S ribosomal subunit (the site for peptidyltransferase activity and the binding site for G protein factors that assist in initiation, elongation, and termination phases of protein synthesis). It was discovered that erythromycin binding sites are composed exclusively of segments of 23S rRNA at the peptidyltransferase cavity and do not involve any interaction of the drugs with ribosomal proteins. Erythromycin does not inhibit the peptidyltransferase activity but prevents the extension of the peptide chain by blocking the polypeptide exit tunnel and provokes the premature release of peptidyl-tRNA.

Table 1.3 Mechanisms of resistance to macrolides in *Streptococcus pneumoniae* and *Streptococcus pyogenes*

Mechanism	Genetic determinants
Ribosomal methylation of 23S rRNA	Erm genes
Active efflux of these antibacterial ribosomal mutations in 23S rRNA and ribosomal L4	*mefA*

Macrolides are bacteriostatic for most bacteria but are bactericidal for group A streptococci and *S. pneumoniae*. It is of note that macrolides also exert an anti-inflammatory effect on lung epithelium and are sometimes regarded as biological response modifiers (effects on neutrophil chemotaxis and interleukin-8 secretion).

Bacterial resistance to macrolides can be due to a variety of changes. The most common mechanisms of resistance are the presence of an efflux pump [mef(A)] and target site modification by methylases (Table 1.3). The mechanism described in the following chapter by Roberts (see chapter 5 in this volume) includes the rRNA methylases. These are enzymes that add 1 or 2 methyl groups to a single adenine (A2058 *E. coli* numbering). Similar to aminoglycosides, more than 30 different genes coding for rRNA methylases have been described. It is important to remember that erythromycin is known to induce their synthesis. Current research suggests that resistance to macrolide antibiotics is increasing among clinical isolates of *Streptococcus pyogenes* and *S. pneumoniae*. However, it must be kept in mind that these worrisome patterns are not always directly related to clinical failure. Unfortunately, many macrolide-resistant strains are also penicillin resistant.

Our studies of the structure of the ribosome have resulted in an intense interest in aminoglycosides and macrolides as therapeutic agents. The lessons learned from the study of the mechanism of action and resistance of these agents will undoubtedly lead us to further development of novel inhibitors of protein synthesis.

References

1. **Daniel, T. M.** 2005. Selman Abraham Waksman and the discovery of streptomycin. *Int. J. Tuberc. Lung Dis.* **9:**120–122.
2. **Francioli, P.** 1995. Antibiotic treatment of streptococcal and enterococcal endocarditis: an overview. *Eur. Heart J.* **16**(Suppl. B):75–79.
3. **Klibanov, O. M., R. H. Raasch, and J. C. Rublein.** 2004. Single versus combined antibiotic therapy for gram-negative infections. *Ann. Pharmacother.* **38:**332–337.
4. **Le, T., and A. S. Bayer.** 2003. Combination antibiotic therapy for infective endocarditis. *Clin. Infect. Dis.* **36:**615–621.

5. **Nguyen, M., and E. P. Chung.** 2005. Telithromycin: the first ketolide antimicrobial. *Clin. Ther.* **27:**1144–1163.

6. **Paul, M., I. Benuri-Silbiger, K. Soares-Weiser, and L. Leibovici.** 2004. Beta lactam monotherapy versus beta lactam-aminoglycoside combination therapy for sepsis in immunocompetent patients: systematic review and meta-analysis of randomised trials. *BMJ* **328:**668.

7. **Vakulenko, S. B., and S. Mobashery.** 2003. Versatility of aminoglycosides and prospects for their future. *Clin. Microbiol. Rev.* **16:**430–450.

Enzyme-Mediated Resistance to Antibiotics: Mechanisms, Dissemination, and Prospects for Inhibition
Edited by Robert A. Bonomo and Marcelo E. Tolmasky

Kanchana Majumder, Lianhu Wei,
Subhash C. Annedi, and Lakshmi P. Kotra

2 Aminoglycoside Antibiotics

INTRODUCTION

"Aminoglycoside antibiotics" is a large group of antibiotics that inhibit protein biosynthesis in bacteria. This group of antibiotics, along with β-lactam antibiotics, played a critical role in the treatment of antimicrobial infections in the 20th century and saved millions of human lives, changing the course of human health and longevity. Aminoglycosides either are naturally produced by bacteria as part of the survival strategies among various species or are semisynthetic derivatives of the naturally occurring ones as improved drugs for the treatment of infections in a clinical setting. Due to the importance of aminoglycosides in the treatment of infections, in the past six decades, there has been intense basic and clinical research focused on this subject. We acknowledge several other reviews and books that were published on this subject in the past decade (33, 61, 64, 65). In this chapter, we focus on the general aspects of aminoglycoside antibiotics and associated resistance factors with particular attention to the literature and developments between 2000 and 2003.

The first aminoglycoside, streptomycin, was isolated from *Streptomyces griseus* in 1944 by Schatz et al., and this was followed by the isolation of neomycin from *Streptomyces fradiae* in 1949 by Waksman and Lechevalier (60, 76). Aminoglycosides gained clinical importance very quickly through the introduction of streptomycin for the treatment of tuberculosis, within one year of this drug's discovery, underlining the clinical effectiveness of these compounds as well as the dire need for clinically useful antibiotics against hard-to-treat infections, especially during the times of World War II.

Aminoglycoside antibiotics are effective against both gram-negative and gram-positive bacteria. Several aminoglycosides are produced by bacteria of the *Actinomycetes* group as part of their microbial biosynthesis process, while some others are derivatives of the naturally occurring compounds (semisynthetic compounds). To distinguish the compounds that are produced by and isolated from *Actinomycetes* species from other compounds (i.e., those produced by other microbial species and semisynthetic compounds), the nomenclature of the compounds bears the suffix "ycin" in the former case (such as streptomycin) and "icin" in the latter case (such as gentamicin).

Structures of Aminoglycosides

Aminoglycoside antibiotics, in general, fall into two different classes: compounds in one group consist of a 2-deoxystreptamine moiety (Fig. 2.1, **1**), and those in the

Kanchana Majumder, Lianhu Wei, Subhash C. Annedi, and Lakshmi P. Kotra, Division of Cell and Molecular Biology, Toronto General Research Institute, University Health Network and Departments of Pharmaceutical Sciences and Chemistry, University of Toronto, Toronto, Ontario, Canada.

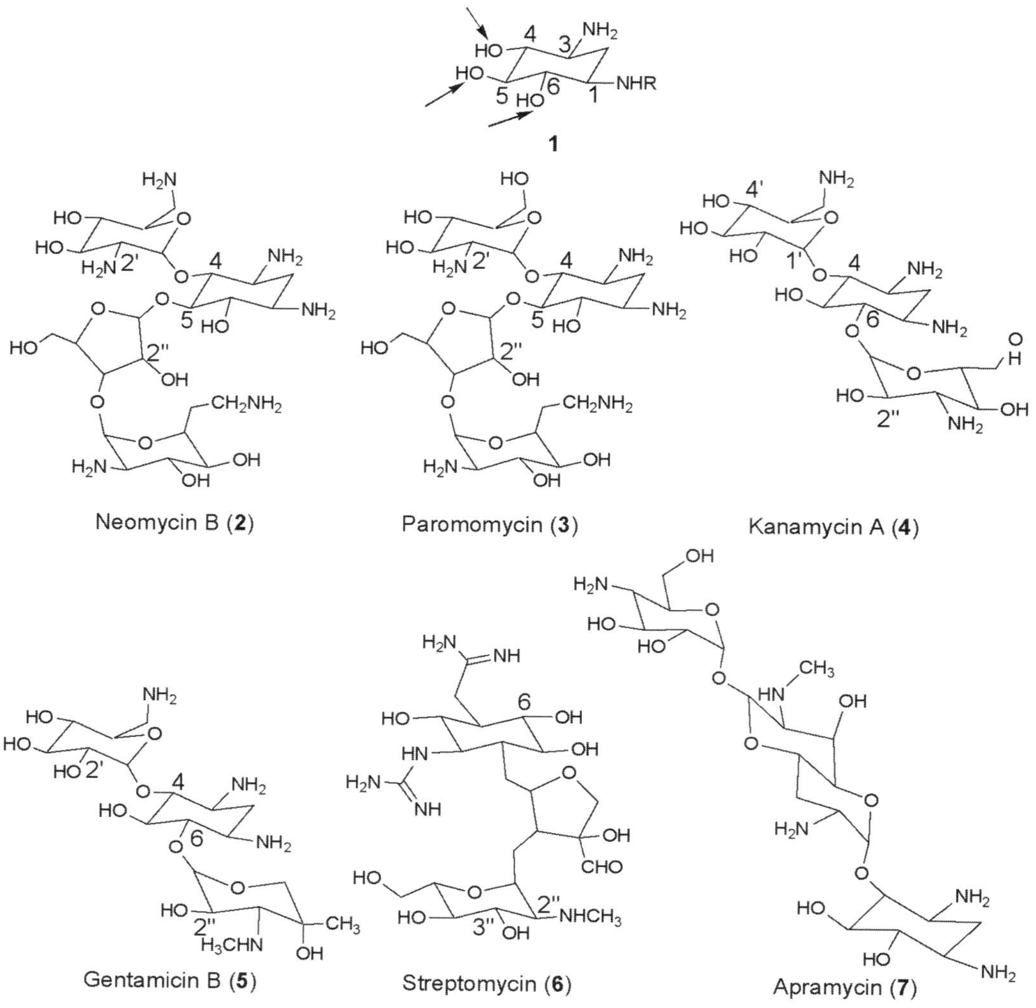

Figure 2.1 Aminoglycoside antibiotics.

other group are devoid of the 2-deoxystreptamine moiety. Popular aminoglycoside antibiotics such as neomycins (2), paromomycin (3), kanamycins (4), gentamicins (5), tobramycin, and amikacin belong to the group of aminoglycosides that contain the 2-deoxystreptamine core moiety. Compounds such as streptomycin (6) and apramycin (7) belong to the latter class of aminoglycosides that lack the 2-deoxystreptamine core (**1**). Among the aminoglycosides that contain the 2-deoxystreptamine core, disubstitutions at positions 4 and 5 or at positions 4 and 6 are common, with a hexose and/or a pentose moiety (see structure **1** for numbering of positions).

Aminoglycosides are water-soluble compounds, and in biological and physiological systems amine groups are mostly protonated. Aminoglycosides are not metabolized by the body, and their antibiotic activity is unchanged by the induction or inhibition of metabolic enzymes. Aminoglycosides are minimally protein bound (ca. 10%) and do not penetrate into the central nervous system or eye due to their high water solubility and hydrophilic character, but they may cross the placenta. Therapeutic concentrations of aminoglycosides are produced only in extracellular fluids. However, relatively high tissue concentrations of the drugs have been found in kidney, the cochlea, and vestibular apparatus with a longer elimination half-life than that from plasma, resulting in their toxic effects.

The hydrophilic sugars on the aminoglycoside antibiotics possess several amino and hydroxyl moieties. Due to the polycationic nature conferred by multiple amine groups, aminoglycosides have binding affinity for nucleic acids, which consist of polyanionic phosphate groups. They have especially high affinity for prokaryotic rRNA (18–20, 51, 52).

Antibiotic Effects of Aminoglycosides

Aminoglycosides in general are effectively used against gram-negative bacteria and are moderately active against certain gram-positive bacteria. They are ineffective against anaerobic bacteria, viruses, and fungi. Some of these antibiotics may be effective against amoeboid or protozoal infections; for example, paromomycin is effective against *Cryptosporidium parvum*. Bactericidal activity of aminoglycosides is principally due to the tight, irreversible binding to ribosomes, but they also interact with other cellular structures and metabolic processes. In addition to the bacterial cellular machinery, aminoglycosides such as neomycin bind to important regions of the genomes in viruses such as the human immunodeficiency virus genome at the *rev* response element, thus blocking the binding of the viral regulatory protein Rev to its response element and at the transactivating response element motifs (77).

These drugs exhibit rapid concentration-dependent bactericidal activity (24, 35). Aminoglycosides have also been shown to possess postantibiotic effects such as persistent suppression of bacterial growth in a dose-dependent manner after a short exposure (24, 35, 74). In vivo, the postantibiotic effect is prolonged by the synergistic effects of the host immune system. It is believed that leukocytes have enhanced phagocytosis and killing activity after exposure to aminoglycosides (13). Side effects due to aminoglycoside treatment include renal toxicity and ototoxicity (vestibular and auditory), and under some rare circumstances, neuromuscular blockade and hypersensitivity reactions may be manifested. Due to these side effects and relatively small therapeutic windows, determination of an optimal dose for these antibiotics is very important for an effective and safe treatment with aminoglycosides.

Drug delivery into the target tissues could be a useful strategy in eliminating the systemic toxicities caused by aminoglycosides. For example, topical application is an effective way to control infections for wounds. Recently, a nebulized formulation of tobramycin solution for inhalation (TSI, TOBI) was introduced into the clinic. This formulation delivers the drug, tobramycin, at a high dose to the lungs of patients with cystic fibrosis carrying *Pseudomonas aeruginosa* (11). This formulation delivers effective amounts of the drug while maintaining low systemic concentrations of the drug and reducing systemic toxicity.

There are several reviews on the activities, structure, resistance, and side effects of aminoglycosides (7, 14, 39). We focus the discussion on the molecular mechanism of action, resistance to aminoglycosides, epidemiology, and clinical importance.

MOLECULAR MECHANISMS

Mechanism of Action of Aminoglycosides

Understanding the mechanisms of action of aminoglycosides not only sheds light on their medicinal properties but also helps explain some of the fundamental cellular processes such as ribosomal protein synthesis. Binding of the aminoglycosides to bacterial 16S rRNA of the 30S ribosomal subunit has been well established (see below). Interactions and binding of aminoglycoside to rRNA inhibit bacterial protein synthesis and result in the misreading of codons and miscoded bacterial proteins. Nuclear magnetic resonance studies reveal their binding to the opened and widened deep grooves of RNA helices at the interface between non-Watson-Crick base-paired regions and hairpin loops. Interactions of aminoglycosides with RNA may have a direct relationship to the number of basic amine groups. These protonated ammonium groups under physiological conditions could potentially interact with the negatively charged metal ion-binding pockets (such as the phosphate groups) on the RNA. Thus, they could affect the folding and conformational changes of the RNA molecules.

Several three-dimensional structures of aminoglycosides that were bound to their target sites on RNA (such as paromomycin and gentamicin C1a bound to a model A-site containing 27 nucleotides, using nuclear magnetic resonance; the complexes of paromomycin-30S RNA in the presence and absence of cognate tRNA and of mRNA, and of gentamicin G bound to an A-site-containing fragment on 16S RNA; and tobramycin and paromomycin bound to a 40-nucleotide A-site RNA fragment, using X-ray crystallography) were determined in the past decade (Fig. 2.2) (9, 18, 44, 49, 70–73, 85). Based on the three-dimensional structures that contain an aminoglycoside complexed at the A-site, the neamine moiety, which is common to tobramycin, paromomycin, gentamicin C1a, and gentamicin G, binds in a similar fashion to the conserved residues at this site, specifically the base pairs U1406-U1495 and C1407-G1494 (numbering of 16S rRNA from *Escherichia coli*). Rings I and II of these aminoglycosides bind at the A site with very close interactions via hydrogen bonding and stacking of the base pairs (Fig. 2.3). The relationship between the binding of these aminoglycosides to the A-site region with interactions involving the bases A1492 and A1493 and the shallow groove and the translation of mRNA by recognizing an appropriate tRNA during the decoding process were discussed by several others in the context of the structure of the ribosome (44, 50, 72, 73). Mutations in the A-site region of rRNA are closely

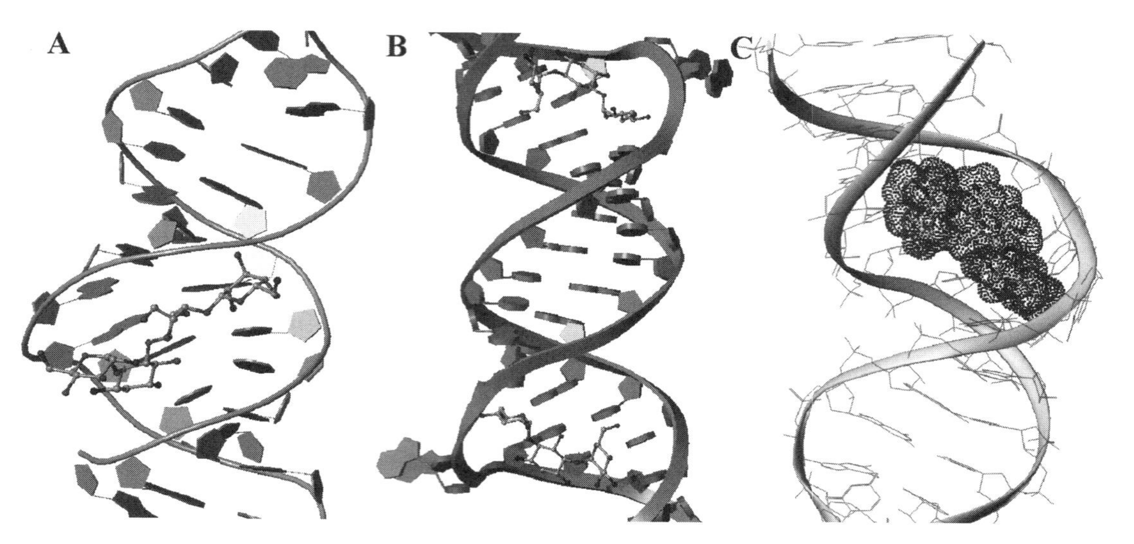

Figure 2.2 Structures of aminoglycosides complexed with rRNAs. (A) Eukaryotic decoding region A-site RNA complexed with paromomycin. (B) Crystal structure of tobramycin bound to the eubacterial 16S rRNA A site. (C) Gentamicin C1A bound at the A-site. The rRNA backbone is represented as a ribbon and the aminoglycoside is shown in ball-and-stick representation (panels A and B) or in space-fill model (panel C).

related to the increase in MICs of the aminoglycosides and can be explained by direct interactions between the aminoglycosides and the compromised interactions with the mutated nucleic base in the A-site or, in other words, by decreased affinity of the aminoglycoside antibiotic for its target site.

Molecular Mechanism of Resistance to Aminoglycosides

Resistance to aminoglycosides is developed in bacteria primarily due to the production and activity of aminoglycoside-modifying enzymes such as acetyltransferases, phosphorylases, and adenyltransferases. For example,

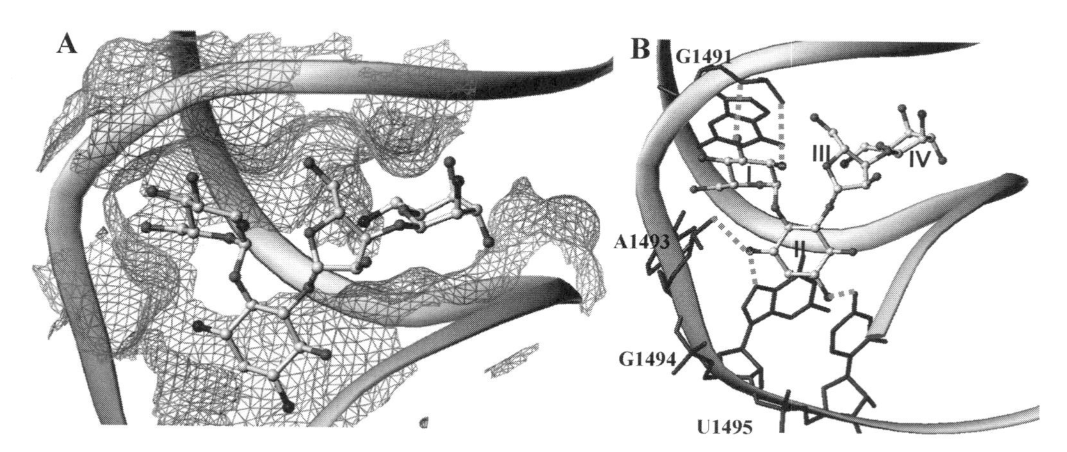

Figure 2.3 Aminoglycoside antibiotic paromomycin bound at the A-site of an RNA oligonucleotide that contains the 30S rRNA. (A) Connolly water-accessible surface for the A-site is shown as chicken mesh, and the drug paromomycin is shown as a capped-stick model. The backbone of the RNA is shown as a ribbon. (B) Rings I and II of paromomycin and their interactions with the A-site nucleotides via hydrogen bonds are shown. The backbone of the RNA is shown as a ribbon, paromomycin is shown as capped-stick, and hydrogen bonds are shown by broken lines.

Pseudomonas aeruginosa and *Burkholderia cepacia* acquire antibiotic resistance by either spontaneous mutation or gene transfer via plasmids or integrins. An aminoglycoside resistance gene, designated *rmtA*, isolated from *P. aeruginosa* AR-2 showed very high-level resistance to various aminoglycosides, including amikacin, tobramycin, isepamicin, arbekacin, kanamycin, and gentamicin (84). This gene encoded a protein, RmtA, which showed similarity to the 16S rRNA methyltransferases of *Actinomycetes* that protect bacterial 16S rRNA from internally produced aminoglycosides (since this mechanism is not intended for true drug resistance in bacteria, this is not covered here in more detail) (84). In addition to the acquired resistance, which is specific to aminoglycosides, bacteria have other modes of survival by forming antibiotic-resistant biofilms around bacterial colonies, creating a very high barrier for the antibiotics to reach the bacterial cell surface. For example, in patients with cystic fibrosis, *P. aeruginosa* forms biofilms within the airways of the patient, creating a different type of resistance to antibiotics, although such resistance is not specific to aminoglycosides (12).

Aminoglycoside-resistant genes are intensively used as genetic markers for eukaryotic and prokaryotic molecular biology, e.g., the neomycin resistance cassette confers resistance to the G-418/gentamicin (toxic to mammalian and bacterial cells), which is widely used in making stable cell lines (e.g., using pcDNA3 vectors). Resistance to aminoglycosides can be divided into two major categories: intrinsic and acquired. From the mechanistic view point, a resistance mechanism could be either a nonenzymatic or an enzyme-mediated process. The nonenzymatic mechanism typically is an intrinsic resistance, and the latter is generally acquired, although not exclusively.

ENZYME-MEDIATED RESISTANCE

The enzymatic mechanism is the most important one due to its prevalence among various pathogenic bacteria. Genes coding for the aminoglycoside-modifying enzymes normally can be disseminated by mobile genetic elements such as plasmids or transposons, and some of these genetic elements are chromosomal in origin.

Aminoglycoside-modifying enzymes can be divided into three groups: (class 1) *O*-nucleotidyltransferase, (class 2) *N*-acetyltransferase, and (class 3) *O*-phosphoryltransferase enzymes. Each enzyme is designated either ANT (for class 1), AAC (for class 2), or APH (for class 3). Genes for these enzymes are designated by lowercase letters, e.g., *ant(6)-Ia* is a gene for the streptomycin nucleotidyltransferase, which modifies position 6 of an aminoglycoside molecule. Each enzyme has several isozymic forms, and each enzyme may have different substrate profiles and associated regiospecificities for inactivation of the aminoglycosides (2). Additionally, the distribution and spread of these enzymes vary depending on the environment and the bacterial species (31). For example, 3′-*O*-phosphotransferase [APH(3′)] is very commonly found in *Pseudomonas* spp., *Klebsiella* spp., *E. coli*, and *Staphylococcus aureus*. Kanamycin is inactivated via phosphorylation, but gentamicin, tobramycin, and amikacin are resistant to this mode of resistance. *Pseudomonas* organisms occasionally have a plasmid coding for a 3′-*N*-acetyltransferase that inactivates gentamicin by acetylation, but not amikacin or tobramycin. *S. aureus* organisms sporadically carry a plasmid that codes for 4′-*O*-adenyltransferase, which links an adenine-ribosyl group to amikacin, tobramycin, and kanamycin. Gentamicin is resistant to this enzyme. Amikacin, a semisynthetic derivative of kanamycin, is resistant to most of the aminoglycoside-modifying enzymes. Dibekacin (3′,4′-dideoxykanamycin B) was the first rationally designed semisynthetic aminoglycoside to overcome the enzymatic resistance.

ANTs

ANTs catalyze the modification of an aminoglycoside (AG) via transfer of a nucleotide monophosphate (NMP) from a corresponding nucleotide triphosphate (NTP) (equation 1).

$$\text{NTP} + \text{AG} \rightarrow \text{pyrophosphate} + \text{NMP-AG conjugate} \quad (1)$$

These enzymes typically are recognized by specifying the modification site on the aminoglycoside. Since a large molecule such as NMP is used to inactivate a single aminoglycoside molecule, the cost of rendering aminoglycosides using one molecule of NTP as well as by the transfer of one NMP is relatively high for bacteria, in comparison to other forms of enzymatic inactivation. Kanamycin nucleotidyltransferase (KNT) was first isolated from *S. aureus* in 1976 as an aminoglycoside resistance enzyme (23). KNT modifies kanamycin at two different positions: at 4′ as well as 4″ hydroxyl groups by transferring NMP. Interestingly, NMP could be transferred from any of ATP, GTP, or UTP onto kanamycin (47).

Among ANTs, ANT(4′) and ANT(2″)-I are the most mechanistically studied ANTs. Typical substrates for ANT(4′) belong to the structural families of kanamycin and neomycin. ANT(4′)-IIb is also found in other clinical isolates such as *P. aeruginosa* BM4492 and confers resistance to tobramycin and amikacin (56). Pedersen et al. and Sakon et al. determined the three-dimensional structure of KNT(4′) complexed with a nonhydrolyzable analog of ATP and kanamycin and compared this structure with a native structure of this enzyme in 1993 (Fig. 2.4) (47, 58).

From the three-dimensional structure of ANT(4′), the active site of the enzyme interacts with rings I and II of the

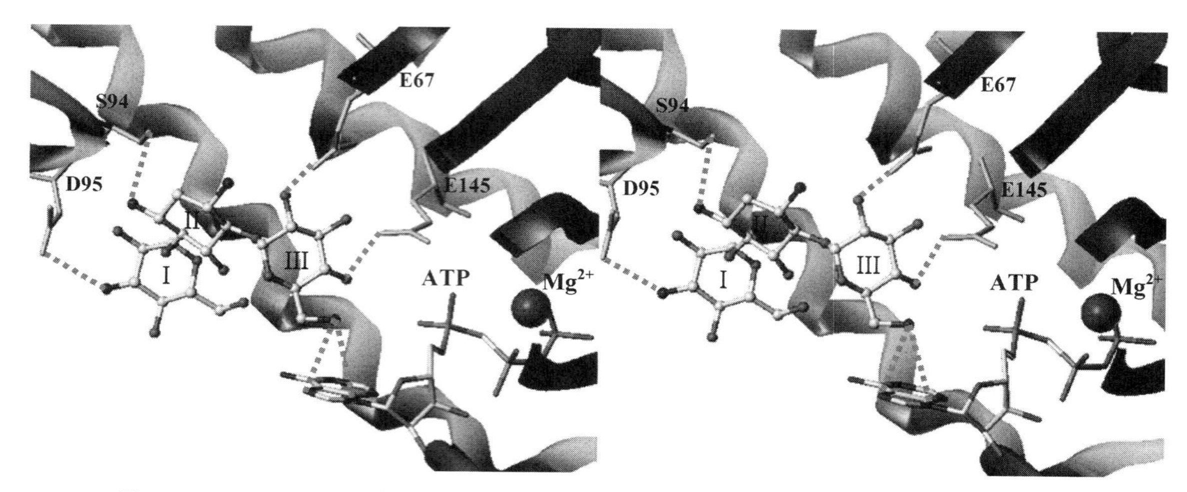

Figure 2.4 Stereo view of X-ray crystal structure of kanamycin-4′-nucleotidyl transferase (KNT) in complex with kanamycin and ATP. Ribbons represent the backbone of dimeric form of KNT. Kanamycin is shown in ball-and-stick representation and ATP is shown as a capped-stick model. Hydrogen bonds between kanamycin and various residues of the enzyme are shown by broken lines. Magnesium ion is shown as a sphere.

aminoglycosides via hydrogen bonds but interactions with the rest of the antibiotic are weaker (Fig. 2.4). Substrate specificity of this enzyme is due to the lack of interactions between the active sites of the enzyme and rings III and IV of the aminoglycoside. Kanamycins and neomycins exhibit similarity in the structure of rings I and II and differ only in ring III. Thus, these two classes of aminoglycosides function as substrates with comparable rates of turnover. The 4′-hydroxyl group of the substrate to KNT is approximately 5 Å from the α-phosphate group on the ATP mimic. The general mechanism of nucleotide transfer suggests that the active site residue, Glu-145, deprotonates 4′-OH acting as a general base. Thus, activated kanamycin attacks the α-phosphorus moiety on the nucleotide, eliminating the Mg^{2+}-pyrophosphate moiety. The orientation of the antibiotic and the nucleotide is appropriately positioned spatially to carry out such a reaction (47).

Another nucleotidyltransferase, ANT(2′)-I, mostly catalyzes a nucleotide transfer onto the aminoglycosides that belong to the groups of gentamicin and kanamycin (22). These aminoglycosides have a common feature in their structure—all of them possess an equatorial 5-OH group. Other ANTs include ANT(6), which modifies streptomycin, and ANT(3), which modifies streptomycin and spectinomycin. Interestingly, ANT(9) and ANT(9)-Ib are found in gram-positive bacteria and these enzymes modify spectinomycin as well. Recently, a bacterial resistance enzyme gene, *aadD2* from *Bacillus clausii*, exhibited approximately 47% identity with *ant(4′)-Ia* from *S. aureus* and could transfer a nucleotide onto kanamycin, tobramycin, and amikacin, thus conferring resistance (6). This novel gene appears to be chromosomal with no possibility for transfer onto other strains. ANT(3″)-Ia, on the other hand, confers resistance to streptomycin and spectinomycin. The gene has been found in association with several transposons (Tn*7*, Tn*21*, etc.) and is ubiquitous among gram-negative bacteria.

ANTs also appear to be part of bifunctional enzymes, i.e., enzymes which would possess two different types of aminoglycoside-modifying activities. Recently, Centron and Roy reported a novel aminoglycoside resistance gene, *ant(3″)-Ii-aac(6′)-IId*, isolated from a multidrug-resistant clinical isolate, *Serratia marcescens* SCH88050909 (10). This gene was characterized for its bifunctional activities of ANT(3″)-I and AAC(6′)-II. Such novel activities suggest ever-changing roles for these resistance enzymes in protecting bacteria from antibacterials such as aminoglycosides and a constant evolutionary selection (46).

AACs

Members of AACs do not have close sequence homologies to the other two classes of aminoglycoside-modifying enzymes, ANTs and APHs. The distribution of AACs among various bacterial species is extensive. The first reported aminoglycoside-modifying enzyme in bacteria was AAC(6′) in 1965, and later a gentamicin acetyltransferase, AAC(3), was reported (45, 79, 80). These enzymes catalyze the transfer of an acetyl group from acetyl coenzyme A (acetyl-CoA) onto an amine on the AG; and coenzyme A (CoA) is released (equation 2).

$$\text{Acetyl-CoA} + \text{AG} \rightarrow \text{CoA} + N\text{-Acetyl-AG} \qquad (2)$$

AACs modify aminoglycosides regiospecifically—acetylation occurs at positions 1 and 3 of the 2-deoxystreptamine ring and at positions 6′ and 2′ of the 6-aminohexose ring. The oldest known enzyme of this class, AAC(6′), is classified into subfamilies I and II, and each subclass is characterized on the basis of substrate specificities. For example, AAC(6′)-I confers resistance to amikacin and gentamicin C1a and C2 but not to gentamicin C1. On the other hand, AAC(6′)-II modifies all gentamicin C isoforms but not amikacin. Additionally, the recent finding that AAC(6′)-Ib is evenly distributed in the cytoplasm in the *E. coli* cell underscores the protective nature of these enzymes for the bacterial rRNA by preventing the unmodified aminoglycoside drug from reaching the target (15). Other AACs such as AAC(2′)-Ic from *Mycobacterium tuberculosis* and AAC(6′)-Iy from *Salmonella enterica* have been studied as well, in the context of their structures and catalytic activities (68, 69). The three-dimensional structures of AAC(3)-Ia complexed with CoA and of AAC(6′)-Ii bound by acetyl-CoA as well as by CoA are available (Fig. 2.5) and reveal a close similarity between these two enzymes in the three-dimensional structure as well as a conserved structural motif with the related superfamily of GCN5-related *N*-acetyltransferases, which were not evident from the low amino acid sequence homology (8, 40, 82, 83). Structures of AAC(6′)-Ii complexed with CoA and acetyl-CoA are similar, and no major differences were observed whether the cofactor was CoA or acetyl-CoA. The three-dimensional structures of AAC(2′)-Ic and AAC(6′)-Iy were also determined, and their enzymatic activities in the context of aminoglycoside modification were discussed by Vetting et al. (69). Structural information on AAC(6′)-Iy is not yet publicly available to allow more comment on this enzyme.

Newer subtypes of AACs, such as AAC(6′)-1z, are emerging from the nosocomial pathogen *Stenotrophomonas maltophilia* and confer resistance to drugs such as tobramycin, netilmicin, and sisomicin (34). Though these are clinically important aminoglycoside-resistant enzymes, little is known about the structure and mechanism of these elements. Some newer variants such as *aac(3)-Ic* from *P. aeruginosa* and *aac(3)-IIa* from *E. coli* are located on the same gene cassette or plasmid as bla(VIM-2) or TEM-1 and were identified in clinical isolates (3, 53).

A new type of enzyme, similar to aminoglycoside resistance enzymes, was discovered recently in *P. aeruginosa*, which possesses very little acetyl-CoA binding ability but binds tightly to aminoglycosides such as tobramycin, dibekacin, kanamycin A, and sisomicin (K_d, <1 μM) and binds weakly to amikacin (K_d, ~ 60 μM) (36). This enzyme, based on sequence homology, is a close relative of AAC-6′ and is anticipated to function "sequestering" of the drug from the solution by binding to the drug very tightly, thus conferring resistance.

APHs

APHs modify most classes of aminoglycosides, and they belong to the kinase superfamily. The substrate specificity of these enzymes varies widely, and to date, there are over 20 known APH enzymes. For example, APH(3′)-II confers resistance to kanamycin, neomycin, paromomycin, ribostamycin, and gentamicin B. Another variant, APH-(3′)-III, confers resistance to kanamycin, neomycin, paromomycin, ribostamycin, lividomycin, and gentamycin B (Belgian Biosafety server, http://www.antibioresistance.be). APH enzymes catalyze the transfer of the γ-phosphoryl group from ATP onto a hydroxyl group on the AG antibiotic (equation 3) and compromise the binding interactions of aminoglycoside with the bacterial rRNA.

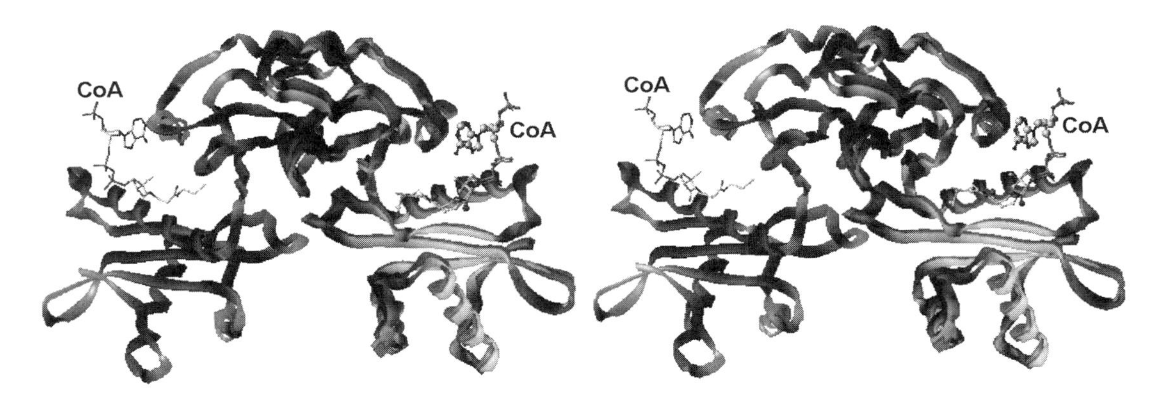

Figure 2.5 Stereo view of overlap of X-ray structures of AAC(2′)-Ic (each monomer is bound to one CoA molecule) and AAC(6′)-Ii (overlapped onto the monomer on the right). Coenzyme A (CoA) in AAC(2′)-Ic is shown as a capped-stick model, one CoA molecule bound to each monomer. CoA in AAC(6′)-Ii is shown in ball-and-stick representation in the binding site of the enzyme and assumes a conformation very close to that in AAC(2′)-Ic.

$$ATP + G \rightarrow ADP + MP\text{-}AG \quad (3)$$

where MP is the γ-phosphoryl group.

The best studied enzyme of this group is APH(3′). APH(3′)-IIIa is found in gram-positive and gram-negative organisms. This enzyme acts on a broad spectrum of 2-deoxystreptamine aminoglycosides such as kanamycin and neomycin. The three-dimensional structures of APH(3′)-IIa in complex with kanamycin and of APH(3′)-IIIa were recently determined (Fig. 2.6). These three-dimensional structures reveal the molecular determinants for the binding of aminoglycosides to these enzymes (17, 25, 42).

The three-dimensional structure of APH(3′)-IIIa, which is normally carried by enterococci and staphylococci, was the first APH structure among this family of resistance enzymes to be determined. Hon et al. observed a close similarity in the three-dimensional structures of this class of enzymes and those of eukaryotic protein kinases (25). There is very little sequence homology between APHs and eukaryotic protein kinases, however. This enzyme was seen as a dimer with two disulfide bonds between the two monomeric units, and the active site of one monomer is facing that of the other with at least a 20-Å separation. Based on the kinetics of this enzyme and the structural features, these two units act as independent enzymes, with no possible cooperation (25). The three-dimensional structure of APH(3′)-IIa complexed with kanamycin distinctly identifies the negatively charged pocket for the aminoglycoside in the binding pocket (42). The structures of APH(3′)-IIa and APH(3′)-IIIa are reportedly very close with 1.1 and 0.9 Å root mean squared deviations upon comparison of the backbone C_α atoms in the N-terminal domain and central cores of the C-terminal domain, respectively. A detailed understanding of the aminoglycoside-phosphorylating enzymes will help to design selective inhibitors against these enzymes.

In addition to the above, APHs are also involved as part of bifunctional enzymes. A bifunctional enzyme consisting of a contiguous sequence representing AAC(6′)-Ie-APH(2″)-Ia and exhibiting acetyltransferase activity as well as phosphoryltransferase activity is becoming a more prevalent enzyme in several bacterial species. Recently, a clinical isolate of a methicillin-resistant *S. aureus* strain exhibited resistance to arbekacin, which was acquired due to the overexpression of AAC(6′)-APH(2″) bifunctional enzyme (37). Substrate specificity for the bifunctional enzyme is very wide and modifies a number of clinically relevant aminoglycosides (5). Biochemical studies indicate that there is no relationship between the acetyltransferase activity and the phosphoryltransferase activity of this enzyme. Several APHs are also used in molecular biology experiments as markers due to their specificity towards aminoglycosides; for example, APH(4) and APH(7″) confer resistance to hygromycin B and are frequently used as a marker in eukaryotic and bacterial cell culture experiments.

NONENZYMATIC RESISTANCE

Mutations in chromosomal genes encoding ribosomal proteins, e.g., *rpsL* (or *strA*), *rpsD* (or *ramA* or sud_2), and *rpsE* (*eps* or *spc* or *spcA*), can cause resistance to streptomycin and spectinomycin. A low-level resistance

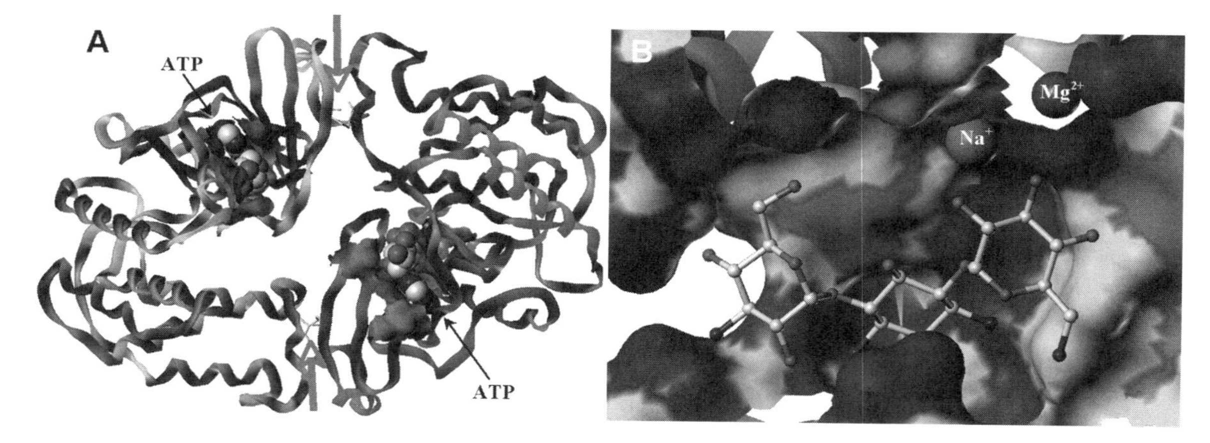

Figure 2.6 (A) X-ray crystal structure of APH(3′)-IIIa shown in ribbon representation. Two monomers are held together via two disulfide bonds (shown by arrows at 6 and 12 o'clock positions). ATP is shown in capped-stick representation. ATP binding sites in each monomer (shown as Connolly water-accessible surfaces) are far from each other. (B) A close view of the kanamycin bound in the active site of APH(3′)-IIa, a negatively charged pocket. The active site is shown as Connolly water-accessible surface. Kanamycin is shown in ball-and-stick representation. Magnesium and sodium ions, shown as spheres, are bound in the active site interacting with the aminoglycoside and the enzyme active site.

against streptomycin can be caused by mutations in *strC* (or *strB*). To understand the molecular mechanism of hygromycin B, Pfister et al. recently found that the specific mutations U1406C, C1496U, and U1498C (*E. coli* numbering) are in close proximity to the binding site of hygromycin B on the helix 44 of 16S rRNA and that any induction of these mutations could compromise the activity of hygromycin B (48). The above specific 16S rRNA residue positions involved in hygromycin B resistance are highly conserved among species, which explains the lack of specificity towards bacteria and the general toxicity of hygromycin B.

Point mutations in rRNA can result in drug resistance; for example, a mutation at C1192 of the rRNA shows spectinomycin resistance in the case of *E. coli* and the MIC for the mutant is >80 μg/ml (usually for the wild type, the MIC is <5 μg/ml). Mutations in other regions of rRNA such as at position C1066 have also been identified in *E. coli* and *Salmonella* spp. and have been found to compromise the activity of this antibiotic (27, 43). In order to understand the influence of bacterial mutations on ribosomal aminoglycoside susceptibility, various MICs were compared with the previously published crystal structures of paromomycin, tobramycin, and geneticin bound to the minimal A site (72). From the structural information, two regions are important for binding of aminoglycosides to the A site: (i) the single adenine residue at position 1408, and (ii) the non-Watson-Crick U1406-U1495 pair (Fig. 2.7). When the structures of the complexes of 30S rRNA with antibiotics such as paromomycin and tobramycin (both bound at the A-site region) are compared, the importance of the U·U base pair is quite evident, and clearly such conserved regions are important in drug design.

The effects of mutations at these positions are modulated by an amino or hydroxy substituent at position 6′

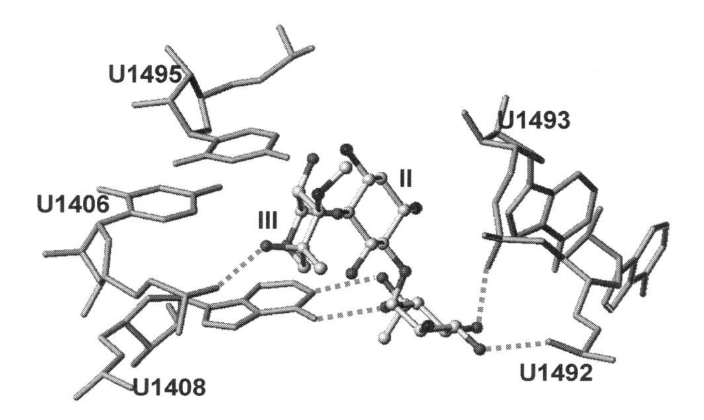

Figure 2.7 Hydrogen bonds between geneticin and various nucleotides in the A-site of the eubacterial 16S rRNA are shown in broken lines. Geneticin is given in capped-stick representation. Critical nucleic bases are shown in capped-stick representation.

(Fig. 2.2), by the number of positive charges on the aminoglycoside antibiotic, and by the type of linkage between rings, either 4, 5 or 4, 6. Since the target site mutations will compromise the affinity of the aminoglycosides, these mutations at the A-site usually compromise the clinical effectiveness and potency of the aminoglycosides.

A high level of streptomycin and spectinomycin resistance can result from mutations in chromosomal genes. An analysis by Vicens and Westhof demonstrated that a high-level resistance to 6′-amine-containing aminoglycosides as a result of the A1408G mutation may be due to the inability of ring I to form a pseudo-base-pair with the mutated G1408 and enter the A-site helix (72). Additionally, they showed that mutations of the uracil residues forming the U1406-U1495 pair either to C or to A residues mostly confer low to moderate levels of drug resistance whereas the U1406C/U1495A double mutation confers high-level resistance (except for neomycin). This underscores the importance of the geometry of the binding site at the U1406-U1495 pair as well as within the A-site for optimal functional activity (33, 72).

An alteration in the aminoglycoside transport system, inadequate membrane potential, or a modification in the lipopolysaccharide phenotype can also result in a cross-resistance to a broad spectrum of aminoglycosides. Transport of aminoglycosides into the cell requires energy derived from the transmembrane electrical potential established by the bacterial electron transport system. Mutations that affect transmembrane electrical potential and modification in the lipopolysaccharide phenotypes show aminoglycoside resistance. Recently, MexXY was identified as an aminoglycoside-inducible multidrug transporter and was shown to contribute to intrinsic or acquired aminoglycoside resistance in laboratory isolates of *P. aeruginosa*. However, the exact molecular determinants for the expression of MexXY are not yet known and remain to be investigated (62).

In addition to the above-discussed mechanisms of resistance, aminoglycosides are facing other types of resistance challenges. Aminoglycosides could be substrates of specialized secretory transporters and decrease the levels of drugs such as arbekacin in rat small intestines. These transporters do not seem to be related to P-glycoprotein transporters (57). Efflux systems in bacteria such as *Pseudomonas* also seem to confer resistance to a number of antibiotics, including aminoglycosides (26). The efflux pumps are located on the outer membranes of the bacterium, and in the future this type of resistance may become more prevalent.

EPIDEMIOLOGY

Resistance to aminoglycosides is a serious problem worldwide and is caused mainly by inactivation using intracellular enzymes in the bacteria. The intensive use of antibiotics results in the emergence of microorganisms that develop resistance against them. Discontinuation of intensively prescribed drugs and replacement of the drugs with newly introduced antibiotics of another class with different structural features (prevalent resistant strains should be susceptible to such new classes) can check this problem; however, experience with such strategies has shown that resistant bacterial strains will evolve in time. A decade ago, according to 1993 estimates, there were more than 50 enzymes so far identified that confer resistance to aminoglycosides (61).

Some frequently found and important aminoglycoside-resistant microorganisms are gram-negative bacilli including *Klebsiella pneumoniae*, *Enterobacter* spp., *Acinetobacter baumannii*, and *P. aeruginosa*, which are also resistant to a broad spectrum of β-lactams and fluoroquinolones. They are found in patients suffering from nosocomial infections (mainly among patients in intensive care). With time, aminoglycosides are being used intensively and the problem of drug resistance is also becoming serious: the number of amikacin-resistant gram-negative bacilli has doubled from 1992 to 1998, and a mostly threefold increase is found among *A. baumannii*/*A. calcoaceticus* isolates from 1997 to 1998 (41).

Only low-level resistance to aminoglycosides is found in all *Enterococcus* spp. due to their anaerobic metabolism. Recently, infective endocarditis has been seen in higher frequency. This is caused by highly aminoglycoside-resistant enterococci. A β-lactam drug is suggested to be used simultaneously with aminoglycoside to help the penetration of the drug into the cell (American Academy of Family Physicians, 1998: http://www.aafp.org/afp/981115ap/gonzalez.html). This endemic and epidemic level of resistance by pathogen is related to the intensive use of antibiotics against which resistance is developed.

Several methods can be used to detect the aminoglycoside resistance enzymes in bacteria, and most useful methods include microbiological assays and PCR. Recently a multiplex PCR methodology was developed to detect a number of aminoglycoside-modifying enzymes simultaneously in enterococci (66). Genes for these resistant factors include *aac(6′)-Ie-aph(2″)-Ia*, *aph(2″)-Ib*, *aph(2″)-Ic*, *aph(2″)-Id*, *aph(3′)-IIIa*, and *ant(4′)-Ia*, spanning all three classes of aminoglycoside-modifying enzymes, i.e., acetyltransferases, phosphoryltransferases and nucleotidyltransferases.

INHIBITION OF AMINOGLYCOSIDE RESISTANCE ENZYMES

New research directions are being targeted to gain insights into the mechanism of action of these drugs against a target resistance enzyme to develop novel aminoglycoside derivatives, which will be resistant to aminoglycoside-modifying enzymes, and potentially have fewer side effects. Dibekacin (3′, 4′-dideoxykanamycin B) (Fig. 2.8, **8**) was the first rationally designed semisynthetic aminoglycoside to overcome the enzymatic resistance (32). Following that, by introducing (*S*)-4-amino-2-hydroxybutyryl, ethyl and (*S*)-3-amino-2-hydroxypropionyl side chains onto the 1-amino group of kanamycin, sisomicin, and gentamicin B, respectively, amikacin, netilmicin, and isepamicin were developed. These side chains are for blocking the access of aminoglycoside-modifying enzymes to their target areas. 7-Hydroxytropolone (**9**), a fermentation product of certain *Pseudomonas* and *Streptomyces* species, and some of its derivatives are potent inhibitors of ANT(4′) and ANT(2′)-I (1, 30, 59).

Some semisynthetic drugs, for example, 5-*epi*-sisomicin (**10**) and 5-*epi*-gentamicin (**11**), are effective against ANT(2′), AAC(3), and AAC(2′) (21, 28, 67, 75). They have an axial 5-hydroxyl group, and this is the reason for the inhibition of the enzyme since ANT(2′) requires an equatorial 5-hydroxyl group for its activity. These results were also supported by in vivo experiments.

A group of antibiotics (tobramycin, gentamicin, amikacin, and dibekacin) are found to affect the growth of *E. coli* that harbors ANT(2′) when combined with 7-HT but not without 7-HT. For AAC(3)-I, a tight-binding bisubstrate analog was synthesized (81). Aminoglycoside derivatives like 5-*epi*-sisomicin and netilmicin (**12**) are resistant to AAC(3)-I (28, 38). 2-Deamino-2-nitro neamine (**13**) and kanamycin B (**14**) are found to inhibit APH(3′) (55). Four bromoacetylated analogs of neamine (**15–18**) are known as inactivators of APH(3′)-IIa (54). Dimers of neamine linked via either amides or 1,2-hydroxyamine moieties are potent inhibitors of APH(2″) of the bifunctional enzyme AAC(6′)-APH(2″) (63). In recent years, several groups have been intensively searching for effective inhibitors of aminoglycoside-modifying enzymes. Some cationic peptides, viz., indolicin and its synthetic analogs, are shown to inhibit both AAC and APH classes of enzymes (4).

CLINICAL RELEVANCE

Aminoglycosides are an important class of antimicrobials in the clinic. An understanding of the mechanism of aminoglycoside activity, their pharmacokinetics, and their physicochemical properties is important to design a next-generation drug with minimum toxicity and an optimal effect against bacteria and the corresponding resistance factors. Aminoglycosides are often combined with a β-lactam drug, mainly in case of *S. aureus* infection, which not only enhances bactericidal activity but also prevents

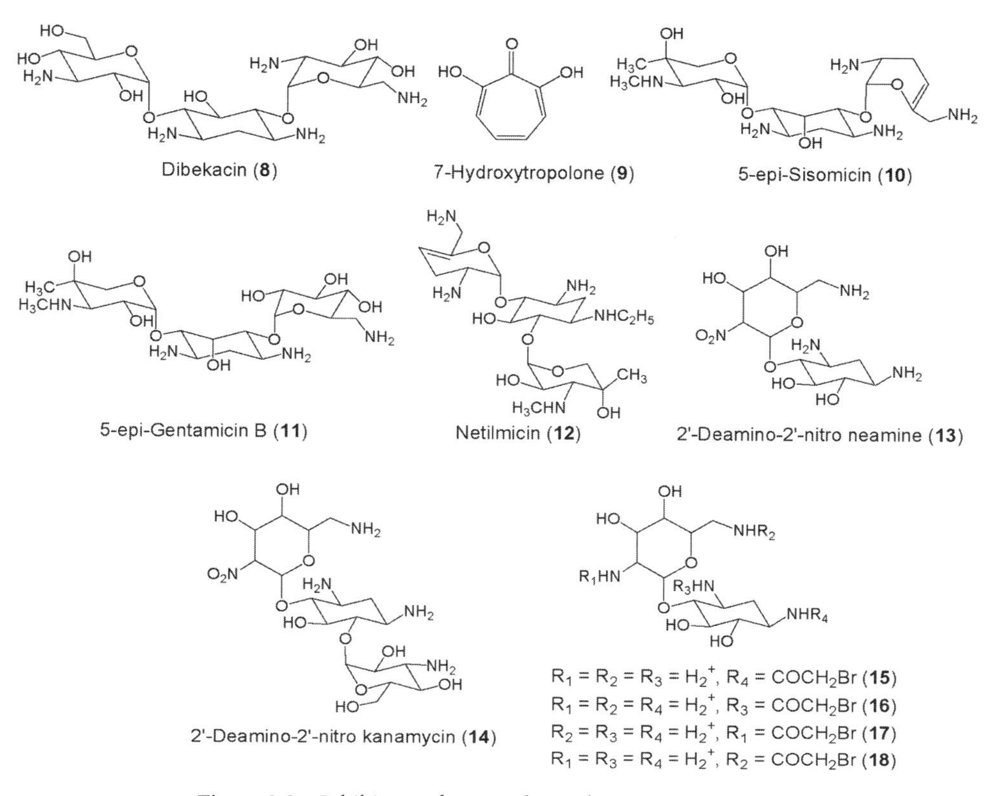

Figure 2.8 Inhibitors of aminoglycoside resistance enzymes.

the generation of resistant staphylococci and clinical relapse after the antibiotic is withdrawn. The challenge in clinical use of aminoglycosides is tolerance and/or severity of side effects and resistant mechanisms acquired by bacteria. Different strategies are employed to minimize the toxicity of aminoglycosides, such as administration of the drug once daily and maintaining an extended interval of dosing.

The emergence of antibiotic-resistant bacteria is a serious problem in the 21st century. The use of antibiotic resistance genes as selection markers in genetically modified organisms is associated with this antibiotic resistance problem—the release of antibiotic resistance genes to sensitive bacterial strains and their transfer to the environment create the problem. The World Health Organization has recognized this drug resistance as a serious problem and is taking several measures (e.g., appropriate antibiotic therapy, hospital hygiene, control of bacterial infections, etc.) in order to address the global antibiotic resistance issue.

As early as 1969, the Swan report in the United Kingdom raised the issue regarding the use of antimicrobials in animals and their subsequent effect on human health (78). In the last decade of the 20th century, effects of antibiotic resistance were seen in the clinic, where multidrug resistance strains to the most effective antibiotics including aminoglycosides were becoming more common. Recent reports support the transmission of antibiotic resistance factors against aminoglycosides such as gentamicin from animals to humans via the food chain in the United States (16). Thus, understanding complex ecological, biochemical, and molecular origins of antimicrobial resistance mechanisms in relation to antibiotics, their development, and their dissemination in addition to developing newer inhibitors is important to continue to use effective drugs such as aminoglycosides in the clinic.

References

1. **Allen, N. E., W. E. Alborn, J. N. Hobbs, and H. A. Kirst.** 1982. 7-Hydroxytropolone: an inhibitor of aminoglycoside-2″-O-adenylyltransferase. *Antimicrob. Agents Chemother.* **22:**824–831.
2. **Azucena, E., and S. Mobashery.** 2000. *Drug Resist. Updat.* **4:**106–117.
3. **Bellaaj, A., C. Bollet, and K. Ben-Mahrez.** 2003. Characterization of the 3-N-aminoglycoside acetyltransferase gene aac(3)-IIa of a clinical isolate of Escherichia coli. *Ann. Microbiol.* **53:**211–217.
4. **Boehr, D. D., K. A. Draker, K. Koteva, M. Bains, R. E. Hancock, and G. D. Wright.** 2003. Broad-spectrum peptide

inhibitors of aminoglycoside antibiotic resistance enzymes. *Chem. Biol.* **10:**189–196.

5. **Boehr, D. D., S. I. Jenkins, and G. D. Wright.** 2003. The molecular basis of the expansive substrate specificity of the antibiotic resistance enzyme aminoglycoside acetyltransferase-6′-aminoglycoside phosphotransferase-2″. The role of Asp-99 as an active site base important for acetyl transfer. *J. Biol. Chem.* **278:**12873–12880.
6. **Bozdogan, B., S. Galopin, G. Gerbaud, P. Courvalin, and R. Leclercq.** 2003. Chromosomal *aadD2* encodes an aminoglycoside nucleotidyltransferase in *Bacillus clausii*. *Antimicrob. Agents Chemother.* **47:**1343–1346.
7. **Brummett, R., and K. Fox.** 1989. Aminoglycoside-induced hearing loss in humans. *Antimicrob. Agents Chemother.* **33:**797–800.
8. **Burk, D. L., N. Ghuman, L. E. Wybenga-Groot, and A. M. Berghuis.** 2003. X-ray structure of the AAC(6′)-Ii antibiotic resistance enzyme at 1.8 angstrom resolution; examination of oligomeric arrangements in GNAT superfamily members. *Protein Sci.* **12:**426–437.
9. **Carter, A. P., W. M. Clemons, D. E. Brodersen, R. J. Morgan-Warren, B. T. Wimberly, and V. Ramakrishnan.** 2000. Functional insights from the structure of the 30S ribosomal subunit and its interactions with antibiotics. *Nature* **407:**340–348.
10. **Centron, D., and P. H. Roy.** 2003. Presence of a group II intron in a multiresistant *Serratia marcescens* strain that harbors three integrons and a novel gene fusion. *Antimicrob. Agents Chemother.* **46:**1402–1409.
11. **Cheer, S. M., J. Waugh, and S. Noble.** 2003. Inhaled tobramycin (TOBI): a review of its use in the management of *Pseudomonas aeruginosa* infections in patients with cystic fibrosis. *Drugs* **63:**2501–2520.
12. **Chernish, R. N., and S. D. Aaron.** 2003. Approach to resistant gram-negative bacterial pulmonary infections in patients with cystic fibrosis. *Curr. Opin. Pulm. Med.* **9:**509–515.
13. **Craig, W. A., and S. Gundmundson.** 1991. Postantibiotic effect, p. 403–431. *In* V. Lorian (ed.), *Antibiotics in Laboratory Medicine,* 3rd ed. Williams and Wilkins, Baltimore, Md.
14. **Davis, B. D.** 1987. Mechanism of bactericidal action of aminoglycosides. *Microbiol. Rev.* **51:**341–350.
15. **Dery, K. J., B. Soballe, M. S. L. Witherspoon, D. Y. Bui, R. Koch, D. J. Sherratt, and M. E. Tolmasky.** 2003. The aminoglycoside 6′-*N*-acetyltransferase type Ib encoded by Tn*1331* is evenly distributed within the cell's cytoplasm. *Antimicrob. Agents Chemother.* **47:**2897–2902.
16. **Donabedian, S. M., L. A. Thal, E. Hershberger, M. B. Perri, J. W. Chow, P. Bartlett, R. Jones, K. Joyce, S. Rossiter, K. Gay, J. Johnson, C. Mackinson, E. Debess, J. Madden, F. Angulo, and M. J. Zervos.** 2003. Molecular characterization of gentamicin-resistant enterococci in the United States: evidence of spread from animals to humans through food. *J. Clin. Microbiol.* **41:**1109–1113.
17. **Fong, D. H., and A. M. Berghuis.** 2002. Substrate promiscuity of an aminoglycoside antibiotic resistance enzyme via target mimicry. *EMBO J.* **21:**2323–2331.
18. **Fourmy, D., M. I. Recht, S. C. Blanchard, and J. D. Puglisi.** 1996. Structure of the A site of *Escherichia coli* 16S ribosomal RNA complexed with an aminoglycoside antibiotic. *Science* **274:**1367–1371.
19. **Fourmy, D., M. I. Recht, and J. D. Puglisi.** 1998. Binding of neomycin-class aminoglycoside antibiotics to the A-site of 16 S rRNA. *J. Mol. Biol.* **277:**347–362.
20. **Fourmy, D., S. Yoshizawa, and J. D. Puglisi.** 1998. Paromomycin binding induces a local conformational change in the A-site of 16 S rRNA. *J. Mol. Biol.* **277:**333–345.
21. **Fu, K. P., and H. C. Neu.** 1978. Activity of 5-episisomicin compared with that of other aminoglycosides. *Antimicrob. Agents Chemother.* **14:**194–200.
22. **Gates, C. A., and D. B. Northrop.** 1988. Substrate specificities and structure-activity relationships for the nucleotidylation of antibiotics catalyzed by aminoglycoside nucleotidyltransferase 2″-I. *Biochemistry* **27:**3820–3825.
23. **Goffic, F. L., B. Baca, C. J. Soussy, A. Bublanchet, and J. Duval.** 1976. ANT(4)-I: a new aminoglycoside nucleotidyltransferase found in "staphylococcus aureus." *Ann. Microbiol.* **127:**391–399.
24. **Hock, R., and R. J. Anderson.** 1995. Prevention of drug-induced nephrotoxicity in the intensive-care unit. *J. Crit. Care* **10:**33–43.
25. **Hon, W. C., G. A. McKay, P. R. Thompson, R. M. Sweet, D. S. C. Yang, G. D. Wright, and A. M. Berghuis.** 1997. Structure of an enzyme required for aminoglycoside resistance reveals homology to eukaryotic protein kinases. *Cell* **89:**887–895.
26. **Jo, J. T. H., F. S. L. Brinkman, and R. E. W. Hancock.** 2003. Aminoglycoside efflux in *Pseudomonas aeruginosa*: involvement of novel outer membrane proteins. *Antimicrob. Agents Chemother.* **47:**1101–1111.
27. **Johanson, U., and D. Hughes.** 1995. A new mutation in 16S rRNA of *Escherichia coli* conferring spectinomycin resistance. *Nucleic Acids Res.* **23:**464–466.
28. **Kabins, S. A., and C. Nathan.** 1978. In vitro activity of 5-episisomicin in bacteria resistant to other aminoglycoside antibiotics. *Antimicrob. Agents Chemother.* **14:**391–397.
29. **Kabins, S. A., C. Nathan, and S. Cohen.** 1976. In vitro comparison of netilmicin, a semisynthetic derivative of sisomicin, and 4 other aminoglycoside antibiotics. *Antimicrob. Agents Chemother.* **10:**139–145.
30. **Kirst, H. A., G. G. Marconi, F. T. Counter, P. W. Ensminger, N. D. Jones, M. O. Chaney, J. E. Toth, and N. E. Allen.** 1982. Synthesis and characterization of a novel inhibitor of an aminoglycoside-inactivating enzyme. *J. Antibiot.* **12:**1651–1657.
31. **Kobayashi, N., M. Alam, Y. Nishimoto, S. Urasawa, N. Uehara, and N. Watanabe.** 2001. Distribution of aminoglycoside resistance genes in recent clinical isolates of *Enterococcus faecalis*, *Enterococcus faecium* and *Enterococcus avium*. *Epidemiol. Infect.* **126:**197–204.
32. **Kondo, S., and K. Hotta.** 1999. Semisynthetic aminoglycoside antibiotics: development and enzymatic modifications. *J. Infect. Chemother.* **5:**1–9.
33. **Kotra, L. P., J. Haddad, and S. Mobashery.** 2000. Aminoglycosides perspectives on mechanisms of action and resistance and strategies to counter resistance. *Antimicrob. Agents Chemother.* **44:**3249–3256.

34. **Li, X. Z., L. Zhang, G. A. McKay, and K. Poole.** 2003. Role of the acetyltransferase AAC(6′)-Iz modifying enzyme in aminoglycoside resistance in *Stenotrophomonas maltophilia*. *J. Antimicrob. Chemother.* **51:**803–811.
35. **Lortholary, O., M. Tod, Y. Cohen, and O. Petitjean.** 1995. Aminoglycosides. *Med. Clin. N. Am.* **79:**761–87.
36. **Magnet, S., T. A. Smith, R. J. Zheng, P. Nordmann, and J. S. Blanchard.** 2003. Aminoglycoside resistance resulting from tight drug binding to an altered aminoglycoside acetyltransferase. *Antimicrob. Agents Chemother.* **47:**1577–1583.
37. **Matsuo, H., M. Kobayashi, T. Kumagai, M. Kuwabara, and M. Sugiyama.** 2003. Molecular mechanism for the enhancement of arbekacin resistance in a methicillin-resistant Staphylococcus aureus. *FEBS Lett.* **546:**401–406.
38. **Miller, G. H., G. Arcieri, M. J. Weinstein, and J. A. Waitz.** 1976. Biological activity of netilmicin, a broad-spectrum semisynthetic aminoglycoside antibiotic. *Antimicrob. Agents Chemother.* **10:**827–836.
39. **Mingeot-Leclercq, M.-P., and P. M. Tulkens.** 1999. Aminoglycosides: nephrotoxicity. *Antimicrob. Agents Chemother.* **43:**1003–1012.
40. **Neuwald, A. F., and D. Landsman.** 1997. GCN5-related histone *N*-acetyltransferases belong to a diverse superfamily that includes the yeast SPT10 protein. *Trends Biochem. Sci.* **22:**154–155.
41. **New Jersey Department of Health and Senior Services.** 1998. *Epidemiology Surveillance System, 1998 Report.* Division of Epidemiology, Environmental and Occupational Health, New Jersey Department of Health and Senior Services. [online.] http://www.state.nj.us/health/cd/ess1998/report.htm.
42. **Nurizzo, D., S. C. Shewry, M. H. Perlin, S. A. Brown, J. N. Dholakia, R. L. Fuchs, T. Deva, E. N. Baker, and C. A. Smith.** 2003. The crystal structure of aminoglycoside-3′-phosphotransferase-IIa, an enzyme responsible for antibiotic resistance. *J. Mol. Biol.* **327:**491–506.
43. **O'Connor, M., and A. E. Dahlberg.** 2002. Isolation of spectinomycin resistance mutations in the 16S rRNA of *Salmonella enterica* serovar Typhimurium and expression in *Escherichia coli* and *Salmonella*. *Curr. Microbiol.* **45:**429–433.
44. **Ogle, J. M., D. E. Brodersen, W. M. Clemons, M. J. Tarry, A. P. Carter, and V. Ramakrishnan.** 2001. Recognition of cognate transfer RNA by the 30S ribosomal subunit. *Science* **292:**897–902.
45. **Okamoto, S., and Y. Suzuki.** 1965. Chloramphenicol-, dihydrostreptomycin-, and kanamycin-inactivating enzymes from multiple drug-resistant *Escherichia coli* carrying episome 'R'. *Nature* **208:**1301–1303.
46. **Over, U., D. Gur, S. Unal, and G. H. Miller.** 2001. The changing nature of aminoglycoside resistance mechanisms and prevalence of newly recognized resistance mechanisms in Turkey. *Clin. Microbiol. Infect.* **7:**470–478.
47. **Pedersen, L. C., M. M. Benning, and H. M. Holden.** 1995. Structural investigation of the antibiotics and ATP-binding sites in kanamycin nucleotidyltransferase. *Biochemistry* **34:**13305–13311.
48. **Pfister, P., A. Risch, D. E. Brodersen, and E. C. Bottger.** 2003. Role of 16S rRNA helix 44 in ribosomal resistance to hygromycin B. *Antimicrob. Agents Chemother.* **47:**1496–1502.
49. **Pfister, P., S. Hobbie, Q. Vicens, E. C. Bottger, and E. Westhof.** 2003. The molecular basis for A-site mutations conferring aminoglycoside resistance: relationship between ribosomal susceptibility and X-ray crystal structures. *Chembiochem* **4:**1078–1088.
50. **Ramakrishnan, V., and P. V. Moore.** 2001. Atomic structures at last: the ribosome in 2000. *Curr. Opin. Struct. Biol.* **11:**144–154.
51. **Recht, M. I., S. Douthwaite, and J. D. Puglisi.** 1999. Basis for prokaryotic specificity of action of aminoglycoside antibiotics. *EMBO J.* **18:**3133–3138.
52. **Recht, M. I., D. Fourmy, S. C. Blanchard, K. D. Dahlquist, and J. D. Puglisi.** 1996. RNA sequence determinants for aminoglycoside binding to an A-site rRNA model oligonucleotide. *J. Mol. Biol.* **262:**421–436.
53. **Riccio, M. L., J. D. Docquier, E. Dell'Amico, F. Luzzaro, G. Amicosante, and G. M. Rossolini.** 2003. Novel 3-*N*-aminoglycoside acetyltransferase gene, *aac(3)-Ic*, from a *Pseudomonas aeruginosa* integron. *Antimicrob. Agents Chemother.* **47:**1746–1748.
54. **Roestamadji, J., and S. Mobashery.** 1998. The use of neamine as a molecular template: inactivation of bacterial antibiotic resistance enzyme aminoglycoside 3′-phosphotransferase type IIa. *Bioorg. Med. Chem. Lett.* **8:**3483–3486.
55. **Roestamadji, J., I. Grapsas, and S. Mobashery.** 1995. Mechanism-based inactivation of bacterial aminoglycoside 3′-phosphotransferases. *J. Am. Chem. Soc.* **117:**80–84.
56. **Sabtcheva, S., M. Galimand, G. Gerbaud, P. Courvalin, and T. Lambert.** 2003. Aminoglycoside resistance gene *ant(4′)-IIb* of *Pseudomonas aeruginosa* BM4492, a clinical isolate from Bulgaria. *Antimicrob. Agents Chemother.* **47:**1584–1588.
57. **Saitoh, H., Y. Arashiki, A. Oka, M. Oda, Y. Hatakeyama, M. Kobayashi, and K. Hosoi.** 2003. Arbekacin is actively secreted in the rat intestine via a different efflux system from P-glycoprotein. *Eur. J. Pharm. Sci.* **19:**133–140.
58. **Sakon, J., H. H. Liao, A. M. Kanikula, M. M. Benning, I. Rayment, and H. M. Holden.** 1993. Molecular structure of kanamycin nucleotidyltransferase determined to 3.0-Å resolution. *Biochemistry* **32:**11977–11984.
59. **Saleh, N. A., A. Zwiefak, W. Peczynska-Czoch, M. Mordarski, and G. Pulverer.** 1988. New inhibitors for aminoglycoside-adenylyltransferase. *Zentbl. Bakteriol. Hyg. A Mikrobiol.* **270:**66–75.
60. **Schatz, A., E. Bugie, and S. A. Waksman.** 1944. Streptomycin, a substance exhibiting antibiotic activity against Gram-positive and Gram-negative bacteria. *Proc. Soc. Exp. Biol. Med.* **55:**66–69.
61. **Shaw, K. J., P. N. Rather, R. S. Hare, and G. H. Miller.** 1993. Molecular genetics of aminoglycoside resistance genes and familiar relationships of the aminoglycoside-modifying enzymes. *Microbiol. Rev.* **57:**138–163.
62. **Sobel, M. L., G. A. McKay, and K. Poole.** 2003. Contribution of the MexXY multidrug transporter to aminoglycoside resistance in *Pseudomonas aeruginosa* clinical isolates. *Antimicrob. Agents Chemother.* **47:**3202–3207.
63. **Sucheck, S. J., A. L. Wong, K. M. Koeller, D. D. Boehr, K. Draker, P. Sears, G. D. Wright, and C. H. Wong.** 2000.

Design of bifunctional antibiotics that target bacterial rRNA and inhibit resistance-causing enzymes. *J. Am. Chem. Soc.* **122:**5230–5231.

64. **Tok, J. B. H., and L. R. Bi.** 2003. Aminoglycoside and its derivatives as ligands to target the ribosome. *Curr. Top Med. Chem.* **3:**1001–1019.
65. **Vakulenko, S. B., and S. Mobashery.** 2003. Versatility of aminoglycosides and prospects for their future. *Clin. Microbiol. Rev.* **16:**430–451.
66. **Vakulenko, S. B., S. M. Donabedian, A. M. Voskresenskiy, M. J. Zervos, S. A. Lerner, and J. W. Chow.** 2003. Multiplex PCR for detection of aminoglycoside resistance genes in enterococci. *Antimicrob. Agents Chemother.* **47:**1423–1426.
67. **Vastola, A. P., J. Altschaefl, and S. Harford.** 1980. 5-Epi-sisomicin and 5-epi-gentamicin B: substrates for aminoglycoside-modifying enzymes that retain activity against aminoglycoside-resistant bacteria. *Antimicrob. Agents Chemother.* **17:**798–802.
68. **Vetting, M., S. L. Roderick, S. Hegde, S. Magnet, and J. S. Blanchard.** 2003. What can structure tell us about in vivo function? The case of aminoglycoside-resistance genes. *Biochem. Soc. Trans.* **31:**520–522.
69. **Vetting, M. W., S. S. Hegde, F. Javid-Magd, J. S. Blanchard, and S. L. Roderick.** 2002. Aminoglycoside 2′-*N*-acetyltransferase from *Mycobacterium tuberculosis* in complex with coenzyme A and aminoglycoside substrates. *Nat. Struct. Biol.* **9:**653–658.
70. **Vicens, Q., and E. Westhof.** 2001. Crystal structure of paromomycin docked into the eubacterial ribosomal decoding A site. *Structure* **9:**647–658.
71. **Vicens, Q., and E. Westhof.** 2002. Crystal structure of a complex between the aminoglycoside tobramycin and an oligonucleotide containing the ribosomal decoding A site. *Chem. Biol.* **9:**747–755.
72. **Vicens, Q., and E. Westhof.** 2003. Crystal structure of geneticin bound to a bacterial 16S ribosomal RNA A site oligonucleotide. *J. Mol. Biol.* **326:**1175–1188.
73. **Vicens, Q., and E. Westhof.** 2003. Molecular recognition of aminoglycoside antibiotics by ribosomal RNA and resistance enzymes: an analysis of x-ray crystal structures. *Biopolymers* **70:**42–57.
74. **Vogelman, B., and W. A. Craig.** 1986. Kinetics of antimicrobial activity. *J. Pediatr.* **108:**835–840.
75. **Waitz, J. A., G. H. Miller, E. Moss, and P. J. S. Chiu.** 1978. Chemotherapeutic evaluation of 5-episisomicin (Sch 22591), a new semisynthetic aminoglycoside. *Antimicrob. Agents Chemother.* **13:**41–48.
76. **Waksman, S. A., and H. A. Lechevalier.** 1949. Neomycin, a new antibiotic active against streptomycin-resistant bacteria, including tuberculosis organism. *Science* **109:**305–307.
77. **Walter, F., Q. Vicens, and E. Westhof.** 1999. Aminoglycoside-RNA interactions. *Curr. Opin. Chem. Biol.* **3:**694–704.
78. **White, D. G., and P. F. McDermott.** 2001. Emergence and transfer of antibacterial resistance. *J. Dairy Sci.* **84**(Suppl. E): E151–E155.
79. **Williams, J. W., and D. B. Northrop.** 1978. Kinetic mechanisms of gentamicin acetyltransferase I. Antibiotic-dependent shift from rapid to nonrapid equilibrium random mechanisms. *J. Biol. Chem.* **253:**5902–5907.
80. **Williams, J. W., and D. B. Northrop.** 1978. Substrate specificity and structure-activity relationships of gentamicin acetyltransferase I. The dependence of antibiotic resistance upon substrate Vmax/Km values. *J. Biol. Chem.* **253:**5908–5914.
81. **Williams, J. W., and D. B. Northrop.** 1979. Synthesis of a tight-binding, multisubstrate analog inhibitor of gentamicin acetyltransferase I. *J. Antibiot.* **32:**1147–1154.
82. **Wolf, E., A. Vassilev, Y. Makino, A. Sali, Y. Nakatani, and S. K. Burley.** 1998. Crystal structure of a GCN5-related *N*-acetyltransferase: *Serratia marcescens* aminoglycoside 3-*N*-acetyltransferase. *Cell* **94:**439–449.
83. **Wybenga-Groot, L. E., K. Draker, G. D. Wright, and A. M. Berghuis.** 1999. Crystal structure of an aminoglycoside 6′-*N*-acetyltransferase: defining the GCN5-related *N*-acetyltransferase superfamily fold. *Structure.* **7:**497–507.
84. **Yokoyama, K., Y. Doi, K. Yamane, H. Kurokawa, N. Shibata, K. Shibayama, T. Yagi, H. Kato, and Y. Arakawa.** 2003. Acquisition of 16S rRNA methylase gene in *Pseudomonas aeruginosa*. *Lancet* **362:**1888–1893.
85. **Yoshizawa, S.** 1998. Structural origins of gentamicin antibiotic action. *EMBO J.* **17:**6437–6448.

Enzyme-Mediated Resistance to Antibiotics: Mechanisms, Dissemination, and Prospects for Inhibition
Edited by Robert A. Bonomo and Marcelo E. Tolmasky

Gerard D. Wright
Albert M. Berghuis

3 Structural Aspects of Aminoglycoside-Modifying Enzymes

The aminoglycoside-aminocyclitol antibiotics (hereafter referred to as aminoglycosides) have been in constant clinical use since the discovery of the first members of this family in the mid-1940s (Table 3.1). They represent a broad group of important agents used for the treatment of infections caused by both gram-positive and gram-negative bacteria and also have found use in the treatment of certain protozoal infections (23). These antibiotics are highly water soluble and cationic, properties that impede their oral availability, and therefore for the most part, aminoglycoside therapy requires parenteral administration with the exception of topical agents such as ocular drops for the treatment of superficial infections of the eye (3).

The net positive charge of the aminoglycosides at neutral pH also has implications in the entry of aminoglycosides into bacterial cells, as they first bind indiscriminately to the negatively charged lipopolysaccharide outer membrane of gram-negative bacteria or the teichoic acid-containing cell wall ultrastructure of gram-positive bacteria (37, 38). The antibiotics then traverse the plasma membrane in an energy-dependent fashion that requires the $\Delta\psi$ component of the proton motive force (37, 38). The primary intracellular target of aminoglycoside antibiotics is the decoding aminoacyl-tRNA recognition site (A-site) of the small, 30S, subunit of the bacterial ribosome (49, 68). Binding to this region of the ribosome results in impairment of translation and specifically a reduced translational fidelity (13, 14). This results in the generation of aberrant proteins, which have been proposed to account for the pleiotropic effects associated with exposure to aminoglycosides including membrane damage and changes in transmembrane ion permeability (16, 17).

All aminoglycosides contain an aminocyclitol ring (carbon ring with appended amino and hydroxyl groups) and can be structurally characterized by the presence or absence of a 2-deoxystreptamine ring. Compounds that lack such a ring include streptomycin and spectinomycin. The 2-deoxystreptamine aminoglycosides can be further subdivided into those in which the 2-deoxystreptamine ring is linked to a hexose at position 4 and a pentose at position 5, such as paromomycin (Fig. 3.1), and those in which the 2-deoxystreptamine ring is linked to hexose rings at positions 4 and 6, such as tobramycin (Fig. 3.1). Biochemical studies including chemical footprinting map the interaction of aminoglycosides within the A-site to the highly conserved 16S rRNA (49, 68). The

Gerard D. Wright, Antimicrobial Research Centre, Department of Biochemistry, McMaster University, Hamilton, ON, Canada L8N 3Z5. **Albert M. Berghuis,** Departments of Biochemistry and Microbiology & Immunology, McGill University, Montreal, PQ, Canada H3A 2B4.

Table 3.1 Aminoglycoside antibiotics in clinical use

Aminoglycoside	Source
Tobramycin	*Streptomyces tenebrarius*
Amikacin	Semisynthetic derivative of kanamycin A
Gentamicin	*Micromonospora purpurea*
Netilmicin	1-*N*-ethyl derivative of sisomicin
Streptomycin	*Streptomyces griseus*
Neomycin	*Streptomyces fradiae*
Spectinomycin	*Streptomyces spectabilis*

2-deoxystreptamine aminoglycosides bind in particular to a region of duplex rRNA that includes a noncanonical base pair between A1408 and A1493 (Fig. 3.2). Aminoglycosides bind to this region of the ribosome with low micromolar dissociation constants (43, 44).

The Puglisi laboratory has used nuclear magnetic resonance (NMR) approaches for the dissection of the molecular details of the interaction of aminoglycosides with their target using an A-site-derived model RNA oligonucleotide. The results of this work revealed that aminoglycosides bind through a series of ionic and hydrogen bond interactions within the major groove of the RNA and in particular through rings I and II (the primed hexose ring and the 2-deoxystreptamine ring, respectively [Fig. 3.1]) (27, 72). Binding of the aminoglycosides results in a conformational change in the 16S rRNA that displaces A1493 (numbering refers to the *Escherichia coli* 16S rRNA) and an unpaired bulge nucleotide, A1492, to form the antibiotic binding site (28). These NMR experiments have been supplemented by the determination of the crystal structure of a similar A-site mimic (with two equivalent sites) and the antibiotics paromomycin and tobramycin (66). These atomic resolution studies demonstrate that these antibiotics, which differ in 2-deoxystreptamine substitution, nonetheless bind to the A-site RNA with rings I and II in similar conformations yet differ in the nature of specific contacts with the RNA; for example, direct contacts between amino or hydroxyl groups in one complex are mediated through water molecules in another. This observation may explain the highly similar magnitude of the dissociation constants among aminoglycosides of divergent structure with the ribosome. For example, the 3′-hydroxyl of gentamicin forms a direct interaction with a phosphate oxygen of A1492, and in the tobramycin structure, which lacks an equivalent 3′-hydroxyl, an intervening water molecule makes the contact with the backbone phosphate of A1493 (66).

These pioneering studies have been supported by X-ray crystallographic structural studies of paromomycin bound to the 30S subunit reported by Carter et al. (8). This structure has revealed that binding of paromomycin to the 30S ribosomal subunit occurs through a series of specific contacts between aminoglycoside hydroxyl and amino groups and 16S rRNA bases and phosphate backbone. Binding of paromomycin to the 16S rRNA results in a conformational change where A1493 and A1493 "flip out" to interact with the minor groove of the codon-anticodon duplex in the A-site. This change is identical to the change when the cognate codon-anticodon duplex is formed. This structure is therefore consistent with the well-established effects of mistranslation that are associated with aminoglycoside action (13–15). The binding of aminoglycoside therefore results in indiscriminate displacement of A1493 and A1493 with the effect of freezing the complex in a

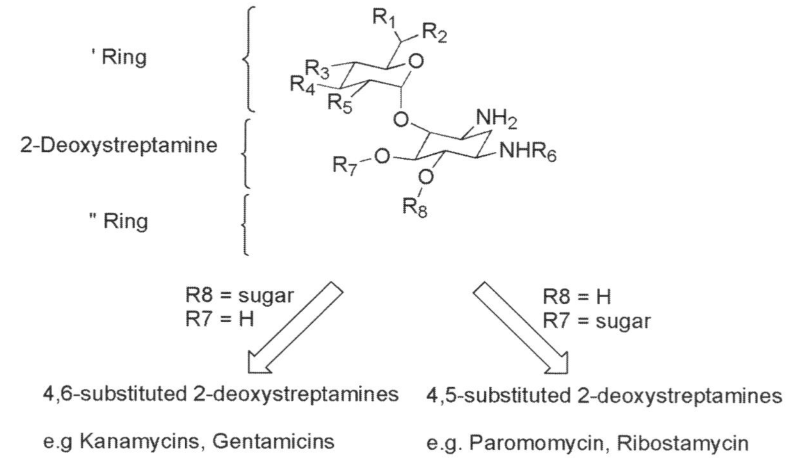

Figure 3.1 Structures and nomenclature of aminoglycoside antibiotics. For a more complete list, see reference 69.

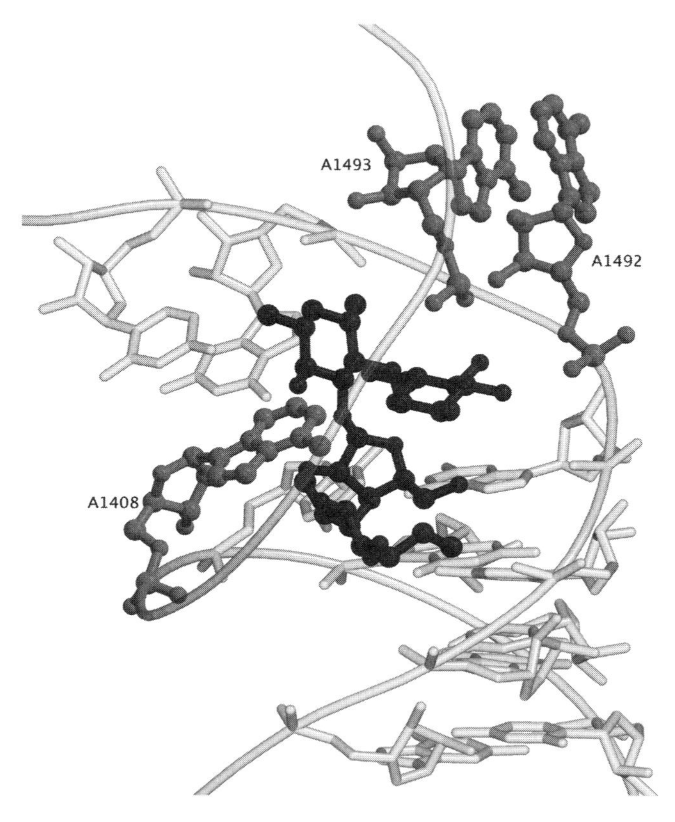

Figure 3.2 Interaction of paromomycin with the A-site 16S rRNA. Paromomycin is shown in black. Nucleotides A1408, A1492, and A1493 are shown in gray. Upon binding of paromomycin, nucleotides A1492 and A1493 are flipped out in translation-competent mode.

translation-competent mode even in the presence of noncognate codon-anticodon pairs, accounting for the decrease in translation fidelity.

Thus, structural studies in the past half-decade have served to decipher the molecular mechanism of aminoglycoside action and can now inform the rational synthesis of new antibiotics. Haddad et al., for example, have developed a series of semisynthetic aminoglycoside derivatives based on molecular modeling of the A-site aminoglycoside interaction, and these show in vitro and in vivo activity, some against aminoglycoside-resistant bacteria (36). In another study, the mirror image of D-neamine, L-neamine, was synthesized and shown to have comparable in vitro binding activity with an A-site mimic (58, 59). Interestingly, even though the unnatural L-isomer was a poorer inhibitor of translation and had fivefold weaker antibacterial activity, it was more effective than the natural D-isomer against aminoglycoside-resistant bacteria. These studies augur well for the leveraging of emerging molecular and atomic information of aminoglycoside action on the development of a new generation of aminoglycoside antibiotics.

OVERVIEW OF AMINOGLYCOSIDE RESISTANCE

Like most other antibiotics, resistance to aminoglycosides can manifest itself in several forms: (i) decreased uptake, (ii) increased efflux, (iii) modification of the target, and (iv) action of inactivating enzymes. Mechanisms of decreased uptake include mutants in the electron transport machinery (51) and various proteins such as the *Pseudomonas aeruginosa* outer membrane protein OprH (73) but are not common. Active efflux of aminoglycosides is gaining increased recognition especially in bacteria of the genera *Pseudomonas* (1) and *Burkholderia* (50). Mutation of the target 16S rRNA, e.g., at position A1408, can be selected for with exposure to a sublethal concentration of drug resulting in resistance (48, 56), but this does not have a significant clinical impact with the exception of ribosomal mutations associated with streptomycin resistance in *Mycobacterium tuberculosis* (25). Enzyme-mediated methylation of A1408 and/or G1405 in bacterial producers of several aminoglycosides can result in high levels of resistance (10, 41), and this mechanism has only very recently been observed in a clinical isolate (29).

By far the most prevalent and clinically relevant mechanisms are enzyme-catalyzed modification of aminoglycosides. Three structurally and functionally unrelated classes of inactivation enzymes are known: the aminoglycoside kinases (APH), the adenylyltransferases (ANT), and the acetyltransferases (AAC). The past decade has seen a dramatic increase in our understanding of the structure and function of members of each of these classes, and each is discussed separately in the sections below.

AMINOGLYCOSIDE-INACTIVATING ENZYMES

Kinases

The three-dimensional structures of two aminoglycoside kinases have been determined; the first was APH(3′)-IIIa, distributed in gram-positive cocci (42), and recently a second enzyme, APH(3′)-IIa, found in *Enterobacteriaceae*, has been reported (52). The first enzyme, APH(3′)-IIIa, has become the most extensively studied aminoglycoside kinase with crystallographic data available for the apoenzyme (7), the enzyme in complex with a nonhydrolyzable analogue of the substrate ATP (7) and the product ADP (42), and two ternary complex structures with the aminoglycoside substrates kanamycin and neomycin (26). Therefore, it is appropriate to describe structural aspects of this enzyme in some detail. The enzyme consists of two lobes, N terminal and C terminal (Fig. 3.3). The N-terminal lobe is composed of one five-stranded antiparallel β-sheet and incorporates a critical helix. The C-terminal lobe is mostly α-helical but also includes a long hairpin loop. The cleft formed between the two lobes is the location of the cofactor-binding pocket. The ATP substrate is located on top of the hairpin loop of the C-terminal lobe and is covered by the N-terminal β-sheet. One side of the pocket is blocked by the N-terminal helix, and the tethering segment that connects the N- and C-terminal lobes blocks a second side. As a consequence, ATP is mostly buried within the protein, leaving principally its phosphate moiety exposed. The location of the aminoglycoside binding site is in front of the exposed phosphate-binding moiety of ATP. Despite the close proximity to the N-terminal lobe, aminoglycosides form interactions only with the C-terminal lobe of the enzyme. Specifically, the C-terminal helix, including its carboxy terminus, and residues located on a flexible loop form most of the interactions with aminoglycoside substrates. This has been validated by a series of site-directed mutagenesis studies (63). Note that the structure of APH(3′)-IIa recently determined by Nurizzo et al. follows closely that of APH(3′)-IIIa with a root mean square difference of only 1.6 Å for 96% of the C-α positions (52).

The availability of crystallographic data for APH(3′)-IIIa in various stages along the reaction coordinate, in conjunction with extensive mechanistic analyses, allows for a detailed examination of the structural foundation for phosphate transfer catalyzed by this enzyme. Extensive biochemical analysis has revealed that APH(3′)-IIIa first binds ATP and two magnesium ions followed by the aminoglycoside substrate (47); phosphate transfer occurs likely through a dissociative-like mechanism, involving a metaphosphate-like transition state (5); this is immediately

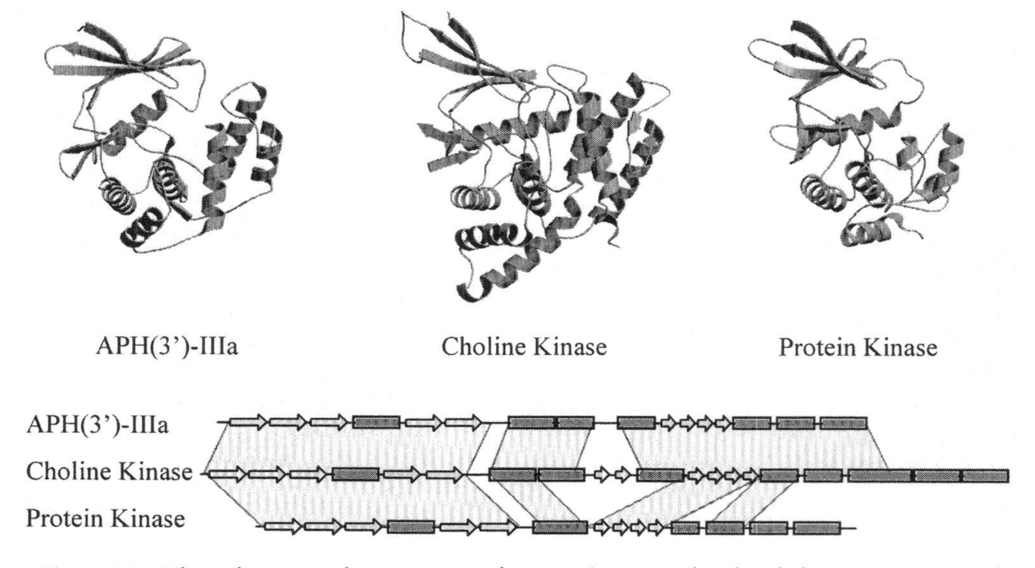

Figure 3.3 Three-dimensional structures of APH(3′)-IIIa and related kinases. Conserved secondary structure elements are shown in the bottom half of the figure.

followed by product release and finally the slow release of ADP and associated magnesium ions (46). The available crystal structures reveal the basis for the ordered reaction in that aminoglycoside binding would physically block the ATP-binding pocket, preventing nucleotide entry. Furthermore, the geometry of the substrate moieties in the protein matrix is fully consistent with a dissociative mechanism.

The various APH(3′)-IIIa crystal structures also reveal that certain regions of the protein are highly flexible and undergo movements during the reaction cycle, while surprisingly other parts of the enzyme are remarkably rigid. Specifically, upon ATP binding and ADP release a loop region located above the tri/diphosphate moiety undergoes movement. This same loop may also be involved in facilitating phosphate transfer as suggested by site-directed mutagenesis studies (62). Similarly, binding of aminoglycosides is also accompanied by movement: a flexible loop folds over the aminoglycoside substrate, thus making several stabilizing interactions, and in fact completing the binding pocket. It is intriguing to note that unlike what is observed in the structurally related protein kinases, no "domain movement" has been observed between the N- and C-terminal lobes in any of the crystal structures (despite the fact that various crystal forms were employed in these studies).

The structure determinations of APH(3′)-IIIa ternary complexes have provided insights into the structural basis for the effectiveness of this enzyme as an antibiotic resistance factor (26). APH(3′)-IIIa is capable of conferring resistance to both 4,5- and 4,6-disubstituted 2-deoxystreptamine aminocyclitol aminoglycosides and has one of the broadest substrate spectrums within the APH class of enzymes. The enzyme is able to bind various structurally diverse aminoglycosides by having the substrate-binding pocket being composed of three subsites (Fig. 3.4). All aminoglycosides detoxified by APH(3′)-IIIa incorporate a neamine moiety, which binds to the A-subsite within the aminoglycoside binding pocket. Various substitutions to the neamine core can either bind to the B-subsite or alternatively to the C-subsite, thus creating a highly promiscuous binding pocket with sufficient specificity to allow effective regiospecific phosphorylation of structurally diverse aminoglycosides.

Thanks to having structural data available for aminoglycosides binding to the bacterial ribosome, it is now also possible to examine how APH(3′)-IIIa vies for substrates. Once an aminoglycoside has reached the cytosol of a resistant bacterium, the drug can either bind to its intended target, the A-site of the 30S ribosomal subunit, or be intercepted by an aminoglycoside-modifying enzyme. While evolutionary selection of antibiotic-producing organisms has resulted in biosynthetic pathways that generate highly effective aminoglycosides, commensurate evolutionary pressures have selected for resistance factors that thwart this binding. Examination of aminoglycoside-macromolecular complexes indicates that aminoglycosides bind in their lowest energy conformation to both the ribosome and APH(3′)-IIIa, thus enhancing binding affinity by curtailing loss of entropy. Moreover, a nearly identical subset of hydrogen bond donor and acceptor groups present in aminoglycosides are used for binding to the ribosome and APH(3′)-IIIa, further explaining the broad substrate spectrum of APH(3′)-IIIa for clinically relevant aminoglycoside antibiotics. Thus, the APH(3′)-IIIa aminoglycoside-binding pocket mimics the ribosomal A-site in all important aspects, therefore efficiently competing for antibiotics. What may actually tip the balance in favor of APH(3′)-IIIa binding is the electrostatic funnel created by the abundance of acidic residues in the active site which attract the invariably positively charged aminoglycosides.

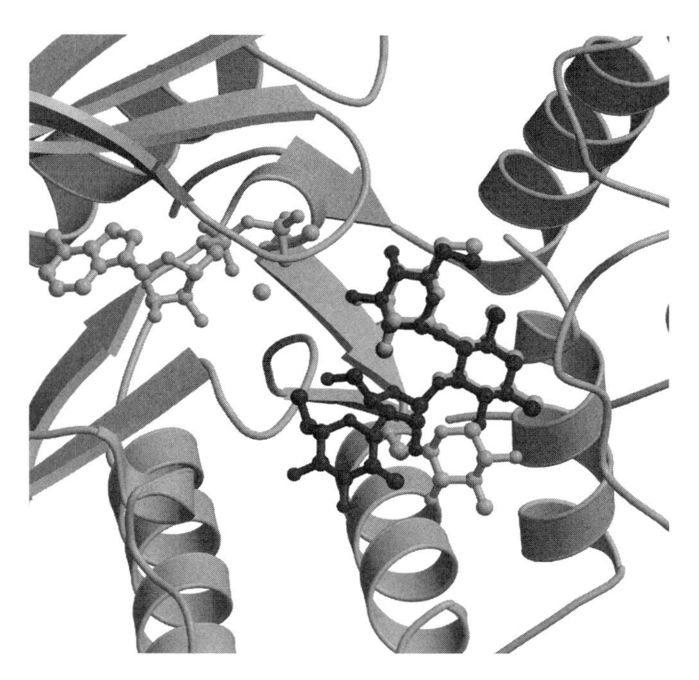

Figure 3.4 Aminoglycoside binding region of APH(3′)-IIIa. The 6-aminohexose (′ ring) and 2-deoxystreptamine rings of neomycin (black) and kanamycin (gray) bind in identical fashion while the double prime (″) rings of both substrates bind in different subsites (26).

In 1988 Wolfgang Piepersberg first suggested that aminoglycoside kinases and protein kinases are related, implying that they have a similar three-dimensional structure (55). Piepersberg based this hypothesis on the observation of similar patterns of conserved amino acid residues in the two enzyme families. However, this theory was mostly overlooked until the crystallographic structure determination of APH(3′)-IIIa, which revealed the extent of structural and functional similarity between these two enzyme families. It was subsequently demonstrated that APHs maintain vestigial protein kinases activity (11). In fact, since the structure determination of APH(3′)-IIIa, which represented the first distant relative of the protein kinases, this superfamily has further expanded and includes now lipid kinases (57) and choline (54) kinases.

The structural core, which is conserved among all currently known members of the protein kinase superfamily, consists of most of the N-terminal lobe of APH(3′) and two segments of the C-terminal lobe, namely the hairpin-like loop and two helices which are positioned below this loop. The residue conservation within the superfamily is minimal to almost nonexistent within this structural core. Only 5 of approximately 100 structurally conserved residues are chemically conserved, and these are predominantly involved in binding to the magnesium-phosphate moiety of ATP. Three additional residues are strongly but not absolutely conserved, and their role is likely in dictating the characteristic protein kinase core motif.

Beyond the core there are a variety of variations in fold within the protein kinase superfamily. For example, the crystal structure of the type II-β phosphatidylinositol phosphate kinase shows an expanded β-sheet in the N-terminal lobe (57). One of the most intriguing differences when comparing Ser/Thr and Tyr kinases with aminoglycoside kinases is the region of the enzymes involved in substrate binding. In protein kinases the segment known as the activation loop is principally involved in interacting with protein substrates; this segment also frequently contains Ser or Thr residues that can undergo autophosphorylation, thereby converting the enzyme from an inactive to an active form. In aminoglycoside kinases a segment that contains the flexible loop, which can form extensive interactions with aminoglycoside substrates, is principally involved in substrate interactions. While the two substrate-interacting segments are in three-dimensional space placed in the same location, in the primary sequence they are far apart. In fact, the activation loop is completely lacking in APH(3′)-IIIa, and vice versa there are no remnants of an aminoglycoside-binding loop segment in protein kinases.

Of the different protein kinase superfamily members, the member most closely related to aminoglycoside kinases is thus far choline kinase. The recent structure determination of the choline kinase from *Caenorhabditis elegans* (CKA-2) reveals that this enzyme also lacks the activation loop present in protein kinases, and it has a homologous segment to the aminoglycoside binding loop present in aminoglycoside kinases (54). Perhaps the most significant difference in fold between choline kinases and aminoglycoside kinases is in the C terminus. In APH(3′)-IIa and APH(3′)-IIIa, the carboxy terminus is intimately involved in interacting with the antibiotic. However, in choline kinase the C terminus is extended and contains two additional helices. This observation is relevant in that while nearly all of the APH(3′) enzymes have their C termini in the same location (based on sequence alignments), aminoglycoside kinases with different regiospecificities such as APH(2″) always have an extended C terminus when compared to APH(3′)-IIIa. Thus, the choline kinase structure may provide insights into how aminoglycoside kinases that, for instance, phosphorylate at the 2″ position may interact with their substrates.

Adenylyltransferases

Aminoglycoside adenylyltransferases (ANTs) catalyze the transfer of AMP to aminoglycoside hydroxyl groups. There are numerous enzymes known; however, only a few are of significant clinical relevance. One of these is ANT(2″)-Ia, which is widespread in gram-negative bacteria and confers resistance to gentamicin, tobramycin, and amikacin. This enzyme has been overexpressed in *E. coli* (24), and its kinetic mechanism has been established (30–32). Recent work by the group of Serpersu has used NMR spectroscopy to probe the conformations of aminoglycosides bound to the enzyme and established that these mirror the conformations of aminoglycosides bound to the target rRNA (24).

ANT(4′)-Ia from *Staphylococcus aureus* is thus far the only member of the ANT family whose structure has been determined. This enzyme is restricted to gram-positive bacteria and confers resistance to a number of aminoglycosides including kanamycin and gentamicin. The regiospecificity of adenylyl transfer has been confirmed by spectroscopic methods (33), and careful isotope effect experiments support a concerted mechanism of adenylyl transfer where bond formation between the 4′-hydroxyl group of kanamycin precedes bond breakage between the α and β phosphates of ATP (34).

The three-dimensional structures of the apoenzyme (60) and in complex with the nonhydrolyzable ATP analogue AMPCPP and kanamycin (53) have been reported. The enzyme is a head-to-tail dimer, consisting of two 2-domain monomers (Fig. 3.5). The N-terminal domain consists of a five-strand antiparallel β-sheet interspersed by five short helices. The C-terminal domain is composed of a bundle of five helices. The enzyme active site is formed at the dimer interface, and both subunits contribute to substrate binding (Fig. 3.6). The N-terminal domain contributes primarily to nucleotide binding, while the kanamycin binding site lies primarily in the C-terminal domain. Like the APHs, the aminoglycoside-binding site is lined with acidic residues

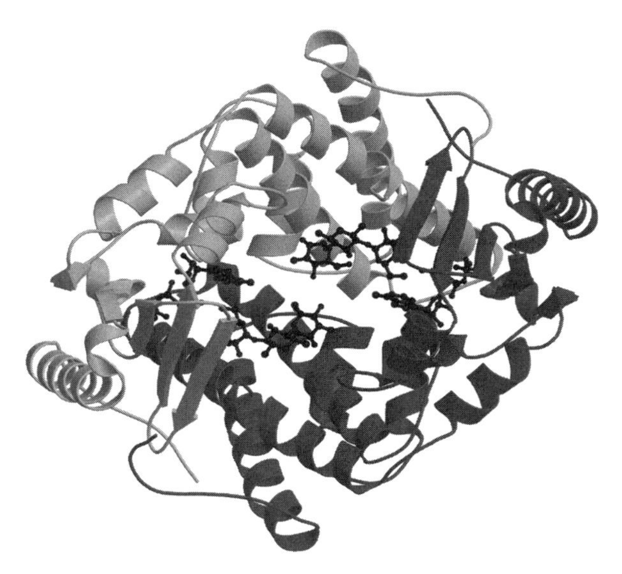

Figure 3.5 Domain structure of ANT(4′). Each monomer in the active site is shown in a different shade of gray. The substrates kanamycin and the nonhydrolyzable ATP analogue AMPCPP are shown in black in the active sites situated at the dimer interface.

for charge compensation with the cationic antibiotic. Both monomers contribute to substrate binding, and there are contacts between both substrates and both subunits. The positioning of the substrates in the active site suggests that Glu145 could act as a general base, deprotonating the aminoglycoside 4′-hydroxyl group.

The ANT enzymes are members of the recently described higher eukaryotes and prokaryotes nucleotide-binding (HEPN) domain family of proteins (35). HEPN proteins of diverse predicted function, including antibiotic resistance, are relatively common in bacteria and were recently identified as a domain in a large human protein, sacsin, which is associated with a hereditary neurodegenerative disease (35). The HEPN domain is found in the C terminus of ANT(4′)-Ia and includes the major contacts for aminoglycoside recognition. The N terminus comprises the nucleotide-binding domain. The fold of this domain is conserved among a number of proteins largely unrelated at the level primary amino acids sequence such as DNA polymerase β (61) and murine terminal deoxynucleotidyltransferase (18).

Acetyltransferases

Aminoglycoside acetyltransferases comprise the largest family of inactivating enzymes. These enzymes are widely distributed in all bacteria and modify aminoglycoside antibiotics at the 1, 3, 2′, and 6′ positions. Over the past few years, four structures of representative enzymes have been determined: AAC(3′)-Ia, widely distributed on plasmids in a number of bacteria (67), and three chromosomally encoded enzymes: AAC(6′)-Ii from *Enterococcus faecium*

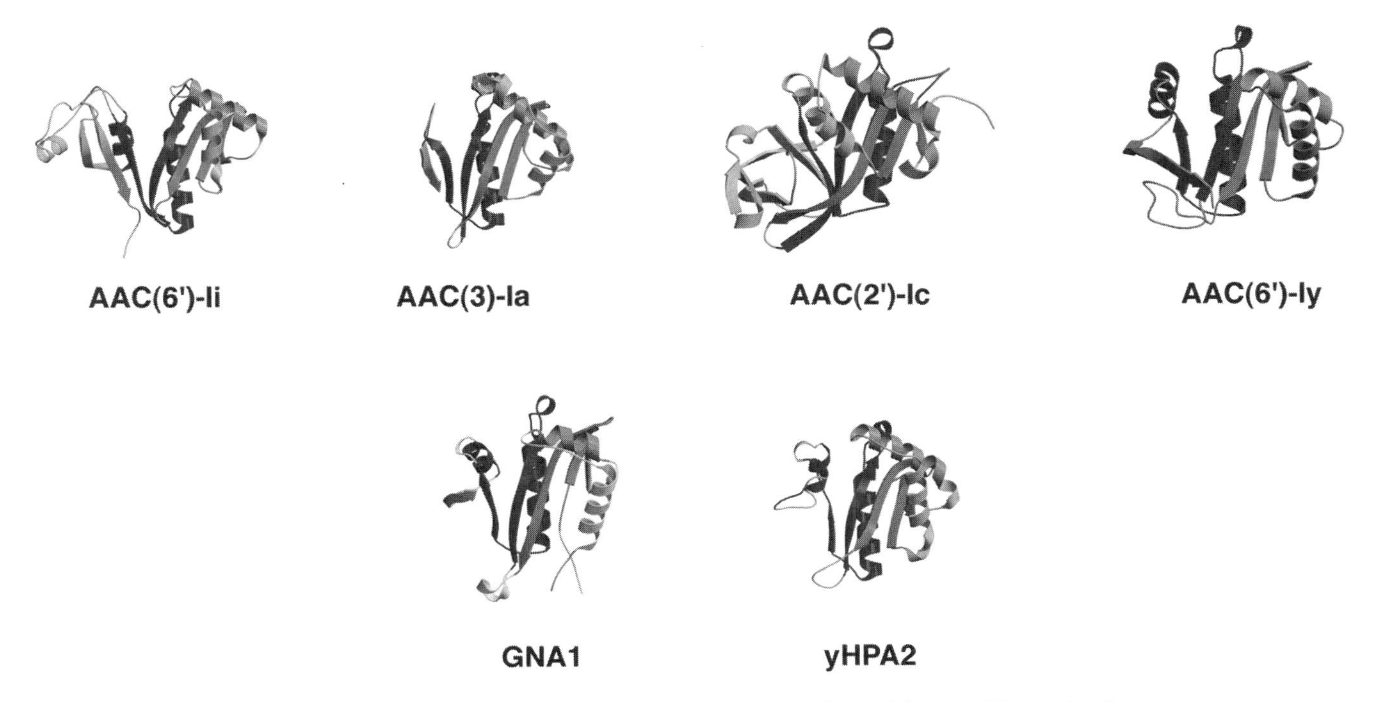

Figure 3.6 Aminoglycoside acetyltransferases are members of the GNAT superfamily.

(71), AAC(2′)-Ic, from *M. tuberculosis* (64), and AAC(6′)-Iy from *Salmonella enterica* (65) (Fig. 3.6). All of these enzymes belong to the larger GCN5 family of *N*-acyltransferases (GNATs) that share a core three-dimensional structure, but not amino acid sequence homology (reviewed in reference 22). Other members of the GNAT family include other small-molecule acyltransferases such as serotonin acetyltransferase and protein acetyltransferases such as the histone acetyltransferases.

Together with the yeast histone acetyltransferase Hat1 (21), AAC(3)-Ia was the first member of the GNAT superfamily whose three-dimensional structure was determined (67). AAC(3)-Ia used for the crystallographic studies was obtained from hospital strains of *Serratia marcescens*, where it is encoded on a multiresistance plasmid. The enzyme represents one of the smallest GNAT structures determined and also represents the only AAC structure whose function is presumably dedicated exclusively to detoxifying aminoglycoside antibiotics (based on its genetic origins). The core GNAT fold, which is present in all members of the GNAT superfamily, occupies approximately 80% of the AAC(3)-Ia structure. The GNAT fold consists of a five-stranded predominantly antiparallel β-sheet, which is flanked by two helices. The acetyl coenzyme A (acetyl-CoA) cofactor is wedged in between strands 4 and 5 and forms hydrogen bond interactions with strand 4, employing the pantothenic acid moiety in a manner reminiscent of an additional β-strand. In addition to the core GNAT fold, AAC(3)-Ia has a small C-terminal extension in the form of a sixth β-strand, which is located adjacent to strand 5. Furthermore, the segment located after the first helix and before the second β-strand, which is highly variable among GNAT superfamily members, is occupied by two small helices in AAC(3)-Ia.

Crystallographic data for AAC(3)-Ia in the CoA-bound state and for the ternary complex with bound spermidine and CoA are available (67). Unfortunately, no structures have yet been determined with a bound aminoglycoside. Analogous to APH and ANT enzymes, the AAC(3)-Ia enzyme has a prominent negatively charged patch that is positioned in the location where one would expect aminoglycoside substrates to bind. Furthermore, this is the location at which the positively charged spermidine binds, presumably mimicking aminoglycoside binding. The putative "gentamicin binding slot" is a narrow cleft, which suggests that aminoglycosides must be inserted into this slot for acetylation to take place. This manner of binding is different from that observed in both APH(3′)-IIIa and ANT(4′)-Ia, where the aminoglycoside binding pocket is more aptly described as a shallow bowl.

Based on the extent of interactions observed between neighboring AAC(3)-Ia molecules in the crystal structure, it can be inferred that the enzyme exists as a dimer under physiological conditions. This suggestion is supported by a comparison with known dimeric members of the GNAT superfamily, most of whom show an analogous manner of quaternary arrangement as seen for AAC(3)-Ia. However, biophysical studies have not yet confirmed this assertion, nor are there data available that shed light on a possible function for dimer formation in AAC(3)-Ia.

AAC(6′)-Ii is found in the human pathogen *E. faecium*, where it is chromosomally encoded (9). The enzyme is responsible for failure of synergistic combination therapy using β-lactams and aminoglycosides, by providing low-level resistance to a broad spectrum of 4,5- and 4,6-disubstituted 2-deoxystreptamine aminocyclitol aminoglycosides. However, several lines of evidence suggest that conferring aminoglycoside resistance is not the primary function of AAC(6′)-Ii and that its actual role within the bacteria might involve protein acetylation (70, 71). First, the steady-state kinetic parameter V_{max} is positively correlated with the MIC of antibiotic, which is suboptimal for a bona fide resistance enzyme (70). Second, AAC(6′)-Ii is, in fact, capable of acetylating proteins and peptides such as histones and poly-L-lysine (71).

The kinetic mechanism for AAC(6′)-Ii has been determined to be an ordered bi-bi ternary complex mechanism with acetyl-CoA binding first, followed by antibiotic; the subsequent step is release of acetylated aminoglycoside; and finally release of CoA occurs (20). In this reaction mechanism, aminoglycoside binding and product release dictate the rate of acetylation, not chemistry (i.e., diffusion control).

Crystallographic studies for AAC(6′)-Ii have been reported for the acetyl-CoA-bound state (6) and the CoA-bound state of the enzyme (71). These studies show that AAC(6′)-Ii also incorporates the core GNAT fold but includes a C-terminal extension, making it larger than AAC(3′)-Ia. Following the fifth β-strand of the core GNAT fold, AAC(6′)-Ii has two additional β-strands which are oriented in an antiparallel fashion. Additionally, the C-terminal extension incorporates two small α-helices and extensive loop structures that connect the β-strands. The variable region between the first helix and the second β-strand is occupied by one helix which does not correspond to either of the two helices present in this location in AAC(3)-Ia. Overall, the additions to the core GNAT fold make AAC(6′)-Ii resemble an embellished letter "V," in which the crevice between the two arms is the location of the cofactor binding site.

Unfortunately, no crystallographic data have yet become available for ternary complexes of AAC(6)-Ii with aminoglycoside substrates or inhibitors. However, based on geometric requirements enforced by the location of acetyl group from the cofactor, an approximate location for the aminoglycoside-binding site can be inferred. This prediction is strengthened by the observation that this putative binding site is extensively lined with acidic residues, providing a negatively charged patch, thus enabling favorably electrostatic interactions with the substrate. Similar to AAC(3)-Ia, the aminoglycoside-binding pocket resembles more a "cleft" than a shallow bowl, suggesting that aminoglycosides have to reach the cofactor edge on. This manner of substrate interaction would be consistent with a primary role for AAC(6′)-Ii in *N*-acetylation of a lysine-containing protein. NMR studies aimed at elucidating the conformation of various aminoglycosides when bound to AAC(6′)-Ii have been performed (19). These studies suggest a certain amount of plasticity in the active site in that two distinct bound conformations were observed for isepamicin, though only one conformation was detected for butirosin A. However, there might be alternative explanations for the observation of multiple bound conformations (e.g., nonproductive complex formation).

The lack of detailed structural information for an AAC(6′)-Ii ternary complex has hampered the examination of the molecular mechanism of acetyl transfer. Nonetheless, the fold of the enzyme, the manner in which the cofactor is bound, and the location of the putative aminoglycoside-binding pocket have provided a rational explanation for the ordered bi-bi reaction mechanism observed for AAC(6′)-Ii. It is evident from the structure that access to the acetyl-CoA binding site will be blocked by aminoglycoside binding, dictating that for catalysis to take place the cofactor must bind first, and after catalysis the acetylated aminoglycoside must leave first.

As is suggested for AAC(3)-Ia, AAC(2′)-Ic, and AAC(6′)-Iy (see below), AAC(6′)-Ii is a dimer under physiological conditions. Due to the particular crystal form of the AAC(6′)-Ii-acetyl-CoA crystals, the quaternary arrangement could initially not be unambiguously identified. The structure of the enzyme in complex with CoA was subsequently determined using an alternate crystal form. This crystallographic study enabled the definitive identification of the physiological dimer. The observed arrangement of the two protomers in the dimer, while having some features in common with that of the AAC(3)-Ia, is unique among dimeric GNAT superfamily members. This suggests that different oligomeric GNAT superfamily members must have employed dissimilar evolutionary pathways for dimer formation. The notion of parallel but dissimilar evolutionary pathways for dimer formations leaves unanswered the question of what the driving force is for oligomerization. Some insights into this question have recently been provided by a detailed kinetic analysis of AAC(6′)-Ii and a monomeric mutant of this enzyme which shows evidence of cooperativity between the two protomers in the physiological dimer (20). Thus, the reason for dimerization of AAC(6′)-Ii, and possibly other oligomeric GNAT superfamily members, may be that it allows for regulation of enzyme activity depending on substrate availability.

AAC(2′)-Ic is a chromosomally encoded aminoglycoside acetyltransferase found in *M. tuberculosis*. Orthologues of this gene are found in other mycobacteria, and it appears that this element is ubiquitous in this genus. The enzyme, when expressed in *E. coli*, confers resistance to a number of aminoglycoside antibiotics, but there is no evidence that it plays a similar role in *M. tuberculosis*, as aminoglycoside resistance in this organism is exclusively associated with mutation in rRNA and ribosomal proteins (2). Purification of the recombinant enzyme has confirmed the regiospecificity of acetyltransfer and explored the kinetic mechanism of the enzyme (40). The crystal structure of AAC(3′)-Ia has been reported to 1.8–1.5 Å and confirms the dimeric quaternary structure of the protein and placement of these enzymes in the GNAT superfamily, despite the lack of amino acid sequence homology with AAC(3), AAC(6′), or other members of the superfamily (Fig. 3.6). The structures of the apoenzyme and ternary complexes with CoA and kanamycin A, tobramycin, and ribostamycin were determined, demonstrating that each monomer in the dimer has distinct active sites (40). Binding of CoA in AAC(2′)-Ic parallels that seen in the other AAC structures and other GNAT enzymes, with this substrate sitting in a long groove and making contacts with a conserved β-sheet structure. The AAC(2′)-Ic structures as well share with the other AAC enzymes a concentration of negatively charged residues in the aminoglycoside-binding region. In the reported ternary structures of 4,6-deoxystreptamine antibiotics kanamycin A, the structurally related tobramycin, and the 4,5-deoxystreptamine aminoglycoside ribostamycin, there are few direct contacts between the substrate and the enzyme other than Asp152, which forms an ionic interaction with the amino group at position 1 of the deoxystreptamine. Instead, the bulk of the interactions occur through intermediary water molecules. This is in direct contrast with the ternary structures of APH and ANT enzymes described above. The structure also does not identify any candidate residues suitable to act as an active site base to promote collapse of the predicted tetrahedral intermediate. However, the C-terminal carboxylate or Glu82 could potentially act in this role remotely through intervening water molecules (40).

The structure of the aminoglycoside-binding region of the enzyme, the knowledge of the regiospecificity of acetyl transfer, and the prediction that aminoglycoside inactivation is not the primary physiological function of AAC(2′)-Ic have raised the possibility that the enzyme may be involved in the acylation of other more biologically relevant amino sugars. One candidate is in mycothiol biosynthesis (40). Mycothiol is the glutathione equivalent in mycobacteria, providing soluble reducing equivalents for general biological use. The relatively small size of the AAC(2′)-Ia active site, especially in comparison with that in the other chromosomal enzyme, AAC(6′)-Ii, lends support to this hypothesis, as the latter provides a larger groove, consistent with a prediction of protein acetylation.

AAC(6′)-Iy is encoded by a chromosomal gene in *S. enterica*. This organism is normally sensitive to aminoglycosides; however, in a clinical strain of *S. enterica*, aminoglycoside resistance was attributed to expression of the cryptic *aac(6′)-Iy* gene following a chromosomal rearrangement that placed the gene downstream from a constitutive promoter (45). The genomic environment of the wild-type *aac(6′)-Iy* gene suggests a role in carbohydrate metabolism (45); however, the gene product has been shown to acylate not only aminoglycosides (39) but also basic proteins such as histones (65). The X-ray crystal structures of AAC(6′)-Iy in the native CoA-bound and CoA-ribostamycin ternary forms have been determined to 2.4- to 2.0-Å resolution (65). The acetyl-CoA and aminoglycoside substrates bind to the enzyme in a random fashion (39) and occupy binding regions on the protein similar to those of the substrates in other AAC structures. Ribostamycin binds to the enzyme through a network of interactions including a likely ionic interaction between N3 of the 2-deoxystreptamine ring and Glu79. The 2-amino group and the 5″OH of the ribose are then able to interact in an intramolecular fashion, positioning the 6′-amino group in line with the thioester of acetyl-CoA, thus facilitating regiospecific acyl transfer. In contrast, an intramolecular interaction between the same 5″OH and the 6′-amino group in AAC(2′)-Ic positions the 2′-amino group for acylation. Thus, slight differences in active-site geometry and binding interactions greatly influence acetyl transfer reactions.

CONCLUSIONS

The availability of three-dimensional structures for all classes of aminoglycoside-inactivating enzymes and, more recently, of additional examples from the aminoglycoside kinase and acetyltransferase group, has transformed the field of aminoglycoside resistance in the past 10 years. We are now in a position to leverage this information to better understand the evolution and molecular basis of resistance. For example, the knowledge of the similarity between aminoglycoside kinases and protein kinases has been used to identify inhibitors of APHs from collections of known protein kinase inhibitors (12). More recently, the observation that while aminoglycoside-inactivating enzymes have no sequence homology among the three different classes, they do share highly anionic aminoglycoside-binding regions, has been exploited with the identification of the first broad-spectrum inhibitors of aminoglycoside-inactivating enzymes (4). In this study, cationic peptides were screened as potential inhibitors of AAC and APH enzymes. Derivatives of the antibacterial cationic peptide indolicidin were found to inhibit several aminoglycoside-inactivating enzymes, and this interaction was further investigated in structure-activity relationship experiments. In principle, inhibitors of aminoglycoside resistance enzymes could find use as leads in therapeutic studies to block the activity of aminoglycoside resistance in the clinic. These inhibitor studies demonstrate the importance of structure-guided investigation of resistance enzyme properties, as knowledge of structure enables creative research in resistance biochemistry. Furthermore, the realization following the determination of enzyme three-dimensional structure that aminoglycoside resistance enzymes are members of larger families of proteins that share similar structures and mechanisms has permitted insight into the origins and evolution of resistance. This information will arm us in the continuing efforts to meet the challenge of resistance at the molecular level and apply this work to the management of infectious disease.

We thank Desiré Fong for help in preparation of the figures. Research in the Wright and Berghuis laboratories in aminoglycoside resistance mechanisms has been funded by the Canadian Institutes for Health Research. G.D.W. is supported by a Canada Research Chair in Antibiotic Biochemistry, and A.M.B. is the recipient of a CIHR/Rx&D-HRF Research Career Award in the Health Sciences and is supported by a Canada Research Chair in Structural Biology.

References

1. **Aires, J. R., T. Kohler, H. Nikaido, and P. Plesiat.** 1999. Involvement of an active efflux system in the natural resistance of *Pseudomonas aeruginosa* to aminoglycosides. *Antimicrob. Agents Chemother.* **43:**2624–2628.
2. **Basso, L. A., and J. S. Blanchard.** 1998. Resistance to antitubercular drugs. *Adv. Exp. Med. Biol.* **456:**115–144.
3. **Boehr, D. D., K.-A. Draker, and G. D. Wright.** 2003. Aminoglycosides and aminocyclitols, p. 155-184. *In* R. G. Finch, D. Greenwood, S. R. Norrby, and R. J. Whitley (ed.), *Antibiotics and Chemotherapy.* Churchill Livingston, Edinburgh, Scotland.
4. **Boehr, D. D., K. Draker, K. Koteva, M. Bains, R. E. Hancock, and G. D. Wright.** 2003. Broad-spectrum peptide inhibitors of aminoglycoside antibiotic resistance enzymes. *Chem. Biol.* **10:**189–196.
5. **Boehr, D. D., P. R. Thompson, and G. D. Wright.** 2001. Molecular mechanism of aminoglycoside antibiotic kinase

APH(3′)-IIIa: roles of conserved active site residues. *J. Biol. Chem.* **276**:23929–23936.

6. **Burk, D. L., N. Ghuman, L. E. Wybenga-Groot, and A. M. Berghuis.** 2003. X-ray structure of the AAC(6′)-Ii antibiotic resistance enzyme at 1.8 Å resolution; examination of oligomeric arrangements in GNAT superfamily members. *Protein Sci.* **12**:426–437.
7. **Burk, D. L., W. C. Hon, A. K. Leung, and A. M. Berghuis.** 2001. Structural analyses of nucleotide binding to an aminoglycoside phosphotransferase. *Biochemistry* **40**: 8756–8764.
8. **Carter, A. P., W. M. Clemons, D. E. Brodersen, R. J. Morgan-Warren, B. T. Wimberly, and V. Ramakrishnan.** 2000. Functional insights from the structure of the 30S ribosomal subunit and its interactions with antibiotics. *Nature* **407**: 340–348.
9. **Costa, Y., M. Galimand, R. Leclercq, J. Duval, and P. Courvalin.** 1993. Characterization of the chromosomal *aac(6′)-Ii* gene specific for *Enterococcus faecium*. *Antimicrob. Agents Chemother.* **37**:1896–1903.
10. **Cundliffe, E.** 1987. On the nature of antibiotic binding sites in ribosomes. *Biochimie* **69**:863–869.
11. **Daigle, D. M., G. A. McKay, P. R. Thompson, and G. D. Wright.** 1998. Aminoglycoside phosphotransferases required for antibiotic resistance are also serine protein kinases. *Chem. Biol.* **6**:11–18.
12. **Daigle, D. M., G. A. McKay, and G. D. Wright.** 1997. Inhibition of aminoglycoside antibiotic resistance enzymes by protein kinase inhibitors. *J. Biol. Chem.* **272**: 24755–24758.
13. **Davies, J., and B. D. Davis.** 1968. Misreading of ribonucleic acid code words induced by aminoglycoside antibiotics. The effect of drug concentration. *J. Biol. Chem.* **243**:3312–3316.
14. **Davies, J., L. Gorini, and B. D. Davis.** 1965. Misreading of RNA codewords induced by aminoglycoside antibiotics. *Mol. Pharmacol.* **1**:93–106.
15. **Davies, J., D. S. Jones, and H. G. Khorana.** 1966. A further study of misreading of codons induced by streptomycin and neomycin using ribopolynucleotides containing two nucleotides in alternating sequence as templates. *J. Mol. Biol.* **18**:48–57.
16. **Davis, B. D.** 1987. Mechanism of action of aminoglycosides. *Microbiol. Rev.* **51**:341–350.
17. **Davis, B. D., L. L. Chen, and P. C. Tai.** 1986. Misread protein creates membrane channels: an essential step in the bactericidal action of aminoglycosides. *Proc. Natl. Acad. Sci. USA* **83**:6164–6168.
18. **Delarue, M., J. B. Boule, J. Lescar, N. Expert-Bezancon, N. Jourdan, N. Sukumar, F. Rougeon, and C. Papanicolaou.** 2002. Crystal structures of a template-independent DNA polymerase: murine terminal deoxynucleotidyltransferase. *EMBO J.* **21**:427–439.
19. **DiGiammarino, E. L., K. A. Draker, G. D. Wright, and E. H. Serpesu.** 1997. Solution studies of isepamicin and conformational comparisons between isepamicin and butirosin A when bound to an aminoglycoside 6-*N*-acetyltransferase determined by NMR spectroscopy. *Biochemistry* **37**: 3638–3644.
20. **Draker, K. A., D. B. Northrop, and G. D. Wright.** 2003. Kinetic mechanism of the GCN5-related chromosomal aminoglycoside acetyltransferase AAC(6′)-Ii from *Enterococcus faecium*: evidence of dimer subunit cooperativity. *Biochemistry* **42**:6565–6574.
21. **Dutnall, R. N., S. T. Tafrov, R. Sternglanz, and V. Ramakrishnan.** 1998. Structure of the histone acetyltransferase Hat1: a paradigm for the GCN5-related *N*-acetyltransferase superfamily. *Cell* **94**:427–438.
22. **Dyda, F., D. C. Klein, and A. B. Hickman.** 2000. GCN5-related N-acetyltransferases: a structural overview. *Annu. Rev. Biophys. Biomol. Struct.* **29**:81–103.
23. **Edson, R. S., and C. L. Terrell.** 1999. The aminoglycosides. *Mayo Clin. Proc.* **74**:519–528.
24. **Ekman, D. R., E. L. DiGiammarino, E. Wright, E. D. Witter, and E. H. Serpersu.** 2001. Cloning, overexpression, and purification of aminoglycoside antibiotic nucleotidyltransferase (2″)-Ia: conformational studies with bound substrates. *Biochemistry* **40**:7017–7024.
25. **Finken, M., P. Kirschner, A. Meier, A. Wrede, and E. C. Bottger.** 1993. Molecular basis of streptomycin resistance in *Mycobacterium tuberculosis*: alterations of the ribosomal protein S12 gene and point mutations within a functional 16S ribosomal RNA pseudoknot. *Mol. Microbiol.* **9**:1239–1246.
26. **Fong, D. H., and A. M. Berghuis.** 2002. Substrate promiscuity of an aminoglycoside antibiotic resistance enzyme via target mimicry. *EMBO J.* **21**:2323–2331.
27. **Fourmy, D., M. I. Recht, S. C. Blanchard, and J. D. Puglisi.** 1996. Structure of the A site of *Escherichia coli* 16S ribosomal RNA complexed with an aminoglycoside antibiotic. *Science* **274**:1367–1371.
28. **Fourmy, D., S. Yoshizawa, and J. D. Puglisi.** 1998. Paromomycin binding induces a local conformational change in the A-site of 16 S rRNA. *J. Mol. Biol.* **277**:333–345.
29. **Galimand, M., P. Courvalin, and T. Lambert.** 2003. Plasmid-mediated high-level resistance to aminoglycosides in *Enterobacteriaceae* due to 16S rRNA methylation. *Antimicrob. Agents Chemother.* **47**:2565–2571.
30. **Gates, C. A., and D. B. Northrop.** 1988. Alternative substrate and inhibition kinetics of aminoglycoside nucleotidyltransferase 2″-I in support of a Theorell-Chance kinetic mechanism. *Biochemistry* **27**:3826–3833.
31. **Gates, C. A., and D. B. Northrop.** 1988. Determination of the rate-limiting segment of aminoglycoside nucleotidyltransferase 2″-I by pH- and viscosity-dependent kinetics. *Biochemistry* **27**:3834–3842.
32. **Gates, C. A., and D. B. Northrop.** 1988. Substrate specificities and structure-activity relationships for the nucleotidylation of antibiotics catalyzed by aminoglycoside nucleotidyltransferase 2″-I. *Biochemistry* **27**:3820–3825.
33. **Gerratana, B., W. W. Cleland, and L. A. Reinhardt.** 2001. Regiospecificity assignment for the reaction of kanamycin nucleotidyltransferase from Staphylococcus aureus. *Biochemistry* **40**:2964–2971.
34. **Gerratana, B., P. A. Frey, and W. W. Cleland.** 2001. Characterization of the transition-state structure of the reaction of kanamycin nucleotidyltransferase by heavy-atom kinetic isotope effects. *Biochemistry* **40**:2972–2977.
35. **Grynberg, M., H. Erlandsen, and A. Godzik.** 2003. HEPN: a common domain in bacterial drug resistance and human neurodegenerative proteins. *Trends Biochem. Sci.* **28**:224–226.

36. **Haddad, J., L. P. Kotra, B. Llano-Sotelo, C. Kim, E. F. Azucena, Jr., M. Liu, S. B. Vakulenko, C. S. Chow, and S. Mobashery.** 2002. Design of novel antibiotics that bind to the ribosomal acyltransfer site. *J. Am. Chem. Soc.* **124:** 3229–3237.
37. **Hancock, R. E.** 1981. Aminoglycoside uptake and mode of action—with special reference to streptomycin and gentamicin. I. Antagonists and mutants. *J. Antimicrob. Chemother.* **8:**249–276.
38. **Hancock, R. E.** 1981. Aminoglycoside uptake and mode of action—with special reference to streptomycin and gentamicin. II. Effects of aminoglycosides on cells. *J. Antimicrob. Chemother.* **8:**429–445.
39. **Hegde, S. S., T. K. Dam, C. F. Brewer, and J. S. Blanchard.** 2002. Thermodynamics of aminoglycoside and acyl-coenzyme A binding to the *Salmonella enterica* AAC(6′)-Iy aminoglycoside N-acetyltransferase. *Biochemistry* **41:** 7519–7527.
40. **Hegde, S. S., F. Javid-Majd, and J. S. Blanchard.** 2001. Overexpression and mechanistic analysis of chromosomally encoded aminoglycoside 2′-N-acetyltransferase (AAC(2′)-Ic) from *Mycobacterium tuberculosis*. *J. Biol. Chem.* **276:** 45876–45881.
41. **Holmes, D. J., and E. Cundliffe.** 1991. Analysis of a ribosomal RNA methylase gene from *Streptomyces tenebrarius* which confers resistance to gentamicin. *Mol. Gen. Genet.* **229:**229–237.
42. **Hon, W. C., G. A. McKay, P. R. Thompson, R. M. Sweet, D. S. C. Yang, G. D. Wright, and A. M. Berghuis.** 1997. Structure of an enzyme required for aminoglycoside resistance reveals homology to eukariotic protein kinases. *Cell* **89:** 887–895.
43. **Le Goffic, F., M. L. Capmau, F. Tangy, and M. Baillarge.** 1979. Mechanism of action of aminoglycoside antibiotics. Binding studies of tobramycin and its 6′-N-acetyl derivative to the bacterial ribosome and its subunits. *Eur. J. Biochem.* **102:**73–81.
44. **Llano-Sotelo, B., E. F. Azucena, Jr., L. P. Kotra, S. Mobashery, and C. S. Chow.** 2002. Aminoglycosides modified by resistance enzymes display diminished binding to the bacterial ribosomal aminoacyl-tRNA site. *Chem. Biol.* **9:**455–463.
45. **Magnet, S., P. Courvalin, and T. Lambert.** 1999. Activation of the cryptic *aac(6′)-Iy* aminoglycoside resistance gene of *Salmonella* by a chromosomal deletion generating a transcriptional fusion. *J. Bacteriol.* **181:**6650–6655.
46. **McKay, G. A., and G. D. Wright.** 1996. Catalytic mechanism of enterococcal kanamycin kinase (APH(3′)-IIIa): viscosity, thio, and solvent isotope effects support a Theorell-Chance mechanism. *Biochemistry* **35:**8680–8685.
47. **McKay, G. A., and G. D. Wright.** 1995. Kinetic mechanism of aminoglycoside phosphotransferase type IIIa: evidence for a Theorell-Chance mechanism. *J. Biol. Chem.* **270:** 24686–24692.
48. **Melancon, P., C. Lemieux, and L. Brakier-Gingras.** 1988. A mutation in the 530 loop of Escherichia coli 16S ribosomal RNA causes resistance to streptomycin. *Nucleic Acids Res.* **16:**9631–9639.
49. **Moazed, D., and H. F. Noller.** 1987. Interaction of antibiotics with functional sites in 16S ribosomal RNA. *Nature* **27:** 389–394.
50. **Moore, R. A., D. DeShazer, S. Reckseidler, A. Weissman, and D. E. Woods.** 1999. Efflux-mediated aminoglycoside and macrolide resistance in *Burkholderia pseudomallei*. *Antimicrob. Agents Chemother.* **43:**465–470.
51. **Muir, M. E., D. R. Hanwell, and B. J. Wallace.** 1981. Characterization of a respiratory mutant of *Escherichia coli* with reduced uptake of aminoglycoside antibiotics. *Biochim. Biophys. Acta* **638:**234–241.
52. **Nurizzo, D., S. C. Shewry, M. H. Perlin, S. A. Brown, J. N. Dholakia, R. L. Fuchs, T. Deva, E. N. Baker, and C. A. Smith.** 2003. The crystal structure of aminoglycoside-3′-phosphotransferase-IIa, an enzyme responsible for antibiotic resistance. *J. Mol. Biol.* **327:**491–506.
53. **Pedersen, L. C., M. M. Benning, and H. M. Holden.** 1995. Structural investigation of the antibiotic and ATP-binding sites in kanamycin nucleotidyltransferase. *Biochemistry* **34:**13305–13311.
54. **Peisach, D., P. Gee, C. Kent, and Z. Xu.** 2003. The crystal structure of choline kinase reveals a eukaryotic protein kinase fold. *Structure* **11:**703–713.
55. **Piepersberg, W., J. Distler, P. Heinzel, and J.-A. Perez-Gonzalez.** 1988. Antibiotic resistance by modification: many resistance genes could be derived from cellular control genes in actinomycetes. A hypothesis. *Actinomycetology* **2:**83–98.
56. **Prammananan, T., P. Sander, B. A. Brown, K. Frischkorn, G. O. Onyi, Y. Zhang, E. C. Bottger, and R. J. Wallace, Jr.** 1998. A single 16S ribosomal RNA substitution is responsible for resistance to amikacin and other 2-deoxystreptamine aminoglycosides in *Mycobacterium abscessus* and *Mycobacterium chelonae*. *J. Infect. Dis.* **177:**1573–1581.
57. **Rao, V. D., S. Misra, I. V. Boronenkov, R. A. Anderson, and J. H. Hurley.** 1998. Structure of type IIbeta phosphatidylinositol phosphate kinase: a protein kinase fold flattened for interfacial phosphorylation. *Cell* **94:**829–839.
58. **Ryu, D. H., A. Litovchick, and R. R. Rando.** 2002. Stereospecificity of aminoglycoside-ribosomal interactions. *Biochemistry* **41:**10499–10509.
59. **Ryu, D. H., C. H. Tan, and R. R. Rando.** 2003. Synthesis of (+),(-)-neamine and their positional isomers as potential antibiotics. *Bioorg. Med. Chem. Lett.* **13:**901–903.
60. **Sakon, J., H. H. Liao, A. M. Kanikula, M. M. Benning, I. Rayment, and H. M. Holden.** 1993. Molecular structure of kanamycin nucleotidyl transferase determined to 3 Å resolution. *Biochemistry* **32:**11977–11984.
61. **Sawaya, M. R., H. Pelletier, A. Kumar, S. H. Wilson, and J. Kraut.** 1994. Crystal structure of rat DNA polymerase β: evidence for a common polymerase mechanism. *Science* **264:**1930–1935.
62. **Thompson, P. R., D. D. Boehr, A. M. Berghuis, and G. D. Wright.** 2002. Mechanism of aminoglycoside antibiotic kinase APH(3′)-IIIa: role of the nucleotide positioning loop. *Biochemistry* **41:**7001–7007.
63. **Thompson, P. R., J. Schwartzenhauer, D. W. Hughes, A. M. Berghuis, and G. D. Wright.** 1999. The COOH terminus of aminoglycoside phosphotransferase (3′)-IIIa is critical for antibiotic recognition and resistance. *J. Biol. Chem.* **274:**30697–30706.
64. **Vetting, M. W., S. S. Hegde, F. Javid-Majd, J. S. Blanchard, and S. L. Roderick.** 2002. Aminoglycoside 2′-N-acetyltransferase from *Mycobacterium tuberculosis* in complex with

coenzyme A and aminoglycoside substrates. *Nat. Struct. Biol.* **9**:653–658.

65. **Vetting, M. W., S. Magnet, E. Nieves, S. L. Roderick, and J. S. Blanchard.** 2004. A bacterial acetyltransferase capable of regioselective *N*-acetylation of antibiotics and histones. *Chem. Biol.* **11**:565–573.

66. **Vicens, Q., and E. Westhof.** 2002. Crystal structure of a complex between the aminoglycoside tobramycin and an oligonucleotide containing the ribosomal decoding a site. *Chem. Biol.* **9**:747–755.

67. **Wolf, E., A. Vassilev, Y. Makino, A. Sali, Y. Nakatani, and S. K. Burley.** 1998. Crystal structure of a GCN5-related N-acetyltransferase: *Serratia marcescens* aminoglycoside 3-*N*-acetyltransferase. *Cell* **94**:439–449.

68. **Woodcock, J., D. Moazed, M. Cannon, J. Davies, and H. F. Noller.** 1991. Interaction of antibiotics with A- and P-site-specific bases in 16S ribosomal RNA. *EMBO J.* **10**: 3099–3103.

69. **Wright, G. D., A. M. Berghuis, and S. Mobashery.** 1998. Aminoglycoside antibiotics: structures, function and resistance, p. 27-69. *In* B. P. Rosen and S. Mobashery (ed.), *Resolving the Antibiotic Paradox: Progress in Drug Design and Resistance*. Plenum Press, New York, N.Y.

70. **Wright, G. D., and P. Ladak.** 1997. Overexpression and characterization of the chromosomal aminoglycoside 6′-*N*-acetyltransferase from *Enterococcus faecium*. *Antimicrob. Agents Chemother.* **41**:956–960.

71. **Wybenga-Groot, L., K. A. Draker, G. D. Wright, and A. M. Berghuis.** 1999. Crystal structure of an aminoglycoside 6′-N-acetyltransferase: defining the GCN5-related *N*-acetyltransferase superfamily fold. *Structure* **7**:497–507.

72. **Yoshizawa, S., D. Fourmy, and J. D. Puglisi.** 1998. Structural origins of gentamicin antibiotic action. *EMBO J.* **17**:6437–6448.

73. **Young, M. L., M. Bains, A. Bell, and R. E. Hancock.** 1992. Role of *Pseudomonas aeruginosa* outer membrane protein OprH in polymyxin and gentamicin resistance: isolation of an OprH-deficient mutant by gene replacement techniques. *Antimicrob. Agents Chemother.* **36**:2566–2568.

Enzyme-Mediated Resistance to Antibiotics: Mechanisms, Dissemination, and Prospects for Inhibition
Edited by Robert A. Bonomo and Marcelo E. Tolmasky

Marcelo E. Tolmasky

4

Aminoglycoside-Modifying Enzymes: Characteristics, Localization, and Dissemination

Aminoglycoside antibiotics are broad-spectrum bactericidal agents that have predictable pharmacokinetics and are used mainly in the treatment of infections caused by gram-negative aerobic bacilli and some gram positives (139). In the latter case they are often used in combination with other antibiotics such as β-lactams or vancomycin, with which they exert a synergistic effect (104, 139). Besides their common utilization to treat infections caused by *Enterobacteriaceae*, *Pseudomonas aeruginosa*, and *Acinetobacter* spp., they have been used to treat tularemia, brucellosis, and tuberculosis and in combination with other antibiotics to treat infections caused by streptococci or enterococci (139). In particular, some aminoglycosides, like amikacin or kanamycin, are used as second-line drugs in the treatment of resistant *Mycobacterium* infections (139). Aminoglycoside antibiotics are commonly administered by intramuscular injection but also intravenously in cases of severe infections. Orally administered aminoglycosides are very poorly absorbed (about 1%), which prevents the utilization of this route of administration in the treatment of systemic infections. However, orally administered aminoglycosides are sometimes used to kill bowel flora before intestinal surgery. Aminoglycoside antibiotics are not metabolized; they are excreted as active compounds, and their half-life in the body is 2 to 3 h as long as the renal function is normal (84). A recent nontraditional application of aminoglycosides is their administration to attempt restoration of dystrophin expression in striated muscles of patients with Duchenne muscular dystrophy caused by a mutation that results in a premature stop codon. Information about the efficacy of this experimental treatment is still contradictory (36, 54, 90). Aminoglycoside antibiotics exist in a variety of formulations including encapsulation in liposomes (31). Their chemical structure consists of an aminocyclitol nucleus, streptidine or 2-deoxystreptamine, and two or more amino sugars linked to the nucleus by glycosidic bonds. Despite their usefulness for treatment of infections, the utilization of these antibiotics is not free of adverse effects. Aminoglycosides have been linked to drug-induced nephrotoxicity, which is usually reversible, and ototoxicity, which in contrast to nephrotoxicity, is irreversible (55, 74, 130).

Streptomycin, the first aminoglycoside, was isolated from *Streptomyces* spp. and introduced in 1944. Its value was quickly established, and other similar compounds were isolated from the same source or from *Micromonospora* spp. To identify the source from which natural aminoglycosides were isolated, they were named with either

Marcelo E. Tolmasky, Department of Biological Science, College of Natural Sciences and Mathematics, California State University—Fullerton, Fullerton, CA 92834-6850.

the suffix "-micin" (*Micromonospora*) or "-mycin" (*Streptomyces*) (139). Their utilization for the treatment of infections was quickly followed by a rise in the number of resistant strains. Companies responded by developing new semisynthetic aminoglycosides (such as amikacin, arbekacin, and netilmicin), which were generated by chemical modification of natural molecules like kanamycin or sisomicin. At first these semisynthetic derivatives were active against most resistant isolates (46, 58, 71). However, new strains resistant to semisynthetic antibiotics were isolated not long after they were introduced.

To exert their biological activity, aminoglycosides, which are RNA binding antibiotics, target the aminoacyl tRNA acceptor site (A-site) of the 16S rRNA that is part of the 30S subunit of the ribosome. This interaction leads to errors in translation, and ultimately bacterial cell death (39, 40). Other metabolic perturbations, probably secondary, include disturbances in DNA and RNA synthesis as well as altered membrane composition, permeability, and cellular ionic concentrations (28). Information about structural aspects of complexes between aminoglycosides and rRNA or oligonucleotides has increased very quickly in recent years, and the physicochemical basis of the interactions between the antibiotic and the RNA molecules is becoming clear (10, 18, 39–41, 60, 65, 126, 127, 131, 141). It was recently determined that the 2-deoxystreptamine moiety makes sequence-specific contact with the major groove of the A-site RNA (66, 140).

Aminoglycosides are thought to follow a three-step process to reach the cytoplasm (110, 121). Initially, the aminoglycoside molecules, which are positively charged, bind negatively charged components of the cell wall, displacing cation bridges between adjacent lipopolysaccharide molecules. This results in an increase in permeability that is followed by the process known as "energy-dependent phase I" consisting of uptake of a small number of aminoglycoside molecules in an energy-dependent fashion (11). The molecules that reached the cytoplasm induce errors in protein synthesis, and the misread membrane proteins are inserted in the cytoplasmic membrane, causing damage to its integrity. This process triggers another energy-dependent phase ("energy-dependent phase II") of uptake, in which a large number of antibiotic molecules penetrate the cell and bind to all ribosomes, leading to cell death. Since the uptake process depends on the respiratory chain, aminoglycosides are active mostly against aerobic bacteria.

The high number of strains that developed resistance has limited the successful use of aminoglycosides in the treatment of serious infections. Several mechanisms by which bacteria resist the action of aminoglycosides have been described (66). They include loss of permeability (81); methylation of 16S rRNA, a mechanism found in most aminoglycoside-producing organisms and recently in clinical strains (6, 33, 43, 87, 138); mutation in the genes encoding the 16S RNA or ribosomal proteins (69); export by efflux pumps (2, 67, 75, 98, 132), one of which has recently been shown to be involved in adaptive resistance (51); enzymatic inactivation of the antibiotic molecule (60); and the recently described sequestration of the drug by tight binding to an acetyltransferase of very low activity (68).

Of all the mechanisms listed above, inactivation mediated by aminoglycoside-modifying enzymes is the most commonly found in the clinical setting (4, 60, 97, 106, 116). The aminoglycoside-modifying enzymes are adenylyltransferases (ANTs), phosphotransferases (APHs), also known as kinases, and acetyltransferases (AACs). Because of the intensive studies on aminoglycoside resistance due to these enzymes and the availability of numerous complete bacterial genomes, the number of aminoglycoside-modifying enzymes identified to date has grown so much and keeps growing so fast that the citations and examples given in this article should be considered representative rather than comprehensive. These enzymes catalyze the modification at –OH or $-NH_2$ groups of the 2-deoxystreptamine nucleus or either of the sugar moieties. They are named using a three-letter identifier of the activity followed by (i) the site of modification between parentheses (class), (ii) a roman number particular to the resistance profile they confer to the host cells (subclass), and (iii) a lowercase letter to provide a unique identifier (106). Thus, AAC(6′)-Ib is an *N*-acetyltransferase that mediates acetylation (indicated by the three-letter identified AAC) at the 6′ position (which corresponds to one of the sugars). It has a resistance profile identical to other AAC(6′)-I enzymes like AAC(6′)-Ic (indicated by the roman numeral I). However, even when the AAC(6′)-Ib substrate profile is identical to that of AAC(6′)-Ic, they are different proteins (identified by the lowercase letters b and c). There is another nomenclature system that is commonly used in which the genes are designated *aac*, *aad*, and *aph* followed by a capital letter that identifies the site of modification (78), e.g., *aacA*, *aacB*, and *aacC* identify aminoglycoside 6′-*N*-acetyltransferase, aminoglycoside 2′-*N*-acetyltransferase, and aminoglycoside 3-*N*-acetyltransferase, respectively. A number is then added to identify different genes; for example, *aacA4* identifies the 6′-*N*-acetyltransferase gene also known as *aac(6′)-Ib*. Neither of these two systems is perfect, and different authors prefer one to the other. It would be desirable to arrive to a consensus and use only one of them. The simultaneous utilization of both systems is a source of confusion and makes it more difficult for people outside the field to follow the quick developments within the aminoglycoside-modifying enzymes field.

FEATURES OF AMINOGLYCOSIDE-MODIFYING ENZYMES

Aminoglycoside O-Phosphotransferases

The hydroxyl positions at which each one of seven known classes of APHs modifies antibiotic molecules by catalyzing the transfer of a phosphate group are shown in Fig. 4.1. The most important class includes APHs that modify the molecule at the 3′ position. Seven resistance profiles have been identified within this group of enzymes [APH(3′)-I through VII] (Fig. 4.1) (121, 136). An important representative of these enzymes, APH(3′)-Ia, was found in several gram negatives and at least one gram positive (121). The *aph(3′)-Ia* gene was originally identified as part of the transposon Tn*903* in *Escherichia coli* (80). This gene has been extensively manipulated, and it is now present in a number of plasmids used as cloning vehicles (24) and tools for molecular biology such as the well-known kanamycin resistance cassette from pUC4K (112). The *aph(3′)-Ia* gene was also found as part of other mobile elements such as Tn*4352* (*Salmonella enterica* serovar Typhimurium) (17), Tn*2680* (*Proteus vulgaris*) (56), Tn*5715* (*Corynebacterium striatum*) (111), and the transposable element from the *Providencia rettgeri* plasmid R391, in which the resistance gene is flanked by two copies of IS*15*-Δ1 in inverse orientation upstream and one copy downstream in inverse orientation to the second element (9). Only one resistance profile has been described for all other six APH classes (Fig. 4.1). The APH(2″)-I enzymes are present mainly in gram positives and in at least one *E. coli* strain (26, 121). A special case within this class of enzymes is that of the APH(2″)-Ia activity found in several *Streptococcus* and *Enterococcus* spp. (121). This activity is part of a bifunctional protein in which the phosphorylating domain is located at the C terminus but that also possesses an acetylating domain [AAC(6′)] at the N terminus (see below) (38). This enzyme eliminates the synergy observed between aminoglycosides and antimicrobials that act at the cell wall level (107, 121). Interestingly, the phosphorylating portion can be separated from the rest of the protein without losing its enzymatic activity (38). APH(3″)-I enzymes are distributed among a wide range of bacteria including the streptomycin-producing *Streptomyces griseus* N2-3-11 and gram negatives. The *aph(3″)-Ib* gene, which was originally found in the plasmid RSF1010, has been found downstream of another *aph* gene, the streptomycin resistance-mediating *aph(6)-Id*, as part of the ~100-kbp *Vibrio cholerae* O93 integrative and conjugative element (ICE) named SXT^{MO10} and its close relative SXT^{ET} (50). ICEs are relatively recently defined mobile genetic elements with plasmid- and phage-like properties (15) that include gram-positive conjugative transposons, certain molecules that have been mislabeled as "conjugative plasmids" and constins (these named after the first letters of conjugative, self transmissible, and integrating; a nomenclature that has not enjoyed widespread adoption) (50, 57). ICEs have the following properties: they

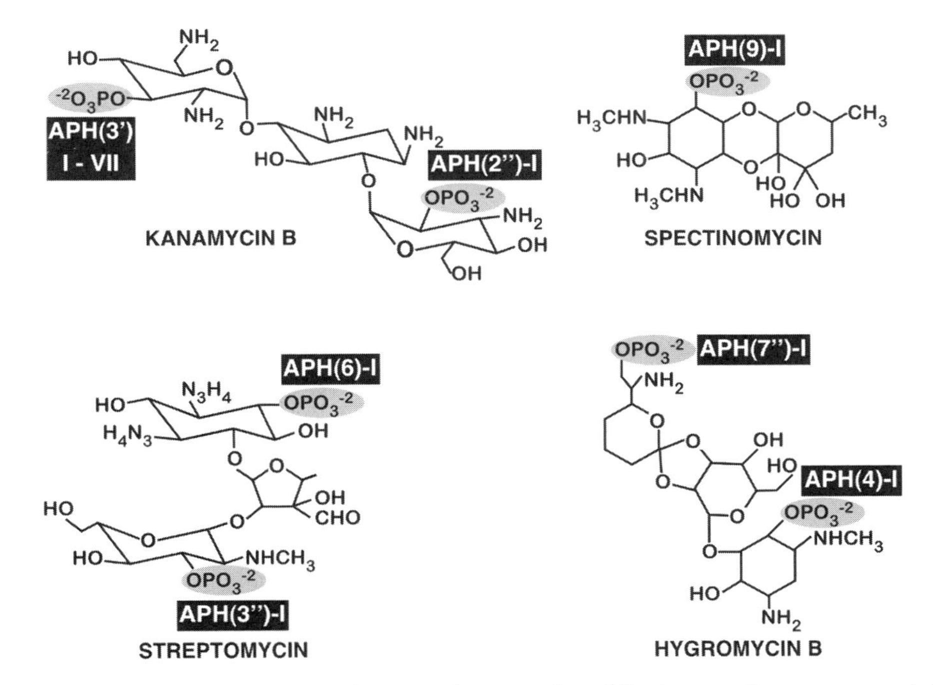

Figure 4.1 Sites of modification of APHs. The sites of modification are shown on one of the substrates of each class of APH. The number of subclasses is indicated.

are of large size, have self-transmissibility and site-specific integration, and are linked to antibiotic resistance genes and maintained integrated within the chromosome of their hosts, which excise from the chromosome by recombination between two specific flanking sequences (*attL* and *attR*) (5, 14, 50, 129). They excise from the chromosome, circularize, and mediate their own transfer to another bacterial cell where they integrate again in a site-specific fashion (14). SXT^{MO10} excises from the chromosome, generating a circular, most probably nonreplicative DNA molecule that seems to be an intermediate for its conjugative transfer between gram-negative bacterial strains (50). At least in one case, transfer of the SXT element was shown to be inhibited by the action of an SXT-encoded repressor known as SetR (5). The SOS response to DNA damage partially interferes with this repression, increasing the transfer frequency (5). Figure 4.2 shows the region that contains the antibiotic resistance genes of SXT^{MO10}. This DNA fragment is thought to have been inserted within the *rumB* gene in the *rumBA* operon of a precursor element. The fragment containing *aph(3″)-Ib* and *aph(6)-Id* is identical to that in the natural plasmid RSF1010 (44).

APH(4)-I and APH(7″)-I enzymes mediate resistance only to hygromycin B, while APH(9)-I enzymes mediate resistance to spectinomycin (106, 121). Analyses of the phylogenetic relationships among APHs and their evolution and mechanisms of action have recently been published (7, 107, 115, 136).

The three-dimensional structures of two aminoglycoside O-phosphotransferases, APH(3′)-IIIa, from gram-positive cocci (52), and APH(3′)-IIa, from *Enterobacteriaceae*, have been determined (79). Details on their structures, mechanisms, and roles of different regions in recognition of substrates, as well as structural and functional relations to protein kinases, are discussed in chapter 3.

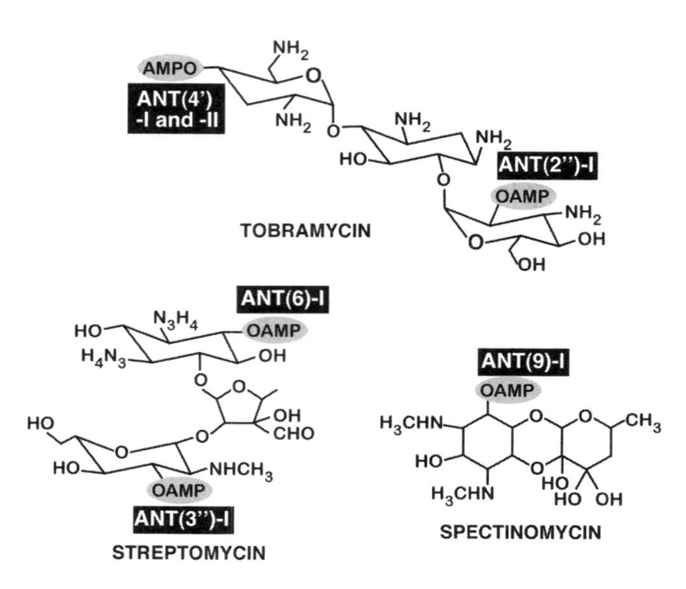

Figure 4.3 Sites of modification of ANTs. The sites of modification are shown on one of the substrates of each class of ANT. The number of subclasses is indicated.

Aminoglycoside O-Nucleotidyltransferases

ANTs catalyze the transfer of an AMP group from the substrate ATP to a hydroxyl group in the aminoglycoside molecule. Five classes of nucleotidyltransferases have been described to date; the positions modified by these enzymes are shown in Fig. 4.3. While two subclasses of ANT(4′) have been found (I and II), only one substrate profile was described for the other four nucleotidyltransferases (106, 121). ANT(2″)-I and ANT(3″)-I enzymes (16, 117), which have very different resistance profiles, have been found in plasmids, transposons, and integrons of gram negatives (121). Genes encoding ANT(3″)-I enzymes are included in Tn*21*, a transposon originally isolated from the

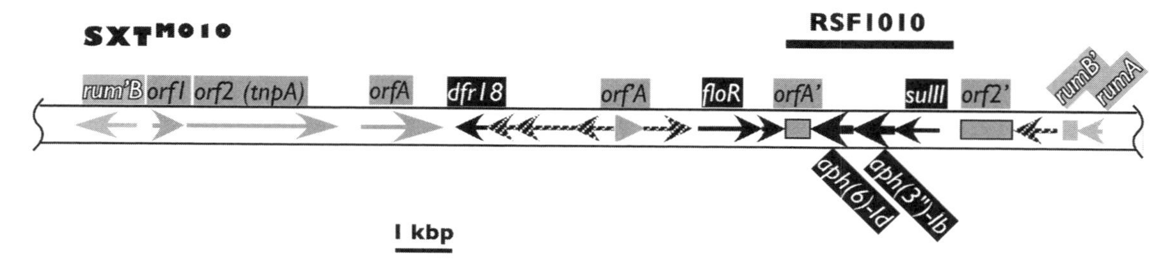

Figure 4.2 Genetic map of the region including resistance genes in the ICE SXT^{MO10}. The region shown is thought to have been inserted within the *rumB* gene of a precursor in a multistep process (50). The *rumA* gene and the fragments of *rumB* are indicated by lighter gray arrows. Transposase related genes (arrows) or truncated genes (boxes) are shown in gray. Truncated genes are indicated with an apostrophe in their names, e.g., *rumB'* means N-terminal coding part of *rumB* and *rum'B* means C-terminal coding part of *rumB*. Antibiotic resistance genes are shown in black. Hatched arrows show genes with other or unknown functions. The DNA fragment identical to RSF1010 is indicated.

conjugative plasmid NR1 (63); in Tn*1331*, a Tn*3*-like transposon with a unique structure within this family (see below) (116, 120); in Tn*7*, a transposon characterized by the utilization of a multiprotein machinery for transposition and target selection (86); and in other transposons, integrons, and plasmids. At least one ANT(3″)-I activity has been found as a fusion protein; it is located at the N-terminal portion of the protein isolated from a *Serratia marcescens* integron with ANT(3″)-I and AAC(6′)-II activities (22) (Fig. 4.4). ANT(4′)-I, ANT(6)-I, and ANT(9)-I enzymes have been isolated from gram-positive bacteria (107). One of them, the *Staphylococcus aureus* ANT(4′)-Ia, is the only enzyme of this kind for which the three-dimensional structure is known (100). ANT(4′)-IIa, which does not share significant homology with ANT(4′)-I enzymes, is present in gram negatives (105).

Aminoglycoside *N*-Acetyltransferases

AACs catalyze the acetylation of -NH_2 groups in the antibiotic molecule using acetyl-coenzyme A as a donor substrate. They belong to the GCN5-related *N*-acetyltransferase superfamily of proteins, which spans all kingdoms of life and comprises several unrelated kinds of enzymes with a wide variety of functions (37, 77). They all transfer the acetyl group from the same acetyl-coenzyme A donor, but they utilize very different acceptor molecules that have a primary amine acceptor group. Proteins in this superfamily have two main loosely conserved motifs (A and B), and several of them share two additional regions of loose conservation (motifs C and D) (77). The longest and most conserved of these regions is motif A, which has been shown to participate in binding of acetyl-coenzyme A and includes amino acids important for the catalytic activity (37, 77). The structure of motif A is similar in those GCN5-related *N*-acetyltransferases for which the three-dimensional structure has been resolved. It includes the β4-strand and the α3-helix, which are the essence of the acetyl-coenzyme A binding site (37). Conversely, the structure of motif B is not that homogeneous. The spatial conformation of this region is different in several GCN5-related *N*-acetyltransferases, and it may play different roles in different proteins in this superfamily. It has even been suggested that the sequence conservation in this motif may be coincidental rather than a consequence of a common structural origin (3). The crystal structures of several representatives of the GCN5-related *N*-acetyltransferase superfamily, including three aminoglycoside *N*-acetyltransferases, AAC(2′)-Ic, AAC(3)-Ia, and AAC(6′)-Ii, have been recently resolved (reviewed in reference 37 and in chapter 3 of this book; also see references 13, 91, 95, 124, 134, and 137). Although within the GCN5-related *N*-acetyltransferase superfamily of enzymes there are monomers and dimers, all three aminoglycoside acetyltransferases for which the crystal structure has been resolved, AAC(2′)-Ic, AAC(3)-Ia, and AAC(6′)-Ii, are of dimeric nature. The *Mycobacterium tuberculosis* AAC(2′)-Ic has been resolved in a ternary

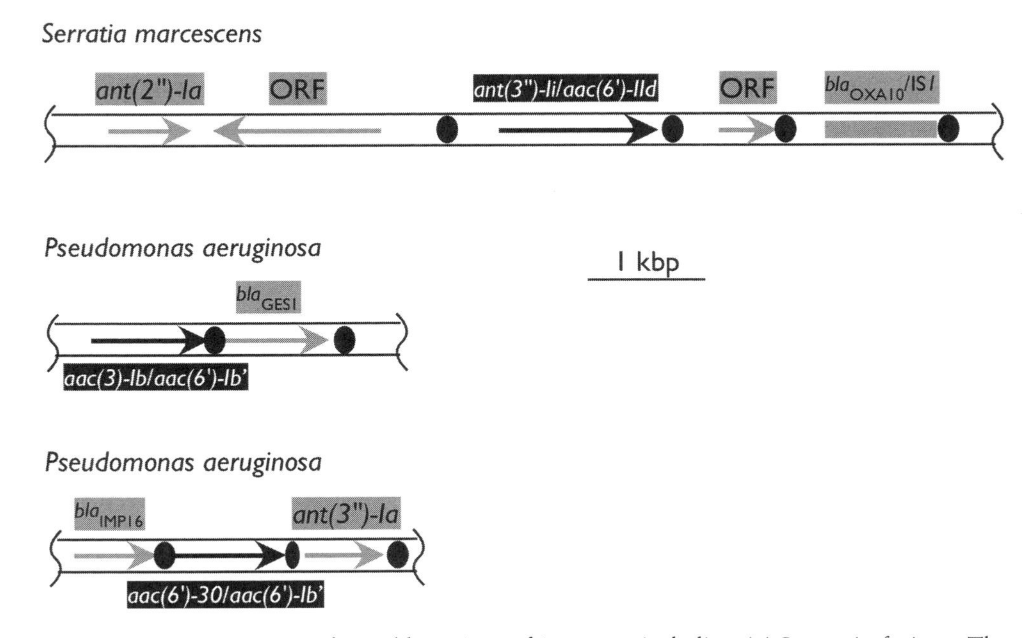

Figure 4.4 Genetic maps of variable regions of integrons including AAC protein fusions. The black dots indicate the *attC* loci in each gene cassette. Gene fusions are shown as black arrows. The genetic maps were redrawn from references 22, 35, and 70.

complex with coenzyme A and either tobramycin, kanamycin, or ribostamycin (125). This complex includes a protein dimer where coenzyme A binds in a groove surrounded by helices α1, α3, and α4 and the β4-strand. The antibiotic molecule binds within a pocket formed by β4, the C-terminal part of the protein, and the loops that join α1-β1, β5-α4, and β8-β9 (125). The resolution of the three-dimensional structure of AAC(6′)-Ii showed that it also forms a dimer but its assembly is different from that in other members of the superfamily (13).

The amino groups that are substituted in reactions mediated by each of the four known classes of aminoglycoside *N*-acetyltransferases are shown in Fig. 4.5. AAC(1), AAC(2′), and AAC(3) enzymes have been detected only in gram negatives (107). The less common acetyltransferase AAC(1) was isolated from a strain of *E. coli* with an unusual resistance phenotype, and it was determined that it mediates acetylation of apramycin, butirosin, lividomycin, and paromomycin (64). In addition, this enzyme catalyzed diacetylation of ribostamycin and neomycin. The second acetylation site has not been unequivocally determined, but it has been suggested to be the 6′-amino group (64).

AAC(2′)-I enzymes are found in the chromosome of *Providencia* [AAC(2′)-Ia] and *Mycobacterium* [AAC(2′)-Ib through e] species (1, 72, 93). AAC(2′)-Ia confers *Providencia stuartii* resistance to several aminoglycosides (72). However, its true function is still a matter of debate. It has been proposed that the ability to mediate acetylation of aminoglycosides is fortuitous and its true role is the *O*-acetylation of the peptidoglycan or peptidoglycan precursors (42). On the other hand, the recent analysis of the three-dimensional structure of another AAC(2′)-I, AAC(2′)-Ic, led the authors to propose that this protein plays a role in acetylation of a biosynthetic intermediary in the synthesis of mycothiol (124). AAC(2′)-I enzymes have been found in all strains of all species of mycobacteria analyzed to date (1). AAC(2′)-Ib through Ie were detected in *Mycobacterium fortuitum* FC1K, *M. tuberculosis* H37Rv, *M. smegmatis* mc^{2}155, and *Mycobacterium leprae*, respectively. Of these enzymes, only AAC(2′)-Id has clearly been associated with aminoglycoside resistance (1).

AAC(3) enzymes are present in a large number of gram negatives, and several subclasses have been described. It is worth mentioning that the subclass AAC(3)-V has been eliminated after confirmation that this enzyme was identical to AAC(3)-II (106). Within the AAC(3)-I subclass, the two best-known enzymes are AAC(3)-Ia and AAC(3)-Ib. They mediate resistance to gentamicin, sisomicin, and fortimicin and are present in about 30% of gram-negative clinical isolates (121). Two other enzymes within this subclass have been recently isolated: AAC(3)-Ic, from a *P. aeruginosa* integron (95), and AAC(3)-Id, from an *S. enterica* serovar Newport gene cluster (34). The subclass AAC(3)-II includes three enzymes, of which AAC(3)-IIa is the most common. This group confers resistance to gentamicin, netilmicin, tobramycin, sisomicin, 2′-*N*-ethylnetilmicin, 6′-*N*-ethylnetilmicin, and dibekacin (106). AAC(3)-III, AAC(3)-IV, and AAC(3)-VI are found in clinical isolates, but they are much less common than subclasses I and II (106, 121). Enzymes belonging to subclasses AAC(3)-VII, AAC(3)-IIX, AAC(3)-IX, and AAC(3)-X have been identified in strains of actinomycetes. The enzyme AAC(3)-X, isolated from *S. griseus* SS-1198PR, mediates acetylation of kanamycin and dibekacin at the 3′-amino group but at the 3″ position of arbekacin and amikacin, becoming the first AAC detected to have also AAC(3″) activity (53). However, while 3″-*N*-acetylamikacin shows loss of activity, 3″-*N*-acetylarbekacin maintains its antibiotic properties.

Two main subclasses of AAC(6′) enzymes have been found, AAC(6′)-I and AAC(6′)-II, coded for by genes that are located in plasmids or chromosomes of gram positives

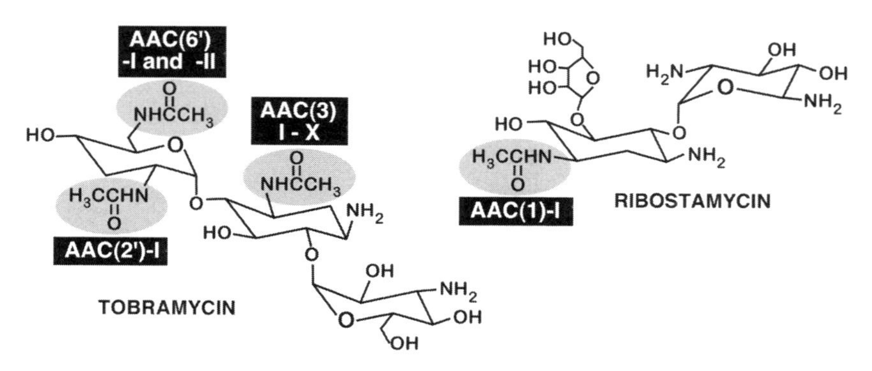

Figure 4.5 Sites of modification of AACs. The sites of modification are shown on one of the substrates of each class of AAC. The number of subclasses is indicated.

and gram negatives, often within transposons or integrons. Both subclasses are capable of conferring resistance to several important aminoglycosides including tobramycin, netilmicin, dibekacin, sisomicin, and kanamycin. They differ in that AAC(6′)-I enzymes confer resistance to amikacin in addition to the above-mentioned antibiotics but have very low gentamicin C1 acetylating activity while AAC(6′)-II enzymes have higher levels of activity when gentamicin C1 is used as a substrate but do not confer resistance to amikacin. About 30 AAC(6′)-I and 4 AAC(6′)-II enzymes have been described. However, it should be noted that the nomenclature used for this class of enzymes is not completely consistent in the literature or GenBank at present. This may be due to the fact that a large number of AAC(6′) enzymes are being discovered and they present different degrees of similarity in sequence and phenotype. In at least one case the simultaneous discovery of two enzymes resulted in the two enzymes being given the same name (122). There have been cases in which an enzyme with AAC(6′)-II resistance profile is much closer to AAC(6′)-I enzymes at the amino acid level, which led to naming them AAC(6′)-I (20, 61). Furthermore, some enzymes have been named using a subscript number because their differences in amino acid sequence are minimal. For example, AAC(6′)-Ib_3, AAC(6′)-Ib_4, AAC(6′)-Ib_6, and AAC(6′)-Ib_7 behave similarly but differ at the N-terminal portion while AAC(6′)-Ib_{11} has a unique combination of amino acids at positions 118 and 119 that results in a different resistance profile (19, 20). In other cases other modifications to the nomenclature were added, as in the case of the *Pseudomonas fluorescens* BM2687 gene that contains a point mutation with respect to the *aac(6′)-Ib* gene present in *P. fluorescens* BM2656. This point mutation, a transition from C to T, leads to a change in phenotype. The enzyme was called AAC(6′)-Ib' even when the phenotype is consistent with an AAC(6′)-II enzyme (61). This protein is identical to the acetyltransferase portion of the recently identified bifunctional protein ANT(3″)-Ii/AAC(6′)-IId (22). In addition, as it is the case for AAC(6′)-29a, AAC(6′)-29b, or AAC(6′)-30 (89), enzymes have been named without the roman number or, as with the enzyme AAC(6′)-I30, with a number following it instead of a lowercase letter (76). AAC(6′)-29a and AAC(6′)-29b are shorter than other AAC(6′)-I enzymes and confer a modified AAC(6′)-I phenotype, and at least one of them shows a feeble acetylating activity (70, 89). A double lowercase letter has also been used, like in AAC(6′)-Iaa, which differs from AAC(6′)-Iy by two amino acids (101), or AAC(6′)Iad (32).

At least four AAC(6′) enzymes exist as fusion proteins. AAC(6′)-Ie is the amino-terminal half of an *Enterococcus faecalis* bifunctional protein with AAC(6′) and APH(2″) activities (38). The AAC(6′)-IId is the carboxy-terminal portion of the protein fusion isolated from an *S. marcescens* integron with ANT(3″)-I and AAC(6′)-II activities mentioned in the above paragraph (22). The variable region of this integron includes a series of unusual features: an uncommon gene cassette containing an insertion of an open reading frame (ORF) coded for by the complementary strands between the *ant(2″)-Ia* gene and its *attC* locus (also known as 59-be) (21, 48) is followed by a gene cassette including *ant(3″)-Ii/aac(6′)-IId*, a gene cassette including an ORF of unknown function, and a gene cassette including the beginning of $bla_{\text{OXA-10}}$ interrupted by a copy of IS*1* (Fig. 4.4) (22). A fusion protein consisting of two AAC(6′)-I activities, AAC(6′)-30/AAC(6′)-Ib′, has recently been identified in a *P. aeruginosa* integron (70). The variable portion of this integron includes gene cassettes containing the genes $bla_{\text{IMP-16}}$, *aac(6′)-30/aac(6′)-Ib′*, and *ant(3″)-Ia*, respectively (Fig. 4.4). An integron isolated from a clinical *P. aeruginosa* strain contained a gene consisting of the *aac(3)-Ib* gene fused with the sequence of *aac(6′)-Ib′* (35). Its variable region includes a gene cassette with the fused genes *aac(3)-Ib/aac(6′)-Ib′* and another gene cassette containing $bla_{\text{GES-1}}$ (Fig. 4.4).

Phylogenetic analyses by some authors led to a division of the AAC(6′) enzymes into three subgroups (47, 106, 108). However, a recent study has proposed that they belong to three groups that are less related than thought before and that acetylation at the 6′ position of aminoglycosides has evolved independently at least three times (101).

Experiments of site-directed mutagenesis identified the amino acid 119 as involved in the difference in resistance profiles conferred by AAC(6′)-Ib (L119) and AAC(6′)-IIa (S119) (92). Later, the AAC(6′)-Ib′ enzyme was isolated from *P. fluorescens* BM2656, which differs from AAC(6′)-Ib in that the equivalent L residue is substituted for an S residue and as a consequence acquired the resistance profile of an AAC(6′)-II enzyme (61). An extended-spectrum AAC(6′)-Ib variant, AAC(6′)-Ib_{11}, was recently identified in an *S. enterica* plasmid (20). This enzyme possesses L118 and S119 as opposed to Q and L or Q and S, respectively, which are the amino acid residues found in all other natural enzymes of this type. As a consequence of this amino acid combination, AAC(6′)-Ib_{11} confers a combination of the resistance profiles of AAC(6′)-I and AAC(6′)-II enzymes, i.e., it confers resistance to both amikacin and gentamicin. The hydrophobicity of the L residue has been proposed as an important factor in the specificity of these enzyme variants. Interestingly, positions 118 and 119 are located within motif A, a region involved in acetyl-coenzyme A binding.

Other amino acids important for specificity and/or activity were detected by random and site-directed mutagenesis of the AAC(6′)-Ib coded for by Tn*1331* (108).

Position 171 in motif B is occupied by an F residue in all AAC(6′)-I enzymes belonging to two of the three phylogenetic subgroups and by a Y in those belonging to the third subgroup [AAC(6′)-Im, AAC(6′)-Iq, AAC(6′)-Ia, and AAC(6′)-Ii]. Mutagenesis at this position provided valuable information. The mutation F171L resulted in a derivative with reduced activity with respect to the wild type when the assays were carried out at 37°C. However, the mutant derivative lost the ability to mediate acetylation of amikacin and netilmicin (which have in common the replacement at the C1 amino group of the deoxystreptamine moiety) at 42°C while the activities when kanamycin or tobramycin were used as substrates were similar at the two temperatures (82). Hence, the substitution F171L resulted in an enzyme that shows thermosensitivity when the substrates are the semisynthetic aminoglycosides amikacin or netilmicin. The substitution F171A resulted in a derivative that lost activity (108), and the substitution F171Y resulted in a derivative with very low activity (25). These results suggested that the side chain of the F or Y at position 171 plays an important role in the activity and, directly or indirectly, specificity of some AAC(6′)-I enzymes. Alanine scanning led to the identification of other important amino acids in motif B. The substitution E167A in AAC(6′)-Ib resulted in a mutant that lost all acetylating activity, and the mutant Y166A resulted in an enzyme that, while capable of conferring as high levels of resistance to kanamycin as the wild type, lost the ability to confer resistance to amikacin, suggesting a role in specificity (108). Although Y is the most common amino acid found at this position among GCN5-related *N*-acetyltransferase enzymes, H residues are also present in some of them (77). Among AAC(6′)-I enzymes, those in two of the phylogenetic subgroups contain a Y residue while the others include an H residue. It has been suggested that the side chain of Y147 [equivalent to Y166 in AAC(6′)-Ib] interacts with acetyl-coenzymeA (137). However, this seems not to be the case for the *S. marcescens* aminoglycoside 3-*N*-acetyltransferase, in which the Y157 is not appropriately positioned for the side chain to be within hydrogen bonding distance of the sulfur atom of acetyl-coenzymeA (134). A mutant derivative of a Y168F serotonin *N*-acetyltransferase led to significant reduction of acetylating activity (49), underscoring the important role of the amino acid at this position in GCN5-related *N*-acetyltransferases.

A very interesting AAC(6′)-Ib variant, AAC(6′)-Ib-cr, has been recently isolated that reduces the activity of the fluoroquinolone ciprofloxacin by acetylating the amino nitrogen on its piperazinyl substituent, resulting in a fourfold-higher MIC (97a). Two amino acid changes occurred in this variant: W102R and D179Y. Mutagenesis experiments showed that in the absence of these two substitutions the enzyme was not associated with resistance to ciprofloxacin. AAC(6′)-Ib-cr is the first aminoglycoside-modifying enzyme to cross boundaries and become able to disable two unrelated kinds of antibiotics (97a).

Deletion at the C terminus of the Tn*1331*-encoded AAC(6′)-Ib enzyme indicated that up to 10 amino acids could be deleted without losing enzymatic activity (108).

Analyses of natural mutants led to identification of other amino acids of importance such as the case of an aminoglycoside-susceptible strain of *Acinetobacter haemolyticus* harboring a derivative of AAC(6′)-Ig in which the natural point mutation M56R resulted in loss of activity (99).

Most versions of the *aac(6′)-Ib* gene, the most frequently identified acetyltransferase gene in several multicenter studies, are part of gene cassettes which are located mainly in class 1 integrons or in a region resembling the variable portion of integrons included in Tn*1331* (62, 117, 119). As a consequence of its high mobility, the gene may have suffered rearrangements that led to the existence of several versions of the enzymes that vary at the N terminus (19, 20). Figure 4.6 shows the N terminus of several AAC(6′)-Ib variants. The fact that all of these enzymes are active demonstrates a high flexibility in the structural requirements at this end of the protein. Casin et al. proposed that this flexibility in requirements may be a factor contributing to the predominance of AAC(6′)-Ib variants among aminoglycoside-resistant *Enterobacteriaceae* (19).

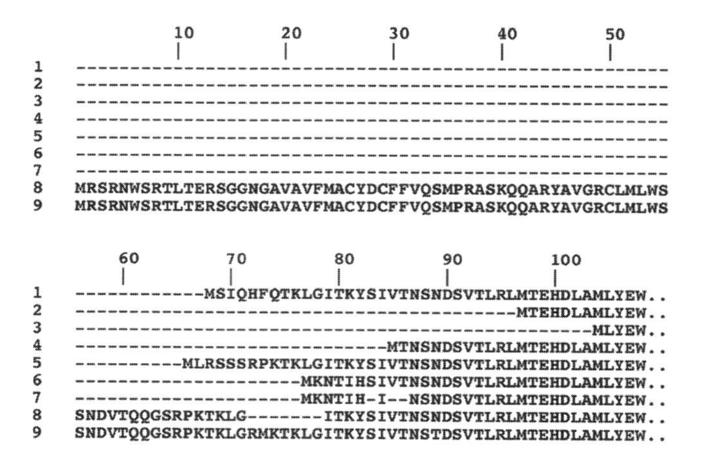

Figure 4.6 N-terminal portions of AAC(6′)-Ib variants. Alignment of the N-terminal amino acid sequences of several AAC(6′)-Ib variants. The accession numbers of the proteins shown are as follows: 1, M55547; 2, AF202035 (S at position 90), L25617 (S90), AF302086 (S90), U90945 (L at position 90), AJ311891 (L90), AF282595 (L90), AF034958 (L90), AJ009820 (L90) (AJ009820 and AF302086 have the last two amino acids different); 3, X60321, AF024602; 4, AF207065, AY033653; 5, AF043381 (S121), U59183 (L121); 6, AY103455; 7, AY136758; 8, Y11948; and 9, Y11947.

Surprisingly, not all proteins belonging to this group confer resistance through enzymatic acetylation of the aminoglycoside molecule. An interesting mechanism of resistance mediated by AAC(6′)-29b has recently been reported. This protein is shorter than most AAC(6′) enzymes and showed very poor acetylating activity (68). However, the mechanism of resistance mediated by this enzyme was shown to be sequestration of the antibiotic as a result of very tight binding rather than chemical modification of the aminoglycoside molecule (68).

SUBCELLULAR LOCATION OF AMINOGLYCOSIDE-MODIFYING ENZYMES

The subcellular location of aminoglycoside-modifying enzymes has been a matter of analysis for some time. Not all studies found these enzymes to be located in the same cellular compartment; some of them indicate that the aminoglycoside-modifying enzyme being studied is located within the cytoplasm (85), while others show a periplasmic location (128). For other enzymes, such as AAC(2′)-Ia, there have been conflicting reports suggesting periplasmic and cytoplasmic locations (42, 121). However, several of these reports were based on sequence analysis and methods of physical fractionation of cellular compartments that may be less than perfect (123). We recently carried out an exhaustive analysis of the location and distribution of the AAC(6′)-Ib enzyme coded for by Tn*1331* (30). This protein was not extracted by spheroplast formation, a technique that includes an osmotic shock step, suggesting that it is located within the cytoplasm of the cells. However, when harsher osmotic shock conditions were used, β-galactosidase, a larger protein known to be located in the cytosol, was not detected in the periplasmic space while a considerable fraction of the total AAC(6′)-Ib was found in this compartment, demonstrating that results from physical separation must be complemented with results obtained by other, more dependable techniques. Operon fusions are powerful tools to determine or confirm location of proteins or protein domains. The wild-type *phoA* gene encodes a protein that includes a signal-sequence that promotes export of alkaline phosphatase into the periplasmic space where it is active. Due to the highly reducing environment, alkaline phosphatase is unable to fold into its native conformation if retained in the cytoplasm of growing cells and therefore is not active in this compartment. Fusion of a gene of interest to a modification of the *phoA* gene in which the signal peptide coding sequence has been removed permits determination of the location of the protein or domain of interest. Cells carrying a functional AAC(6′)-Ib-PhoA fusion protein did not produce detectable alkaline phosphatase activity, indicating that AAC(6′)-Ib is cytoplasmic. These results were further confirmed by analyzing the fluorescence of cells possessing a functional AAC(6′)-Ib-cyan fluorescent protein fusion. Figure 4.7 shows that the fluorescence was observed within the cytoplasmic compartment. Controls consisting of *E. coli* cells harboring green fluorescent protein fusions that direct the protein to the periplasmic space or to a polar location within the periplasm are included (Fig. 4.7). Additionally, determination of this AAC(6′)-Ib protein N terminus was consistent with a mature protein that did not undergo processing after translation. These results, taken together, unequivocally demonstrated that the AAC(6′)-Ib enzyme coded for by Tn*1331* is evenly distributed within the cytoplasmic compartment of *E. coli* (30).

DISSEMINATION OF GENES CODING FOR AMINOGLYCOSIDE-MODIFYING ENZYMES

Two main, not necessarily exclusive, theories have been postulated to explain the origin of genes coding for aminoglycoside-modifying enzymes: (i) they have originated in aminoglycoside-producing organisms that needed them for protection against the antibiotics they produced, and (ii) the selective pressure of aminoglycoside use selected evolutionary products of genes coding for activities that participate in cellular functions (106). Regardless of their origin,

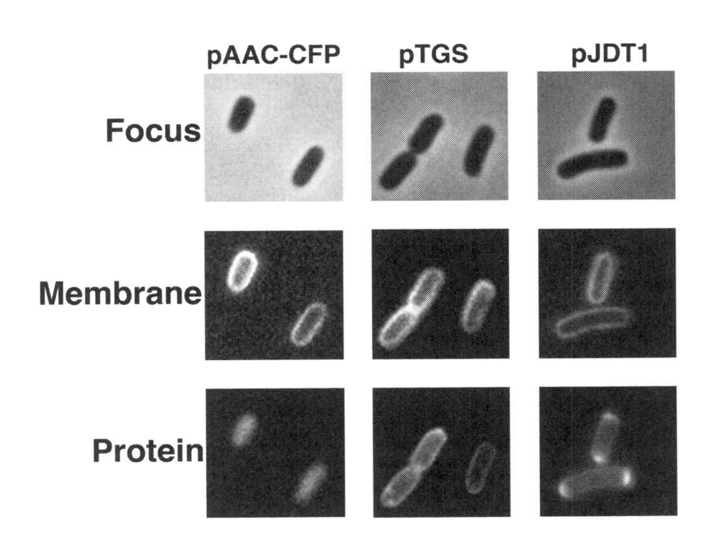

Figure 4.7 Visualization of AAC(6′)-Ib-cyan fluorescent protein. Cells containing plasmid pAAC-CFP were stained with the membrane dye FM5-95 and examined on a fluorescent microscope using the appropriate filters to detect FM5-95 or cyan fluorescent protein (30). Control experiments were done using plasmid pTGS, which encodes a periplasmic protein fused to green fluorescent protein (27), and pJDT1, which encodes a periplasmic protein fused to green fluorescent protein that accumulates at the poles of the cell (114).

a large number of these genes found ways to disseminate. They are found associated to chromosome and extrachromosomal elements including plasmids with diverse characteristics, transposable elements, ICEs, or integrons; and many of them exist as gene cassettes (22). The presence of genes coding for aminoglycoside-modifying enzymes within many of these elements permitted them to disseminate at the molecular as well as at the cellular level, contributing to the rise of multiresistant bacteria. The transfer of genes has been shown to occur not only among gram negatives and gram positives but also between the two groups of bacteria (113). An interesting paradigm of a complex transposable element with unique structural features and that includes genes coding for two aminoglycoside-modifying enzymes, *aac(6')-Ib* and *ant(3")-Ia*, is Tn*1331* (120). This transposon has been originally found in pJHCMW1, a small *Klebsiella pneumoniae* 11,354-bp plasmid (135), of which 7,993 bp correspond to Tn*1331* while the rest corresponds to the maintenance functions including the replication region, a functional *oriT*, and *mwr*, a target site for Xer recombination (Fig. 4.8a) (29, 88, 103). Although in a molecular epidemiology study pJHCMW1 was not prevalent in multiresistant *K. pneumoniae* clinical isolates, a large number of strains carried plasmids that include Tn*1331* or its close relative Tn*1331.2* (23, 118). Tn*1331* belongs to the Tn*3* family of transposons, a group of related mobile elements in which replicative transposition is mediated by transposase and resolvase, the products of *tnpA* and *tnpR*, respectively. Upstream of *tnpA* there is a site-specific recombination target of resolvase, *res*, at which resolution of the cointegrate intermediary occurs (63). There are two Tn*3* subfamilies, one known as the Tn*21*subfamily, characterized by having the *tnpA* and *tnpR* genes transcribed in the same direction, and the other known as the Tn*3* subfamily, in which *tnpA* and *tnpR* are divergently transcribed. While there are several transposons within the Tn*21* subfamily (63), the Tn*3* subfamily is much smaller. One of its representatives is Tn*1331*, a transposon that seems to be an evolutionary product of Tn*3* in which it acquired a DNA region that has a structure similar to that of the variable portion of the integrons (Fig. 4.8a) (117). The inverted repeats at the Tn*1331* ends as well as the *tnpR* gene are identical to those of Tn*3*. The Tn*1331 tnpA* nucleotide sequence differs from that of Tn*3*; it lacks 9 nucleotides, which are repeated in Tn*3*. As a result, while in Tn*3* there is an amino acid sequence GFHGFH, the Tn*1331* transposase carries the amino acid sequence GFH (103). However, both transposases mediate transposition of both Tn*3* and Tn*1331* (116, 120). A comparison of the genetic maps of Tn*3* and Tn*1331* shows that in the latter transposon there is an insertion of a sequence similar to the variable portion of the integrons flanked by 520-bp direct repeats (Fig. 4.8a and b). However, in this case, expression of the genes is not driven by the promoter present within the 5'-conserved region of integrons (94), but by a promoter identical to the $bla_{\text{TEM-1}}$ promoter, which is located upstream of the structural *aac(6')-Ib* gene due to the duplication present in Tn*1331* (Fig. 4.8a and b). This duplication could have taken place during the process of genesis of Tn*1331*, and as a consequence a gene fusion was formed originating the Tn*1331* version of *aac(6')-Ib* (Fig. 4.8a and b) (117). This version of the gene codes for a protein with an N-terminal amino acid sequence SIQHF, identical to the first five amino acids of the leader peptide of the TEM β-lactamase (30) (Fig. 4.8). At least two mechanisms for the genesis of Tn*1331* have been postulated. One of the theories proposes a defective transposition event of Tn*3* into the appropriate DNA site. After formation of the cointegrate, deletion of two DNA fragments, one of them including one of the *res* sites, makes resolution impossible, freezing the structure of the deleted cointegrate and defining a new transposon between the inverted repeats (116, 117). The other theory considers an event of duplication-deletion after a normal Tn*3* transposition event into an appropriate DNA site. The duplication flanks a DNA fragment that becomes part of the newly formed transposon (116, 117).

The genetic organization of the region flanked by the 520-bp direct repeats shows that, as it is the case in the variable portion of the integrons, *aac(6')-Ib* is part of a gene cassette with the common structure *aac(6')-Ib-attC*. The other two genes, however, present a more unusual structure containing a gene cassette that includes two resistance genes. Unlike the common situation in which *ant(3")-Ia* is followed by an *attC* site (96), in Tn*1331* this gene is followed by a sequence that seems to be a remnant of an *attI* site (*attI**) and then by $bla_{\text{OXA-9}}$ and *attC* generating the arrangement *ant(3")-Ia-attI**-$bla_{\text{OXA-9}}$-*attC* (103, 119). This gene cassette could have been formed by illegitimate recombination between two integrons (Fig. 4.9). In one of the integrons the gene cassette adjacent to the 5' conserved region was that containing $bla_{\text{OXA-9}}$. An illegitimate recombination event may have taken place between the *attI* 5' of $bla_{\text{OXA-9}}$ and *attC* 3' of *ant(3")-Ia* in the second integron, resulting in the generation of *attI** and the loss of *attC* (Fig. 4.9) (103). The unusual structure of a gene cassette containing two genes is not unique to Tn*1331*, analysis of *Pasteurella* and *Mannheimia* strains carrying the *sulII* and *strA* genes indicated that a large percentage of them contained a copy of the *catAIII* gene, which was inserted by illegitimate recombination (59). In another case, a gene cassette present in Tn*2424* includes the gene *aac(6')-Ia* followed by an ORF that includes a Shine-Dalgarno sequence (83).

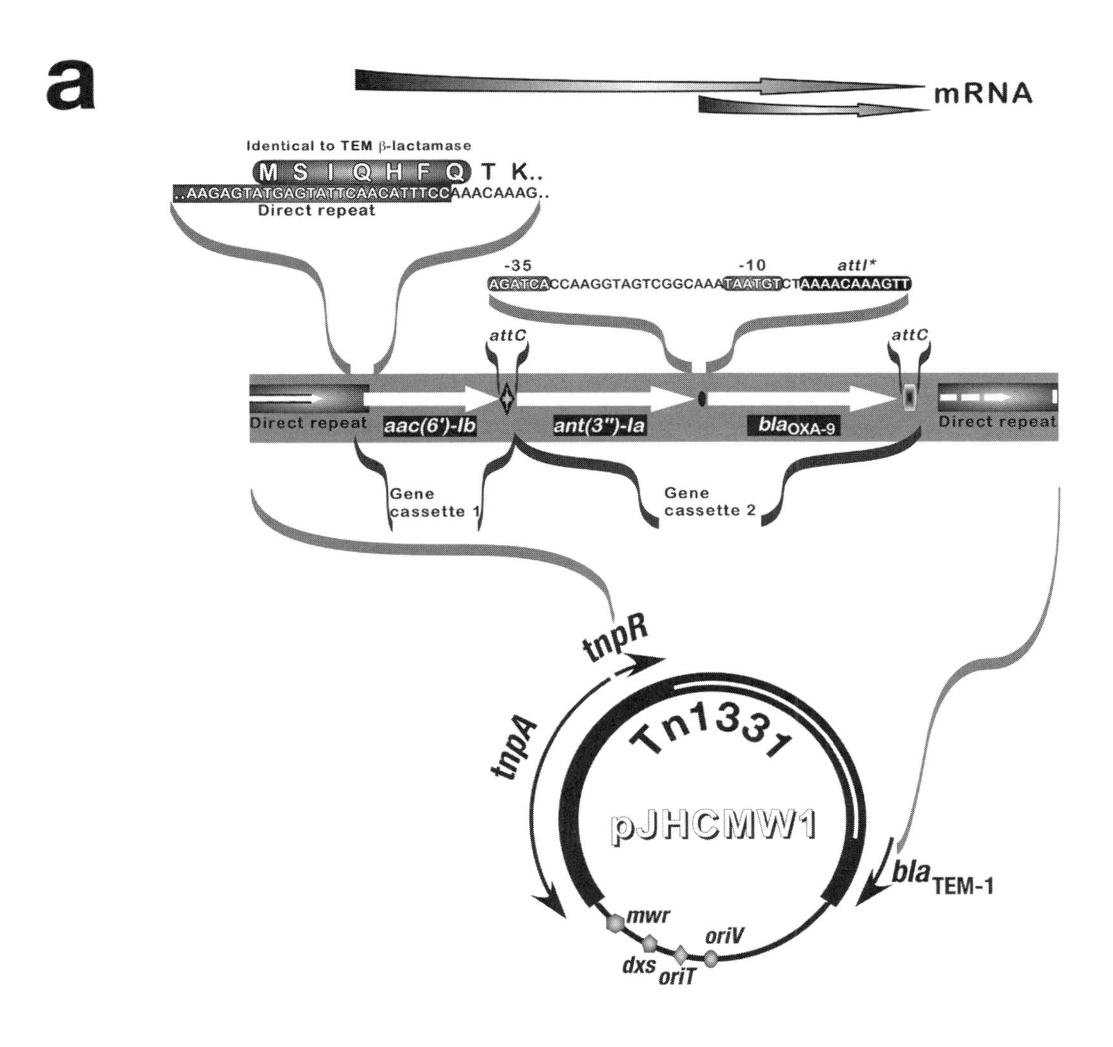

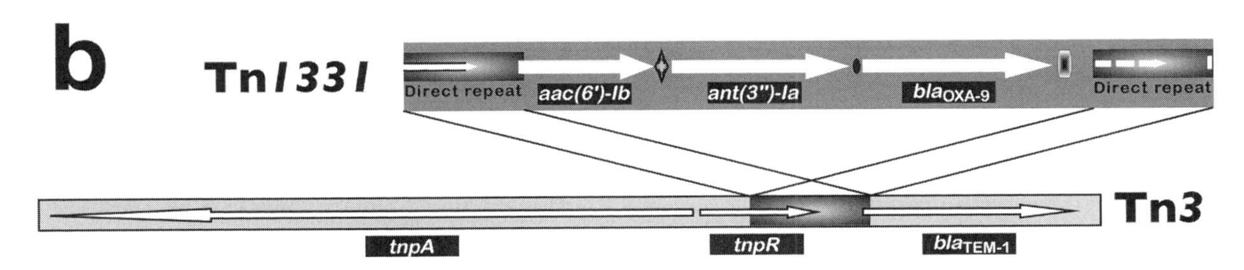

Figure 4.8 Genetic map of pJHCMW1 and Tn*1331*. (a) The map shows important loci of pJHCMW1: *mwr*, a target for Xer site-specific recombination (88); *dxs*, a weak XerD binding site of unknown function (103); *oriT* and *oriV* (29); and Tn*1331*. Tn*1331* can be considered to be Tn*3* with the addition of a DNA region harboring three resistance genes. The genetic map of this region is shown in detail including the N terminus of AAC(6')-Ib, the *attC* sites, the gene cassettes, and the nucleotide sequence resulting from the formation of a gene cassette including both *ant(3")-Ia* and bla_{OXA-9}. The arrows on top represent mRNA species. (b) Diagram showing Tn*3* and the region duplicated (boxes) flanking the fragment containing *aac(6')-Ib*, *ant(3")-Ia*, and bla_{OXA-9} in Tn*1331*. The dotted arrow inside the box to the right indicates that only a portion of *tnpR* is present on this side.

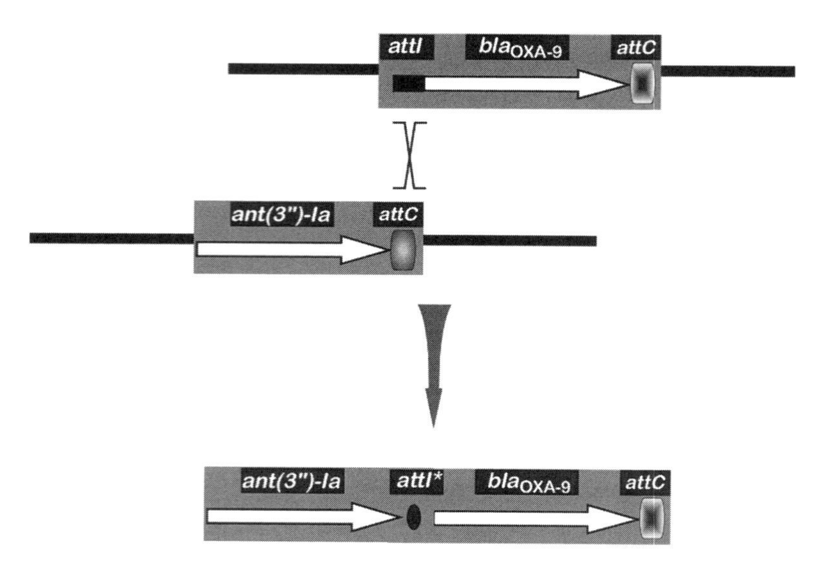

Figure 4.9 Possible mechanism of generation of the cassette including *ant(3")-Ia-attI*-bla*$_{OXA\text{-}9}$-*attC*: illegitimate recombination between an integron containing the *bla*$_{OXA\text{-}9}$-*attC* gene cassette immediately following *attI* and another one containing an *ant(3")-Ia-attC* gene cassette (103, 119).

Transcription of the genes included in the Tn*1331* two-gene cassettes was studied using a variety of techniques. The results indicated that there is polycistronic mRNA that includes information for translation of all three genes, *aac(6')-Ib*, *ant(3")-Ia*, and *bla*$_{OXA\text{-}9}$. In addition the *bla*$_{OXA\text{-}9}$ gene is transcribed from a promoter located immediately upstream of this gene (Fig. 4.8a).

STRATEGIES TO OVERCOME THE ACTION OF AMINOGLYCOSIDE-MODIFYING ENZYMES

Various approaches have been followed to overcome the action of aminoglycoside-modifying enzymes. New antibiotics that are not substrates of some modifying enzymes were successfully obtained. Dibekacin (3',4'-dideoxykanamycin B) as well as 3'-dideoxykanamycin A, which could not be modified by some APH and ANT enzymes, were among the first derivatives active against resistant *E. coli* or *P. aeruginosa* (73). Modification of the C_1 amino group of the deoxystreptamine moiety of aminoglycosides originated the semisynthetic amikacin and netilmicin which are refractory to most modifying enzymes (106). Recently new aminoglycosides that showed killing activity against bacteria containing modifying enzymes were designed based on the structure of neamine bound to the A-site of the bacterial rRNA (45). Cross-linked dimers of neamine proved to be bactericidal by binding the A-site of the rRNA with high affinity while being poor substrates of aminoglycoside-modifying enzymes (109). Another approach consisted of the development of enzymatic inhibitors. Cationic peptides with inhibitory effect on AAC(6')-Ii and two phosphotransferases were synthesized (8). Another approach consisted of taking advantage of the structural relation found between eukaryotic protein kinases and APHs. Protein kinase inhibitors that disable aminoglycoside-modifying enzymes by targeting the ATP-binding site are showing very promising prospects for their use in combination with aminoglycosides (12).

Antisense oligonucleotide-mediated inhibition of expression of genes coding for aminoglycoside-modifying enzymes shows promise for development as another alternative against resistance. Regions accessible for interaction with antisense oligodeoxynucleotides in the *aac(6')-Ib* mRNA were identified by RNase H mapping in combination with computer-generated secondary structures (Fig. 4.10) (102). A series of oligodeoxynucleotides were synthesized and assayed to determine their ability to bind the mRNA, induce in vitro RNAse H-mediated digestion of the mRNA, inhibition of in vitro synthesis of AAC(6')-Ib, and in vivo interference with amikacin activity. At least three oligodeoxynucleotides were identified that induced in vitro degradation of mRNA, inhibit in vitro synthesis of the enzyme, and significantly reduced the number of cells surviving after exposure to amikacin (Fig. 4.10) (102). A similar strategy was successfully used to design antisense compounds that inhibit expression of the *mar* operon in *E. coli* (133). While these experiments show potential for

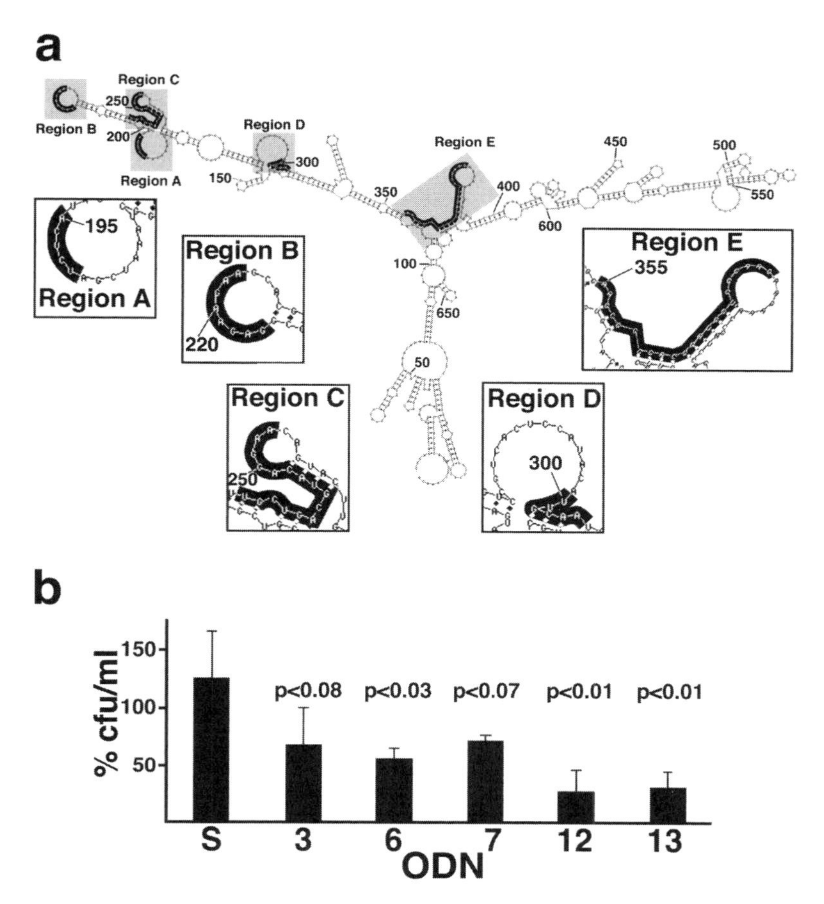

Figure 4.10 Determination of *aac(6')-Ib* mRNA region accessible to antisense oligodeoxynucleotides and reduction of resistance levels by their action. (a) Regions accessible to interaction with antisense oligodeoxynucleotides were determined by a combination of RNase H mapping and computer-generated secondary structure of *aac(6')-Ib* mRNA. The nucleotides over a black background indicate the regions identified by RNase H mapping (enlarged in the insets) (102). (b) Selected antisense oligodeoxynucleotides were introduced into cells by electroporation before being exposed to amikacin. The percent CFU per milliliter values represent the fraction of surviving cells compared to the sample that was subjected to electroporation without adding oligodeoxynucleotides (102). S, sense oligodeoxynucleotides. The sequences of the oligodeoxynucleotides 3, 6, 7, 12, and 13 can be found in reference 102. *P* values ($n = 3$) were calculated with respect to the control sense oligodeoxynucleotides (ODN). *P* values of < 0.05 are statistically significant.

antisense oligonucleotides to be developed as a way to inhibit the action of aminoglycoside modifying enzymes, many problems remain to be solved. Some of these problems are that (i) an antisense oligodeoxynucleotide may be effective against only one or a very limited number of genes, (ii) the presence of aminoglycoside-modifying enzyme genes as part of high-copy-number plasmids and/or under the control of very strong promoters may result in inhibition levels that are not enough for phenotypic conversion to susceptibility, and (iii) methods to ensure that the compounds penetrate the membranes and reach the cell's cytoplasm are still to be developed. Research is presently under way to overcome these problems.

The work carried out in M. E. Tolmasky's laboratory was supported by Public Health grant AI47115-02 and a grant from CSUPERB.

References

1. **Ainsa, J. A., E. Perez, V. Pelicic, F. X. Berthet, B. Gicquel, and C. Martin.** 1997. Aminoglycoside 2'-*N*-acetyltransferase genes are universally present in mycobacteria: characterization of the *aac(2')-Ic* gene from *Mycobacterium tuberculosis* and the *aac(2')-Id* gene from *Mycobacterium smegmatis*. *Mol. Microbiol.* **24:**431–441.
2. **Aires, J. R., T. Kohler, H. Nikaido, and P. Plesiat.** 1999. Involvement of an active efflux system in the natural

resistance of *Pseudomonas aeruginosa* to aminoglycosides. *Antimicrob. Agents Chemother.* **43:**2624–2628.

3. **Angus-Hill, M. L., R. N. Dutnall, S. T. Tafrov, R. Sternglanz, and V. Ramakrishnan.** 1999. Crystal structure of the histone acetyltransferase Hpa2: a tetrameric member of the Gcn5-related N-acetyltransferase superfamily. *J. Mol. Biol.* **294:**1311–1325.
4. **Azucena, E., and S. Mobashery.** 2001. Aminoglycoside-modifying enzymes: mechanisms of catalytic processes and inhibition. *Drug. Resist. Updat.* **4:**106–117.
5. **Beaber, J., B. Hochhut, and M. K. Waldor.** 2004. SOS response promotes horizontal dissemination of antibiotic resistance genes. *Nature* **427:**72–74.
6. **Beauclerk, A. A., and E. Cundliffe.** 1987. Sites of action of two ribosomal RNA methylases responsible for resistance to aminoglycosides. *J. Mol. Biol.* **193:**661–671.
7. **Boehr, D. D., A. R. Farley, G. D. Wright, and J. R. Cox.** 2002. Analysis of the pi-pi stacking interactions between the aminoglycoside antibiotic kinase APH(3′)-IIIa and its nucleotide ligands. *Chem. Biol.* **9:**1209–1217.
8. **Boehr, D. D., S. I. Jenkins, and G. D. Wright.** 2003. The molecular basis of the expansive substrate specificity of the antibiotic resistance enzyme aminoglycoside acetyltransferase-6′-aminoglycoside phosphotransferase-2″. The role of ASP-99 as an active site base important for acetyl transfer. *J. Biol. Chem.* **278:**12873–12880.
9. **Boltner, D., C. MacMahon, J. T. Pembroke, P. Strike, and A. M. Osborn.** 2002. R391: a conjugative integrating mosaic comprised of phage, plasmid, and transposon elements. *J. Bacteriol.* **184:**5158–5169.
10. **Brodersen, D. E., W. M. Clemons, Jr., A. P. Carter, R. J. Morgan-Warren, B. T. Wimberly, and V. Ramakrishnan.** 2000. The structural basis for the action of the antibiotics tetracycline, pactamycin, and hygromycin B on the 30S ribosomal subunit. *Cell* **103:**1143–1154.
11. **Bryan, L., and H. van der Elzen.** 1977. Effects of membrane-energy mutations and cations on streptomycin and gentamicin accumulation by bacteria: a model for entry of streptomycin and gentamicin in susceptible and resistant bacteria. *Antimicrob. Agents Chemother.* **12:**163–177.
12. **Burk, D. L., and A. M. Berghuis.** 2002. Protein kinase inhibitors and antibiotic resistance. *Pharmacol. Ther.* **93:** 283–292.
13. **Burk, D. L., N. Ghuman, L. E. Wybenga-Groot, and A. M. Berghuis.** 2003. X-ray structure of the AAC(6′)-Ii antibiotic resistance enzyme at 1.8 A resolution; examination of oligomeric arrangements in GNAT superfamily members. *Protein Sci.* **12:**426–437.
14. **Burrus, V., and M. K. Waldor.** 2004. Formation of SXT tandem arrays and SXT-R391 hybrids. *J. Bacteriol.* **186:** 2636–2645.
15. **Burrus, V., and M. K. Waldor.** 2004. Shaping bacterial genomes with integrative and conjugative elements. *Res. Microbiol.* **155:**376–386.
16. **Cameron, F. H., D. J. Groot Obbink, V. P. Ackerman, and R. M. Hall.** 1986. Nucleotide sequence of the AAD(2″) aminoglycoside adenylyltransferase determinant *aadB*. Evolutionary relationship of this region with those surrounding *aadA* in R538-1 and *dhfrII* in R388. *Nucleic Acids Res.* **14:**8625–8635.
17. **Cannon, P. M., and P. Strike.** 1992. Complete nucleotide sequence and gene organization of plasmid NTP16. *Plasmid* **27:**220–230.
18. **Carter, A. P., W. M. Clemons, D. E. Brodersen, R. J. Morgan-Warren, B. T. Wimberly, and V. Ramakrishnan.** 2000. Functional insights from the structure of the 30S ribosomal subunit and its interactions with antibiotics. *Nature* **407:**340–348.
19. **Casin, I., F. Bordon, P. Bertin, A. Coutrot, I. Podglajen, R. Brasseur, and E. Collatz.** 1998. Aminoglycoside 6′-*N*-acetyltransferase variants of the Ib type with altered substrate profile in clinical isolates of *Enterobacter cloacae* and *Citrobacter freundii*. *Antimicrob. Agents Chemother.* **42:** 209–215.
20. **Casin, I., B. Hanau-Bercot, I. Podglajen, H. Vahaboglu, and E. Collatz.** 2003. *Salmonella enterica* serovar Typhimurium *bla*(PER-1)-carrying plasmid pSTI1 encodes an extended-spectrum aminoglycoside 6′-*N*-acetyltransferase of type Ib. *Antimicrob. Agents Chemother.* **47:**697–703.
21. **Centron, D., and P. H. Roy.** 1998. Characterization of the 6′-*N*-aminoglycoside acetyltransferase gene *aac(6′)-Iq* from the integron of a natural multiresistance plasmid. *Antimicrob. Agents Chemother.* **42:**1506–1508.
22. **Centron, D., and P. H. Roy.** 2002. Presence of a group II intron in a multiresistant *Serratia marcescens* strain that harbors three integrons and a novel gene fusion. *Antimicrob. Agents Chemother.* **46:**1402–1409.
23. **Chamorro, R. M., L. A. Actis, J. H. Crosa, and M. E. Tolmasky.** 1990. Dissemination of plasmid-mediated amikacin resistance among pathogenic *Klebsiella pneumoniae*. *Medicina* (Buenos Aires) **50:**543–547.
24. **Chang, A. C. Y., and S. N. Cohen.** 1978. Construction and characterization of amplifiable multicopy DNA cloning vehicles derived from the P15A cryptic miniplasmid. *J. Bacteriol.* **134:**1141–1156.
25. **Chavideh, R., S. Sholly, D. Panaite, and M. E. Tolmasky.** 1999. Effects of F171 mutations in the 6′-*N*-acetyltransferase type Ib [AAC(6′)-Ib] enzyme on susceptibility to aminoglycosides. *Antimicrob. Agents Chemother.* **43:**2811–2812.
26. **Chow, J. W., V. Kak, I. You, S. J. Kao, J. Petrin, D. B. Clewell, S. A. Lerner, G. H. Miller, and K. J. Shaw.** 2001. Aminoglycoside resistance genes *aph(2″)-Ib* and *aac(6′)-Im* detected together in strains of both *Escherichia coli* and *Enterococcus faecium*. *Antimicrob. Agents Chemother.* **45:**2691–2694.
27. **Danese, P. N., and T. J. Silhavy.** 1998. Targeting and assembly of periplasmic and outer-membrane proteins in *Escherichia coli*. *Annu. Rev. Genet.* **32:**59–94.
28. **Davis, B. D.** 1987. Mechanism of bactericidal action of aminoglycosides. *Microbiol. Rev.* **51:**341–350.
29. **Dery, K. J., R. Chavideh, V. Waters, R. Chamorro, L. S. Tolmasky, and M. E. Tolmasky.** 1997. Characterization of the replication and mobilization regions of the multiresistance *Klebsiella pneumoniae* plasmid pJHCMW1. *Plasmid* **38:**97–105.
30. **Dery, K. J., B. Soballe, M. S. Witherspoon, D. Bui, R. Koch, D. J. Sherratt, and M. E. Tolmasky.** 2003. The aminoglycoside 6′-*N*-acetyltransferase type Ib encoded by Tn*1331* is evenly distributed within the cell's cytoplasm. *Antimicrob. Agents Chemother.* **47:**2897–2902.

31. **Dhillon, J., R. Fielding, J. Adler-Moore, R. L. Goodall, and D. Mitchison.** 2001. The activity of low-clearance liposomal amikacin in experimental murine tuberculosis. *J. Antimicrob. Chemother.* **48:**869–876.

32. **Doi, Y., J. Wachino, K. Yamane, N. Shibata, T. Yagi, K. Shibayama, H. Kato, and Y. Arakawa.** 2004. Spread of novel aminoglycoside resistance gene *aac(6')-Iad* among *Acinetobacter* clinical isolates in Japan. *Antimicrob. Agents Chemother.* **48:**2075–2080.

33. **Doi, Y., K. Yokoyama, K. Yamane, J. Wachino, N. Shibata, T. Yagi, K. Shibayama, H. Kato, and Y. Arakawa.** 2004. Plasmid-mediated 16S rRNA methylase in *Serratia marcescens* conferring high-level resistance to aminoglycosides. *Antimicrob. Agents Chemother.* **48:**491–496.

34. **Doublet, B., F. Weill, L. Fabre, E. Chaslus-Dancla, and A. Cloeckaert.** 2004. Variant *Salmonella* genomic island 1 antibiotic resistance gene cluster containing a novel 3′-*N*-aminoglycoside acetyltransferase gene cassette, *aac(3)-Id*, in *Salmonella enterica* serovar Newport. *Antimicrob. Agents Chemother.* **48:**3806–3812.

35. **Dubois, V., L. Poirel, C. Marie, C. Arpin, P. Nordmann, and C. Quentin.** 2002. Molecular characterization of a novel class 1 integron containing *bla*(GES-1) and a fused product of *aac3-Ib/aac6'-Ib'* gene cassettes in *Pseudomonas aeruginosa*. *Antimicrob. Agents Chemother.* **46:**638–645.

36. **Dunant, P., M. C. Walter, G. Karpati, and H. Lochmuller.** 2003. Gentamicin fails to increase dystrophin expression in dystrophin-deficient muscle. *Muscle Nerve* **27:**624–627.

37. **Dyda, F., D. C. Klein, and A. B. Hickman.** 2000. GCN5-related N-acetyltransferases: a structural overview. *Annu. Rev. Biophys. Biomol. Struct.* **29:**81–103.

38. **Ferretti, J. J., K. S. Gilmore, and P. Courvalin.** 1986. Nucleotide sequence analysis of the gene specifying the bifunctional 6′-aminoglycoside acetyltransferase 2″-aminoglycoside phosphotransferase enzyme in *Streptococcus faecalis* and identification and cloning of gene regions specifying the two activities. *J. Bacteriol.* **167:**631–638.

39. **Fourmy, D., M. I. Recht, S. C. Blanchard, and J. D. Puglisi.** 1996. Structure of the A site of Escherichia coli 16S ribosomal RNA complexed with an aminoglycoside antibiotic. *Science* **274:**1367–1371.

40. **Fourmy, D., M. I. Recht, and J. D. Puglisi.** 1998. Binding of neomycin-class aminoglycoside antibiotics to the A-site of 16 S rRNA. *J. Mol. Biol.* **277:**347–362.

41. **Fourmy, D., S. Yoshizawa, and J. D. Puglisi.** 1998. Paromomycin binding induces a local conformational change in the A-site of 16 S rRNA. *J. Mol. Biol.* **277:**333–345.

42. **Franklin, K., and A. J. Clarke.** 2001. Overexpression and characterization of the chromosomal aminoglycoside 2′-*N*-acetyltransferase of *Providencia stuartii*. *Antimicrob. Agents Chemother.* **45:**2238–2244.

43. **Galimand, M., P. Courvalin, and T. Lambert.** 2003. Plasmid-mediated high-level resistance to aminoglycosides in *Enterobacteriaceae* due to 16S rRNA methylation. *Antimicrob. Agents Chemother.* **47:**2565–2571.

44. **Guerry, P., J. van Embden, and S. Falkow.** 1974. Molecular nature of two nonconjugative plasmids carrying drug resistance genes. *J. Bacteriol.* **117:**619–630.

45. **Haddad, J., L. P. Kotra, B. Llano-Sotelo, C. Kim, E. F. Azucena, Jr., M. Liu, S. B. Vakulenko, C. S. Chow, and S. Mobashery.** 2002. Design of novel antibiotics that bind to the ribosomal acyltransfer site. *J. Am. Chem. Soc.* **124:**3229–3237.

46. **Hamilton-Miller, J. M., and S. Shah.** 1995. Activity of the semi-synthetic kanamycin B derivative, arbekacin against methicillin-resistant *Staphylococcus aureus*. *J. Antimicrob. Chemother.* **35:**865–868.

47. **Hannecart-Pokorni, E., F. Depuydt, L. de Wit, E. van Bossuyt, J. Content, and R. Vanhoof.** 1997. Characterization of the 6′-*N*-aminoglycoside acetyltransferase gene *aac(6')-Im* [corrected] associated with a sulI-type integron. *Antimicrob. Agents Chemother.* **41:**314–318.

48. **Hansson, K., O. Skold, and L. Sundstrom.** 1997. Non-palindromic attl sites of integrons are capable of site-specific recombination with one another and with secondary targets. *Mol. Microbiol.* **26:**441–453.

49. **Hickman, A. B., M. A. Namboodiri, D. C. Klein, and F. Dyda.** 1999. The structural basis of ordered substrate binding by serotonin *N*-acetyltransferase: enzyme complex at 1.8 A resolution with a bisubstrate analog. *Cell* **97:**361–369.

50. **Hochhut, B., Y. Lotfi, D. Mazel, S. M. Faruque, R. Woodgate, and M. K. Waldor.** 2001. Molecular analysis of antibiotic resistance gene clusters in *Vibrio cholerae* O139 and O1 SXT constins. *Antimicrob. Agents Chemother.* **45:**2991–3000.

51. **Hocquet, D., C. Vogne, F. El Garch, A. Vejux, N. Gotoh, A. Lee, O. Lomovskaya, and P. Plesiat.** 2003. MexXY-OprM efflux pump is necessary for adaptive resistance of *Pseudomonas aeruginosa* to aminoglycosides. *Antimicrob. Agents Chemother.* **47:**1371–1375.

52. **Hoshi, H., S. Aburaki, S. Iimura, T. Yamasaki, T. Naito, and H. Kawaguchi.** 1990. Amikacin analogs with a fluorinated amino acid side chain. *J. Antibiot.* (Tokyo) **43:**858–872.

53. **Hotta, K., A. Sunada, J. Ishikawa, S. Mizuno, Y. Ikeda, and S. Kondo.** 1998. The novel enzymatic 3″-*N*-acetylation of arbekacin by an aminoglycoside 3-*N*-acetyltransferase of *Streptomyces* origin and the resulting activity. *J. Antibiot.* (Tokyo) **51:**735–742.

54. **Howard, M., C. Anderson, U. Fass, S. Khatri, R. Gesteland, J. Atkins, and K. Flanigan.** 2004. Readthrough of dystrophin stop codon mutations induced by aminoglycosides. *Ann. Neurol.* **55:**422–426.

55. **Hutchin, T., and G. Cortopassi.** 1994. Proposed molecular and cellular mechanism for aminoglycoside ototoxicity. *Antimicrob. Agents Chemother.* **38:**2517–2520.

56. **Iida, S., J. Meyer, P. Linder, N. Goto, R. Nakaya, H. J. Reif, and W. Arber.** 1982. The kanamycin resistance transposon Tn*2680* derived from the R plasmid Rts1 and carried by phage P1Km has flanking 0.8-kb-long direct repeats. *Plasmid* **8:**187–198.

57. **Iwanaga, M., C. Toma, T. Miyazato, S. Insisiengmay, N. Nakasone, and M. Ehara.** 2004. Antibiotic resistance conferred by a class I integron and SXT constin in *Vibrio cholerae* O1 strains isolated in Laos. *Antimicrob. Agents Chemother.* **48:**2364–2369.

58. **Kawaguchi, H.** 1976. Discovery, chemistry, and activity of amikacin. *J. Infect. Dis.* **134**(Suppl.):S242–S248.

59. **Kehrenberg, C., and S. Schwarz.** 2002. Nucleotide sequence and organization of plasmid pMVSCS1 from *Mannheimia varigena*: identification of a multiresistance gene cluster. *J. Antimicrob. Chemother.* **49:**383–386.

60. **Kotra, L. P., J. Haddad, and S. Mobashery.** 2000. Aminoglycosides: perspectives on mechanisms of action and resistance and strategies to counter resistance. *Antimicrob. Agents Chemother.* **44:**3249–3256.

61. **Lambert, T., M. C. Ploy, and P. Courvalin.** 1994. A spontaneous point mutation in the *aac(6')-Ib'* gene results in altered substrate specificity of aminoglycoside 6′-*N*-acetyltransferase of a *Pseudomonas fluorescens* strain. *FEMS Microbiol. Lett.* **115:**297–304.

62. **Levesque, C., L. Piche, C. Larose, and P. H. Roy.** 1995. PCR mapping of integrons reveals several novel combinations of resistance genes. *Antimicrob. Agents Chemother.* **39:**185–191.

63. **Liebert, C. A., R. M. Hall, and A. O. Summers.** 1999. Transposon Tn*21*, flagship of the floating genome. *Microbiol. Mol. Biol. Rev.* **63:**507–522.

64. **Lovering, A. M., L. O. White, and D. S. Reeves.** 1987. AAC(1): a new aminoglycoside-acetylating enzyme modifying the Cl aminogroup of apramycin. *J. Antimicrob. Chemother.* **20:**803–813.

65. **Lynch, S., R. Gonzalez, and J. D. Puglisi.** 2003. Comparison of X-ray crystal structure of the 30S subunit-antibiotic complex with NMR structure of decoding site oligonucleotide-paromomycin complex. *Structure* **11:**43–53.

66. **Magnet, S., and J. S. Blanchard.** 2005. Molecular insights into aminoglycoside action and resistance. *Chem. Rev.* **105:** 477–498.

67. **Magnet, S., P. Courvalin, and T. Lambert.** 2001. Resistance-nodulation-cell division-type efflux pump involved in aminoglycoside resistance in *Acinetobacter baumannii* strain BM4454. *Antimicrob. Agents Chemother.* **45:**3375–3380.

68. **Magnet, S., T. A. Smith, R. Zheng, P. Nordmann, and J. S. Blanchard.** 2003. Aminoglycoside resistance resulting from tight drug binding to an altered aminoglycoside acetyltransferase. *Antimicrob. Agents Chemother.* **47:**1577–1583.

69. **Meier, A., P. Kirschner, F. C. Bange, U. Vogel, and E. C. Bottger.** 1994. Genetic alterations in streptomycin-resistant *Mycobacterium tuberculosis*: mapping of mutations conferring resistance. *Antimicrob. Agents Chemother.* **38:**228–233.

70. **Mendes, R., M. Toleman, J. Ribeiro, H. Sader, R. Jones, and T. Walsh.** 2004. Integron carrying a novel metallo-β-lactamase gene, bla_{IMP-16}, and a fused form of aminoglycoside-resistance gene *aac(6')-30/aac(6')-Ib*: report from the SENTRY antimicrobial surveillance program. *Antimicrob. Agents Chemother.* **48:**4693–4702.

71. **Miller, G. H., G. Arcieri, M. J. Weinstein, and J. A. Waitz.** 1976. Biological activity of netilmicin, a broad-spectrum semisynthetic aminoglycoside antibiotic. *Antimicrob. Agents Chemother.* **10:**827–836.

72. **Miller, G. H., F. J. Sabatelli, R. S. Hare, Y. Glupczynski, P. Mackey, D. Shlaes, K. Shimizu, K. J. Shaw, et al.** 1997. The most frequent aminoglycoside resistance mechanisms—changes with time and geographic area: a reflection of aminoglycoside usage patterns? *Clin. Infect. Dis.* **24**(Suppl. 1): S46–S62.

73. **Mingeot-Leclercq, M. P., Y. Glupczynski, and P. M. Tulkens.** 1999. Aminoglycosides: activity and resistance. *Antimicrob. Agents Chemother.* **43:**727–737.

74. **Mingeot-Leclercq, M. P., and P. M. Tulkens.** 1999. Aminoglycosides: nephrotoxicity. *Antimicrob. Agents Chemother.* **43:**1003–1012.

75. **Moore, R. A., D. DeShazer, S. Reckseidler, A. Weissman, and D. E. Woods.** 1999. Efflux-mediated aminoglycoside and macrolide resistance in *Burkholderia pseudomallei*. *Antimicrob. Agents Chemother.* **43:**465–470.

76. **Mulvey, M., D. Boyd, L. Baker, O. Mykytczuk, E. Resi, M. Asensi, D. Rodrigues, and L. Ng.** 2004. Characterization of a *Salmonella enterica* serovar Agona strain harbouring a class 1 integron containing novel OXA-type β-lactamase (bla_{OXA-53}) and 6′-*N*-aminoglycoside acetyltransferase genes *[aac(6')-I30]*. *J. Antimicrob. Chemother.* **54:**354–359.

77. **Neuwald, A. F., and D. Landsman.** 1997. GCN5-related histone *N*-acetyltransferases belong to a diverse superfamily that includes the yeast SPT10 protein. *Trends Biochem. Sci.* **22:**154–155.

78. **Novick, R. P., R. C. Clowes, S. N. Cohen, R. Curtiss III, N. Datta, and S. Falkow.** 1976. Uniform nomenclature for bacterial plasmids: a proposal. *Bacteriol. Rev.* **40:**168–189.

79. **Nurizzo, D., S. C. Shewry, M. H. Perlin, S. A. Brown, J. N. Dholakia, R. L. Fuchs, T. Deva, E. N. Baker, and C. A. Smith.** 2003. The crystal structure of aminoglycoside-3′-phosphotransferase-IIa, an enzyme responsible for antibiotic resistance. *J. Mol. Biol.* **327:**491–506.

80. **Oka, A., H. Sugisaki, and M. Takanami.** 1981. Nucleotide sequence of the kanamycin resistance transposon Tn*903*. *J. Mol. Biol.* **147:**217–226.

81. **Over, U., D. Gur, S. Unal, and G. H. Miller.** 2001. The changing nature of aminoglycoside resistance mechanisms and prevalence of newly recognized resistance mechanisms in Turkey. *Clin. Microbiol. Infect.* **7:**470–478.

82. **Panaite, D. M., and M. E. Tolmasky.** 1998. Characterization of mutants of the 6′-*N*-acetyltransferase encoded by the multiresistance transposon Tn*1331*: effect of Phen171-to-Leu171 and Tyr80-to-Cys80 substitutions. *Plasmid* **39:** 123–133.

83. **Parent, R., and P. H. Roy.** 1992. The chloramphenicol acetyltransferase gene of Tn*2424*: a new breed of cat. *J. Bacteriol.* **174:**2891–2897.

84. **Pechere, J. C., and R. Dugal.** 1979. Clinical pharmacokinetics of aminoglycoside antibiotics. *Clin. Pharmacokinet.* **4:**170–199.

85. **Perlin, M. H., and S. A. Lerner.** 1981. Localization of an amikacin 3′-phosphotransferase in *Escherichia coli*. *J. Bacteriol.* **147:**320–325.

86. **Peters, J. E., and N. L. Craig.** 2001. Tn*7*: smarter than we thought. *Nat. Rev. Mol. Cell Biol.* **2:**806–814.

87. **Pfister, P., M. Risch, D. E. Brodersen, and E. C. Bottger.** 2003. Role of 16S rRNA helix 44 in ribosomal resistance to hygromycin B. *Antimicrob. Agents Chemother.* **47:**1496–1502.

88. **Pham, H., K. J. Dery, D. J. Sherratt, and M. E. Tolmasky.** 2002. Osmoregulation of dimer resolution at the plasmid pJHCMW1 *mwr* locus by *Escherichia coli* XerCD recombination. *J. Bacteriol.* **184:**1607–1616.

89. **Poirel, L., T. Lambert, S. Turkoglu, E. Ronco, J. Gaillard, and P. Nordmann.** 2001. Characterization of Class 1 integrons from *Pseudomonas aeruginosa* that contain the *bla*(VIM-2) carbapenem-hydrolyzing β-lactamase gene and of two novel aminoglycoside resistance gene cassettes. *Antimicrob. Agents Chemother.* **45:**546–552.

90. **Politano, L., G. Nigro, V. Nigro, G. Piluso, S. Papparella, O. Paciello, and L. I. Comi.** 2003. Gentamicin administration in Duchenne patients with premature stop codon. Preliminary results. *Acta Myol.* **22:**15–21.

91. **Poux, A. N., M. Cebrat, C. M. Kim, P. A. Cole, and R. Marmorstein.** 2002. Structure of the GCN5 histone acetyltransferase bound to a bisubstrate inhibitor. *Proc. Natl. Acad. Sci. USA* **99:**14065–14070.

92. **Rather, P. N., H. Munayyer, P. A. Mann, R. S. Hare, G. H. Miller, and K. J. Shaw.** 1992. Genetic analysis of bacterial acetyltransferases: identification of amino acids determining the specificities of the aminoglycoside 6′-*N*-acetyltransferase Ib and IIa proteins. *J. Bacteriol.* **174:**3196–3203.

93. **Rather, P. N., E. Orosz, K. J. Shaw, R. Hare, and G. Miller.** 1993. Characterization and transcriptional regulation of the 2′-*N*-acetyltransferase gene from *Providencia stuartii*. *J. Bacteriol.* **175:**6492–6498.

94. **Recchia, G. D., and R. M. Hall.** 1995. Gene cassettes: a new class of mobile element. *Microbiology* **141**(Pt. 12):3015–3027.

95. **Riccio, M. L., J. D. Docquier, E. Dell'Amico, F. Luzzaro, G. Amicosante, and G. M. Rossolini.** 2003. Novel 3-*N*-aminoglycoside acetyltransferase gene, *aac(3)-Ic*, from a *Pseudomonas aeruginosa* integron. *Antimicrob. Agents Chemother.* **47:**1746–1748.

96. **Riccio, M. L., N. Franceschini, L. Boschi, B. Caravelli, G. Cornaglia, R. Fontana, G. Amicosante, and G. M. Rossolini.** 2000. Characterization of the metallo-β-lactamase determinant of *Acinetobacter baumannii* AC-54/97 reveals the existence of *bla*(IMP) allelic variants carried by gene cassettes of different phylogeny. *Antimicrob. Agents Chemother.* **44:**1229–1235.

97. **Rice, L., D. Sahm, and R. Bonomo.** 2003. Mechanisms of resistance to antibacterial agents, p. 1074–1101. *In* P. Murray, E. Baron, J. Jorgensen, M. Pfaller, and R. Yolken (ed.), *Manual of Clinical Microbiology*, vol. 1. ASM Press, Washington, D.C.

97a. **Robicsek, A., J. Strahilevitz, G. Jacoby, M. Macielag, D. Abbanat, C. Park, K. Bush, and D. Hooper.** Fluoroquinolone-modifying enzyme: a new adaptation of a common aminoglycoside acetyltransferase. *Nat. Med.* **12:**83–87.

98. **Rosenberg, E. Y., D. Ma, and H. Nikaido.** 2000. AcrD of *Escherichia coli* is an aminoglycoside efflux pump. *J. Bacteriol.* **182:**1754–1756.

99. **Rudant, E., P. Courvalin, and T. Lambert.** 1997. Loss of intrinsic aminoglycoside resistance in *Acinetobacter haemolyticus* as a result of three distinct types of alterations in the *aac(6′)-Ig* gene, including insertion of IS*17*. *Antimicrob. Agents Chemother.* **41:**2646–2651.

100. **Sakon, J., H. H. Liao, A. M. Kanikula, M. M. Benning, I. Rayment, and H. M. Holden.** 1993. Molecular structure of kanamycin nucleotidyltransferase determined to 3.0-A resolution. *Biochemistry* **32:**11977–11984.

101. **Salipante, S. J., and B. G. Hall.** 2003. Determining the limits of the evolutionary potential of an antibiotic resistance gene. *Mol. Biol. Evol.* **20:**653–659.

102. **Sarno, R., H. Ha, N. Weinsetel, and M. E. Tolmasky.** 2003. Inhibition of aminoglycoside 6′-*N*-acetyltransferase type Ib-mediated amikacin resistance by antisense oligodeoxynucleotides. *Antimicrob. Agents Chemother.* **47:**3296–3304.

103. **Sarno, R., G. McGillivary, D. J. Sherratt, L. A. Actis, and M. E. Tolmasky.** 2002. Complete nucleotide sequence of *Klebsiella pneumoniae* multiresistance plasmid pJHCMW1. *Antimicrob. Agents Chemother.* **46:**3422–3427.

104. **Scaglione, F., S. Dugnani, G. Demartini, M. M. Arcidiacono, C. E. Cocuzza, and F. Fraschini.** 1995. Bactericidal kinetics of an in vitro infection model of once-daily ceftriaxone plus amikacin against gram-positive and gram-negative bacteria. *Chemotherapy* **41:**239–246.

105. **Shaw, K. J., H. Munayyer, P. N. Rather, R. S. Hare, and G. H. Miller.** 1993. Nucleotide sequence analysis and DNA hybridization studies of the *ant(4′)-IIa* gene from *Pseudomonas aeruginosa*. *Antimicrob. Agents Chemother.* **37:**708–714.

106. **Shaw, K. J., P. N. Rather, R. S. Hare, and G. H. Miller.** 1993. Molecular genetics of aminoglycoside resistance genes and familial relationships of the aminoglycoside-modifying enzymes. *Microbiol. Rev.* **57:**138–163.

107. **Shaw, K. J., and G. D. Wright.** 2000. Aminoglycoside resistance in gram-positive bacteria, p. 635–646. *In* V. Fischetti, R. Novick, J. Ferretti, D. Portnoy, and J. Rood (ed.), *Gram-Positive Pathogens*. ASM Press, Washington, D.C.

108. **Shmara, A., N. Weinsetel, K. J. Dery, R. Chavideh, and M. E. Tolmasky.** 2001. Systematic analysis of a conserved region of the aminoglycoside 6′-*N*-acetyltransferase type Ib. *Antimicrob. Agents Chemother.* **45:**3287–3292.

109. **Sucheck, S., A. Wong, K. Koeller, D. Boher, K. Draker, P. Sears, and G. Wright.** 2000. Design of bifunctional antibiotics that target bacterial rRNA and inhibit resistance-causing enzymes. *J. Am. Chem. soc.* **122:**5230–5231.

110. **Taber, H. W., J. P. Mueller, P. F. Miller, and A. S. Arrow.** 1987. Bacterial uptake of aminoglycoside antibiotics. *Microbiol. Rev.* **51:**439–457.

111. **Tauch, A., S. Krieft, J. Kalinowski, and A. Puhler.** 2000. The 51,409-bp R-plasmid pTP10 from the multiresistant clinical isolate *Corynebacterium striatum* M82B is composed of DNA segments initially identified in soil bacteria and in plant, animal, and human pathogens. *Mol. Gen. Genet.* **263:**1–11.

112. **Taylor, L. A., and R. E. Rose.** 1988. A correction in the nucleotide sequence of the Tn*903* kanamycin resistance determinant in pUC4K. *Nucleic Acids Res.* **16:**358.

113. **Tenover, F. C.** 2001. Development and spread of bacterial resistance to antimicrobial agents: an overview. *Clin. Infect. Dis.* **33**(Suppl. 3):S108–S115.

114. **Thomas, J. D., R. A. Daniel, J. Errington, and C. Robinson.** 2001. Export of active green fluorescent protein to the periplasm by the twin-arginine translocase (Tat) pathway in *Escherichia coli*. *Mol. Microbiol.* **39:**47–53.

115. **Thompson, P. R., D. D. Boehr, A. M. Berghuis, and G. D. Wright.** 2002. Mechanism of aminoglycoside antibiotic kinase APH(3′)-IIIa: role of the nucleotide positioning loop. *Biochemistry* **41:**7001–7007.

116. **Tolmasky, M. E.** 2000. Bacterial resistance to aminoglycosides and β-lactams: the Tn*1331* transposon paradigm. *Front. Biosci.* **5:**D20–D29.

117. **Tolmasky, M. E.** 1990. Sequencing and expression of *aadA*, *bla*, and *tnpR* from the multiresistance transposon Tn*1331*. *Plasmid* **24:**218–226.

118. **Tolmasky, M. E., R. M. Chamorro, J. H. Crosa, and P. M. Marini.** 1988. Transposon-mediated amikacin resistance in *Klebsiella pneumoniae*. *Antimicrob. Agents Chemother.* **32:**1416–1420.

119. **Tolmasky, M. E., and J. H. Crosa.** 1993. Genetic organization of antibiotic resistance genes (*aac(6′)-Ib*, *aadA*, and *oxa9*) in the multiresistance transposon Tn*1331*. *Plasmid* **29:**31–40.

120. **Tolmasky, M. E., and J. H. Crosa.** 1987. Tn*1331*, a novel multiresistance transposon encoding resistance to amikacin and ampicillin in *Klebsiella pneumoniae*. *Antimicrob. Agents Chemother.* **31:**1955–1960.

121. **Vakulenko, S. B., and S. Mobashery.** 2003. Versatility of aminoglycosides and prospects for their future. *Clin. Microbiol. Rev.* **16:**430–450.

122. **Vanhoof, R., E. Hannecart-Pokorni, and J. Content.** 1998. Nomenclature of genes encoding aminoglycoside-modifying enzymes. *Antimicrob. Agents Chemother.* **42:** 483.

123. **Vazquez-Laslop, N., H. Lee, R. Hu, and A. A. Neyfakh.** 2001. Molecular sieve mechanism of selective release of cytoplasmic proteins by osmotically shocked *Escherichia coli*. *J. Bacteriol.* **183:**2399–2404.

124. **Vetting, M., S. L. Roderick, S. Hegde, S. Magnet, and J. S. Blanchard.** 2003. What can structure tell us about in vivo function? The case of aminoglycoside-resistance genes. *Biochem. Soc. Trans.* **31:**520–522.

125. **Vetting, M. W., S. S. Hegde, F. Javid-Majd, J. S. Blanchard, and S. L. Roderick.** 2002. Aminoglycoside 2′-*N*-acetyltransferase from *Mycobacterium tuberculosis* in complex with coenzyme A and aminoglycoside substrates. *Nat. Struct. Biol.* **9:**653–658.

126. **Vicens, Q., and E. Westhof.** 2001. Crystal structure of paromomycin docked into the eubacterial ribosomal decoding A site. *Structure* **9:**647–658.

127. **Vicens, Q., and E. Westhof.** 2003. Molecular recognition of aminoglycoside antibiotics by ribosomal RNA and resistance enzymes: an analysis of x-ray crystal structures. *Biopolymers* **70:**42–57.

128. **Vliegenthart, J. S., P. A. Ketelaar-van Gaalen, J. Eelhart, and J. A. van de Klundert.** 1991. Localisation of the aminoglycoside-(3)-*N*-acetyltransferase isoenzyme II in *Escherichia coli*. *FEMS Microbiol. Lett.* **65:**101–105.

129. **Waldor, M. K., H. Tschape, and J. J. Mekalanos.** 1996. A new type of conjugative transposon encodes resistance to sulfamethoxazole, trimethoprim, and streptomycin in *Vibrio cholerae* O139. *J. Bacteriol.* **178:**4157–4165.

130. **Walker, R. J., and G. G. Duggin.** 1988. Drug nephrotoxicity. *Annu. Rev. Pharmacol. Toxicol.* **28:**331–345.

131. **Walter, F., Q. Vicens, and E. Westhof.** 1999. Aminoglycoside-RNA interactions. *Curr. Opin. Chem. Biol.* **3:**694–704.

132. **Westbrock-Wadman, S., D. R. Sherman, M. J. Hickey, S. N. Coulter, Y. Q. Zhu, P. Warrener, L. Y. Nguyen, R. M. Shawar, K. R. Folger, and C. K. Stover.** 1999. Characterization of a *Pseudomonas aeruginosa* efflux pump contributing to aminoglycoside impermeability. *Antimicrob. Agents Chemother.* **43:**2975–2983.

133. **White, D. G., K. Maneewannakul, E. von Hofe, M. Zillman, W. Eisenberg, A. K. Field, and S. B. Levy.** 1997. Inhibition of the multiple antibiotic resistance (*mar*) operon in *Escherichia coli* by antisense DNA analogs. *Antimicrob. Agents Chemother.* **41:**2699–2704.

134. **Wolf, E., A. Vassilev, Y. Makino, A. Sali, Y. Nakatani, and S. K. Burley.** 1998. Crystal structure of a GCN5-related N-acetyltransferase: *Serratia marcescens* aminoglycoside 3-*N*-acetyltransferase. *Cell* **94:**439–449.

135. **Woloj, M., M. E. Tolmasky, M. C. Roberts, and J. H. Crosa.** 1986. Plasmid-encoded amikacin resistance in multiresistant strains of *Klebsiella pneumoniae* isolated from neonates with meningitis. *Antimicrob. Agents Chemother.* **29:**315–319.

136. **Wright, G. D., and P. R. Thompson.** 1999. Aminoglycoside phosphotransferases: proteins, structure, and mechanism. *Front. Biosci.* **4:**D9–D21.

137. **Wybenga-Groot, L. E., K. Draker, G. D. Wright, and A. M. Berghuis.** 1999. Crystal structure of an aminoglycoside 6′-*N*-acetyltransferase: defining the GCN5-related *N*-acetyltransferase superfamily fold. *Structure* **7:**497–507.

138. **Yamane, K., Y. Doi, K. Yokoyama, T. Yagi, H. Kurokawa, N. Shibata, K. Shibayama, H. Kato, and Y. Arakawa.** 2004. Genetic environments of the *rmtA* gene in *Pseudomonas aeruginosa* clinical isolates. *Antimicrob. Agents Chemother.* **48:**2069–2074.

139. **Yao, J., and R. Moellering.** 2003. Antibacterial agents, p. 1031–1073. *In* P. Murray, E. Baron, J. Jorgensen, M. Pfaller, and R. Yolken (ed.), *Manual of Clinical Microbiology*, vol. 1. ASM Press, Washington, D.C.

140. **Yoshizawa, S., D. Fourmy, R. G. Eason, and J. D. Puglisi.** 2002. Sequence-specific recognition of the major groove of RNA by deoxystreptamine. *Biochemistry* **41:**6263–6270.

141. **Yoshizawa, S., D. Fourmy, and J. D. Puglisi.** 1998. Structural origins of gentamicin antibiotic action. *EMBO J.* **17:**6437–6448.

Enzyme-Mediated Resistance to Antibiotics: Mechanisms, Dissemination, and Prospects for Inhibition
Edited by Robert A. Bonomo and Marcelo E. Tolmasky

Marilyn C. Roberts

5

rRNA Methylases and Resistance to Macrolide, Lincosamide, Streptogramin, Ketolide, and Oxazolidinone (MLSKO) Antibiotics

CLINICAL USE

Macrolides, lincosamides, streptogramins, ketolides, and oxazolidinones (MLSKO), though chemically distinct, are usually considered together because they all act at the 50S subunit of the ribosome (26). Bacteria acquire resistance to more than one of these antibiotic classes with the acquisition of a single gene or by mutations (65, 77). Erythromycin was the first to be used clinically in humans, while tylosin, another macrolide, has been used extensively in animals. Both were introduced over 50 years ago. Macrolides have good activity against a variety of gram-positive cocci, *Mycoplasma pneumoniae*, *Legionella pneumophila*, *Corynebacterium diphtheriae*, *Listeria monocytogenes*, *Neisseria* spp., *Bordetella pertussis*, *Chlamydia*, and the parasite *Entamoeba histolytica* (26, 65). To improve on the activity of macrolides, semisynthetic derivatives of erythromycin (roxithromycin, clarithromycin, azithromycin, and dirithromycin) were developed in the 1980s. These were designed to overcome instability in acidic environments and to provide improved pharmacokinetics. They had a broader spectrum of activities and have been used extensively for treatment of community-acquired pneumonia and infections with chlamydia, *Mycobacterium avium-intracellulare*, other rapid-growing *Mycobacterium* spp., *Moraxella catarrhalis*, *Borrelia burgdorferi*, and the parasites *Plasmodium* spp., *Toxoplasma gondii*, and *Babesia microti* (26). These semisynthetic derivatives were more stable, could be better absorbed, had improved intracellular and tissue penetration and improved pharmacodynamics, and required fewer doses per day compared with erythromycin. Unfortunately, they do share cross-resistance with erythromycin A. The semisynthetic derivatives became available just as the human immunodeficiency virus era began, which, along with their broader spectrum of activity, helped encourage the increased use of macrolides in the last 15 years.

More recently, clinical efficacy of macrolide treatment in chronic *Pseudomonas* airway infections has been described, and it is becoming the recommended therapy for cystic fibrosis patients (9, 27, 31, 66, 67). Macrolide therapy has also been used for other gram-negative infections such as typhoid fever (17). In addition, discovery and development of novel macrolides still occur in the hope of improving therapeutic potential.

Marilyn C. Roberts, Department of Pathobiology, Box 357238, School of Public Health and Community Medicine, University of Washington, Seattle, WA 98195.

Unfortunately, macrolide-resistant bacteria have become a world wide concern. Many have speculated on why the bacterial resistance has increased. A study by a Finnish group (71) supports the hypothesis of a linkage between the use of erythromycin therapy and increased resistance in the pathogen being treated. In Finland it was noted that there was an increased use of macrolides for treatment of *Streptococcus pyogenes* along with an increase in erythromycin-resistant *S. pyogenes* (71). A change in recommendations was issued that resulted in the reduction in the use of macrolides in Finland. After this therapy change, a steady decrease in the frequency of erythromycin-resistant group A streptococci was found.

In response to increased bacterial resistance to erythromycin in gram-positive cocci, a new series of compounds (the ketolides) were developed. These are semisynthetic derivatives of erythromycin A with a 3-keto group instead of an L-cladinose moiety (6). The ketolides penetrate a variety of cells and achieve high intracellular concentrations, which may account for their activity against intracellular pathogens (6). The first ketolide, telithromycin, is two to four times more active than clarithromycin on susceptible gram-positive cocci. Telithromycin is active against macrolide-resistant gram-positive cocci carrying an inducibly regulated rRNA methylase and/or macrolide efflux mechanisms (see below). Unfortunately, telithromycin-resistant *S. pyogenes* strains have already been identified in places like Taiwan, where this drug is currently not available (24).

Lincosamides (clindamycin and lincomycin) have activity against gram-positive cocci and have been important for the treatment of a variety of gram-positive and gram-negative anaerobic diseases, as well as for the parasite *T. gondii* (8, 14, 33, 43, 65). Clindamycin, a semisynthetic derivative of lincomycin, is used in the treatment of human disease and some animal infections, while lincomycin, a natural product, is used exclusively for a variety of animal diseases.

Streptogramins are made of two chemically distinct components, streptogramin A and streptogramin B. The current commercial product used for humans (Synercid) has 30% streptogramin A (quinupristin) and 70% streptogramin B (dalfopristin). The activity of the combination drug is about 10 times higher than the sum of the activities of the single components. One member of this group is virginiamycin, which has been used as a feed additive in poultry, swine, and cattle for 20 years prior to its use in humans. Synercid resistance has been found among methicillin-resistant *Staphylococcus aureus*, coagulase-negative staphylococci, *Streptococcus pneumoniae*, viridans group streptococci, vancomycin-resistant *Enterococcus* spp., *Leuconostoc* spp., *Lactobacillus* spp., and *Pediococcus* spp. in Taiwan (40) and other locations (18, 62). Luh et al. (40) suggested that previous extensive animal use of virginiamycin may have contributed to the high rates of resistance to Synercid shortly after the drug was licensed for human use.

Oxazolidinones were discovered in the late 1980s, but linezolid, the first analogue to be marketed, was discovered only when this family of agents was examined in the 1990s (39). Linezolid is well absorbed into the cerebrospinal fluid, bone, fat, and muscle and can be taken orally or intravenously (42). Linezolid has good activity against most gram-positive bacteria that cause community-acquired pneumonia, nosocomial pneumonia, bacteremia, and skin and soft tissue infections, methicillin-resistant *S. aureus*, penicillin-resistant *S. pneumoniae*, vancomycin-resistant enterococci, and *Nocardia brasiliensis* (15, 28, 39, 42, 83, 90). They may also have potential against anaerobic bacteria (91).

MECHANISM OF ACTION

All the MLSKO antibiotics inhibit protein synthesis by binding to the 50S bacterial ribosomal subunit, close to the peptidyltransferase center (6, 39, 57, 77, 79). The erythromycin binding site overlaps the binding site of the newer macrolides, lincosamides, and streptogramin B. When these antibiotics bind to the ribosome, they cause a dissociation of the peptidyl-tRNAs from the ribosome, with the result that protein synthesis is stopped. Streptogramin A binds at a different part of the subunit which inactivates the donor and acceptor sites of the peptidyltransferase. One hypothesis for the synergistic action of streptogramin A and B is that the binding of streptogramin A to protein L24 leads to a conformational change in the ribosome. This change enhances the binding capacity for streptogramin B, which binds to proteins L10 and L11. The three proteins are part of the exit channel in the ribosome, and binding of both antibiotics leads to an irreversible constriction of the channel which prevents the nascent peptide from exiting (4). More recent experiments based on rRNA footprinting suggest that the two antibiotics may actually have overlapping binding sites which allow for direct interaction between both drugs and the ribosome (58).

Linezolid differs from macrolides and lincosamides because its target does not overlap with the antibiotics described above. When linezolid binds to the 50S subunit, it prevents the initiation of protein synthesis (39). Its activity is not affected by methylation of the 23S rRNA which blocks the binding of macrolides, lincosamides, and group B streptogramins (MLS) to the ribosomes (39). Oxazolidinones bind to the 50S subunit and prevent the initiation of protein synthesis (39). Mutations in the 23S rRNA can affect both resistance to the MLS antibiotics and susceptiblity to oxazolidinones (13, 18, 32, 35, 39, 50, 55, 57, 77, 78, 84), or even the rRNA methylase (69). More details on the bacterial ribosome can be found in references 18 and 48.

MECHANISMS OF RESISTANCE

Macrolide resistance in the United States and Europe has doubled between 1992 and 1995 for *S. pneumoniae* (73). A 2002 review (33) has a table which lists erythromycin resistance levels for *S. pneumoniae* internationally over various time periods and illustrates that different bacterial populations vary in their level of macrolide resistance. The lowest levels were 1 to 2%, and the highest levels were >60% resistance.

Bacterial resistance to macrolides can be due to the acquisition of a resistant gene, usually associated with a mobile element, or by mutation with changes in the 23S rRNA at either the A2058 or A2059 position (*Escherichia coli* numbering) and/or ribosomal proteins L4 and/or L22 (33–35, 50, 55, 57, 58, 65, 77, 78, 84). Originally these mutations were found in pathogens that had one or two copies of the 23S rRNA like *Mycobacterium* or *Helicobacter* (50, 78), but similar mutations can now be found in species with 4 or more copies of the 23S rRNA such as *S. pneumoniae* (77). Mutations account for a relatively small percentage (1 to 3%) of macrolide-resistant *S. pneumoniae* and *Enterococcus* spp. (13, 24, 28, 57, 65). Mutations in innate efflux pumps can also alter susceptibility to the MLSKO antibiotics (15).

Recently an *S. aureus* strain carrying an inducibly expressed *erm*(A) gene was used to select for constitutive mutants using various MSLKO antibiotics. All the mutants had structural alterations in the translational attenuator region. These mutations prevented the formation of the mRNA secondary structure in the *erm*(A) regulatory region, creating constitutively expressed variants (69). This work suggests that acquired resistance genes may provide additional target sites for mutations, with the potential to change regulation of the gene as well as provide increased resistance to the bacteria.

A variety of other acquired genes which confer MLSKO resistance have been described and are listed in Table 5.1. These include genes for inactivating enzymes and 14 efflux genes coding proteins which pump the antibiotics out of the cell (2, 30, 41, 44, 49, 53, 54, 56, 60, 62, 74, 77, 80, 88). Among the newest efflux genes is the previously identified *mel* gene, which has recently been shown to be functional when cloned away from the other efflux genes and has been named *msr*(D) (12). The *msr*(D) gene has always been found with the *mef*(A) gene, and both genes have host ranges which include a number of gram-positive and gram-negative genera (54, 62). This is important because the *mef*(A) gene is now predominant in macrolide-resistant *S. pneumoniae* from North America and other locations rather than the *erm*(B) gene (1, 33, 34, 73, 81, 89). In addition, the *mef*(A)-*msr*(D) genes were the most commonly found of the seven acquired genes screened from randomly selected gram-negative commensal isolates (54).

The last group of acquired genes, and the focus of this chapter, are those that encode an adenine-N^6-methyltransferase (rRNA methylase) (33, 34, 62, 65, 70, 77). Currently, 32 different genes coding for enzymes that modify the 23S rRNA have been described (Table 5.1). This modification blocks the binding of the MLSK group of antibiotics and allows the bacterial ribosomes to continue to produce protein in the presence of macrolides, lincosamides, and streptogramin B.

GEOGRAPHIC DISTRIBUTION

The rRNA methylases add 1 or 2 methyl groups to a single adenine (A2058 *E. coli* numbering) in the 23S rRNA molecule (33, 34, 65, 77). This adenine is critical to the function of the ribosome, and methylation or change in the adenine to another base, either naturally or by mutation, changes the bacteria's resistance to the MLSKO antibiotics. Of the 32 characterized adenine-N^6-methyltransferase genes, 12 (38%) have been found exclusively in the *Streptomyces* spp., which are producers of macrolides. These are innate genes which in the *Streptomyces* spp. are often linked to biosynthetic pathways for macrolides production. Unfortunately, many of the *Streptomyces* genes are no longer available to the field, making it unlikely that further work will be done with them. Another 12 (41%) rRNA methylases have been found in only a single genus. These include four innate *Mycobacterium* rRNA methylases [encoded by *erm*(37), *erm*(38), *erm*(39), and *erm*(40)] from four different species (Table 5.2) (50–52). There is speculation that many of the different *Mycobacterium* spp. have different innate rRNA methylases (K. Nash, personal communication). The eight other methylases [encoded by *erm*(D), *erm*(34), *erm*(R), *erm*(W), *erm*(Y), *erm*(33), *erm*(35), and *erm*(36)] identified in a single genus have not had their host ranges extensively examined. Therefore, they may be in more genera than depicted in Table 5.2. The remaining eight of the methylases [encoded by *erm*(A), *erm*(B), *erm*(C), *erm*(F), *erm*(G), *erm*(Q), *erm*(T), and *erm*(X)] have been identified in multiple genera [2 to 32 genera] (Table 5.2). All of these *erm* genes except *erm*(X) have low G+C content (<35% G+C) and are thought to have originated in gram-positive bacteria, even if they were first identified in gram-negative species, as was the case with *erm*(F) (23). These six *erm* genes are also found in both gram-positive and gram-negative genera (Table 5.2).

Table 5.1. Mechanism of MLS resistance[a]

rRNA methylase	Efflux genes	Inactivating enzymes			
32 rRNA methylases (*erm* genes): (A), (B), (C), (D), (E), (F), (G), (H), (I), (N), (O), (Q), (R), (S), (T), (U) (V), (W), (X), (Y), (Z), (30), (31), (32), (33), (34), (35), (36), (37), (38), (39), (40)	14 efflux genes: *mef*(A), *msr*(A), (C), (D); *car*(A); *lmr*(A); *ole*(B), (C); *srm*(B); *tlr*(C); *vga*(A), (B); *lsa*(A), (B)	2 esterases: *ere*(A), (B)	2 lyases: *vgb*(A), (B)	10 transferases: *lnu*(A), (B), (C), (F); *vat*(A), (B), (C), (D), (E), (F)	4 phosphorylases: *mph*(A), (B), (C), (D)

[a]Data from reference 62.

The *erm*(33) gene sequence suggests that this gene was created by a recombination event between an *erm*(C) and an *erm*(A) gene. The first 284 bp at the 5′ end of the gene shows a high degree of homology to the *erm*(C) gene sequence, while the 403 bp at the 3′ end is indistinguishable from the *erm*(A) gene. The complete *erm*(33) gene shows overall homology to the both the *erm*(C) and *erm*(A) genes of 62% at the DNA level and 58% at the amino acid level and was thus given a new name (69). A recombination event between two tetracycline resistance genes [*tet*(O) and *tet*(W)] has also been recently identified (76), suggesting that the creation of the *erm*(33) gene may not represent a unique event. How frequently these types of recombination events occur is not known. However, one can speculate that other recombinant genes may be identified in the future.

Previous work suggests that the distribution of the various *erm* genes, even among gram-positive cocci (streptococci, staphylococci, and enterococci), varies by genus and sometimes by species. Studies have shown that the

Table 5.2 Location of rRNA methylases in the published literature[a]

Gene	No. of genera	Genera
erm(A)	7	*Actinobacillus, Bacteroides, Prevotella, Peptostreptococcus, Staphylococcus, Streptococcus, Helcococcus*
erm(B)	31	*Acinetobacter, Actinobacillus, Aerococcus, Arcanobacterium, Bacillus, Bacteroides, Citrobacter, Catenibacterium, Staphylococcus Corynebacterium, Clostridium, Enterobacter, Enterococcus, Escherichia, Eubacterium, Fusobacterium, Haemophilus, Klebsiella, Lactobacillus, Micrococcus, Neisseria, Pantoea, Pediococcus, Peptostreptococcus, Porphyromonas, Proteus, Pseudomonas, Serratia, Rothia, Staphylococcus, Streptococcus, Treponema, Wolinella*
erm(C)	16	*Actinobacillus, Actinomyces, Bacillus, Bacteroides, Corynebacterium, Enterococcus, Eubacterium, Haemophilus, Lactobacillus, Micrococcus, Neisseria, Peptostreptococcus, Prevotella, Staphylococcus, Streptococcus, Wolinella*
erm(D), (34)	1	*Bacillus*
erm(F)	22	*Actinobacillus, Actinomyces, Bacteroides, Clostridium, Corynebacterium, Enterococcus, Eubacterium, Fusobacterium, Gardnerella, Haemophilus, Lactobacillus, Mobiluncus, Neisseria, Porphyromonas, Prevotella, Peptostreptococcus, Selenomonas, Staphylococcus, Streptococcus, Treponema, Veillonella, Wolinella*
erm(G)	7	*Bacillus, Bacteroides, Lactobacillus, Prevotella, Porphyromonas, Catenibacterium, Staphylococcus*
erm(E), (H), (I), (N), (O), (S), (U), (V), (Z), (30), (31), (32)	1	*Streptomyces*
erm(Q)	6	*Actinobacillus, Bacteroides, Clostridium, Staphylococcus, Streptococcus, Wolinella*
erm(R)	1	*Arthrobacter*
erm(T)	2	*Lactobacillus, Streptococcus*
erm(W)	1	*Micromonospora*
erm(X)	3	*Arcanobacterium, Corynebacterium, Propionibacterium*
erm(Y)	1	*Staphylococcus*
erm(33)	1	*Staphylococcus*
erm(35)	1	*Bacteroides*
erm(36)	1	*Micrococcus*
erm(37), (38), (39), (40)	1	*Mycobacterium* (innate)

[a]*erm*(B) is the only rRNA methylase gene screened in aerobic and facultative gram-negative bacteria studies. Thus, the distribution of other *erm* genes in aerobic gram-negative genera is most likely underrepresented in the table. Data from references 2, 3, 5, 10 through and 12, 20 through 23, 29, 33, 34, 36, 37, 41, 43 through 47, 51, 52, 54, 59, 62 though 65, 69, 70, 73, 77, and 85 through 89).

erm(B) gene predominates in *Enterococcus* spp., oral viridans group streptococci, and *S. pneumoniae*, while *Staphylococcus* spp. commonly carry *erm*(B), *erm*(C), and *erm*(F) genes in the same population (1, 25, 34, 37, 41, 47, 62, 63, 68, 73, 81). We have expanded our recent work (41) to include 396 gram-positive oral isolates, of which 75% were streptococci. In this group, 30% of the isolates carried the *erm*(B) gene, 29% carried the *erm*(C) gene, and 16% carried the *erm*(F) gene, while <1% carried the *erm*(A) gene. We also examined 237 gram-positive urine isolates, of which 80% were staphylococci. In this group, 25% carried the *erm*(B) gene, 25% carried the *erm*(C) gene, 12% carried the *erm*(F) gene, and 3% carried the *erm*(A) gene (M. C. Roberts, unpublished data).

Less work has been done on the host range of *erm* genes in gram-negative species (Table 5.2) (2, 11, 54, 87). Previously, we have found the *erm*(B), *erm*(C), and *erm*(F) genes in *Neisseria* spp., *Actinobacillus pleuropneumoniae*, *Treponema denticola*, and anaerobic gram-negative bacteria (Table 5.2) (10, 11, 41, 62, 64, 87). Yet in most studies of other gram-negative bacteria only the presence of the *erm*(B) gene has been examined (2, 62). However, we have recently examined 176 randomly selected commensal gram-negative oral and urine bacterial strains for the presence of seven macrolide resistance genes. Seventy-nine (92%) oral and 59 (67%) urine isolates carried one or more acquired macrolide resistance genes (54). The *erm*(B) gene was found in 16 of 72 oral isolates and 9 of 54 urine isolates with known macrolide resistance genes. No other *erm* gene was screened, and the most common acquired gene was *mef*(A). Therefore, Table 5.2 may underestimate the host range of the *erm* genes.

MOBILE ELEMENTS

Lateral gene exchange has been recognized as an important vehicle for transfer of antibiotic resistance genes for almost 50 years. Mobile elements often carry multiple different genes which confer antibiotic resistance to a variety of different antibiotic classes. Association of the *erm* genes with mobile elements (plasmids, conjugative transposons, and transposons) provides the potential to move from one species to another, one genus to another, and one ecosystem to another, between bacterial hosts, and from food to man and the environment. Linkage between antibiotic resistance genes on mobile elements may be a mechanism, for nonfunctional genes, for stability in a bacterial population without selection pressure (Table 5.3). The *erm*(F) gene is linked to *tet*(X) gene, which codes for an enzyme that breaks down tetracycline in the presence of oxygen but is nonfunctional in its *Bacteroides* host. People have speculated that the *erm*(F) and *tet*(X) genes entered the *Bacteroides* on a unit from an aerobic ancestor. Because the *erm*(F) gene functions and is beneficial to the bacteria during clindamycin therapy, both genes have been maintained, even though the *tet*(X) is not functional in *Bacteroides*. This argument has been used to explain the presence of antibiotic resistance genes in recent isolates of other bacteria, which either are not functional, as in the case of *tet*(X), or code for resistance to antibiotics no longer or rarely used, such as aminoglycosides and/or chloramphenicol. Various previously described linkages between specific *erm* genes and other antibiotic resistance genes are listed in Table 5.3. In addition to antibiotic resistance genes a variety of other genes may also be found on these mobile elements (23, 61, 82). These other genes may also influence carriage of the entire mobile unit and help to maintain genes in a bacterial population.

Table 5.3 Linkage between rRNA methylase and other antibiotic resistance genes[a]

rRNA methylase gene	Other resistance gene [antibiotic or element]	Reference(s)
erm(A)	*lun*(A) [lincomycin]	29, 37
	msr(A) [erythromycin and streptogramin B]	29, 37
	vga(A) [streptogramin A]	29, 37
	vgb(A) [streptogramin B]	29, 37
erm(B)	*vat*(E) [streptogramin A]	29
	tet(M) [tetracycline]	61
	*cat*221 [chloramphenicol]	61
	aphA-3 [kanamycin]	61
	vat(B) [streptogramin A]	29, 37
	lun(A) [lincomycin]	29, 37
	msr(A) [erythromycin and streptogramin B]	29, 37
	vga(A) [streptogramin A]	29, 37
	vgb(A) [streptogramin B]	29, 37
	tcrB [copper resistance]	22
erm(C)	*lun*(A) [lincomycin]	29, 37
	msr(A) [erythromycin and streptogramin B]	29, 37
	vga(A) [streptogramin A]	29, 37
	vgb(A) [streptogramin B]	29, 37
erm(F)	*tet*(X) [tetracycline]	23, 72
	tet(Q) [tetracycline]	23, 72

[a]Data from references 10, 21, 44, and 47.

The *erm*(A), *erm*(B), *erm*(C), *erm*(F), and *erm*(Q) genes have all been found on conjugative elements in chromosomes (10, 11, 23, 29, 41, 61, 65, 82). Some of these genes have also been found on plasmids (2, 29, 44). For example, the *erm*(B) gene in streptococci is usually found on conjugative transposons in the chromosome, while the *erm*(B) gene can be found on plasmids in enterococci (29).

Similarly, the *erm*(B) gene in gram-negative bacteria may be located on a plasmid or a mobile element that acts like a conjugative transposon (2, 54; Roberts, unpublished). The location of the *erm*(B) gene is important when considering gene exchange in nature because the location of the gene directly influences what types of recipients it can be conjugally transferred to. This is especially true when crossing genus barriers or between gram-positive and gram-negative pairs. If a gram-negative donor carries the *erm*(B) gene on a plasmid, this may be transferred by conjugation and/or transformation to an *E. coli* recipient, but the *erm*(B) gene is unlikely to be transferred to a gram-positive *Enterococcus faecalis* recipient. In contrast, if the *erm*(B) gene is on a mobile element in the chromosome, then conjugal transfer to an *E. faecalis* recipient can normally occur and the *erm*(B) gene may also be transferred to an *E. coli* recipient by conjugation (Roberts, unpublished).

The *erm*(B) gene is often associated with Tn*916*-like or other conjugative transposons (61). These elements have a wide host range (61). The *erm*(F) gene is associated with conjugative chromosomal elements, such as Tc^r Em^r DOT, Tc^r Em^r 12256, or Tc^r Em^r CEST in *Bacteroides* (23, 72). These elements have been shown to mobilize *Bacteroides* nonconjugative coresident plasmids in the laboratory and may do this in nature as well (23). These elements have been found in other genera (10), but it is unknown whether they can mobilize nonconjugative coresident plasmids in these bacterial hosts. It is also not know if *erm* genes associated with Tn*916*-like or other conjugative transposons have similar capacities to move nonconjugative coresident plasmids or if this is unique to the *Bacteroides* elements in *Bacteroides*.

AGE OF THE rRNA METHYLASES

The innate methylases found in *Streptomyces* spp. and perhaps *Mycobacterium* spp. may have been in these organisms for a very long time. In *Streptomyces* spp. they are thought to be protective to the host. Unfortunately, we do not have ready access to ancient strains. On the other hand it is clear that most acquired resistance genes have occurred in the last 50 years. There are some isolates which do exist from the 1950s and 1960s (3, 11). A few have been screened for the presence of *erm* genes. An *erm*(B) gene was identified from an enterococcus isolated in the 1950s (3). We have recently identified a 1955 *Neisseria gonorrhoeae isolate* carrying both the *erm*(B) and *erm*(F) genes and a 1963 *Neisseria meningitidis* isolate carrying an *erm*(F) gene (11). Previously, we have shown that some of the ATCC anaerobic gram-negative isolates from the 1950s carry a variety of *erm* genes including *erm*(A), *erm*(B), *erm*(C), *erm*(F), and/or *erm*(Q) genes (10). The first erythromycin-resistant *S. pneumoniae* isolates described by Dixon (14) in 1967 have been shown to carry an *erm*(B) gene. Together these data suggest that the *erm*(A), *erm*(B), *erm*(C), *erm*(F), and *erm*(Q) genes have been dispersed to a variety of gram-positive and gram-negative bacteria for about 50 years. These genes are also the ones with the widest host range (Table 5.2).

MOLECULAR MECHANISM

The first and most common mechanism of MLS_B resistance is due to the acquisition of rRNA methylases. These enzymes add one or two methyl groups to a single adenine (A2058 in *E. coli*) in the 23S rRNA moiety (38). This gives the host bacteria resistance to MLS_B antibiotics. The methyl groups encroach into the MLS_B binding site within the tunnel, where the nascent peptide is formed (68). Two methyl groups at the N(6) of the A2058 are needed to effectively block the MLS_B antibiotics including the ketolides from binding (38). Though we tend to lump all of the rRNA methylases into one group, they do not confer identical susceptibility to the MLSKO group of antibiotics. Part of this is due to the fact that the rRNA methylase genes can be expressed constitutively or inducibly (33, 65, 69, 77). Inducibly regulated genes are turned on only in the presence of an inducer. In inducibly regulated hosts, the methylase is not produced from the mRNA unless there is an inducer, such as when erythromycin is present. Induction occurs because the inducer and the leader region of the mRNA interact with each other, which leads to rearrangement of the mRNA and production of the methylase. The bacterial host also is important in the specificity of induction. Erythromycin is usually a good inducer for these genes, while lincosamides and streptogramins generally are not. Traditionally it was thought that in a strain carrying an inducibly regulated *erm* gene, the bacterial host was resistant only to inducers such as erythromycin and not resistant to noninducers like clindamycin and ketolides. However, in some animal strains lincomycin rather than erythromycin was the inducer (63). This may be due to the extensive use of lincomycin in animals. A variety of structural alterations of the translational attenuator have been found in nature. Gene expression may vary with the structure of the regulatory region. Constitutive clinical strains carrying tandem duplications of 25 bp in the *erm*(A) gene and/or a deletion of 107 bp in *erm*(C) have been described (69).

In constitutively regulated genes, an active methylase gene is produced from the mRNA in the absence of an inducer (33). The host bacteria can also influence susceptibility profiles in constitutively regulated genes. For example in *S. pyogenes*, the inducibly regulated *erm*(A) and *erm*(B) genes differ in their MICs to telithromycin, a new ketolide, and to clindamycin compared to constitutively regulated genes (28). Isolates carrying inducible *erm*(A) genes often have lower MIC_{50} and MIC_{90} for erythromycin than do isolates carrying inducibly regulated *erm*(B) genes (28). Monomethylation of the A2058 confers low to intermediate resistance to macrolide and streptogramin B antibiotics, low-level resistance to telithromycin (≤5 μg/ml), and high-level resistance to lincosamides in streptococci (28, 38). Jalava et al. (28) identified six *S. pyogenes* isolates with telithromycin MICs ranging from 4 to 64 μg/ml. All carried the constitutive *erm*(B) gene, while an *S. pyogenes* constitutive *erm*(A) gene had a low telithromycin MIC (0.25 μg/ml). However, constitutive expression of the *erm*(B) gene in a related bacteriaum, *S. pneumoniae*, gave lower telithromycin MICs (≤0.008 to 0.25 μg/ml) than found for the *erm*(B)-carrying *S. pyogenes*. Liu and Douthwaite (38) have hypothesized that the difference in levels of protection between *S. pyogenes* and *S. pneumoniae*, both carrying constitutively expressed *erm*(B) genes, is due to the effectiveness of methylation, with the *S. pyogenes* primarily having a dimethylated A2058 nucleotide versus the monomethylation found in *S. pneumoniae*. The effect could be reproduced for erythromycin, clarithromycin, and telithromycin in an *E. coli* strain. Differences in monomethylation versus dimethylation may also explain the different levels of resistance found between *S. pyogenes* strains carrying inducible *erm*(A) and those carrying *erm*(B) genes (38).

LABORATORY DETECTION

There are well-established phenotypic methods for distinguishing inducible versus constitutive expression of *erm* resistance genes, which can be done using disk diffusion. A clindamycin disk is placed close to the erythromycin disk in a standard disk diffusion assay (85). A flattening of the clindamycin zone of inhibition on the erythromycin side is the classical way to determine inducibly regulated *erm* genes (85). However, considerable variation can be found with different strains and species. We have found that gram-positive species which are borderline intermediate or resistant to erythromycin and/or clindamycin most likely carry an inducible *erm* gene. Another method for distinguishing inducible versus constitutive expression is to compare susceptibility with and without exposure to a low dose of erythromycin (0.5 μg). If the strain is inducible, the clindamycin susceptibility will be different for the two assays. This method is used primarily in research.

The use of molecular tests to determine carriage of specific antibiotic resistance genes began as a research tool approximately 20 years ago. Flutt et al. (16) suggest that identification of MLS resistance genes for diagnostic purposes should be currently restricted to epidemiology and basic research and suggested that there was limited clinical value in distinguishing which gene(s) is present because different genes may give similar susceptibility profiles. They suggested that there might be clinical value in determining whether *S. pneumoniae* or other species carried a *mef*(A) gene versus an *erm* gene(s), because the *mef*(A) gene confers resistance only to macrolides, while the *erm* genes confers resistance to MLS. If a strain could be characterized quickly and shown to carry the *mef*(A) gene, then clindamycin therapy could be used.

With the introduction of molecular methods an issue has occurred which has not been adequately addressed, i.e., how to classify a strain which is genotypically positive by a DNA test but is phenotypically negative. Such a strain could carry an inducibly regulated resistance gene or carry a nonfunctional gene. It is unclear how these strains would be reported and whether macrolides could be used for therapy.

Currently, a number of surveillance studies are determining the genotype of the MLS genes carried (24, 32). Detection can be done using DNA-DNA hybridization or PCR assays, either individually or in multiplex. Usually, short (≤30-bp) oligonucleotide probes are used for both types of assays. Often the same probes can be used for DNA-DNA hybridization and PCR assays. Templates for either assay can be purified DNA, proteinase K-treated whole bacteria, or direct samples from any sample that contains bacteria. The *erm*(33) gene could give false answers with either assay because it appears to be a hybrid created from an *erm*(A) and *erm*(C) crossover (70). Unless sequencing is done, it may be difficult to distinguish by DNA-DNA hybridization whether the isolate carries an *erm*(33) gene or both *erm*(A) and *erm*(C) genes (70). This gene can appear to be either an *erm*(C) or an *erm*(A) gene depending on the probes used. Only if multiple probes for both sides of the molecule and/or PCR products, which bridge the two different regions, are used would the gene be correctly identified as *erm*(33). Whether there are other composite *erm* genes in bacterial populations is unknown.

If DNA-DNA hybridization is used, confirmation by a PCR assay is advisable. Similarly, showing that a test organism gives the same-size PCR product in a gel is not adequate. To reduce false-positive results in the PCR assay, we use an internally labeled oligonucleotide probe that is hybridized to PCR products. This allows one to verify that the PCR product visualized is the correct one. Alternatively, the PCR product should be sequenced. Either method ensures that the PCR product obtained contains the correct sequence.

The introduction of DNA microarray technology will someday make it possible to screen a single isolate for large numbers of acquired antibiotic resistance genes. This is just now being explored in the research setting (7). However, the costs need to be reduced, more experience must be gained, and the availability of kits is required before this technology moves from a research setting to the clinical laboratory. Eventually even common mutational changes in the normal chromosome may be detected using the DNA array technology in which a microchip containing gel-immobilized oligonucleotides is used to hybridize directly with test DNA, or as a base for PCR amplification and hybridization of the PCR products in the clinical laboratory. It is hoped that in the future microarrays will be available to allow for screenings of large numbers of the different antibiotic resistance genes. This could be of great help to groups doing large surveillance studies and to large clinical labs or centers.

Real-time PCR is another tool which has potential for use in the detection of antibiotic resistance genes. This system, which analyzes the results in real time, is also being used for analysis in the research setting. Whether this will be adapted for detection of antibiotic resistance genes in a clinical setting is not known.

CONCLUSION

At least two of the *erm* genes, *erm*(B) and *erm*(F), have been identified from a variety of different isolates from the 1950s and 1960s, suggesting that these genes have been present in both gram-positive and gram-negative hosts for 50 years. Papers have shown a linkage between macrolide use and the increased resistance found in pathogenic bacterial populations. One can speculate that as the newer macrolide derivatives became available in the 1980s this influenced the increase in carriage of macrolide resistance genes. How the introduction of the ketolides and linezolids into therapeutic use will influence the continuing evolution in these bacteria is not clear. However, one can certainly state that this field will continue to evolve in the years to come. How these changes will ultimately affect treatment and antibiotic prescribing is unknown. Whether new tools like real-time PCR will become a standard piece of equipment in clinical laboratories is unknown. However, one can predict that the number of specific acquired MLS genes will continue to increase, especially as studies begin to examine environmental bacterial populations. We may also find new innate genes, which when mutated provide increased levels of resistance to all the MLSKO antibiotics. It is even possible that new mechanisms of MLS resistance will be described in the coming years. This field is not stagnant, and change may occur faster now than it has in the past 50 years.

References

1. **Appelbaum, P. C.** 1992. Antimicrobial resistance in *Streptococcus pneumoniae*: an overview. *Clin. Infect. Dis.* **15:**77–83.
2. **Arthur, M., A. Andremont, and P. Courvalin.** 1987. Distribution of erythromycin esterase and rRNA methylase genes in members of the family *Enterobacteriaceae* highly resistant to erythromycin. *Antimicrob. Agents Chemother.* **31:**404–409.
3. **Atkinson, B. A., A. Abu-Al-Jaibat, and D. J. LeBlanc.** 1997. Antibiotic resistance among enterococci isolated from clinical specimens between 1953 and 1954. *Antimicrob. Agents Chemother.* **41:**1598–1600.
4. **Aumercier, M., S. Bouchallab, M. L. Capmau, and F. Le Goffic.** 1992. RP59500: a proposed mechanism for its bacterial activity. *J. Antimicrob. Chemother.* **30:**9–14.
5. **Bozdogan, B., S. Galopin, and R. Leclercq.** 2004. Characterization of a new *erm*-related macrolide resistance gene present in probiotic strains of *Bacillus clausii*. *Appl. Environ. Microbiolo.* **70:**280–284.
6. **Bryskier, A., and A. Denis.** 2001. Ketolides: novel antibacterial agents designed to overcome resistance to erythromycin A within gram-positive cocci, p. 97–140. In W. Schonfeld and H. A. Kirst. (ed.), *Macrolide Antibiotics*. Birkhauser Verlag, Basel, Switzerland.
7. **Call, D. R., M. K. Bakko, M. J. Krug, and M. C. Roberts.** 2003. Identifying antimicrobial resistance genes using DNA microarrays. *Antimicrob. Agents Chemother.* **47:**3290–3295.
8. **Camps, M., G. Arrizabalaga, and J. Boothroyd.** 2002. An rRNA mutation identifies the apicoplast as the target for clindamycin in *Toxoplasma gondii*. *Mol. Microbiol.* **43:**1309–1328.
9. **Carfartan, G., P. Gerardin, D. Turck, and M.-O. Husson.** 2004. Effect of subinhibitory concentrations of azithromycin on adherence of *Pseudomonas aeruginosa* to bronchial mucins collected from cystic fibrosis patients. *J. Antimicrob. Chemother.* **53:**686–688.
10. **Chung, W. O., C. Werckenthin, S. Schwarz, and M. C. Roberts.** 1999. Host range of the *ermF* rRNA methylase gene in human and animal bacteria. *J. Antimicrob. Chemother.* **43:**5–14.
11. **Cousin, S. L., Jr., W. L. Whittington, and M. C. Roberts.** 2003. Acquired macrolide resistance genes in pathogenic *Neisseria* spp. isolated between 1940 and 1987. *Antimicrob. Agents Chemother.* **47:**3877–3880.
12. **Daly, M., S. Doktor, R. Flamm, and V. Shortridge.** 2004. Characterization and prevalence of MefA, MefE, and the associated *msr*(D) in *Streptococcus pneumoniae* clinical isolates. *J. Clin. Microbiol.* **42:**3570–3574.
13. **Depardieu, F., and P. Courvalin.** 2001. Mutation in 23S rRNA responsible for resistance to 16-membered macrolides and streptogramins in *Streptococcus pneumoniae*. *Antimicrob. Agents Chemother.* **45:**319–323.
14. **Dixon, J.** 1967. Pneumococcus resistant to erythromycin and lincomycin. *Lancet* **1:**573.
15. **Edlund, C., E. Sillerstrom, E. Wahlund, and C. E. Nord.** 1998. *In vitro* activity of HMR 3647 against anaerobic bacteria. *J. Chemother.* **10:**280–284.
16. **Flutt, A. C., M. R. Visser, and F.-J. Schmitz.** 2001. Molecular detection of antimicrobial resistance. *Clin. Microbiolo Rev.* **14:**836–871.

17. **Frenck, R. W., Jr., A. Mansour, I. Nakhla, Y. Sultan, S. Putam, T. Wierzba, M. Morsy, and C. Knirshc.** 2004. Short-course azithromycin for the treatment of uncomplicated typhoid fever in children and adolescents. *Clin. Infect. Dis.* **38:**951–957.

18. **Garza-Ramos, G., L. Xiong, P. Zhong, and A. Mankin.** 2001. Binding site of macrolide antibiotics on the ribosome: new resistance mutation identifies a specific interaction of ketolides with rRNA. *J. Bacteriol.* **184:**6898–6907.

19. **Grkovic, S. G., M. H. Brown, and R. A. Skurray.** 2002. Regulation of bacterial drug export systems. *Microbiol. Mol. Biol. Rev.* **66:**671–701.

20. **Gupta, A., H. Vlamakis, N. Shoemaker, and A. A. Salyers.** 2003. A new *Bacteroides* conjugative transposon that carries an *ermB* gene. *Appl. Environ. Microbiol.* **69:**6455–6463.

21. **Hammerum, A. M., S. E. Flannagan, D. B. Clewell, and L. B. Jensen.** 2001. Indication of transposition of a mobile DNA element containing the *vat*(D) and *erm*(B) genes in *Enterococcus faecium*. *Antimicrob. Agents Chemother.* **45:** 3223–3225.

22. **Hasman, H., and F. M. Aarestrup.** 2005. Relationship between copper, glycopeptide, and macrolide resistance among *Enterococcus faecium* strains isolated from pigs in Denmark between 1997 and 2003. *Antimicrob. Agents Chemother.* **49:**454–456.

23. **Hecht, D. W., J. S. Thompson, and M. H. Malamy.** 1989. Characterization of the termini and transposition products of Tn*4399*, a conjugal mobilizing transposon of *Bacteroides fragilis*. *Proc. Natl. Acad. Sci. USA* **86:**5340–5344.

24. **Hsueh, P.-R., L.-J. Teng, C.-M. Lee, W.-K. Huang, T.-L. Wu, J.-H. Wan, D. Yang, J.-M. Shyr, Y.-C. Chuang, J.-J. Yan, J.-J. Lu, J.-J. Wu, W.-C. Ko, F.-Y. Chang, Y.-C. Yang, Y.-J. Lau, Y.-C. Liu, H.-S. Leu, C.-Y. Liu, and K.-T. Luh.** 2003. Telithromycin and quinupristin-dalfopristin resistance in clinical isolates of *Streptococcus pyogenes*: SMART Program 2001 Data. *Antimicrob. Agents Chemother.* **47:** 2152–2157.

25. **Hyde, T. B., K. Gay, D. S. Stephens, D. J. Vugia, M. Pass, S. Johnson, N. L. Barrett, W. Schaffner, P. R. Cieslak, P. S. Maupin, E. R. Zell, J. H. Jorgensen, R. R. Facklam, and C. G. Whitney.** 2001. Macrolide resistance among invasive *Streptococcus pneumoniae* isolates. *JAMA* **286:**1857–1862.

26. **Iacoviello, V. R., and S. H. Zinner.** 2001. Macrolides: a clinical overview, p. 15–24. *In* W. Schonfeld and H. A. Kirst (ed.), *Macrolide Antibiotics*. Birkhauser Verlag, Basel, Switzerland.

27. **Jaffe, A., F. Francis, M. Rosenthal, and M. Bush.** 1998. Long-term azithromycin may improve lung function in children with cystic fibrosis. *Lancet* **351:**420.

28. **Jalava, J., J. Kataja, H. Seppala, and P. Huovinen.** 2001. In vitro activities of the novel ketolide telithromycin (HMR 3647) against erythromycin-resistant *Streptococcus* species. *Antimicrob. Agents Chemother.* **45:**789–793.

29. **Jensen, L. B., A. M. Hammerum, and R. M. Aarestrup.** 2000. Linkage of *vat*(E) and *erm*(B) in streptogramin-resistant *Enterococcus faecium* isolates from Europe. *Antimicrob. Agents Chemother.* **44:**2231–2232.

30. **Kim, Y.-H., C.-J. Cha, and C.-E. Cerniglia.** 2002. Purification and characterization of an erythromycin esterase from an erythromycin-resistant *Pseudomonas* sp. *FEMS Microbiol. Lett.* **210:**239–244.

31. **Kudoh, S., A. Azuma, M. Yamamoto, T. Izumin, and M. Ando.** 1998. Improvement of survival in patients with diffuse panbronchiolitis treated with low-dose erythromycin. *Am. J. Resp. Crit. Care Med.* **157:**1829–1832.

32. **Kugler, K. C., G. A. Denys, M. L. Wilson, and R. N. Jones.** 2000. Serious streptococcal infections produced by isolates resistant to streptogramins (quinupristin-dalfopristin): case reports from the SENTRY antimicrobial surveillance program. *Diagn. Microbiol. Infect. Dis.* **36:**269–272.

33. **Leclercq, R.** 2002. Mechanisms of resistance to macrolides and lincosamides: nature of the resistance elements and their clinical implications. *Clin. Infect. Dis.* **34:**482–492.

34. **Leclercq, R., and P. Courvalin.** 2002. Resistance to macrolides and related antibiotics in *Streptococcus pneumoniae*. *Antimicrob. Agents Chemother.* **46:**2727–2734.

35. **Lee, S. Y., Y. Ning, and J. C. Fenno.** 2002. 23S rRNA point mutation associated with erythromycin resistance in *Treponema denticola*. *FEMS Microbiol Lett.* **207:**39–42.

36. **Liebl, W., W. E. Kloos, and W. Ludwig.** 2002. Plasmid-borne macrolide resistance in *Micrococcus luteus*. *Microbiology* **148:**2479–2487.

37. **Lina, G., A. Quaglia, M.-E. Reverdy, R. Leclercq, R. Vandenesch, and J. Etienne.** 1999. Distribution of genes encoding resistance to macrolides, lincosamides, and streptogramins among staphylococci. *Antimicrob. Agents Chemother.* **43:** 1062–1066.

38. **Liu, M., and D. S. Douthwaite.** 2002. Activity of the ketolide telithromycin is refractory to *erm* monomethylation of bacterial rRNA. *Antimicrob. Agents Chemother.* **46:**1629–1633.

39. **Livermore, D. M.** 2003. Linezolid *in vitro*: mechanism and antibacterial spectrum. *J. Antimicrob. Chemother.* **51:** ii9–ii16.

40. **Luh, K.-T., P.-R. Hsueh., L.-J. Teng, H.-J, Pan, Y.-C. Chen, J.-J. Lu, J.-J. Wu, and S.-W. Ho.** 2000. Quinupristin-dalfopristin resistance among gram-positive bacteria in Taiwan. *Antimicrob. Agents Chemother.* **44:**3374–3380.

41. **Luna, V. A., M. Heiken, K. Judge, C. Uelp, N. Van Kirk, H. Luis, M. Bernardo, J. Leitao, and M. C. Roberts.** 2002. Distribution of the *mef*(A) gene in gram-positive bacteria from healthy Portuguese children. *Antimicrob. Agents Chemother.* **46:**2513–2517.

42. **MacGowan, A. P.** 2003. Pharmacokinetic and pharmacodynamic profile of linezolid in healthy volunteers and patients with gram-positive infections. *J. Antimicrob. Chemother.* **51:** ii17–ii25.

43. **Martel, A., V. Meulenaere, L. A. Devriese, A. Decostere, and F. Haesebrouck.** 2003. Macrolide- and lincosamide-resistance in the Gram-positive nasal and tonsillar flora of pigs. *Microb. Drug Resist.* **9:**293–297.

44. **Matsuoka, M., K. Endou, H. Kobayashi, M. Inoue, and Y. Nakajima, Y.** 1997. A dyadic plasmid that shows MLS and PMS resistance in *Staphylococcus aureus*. *FEMS Microbiol. Lett.* **148:**91–96.

45. **Matsuoka, M., M. Inoue, Y. Nakajima, and K. Endou.** 2002. New *erm* gene in *Staphylococcus aureus* clinical isolates. *Antimicrob. Agents Chemother.* **46:**211–215.

46. **McDougal, L. K., F. C. Tenover, L. N. Lee, J. K. Rasheed, J. E. Patterson, J. H. Jorgensen, and D. J. LeBlanc.** 1998.

Detection of Tn*917*-like sequences within a Tn*916*-like conjugative transposon (Tn*3872*) in erythromycin-resistant isolates of *Streptococcus pneumoniae*. *Antimicrob. Agents Chemother.* **42:**2312–2318.

47. **Montanari, M. P., I. Cochetti, M. Mingoia, and P. E. Varaldo.** 2003. Phenotypic and molecular characterization of tetracycline- and erythromycin-resistant strains of *Streptococcus pneumoniae*. *Antimicrob. Agents Chemother.* **47:**2236–2241.

48. **Moore, P. B., and T. A. Steitz.** 2003. The structural basis of large ribosomal subunit function. *Annu. Rev. Biochem.* **72:**813–850.

49. **Nakamura, A., I. Miyakozawa, K. Nakazawa, K. O'Hara, and T. Sawai.** 2000. Detection and characterization of a macrolide 2′-phosphotransferase from a *Pseudomonas aeruginosa* clinical isolate. *Antimicrob. Agents Chemother.* **44:**3241–3242.

50. **Nash, K. A., and C. B. Inderlied.** 1996. Rapid detection of mutations associated with macrolide resistance in *Mycobacterium avium* complex. *Antimicrob. Agents Chemother.* **40:**1748–1750.

51. **Nash, K. A.** 2003. Intrinsic macrolide resistance in *Mycobacterium smegmatis* is conferred by a novel *erm* gene, *erm*(38). *Antimicrob. Agents Chemother.* **47:**3053–3060.

52. **Nash, K. A., Y. Zhang, B. A. Brown-Elliott, and R. J. Wallace, Jr.** 2005. Molecular basis of intrinsic macrolide resistance in clinical isolates of *Mycobacterium fortuitum*. *J. Antimicrob. Chemother.* **55:**170–177.

53. **Noguchi, N., A. Emura, H. Matsuyama, K. O'Hara, M. Sasatsum, and M. Kono.** 1995. Nucleotide sequence and characterization of erythromycin resistance determinant that encodes macrolide 2′-phosphotransferase I in *Escherichia coli*. *Antimicrob. Agents Chemother.* **39:**2359–2363.

54. **Ojo, K. K., C. Ulep, N. Van Kirk, H. Luis, M. Bernardo, J. Leitao, and M. C. Roberts.** 2004. The *mef*(A) gene predominates among seven macrolide resistant genes identified in 13 gram-negative genera from healthy Portuguese children. *Antimicrob. Agents Chemother.* **48:**3451–3456.

55. **Pereyre, S., P. Gonzalez, B. de Barbeyrac, A. Darnige, H. Renaudin, A. Charron, S. Raherison, C. Bebear, and C. M. Bebear.** 2002. Mutations in 23S rRNA account for intrinsic resistance to macrolides in *Mycoplasma hominis* and *Mycoplasma fermentans* and for acquired resistance to macrolides in *M. hominis*. *Antimicrob. Agents Chemother.* **46:** 3142–3150.

56. **Plante, I., D. Centron, and P. H. Roy.** 2003. An integron cassette encoding erythromycin esterase, *ere*(A), from *Providencia stuartii*. *Antimicrob. Agents Chemother.* **51:**787–790.

57. **Poehlsgaard, J., and S. Douthwaite.** 2003. Macrolide antibiotic interaction and resistance on the bacterial ribosome. *Curr. Opin. Investig. Drugs* **4:**140–148.

58. **Porse, B. T., and R. A. Garrett.** 1999. Sites of interaction of streptogramin A and B antibiotics in the peptidyl transferase loop of 23S rRNA and the synergism of their inhibitory mechanisms. *J. Mol. Biol.* **286:**275–387.

59. **Reig, M., J.-C. Galan, F. Bazuero, and J. C. Perez-Diaz.** 2001. Macrolide resistance in *Peptostreptococcus* spp. mediated by *ermTR*: possible source of macrolide-lincosamide-streptogramin B resistance in *Streptococcus pyogenes*. *Antimicrob. Agents Chemother.* **45:**630–632.

60. **Reynolds, E., J. I. Ross, and J. H. Cove.** 2003. *msr*(A) and related macrolide/streptogramin resistance determinants: incomplete transporters? *Int. J. Antimicrob. Agents* **22:**228–236.

61. **Rice, L. B.** 1998. Tn*916* family conjugative transposons and dissemination of antimicrobial resistance determinants. *Antimicrob. Agents Chemother.* **42:**1871–1877.

62. **Roberts, M. C.** Updated January 2007. 2000 rRNA Methylase Genes Table; Location of the Various Genes Table. [Online.] http://faculty.washington. edu/marilynr/.

63. **Roberts, M. C., and M. B. Brown.** 1994. Macrolide-lincosamide resistance determinants in streptococcal species isolated from the bovine mammary gland. *Vet. Microbiol.* **40:**253–261.

64. **Roberts, M. C., W. Chung, and D. E. Roe.** 1996. Characterization of tetracycline and erythromycin determinants in *Treponema denticola*. *Antimicrob. Agents Chemother.* **40:**1690–1694.

65. **Roberts, M. C., J. Sutcliffe, P. Courvalin, L. B. Jensen, J. Rood, and H. Seppala.** 1999. Nomenclature for macrolide and macrolide-lincosamide streptogramin B antibiotic resistance determinants. *Antimicrob. Agents Chemother.* **43:** 2823–2830.

66. **Saiman, L., Y. Chen, P. San Gabriel, and C. Knirsch.** 2002. Synergistic activities of macrolide antibiotics against *Pseudomonas aeruginosa*, *Burkholderia cepacia*, *Stenotrophomonas maltophilia*, and *Alcaligenes xylosoxidans* isolated from patients with cystic fibrosis. *Antimicrob. Agents Chemother.* **46:**1105–1107.

67. **Saiman, L., B. C. Marshall, N. Mayer-Hamblett, J. L. Burns, A. L. Quittner, D. A. Cibene, S. Coquillette, A. Y. Fieberg, F. J. Accurso, and P. W. Campbell III.** 2003. Azithromycin in patients with cystic fibrosis chronically infected with *Pseudomonas aeruginosa*: a randomized controlled trial. *JAMA* **290:**1749–1756.

68. **Schmitz, F.-J., W. Witte, G. Werner, J. Petridou, A. C. Fluit, and S. Schwarz.** 2001. Characterization of the translational attenuator of 20 methicillin-resistant, quinupristin/dalfopristin-resistant *Staphylococcus aureus* isolates with reduced susceptibility to glycopeptides. *J. Antimicrob. Chemother.* **48:**939–941.

69. **Schmitz, F.-J., J. Petridou, H. Jagusch, N. Astfalk, S. Scheuring, and S. Schwarz.** 2002. Molecular characterization of ketolide-resistance *erm*(A)-carrying *Staphylococcus aureus* isolates selected in vitro by telithromycin, ABT-773, quinupristin and clindamycin. *J. Antimicrob. Chemother.* **49:** 611–617.

70. **Schwarz, S., C. Kehrenberg , and K. K. Ojo.** 2002. *Staphylococcus sciuri* gene *erm*(33), encoding inducible resistance to macrolides, lincosamides, and streptogramin B antibiotics, is a product of recombination between *erm*(C) and *erm*(A). *Antimicrob. Agents Chemother.* **46:**3621–3623.

71. **Seppala, H., T. Klaukka, J. Vuopio-Varkila, A. Muotiala, H. Helenius, K. Lager, and P. Huovinen.** 1997. The effect of changes in the consumption of macrolide antibiotics on erythromycin resistance in Group A streptococci in Finland. *N. Engl. J. Med.* **337:**441–446.

72. **Shoemaker, N. B., H. Vlamikis, K. Hayes, and A. A. Salyers.** 2001. Evidence for extensive resistance gene transfer among *Bacteroides* spp. and between *Bacteroides* and other genera in the human colon. *Appl. Environ. Microbiol.* **67:**561–568.

73. **Shortridge, V. D., G. V. Doern, A. B. Brueggmann, J. M. Beyer, and R. K. Flamm.** 1999. Prevalence of macrolide resistance mechanism in *Streptococcus pneumoniae* isolates from a multi-center antibiotic resistance surveillance study conducted in the United States in 1994–1995. *Clin. Infect. Dis.* **29:**1186–1188.

74. **Singh, K. V., K. Malathum, and B. E. Murray.** 2001. Disruption of an *Enterococcus faecium* species-specific gene, a homologue of acquired macrolide resistance genes of staphylococci, is associated with an increase in macrolide susceptiblity. *Antimicrob. Agents Chemother.* **45:**3672–3673.

75. **Speciale, A., K. La Ferla, F. Caccamo, and G. Nicoletti.** 1999. Antimicrobial activity of quinupristin/dalfopristin, a new injectable streptogramin with wide Gram-positive spectrum. *Int. J. Antimicrob. Agents* **13:**21–28.

76. **Stanton, T. B., and S. B. Humphrey.** 2003. Isolation of tetracycline-resistant *Megasphaera elsdenii* strains with novel mosaic gene combinations of *tet*(O) and *tet*(W) from swine. *Appl. Environ. Microbiol.* **69:**3874–3882.

77. **Sutcliffe, J. A., and R. Leclercq.** 2003. Mechanisms of resistance to macrolides, lincosamides and ketolides, p. 281–317. *In* W. Schonfeld and H. A. Kirst (ed.), *Macrolide Antibiotics*. Birkhauser Verlag, Basel, Switzerland.

78. **Taylor, D. E., Z. Ge, D. Purych, T. Lo, and K. Hiratsuka.** 1997. Cloning and sequence analysis of two copies of a 23S rRNA gene from *Helicobacter pylori* and association of clarithromycin resistance with 23S rRNA mutations. *Antimicrob. Agents Chemother.* **43:**2621–2628.

79. **Tenson, T., M. Lovmar, and M. Ehrenbert.** 2003. The mechanism of action of macrolides, lincosamides and streptogramin B reveals the nascent peptide exit path in the ribosome. *J. Mol. Biol.* **330:**1005–1014.

80. **Thungapathra, M., Amita, K. K. Sinha, S. R. Chaudhuri, P. Garg, T. Ramamurthy, G. B. Nair, and A. Ghosh.** 2002. Occurrence of antibiotic resistance gene cassettes *aac(6')-Ib*, *dfrA5*, *dfrA12*, and *ereA2* in Class I integrons in non-O1, non-O139 *Vibrio cholerae* strains in India. *Antimicrob. Agents Chemother.* **46:**2948–2955.

81. **Tomasz, A.** 1999. New faces of an old pathogen: emergence and spread of a multidrug-resistant *Streptococcus pneumoniae*. *Am. J. Med.* **107:**55S–66S.

82. **Valentine, P. J., N. B. Shoemaker, and A. A. Salyers.** 1988. Mobilization of *Bacteroides* plasmids by *Bacteroides* conjugal elements. *J. Bacteriol.* **170:**1319–1324.

83. **Vera-Cabrera, L., A. Gomez-Flores, W. G. Escalante-Fuentes, and O. Welsh.** 2001. In vitro activity of PNU-100766 (linezolid), a new oxazolidinone antimicrobial, against *Norcardia brasiliensis*. *Antimicrob. Agents Chemother.* **45:**3629–3630.

84. **Vester, B., and S. Douthwaite.** 2001 Macrolide resistance conferred by base substitutions in 23S rRNA. *Antimicrob. Agents Chemother.* **45:**1–12.

85. **Waites, K., C. Johnson, B. Gray, K. Edwards, M. Crain, and W. Benjamin, Jr.** 2000. Use of clindamycin disks to detect macrolide resistance mediated by *ermB* and *mefE* in *Streptococcus pneumoniae* isolates from adults and children. *J. Clin. Microbiol.* **38:**1731–1734.

86. **Wang, Y., G. R., Wang, A. Shelby, N. B. Shoemaker, and A. A. Salyers.** 2003. A newly discovered *Bacteroides* conjugative transposon, CTnGERM1, contains genes also found in gram-positive bacteria. *Appl. Environ. Microbiol.* **69:**4594–4603.

87. **Wasteson, Y., D. E. Roe, K. Falk, and M. C. Roberts. 1994.** Characterization of antibiotic resistance in *Actinobacillus pleuropneumoniae*. *Vet. Microbiol.* **48:**41–50.

88. **Werner, G., I. Klare, and W. Witte. 2002.** Molecular analysis of streptogramin resistance in enterococci. *Int. J. Med. Microbiol.* **292:**81–94.

89. **Widdowson, C. A., and L. P. Klugman.** 1998. Emergence of M phenotype of erythromycin-resistance pneumococci in South Africa. *Emerg. Infect. Dis.* **4:**277–281.

90. **Wilcox, M. H. 2003.** Efficacy of linezolid versus comparator therapies in Gram-positive infections. *J. Antimicrob. Chemother.* **51**(Suppl. S2): ii27–ii35.

91. **Yagi, B. H., and G. E. Zurenko. 2003.** An in vitro time-kill assessment of linezolid and anaerobic bacteria. *Anaerobe* **9:**1–3.

Enzymes in Defense of the Bacterial Cell Wall

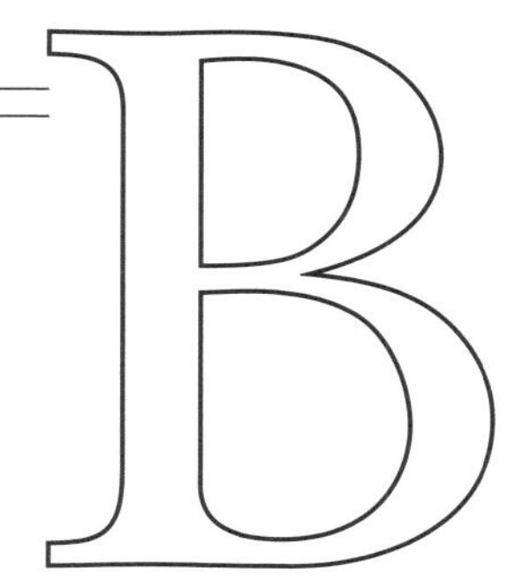

Enzyme-Mediated Resistance to Antibiotics: Mechanisms, Dissemination, and Prospects for Inhibition
Edited by Robert A. Bonomo and Marcelo E. Tolmasky

Karen Bush
Patricia A. Bradford

6 β-Lactamases: Historical Perspectives

β-Lactamases are arguably the most well-studied antibiotic resistance determinants that have been described (14, 20, 69, 80), with over 450 unique enzymes currently identified (21). These enzymes serve to hydrolyze the β-lactam bonds found in penicillins, cephalosporins, and other β-lactam-containing antibacterial agents, thereby rendering these drugs inactive as antibacterial agents (52, 77). Abraham and Chain provided the first description of these enzymes well before penicillin or its subsequent offspring were widely used clinically (1), thereby establishing their existence prior to the introduction of selective pressure from antibiotic usage.

Molecular relationships, as well as functional relationships, exist between the β-lactamases and penicillin-binding proteins (PBPs), the essential enzymes involved in the terminal stages of bacterial cell wall synthesis (76). Both enzymes recognize β-lactams as substrates or inhibitors, resulting in eventual hydrolysis of the inactivated β-lactams. However, release of the hydrolyzed β-lactam can be measured in hours for PBPs (43), whereby some β-lactamases can release hydrolyzed product at rates exceeding 1,000 molecules per s (77). Notably, this variability in hydrolysis rates and diversity of function also can be seen among the β-lactamases themselves, providing the basis for many of the studies of this enticing set of enzymes.

CLASSIFICATIONS

Early descriptions of β-lactamases were based on the characteristics of the producing organisms, focusing initially on the penicillin-hydrolyzing enzymes from gram-positive bacteria. Cell-free "penicillin inactivators" were first described as species-specific enzymes, when it was discovered in the 1940s that *Staphylococcus aureus* (65), *Bacillus anthracis* (6), *Bacillus cereus* (11), *Bacillus licheniformis* (38, 108), and *Bacillus subtilis* (37, 42) produced different but related sets of β-lactamases that all destroyed the microbiological activity of penicillin. Because penicillin with its gram-positive microbiological spectrum was the only β-lactam in clinical use at that time, the original report of a penicillin inactivator from the gram-negative *Bacillus coli* (e.g., *Escherichia coli*) (1) was thought to be clinically unimportant.

By the mid-1950s multiple enzymes were known to exist in *B. cereus* isolates (94), leading to the realization that two very different β-lactamases existed in nature, both of which were capable of hydrolyzing β-lactam antibiotics (116). When Sabath and Abraham recognized that β-lactamase II from *B. cereus* required the zinc ion for hydrolytic activity in the 1960s (117), they succeeded in identifying the two major structural classes of β-lactamases, one of which is the metallo-enzyme class, the other one being the set of

Karen Bush, Johnson & Johnson Pharmaceutical Research & Development, Raritan, NJ 08869. **Patricia A. Bradford,** Wyeth Research, Pearl River, NY 10965.

enzymes that utilize serine at the active site. The enzymes utilizing serine at the active site hydrolyze their substrates by forming acyl enzymes that are then attacked by an active-site water molecule, resulting in the breaking of the β-lactam bond and subsequent release of hydrolyzed product (123). Although the precise hydrolysis mechanism for metallo-β-lactamases is still a matter of investigation, one hypothesis is that a tetrahedral intermediate is stabilized through at least one Zn^{2+} active-site atom (99), whereas other investigators believe that binuclear zinc complexes are critical for hydrolysis to occur (132).

Two sets of β-lactamase classification schemes have evolved, based on either structural (4, 56, 59) or functional descriptions (22) of the enzymes. Fortuitously, the two schemes fit together according to major divisions but show some diversity among various subgroups for each of the divisions as shown in Table 6.1.

METHODOLOGY

Microbiological Assays

Microbiological methods have traditionally indicated the potential for β-lactamase production, based on susceptibility profiles for β-lactam-containing agents. If inhibitory concentrations of β-lactams are elevated when testing bacteria that are assumed to be susceptible to these agents, such as *E. coli* or *Haemophilus influenzae*, a first assumption is made that resistance may be due to β-lactamase production. This is frequently an accurate assessment for gram-negative bacteria, in which expression of β-lactamases represents the major resistance mechanism for β-lactam-containing agents. This resistance is often inoculum dependent, with higher MICs observed at higher bacterial inocula (111).

Gram-positive bacteria generally rely less extensively on β-lactamase production for resistance to these agents and instead exhibit major changes in the PBPs, particularly in the streptococci (34), which produce no intrinsic β-lactamase and have never been reported to acquire a gene for their production. Notable exceptions include the β-lactamase-producing gram-positive bacilli, as noted above, and *S. aureus*, in which β-lactamase production preceded the development of β-lactam resistance via acquisition of a new PBP (18, 65). Thus, elevated MICs or smaller zones of inhibition in agar dilution assays may suggest the presence of a β-lactamase but are not always predictive of β-lactamase production, especially in gram-positive cocci.

Organisms producing β-lactamases with specific substrate profiles have been examined in detail to try to devise standardized testing procedures to distinguish β-lactamase producers. A prime example of this is the recently introduced CLSI (formerly NCCLS) methodology for detection of extended-spectrum β-lactamase (ESBL) production in *E. coli* and *Klebsiella pneumoniae* (92, 93). This testing includes both a screening test and a confirmation test for ESBL production. In these tests, MICs (≥ 2 μg/ml for extended-spectrum cephalosporins or aztreonam or ≥ 8 μg/ml for cefpodoxime) are first used as a screen, with confirmation related to the ability of the β-lactamase inhibitor clavulanic acid to lower the cephalosporin MICs at least three doubling dilutions in MIC testing. Similar principles are employed for disk diffusion testing. However, this testing may identify both false positives and false negatives (111) and has not been as widely adopted as had been hoped. It is also limited to the two organisms listed above because the presence of chromosomal or plasmid-encoded AmpC β-lactamases confounds the picture in other *Enterobacteriaceae* (119).

Enzymatic Assays

Hydrolysis activity in the presence of an appropriate substrate is the most reliable way to determine the production of β-lactamase. The most widely used substrate for general β-lactamase screening is nitrocefin, a cephalosporin that is yellow when intact at neutral pH but turns to a bright red color after hydrolysis of the β-lactam ring (96). This substrate is hydrolyzed by almost all β-lactamases, with the notable exception of some metallo-β-lactamases from

Table 6.1 Summary of β-lactamase classification schemes

Functional group[a]	Functional characteristics	Active site	Structural class[b]
1	Cephalosporinases poorly inhibited by clavulanic acid	Serine	C
2	Penicillinases and/or cephalosporinases with diverse hydrolytic properties according to subdivisions 2a, 2b, 2be, 2br, 2c, 2d, 2e, and 2f, all of which are inhibited to some extent by clavulanic acid	Serine	A and D
3	Broad spectrum, carbapenem-hydrolyzing enzymes not inhibited by clavulanic acid, but inhibited by EDTA	Metallo (Zn^{2+})	B
4	Miscellaneous β-lactamases that do not fit other functional profiles	Not determined	Not assigned

[a] Data from reference 22.
[b] Data from references 4, 56, and 59.

Aeromonas spp. (50, 120). Nitrocefin at a concentration of 100 μg/ml can be used to detect β-lactamase production in whole cells, in cell-free extracts, during β-lactamase purification, and on isoelectric focusing gels.

In addition to using nitrocefin for β-lactamase quantitative studies, all β-lactams exhibit other chemical characteristics that allow for quantitative assays using iodometric (115), acidometric (54), or spectrophotometric techniques (115), with details outlined in the cited publications. Of these, the most widely accepted methodology has been the use of spectrophotometric assays for studying the kinetics of hydrolysis by purified enzymes (23).

Isoelectric Focusing

Isoelectric focusing, using nitrocefin to overlay acrylamide gels following electrophoresis of cell-free bacterial extracts, was developed as an early tool to differentiate individual β-lactamases. When introduced in the mid-1970s by Matthew et al. (78, 79), it became the "gold standard" for identifying specific β-lactamases, either when produced as single enzymes or when multiple enzymes were produced by a single organism. Its utility has changed in the past few years due to the large numbers of unique β-lactamases with identical isoelectric points (G. Jacoby and K. Bush, http://www.lahey.org/Studies/webt.stm). However, the technique is still valuable in determining the minimum number of enzymes produced in a strain, similarities of enzymes and enzyme patterns within an outbreak, and the identity of enzymes in wild-type strains and transconjugants produced during molecular characterization.

MAJOR β-LACTAMASES AND THEIR FAMILIES

After the penicillinases from staphylococci and bacilli became well recognized as the prototypical β-lactamases in the 1950s, β-lactamases with a broader substrate specificity began to become more important clinically. The introduction of cephalosporins and penicillinase-stable penicillins to the clinic in the late 1950s and 1960s provided the selective pressure for the emergence of gram-negative pathogens, organisms that produced a variety of cephalosporin-hydrolyzing β-lactamases (19, 81). Group 1 (class C) cephalosporinases were observed as endogenous enzymes in most *Enterobacteriaceae* and could appear either as inducible enzymes or as constitutive β-lactamases with high periplasmic concentrations.

By the early 1970s, plasmid-encoded β-lactamases were identified in organisms that could transfer β-lactam resistance determinants to pathogens of other species of *Enterobacteriaceae* (51). At this time phenotypic behavior and isoelectric focusing profiles formed the basis for β-lactamase identification (78). In multiple surveys conducted during the late 1970s the most commonly identified plasmid-encoded β-lactamases among the *Enterobacteriaceae* were the broad-spectrum (group 2b) SHV-1 and TEM-1 or TEM-2 β-lactamase, followed by the oxacillin-hydrolyzing OXA-1 enzyme (19, 81). These enzymes served as the target for novel β-lactam drug discovery efforts, resulting in the extended-spectrum cephalosporins and monobactams that were stable to hydrolysis by the SHV and TEM enzymes of the early 1980s (124).

Unfortunately, the widespread use of these β-lactamase-stable β-lactams soon selected for variants of the most common, transferable β-lactamases, resulting in the proliferation of ESBLs that now exists (14, 69). With the advent of rapid and inexpensive nucleotide sequencing, hundreds of TEM-, SHV-, and OXA-derived ESBLs have been identified with minor amino acid variations from their parents, leading to an explosion of new β-lactamases by the end of the 20th century.

NEW PHENOTYPES

In the current age of rapid nucleotide sequencing, molecular and epidemiological characterizations of numerous β-lactamase genes in a wide variety of organisms have been reported. The following narrative describes important new enzymes characterized in the published literature since the year 2000.

Plasmid-Encoded Class A ESBLs

Many of the ESBLs that are isolated worldwide belong to three families of enzymes, TEM, SHV, and CTX-M, groups of β-lactamases which continue to grow rapidly. The most recent information regarding these families of β-lactamases can be found at http://www.lahey.org/studies/webt.stm.

There are, however some ESBLs that do not fit in to any of the previously described families. Two very closely related ESBLs have been described in two European countries. A novel ESBL, designated IBC-1, discovered in an isolate of *Enterobacter cloacae* in Greece (46), was reported to hydrolyze ceftazidime and cefotaxime and was inhibited by imipenem, tazobactam, and to a lesser extent by clavulanate. *E. cloacae* strains expressing this enzyme were later found to be the cause of an outbreak of multidrug-resistant infections among premature infants in a neonatal intensive care unit (63). The IBC-1 β-lactamase is harbored in the In111 integron that also contains gene cassettes that confer resistance to aminoglycosides and trimethoprim-sulfamethoxazole (129).

The GES-1 β-lactamase, isolated from a strain of *K. pneumoniae* in France, differed from IBC-1 by only two amino acids (106). GES-1 was later found in clinical isolates of *Pseudomonas aeruginosa* (36). Like IBC-1, GES-1 was also part of an integron that contained several aminoglycoside-modifying enzymes in gene cassettes. GES-1 was also detected in an outbreak of multidrug-resistant *K. pneumoniae* in a hospital in Portugal (35). Subsequently, a derivative named GES-2, which differed from GES-1 by a single amino acid substitution, was found in a clinical isolate of *P. aeruginosa* in South Africa (107). This small change in the GES-2 protein confers increased hydrolytic activity against imipenem, although not enough to confer resistance. The IBC-1 and GES-1 enzymes were found to be related to the chromosomally encoded class A penicillinases found in *Proteus mirabilis* and *Yersinia enterocolitica* (46, 106).

Plasmid-Encoded Class C Cephalosporinases

New plasmid-encoded AmpC-type β-lactamases are being described with increasing frequency. Many of these new enzymes are members of existing families of β-lactamases including FOX, MOX, and CMY β-lactamases (33, 112, 114). Interestingly, the CMY-2 β-lactamase is often found in or associated with isolates of veterinary origin (17, 118, 133, 134, 142). In addition, several novel AmpC-type enzymes have been described. Among these is ABA-1, identified in a clinical isolate of *Oligella urethralis* that is unusually resistant to penicillins, narrow-spectrum cephalosporins, and cefoxitin (75). The gene encoding ABA-1 had 98% amino acid identity with the chromosomally encoded AmpC β-lactamase from *Acinetobacter baumannii*.

Class D Enzymes

A number of new OXA-type β-lactamases have been characterized in strains of *P. aeruginosa* and *A. baumannii*. Some of these new OXA variants provide an extended-spectrum β-lactamase in that they show some hydrolytic activity against extended-spectrum cephalosporins (102, 127). Other variants have weak hydrolytic activity against carbapenems (2, 28, 70). Both of these groups of enzymes serve to increase the spectrum of resistance caused by the OXA family of β-lactamases, classically known for their ability to hydrolyze penicillinase-stable penicillins such as cloxacillin.

Chromosomally Encoded Enzymes

Rapid sequencing techniques and bioinformatics technology have also aided in the discovery of β-lactamase genes in many species of bacteria that had previously been uncharacterized (Table 6.2). These investigations have led to the discovery of new β-lactamases in all four of the molecular classes and they have been described in both aerobic and anaerobic bacteria. Many of these β-lactamases are interesting enzymes by themselves. However, these sequences have also given insight into the origin of some of the plasmid-encoded β-lactamases that have been disseminated throughout the world. For example, new AmpC genes have been found that appear to be progenitors for some of the plasmid-encoded enzymes that have invaded *E. coli* and *K. pneumoniae*. Sequence analysis revealed that the chromosomally encoded CAV-1 β-lactamase of *Aeromonas caviae* had >96% homology to the FOX-1 β-lactamase and related enzymes (41).

β-Lactamase-Mediated Carbapenem Resistance

Resistance to the carbapenems is on the increase. There are a variety of mechanisms by which this resistance is occurring. Although the overexpression of an AmpC β-lactamase combined with the loss of an outer membrane protein porin has been documented and well characterized for a number of years in *Enterobacter* spp., *Citrobacter* spp., and *K. pneumoniae* (16, 73, 138), the spread of plasmid-encoded AmpC β-lactamases into a variety of *Enterobacteriaceae* has resulted in this mechanism occurring in additional species. Recently, a strain of *Salmonella enterica* serovar Wein that was resistant to imipenem and multiple other antibiotics was isolated in Tunisia (5). This strain possessed multiple β-lactamases including TEM-1, TEM-2, CTX-M-3, and CMY-4 and was also noted to be lacking the OmpF porin protein.

The number of plasmid-encoded metallo-β-lactamases is also increasing, most notably members of the IMP and VIM-type enzyme families. Originally, the IMP-1 β-lactamase was found in Japan in isolates of *P. aeruginosa* and *Serratia marcescens* (98). However, in recent years, the IMP-type β-lactamases have been found in a variety of species of gram-negative bacteria in areas of the world distant from Japan (see below).

WORLDWIDE SPREAD AND DIFFERENTIAL PATTERNS OF ENZYME ISOLATION

Extended-spectrum β-lactamases have now been found in virtually every corner of the world (Table 6.3). The reported incidence of these β-lactamases depends, in part, on the laboratory technique used for detection by the clinical laboratory (14, 97, 125), and the type of β-lactamase varies by country and by region. For example, the CTX-M-type β-lactamases are the most commonly isolated ESBLs in Argentina and TEM-type ESBLs are rarely found there, whereas the converse is true for North America (Table 6.3).

The increased incidence of plasmid-encoded metallo-β-lactamases that has occurred over the last five years is a good example of the worldwide spread of a specific resistance mechanism. The first plasmid-encoded metallo-β-lactamase, IMP-1, was first discovered in 1994 in Japan

Table 6.2 Chromosomally encoded β-lactamase genes discovered since 2000

Class A—groups 2b and 2e		Class A—groups 2be (ESBL)		Class B		Class C		Class D	
β-Lactamase	Organism (reference)	β-Lactamase	Organism (reference)	β-Lactamase	Organism (reference)	β-Lactamase	Organism (reference)	β-Lactamase	Organism (reference)
HER-1	*Escherichia hermannii* (12)	DES-1	*Desulfovibrio desulfuricans* (88)	EBR-1	*Empedobacter brevis* (8)	AmpC	*Citrobacter braakii* (89)	?	*Burkholderia pseudomallei* (95)
CfxA2	*Prevotella intermedia* (72)	Sed-1	*Citrobacter sedlakii* (100)	GOB	*Chryseobacterium meningosepticum* (7, 128)	AmpC	*Citrobacter murliniae* (89)	CARB-7	*Vibrio cholerae* (83)
HugA	*Proteus penneri* (68)	RAHN-1	*Rahnella aquatilis* (10)	IND-1	*Chryseobacterium indologenes* (8)	AmpC	*Citrobacter werkmanii* (89)		
		OXY-3	*Klebsiella oxytoca* (49)	CGB-1	*Chryseobacterium gleum* (9)	AmpC	*Enterobacter cancerogenus* (89)		
		OXY-4	*Klebsiella oxytoca* (49)	JOHN-1	*Flavobacterium johnsoniae* (90)	AmpC	*Escherichia fergusonii* (89)		
		KLUA-1	*Kluyvera ascorbata* (55)	CAU-1, Mb11b	*Caulobacter crescentus* (31, 121)	BUT-1	*Buttiauxella* spp. (40)		
		KLUC-1	*Kluyvera cryocrescens* (30)			CAV-1	*Aeromonas caviae* (41)		
		KLUG-1	*Kluyvera georgiana* (103)			OCH-1	*Ochrobactrum anthropi* (91)		

Table 6.3 Diversity of plasmid-mediated β-lactamases in various regions of the world

Region	Country	Date of isolation or outbreak	Organism(s)	ESBL	IRT[b]	Plasmid-mediated AmpC	Class A carbapenemase	Metallo-β-lactamases	Reference
North America	United States	NR[a]	*K. pneumoniae*				KPC-1		139
	United States	1996–2000	*K. pneumoniae*			ACT-1, CMY-2, DHA-1, FOX-5	KPC-2		86
	United States	1998	*S. enterica* serovar Cubana				KPC-2		84
	United States	1998–1999	*K. pneumoniae*	SHV-5			KPC-2		87
	United States	1997–2002	*K. pneumoniae*	SHV-7, SHV-12	TEM-30		KPC-2		15
	Canada	1995–1999	*P. aeruginosa*					IMP-7	48
Latin America	Argentina	NR	*K. pneumoniae*	CTX-M-2, PER-2					82
	Argentina	1992–1998	*V. cholerae*	CTX-M-2, PER-2					101
	Argentina	2000	*E. coli, K. pneumoniae, P. mirabilis, E. aerogenes, S. marcescens, Providencia* spp.	SHV-type, CTX-M-2, CTX-M-31, PER-2					113
	Brazil	1999	*K. pneumoniae*	SHV-27					26
	Brazil	2002	*P. aeruginosa*					SPM-1	44
Western Europe	Spain	1996–1998	*E. coli*		TEM-30, TEM-31, TEM-33, TEM-34, TEM-37, TEM-40, TEM-51, TEM-54				85
	Spain	2001	*E. coli, K. pneumoniae*	CTX-M-14					13
	France	1997–1998	*P. aeruginosa*					VIM-2	104
	France	2001	*A. baumannii*	VEB-1					105
	Portugal	1998	*A. baumannii*					IMP-5	28
	Italy	1997–1998	*P. aeruginosa*					VIM-1	27
	Italy	1998–1999	*P. aeruginosa*	PER-1					71
	Italy	1999	*E. coli, K. pneumoniae, K. oxytoca, P. mirabilis, P. stuartii, E. aerogenes, E. cloacae, C. freundii, C. koserii, S. marcescens*	TEM-type, ESBL, SHV-type, ESBL					122
	Italy	2000	*P. putida*					IMP-12	32
	Italy	2001	*P. aeruginosa*					IMP-13	126
	Greece	2001	*P. aeruginosa*					VIM-4	109
	Greece	2001	*P. aeruginosa*					VIM-2	45
	Greece	2002	*K. pneumoniae*					VIM-1	47

Eastern Europe	Hungary	1996	*K. pneumoniae*	SHV-2, SHV-5			110
	Poland		*P. aeruginosa*			VIM-2	130
Middle East							
Asia	India	1999	*E. coli*, *K. pneumoniae*, *E. aerogenes*	CTX-M-3	CMY-1		61
	Korea	1993–1998	*E. coli*, *K. pneumoniae*	TEM-15, TEM-52, TEM-88, SHV-2a, SHV-12, CTX-M-14			64
	Korea	1995–1997	*P. aeruginosa*, *P. putida*			VIM-7	67
	Korea	2000	*E. cloacae*			VIM-2	60
	Korea	2000–2001	*Acinetobacter* spp.	PER-1			141
	Hong Kong	1994–1998	*Acinetobacter* spp.			IMP-4	25
	Japan	1996	*S. marcescens*			IMP-6	137
	Japan	1997	*P. putida*			IMP-1	140
	Japan	1997–2000	*P. aeruginosa*, *A. xylosoxidans*			IMP-10	58
	Japan	1998–2000	*E. coli*, *K. pneumoniae*, *E. cloacae*, *S. marcescens*	SHV-12, CTX-M-2, CTX-M-3		IMP-1	135
	Malaysia	1999	*P. aeruginosa*			IMP-7	53
	China	1999	*E. coli*, *K. pneumoniae*, *E. cloacae*, *C. freundii*	SHV-43, CTX-M-3			131
	Taiwan	1997–2000	*P. aeruginosa*, *P. putida*, *P. stutzeri*, *P. fluorescens*			IMP-1, IMP-2, VIM-1, VIM-2, VIM-3	136
	Vietnam	2000–2001	*K. pneumoniae*, *P. mirabilis*	SHV-2, CTX-M-14, CTX-M-17, VEB-1			24
Africa	South Africa	1994–1996	*K. pneumoniae*	TEM-53, TEM-63, SHV-2a, SHV-5 SHV-19, SHV-20, SHV-21, SHV-22			39
	Tunisia	1995–1999	*S. enterica* serovar Mbandaka	TEM-4, SHV-2a	ACC-1		74
	Kenya	1999–2000	*K. pneumoniae*	CTX-M-12			62
	Nigeria	2001	*E. aerogenes*, *E. cloacae*	TEM-type ESBL, SHV-type ESBL			3

[a]NR, not reported
[b]IRT, inhibitor-resistant TEM(s).

in clinical isolates of *S. marcescens* and then spread to *P. aeruginosa* (57, 66, 98). For a number of years, this enzyme was contained within that country. However, in recent years, the IMP-1 β-lactamase and at least 16 related enzymes (http://www.lahey.org/Studies.webt.stm) have been identified, first in other Asian countries (25, 53, 136) (Table 6.3) but now in Europe (29, 32, 126) as well as in North America (48). Interestingly, some of the more recent reports of the IMP-type of metallo-β-lactamases have involved isolates of *Acinetobacter* spp. (25, 29), *Pseudomonas putida* (32, 136, 140), and *Alcaligenes xylosoxidans* (58).

Another growing family of plasmid-encoded metallo-β-lactamases are the VIM-type enzymes. There are now at least 10 derivatives of VIM-1 (http://www.lahey.org/Studies.webt.stm). Unlike the IMP-type enzymes, the VIM-type enzymes were first discovered in Europe, in Italy in 1997 (67), and have been spreading eastward. In the last few years VIM-type enzymes have been reported throughout Europe (27, 45, 47, 104, 109, 130) as well as South Korea (60, 67) and Taiwan (136) in strains of *P. aeruginosa*, *K. pneumoniae*, and *E. cloacae*. Both of the IMP- and VIM-type enzymes are often harbored in gene cassettes and are associated with integrons (45).

CONCLUSIONS

Identification of new β-lactamases can be expected to increase due to the rising number of multidrug-resistant *Enterobacteriaceae* and the ease with which resistance determinants are being exchanged. In addition, the numbers of these enzymes will multiply with the propensity of β-lactamase experts to name every discrete variant of enzyme families based on one or two amino acid modifications. Finally, the search for new β-lactamases in environmental isolates will serve to expand this important family of enzymes that are expected to continue as a major chapter in the story on antibiotic resistance.

References

1. **Abraham, E. P., and E. Chain.** 1940. An enyzme from bacteria able to destroy penicillin. *Nature* **146:**837.
2. **Afazal-Shah, M., N. Woodford, and D. M. Livermore.** 2001. Characterization of OXA-25, OXA-26, and OXA-27, molecular class D β-lactamases associated with carbapenem resistance in clinical isolates of *Acinetobacter baumannii*. *Antimicrob. Agents Chemother.* **45:**583–588.
3. **Aibinu, I. E., V. C. Ohaegbulam, E. A. Adenipekun, F. T. Ogunsola, T. O. Odugbemi, and B. J. Mee.** 2003. Extended-spectrum β-lactamase enzymes in clinical isolates of *Enterobacter* species from Lagos, Nigeria. *J. Clin. Microbiol.* **41:** 2197–2200.
4. **Ambler, R. P.** 1980. The structure of β-lactamases. *Philos. Trans. R. Soc. Lond. B* **289:**321–331.
5. **Armand-Lefèvre, L., V. Leflon-Guibout, J. Bredin, F. Barguellil, A. Amor, J. M. Pagès, and M.-H. Nicolas-Chanoine.** 2003. Imipenem resistance in *Salmonella enterica* serovar Wein related to porin loss and CMY-4 β-lactamase production. *Antimicrob. Agents Chemother.* **47:**1165–1168.
6. **Barnes, J. M.** 1947. Penicillin and *B. anthracis*. *J. Pathol. Bacteriol.* **LIX:**113–125.
7. **Bellais, S., D. Aubert, T. Naas, and P. Nordmann.** 2000. Molecular and biochemical heterogeneity of class B carbapenem-hydrolyzing β-lactamases in *Chryseobacterium meningosepticum*. *Antimicrob. Agents Chemother.* **44:** 1878–1886.
8. **Bellais, S., D. Girlich, A. Karim, and P. Nordmann.** 2002. EBR-1, a novel Ambler subclass B1 β-lactamase from *Empedobacter brevis*. *Antimicrob. Agents Chemother.* **46:** 3223–3227.
9. **Bellais, S., T. Naas, and P. Nordmann.** 2002. Genetic and biochemical characterization of CGB-1, an Ambler class B carbapenem-hydrolyzing β-lactamase from *Chryseobacterium gleum*. *Antimicrob. Agents Chemother.* **46:**2791–2796.
10. **Bellais, S., L. Poirel, N. Fortineau, J. W. Decousser, and P. Nordmann.** 2001. Biochemical-genetic characterization of the chromosomally encoded extended-spectrum class A β-lactamase from *Rahnella aquatilis*. *Antimicrob. Agents Chemother.* **45:**2965–2968.
11. **Benedict, R. G., W. H. Schmidt, and R. D. Coghill.** 1945. Penicillin. VII. Penicillinase. *Arch. Biochem.* **8:**377–384.
12. **Beuchef-Havard, A., G. Arlet, V. Gautier, R. Labia, P. Grimont, and A. Philippon.** 2003. Molecular and biochemical characterization of a novel class A β-lactamase (HER-1) from *Escherichia hermannii*. *Antimicrob. Agents Chemother.* **47:**2669–2673.
13. **Bou, G., M. Cartelle, M. Tomas, D. Canele, F. Molina, R. Moure, J. M. Eiros, and A. Guerrero.** 2002. Identification and broad dissemination of the CTX-M-14 β-lactamase in different *Escherichia coli* strains in the Northwest area of Spain. *J. Clin. Microbiol.* **40:**4030–4036.
14. **Bradford, P. A.** 2001. Extended-spectrum β-lactamases in the 21st century: characterization, epidemiology, and detection of this important resistance threat. *Clin. Microbiol. Rev.* **14:**933–951.
15. **Bradford, P. A., S. Bratu, C. Urban, M. Visalli, N. Mariano, D. L. Landman, J. J. Rahal, S. Brooks, S. Cebular, and J. Quale.** 2004. Emergence of carbapenem-resistant *Klebsiella* spp. possessing the class A carbapenem-hydrolyzing KPC-2 and inhibitor-resistant TEM-30 β-lactamases in New York City. *Clin. Infect. Dis.* **39:**55–60.
16. **Bradford, P. A., C. Urban, N. Mariano, S. J. Projan, J. J. Rahal, and K. Bush.** 1997. Imipenem resistance in *Klebsiella pneumoniae* is associated with the combination of ACT-1, a plasmid-mediated AmpC β-lactamase and the loss of an outer membrane protein. *Antimicrob. Agents Chemother.* **41:**563–569.
17. **Brinas, L., M. A. Moreno, M. Zarazaga, C. Porerro, Y. Sáenz, M. García, L. Dominguez, and C. Torres.** 2003. Detection of CMY-2, CTX-M-14, and SHV-12 β-lactamases

in *Escherichia coli* fecal-sample isolates from healthy chickens. *Antimicrob. Agents Chemother.* **47:**2056–2058.

18. **Brown, D. F. J., and P. E. Reynolds.** 1980. Intrinsic resistance to beta-lactam antibiotics in *Staphylococcus aureus*. *FEBS Lett.* **122:**275–278.
19. **Bush, K.** 1997. The evolution of β-lactamases, p. 152–166. *In* D. J. Chadwick and J. Goode (ed.), *Antibiotic Resistance: Origins, Evolution, Selection and Spread*, vol. 207. John Wiley & Sons, Chichester, United Kingdom.
20. **Bush, K.** 2002. The impact of beta-lactamases on the development of novel antimicrobial agents. *Curr. Opin. Investig. Drugs* **3:**1284–1290.
21. **Bush, K.** 2003. Structural basis of resistance to β-lactamase inhibitors, abstr. 1790, p. 531. *In Program and Abstracts of 43rd ICAAC*, 14–17 September 2003, Chicago, IL.
22. **Bush, K., G. A. Jacoby, and A. A. Medeiros.** 1995. A functional classification scheme for β-lactamases and its correlation with molecular structure. *Antimicrob. Agents Chemother.* **39:**1211–1233.
23. **Bush, K., and R. B. Sykes.** 1986. Methodology for the study of β-lactamases. *Antimicrob. Agents Chemother.* **30:**6–10.
24. **Cao, V., T. Lambert, D. Q. Nhu, H. K. Loan, N. K. Hoang, G. Arlet, and P. Courvalin.** 2002. Distribution of extended-spectrum β-lactamases in clinical isolates of *Enterobacteriaceae* in Vietnam. *Antimicrob. Agents Chemother.* **46:**3739–3743.
25. **Chu, Y.-W., M. Afzal-Shah, E. T. S. Houang, M.-F. I. Palepou, D. J. Lyon, N. Woodford, and D. M. Livermore.** 2001. IMP-4, a novel metallo-β-lactamase from nosocomial *Acinetobacter* spp. collected in Hong Kong between 1994 and 1998. *Antimicrob. Agents Chemother.* **45:**710–714.
26. **Corkill, J. E., L. E. Cuevas, R. Q. Gurgel, J. Greensill, and C. A. Hart.** 2001. SHV-27, a novel cefotaxime-hydrolysing β-lactamase, identified in *Klebsiella pneumoniae* isolates from a Brazilian hospital. *J. Antimicrob. Chemother.* **47:**463–465.
27. **Cornaglia, G., A. Mazzariol, L. Lauretti, G. M. Rossolini, and R. Fontana.** 2000. Hospital outbreak of carbapenem-resistant *Pseudomonas aeruginosa* producing VIM-1, a novel transferable metallo-β-lactamase. *Clin. Infect. Dis.* **31:**1119–1125.
28. **Dalla-Costa, L. M., J. M. Coelho, H. A. P. H. M. Souza, M. E. S. Castro, C. J. N. Stier, K. L. Bragagnolo, A. Rea-Neto, S. R. Penteado-Filho, D. M. Livermore, and N. Woodford.** 2003. Outbreak of carbapenem-resistant *Acinetobacter baumannii* producing the OXA-23 enzyme in Curitiba, Brazil. *J. Clin. Microbiol.* **41:**3403–3406.
29. **Da Silva, G. J., M. Correia, C. Vital, G. Ribeiro, J. C. Sousa, R. Leitão, L. Peixe, and A. Duarte.** 2002. Molecular characterization of bla_{IMP-5}, a new integron-borne metallo-β-lactamase gene from an *Acinetobacter baumannii* nosocomial isolate from Portugal. *FEMS Microbiol. Lett.* **215:**33–39.
30. **Decousser, J.-W., L. Poirel, and P. Nordmann.** 2001. Characterization of a chromosomally encoded extended-spectrum class A β-lactamase from *Kluyvera cryocrescens*. *Antimicrob. Agents Chemother.* **45:**3595–3598.
31. **Docquier, J.-D., F. Pantanella, F. Giuliani, M. C. Thaller, G. Amicosante, M. Galleni, J.-M. Frère, K. Bush, and G. M. Rossolini.** 2002. CAU-1, a subclass B3 metallo-β-lactamase of low substrate affinity encoded by an ortholog present in the *Caulobacter crescentus* chromosome. *Antimicrob. Agents Chemother.* **46:**1823–1830.
32. **Docquier, J.-D., M. L. Riccio, C. Mugnaioli, F. Luzzaro, A. Endimiani, A. Toniolo, G. Amicosante, and G. M. Rossolini.** 2003. IMP-12, a new plasmid-encoded metallo-β-lactamase from a *Pseudomonas putida* clinical isolate. *Antimicrob. Agents Chemother.* **47:**1522–1528.
33. **Doi, Y., N. Shibata, K. Shibayama, K. Kamachi, H. Kurkawa, K. Yokoyama, T. Yagi, and Y. Arakawa.** 2002. Characterization of a novel plasmid-mediated cephalosporinase (CMY-9) and its genetic environment in an *Escherichia coli* clinical isolate. *Antimicrob. Agents Chemother.* **46:**2427–2434.
34. **Dowson, C. G., A. Hutchison, J. A. Brannigan, R. C. George, D. Hansman, J. Linares, A. Tomasz, J. M. Smith, and B. G. Spratt.** 1989. Horizontal transfer of penicillin-binding protein genes in penicillin-resistant clinical isolates of *Streptococcus pneumoniae*. *Proc. Natl. Acad. Sci. USA* **86:**8842–8846.
35. **Duarte, A., F. Boavida, F. Grosso, M. Correia, L. M. Lito, J. Melo Cristino, and M. J. Salgado.** 2003. Outbreak of GES-1 β-lactamase-producing multidrug-resistant *Klebsiella pneumoniae* in a university hospital in Lisbon, Portugal. *Antimicrob. Agents Chemother.* **47:**1481–1482.
36. **Dubois, V., L. Poirel, C. Marie, C. Arpin, P. Nordmann, and C. Quentin.** 2002. Molecular characterization of a novel class 1 integron containing bla_{GES-1} and a fused product of *aac(3)-Ib/aac(6′)-I-b′* gene cassettes in *Pseudomonas aeruginosa*. *Antimicrob. Agents Chemother.* **46:**638–645.
37. **Duthie, E. C.** 1944. The production of penicillinase by organisms of the *subtilis* group. *Br. J. Exp. Pathol.* **25:**96–100.
38. **Duthie, E. S.** 1947. The production of stable potent preparations of penicillinase. *J. Gen. Microbiol.* **1:**370–377.
39. **Essack, S. Y., L. M. C. Hall, D. G. Pillay, M. L. McFaydyen, and D. M. Livermore.** 2001. Complexity and diversity of *Klebsiella pneumoniae* strains with extended-spectrum β-lactamases isolated in 1994 and 1996 at a teaching hospital in Durban, South Africa. *Antimicrob. Agents Chemother.* **45:**88–95.
40. **Fihman, V., M. Rottman, Y. Benzerara, F. Delisle, R. Labia, A. Philippon, and G. Arlet.** 2002. BUT-1: a new member in the chromosomal inducible class C β-lactamases family from a clinical isolate of *Buttiauxella* sp. *FEMS Microbiol. Lett.* **213:**103–111.
41. **Fosse, T., C. Giraud-Morin, I. Madinier, and R. Labia.** 2003. Sequence analysis and biochemical characterization of chromosomal CAV-1 (*Aeromonas caviae*), the parental cephalosporinase of plasmid-mediated AmpC 'FOX' cluster. *FEMS Microbiol. Lett.* **222:**93–98.
42. **Foster, J. W.** 1945. Acid formation from penicillin during enzymatic inactivation. *Science* **101:**205.
43. **Frère, J.-M., and B. Joris.** 1985. Penicillin-sensitive enzymes in peptidoglycan biosynthesis. *CRC Crit. Rev. Microbiol.* **11:**299–396.
44. **Gales, A. C., L. C. Menezes, S. Silbert, and H. S. Sader.** 2003. Dissemination in distinct Brazilian regions of an epidemic carbapenem-resistant *Pseudomonas aeruginosa* producing SPM metallo-β-lactamase. *J. Antimicrob. Chemother.* **52:**699–702.

45. **Giakkoupi, P., G. Petrikkos, L. S. Tzouvelekis, S. Tsonas, The WHONET GREECE Study Group, N. J. Legakis, and A. C. Vatopoulos.** 2003. Spread of integron-associated VIM-type metallo-β-lactamase genes among imipenem-nonsusceptible *Pseudomonas aeruginosa* strains in Greek hospitals. *J. Clin. Microbiol.* **41:**822–825.

46. **Giakkoupi, P., L. S. Tzouvelekis, A. Tsakris, V. Loukova, D. Sofianou, and E. Tzelepi.** 2000. IBC-1, a novel integron-associated class A β-lactamase with extended-spectrum properties produced by an *Enterobacter cloacae* clinical strain. *Antimicrob. Agents Chemother.* **44:**2247–2253.

47. **Giakkoupi, P., A. Xanthaki, M. Kanelopoulou, A. Vlahaki, V. Miragou, S. Kontou, E. Papfraggas, H. Malamou-Lada, L. S. Tzouvelekis, N. J. Legakis, and A. C. Vatopoulos.** 2003. VIM-1 metallo-β-lactamase-producing *Klebsiella pneumoniae* strains in Greek hospitals. *J. Clin. Microbiol.* **41:**3893–3896.

48. **Gibb, A. P., C. Tribuddharat, R. A. Moore, T. J. Louie, W. Krulicki, D. M. Livermore, M.-F. I. Palepou, and N. Woodford.** 2002. Nosocomial outbreak of carbapenem-resistant *Pseudomonas aeruginosa* with a new bla_{IMP} allele, bla_{IMP-7}. *Antimicrob. Agents Chemother.* **46:**255–258.

49. **Granier, S. A., V. Leflon-Guibout, F. W. Goldstein, and M.-H. Nicolas-Chanoine.** 2003. New *Klebsiella oxytoca* β-lactamase genes bla_{OXY-3} and bla_{OXY-4} and a third genetic group of *K. oxytoca* based on bla_{OXY-3}. *Antimicrob. Agents Chemother.* **47:**2922–2928.

50. **Hayes, M. V., C. J. Thomson, and S. G. B. Amyes.** 1994. Three beta-lactamases isolated from *Aeromonas salmonicida*, including a carbapenemase not detectable by conventional methods. *Eur. J. Clin. Microbiol. Infect. Dis.* **13:**805–811.

51. **Hedges, R. W., N. Datta, P. Kontomichalou, and J. T. Smith.** 1974. Molecular specificities of R factor-determined beta-lactamases: correlation with plasmid compatibility. *J. Bacteriol.* **117:**56–62.

52. **Henry, R. J., and R. D. Housewright.** 1947. Studies on penicillinase. II. Manometric method of assaying penicillinase and penicillin, kinetics of the penicillin-penicillinase reaction, and the effects of inhibitors on penicillinase. *J. Biol. Chem.* **167:**559–571.

53. **Ho, S. E., G. Subramaniam, S. Palasubramaniam, and P. Navaratnam.** 2002. Carbapenem-resistant *Pseudomonas aeruginosa* in Malaysia producing IMP-7 β-lactamase. *Antimicrob. Agents Chemother.* **46:**3286–3287.

54. **Hou, J. P., and J. W. Poole.** 1972. Measurement of beta-lactamase activity and rate of inactivation of penicillins by a pH-stat alkalimetric titration method. *J. Pharma. Sci.* **61:**1594–1598.

55. **Humeniuk, G., G. Arlet, R. Labia, P. Grimont, and A. Philippon.** 2000. Presented at the Reunion Interdisciplinaire de Chimiotherapie Anti-Infectieuse, Paris, France.

56. **Huovinen, P., S. Huovinen, and G. A. Jacoby.** 1988. Sequence of PSE-2 beta-lactamase. *Antimicrob. Agents Chemother.* **32:**134–136.

57. **Ito, H., Y. Arakawa, S. Ohsuka, R. Wacharotayankun, N. Kato, and M. Ohta.** 1995. Plasmid-mediated dissemination of the metallo-β-lactamase gene bla_{IMP} among clinically isolated strains of *Serratia marcescens. Antimicrob. Agents Chemother.* **39:**824–829.

58. **Iyobe, S., H. Kusadokoro, A. Takahashi, S. Yomoda, T. Okubo, A. Nakamura, and K. O'Hara.** 2002. Detection of a variant metallo-β-lactamase, IMP-10, from two unrelated strains of *Pseudomonas aeruginosa* and an *Alcaligenes xylosoxidans* strain. *Antimicrob. Agents Chemother.* **46:**2014–2016.

59. **Jaurin, B., and T. Grundstrom.** 1981. *ampC* cephalosporinase of *Escherichia coli* K-12 has a different evolutionary origin from that of β-lactamases of the penicillinase type. *Proc. Natl. Acad. Sci. USA* **78:**4897–4901.

60. **Jeong, S. H., K. Lee, Y. Chong, J. H. Yum, S. H. Lee, H. J. Choi, J. M. Kim, K. H. Park, B. H. Han, S. W. Lee, and T. S. Jeong.** 2003. Characterization of a new integron containing VIM-2, a metallo-β-lactamase gene cassette, in a clinical isolate of *Enterobacter cloacae. J. Antimicrob. Chemother.* **51:**397–400.

61. **Karim, A., L. Poirel, S. Nagarajan, and P. Nordmann** 2001. Plasmid-mediated extended-spectrum β-lactamase (CTX-M-3 like) from India and gene association with insertion sequence IS*Ecp1*. *FEMS Microbiol. Lett.* **201:**237–241.

62. **Kariuki, S., J. E. Corkill, G. Revathi, R. Musoke, and C. A. Hart.** 2001. Molecular characterization of a novel plasmid-encoded cefotaximase (CTX-M-12) found in clinical *Klebsiella pneumoniae* isolates from Kenya. *Antimicrob. Agents Chemother.* **45:**2141–2143.

63. **Kartali, G., E. Tzelepi, S. Pournaras, C. Kontopoulou, F. Kontos, D. Sofianou, A. N. Maniatis, and A. Tsakris.** 2002. Outbreak of infections caused by *Enterobacter cloacae* producing the integron-associated β-lactamase IBC in a neonatal intensive care unit of a Greek hospital. *Antimicrob. Agents Chemother.* **46:**1577–1580.

64. **Kim, Y.-K., H. Pai, H.-J. Lee, S.-E. Park, E.-H. Choi, J. Kim, J.-H. Kim, and E.-C. Kim.** 2002. Bloodstream infections by extended-spectrum β-lactamase-producing *Escherichia coli* and *Klebsiella pneumoniae* in children: epidemiology and clinical outcome. *Antimicrob. Agents Chemother.* **46:**1481–1491.

65. **Kirby, W. M. N.** 1944. Extraction of a highly potent penicillin inactivator from penicillin resistant staphylococci. *Science* **99:**452–453.

66. **Laraki, N., M. Galleni, I. Thamm, M. L. Riccio, G. Amicosante, J.-M. Frere, and G. M. Rossolini.** 1999. Structure of In31, a bla_{IMP}-containing *Pseudomonas aeruginosa* integron phyletically related to In5, which carries an unusual array of gene cassettes. *Antimicrob. Agents Chemother.* **43:**890–901.

67. **Lee, K., J. B. Lim, J. H. Yum, D. Yong, Y. Chong, J. M. Kim, and D. M. Livermore.** 2002. bla_{VIM-2} cassette-containing novel integrons in metallo-β-lactamase-producing *Pseudomonas aeruginosa* and *Pseudomonas putida* isolates disseminated in a Korean hospital. *Antimicrob. Agents Chemother.* **46:**1053–1058.

68. **Liassine, N., S. Madec, B. Ninet, C. Metral, M. Fouchereau-Peron, R. Labia, and R. Auckenthaler.** 2002. Postneurosurgical meningitis due to *Proteus penneri* with selection of a ceftriaxone-resistant isolate: analysis of chromosomal class A β-lactamase HugA and its LysR-type regulatory protein HugR. *Antimicrob. Agents Chemother.* **46:**216–219.

69. **Livermore, D. M.** 1998. β-lactamase-mediated resistance and opportunities for its control. *J. Antimicrob. Chemother.* **41:**25–43.

70. **Lopez-Otsoa, F., L. Gallego, K. J. Towner, L. Tysall, N. Woodford, and D. M. Livermore.** 2002. Endemic carbapenem resistance associated with OXA-40 carbapenemase among *Acinetobacter baumannii* isolates from a hospital in Northern Spain. *J. Clin. Microbiol.* **40:**4741–4743.

71. **Luzzaro, F., E. Mantengoli, M. Perilli, G. Lombardi, V. Orlandi, A. Orsatti, G. Amicosante, G. M. Rossolini, and A. Toniolo.** 2001. Dynamics of a nosocomial outbreak of multidrug-resistant *Pseudomonas aeruginosa* producing PER-1 extended-spectrum β-lactamase. *J. Clin. Microbiol.* **39:**1865–1870.

72. **Madinier, I., T. Fosse, J. Giudicelli, and R. Labia.** 2001. Cloning and biochemical characterization of a class A β-lactamase from *Prevotella intermedia. Antimicrob. Agents Chemother.* **45:**2386–2389.

73. **Mainardi, J.-L., P. Mugnier, A. Coutrot, A. Buu-Hoi, E. Collatz, and L. Gutmann.** 1997. Carbapenem resistance in a clinical isolate of *Citrobacter freundii. Antimicrob. Agents Chemother.* **41:**2352–2354.

74. **Makanera, A., G. Arlet, V. Gautier, and M. Manai.** 2003. Molecular epidemiology and characterization of plasmid-encoded β-lactamases produced by Tunisian clinical isolates of *Salmonella enterica* serotype Mbandaka resistant to broad-spectrum cephalosporins. *J. Clin. Microbiol.* **41:** 2940–2945.

75. **Mammeri, H., L. Poirel, N. Mangeny, and P. Nordmann.** 2003. Chromosomal integration of a cephalosporinase gene from *Acinetobacter baumannii* into *Oligella urethralis* as a source of acquired resistance to β-lactams. *Antimicrob. Agents Chemother.* **47:**1536–1542.

76. **Massova, I., and S. Mobashery.** 1998. Kinship and diversification of bacterial penicillin-binding proteins and β-lactamases. *Antimicrob. Agents Chemother.* **42:**1–17.

77. **Matagne, A., A. Dubus, M. Galleni, and J.-M. Frere.** 1999. The beta-lactamase cycle: a tale of selective pressure and bacterial ingenuity. *Nat. Prod. Rep.* **16:**1–19.

78. **Matthew, M., and A. M. Harris.** 1976. Identification of β-lactamases by analytical isoelectric focusing: correlation with bacterial taxonomy. *J. Gen. Microbiol.* **94:**55–67.

79. **Matthew, M., A. M. Harris, M. J. Marshall, and G. W. Ross.** 1975. The use of isoelectric focusing for detection and identification of beta-lactamases. *J. Gen. Microbiol.* **88:** 169–178.

80. **Medeiros, A. A.** 2000. Cooperative evolution of mechanisms of beta-lactam resistance. *Clin. Microbiol. Infect.* **6** (Suppl. 3):3–5.

81. **Medeiros, A. A.** 1997. Evolution and dissemination of β-lactamases accelerated by generations of β-lactam antibiotics. *Clin. Infect. Dis.* **24:**S19–S45.

82. **Melano, R., A. Corso, A. Petroni, D. Centrón, B. Orman, A. Pereyra, N. Moreno, and M. Galas.** 2003. Multiple antibiotic-resistance mechanisms including a novel combination of extended-spectrum β-lactamase in a *Klebsiella pneumoniae* clinical strain isolated in Argentina. *J. Antimicrob. Chemother.* **52:**36–42.

83. **Melano, R., A. Petroni, A. Garutti, H. A. Saka, L. Mange, F. Pasterán, M. Rapoport, A. Rossi, and M. Galas.** 2002. New carbenicillin-hydrolyzing β-lactamase (CARB-7) from *Vibrio cholerae* non-O1, non-O139 strains encoded by the VCR region of the *V. cholerae* genome. *Antimicrob. Agents Chemother.* **46:**2162–2168.

84. **Miragou, V., L. S. Tzouvelekis, S. Rossiter, E. Tzelepi, F. J. Angulo, and J. M. Whichard.** 2003. Imipenem resistance in a Salmonella clinical strain due to plasmid-mediated class A carbapenemase KPC-2. *Antimicrob. Agents Chemother.* **47:**1297–1300.

85. **Miró, E., F. Navarro, B. Mirelis, M. Sabaté, A. Rivera, P. Coll, and G. Prats.** 2002. Prevalence of clinical isolates of *Escherichia coli* producing inhibitor-resistant β-lactamases at a university hospital in Barcelona, Spain, over a 3-year period. *Antimicrob. Agents Chemother.* **46:**3991–3994.

86. **Moland, E. S., J. A. Black, J. Ourada, M. D. Reisbig, N. D. Hanson, and K. S. Thomson.** 2002. Occurrence of newer β-lactamases in *Klebsiella pneumoniae* isolates from 24 U.S. hospitals. *Antimicrob. Agents Chemother.* **46:**3837–3842.

87. **Moland, E. S., N. D. Hanson, V. Herrera, J. A. Black, T. J. Lockhart, A. Hossain, J. A. Johnson, R. V. Goering, and K. S. Thomson.** 2003. Plasmid-mediated, carbapenem-hydrolysing β-lactamase, KPC-2 in *Klebsiella pneumoniae* isolates. *J. Antimicrob. Chemother.* **51:**711–714.

88. **Morin, A.-S., L. Poirel, F. Mory, R. Labia, and P. Nordmann.** 2002. Biochemical-genetic analysis and distribution of DES-1, an Ambler class A extended-spectrum β-lactamase from *Desulfovibrio desulfuricans. Antimicrob. Agents Chemother.* **46:**3215–3222.

89. **Naas, T., D. Aubert, N. Fortineau, and P. Nordmann.** 2002. Cloning and sequencing of the β-lactamase gene and surrounding DNA sequences of *Citrobacter braakii, Citrobacter murliniae, Citrobacter werkmanii, Escherichia fergusonii* and *Enterobacter cancerogenus. FEMS Microbiol. Lett.* **215:**81–87.

90. **Naas, T., S. Bellais, and P. Nordmann.** 2003. Molecular and biochemical characterization of a carbapenem-hydrolysing β-lactamase from *Flavobacterium johnsoniae. J. Antimicrob. Chemother.* **51:**267–273.

91. **Nadjar, D., R. Labia, C. Cerceau, C. Bizet, A. Philippon, and G. Arlet.** 2001. Molecular characterization of chromosomal class C β-lactamase and its regulatory gene in *Ochrobactrum anthropi. Antimicrob. Agents Chemother.* **5:**2324–2330.

92. **National Committee for Clinical Laboratory Standards.** 2003. *Performance Standards for Antimicrobial Susceptibility Testing. NCCLS Approved Standard M100-S13 (M2).* National Committee for Clinical Laboratory Standards, Wayne, Pa.

93. **National Committee for Clinical Laboratory Standards.** 2003. *Performance Standards for Antimicrobial Susceptibility Testing. NCCLS Approved Standard M100-S13 (M7).* National Committee for Clinical Laboratory Standards, Wayne, Pa.

94. **Newton, G. G. F., and E. P. Abraham.** 1956. Isolation of cephalosporin C, a penicillin-like antibiotic containing D-alpha aminoadipic acid. *Biochem. J.* **62:**651–658.

95. **Niumsup, P., and V. Wuthiekanun.** 2002. Cloning of the class D β-lactamase gene from *Burkholderia pseudomallei* and studies on its expression in ceftazidime-susceptible and -resistant strains. *J. Antimicrob. Chemother.* **50:**445–455.

96. **O'Callaghan, C., A. Morris, S. M. Kirby, and A. H. Shingler.** 1972. Novel method for detection of β-lactamases by using a chromogenic cephalosporin substrate. *Antimicrob. Agents Chemother.* **1:**283–288.

97. **Oliver, A., J. C. Pérez-Díaz, T. M. Coque, F. Baquero, and R. Cantón.** 2000. Presented at the 40th Interscience Conference on Antimicrobial Agents and Chemotherapy, Toronto, Canada.
98. **Osano, E., Y. Arakawa, R. Wacharotayankun, M. Ohta, T. Horii, H. Ito, F. Yosimura, and N. Kato.** 1994. Molecular characterization of an enterobacterial metallo β-lactamase found in a clinical isolate of *Serratia marcescens* that shows imipenem resistance. *Antimicrob. Agents Chemother.* **38:**71–78.
99. **Page, M. I.** 1999. The reactivity of beta-lactams, the mechanism of catalysis and the inhibition of beta-lactamases. *Curr. Pharm. Des.* **5:**895–913.
100. **Petrella, S., D. Clermont, I. Casin, V. Jarlier, and W. Sougakoff.** 2001. Novel class A β-lactamase Sed-1 from *Citrobacter sedlakii*: genetic diversity of β-lactamases within the *Citrobacter* genus. *Antimicrob. Agents Chemother.* **45:**2287–2298.
101. **Petroni, A., A. Corso, R. Melano, M. L. Cacace, A. M. Bru, A. Rossi, and M. Galas.** 2002. Plasmidic extended-spectrum β-lactamases in *Vibrio cholerae* O1 El Tor isolates in Argentina. *Antimicrob. Agents Chemother.* **46:**1462–1486.
102. **Poirel, L., D. Girlich, T. Naas, and P. Nordmann.** 2001. OXA-28, an extended-spectrum variant of OXA-10, β-lactamase from *Pseudomonas aeruginosa* and its plasmid- and integron-located gene. *Antimicrob. Agents Chemother.* **45:**447–453.
103. **Poirel, L., P. Kampfer, and P. Nordmann.** 2002. Chromosome-encoded Ambler class A β-lactamase of *Kluyvera georgiana*, a probable progenitor of a subgroup of CTX-M extended-spectrum β-lactamases. *Antimicrob. Agents Chemother.* **46:**4038–4040.
104. **Poirel, L., T. Lambert, S. Türkoglü, E. Ronco, J.-L. Gaillard, and P. Nordmann.** 2001. Characterization of class I integrons from *Pseudomonas aeruginosa* that contain the bla_{VIM-2} carbapenem-hydrolyzing β-lactamase gene and of two novel aminoglycoside resistance gene cassettes. *Antimicrob. Agents Chemother.* **45:**456–552.
105. **Poirel, L., O. Menuteau, N. Agoli, C. Cattoen, and P. Nordmann.** 2003. Outbreak of extended-spectrum β-lactamase VEB-1-producing isolates of *Acinetobacter baumannii* in a French hospital. *J. Clin. Microbiol.* **41:**3542–3547.
106. **Poirel, L., I. L. Thomas, T. Naas, A. Karim, and P. Nordman.** 2000. Biochemical sequence analyses of GES-1, a novel class A extended-spectrum β-lactamase, and the class 1 integron IN52 from *Klebsiella pneumoniae*. *Antimicrob. Agents Chemother.* **44:**622–632.
107. **Poirel, L., G. F. Weldhagen, T. Naas, C. D. Champs, M. G. Dove, and P. Nordmann.** 2001. GES-2, a class A β-lactamase from *Pseudomonas aeruginosa* with increased hydrolysis of imipenem. *Antimicrob. Agents Chemother.* **45:**2598–2603.
108. **Pollock, M. R.** 1965. Purification and properties of penicillinases from two strains of *Bacillus licheniformis*: a chemical physicochemical and physiological comparison. *Biochem. J.* **94:**666–675.
109. **Pournaras, S., A. Tsakris, M. Maniati, L. S. Tzouvelekis, and A. N. Maniatis.** 2002. Novel variant (bla_{VIM-4}) of the metallo-β-lactamase gene bla_{VIM-1} in a clinical strain of *Pseudomonas aeruginosa*. *Antimicrob. Agents Chemother.* **46:**4026–4028.
110. **Pragai, Z., Z. Koczian, and E. Nagy.** 1998. Characterization of the extended-spectrum β-lactamases and determination of the antibiotic susceptibilities of *Klebsiella pneumoniae* isolates in Hungary. *J. Antimicrob. Chemother.* **42:**401–403.
111. **Queenan, A. M., B. Foleno, C. Gownley, E. Wira, and K. Bush.** 2004. Effects of inoculum and beta-lactamase activity in AmpC- and extended-spectrum beta-lactamase (ESBL)-producing *Escherichia coli* and *Klebsiella pneumoniae* clinical isolates tested by using NCCLS ESBL methodology. *J. Clin. Microbiol.* **42:**269.
112. **Queenan, A. M., S. Jenkins, and K. Bush.** 2001. Cloning and biochemical characterization of FOX-5, an AmpC-type plasmid-encoded β-lactamase from a New York City *Klebsiella pneumoniae* isolate. *Antimicrob. Agents Chemother.* **45:**3189–3194.
113. **Quinteros, M., M. Radice, N. Gardella, M. M. Rodriguez, N. Costa, D. Korbenfeld, E. Couto, G. Gutkind, and Microbiology Study Group.** 2003. Extended-spectrum β-lactamases in *Enterobacteriaceae* in Buenos Aires, Argentina, public hospitals. *Antimicrob. Agents Chemother.* **47:**2864–2867.
114. **Raskine, L., I. Borrel, G. Barnaud, S. Boyer, B. Hanau-Bercot, J. Gravisse, R. Labia, G. Arlet, and M.-J. Sanson-Le-Pors.** 2002. Novel plasmid-encoded class C β-lactamase (MOX-2) in *Klebsiella pneumoniae* from Greece. *Antimicrob. Agents Chemother.* **46:**2262–2265.
115. **Ross, G. W., K. V. Chanter, A. M. Harris, S. M. Kirby, M. J. Marshall, and C. H. O'Callaghan.** 1973. Comparison of assay techniques for beta-lactamase activity. *Anal. Biochem.* **54:**9–16.
116. **Sabath, L. D., and E. P. Abraham.** 1965. Cephalosporinase and penicillinase activity of *Bacillus cereus*. *Antimicrob. Agents Chemother.* **1:**392–397.
117. **Sabath, L. D., and E. P. Abraham.** 1966. Zinc as a cofactor for cephalosporinase from *Bacillus cereus* 569. *Biochem. J.* **98:**11c–13c.
118. **Sanchez, S., M. A. McCrackin Stevenson, C. R. Hudson, M. Maier, T. Buffington, Q. Dam, and J. J. Maurer.** 2002. Characterization of multidrug-resistant *Escherichia coli* isolates associated with nosocomial infections in dogs. *J. Clin. Microbiol.* **40:**3586–3595.
119. **Schwaber, M. J., P. M. Raney, J. K. Rasheed, J. W. Biddle, P. Williams, J. E. McGowan, Jr., and F. C. Tenover.** 2004. Utility of NCCLS guidelines for identifying extended-spectrum beta-lactamases in non-*Escherichia coli* and non-*Klebsiella* spp. of *Enterobacteriaceae*. *J. Clin. Microbiol.* **42:**294.
120. **Shannon, K., A. King, and I. Phillips.** 1986. β-Lactamases with high activity against imipenem and Sch 34343 from *Aeromonas hydrophila*. *J. Antimicrob. Chemother.* **17:**45–50.
121. **Simm, A. M., C. S. Higgins, S. T. Pullan, M. B. Avison, P. Niumsup, O. Erdozain, P. M. Bennett, and T. R. Walsh.** 2001. A novel metallo-β-lactamase, Mb11b, produced by the environmental bacterium *Caulobacter crescentus*. *FEBS Lett.* **509:**350–354.
122. **Spanu, T., F. Luzzaro, M. Perilli, G. Amicosante, A. Toniolo, G. Fadda, and The Italian ESBL Study Group.** 2002. Occurrence of extended-spectrum β-lactamases in members of the family *Enterobacteriaceae* in Italy: implications

for resistance to β-lactams and other antimicrobial drugs. *Antimicrob. Agents Chemother.* **46:**196–202.

123. **Strydnadka, N. C. J., H. Adachi, S. E. Jensen, K. Johns, A. Sielecke, C. Betzel, K. Sutoh, and M. N. G. James.** 1992. Molecular structure of the acyl-enzyme intermediate in β-lactam hydrolysis at 1.7 A resolution. *Nature* **359:**700–705.
124. **Sykes, R. B., D. P. Bonner, K. Bush, and N. H. Georgopapadakou.** 1982. Azthreonam (SQ 26,776), a synthetic monobactam specifically active against aerobic gram-negative bacteria. *Antimicrob. Agents Chemother.* **21:**85–92.
125. **Tenover, F. C., P. M. Raney, P. P. Williams, J. K. Rasheed, J. W. Biddle, A. Oliver, S. K. Fridkin, L. Jevitt, and J. E. McGowan, Jr.** 2003. Evaluation of the NCCLS extended-spectrum β-lactamase confirmation methods for *Escherichia coli* with isolates collected during project ICARE. *J. Clin. Microbiol.* **41:**3142–3146.
126. **Toleman, M. A., D. Biedenbach, D. Bennett, R. N. Jones, and T. R. Walsh.** 2003. Genetic characterization of a novel metallo-β-lactamase gene bla_{IMP-13}, harboured by a novel Tn*5051*-type transposon disseminating carbapenemase genes in Europe: report from the SENTRY worldwide antimicrobial surveillance programme. *J. Antimicrob. Chemother.* **52:**583–590.
127. **Toleman, M. A., K. Rolston, R. N. Jones, and T. R. Walsh.** 2003. Molecular and biochemical characterization of OXA-45, an extended-spectrum class 2d′ β-lactamase in *Pseudomonas aeruginosa. Antimicrob. Agents Chemother.* **47:**2859–2863.
128. **Vessillier, S., J.-D. Docquier, S. Rival, J.-M. Frere, M. Galleni, G. Amicosante, G. M. Rossolini, and N. Franceschini.** 2002. Overproduction and biochemical characterization of the *Chryseobacterium meningosepticum* BlaB metallo-β-lactamase. *Antimicrob. Agents Chemother.* **46:**1921–1927.
129. **Vourli, S., L. S. Tzouvelekis, E. Tzelepi, E. Lebessi, N. J. Legakis, and V. Miragou.** 2003. Characterization of In111, a class 1 integron that carries the extended-spectrum β-lactamase gene bla_{IBC-1}. *FEMS Microbiol. Lett.* **225:**149–153.
130. **Walsh, T. R., M. A. Toleman, W. Hryniewicz, P. M. Bennett, and R. N. Jones.** 2002. Evolution of an integron carrying bla_{VIM-2} in Eastern Europe: report from the SENTRY antimicrobial surveillance program. *J. Antimicrob. Chemother.* **52:**116–119.
131. **Wang, H., S. Kelkar, W. Wu, M. Chen, and J. P. Quinn.** 2003. Clinical isolates of *Enterobacteriaceae* producing extended-spectrum β-lactamases: prevalence of CTX-M-3 at a hospital in China. *Antimicrob. Agents Chemother.* **47:**790–793.
132. **Wang, Z., W. Fast, and S. J. Benkovic.** 1999. On the mechanism of the metallo-beta-lactamase from *Bacteroides fragilis. Biochemistry* **38:**10013–10023.
133. **Winokur, P. L., A. Brueggemann, D. L. DeSalvo, L. Hoffmann, M. D. Apley, E. K. Uhlenhopp, M. A. Pfaller, and G. V. Doern.** 2000. Animal and human multidrug-resistant, cephalosporin-resistant salmonella isolates expressing a plasmid-mediated CMY-2 AmpC beta-lactamase. *Antimicrob. Agents Chemother.* **44:**2777–2783.
134. **Winokur, P. L., D. L. Vonstein, L. J. Hoffman, E. K. Uhlenhopp, and G. V. Doern.** 2001. Evidence for transfer of CMY-2 AmpC beta-lactamase plasmids between *Escherichia coli* and *Salmonella* isolates from food animals and humans. *Antimicrob. Agents Chemother.* **45:**2716–2722.
135. **Yamasaki, K., M. Komatsu, T. Yamashita, K. Shimakawa, T. Ura, H. Nishio, K. Satoh, R. Washidu, S. Kinoshita, and M. Aihara.** 2003. Production of CTX-M-3 extended-spectrum β-lactamase and IMP-1 metallo β-lactamase by five Gram-negative bacilli: survey of clinical isolates from seven laboratories collected in 1998 and 2000, in the Kinki region of Japan. *J. Antimicrob. Chemother.* **51:**631–638.
136. **Yan, J.-J., P.-R. Hsueh, W.-C. Ko, K.-T. Luh, S.-H. Tsai, H.-M. Wu, and J.-J. Wu.** 2001. Metallo-β-lactamases in clinical *Pseudomonas* isolates in Taiwan and identification of VIM-3, a novel variant of the VIM-2 enzyme. *Antimicrob. Agents Chemother.* **45:**2224–2228.
137. **Yano, H., A. Kuga, R. Okamoto, H. Kitasato, T. Kobayashi, and M. Inoue.** 2001. Plasmid-encoded metallo-β-lactamase (IMP-6) conferring resistance to carbapenems, especially meropenem. *Antimicrob. Agents Chemother.* **45:** 1343–1348.
138. **Yigit, H., G. J. Anderson, J. W. Biddle, C. D. Steward, J. K. Rasheed, L. L. Valera, J. E. McGowan, Jr., and F. C. Tenover.** 2002. Carbapenem resistance in a clinical isolate of *Enterobacter aerogenes* is associated with decreased expression of OmpF and OmpC porin analogs. *Antimicrob. Agents Chemother.* **46:**3817–3822.
139. **Yigit, H., A. M. Queenan, G. J. Anderson, A. Domenech-Sanchez, J. W. Biddle, C. D. Steward, S. Alberti, K. Bush, and F. C. Tenover.** 2001. Novel carbapenem-hydrolyzing β-lactamase, KPC-1 from a carbapenem-resistant strain of *Klebsiella pneumoniae. Antimicrob. Agents Chemother.* **45:** 1151–1161.
140. **Yomoda, S., T. Okubo, A. Takahashi, M. Murakami, and S. Iyobe.** 2003. Presence of *Pseudomonas putida* strains harboring plasmids bearing the metallo-β-lactamase gene bla_{IMP} in a hospital in Japan. *J. Clin. Microbiol.* **41:**4246–4251.
141. **Yong, D., J. H. Shin, S. Kim, Y. Lim, J. H. Yum, K. Lee, Y. Chong, and A. Bauernfeind.** 2003. High prevalence of PER-1 extended-spectrum β-lactamase-producing *Acinetobacter* spp. in Korea. *Antimicrob. Agents Chemother.* **47:** 1749–1751.
142. **Zhao, S., D. G. White, P. F. McDermott, S. Friedman, L. English, S. Ayers, J. Meng, J. J. Maurer, R. Holland, and R. D. Walker.** 2001. Identification and expression of cephamycinase bla_{CMY} genes in *Escherichia coli* and *Salmonella* isolates from food animals and ground meat. *Antimicrob. Agents Chemother.* **45:**3647–3650.

Enzyme-Mediated Resistance to Antibiotics: Mechanisms, Dissemination, and Prospects for Inhibition
Edited by Robert A. Bonomo and Marcelo E. Tolmasky

Malcolm G. P. Page

7 Resistance Mediated by Penicillin-Binding Proteins

INTRODUCTION

The cell wall imparts strength and shape to bacteria. It is made up of polymeric glycan chains cross-linked by peptide branches. The cross-linking reaction, catalyzed by transpeptidases, is one of the last steps in cell wall biosynthesis. The transpeptidases are members of the family of penicillin-binding proteins (PBPs), which have become known as the targets of β-lactam antibiotics. PBPs perform a variety of critical functions for the bacterial cell. They alert the cell to the presence of penicillin, they make the cross-links essential for structural integrity of the cell wall, they are drug targets, and they mediate resistance to β-lactam antibiotics. Some PBPs are multidomain proteins and are thus functionally interrelated within the family and with other protein families. The functions of the non-penicillin-binding domains (nPBDs) are manifold and, in many cases, uncertain. They may be involved in folding, signal transduction, other enzymic activities, substrate binding, and protein-protein interactions. Dissecting the role of each PBP, determining which are essential, and elucidating the contributions of the nPBD and PBD to structure and function will all improve our understanding of drug resistance and may suggest new targets for drug development.

Nomenclature

Perhaps unfortunately, it is customary to identify PBPs by their relative position on sodium dodecyl sulfate-polyacrylamide gel electrophoresis gels and not, for example, by the apparent molecular mass deduced from the mobility in that gel system (Fig. 7.1). This results in unrelated proteins in different organisms receiving the same identifier, and considerable confusion springs therefrom. For example, PBP5 of *Escherichia coli* has an apparent molecular mass of 40 kDa, is primarily a D-Ala-D-Ala carboxypeptidase with negligible transpeptidase activity, and has little to do with β-lactam resistance, whereas PBP5 of *Enterococcus faecium* has an apparent molecular mass of 75 kDa, is primarily a transpeptidase with negligible carboxypeptidase activity, and is a major determinant of β-lactam resistance in this organism (Table 7.1). Further, for one of the clinically most important proteins, the MecA protein of staphylococci, there are two nomenclatures (PBP2′ and PBP2a) in use and typographical errors may result in loss of the last character, which separates this highly resistant protein from the remotely related, sensitive PBP2 found in the same organism. This being said, neither the apparent molecular mass nor the assigned gene name is a more reliable indicator of affinity and one can only advocate extreme

Malcolm G. P. Page, Basilea Pharmaceutica AG, Grenzacherstrasse 487, CH-4005 Basel, Switzerland.

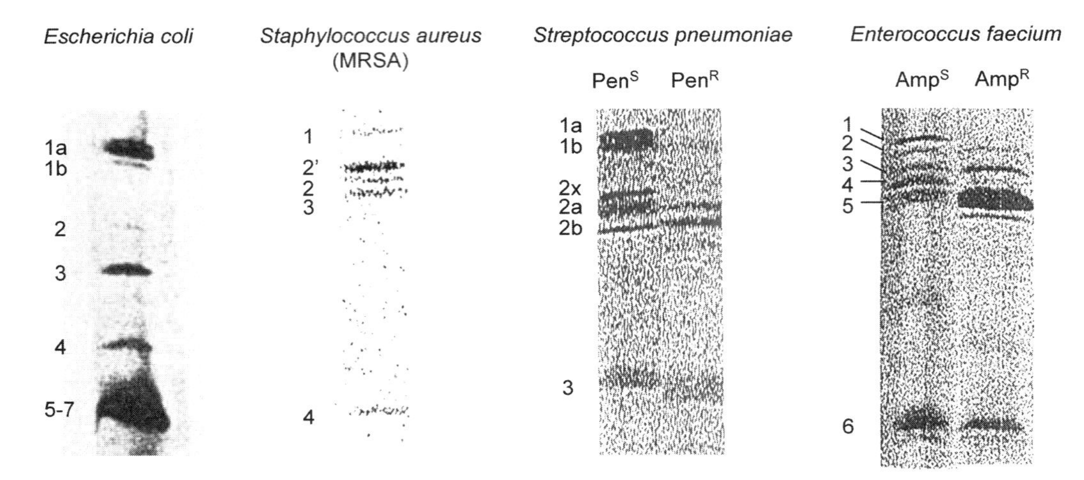

Figure 7.1 Comparison of PBP profiles of selected organisms and pattern alterations commonly encountered in resistant organisms. The PBPs have been labeled by incubation of membranes derived by sonication of cultures of the indicated organisms with radiolabeled benzylpenicillin and then resolution of the proteins by standard sodium dodecyl sulfate-polyacrylamide gel electrophoresis. From left to right: *E. coli*, MRSA, *Streptococcus pneumoniae*, penicillin susceptible (Pen^S) and an example of a penicillin-resistant (Pen^R) isolate where loss of affinity makes some PBPs disappear, and *Enterococcus faecium*, ampicillin-susceptible (Amp^S) and an example of an ampicillin-resistant (Amp^R) isolate where loss of affinity of PBP1 and PBP4 as well as overexpression of PBP5 are evident.

caution when comparing results from one organism to another.

With the availability of the complete genomes of many bacteria it is now possible to cluster the PBPs by homology of the primary sequence data. Ghuysen and coworkers (34, 39, 40) have proposed a classification that divides the high-molecular-weight PBPs into type A PBPs, which have an integral glycosyltransferase domain as well as a transpeptidase domain, and type B PBPs, which have an N-terminal β-sheet domain but no equivalent glycosyltransferase domain. Types A and B were divided into a number of subclasses according to sequence homology and the presence, in the type B subfamily, of further domains of unknown function.

Role in Resistance

The PBPs that comprise the normal complement of β-lactam-susceptible organisms differ considerably in their natural susceptibility towards inhibition by these agents (40). Benzyl penicillin, a natural PBP inhibitor, works well and evenly against the PBPs of many organisms and, indeed, has been suggested to be optimized during the course of evolution for exactly this purpose (16). In contrast, the semisynthetic penicillin mecillinam works very potently against *E. coli* PBP2, but not against any of the other PBPs of this organism (Table 7.2).

Low-affinity PBPs that contribute to the β-lactam resistance of several pathogenic bacteria have different origins (Table 7.3). Natural transformation and recombination events with DNA acquired from neighboring intrinsically resistant organisms are responsible for the appearance of mosaic genes encoding low-affinity PBPs in highly resistant strains of transformable microorganisms such as *Neisseria* and *Streptococcus*. Methicillin-resistant *Staphylococcus aureus* (MRSA) and some coagulase-negative staphylococci possess an extra PBP (PBP2′), which probably evolved within the *Staphylococcus* genus. Enterococci have a natural low susceptibility to β-lactams due to an intrinsic low-affinity PBP that is homologous to the staphylococcal PBP2′. Highly resistant enterococcal strains overexpress this PBP and may also collect point mutations that further reduce its affinity. PBP-mediated resistance is less common in gram-negative pathogens apart from the *Neisseria*, but mutant PBPs with decreased susceptibility towards β-lactams have been noted in several genera, especially *Haemophilus*, *Pseudomonas*, and *Acinetobacter*.

EPIDEMIOLOGY AND CLINICAL IMPORTANCE

PBPs are found in all pathogenic bacteria except those of the genus *Mycoplasma*, which do not have a cell wall. Sensitivity of PBPs towards β-lactams is frequently noted as a contributory factor to the susceptibility of the organism to these antibiotics.

Table 7.1 Comparison of the PBPs in *E. coli*, *P. aeruginosa*, *S. aureus*, *S. pneumoniae*, and *E. faecalis*[a]

	E. coli				*P. aeruginosa*				*S. aureus*				*S. pneumoniae*				*E. faecalis*			
Class	PBP no.	Gene name	No. of residues	Predicted kDa	PBP no.	Gene name	No. of residues	Predicted kDa	PBP no.	Gene name	No. of residues	Predicted kDa	PBP no.	Gene name	No. of residues	Predicted kDa	PBP no.	Gene name	No. of residues	Predicted kDa
Transglycosylase transpeptidase																				
A1	1a	*ponA mrcA*	850	93.6	1a	*ponA mrcA*	822	91.1												
A2	1b	*ponB mrcB*	844	94.3	1b	*ponB*	774	85.5												
A3									2	*pbp2*	727, 716	80.5, 79.2	1a	*ponA*	719	79.8	1a	*ponA*	778	85.3
A4													2a	*pbp2a*	731	80.8	2a	*pbp2a*	728	79.5
A5													1b	*ponB*	821	89.5	1b	*ponB*	803	88.5
A?	1c	*pbpC*	770	85.1																
Transpeptidases																				
B1									2′, 2a	*mecA*	668	76.2					4	*pbp4*	680	74
B2	2	*mrdA pbpA*	633	70.9	2	*pbpA*	646	72.2												
B3	3	*ftsI pbpB*	588	63.9	3	*pbpB*	579	62.9												
					3x	*pbpC*	565	61.1												
B4									1	*pbpA*	744	82.7	2x	*pbpX*	750	82.3		*pbpC*	742	81.7
B5									3, 2B	*pbpF*	691	77.18	2b	*penA*	680	73.9	2b	*pbp2b*	711	77.9
Carboxypeptidases																				
C1	4	*dacB*	477	51.8	4		476	51.9												
C2									4	*pdp4*	431	48.3								
													3	*dacA*	413	45.2		*dacA*	493	53.9
C3	5	*dacA*	403	44.4																
	6	*dacC*	400	43.6	5	*dacC*	386	42.5												
	6b	*dacD*	388	43.4																
	7	*pbpG*	313	34.3	6	*pbpG*	310	34												
	8	Fragment of PBP7																		

[a]The type definition follows that of Ghuysen and coworkers (38, 39).

Table 7.2 Comparison of susceptibilities of the PBPs of *E. coli* and *S. aureus* to different β-lactam classes

Compound	Approx concn (mg/liter) required for complete inhibition[a]										
	E. coli						MRSA				
	1a	1b	2	3	4	5/6	1	2′	2	3	4
Benzylpenicillin	0.5	5	0.5	0.5	1	10	0.001	10	0.05	0.05	1
Mecillinam	>10	>10	0.1	>100	>100	>100	>100	>100	>100	>100	>100
Piperacillin	1	5	1	1	5	10	50	>100	50	100	>100
Cephalothin	0.5	5	50	1	50	10	1	>100	0.5	0.1	0.1
Ceftriaxone	0.1	1	1	0.05	50	10	0.5	>100	0.5	0.5	0.1
Ceftobiprole	100	1	1	5	10	50	10	1	10	0.01	10
Imipenem	0.5	1	0.05	10	100	0.5	0.1	>100	5	0.5	0.5
Meropenem	1	5	0.05	1	10	10	0.05	>100	1	1	0.05

[a]Inhibition of binding of radiolabeled benzylpenicillin to membrane-bound PBP after 15 min of incubation.

Sphingomonadales

Zymomonas

Karibian and Starka (68) reported that *Zymomonas mobilis* had three PBPs with apparent molecular masses of 90, 60, and 44 kDa. PBP1 (90 kDa, probably a type A PBP) and PBP2 (60 kDa, probably a type B PBP) were half-saturated at the MIC required to prevent bacterial growth, showing that they are probably killing targets. Both PBPs had low affinity for mecillinam, to which *Z. mobilis* was relatively resistant.

Neisseriales

Neisseria

Acquired resistance has been demonstrated in clinical isolates, involving mutation of two of the high-molecular-mass proteins. Mutations in PBP1 (type A1) have been identified in *Neisseria gonorrhoeae* (21, 22, 118, 119), *N. denitrificans* (87), and *N. meningitidis* (106). Mutations in, or alteration by genetic exchange between related genes of, PBP2 (type B3) have been identified in *N. gonorrhoeae* (1, 13, 20–23, 130, 132), *N. denitrificans* (87), *N. meningitidis* (2–15, 92, 131, 157), *N. lactamica* (85, 86), and *N. perflava/sicca* (113). Genetic exchange between the species pathogenic to humans (*N. meningitidis* and *N. gonorrhoeae*) and human commensal species (including *N. flavescens* and *N. cinerea*) can be induced to occur under laboratory conditions (10, 105, 106).

Several authors have found that PBP2 from penicillin-resistant strains have an additional residue (Asp345) that is absent from PBP2 of penicillin-sensitive strains (11, 139, 140). The role of the additional aspartic acid residue in the decreased affinity of PBP2 is unclear.

Pseudomonales

Pseudomonas

Alterations in the apparent affinity and expression levels of PBPs have been noted in clinical isolates of *Pseudomonas aeruginosa*, although a causal relation with clinical resistance has yet to be demonstrated (107, 133, 156).

Mutations in the bifunctional glycosyltransferase/transpeptidases PBP1a (type A1) and PBP1b (type A2) have been observed (38, 94, 116). Further, decreased affinity has been reported for PBP2 (type B2) (113) and PBP3 (type B3) (37, 38, 42, 94). Liao and Hancock also noted the appearance of an additional protein, PBP3x (Type B3), with growth-phase-dependent expression (76, 77).

Several reports have implicated altered levels of expression of PBPs from the D,D-carboxypeptidase families in the appearance of resistance in *Pseudomonas* strains. They include members of the PBP C1 (7, 81, 117) and PBP C4 groups (37, 38, 81).

Acinetobacter

Reduced expression of PBP2 (type B, assignment not certain) is one of the most frequently observed mechanisms of resistance to carbapenems (26). Decreased labeling of several PBPs in imipenem-resistant strains has been reported (31, 146). Increased labeling of a 24-kDa PBP (presumably a D,D-carboxypeptidase, group PBP C4) has been observed (32). Laboratory mutants of *Acinetobacter calcoaceticus* resistant to cefoxitin, cefoperazone, or ceftazidime were selected from a strain producing a low level of β-lactamase. The PBPs of the resistant isolates exhibited altered expression and/or affinity for β-lactams (102).

Table 7.3 Distribution of PBP types among the bacteria and their contribution to β-lactam resistance[a]

Taxonomic group	PBP class																
	A1 (*E. coli* PBP 1a)	A1* (weak *E. coli* PBP1a homologues)	A2 (*E. coli* PBP 1b)	A3 (*Spn* PBP 1a)	A4 (*Spn* PBP 2a)	A5 (*Spn* PBP 1b)	A6 (*E. coli* 1c)	B1 (*S. aureus* PBP 2′)	B2 (*E. coli* PBP 2)	B3 (*E. coli* PBP 3)	B4 (*Spn* PBP 2x)	B5 (*Spn* PBP 2b)	C1 (*E. coli* PBP 4)	C2 (*S. aureus* PBP 4)	C3 (*E. coli* PBP 5)	C4 (*Spn* PBP3)	C5 (*B. subtilis* DacF)
Aquificae		X							X								
Thermotogae		X							X								
Deinococcus		X															
Cyanobacteria		X	X														
Chlorobi									X	X							
Proteobacteria										X							
Alpha																	
Rickettsiales	X						X		X	X							
Magnetococcales	X								X								
Rhodobacterales	X						X		X								
Caulobacterales	X						X		X								
Rhizobiales	X	X					X										
Beta																	
Burkholderiales	X						X		X	X			X		X		
Neisseriales	XX								X	XX			X		X		
Nitrosomonadales	X								X	X			X		X		
Gamma																	
"*Xanthomonadales*"	X		X				X		X	X					X		
Methylococcales	X		X				X		X	X					X		
Pseudomonadales	XX		XX				X		X	XX			X		X		
"*Alteromonadales*"	X		X				X		X	X			X		X		
Vibrionales	X		X						X	X					X		
Enterobacteriales	X		X				X		X	X			X		X		
Pasteurales	X		X				X		X	XX			X		X		
Delta																	
Desulfovibrionales		X							X								
Desulfobacterales		X											X				
Desulfuromonadales		X							X	X			X				
Bdellovibrionales		X	X				X			X			X				
Myxococcales		X					X			X			X				
Epsilon																	
Campylobacteriales		XX	X				X		X								
Firmicutes																	
Clostridia																	
Clostridiales					XX						XX						X
Thermoanaerobacterales				X	X						X						X
Bacilli																	
Bacilliales				X	XX			XX			XX	XX		X		X	X
Lactobacilliales				X	X	X		XX			XX	XX				X	
Actinobacteria																	
Actinomycetales		X															
Bifidobacteriales		X															
Chlamydiae									X	X							
Spirochaetes		X					X		X	X							
Bacteroidetes		XX					X										
Fusobacteria		X					X										

[a]The type definition follows that of Ghuysen and coworkers (38, 39). X, presence of a particular type of PBP; XX, presence of a resistant PBP or of one that has been implicated in acquired resistance. *Spn*, *S. pneumoniae*.

The *Acinetobacter* species are unusual in that several types of β-lactamase inhibitor have antibacterial activity (91). Kawahata et al. examined the antibacterial activity of β-lactams against *A. calcoaceticus*, which they found to decrease in the order of N-formimidoylthienamycin > sulbactam > clavulanic acid > 6-aminopenicillanic acid > 6-substituted penicillins (70). Six PBPs were recognized in *A. calcoaceticus*, and it was observed that 6-aminopenicillanic acid bound to PBP2, sulbactam bound to PBP3, and N-formimidoylthienamycin bound to all PBPs.

Vibrionales

Vibrio

Prolonged treatment with β-lactam antibiotics produced resistant cells that exhibited significant differences in the relative expression levels of low-molecular-mass PBPs (type C) (127). The PBPs from resistant cells were reported to form more-stable complexes with penicillin than those from the parental strain.

Enterobacteriales

Proteus

Resistance of *Proteus mirabilis* to carbapenems, in particular imipenem, has been associated with changes in the sensitivity of PBP2 (type B) (101, 150). Decreased sensitivity towards penicillin *P. mirabilis*, with growth as L-form, depends on the continuing faction of PBP5 (type C) as a D,D-carboxypeptidase and transpeptidase in the presence of the antibiotic (125).

Pasteurellales

Haemophilus

Non-lactamase-mediated resistance to β-lactams in *Haemophilus* has been associated with the accumulation of multiple mutations in PBP3 (type B3) (93, 144, 145). Further, changes in expression of low-molecular-mass PBPs from the D,D-carboxypeptidase families type C4 and type C5 have been implicated in resistance (93, 145). Alterations in PBP binding properties appear to be a more common resistance mechanism in Japan than it does in western countries, where β-lactamase-mediated resistance appears to be more prevalent.

Campylobacterales

Campylobacter

Okada (103) reported that clinical isolates of *Campylobacter jejuni*, *C. coli*, and *C. fetus* were resistant to ampicillin and cefazolin, which correlated with a poor affinity of these two antibiotics for the PBPs of these organisms.

Helicobacter

Resistance to ampicillin has been associated with acquisition of mutations in the type A2 PBP of *Helicobacter pylori* (33, 74, 105, 129). Further, changes in expression or affinity of a type C PBP with an apparent molecular mass of 30 kDa have been reported (19).

Clostridiales

Clostridium

Surveys conducted with South African *Clostridium perfringens* isolates have identified alterations in the labeling of several PBPs with β-lactam resistance (12). In particular PBP1 (type A4), PBPs 2 and 2B (type B4) and 4B (type C) showed low binding affinities for penicillin.

Bacillales

Listeria

The natural resistance of the *Listeria* genus to certain β-lactam compounds appears to correlate with the sensitivity of PBPs towards the agents: for example, cefotaxime and cefoxitin have a weak activity against *L. monocytogenes* and exhibit poor inhibition of penicillin binding to the PBPs of this organism (148). PBP2 (type B4) and PBP3 (type B5) have been specifically implicated in resistance towards β-lactams (46, 115, 149).

Staphylococcus

β-Lactam resistance has been associated with mutations leading to altered affinity of PBP1 (type B4) and PBP2 (type A4) of *S. aureus* (14, 47). Increased expression of both PBPs, with consequent hyperproduction of peptidoglycan, has been implicated in decreased sensitivity towards glycopeptides such as vancomycin (95). Increased expression of PBP4 (type C2) has been implicated in resistance to both β-lactams (14, 58, 59) and glycopeptides (95).

Broad, high-level β-lactam resistance in the staphylococci is usually associated with the presence of an additional PBP, identified as PBP2′ or PBP2a (type B1) (8, 28, 30, 51, 52, 59, 97, 135, 147). Mutants of *S. aureus* PBP2′ that confer high-level resistance to anti-MRSA carbapenems (see "Novel β-Lactams" below) have been described (69).

Lactobacillales

Enterococcus

Broad, high-level β-lactam resistance in the enterococci is usually associated with the overexpression and eventual mutation of a type B1 PBP that is part of the normal genome of these organisms. Resistance mediated by the type B1 homologues of enterococci has been recorded for *E. faecium* PBP5 (15, 27, 78, 121, 123, 158), *E. faecalis* PBP4 (15), and *E. hirae* S185 PBP3r and *E. hirae* ATCC 9790 PBP5 (78, 114).

Chen and Williams (15) also implicated type B4 PBPs of *E. faecium* (PBP4) and *E. faecalis* (PBP3) in resistance. Grayson et al. (44) implicated *E. raffinosus* PBP7, a PBP with an apparent molecular mass of 52 kDa, in resistance of this species to β-lactams.

Streptococcus

Acquired resistance involving mutation of one or more of the three type A bifunctional proteins found in streptococci has been demonstrated in clinical isolates. Mutation of PBP1a (type A3) has been recorded for *Streptococcus pneumoniae* (6, 48, 49, 64, 99) and mutations in PBP1b (type A5) have been recorded for *S. pneumoniae* (49) and *S. mitis* (100). Altered affinity has been reported for *S. pneumoniae* PBP2a (type A4) (49, 64).

Both of the monofunctional type B proteins of streptococci have been implicated in acquired resistance, either through accumulation of point mutations or natural recombination with related genes of commensal species. Decreased susceptibility of PBP2b (type B5) has been observed in *S. pneumoniae* (6, 45, 48, 49, 64, 99) and *S. mitis* (100). Decreased susceptibility of PBP2x (type B4) has been observed in *S. pneumoniae* (6, 45, 49, 97) and *S. mitis* (100).

Hakenbeck et al. (48) implicated the D,D-carboxypeptidase *S. pneumoniae* PBP3 (type C3) in resistance.

Actinomycetales

Mycobacterium

The PBPs of the mycobacteria and other actinomycetes are only distantly related to those found in other pathogenic bacteria, and several appear to have a rather low intrinsic susceptibility to β-lactams (40). The type A proteins (PBPA4 and PBPA5 groups) appear to have a high intrinsic affinity and have not been implicated in clinical resistance (9, 75, 89).

Chlamydiales

Chlamydia

The chlamydias have three PBPs that are distantly related to the PBPs found in free-living organisms (35). Mecillinam selectively inhibited PBP1 (type B), with a 50% inhibitory concentration (IC_{50}) for PBP1 binding similar to the minimum bactericidal concentration (0.25 μg/ml). Although the other β-lactams inhibited a wider range of PBPs than mecillinam, their antichlamydial activities were inferior to that of mecillinam (137).

Bacteroidales

Bacteroides

The PBPs of the anaerobes belonging to the *Bacteroidales* appear to belong to discrete groups only distantly related to their equivalents in other families of bacteria. Changes in the PBPs of *B. fragilis* have been identified as a cause of β-lactam resistance (57). These include PBP1 (type A) (56, 151), PBP3 (type B) of *B. fragilis* (155), and PBP4 (type B) of *B. uniformis* (56). The expression of an additional type C PBP that was not seen in fully sensitive strains of *B. fragilis* was detected in strains in which reduced susceptibility to imipenem was not associated with metallo-β-lactamase activity (25).

LABORATORY DETECTION

PCR-based methods for the detection of antibiotic resistance are becoming increasingly important with the expanding use of molecular techniques for bacteriological diagnosis. Several such tests based on PBP sequences have been investigated for the identification of bacterial strains and diagnosis of resistance mechanism. These include PCR-based tests for *N. meningitidis* (88, 136), *S. pneumoniae* (25), and methicillin-resistant staphylococci (142). Antibody-based tests have also been investigated for detection of methicillin resistance in staphylococci (31, 61).

MOLECULAR MECHANISM

Kinetic Mechanism

The reaction between β-lactams and PBPs is frequently represented as a three-step mechanism (see Goffin and Ghuysen [40]), and references therein):

$$E + L \overset{K_S}{\leftrightarrow} E \times L \overset{k_{ac}}{\rightarrow} E\text{-}I \overset{k_{dc}}{\rightarrow} E + P$$

where E is the PBP, L is the β-lactam, E-I is the covalent acyl enzyme intermediate, and P is the products (hydrolyzed β-lactam and other potential breakdown products). The first step is presumed to be a rapid equilibrium, defined by a dissociation constant, K_S (also known as K_d or K_I); the second step is an irreversible acylation defined by a limiting acylation rate k_{ac} (also known as k_2), and the third step is the slow recovery of free PBP defined by a deacylation rate k_{dc} (also known as k_3). Conformation changes indicated by changes in intrinsic protein fluorescence have been identified with the acylation step in *S. pneumoniae* PBP2x (65) and *S. aureus* PBP2′ (43). Chittock et al. used time-resolved Fourier transform infrared spectroscopy studies to provide further evidence for protein conformation alteration accompanying or dependent on the acylation reaction of *S. pneumoniae* PBP2x (16). Page (108) discussed the possibility of a conformation change during the reaction leading to a stable acyl enzyme complex between cephalosporins and a soluble construct of *E. coli* PBP1b. All of the PBPs that have been studied are characterized by low affinity and low rates of deacylation: in consequence the second-order acylation constant k_{ac}/K_S is frequently used to compare PBPs (Table 7.4).

PBPA and Resistant Mutants

The structures of two bifunctional type A PBPs, *S. pneumoniae* PBP1a (17) and PBP1b (82, 90), have recently been solved. The structures of the apoenzyme and of β-lactam complexes of PBP1b indicate that a significant conformation change has to occur during formation of the acyl enzyme complex (82, 90), which is consistent with the kinetic mechanism suggested for *E. coli* PBP1b (108).

A number of residues have been identified in clinical or laboratory mutants. These include Thr-371, mutated to Ala or Ser, in *S. pneumoniae* PBP1a (type A3) (6, 99); Ser-569, mutated to Ala in *S. aureus* PBP 2 (type A4) (47); Ser-414, mutated to Arg in *H. pylori* PBP1a (type A2) (33); and Leu-421, mutated in *N. meningitidis* PBP1a (type A1) (118). Thr-371 occurs immediately after the catalytic Ser residue, whereas Ser-414 and Leu-421 are located in positions that are likely to be somewhat removed from the catalytic center, being respectively 45 residues after and 40 residues before the catalytic Ser residue in their respective proteins (Fig. 7.2). The effect of the Thr-to-Ser mutation in *S. pneumoniae* PBP1a must be one that is specific to the structure of that protein, since Ser occurs in this position in other PBP1a homologues. The crystal structure of *S. pneumoniae* PBP1a indicates that the mutations in β-lactam-resistant variants cluster at the entrance of the catalytic cleft and modify the polarity and accessibility of the active site (17).

PBP Group B1: Intrinsic Resistance and Hyperresistant Mutants

Detailed kinetic studies have been performed with soluble constructs of *S. aureus* PBP2′ (29, 43, 84, 111, 120) and *E. faecium* PBP5 (60). Both proteins are characterized by very low apparent affinity and low rates of acylation, with second-order acylation constants (k_{ac}/K_S) in the range of 1 to 50 $M^{-1}s^{-1}$ for many inhibitors.

We are fortunate to have two structures of highly resistant PBPs from this group (Color Plate 1): *S. aureus* PBP2′ (79) and *E. faecium* PBP5 (124). The crystal structure of a soluble construct of PBP2′ was determined to 1.8-Å resolution as the apoenzyme and as acyl enzyme complexes with nitrocefin (Color Plate 2), penicillin G, and methicillin. The crystal structure of the acyl enzyme complex of *E. faecium* PBP5 with benzylpenicillin was determined at a resolution of 2.4 Å (124).

The active site in the apoenzyme of *S. aureus* PBP2′ has a closed conformation, which denies access of the β-lactam to the catalytic center and distorts the arrangement of residues such that the Ser residue will not be activated for attack. The binding pocket that accommodates the 3′ side chain of cephalosporins and the equivalent in carbapenems is a long groove with somewhat hydrophobic character (Color Plate 3), whereas it is more open in *E. faecium* PBP5 and *S. pneumoniae* PBP2x (type B4, see below). This may explain the inhibitor selectivity of the staphylococcal enzyme ("Novel β-Lactams" below).

High-level ampicillin resistance in *E. faecium* has been shown to be associated with the synthesis of a modified PBP5 which had been mutated to further decrease its susceptibility to β-lactams (79, 121, 158). The mutations that have been identified with this increased resistance (116) are located away from the active site (Color Plate 3). Mutations of *S. aureus* PBP2′ that lead to high-level resistance towards anti-MRSA carbapenems (see "Novel β-Lactams" below) map in equivalent regions of PBP2′ (Fig. 7.2) (69).

Table 7.4 Kinetic parameters describing the reaction of soluble constructs of PBPs

Class	Protein	Inhibitor	K_S (mM)	k_{ac} s^{-1}	k_{dc} 10^{-6} s^{-1}	k_{ac}/K_S $M^{-1}\cdot s^{-1}$	Reference(s)
PBPA2	*E. coli* PBP1b	Benzylpenicillin	0.005	0.18	12	36,000	105
		Nitrocefin	0.028	0.39	18	13,900	105
		Cephalothin	0.008	0.25	22	32,000	105
PBPB1	*S. aureus* PBP 2′	Benzylpenicillin	1.5	0.007	70	4.5	28, 107
		Nitrocefin	0.20	0.009	120	43	28, 107
		Cephalothin	15	0.018	180	1.2	107
PBPB4	*S. pneumoniae* PBP 2x	Benzylpenicillin	0.9	180	8	200,000	81
	S. pneumoniae PBP 2x Pen^R	Benzylpenicillin	4	0.56	570	137	81

```
PBP A
Nme: 413 PLLQGALVSLDAKTGAVRALVGG-YDFHSKTF--NRAVQAMRQPGSTFKPFVYSAALSKGMTASTVVNDAP--ISLPGKGPNGS----VWTPKNSDGRYSGYITLRQALTASKN
Hpy: 326 DNLNASMIVTDTSTGKILALVGG-IDYKKSAF--NRATQAKRQFGSAIKPFVYQIAFDNGYSTTSKIPDTARNFEN-GNYSKNSEQNHAWHPSNYSRKFLGLVTLQEALSHSLN
Spn: 319 DELQVASTIVDVSNGKVIAQLGARHQSSNVSFGINQAVETNRDWGSTMKPITDYAPALEYGVYESTATIVHDEPYNY----PGT----NTPVYNWDRGYFGNITLQYALQQSRN

PBP B1
           ↓
Efc: 464 RVSDVSHVDLKTALIYSDNIYMAQETLKMGEKNFRTGLDKFIFGEDLDLPISMNPAQISNEESFNSDILLADTGYGQGELLINPIQQAAMYSVFANNGTLVYPKLIADKETKDKKN 580
Sau: 446 YEVVNGNIDLKQAIESSDNIFFARVALELGSKKFEKGMKKLGVGEDIPSDYPFYNAQISNKNLDN-EILLADSGYGQGEILINPVQILSIYSALENNGNINAPHLLK-DTKNK

Efc: 600 NGTAHSLSALGIPLAAKTGTAEIKEKQDEKVKENSFLFAFNPDNQGYMMVSMLENKEDDDSATKRASELL 670
Sau: 587 KEDIYRSYANLI---GKSGTAELKM-KQGETGRQIGWFISYDKDNPNMMMAINVKDVQDKGMASYNAKIS 649

PBP B3 B4 B5
                                                                                  ↓
Nme 2:   303 TDMIEPGSAIKPFVIAKALDA-----GKTDL-NERLNTQPYKIGPSPVRDTHVYPSLDVRGIMQKSSNVGTSKLSARFGAEEMYDFYHELG 386
Hin 3:   314 TDTFEPGSTVKPFVVLTALQR-----GVVKR-DEIIDTTSFKLSGKEIVDVAPRAQQTLDEILMNSSNRGVSRLALRMPPSALMETYQNAG 404
Spn 2x:  330 QSNYEPGSTMKVMMLAAAIDNNTFPGGEVFNSSELKIADATIRDWDVNEGLTGGRTMTFSQGFAHSSNVGMTLLEQKMGDATWLDYLNRFK 420
Spn 2b:  388 TNVFVPGSVVKAATISSGWEN-GVLSGNQTLTDQSIVFQGSAPINSWYTQAYGSFPITAVQALEYSSNTYMVQTALGLMGQTYQPNMFVGT 468

Spn 2x:  541 GQNVALKSGTAQIADEKNGGYLVGLTDYIFSAVSMSPAENPDFILYVTVQQPEHYSGIQL 600
Hin 3:   506 GYRVGVKTGTARKIENGHYVNKYVAFTAGIAPISDPRYALVVLINDPKAGEYYGGA 561
```

Figure 7.2 Location of mutations in PBPs conferring resistance to β-lactams. Regions from the sequences of susceptible proteins are shown, with the residues that are mutated in resistant isolates shown in bold and boxed. The highly conserved residues that comprise parts of the active site (38, 39) are shown in bold. The organisms denoted are *E. faecium* (Efc), *H. influenzae* (Hin), *H. pylori* (Hpy), *N. meningitidis* (Nme), *S. aureus* (Sau), and *S. pneumoniae* (Spn).

Table 7.5 Representative anti-MRSA cephalosporins

Compound	Company of origin	Comments	Reference
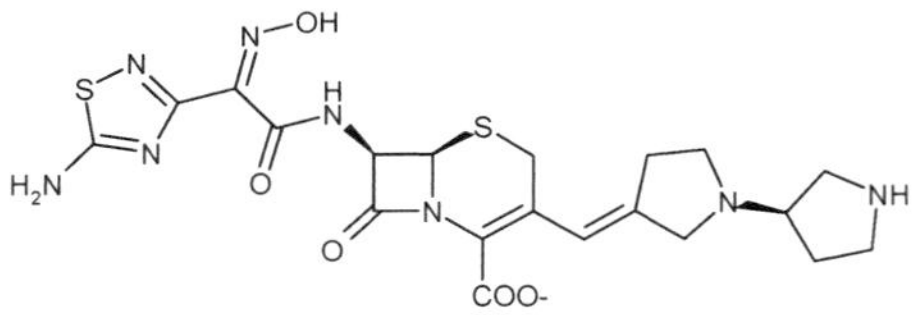	Roche	Ceftobiprole. In clinical development by Basilea Pharmaceutica and Johnson & Johnson	54
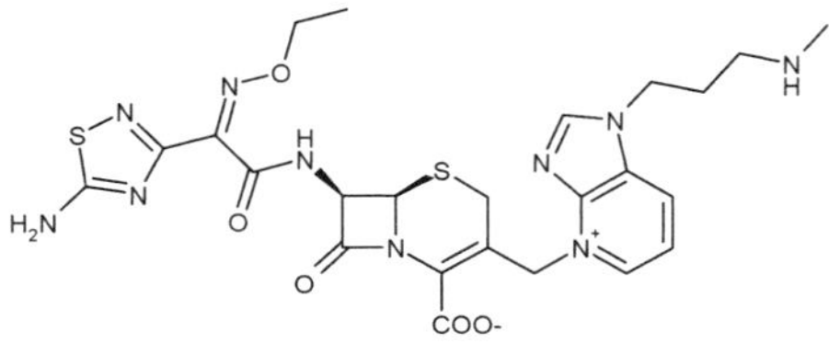	Shionogi	In clinical development by Shionogi	150
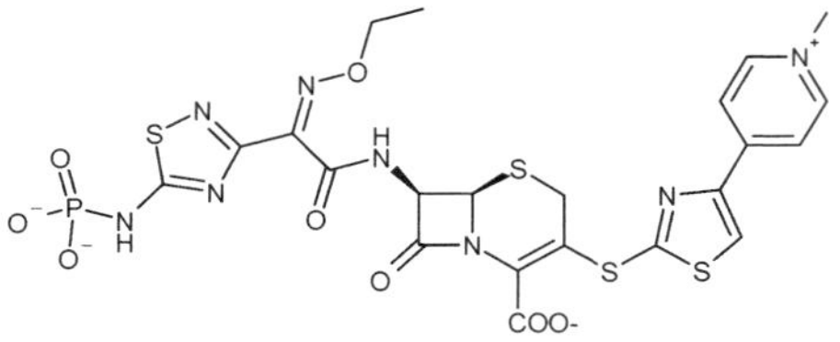	Takeda	In clinical development by Peninsula Pharmaceuticals	62
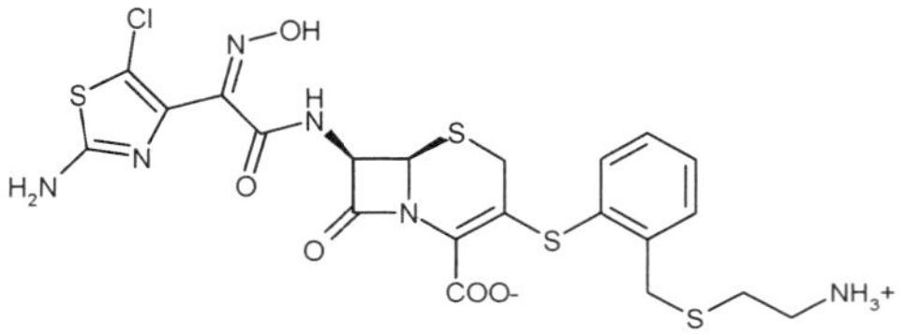	Microcide		35
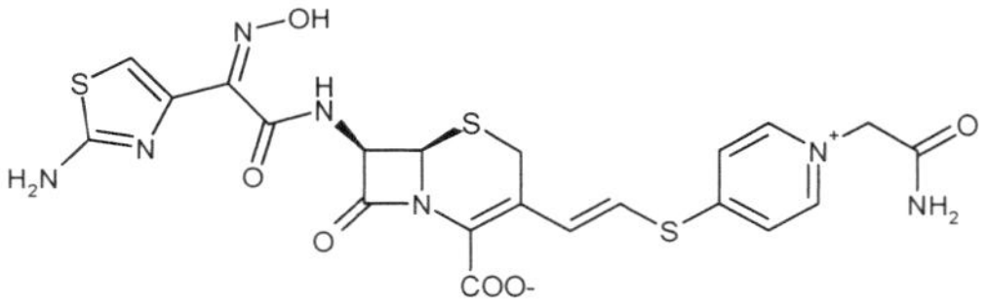	Taiho Pharmaceuticals		49
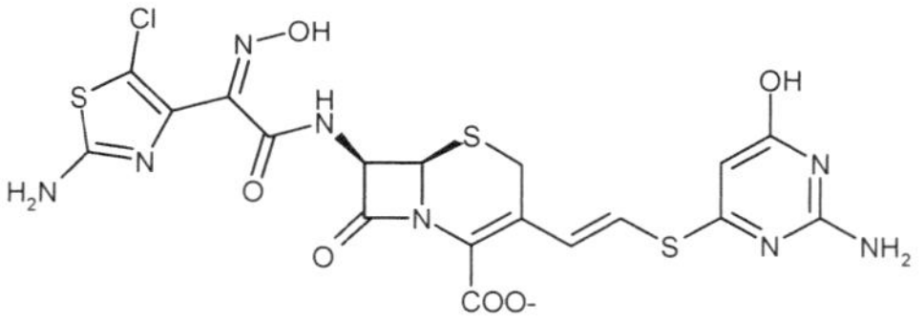	LG Life Sciences		147
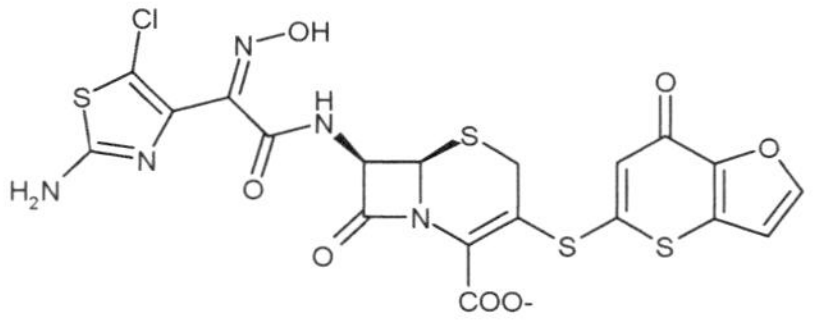	Zenyaku		149
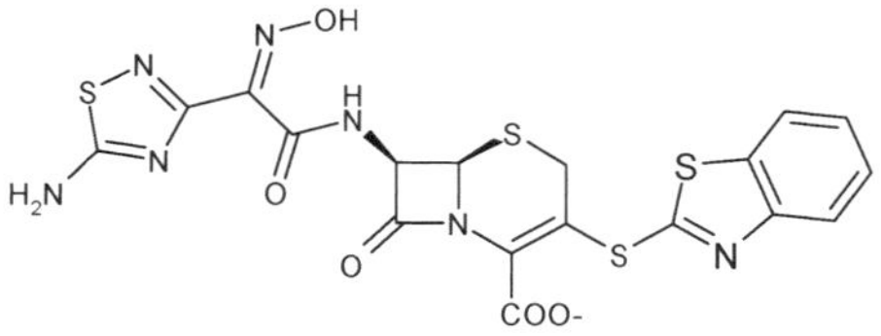	Meiji Seika Kaisha		139

Table 7.6 Anti-MRSA carbapenems

Compound	Company	Comments	Reference
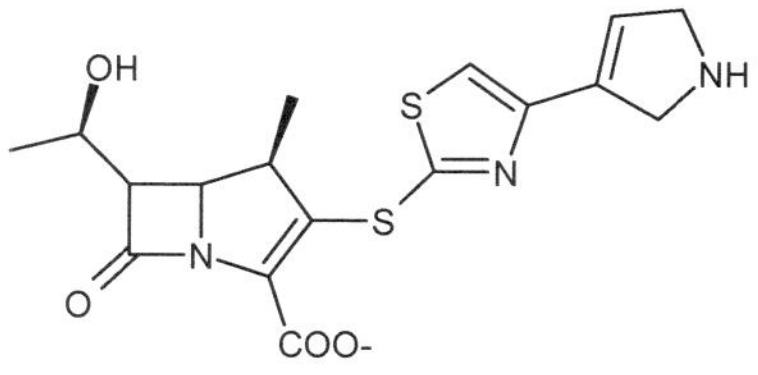	Sumitomo	Experimental	134
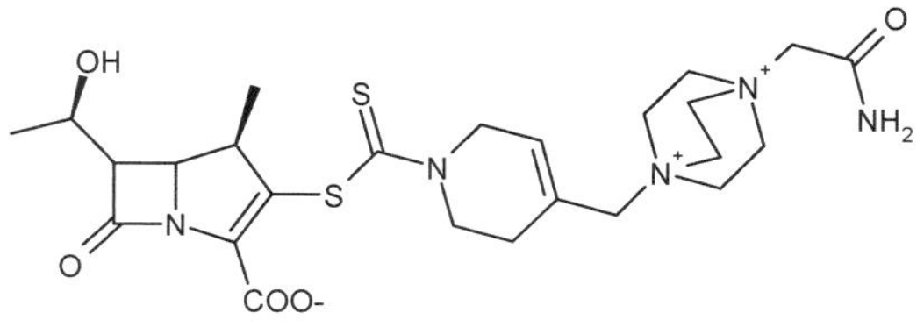	Banyu	Experimental	61
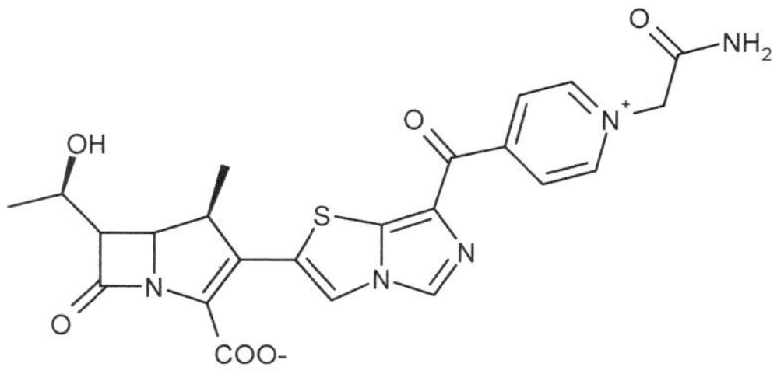	Meiji Saika Kaisha	Potent inhibition of *S. aureus* PBP2′ (IC_{50}, 1.5 μM)	66
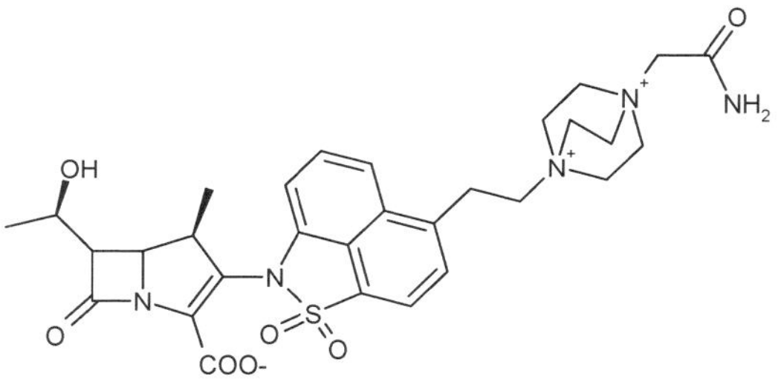	Merck	Discontinued	65
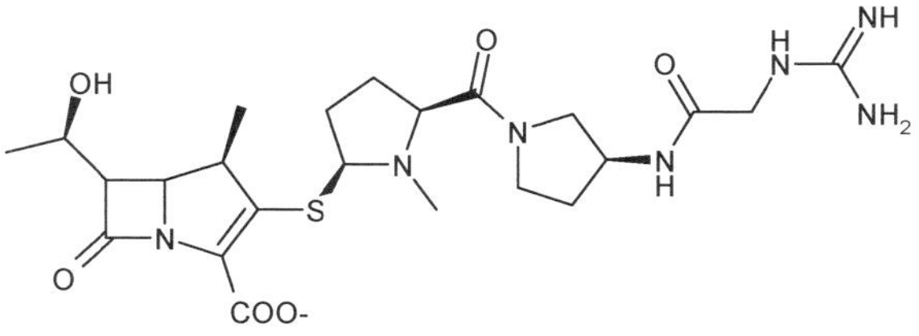	Sankyo	In clinical development by Roche. Weak inhibition of *S. aureus* PBP2′ (IC_{50} 1.5 mM). Strains with elevated MIC already described.	70

PBP Groups B3, B4, and B5 Resistant Mutants

The PBP B3 homologues found in the gram-negative bacteria belonging to *Pseudomonas, Haemophilus*, and *Neisseria*, as well as proteins belonging to the PBP B4 and B5 groups of gram-positive bacteria, have been implicated in acquired resistance through mutation or genetic exchange. The point mutations that have been identified to date in these proteins cluster around the active-site components (Fig. 7.2) (23, 45, 85, 86, 96, 99, 126, 131, 145, 157). Where it has been investigated, the mutations have been found to affect all steps in the reaction (83).

The crystal structure of *S. pneumoniae* PBP2x (Color Plate 1) has been solved as the apoenzyme (108) and as the complex with cefuroxime (41). Surprisingly, two antibiotic molecules were observed, one as a covalent complex with the active-site serine residue, and a second one between the C-terminal and the transpeptidase domains. Further, the structure of PBP2x from a highly penicillin-resistant clinical isolate of *S. pneumoniae* has also been solved, which provides the opportunity to examine the effects that some of the point mutations implicated in resistance have on the mechanism of the protein. In the proximity of the active

Table 7.7 Modulators of PBP2′-mediated resistance in staphylococci

Compound	Comments	Reference(s)
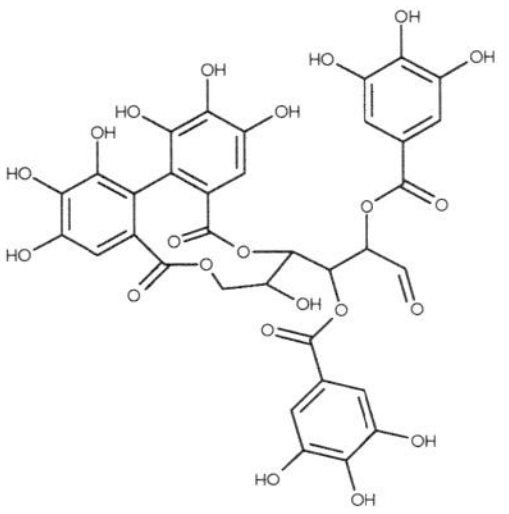	Tellimagrandin I from dog rose (*Rosa canina*)	124
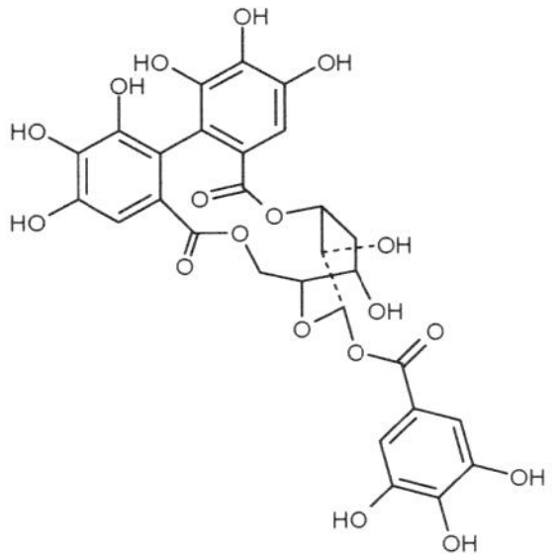	Corilagin from bearberry (*Arctostaphylos uvaursi*)	124
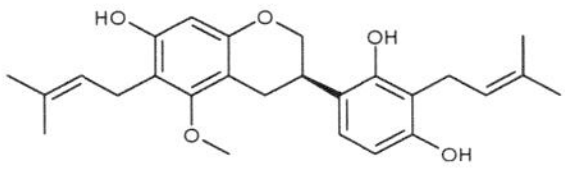	Licoricidin from licorice (*Glycyrrhiza glabra*)	52
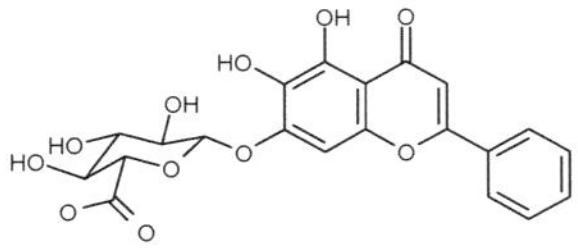	Baicalin from Xi-nan Huangqin (*Scutellaria amoena*)	79
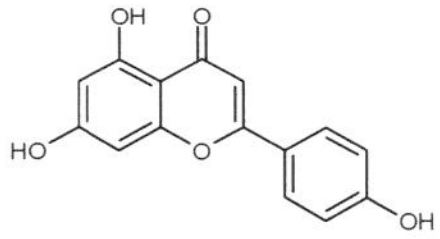	Flavanoids (e.g., apigenin) from many plants	118
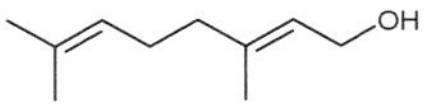	Geraniol from Sichuan pepper (*Zanthoxylum piperitum*)	53
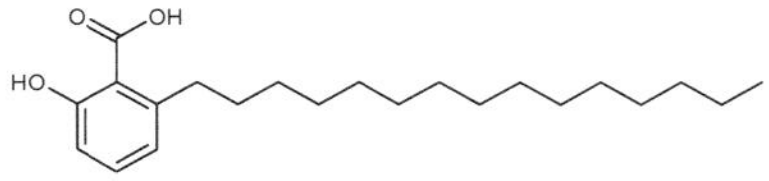	Anacardic acids from cashew (*Anacardium occidentale*)	72, 95
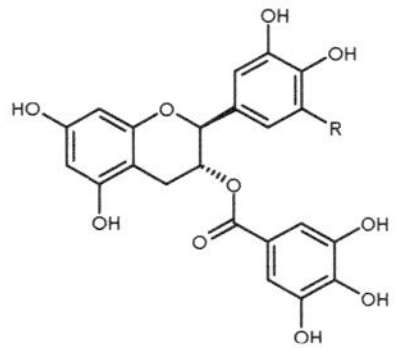	(−)-*Epi*(gallo)catechin gallates from tea (*Camellia sinensis*)	130, 137
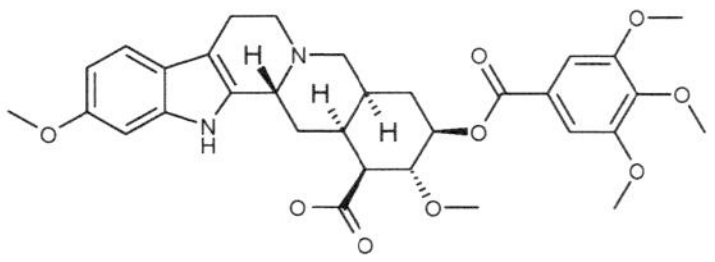	Reserpine	71
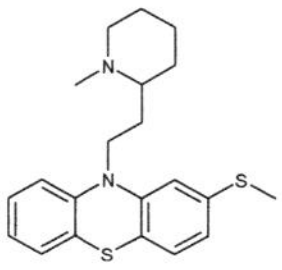	Phenothiazines	71

site, the Thr338→Ala mutation weakens the local hydrogen-bonding network, thus abrogating the stabilization of a crucial buried water molecule. Further, the Ser389→Leu and Asn514→His mutations produce a destabilizing effect that generates an "open" active site (18), while Gln552 and Thr550, which belong to strand β3, are in direct contact with the cephalosporin (94).

PROSPECTS FOR INHIBITION

There have been considerable efforts devoted to the discovery of new agents that can overcome PBP-mediated resistance either through increasing the reactivity of β-lactams towards the resistant PBPs (see "Novel β-Lactams" below) or by suppressing the activity of the resistant PBP (see "Modulators of PBP Expression and Activity" below).

Novel β-Lactams

There are several experimental β-lactams (36, 50, 55, 138, 143, 153, 154) now known to be potent inhibitors of the staphylococcal type B1 PBP that is the primary determinant of β-lactam resistance in these organisms, of which ceftobiprole is the most advanced in clinical development (109, 110). Many of these molecules have aminothiazolylhydroxyimino 7-acyl side chains typical of the broad-spectrum cephalosporins, while on the 3′ side chain, many have a basic, or positively charged group (Table 7.5). Carbapenems with activity against methicillin-resistant staphylococci have also been described (Table 7.6). Like the cephalosporins, many of these carbapenems are characterized by long side chains carrying basic or positively charged substituents.

Modulators of PBP Expression and Activity

An understanding of the mechanism of methicillin resistance has led to the discovery of accessory factors that influence the level and nature of methicillin resistance (73, 80, 98, 122, 128, 141). Accessory factors, such as Fem factors, provide possible new targets, while compounds that modulate methicillin resistance (Table 7.7) such as epicatechin gallate, derived from green tea, and corilagin, derived from bearberry extract, provide possible lead compounds for development of inhibitors that could be used in combination with established antistaphylococcal β-lactams in order to promote their activity against methicillin-resistant strains (135).

References

1. **Ameyama, S., S. Onodera, M. Takahata, S. Minami, N. Maki, K. Endo, H. Goto, H. Suzuki, and Y.Oishi.** 2002. Mosaic-like structure of penicillin-binding protein 2 gene (*penA*) in clinical isolates of *Neisseria gonorrhoeae* with reduced susceptibility to cefixime. *Antimicrob. Agents Chemother.* **46:**3744–3749.
2. **Antignac, A., J.-M. Alonso, and M.-K. Taha.** 2000. Update on the resistance of *Neisseria meningitidis* to antibiotics, based on the strains tested at the French National Reference Center in 1998. *Antibiotiques* **2:**241–245.
3. **Antignac, A., I. G. Boneca, J.-C. Rousselle, A. Namane, J.-P. Carlier, J. A. Vazquez, A. Fox, J.-M. Alonso, and M.-K. Taha.** 2003. Correlation between alterations of the penicillin-binding protein 2 and modifications of the peptidoglycan structure in *Neisseria meningitidis* with reduced susceptibility to penicillin G. *J. Biol. Chem.* **278:**31529–31535.
4. **Antignac, A., P. Kriz, G. Tzanakaki, J.-M. Alonso, and M.-K. Taha.** 2001. Polymorphism of *Neisseria meningitidis penA* gene associated with reduced susceptibility to penicillin. *J. Antimicrob. Chemother.* **47:**285–296.
5. **Arreaza, L., B. Alcala, C. Salcedo, L. de la Fuente, and J. A.Vazquez.** 2003. Dynamics of the *penA* gene in serogroup C meningococcal strains. *J. Infect. Dis.* **187:**1010–1014.
6. **Asahi, Y., and K. Ubukata.**1998. Association of a Thr-371 substitution in a conserved amino acid motif of penicillin-binding protein 1A with penicillin resistance of *Streptococcus pneumoniae. Antimicrob. Agents Chemother.* **42:**2267–2273.
7. **Bellido, F., C. Veuthey, J. Blaser, A. Bauernfeind, and J. C. Pechere.** 1990. Novel resistance to imipenem associated with an altered PBP-4 in a *Pseudomonas aeruginosa* clinical isolate. *J. Antimicrob. Chemother.* **25:**57–68.
8. **Berger-Bachi, B., and S. Rohrer.** 2002. Factors influencing methicillin resistance in staphylococci. *Arch. Microbiol.* **178:**165–171.
9. **Bhakta, S., and J. Basu.** 2002. Overexpression, purification and biochemical characterization of a class A high-molecular-mass penicillin-binding protein (PBP), PBP1* and its soluble derivative from *Mycobacterium tuberculosis. Biochem. J.* **361:**635–639.
10. **Bowler, L. E., Q. Y. Zhang, J. Y. Riou, and B. G. Spratt.** 1994. Interspecies recombination between the *penA* genes of *Neisseria meningitidis* and commensal *Neisseria* species during the emergence of penicillin resistance in *N. meningitidis*: natural events and laboratory simulation. *J. Bacteriol.* **176:**333–337.
11. **Brannigan, J. A., I. A. Tirodimos, Q. Y. Zhang, C. G. Dowson, and B. G. Spratt.** 1990. Insertion of an extra amino acid is the main cause of the low affinity of penicillin-binding protein 2 in penicillin-resistant strains of *Neisseria gonorrhoeae. Mol. Microbiol.* **4:**913–919.
12. **Chalkley, L., and I. Van der Westhuyzen.** 1993. Penicillin-binding proteins of clostridia. *Curr. Microbiol.* **26:**109–112.
13. **Chalkley, L. J., S. van Vuuren, R. C. Ballard, and P. L Botha.** 1995. Characterization of *penA* and *tetM* resistance genes of *Neisseria gonorrhoeae* isolated in southern Africa—epidemiological monitoring and resistance development. *S. Afr. Med. J.* **85:**775–780.
14. **Chambers, H. F., M. J. Sachdeva, and C. J. Hackbarth.** 1994. Kinetics of penicillin binding to penicillin-binding proteins of *Staphylococcus aureus. Biochem. J.* **301:**139–144.

15. **Chen, H. Y., and J. D. Williams.** 1987. Penicillin-binding proteins in *Streptococcus faecalis* and *S. faecium*. *J. Med. Microbiol.* **23:**141–147.

16. **Chittock, R. S., S. Ward, A.-S. Wilkinson, P. Caspers, B. Mensch, M.G. P. Page, and C. W. Wharton.** 1999. Hydrogen bonding and protein perturbation in β-lactam acyl-enzymes of *Streptococcus pneumoniae* penicillin-binding protein PBP2x. *Biochem. J.* **338:**153–159.

17. **Contreras-Martel, C., V. Job, A. M. Di Guilmi, T. Vernet, O. Dideberg, and A. Dessen.** 2006. Crystal structure of penicillin-binding protein 1a (PBP1a) reveals a mutational hotspot implicated in β-lactam resistance in *Streptococcus pneumoniae*. *J. Mol. Biol.* **355:**684–696.

18. **Dessen, A., N. Mouz, E. Gordon, J. Hopkins, and O. Dideberg.** 2001. Crystal structure of PBP2x from a highly penicillin-resistant *Streptococcus pneumoniae* clinical isolate. A mosaic framework containing 83 mutations. *J. Biol. Chem.* **276:**45106–45112.

19. **Dore, M. P., D. Y. Graham, and A. R. Sepulveda.** 1999. Different penicillin-binding protein profiles in amoxicillin-resistant *Helicobacter pylori*. *Helicobacter* **4:**154–161.

20. **Dougherty, T. J.** 1983. Peptidoglycan biosynthesis in *Neisseria gonorrhoeae* strains sensitive and intrinsically resistant to β-lactam antibiotics. *J. Bacteriol.* **153:**429–435.

21. **Dougherty, T. J.** 1986. Genetic analysis and penicillin-binding protein alterations in *Neisseria gonorrhoeae* with chromosomally mediated resistance. *Antimicrob. Agents Chemother.* **30:**649–652.

22. **Dougherty, T. J., A. E. Koller, and A.Tomasz.** 1980. Penicillin-binding proteins of penicillin-susceptible and intrinsically resistant *Neisseria gonorrhoeae*. *Antimicrob. Agents Chemother.* **18:**730–737.

23. **Dowson, C. G., A. E. Jephcott, K. R. Gough, and B. G. Spratt.** 1989. Penicillin-binding protein 2 genes of non-β-lactamase-producing, penicillin-resistant strains of *Neisseria gonorrhoeae*. *Mol. Microbiol.* **3:**35–41.

24. **du Plessis, M., A. M. Smith, and K. P. Klugman.** 1999. Application of pbp1A PCR in identification of penicillin-resistant *Streptococcus pneumoniae*. *J. Clin. Microbiol.* **37:**628–632.

25. **Edwards, R., and D. Greenwood.** 1996. Mechanisms responsible for reduced susceptibility to imipenem in *Bacteroides fragilis*. *J. Antimicrob. Chemother.* **38:**941–951.

26. **Fernandez-Cuenca, F., L. Martinez-Martinez, M. C. Conejo, J. A. Ayala, E. J. Perea, and A. Pascual.** 2003. Relationship between β-lactamase production, outer membrane protein and penicillin-binding protein profiles on the activity of carbapenems against clinical isolates of *Acinetobacter baumannii*. *J. Antimicrob. Chemother.* **51:**565–574.

27. **Fontana, R., A. Grossato, L. Rossi, Y. R Cheng, and G. Satta.** 1985. Transition from resistance to hypersusceptibility to β-lactam antibiotics associated with loss of a low-affinity penicillin-binding protein in a *Streptococcus faecium* mutant highly resistant to penicillin. *Antimicrob. Agents Chemother.* **28:**678–683.

28. **Fontana, R., L. Rossi, Y. C. Rong, and E. Tonin.** 1985. Penicillin-binding proteins and resistance to β-lactam antibiotics in *Staphylococcus aureus*. *Chemioterapia* **4:**53–55.

29. **Fuda, C., M. Suvorov, S. B. Vakulenko, and S. Mobashery.** 2004. The basis for resistance to β-lactam antibiotics by penicillin-binding protein 2a of methicillin-resistant *Staphylococcus aureus*. *J. Biol. Chem.* **279:**40802–40806.

30. **Gaisford, W. C., and P. E. Reynolds.** 1989. Methicillin resistance in *Staphylococcus epidermidis*. Relationship between the additional penicillin-binding protein and an attachment transpeptidase. *Eur. J. Biochem.* **185:**211–218.

31. **Gehrlein, M., H. Leying, W. Cullmann, S. Wendt, and W. Opferkuch.** 1991. Imipenem resistance in *Acinetobacter baumannii* is due to altered penicillin-binding proteins. *Chemotherapy* **37:**405–412.

32. **Gerberding, J. L., C. Miick, H. H. Liu, and H. F. Chambers.** 1991. Comparison of conventional susceptibility tests with direct detection of penicillin-binding protein 2a in borderline oxacillin-resistant strains of *Staphylococcus aureus*. *Antimicrob. Agents Chemother.* **35:**2574–2579.

33. **Gerrits, M. M., D. Schuijffel, A. A. Van Zwet, E. J. Kuipers, C. M. J. E. Vandenbroucke-Grauls, and J. G. Kusters.** 2002. Alterations in penicillin-binding protein 1A confer resistance to β-lactam antibiotics in *Helicobacter pylori*. *Antimicrob. Agents Chemother.* **46:**2229–2233.

34. **Ghuysen, J. M.** 1991. Serine β-lactamases and penicillin-binding proteins. *Annu. Rev. Microbiol* **45:**37–67.

35. **Ghuysen, J. M., and C. Goffin.** 1999. Lack of cell wall peptidoglycan versus penicillin sensitivity: new insights into the chlamydial anomaly. *Antimicrob. Agents Chemother.* **43:**2339–2344.

36. **Glinka, T. W., A. Cho, Z. J. Zhang, M. Ludwikow, D. Griffith, K. Huie, S. J. Hecker, M. N. Dudley, V. J. Lee, and S. Chamberland.** 2000. SAR studies of anti-MRSA non-zwitterionic 3-heteroarylthiocephems. *J. Antibiot.* **53:**1045–1052.

37. **Godfrey, A. J., L. E. Bryan, and H. R. Rabin.** 1981. β-Lactam-resistant *Pseudomonas aeruginosa* with modified penicillin-binding proteins emerging during cystic fibrosis treatment. *Antimicrob. Agents Chemother.* **19:**705–711.

38. **Godfrey, A. J., and L. E. Bryan.** 1982. Mutation of *Pseudomonas aeruginosa* specifying reduced affinity for penicillin G. *Antimicrob. Agents Chemother.* **21:**216–223.

39. **Goffin C., and J. M. Ghuysen.** 1998. Multimodular penicillin-binding proteins: an enigmatic family of orthologs and paralogs. *Microbiol. Mol. Biol. Rev.* **62:**1079–1093.

40. **Goffin, C., and J.-M. Ghuysen.** 2002. Biochemistry and comparative genomics of SxxK superfamily acyltransferases offer a clue to the mycobacterial paradox: presence of penicillin-susceptible target proteins versus lack of efficiency of penicillin as therapeutic agent. *Microbiol. Mol. Biol. Rev.* **66:**702–738.

41. **Gordon, E., N. Mouz, E. Duee, and O. Dideberg.** 2000. The crystal structure of the penicillin-binding protein 2x from *Streptococcus pneumoniae* and its acyl-enzyme form: implication in drug resistance. *J. Mol. Biol.* **299:**477–485.

42. **Gotoh, N., K. Nunomura, and T. Nishino.** 1990. Resistance of *Pseudomonas aeruginosa* to cefsulodin: modification of penicillin-binding protein 3 and mapping of its chromosomal gene. *J. Antimicrob. Chemother.* **25:**513–523.

43. **Graves-Woodward, K., and R. F. Pratt.** 1998. Reaction of soluble penicillin-binding protein 2a of methicillin-resistant *Staphylococcus aureus* with β-lactams and acyclic substrates: kinetics in homogeneous solution. *Biochem. J.* **332:**755–761.

44. **Grayson, M. L., G. M. Eliopoulos, C. B. Wennersten, K. L. Ruoff, K. Klimm, F. L. Sapico, A. S. Bayer, and R. C. Moellering, Jr.** 1991. Comparison of *Enterococcus raffinosus* with *Enterococcus avium* on the basis of penicillin susceptibility, penicillin-binding protein analysis, and high-level aminoglycoside resistance. *Antimicrob. Agents Chemother.* **35:**1408–1412.

45. **Grebe, T., and R. Hakenbeck.** 1996. Penicillin-binding proteins 2b and 2x of *Streptococcus pneumoniae* are primary resistance determinants for different classes of β-lactam antibiotics. *Antimicrob. Agents Chemother.* **40:**829–834.

46. **Gutkind, G. O., S. B. Ogueta, A. C. De Urtiaga, M. E. Mollerach, and R. A. De Torres.** 1990. Participation of PBP 3 in the acquisition of dicloxacillin resistance in *Listeria monocytogenes*. *J. Antimicrob. Chemother.* **25:**751–758.

47. **Hackbarth, C. J., T. Kocagoz, S. Kocagoz, and H. F. Chambers.** 1995. Point mutations in *Staphylococcus aureus* PBP 2 gene affect penicillin-binding kinetics and are associated with resistance. *Antimicrob. Agents Chemother.* **39:** 103–106.

48. **Hakenbeck, R., H. Ellerbrok, T. Briese, S. Handwerger, and A. Tomasz.** 1986. Penicillin-binding proteins of penicillin-susceptible and -resistant pneumococci: immunological relatedness of altered proteins and changes in peptides carrying the β-lactam binding site. *Antimicrob. Agents Chemother.* **30:**553–558.

49. **Hakenbeck, R., A. Konig, I. Kern, M. Van Der Linden, W. Keck, D. Billot-Klein, R. Legrand, B. Schoot, and L. Gutmann.** 1998. Acquisition of five high-Mr penicillin-binding protein variants during transfer of high-level β-lactam resistance from *Streptococcus mitis* to *Streptococcus pneumoniae*. *J. Bacteriol.* **180:**1831–1840.

50. **Hanaki, H., H. Akagi, Y. Masaru, T. Otani, A. Hyodo, and K. Hiramatsu.** 1995. TOC-39, a novel parenteral broad-spectrum cephalosporin with excellent activity against methicillin-resistant *Staphylococcus aureus*. *Antimicrob. Agents Chemother.* **39:**1120–1126.

51. **Hartman, B. J., and A. Tomasz.** 1984. Low-affinity penicillin-binding protein associated with β-lactam resistance in *Staphylococcus aureus*. *J. Bacteriol.* **158:**513–516.

52. **Hartman, B. J., and A. Tomasz.** 1986. Expression of methicillin resistance in heterogeneous strains of *Staphylococcus aureus*. *Antimicrob. Agents Chemother.* **29:**85–92.

53. **Hatano, T., Y. Shintani, Y. Aga, S. Shiota, T. Tsuchiya, and T. Yoshida.** 2000. Phenolic constituents of licorice. VIII. Structures of glicophenone and glicoisoflavanone, and effects of licorice phenolics on methicillin-resistant *Staphylococcus aureus*. *Chem. Pharm. Bull.* **48:**1286–1292.

54. **Hatano, T., and T. Yoshida.** 2003. Constituents of *Zanthoxylum* fruits and the other herbs and the spices effective on methicillin-resistant *Staphylococcus aureus* (MRSA). *Aroma Res.* **4:**384–388.

55. **Hebeisen, P., I. Heinze-Krauss, P. Angehrn, P. Hohl, M. G. P. Page, and R. L. Then.** 2001. In vitro and in vivo properties of Ro 63-9141, a novel broad-spectrum cephalosporin with activity against methicillin-resistant staphylococci. *Antimicrob. Agents Chemother.* **45:**825–836.

56. **Hedberg, M., K. Bush, P. A. Bradford, N. Bhachech, C. Edlund, K. Tuner, and C. E. Nord.** 1996. The role of penicillin-binding proteins for β-lactam resistance in a β-lactamase-producing *Bacteroides uniformis* strain. *Anaerobe* **2:**111–115.

57. **Hedberg, M., E. Nagy, and C. E. Nord.** 1997. Role of penicillin-binding proteins in resistance of *Bacteroides fragilis* group species to β-lactam drugs. *Clin. Infect. Dis.* **25**(Suppl. 2):S270–S271.

58. **Henze, U. U., and B. Berger-Baechi.** 1996. Penicillin-binding protein 4 overproduction increases β-lactam resistance in *Staphylococcus aureus*. *Antimicrob. Agents Chemother.* **40:**2121–2125.

59. **Higashi, Y., A. Wakabayashi, Y. Matsumoto, Y. Watanabe, and A. Ohno.** 1999. Role of inhibition of penicillin binding proteins and cell wall crosslinking by beta-lactam antibiotics in low- and high-level methicillin resistance of *Staphylococcus aureus*. *Chemotherapy* **45:**37–47.

60. **Hujer, A. M., M. Kania, T. Gerken, V. E. Anderson, J. Buynak, X. Ge, P. Caspers, M. G. P. Page, L. B. Rice, and R. A. Bonomo.** 2005. Structure-activity relationships of different β-lactam antibiotics against a soluble form of *Enterococcus faecium* PBP5, a type II bacterial transpeptidase. *Antimicrob. Agents Chemother* **49:**612–618.

61. **Hussain, Z., L. Stoakes, S. Garrow, S. Longo, V. Fitzgerald, and R. Lannigan.** 2000. Rapid detection of *mecA*-positive and *mecA*-negative coagulase-negative staphylococci by an anti-penicillin binding protein 2a slide latex agglutination test. *J. Clin. Microbiol.* **38:**2051–2054.

62. **Imamura, H., N. Ohtake, H. Jona, A. Shimizu, M. Moriya, H. Sato, Y. Sugimoto, C. Ikeura, H. Kiyonaga, M. Nakano, R. Nagano, S. Abe, K. Yamada, T. Hashizume, and H. Morishima.** 2001. Dicationic dithiocarbamate carbapenems with anti-MRSA activity. *Bioorg. Med. Chem.* **9:**1571–1578.

63. **Ishikawa, T., N. Matsunaga, H. Tawada, N. Kuroda, Y. Nakayama, Y. Ishibashi, M. Tomimoto, Y. Ikeda, Y. Tagawa, Y. Iizawa, K. Okonogi, S. Hashiguchi, and A. Miyake.** 2003. TAK-599, a novel N-phosphono type prodrug of anti-MRSA cephalosporin T-91825: synthesis, physicochemical and pharmacological properties. *Bioorg. Med. Chem.* **11:**2427–2437.

64. **Jabes, D., S. Nachman, and A. Tomasz.** 1989. Penicillin-binding protein families: evidence for the clonal nature of penicillin resistance in clinical isolates of pneumococci. *J. Infect. Dis.* **159:**16–25.

65. **Jamin, M., C. Damblon, S. Millier, R. Hakenbeck, and J.-M. Frère.** 1993. Penicillin-binding protein 2x of *Streptococcus pneumoniae*: enzymic activities and interactions with beta-lactams. *Biochem. J.* **292:**735–741.

66. **Jensen, M. S., C. Yang, Y. Hsiao, N. Rivera, K. M. Wells, J. Y. L. Chung, N. Yasuda, D. L. Hughes, and P. J. Reider.** 2000. Synthesis of an anti-methicillin-resistant *Staphylococcus aureus* (MRSA) carbapenem via stannatrane-mediated Stille coupling. *Org. Lett.* **2:**1081–1084.

67. **Kano, Y., T. Maruyama, Y. Sambongi, K. Aihara, K. Atsumi, K. Iwamatsu, and T. Ida.** 2001. Preparation of novel carbapenem derivatives as antimicrobial agents. *PCT Int. Appl.* Patent no. WO 2001055154.

68. **Karibian, D., and G. Starka.** 1987. The penicillin-binding proteins of *Zymomonas mobilis* Zm4. *FEMS Microbiol. Lett.* **41:**121–125

69. **Katayama, Y., H.-Z. Zhang, and H. F. Chambers.** 2004. PBP 2a mutations producing very-high-level resistance to beta-lactams. *Antimicrob. Agents Chemother.* **48:**453–459.

70. **Kawahata, Y., S. Tomida, T. Nishino, and T. Tanino.** 1983. Studies on antibacterial activity of β-lactam antibiotics against *Acinetobacter calcoaceticus*. *In* K. H. Spitzy and K. Karrer (ed.), *Proceedings of the 13th International Congress of Chemotherapy*. 2 88/58–88/62. Verlag H. Egermann, Vienna, Austria.

71. **Kawamoto, I., Y. Shimoji, O. Kanno, K. Kojima, K. Ishikawa, E. Matsuyama, Y. Ashida, T. Shibayama, T. Fukuoka, and S. Ohya.** 2003. Synthesis and structure-activity relationships of novel parenteral carbapenems, CS-023 (R-115685) and related compounds containing an amidine moiety. *J. Antibiot.* **56:**565–579.

72. **Kristiansen, M. M., C. Leandro, D. Ordway, M. Martins, M. Viveiros, T. Pacheco, J. E. Kristiansen, and L. Amaral.** 2003. Phenothiazines alter resistance of methicillin-resistant strains of *Staphylococcus aureus* (MRSA) to oxacillin in vitro. *Int. J. Antimicrob. Agents* **22:**250–253.

73. **Kubo, I., K. Nihei, and K. Tsujimoto.** 2003. Antibacterial action of anacardic acids against methicillin resistant *Staphylococcus aureus* (MRSA). *J. Agric. Food Chem.* **51:**7624–7628.

74. **Kwon, D. H., M. P. Dore, J. J. Kim, M. Kato, M. Lee, J. Y. Wu, and D. Y. Graham.** 2003. High-level β-lactam resistance associated with acquired multidrug resistance in *Helicobacter pylori*. *Antimicrob. Agents Chemother.* **47:**2169–2178.

75. **Lepage, S., P. Dubois, T. K. Ghosh, B. Joris, S. Mahapatra, M. Kundu, J. Basu, P. Chakrabarti, S. T. Cole, M. Nguyen-Disteche, and J.-M. Ghuysen.** 1997. Dual multimodular class A penicillin-binding proteins in *Mycobacterium leprae*. *J. Bacteriol.* **179:**4627–4630.

76. **Liao, X., and R. E. Hancock.** 1997. Identification of a penicillin-binding protein 3 homolog, PBP3x, in *Pseudomonas aeruginosa*: gene cloning and growth phase-dependent expression. *J. Bacteriol.* **179:**1490–1496.

77. **Liao, X., and R. E. W. Hancock.** 1997. Susceptibility to β-lactam antibiotics of *Pseudomonas aeruginosa* overproducing penicillin-binding protein 3. *Antimicrob. Agents Chemother.* **41:**1158–1161.

78. **Ligozzi, M., F. Pittaluga, and R. Fontana.** 1996. Modification of penicillin-binding protein 5 associated with high-level ampicillin resistance in *Enterococcus faecium*. *Antimicrob. Agents Chemother.* **40:**354–357.

79. **Lim, D., and N. C. J. Strynadka.** 2002. Structural basis for the β lactam resistance of PBP2a from methicillin-resistant *Staphylococcus aureus*. *Nat. Struct. Biol.* **9:**870–876.

80. **Liu, I. X., D. G. Durham, and R. M. E. Richards.** 2000. Baicalin synergy with β-lactam antibiotics against methicillin-resistant *Staphylococcus aureus* and other β-lactam-resistant strains of *S. aureus*. *J. Pharm. Pharmacol.* **52:**361–366.

81. **Livermore, D. M.** 1987. Radiolabeling of penicillin-binding proteins (PBPs) in intact *Pseudomonas aeruginosa* cells: consequences of β-lactamase activity by PBP-5. *J. Antimicrob. Chemother.* **19:**733–742.

82. **Lovering, A. L., L. De Castro, D. Lim, and N. C. J. Strynadka.** 2006. Structural analysis of an "open" form of PBP1B from *Streptococcus pneumoniae*. *Protein Sci.* **15:**1701–1709.

83. **Lu, W.-P., E. Kincaid, Y. Sun, and M. D. Bauer.** 2001. Kinetics of β-lactam interactions with penicillin-susceptible and -resistant penicillin-binding protein 2x proteins from *Streptococcus pneumoniae*: involvement of acylation and deacylation in β-lactam resistance. *J. Biol. Chem.* **276:**31494–31501.

84. **Lu, W.-P., Y. Sun, M. D. Bauer, S. Paule, P. M. Koenigs, and W. G. Kraft.** 1999. Penicillin-binding protein 2a from methicillin-resistant *Staphylococcus aureus*: kinetic characterization of its interactions with β-lactams using electrospray mass spectrometry. *Biochemistry* **38:**6537–6546.

85. **Lujan, R., J. A. Saez-Nieto, J. V. Martinez-Suarez, B. G. Spratt, L. Bowler, and Q. Y. Zhang.** 1991. Nucleotide sequences and genetic diversity of the *penA* genes from penicillin sensitive and moderately penicillin resistant strains of *Neisseria lactamica*, p. 93–98. *In* M. Achtman (ed.), *Neisseriae 1990, Proceedings of the 7th International Pathogenic Neisseria Conference 1990*. Walter de Gruyter, Berlin, Germany.

86. **Lujan, R., Q. Y. Zhang, J. A. Saez Nieto, D. M. Jones, and B. G. Spratt.** 1991. Penicillin-resistant isolates of *Neisseria lactamica* produce altered forms of penicillin-binding protein 2 that arose by interspecies horizontal gene transfer. *Antimicrob. Agents Chemother.* **35:**300–304.

87. **MacKenzie, C. R., I. J. McDonald, and K. G. Johnson.** 1980. Antibiotic resistance in *Neisseria denitrificans*. *Antimicrob. Agents Chemother.* **17:**789–797.

88. **Maggs, A. F., J. M. J. Logan, P. E. Carter, and T. H. Pennington.** 1998. The detection of penicillin insensitivity in *Neisseria meningitidis* by polymerase chain reaction. *J. Antimicrob. Chemother.* **42:**303–307.

89. **Mahapatra, S., S. Bhakta, J. Ahamed, and J. Basu.** 2000. Characterization of derivatives of the high-Mol.-mass penicillin-binding protein (PBP) 1 of *Mycobacterium leprae*. *Biochem. J.* **350:**75–80.

90. **Macheboeuf, P., A. M. Di Guilmi, V. Job, T. Vernet, O. Dideberg, and A. Dessen.** 2005. Active site restructuring regulates ligand recognition in class A penicillin-binding proteins. *Proc. Natl. Acad. Sci. USA* **102:**577–582.

91. **Masson, J. M., A. Kazmierczak, and R. Labia.** 1983. Interactions of clavulanic acid and sulbactam with penicillin-binding proteins. *Drugs Exp. Clin. Res.* **9:**513–518.

92. **Mendelman, P. M., J. Campos, D. O. Chaffin, D. A. Serfass, A. L. Smith, and J. A. Saez-Nieto.** 1988. Relative penicillin G resistance in *Neisseria meningitidis* and reduced affinity of penicillin-binding protein 3. *Antimicrob. Agents Chemother.* **32:**706–709.

93. **Mendelman, P. M., D. O. Chaffin, and G. Kalaitzoglou.** 1990. Penicillin-binding proteins and ampicillin resistance in *Haemophilus influenzae*. *J. Antimicrob. Chemother.* **25:**525–534.

94. **Mirelman, D., Y. Nuchamowitz, and E. Rubinstein.** 1981. Insensitivity of peptidoglycan biosynthetic reactions to β-lactam antibiotics in a clinical isolate of *Pseudomonas aeruginosa*. *Antimicrob. Agents Chemother.* **19:**687–695.

95. **Moreira, B., S. Boyle-Vavra, B. L. M. Dejonge, and R. S. Daum.** 1997. Increased production of penicillin-binding protein 2, increased detection of other penicillin-binding

proteins, and decreased coagulase activity associated with glycopeptide resistance in *Staphylococcus aureus*. *Antimicrob. Agents Chemother.* **41:**1788–1793.

96. **Mouz, N., A. M. Di Guilmi, E. Gordon, R. Hakenbecki, O. Dideberg, and T. Vernet.** 1999. Mutations in the active site of penicillin-binding protein PBP2x from *Streptococcus pneumoniae*. Role in the specificity for β-lactam antibiotics. *J. Biol. Chem.* **274:**19175–19180.
97. **Murakami, K., K. Nomura, M. Doi, and T. Yoshida.** 1987. Production of low-affinity penicillin-binding protein by low- and high-resistance groups of methicillin-resistant *Staphylococcus aureus*. *Antimicrob. Agents Chemother.* **31:**1307–1311.
98. **Muroi, H., and I. Kubo.** 1996. Antibacterial activity of anacardic acid and totarol, alone and in combination with methicillin, against methicillin-resistant *Staphylococcus aureus*. *J. Appl. Bacteriol.* **80:**387–394.
99. **Nagai, K., T. A. Davies, M. R. Jacobs, and P. C. Appelbaum.** 2002. Effects of amino acid alterations in penicillin-binding proteins (PBPs) 1a, 2b, and 2x on PBP affinities of penicillin, ampicillin, amoxicillin, cefditoren, cefuroxime, cefprozil, and cefaclor in 18 clinical isolates of penicillin-susceptible, -intermediate, and -resistant pneumococci. *Antimicrob. Agents Chemother.* **46:**1273–1280.
100. **Nakayama, A., and A. Takao.** 2003. Beta-lactam resistance in *Streptococcus mitis* isolated from saliva of healthy subjects. *J. Infect. Chemother.* **9:**321–327.
101. **Neuwirth, C., E. Siebor, J.-M. Duez, A. Pechinot, and A. Kazmierczak.** 1995. Imipenem resistance in clinical isolates of *Proteus mirabilis* associated with alterations in penicillin-binding proteins. *J. Antimicrob. Chemother.* **36:**335–342.
102. **Obara, M., and T. Nakae.** 1991. Mechanisms of resistance to β-lactam antibiotics in *Acinetobacter calcoaceticus*. *J. Antimicrob. Chemother.* **28:**791–800.
103. **Okada, M.** 1991. Resistance mechanisms of *Campylobacter* to β-lactam antibiotics. *Hiroshima Daigaku Shigaku Zasshi* **23:**18–30.
104. **Okamoto, T., H. Yoshiyama, T. Nakazawa, I.-D. Park, M.-W. Chang, H. Yanai, K. Okita, and M. Shirai.** 2002. A change in PBP1 is involved in amoxicillin resistance of clinical isolates of *Helicobacter pylori*. *J. Antimicrob. Chemother.* **50:**849–856.
105. **Orus, P., and M. Vinas.** 2000. Transfer of penicillin resistance between Neisseriae in microcosm. *Microb. Drug Res.* **6:**99–104.
106. **Orus, P., and M. Vinas.** 2001. Mechanisms other than penicillin-binding protein-2 alterations may contribute to moderate penicillin resistance in *Neisseria meningitidis*. *Int. J. Antimicrob. Agents* **18:**113–119.
107. **Pagani, L., M. Debiaggi, R. Tenni, P. M. Cereda, P. Landini, and E. Romero.** 1988. β-Lactam resistant *Pseudomonas aeruginosa* strains emerging during therapy: synergistic resistance mechanisms. *Microbiologica* **11:**47–53.
108. **Page, M. G. P.** 1994. The reaction of cephalosporins with penicillin-binding protein 1bg from *Escherichia coli*. *Biochim. Biophys. Acta* **1205:**199–206.
109. **Page, M. G. P.** 2004. Cephalosporins in development. *Expert Opin. Investig. Drugs* **13:**973–985.
110. **Page, M. G. P.** 2006. Anti-MRSA β-lactams in development. *Curr. Opin. Pharmacol.* **6:**480–485.
111. **Page, M., D. Bur, F. Danel, I. Heinze-Krauss, M. Kania, B. Mensch, V. Runtz, U. Weiss, and F. Winkler.** 1998. Inhibition of the penicillin-binding proteins of methicillin-resistant staphylococci by pyrrolidinone-3-ylidenemethyl cephems, Poster F-022. Presented at the 38th ICAAC Intersic. Conf. Antimicrob. Agents Chemother.
112. **Pares, S., N. Mouz, Y. Petillot, R. Hakenbeck, and O. Dideberg.** 1996. X-ray structure of *Streptococcus pneumoniae* PBP2x, a primary penicillin target enzyme. *Nat. Struct. Biol.* **3:**284–289.
113. **Perez-Castillo, A., A. M. Perez-Castillo, and J. A. Saez-Nieto.** 1994. Sequence of the penicillin-binding protein 2-encoding gene (penA) of *Neisseria perflava/sicca*. *Gene* **146:**91–93.
114. **Pierre, J., A. Boisivon, and L. Gutmann.** 1990. Alteration of PBP 3 entails resistance to imipenem in *Listeria monocytogenes*. *Antimicrob. Agents Chemother.* **34:**1695–1698.
115. **Piras, G., A. El Kharroubi, J. Van Beeumen, E. Coeme, J. Coyette, and J.-M. Ghuysen,** 1990. Characterization of an *Enterococcus hirae* penicillin-binding protein 3 with low penicillin affinity. *J. Bacteriol.* **172:**6856–6862.
116. **Rice, L. B., S. Bellais, L. L. Carias, R. Hutton-Thomas, R. A. Bonomo, P. Caspers, M. G. Page, and L. Gutmann.** 2004. Impact of specific *pbp5* mutations on expression of β-lactam resistance in *Enterococcus faecium*. *Antimicrob. Agents Chemother.* **48:**3028–3032.
117. **Rodriguez-Tebar, A., F. Rojo, D. Damaso, and D. Vazquez.** 1982. Carbenicillin resistance of *Pseudomonas aeruginosa*. *Antimicrob. Agents Chemother.* **22:**255–261.
118. **Ropp, P. A., M. Hu, M. Olesky, and R. A. Nicholas.** 2002. Mutations in *ponA*, the gene encoding penicillin-binding protein 1, and a novel locus, *penC*, are required for high-level chromosomally mediated penicillin resistance in *Neisseria gonorrhoeae*. *Antimicrob. Agents Chemother.* **46:** 769–777.
119. **Ropp, P. A., and R. A. Nicholas.** 1997. Cloning and characterization of the *ponA* gene encoding penicillin-binding protein 1 from *Neisseria gonorrhoeae* and *Neisseria meningitidis*. *J. Bacteriol.* **179:**2783–2787.
120. **Roychoudhury, S., R. E. Kaiser, D. N. Brems, and W.-K. Yeh.** 1996. Specific interaction between β-lactams and soluble penicillin-binding protein 2a from methicillin-resistant *Staphylococcus aureus*: development of a chromogenic assay. *Antimicrob. Agents Chemother.* **40:**2075–2079.
121. **Rybkine, T., J.-L. Mainardi, W. Sougakoff, E. Collatz, and L. Gutmann.** 1998. Penicillin-binding protein 5 sequence alterations in clinical isolates of *Enterococcus faecium* with different levels of β-lactam resistance. *J. Infect. Dis.* **178:**159–163.
122. **Sato, Y., H. Shibata, T. Arai, A.Yamamoto, Y. Okimura, N. Arakaki, and T. Higuti.** 2004. Variation in synergistic activity by flavone and its related compounds on the increased susceptibility of various strains of methicillin-resistant *Staphylococcus aureus* to β-lactam antibiotics. *Int. J. Antimicrob. Agents* **24:**226–233.
123. **Satta, G., P. Canepari, R. Maurici, R. Pompei, and M. A. Marcialis.** 1985. Shifting of the penicillin-binding proteins that are the target for inhibition by β-lactams as a likely mechanism of resistance to antibiotics during therapy. *Chemioterapia* **4:**113–115.

124. **Sauvage, E., F. Kerff, E. Fonze, R. Herman, B. Schoot, J.-P. Marquette, Y. Taburet, D. Prevost, J. Dumas, G. Leonard, P. Stefanic, J. Coyette, and P. Charlier.** 2002. The 2.4 Å crystal structure of the penicillin-resistant penicillin-binding protein PBP5fm from *Enterococcus faecium* in complex with benzylpenicillin. *Cell. Mol. Life Sci.* **59:**1223–1232.

125. **Schilf, W., and H. H. Martin.** 1980. Purification of two DD-carboxypeptidases /transpeptidases with different penicillin sensitivities from *Proteus mirabilis. Eur. J. Biochem.* **105:**361–370.

126. **Schultz, D. E., B. G. Spratt, and R. A. Nicholas.** 1991. Expression and purification of a soluble form of penicillin-binding protein 2 from both penicillin-susceptible and penicillin-resistant *Neisseria gonorrhoeae. Protein Expr. Purif.* **2:**339–349.

127. **Sengupta, T. K., K. Chaudhuri, S. Majumdar, A. Lohia, A. N. Chatterjee, and J. Das.** 1992. Interaction of *Vibrio cholerae* cells with β-lactam antibiotics: emergence of resistant cells at a high frequency. *Antimicrob. Agents Chemother.* **36:**788–795.

128. **Shiota, S., M. Shimizu, J. Sugiyama, Y. Morita, T. Mizushima, and T. Tsuchiya.** 2004. Mechanisms of action of corilagin and tellimagrandin I that remarkably potentiate the activity of β-lactams against methicillin-resistant *Staphylococcus aureus. Microbiol. Immunol.* **48:**67–73.

129. **Shirai, M., A. Nakazawa, K. Okita, K. Okamoto, and H. Kichiyama.** 2004. Mutation in *Helicobacter pylori* gene pbp1 for amoxicillin resistance and the use of microbial for drug screening. *Jpn. Kokai Tokkyo Koho* Patent no. JP2004121141.

130. **Spratt, B. G.** 1988. Hybrid penicillin-binding proteins in penicillin-resistant strains of *Neisseria gonorrhoeae. Nature* **332:**173–176.

131. **Spratt, B. G., Q. Y. Zhang, D. M. Jones, A. Hutchison, J. A. Brannigan, and C. G. Dowson.** 1989. Recruitment of a penicillin-binding protein gene from *Neisseria flavescens* during the emergence of penicillin resistance in *Neisseria meningitidis. Proc. Natl. Acad. Sci. USA* **86:**8988–8992.

132. **Spratt, B. G., L. D. Bowler, Q. Y. Zhang, J. Zhou, and J. M. Smith.** 1992. Role of interspecies transfer of chromosomal genes in the evolution of penicillin resistance in pathogenic and commensal *Neisseria* species. *J. Mol. Evol.* **34:**115–125.

133. **Srikumar, R., E. Tsang, and K. Poole.** 1999. Contribution of the MexAB-OprM multidrug efflux system to the β-lactam resistance of penicillin-binding protein and β-lactamase-derepressed mutants of *Pseudomonas aeruginosa. J. Antimicrob. Chemother.* **44:**537–540.

134. **Stapleton, P. D., S. Shah, J. C. Anderson, Y. Hara, J. M. T. Hamilton-Miller, and P. W. Taylor.** 2004. Modulation of β-lactam resistance in *Staphylococcus aureus* by catechins and gallates. *Int. J. Antimicrob. Agents* **23:**462–467.

135. **Stapleton, P. D., and P. W. Taylor.** 2002. Methicillin resistance in *Staphylococcus aureus*: mechanisms and modulation. *Sci. Prog.* **85:**57–72.

136. **Stefanelli, P., A. Carattoli, A. Neri, C. Fazio, and P. Mastrantonio.** 2003. Prediction of decreased susceptibility to penicillin of *Neisseria meningitidis* strains by real-time PCR. *J. Clin. Microbiol.* **41:**4666–4670.

137. **Storey, C., and I. Chopra.** 2001. Affinities of β-lactams for penicillin binding proteins of *Chlamydia trachomatis* and their antichlamydial activities. *Antimicrob. Agents Chemother.* **45:**303–305.

138. **Sunagawa, M., M. Itoh, K. Kubota, A. Sasaki, Y. Ueda, P. Angehrn, A. Bourson, E. Goetschi, P. Hebeisen, and R. L. Then.** 2002. New anti-MRSA and anti-VRE carbapenems; synthesis and structure-activity relationships of 1 β-methyl-2-(thiazol-2-ylthio)carbapenems. *J. Antibiot.* **55:**722–757.

139. **Tirodimos, I., E. Tzelepi, and V. C. Katsougiannopoulos.** 1993. Penicillin-binding protein 2 genes of chromosomally-mediated penicillin-resistant *Neisseria gonorrhoeae* from Greece: screening for codon Asp-345A. *J. Antimicrob. Chemother.* **32:**677–684.

140. **Tirodimos, I., E. Tzelepi, N. Vavatsi, K. Delidou, and J. Doubogias.** 1993. Presence of mutation in the penicillin-binding protein 2 (PBP 2) genes of non-β-lactamase producing penicillin-resistant strains of *Neisseria gonorrhoeae* isolated in Greece. *Delt. Hell. Mikrobiol. Hetair.* **38:**331–341.

141. **Toda, M., S. Okubo, Y. Hara, and T. Shimamura.** 1991. Antibacterial and bactericidal activities of tea extracts and catechins against methicillin resistant *Staphylococcus aureus. Nippon Saikingaku Zasshi.* **46:**839–845.

142. **Tokue, Y., S. Shoji, K. Satoh, A. Watanabe, and M. Motomiya.** 1992. Comparison of a polymerase chain reaction assay and a conventional microbiologic method for detection of methicillin-resistant *Staphylococcus aureus. Antimicrob. Agents Chemother.* **36:**6–9.

143. **Tsushima, M., K. Iwamatsu, A. Tamura, and S. Shibahara.** 1998. Novel cephalosporin derivatives possessing a bicyclic heterocycle at the 3-position. I. Synthesis and biological activities of 3-(benzothiazol-2-yl)thiocephalosporin derivatives, CP0467 and related compounds. *Bioorg. Med. Chem.* **6:**1009–1017.

144. **Ubukata, K., N. Chiba, N. Nakayama, and M. Konno.** 1999. Drug-resistance mechanism of β-lactamase nonproducing ampicillin-resistant strains of *Haemophilus influenzae. Nippon Rinsho Biseibutsugaku Zasshi* **9:**22–29.

145. **Ubukata, K., Y. Shibasaki, K. Yamamoto, N. Chiba, K. Hasegawa, Y. Takeuchi, K. Sunakawa, M. Inoue, and M. Konno.** 2001. Association of amino acid substitutions in penicillin-binding protein 3 with beta-lactam resistance in beta-lactamase-negative ampicillin-resistant *Haemophilus influenzae. Antimicrob. Agents Chemother.* **45:**1693–1699.

146. **Urban, C., E. Go, N. Mariano, and J. J. Rahal.** 1995. Interaction of sulbactam, clavulanic acid and tazobactam with penicillin-binding proteins of imipenem-resistant and -susceptible *Acinetobacter baumannii. FEMS Microbiol. Lett.* **125:**193–198.

147. **Utsui, Y., M. Tajima, R. Sekiguchi, E. Suzuki, and T. Yokota.** 1983. Role of an altered penicillin-binding protein (PBP) and membrane-bound penicillinase in cephem-resistant *Staphylococcus aureus. In* K. H. Spitzy and K. Karrer (ed.), *Proceedings of the 13th International Congress of Chemotherapy*. Verlag H. Egermann, Vienna, Austria.

148. **Vicente, M. F., J. Berenguer, M. A. De Pedro, J. C. Perez-Diaz, and F. Baquero.** 1990. Penicillin-binding proteins in *Listeria monocytogenes. Acta Microbiol. Hung.* **37:**227–231.

149. **Vicente, M. F., J. C. Perez-Diaz, F. Baquero, M. A. De Pedro, and J. Berenguer.** 1990. Penicillin-binding protein 3

of *Listeria monocytogenes* as the primary lethal target for β-lactams. *Antimicrob. Agents Chemother.* **34**:539–542.

150. **Villar, H. E., F. Danel, and D. M. Livermore.** 1997. Permeability to carbapenems of *Proteus mirabilis* mutants selected for resistance to imipenem or other β-lactams. *J. Antimicrob. Chemother.* **40**:365–370.

151. **Vouillamoz, J., J. M. Entenza, P. Hohl, and P. Moreillon.** 2004. LB11058, a new cephalosporin with high penicillin-binding protein 2a affinity and activity in experimental endocarditis due to homogeneously methicillin-resistant *Staphylococcus aureus. Antimicrob. Agents Chemother.* **48**:4322–4327.

152. **Wexler, H. M., and S. Halebian.** 1990. Alterations to the penicillin-binding proteins in the *Bacteroides fragilis* group: a mechanism for non-β-lactamase mediated cefoxitin resistance. *J. Antimicrob. Chemother.* **26**:7–20.

153. **Yamazaki, H., Y. Tsuchida, H. Satoh, S. Kawashima, H. Hanaki, and K. Hiramatsu.** 2000. Novel cephalosporins 2. Synthesis of 3-heterocyclic-fused thiopyranylthiovinyl cephalosporins and antibacterial activity against methicillin-resistant *Staphylococcus aureus* and vancomycin-resistant *Enterococcus faecalis. J. Antibiot.* **53**:551–555.

154. **Yoshizawa, H., H.Itani, K. Ishikura, T. Irie, K. Yokoo, T. Kubota, K. Minami, T. Iwaki, H. Miwa, and Y. Nishitani.** 2002. S-3578, a new broad spectrum parenteral cephalosporin exhibiting potent activity against both methicillin-resistant *Staphylococcus aureus* (MRSA) and *Pseudomonas aeruginosa.* Synthesis and structure-activity relationships. *J. Antibiot.* **55**:975–992.

155. **Yotsuji, A., J. Mitsuyama, R. Hori, T. Yasuda, I. Saikawa, M. Inoue, and S. Mitsuhashi.** 1988. Mechanism of action of cephalosporins and resistance caused by decreased affinity for penicillin-binding proteins in *Bacteroides fragilis. Antimicrob. Agents Chemother.* **32**:1848–1853.

156. **Zhang, F.-K., M. G. P. Page, and S.-H. Jin.** 2000. Factors determining the resistance of *Pseudomonas aeruginosa* to β-lactam antibiotics. *Zhongguo Kang Sheng Su Za Zhi* **25**:362–367.

157. **Zhang, Q. Y., D. M. Jones, J. A. Saez Nieto, E. P. Trallero, and B. G. Spratt.** 1990. Genetic diversity of penicillin-binding protein 2 genes of penicillin-resistant strains of *Neisseria meningitidis* revealed by fingerprinting of amplified DNA. *Antimicrob. Agents Chemother* **34**:1523–1528.

158. **Zorzi, W., X. Y. Zhou, O. Dardenne, J. Lamotte, D. Raze, J. Pierre, L. Gutmann, and J. Coyette.** 1996. Structure of the low-affinity penicillin-binding protein 5 PBP5fm in wild-type and highly penicillin-resistant strains of *Enterococcus faecium. J. Bacteriol.* **178**:4948–4957.

Enzyme-Mediated Resistance to Antibiotics: Mechanisms, Dissemination, and Prospects for Inhibition
Edited by Robert A. Bonomo and Marcelo E. Tolmasky

Samy O. Meroueh
Jooyoung Cha
Shahriar Mobashery

8

Inhibition of Class A β-Lactamases

The catalytic function of β-lactamases is the primary mechanism of bacterial resistance to β-lactam antibiotics (penicillins, cephalosporins, carbapenems, etc.). β-Lactamases hydrolyze the β-lactam bond of these antibiotics, a structure modification that abrogates the antibacterial activity. Four classes of these enzymes, classes A through D, are known. Enzymes of classes A, C, and D undergo a two-step catalytic process involving intermediary acyl-enzyme species. The antibiotic acylates an active-site serine in these enzymes, and this species undergoes hydrolysis in the second step. Enzymes of class B are zinc dependent and operate by a distinct mechanism that does not involve covalent chemistry with the enzyme. Classes A and C of β-lactamases are the most and second most prevalent, respectively, among pathogens presently. Several excellent reviews have addressed the issue of inhibition of β-lactamases (22, 37, 45, 51, 54). The focus of this review is on recent developments in inhibition of class A β-lactamases, but key features of mechanisms of earlier inhibitors are complemented with new insights into their mechanistic details.

Among class A β-lactamases, three subclasses appear to dominate: the (historically gram-negative plasmid penicillinase) TEM/SHV, the *Pseudomonas aeruginosa* PER/OXA/TOHO cephalosporinases, and the CTX-M (NMC-A) carbapenemase subclasses. The TEM and SHV enzymes are the most prominent members of the class A β-lactamase family and have been extensively studied. As of September 2004, 133 variants of TEM and 54 variants of SHV have been reported. A glimpse into their three-dimensional structures became possible in the late 1980s and early 1990s, when the structure of class A β-lactamases, among them TEM-1 from *Escherichia coli*, was solved by X-ray diffraction (20, 23, 48). The structure of TEM-1 consists of two domains, one helical, and the other composed of a mixed α/β domain where the α helices flank the β-sheet structure on both sides (Fig. 8.1A). The active site is sequestered between the two domains (Fig. 8.1B), which contains the serine residue that is acylated at the N terminus of an α-helix (Fig. 8.1C). This position has been proposed to be important in modulating the pK_a of this residue and facilitating its activation for catalysis (31).

The interest in inhibitors of class A β-lactamases stems from the recognition that abrogation of the activities of these deleterious enzymes would preserve the antibiotic activity of β-lactams, such as benzylpenicillin (**1**) (numbers

Samy O. Meroueh, Jooyoung Cha, and Shahriar Mobashery, Department of Chemistry and Biochemistry, University of Notre Dame, Notre Dame, IN 46556.

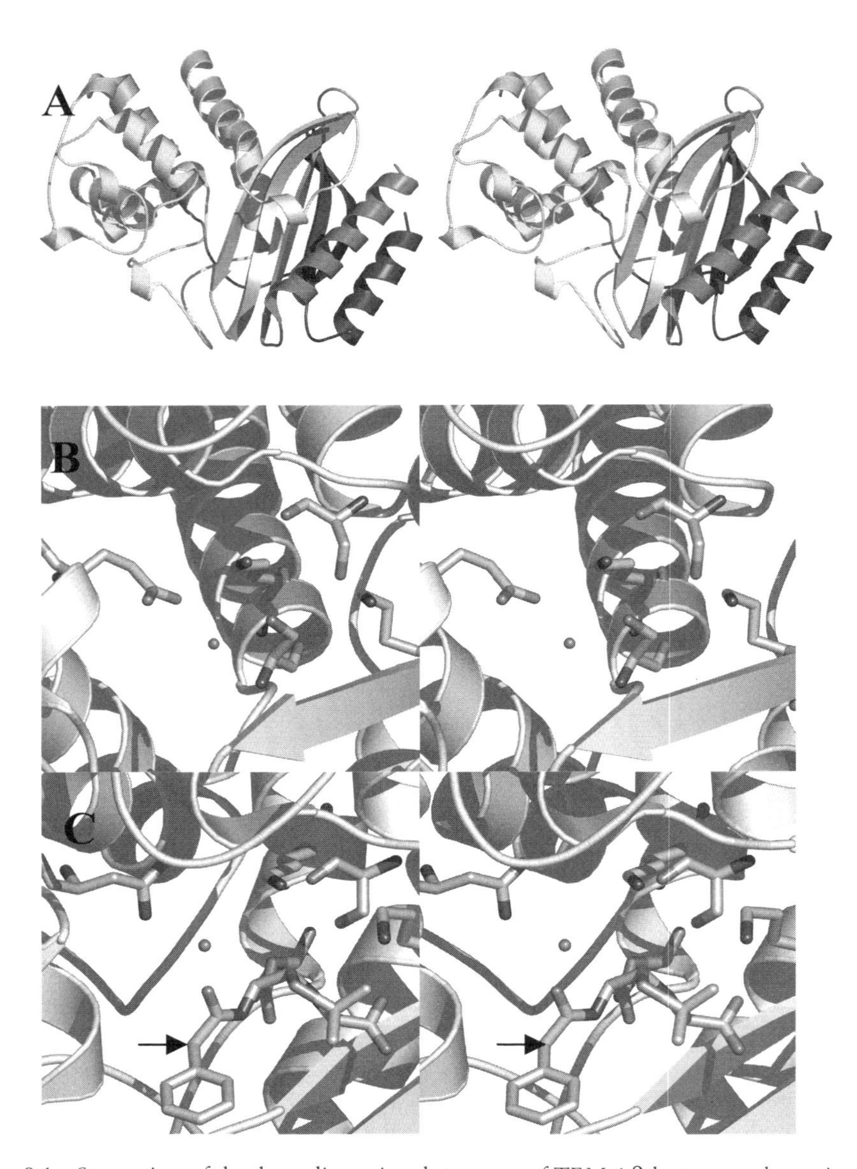

Figure 8.1 Stereoview of the three-dimensional structure of TEM-1 β-lactamase shown in ribbon representation (Protein Data Bank [PDB] code 1BTL). (B) Stereoview of a close-up of the active site of TEM-1 showing Glu-166 (9 o'clock), Lys-73 (11 o'clock), Ser-130 (12 o'clock), Lys-234 (2 o'clock), and Ser-70 (center). The enzyme is shown in ribbon representation, and residues are shown in capped-sticks representation. The conserved water molecule is shown as a sphere. (C) Stereoview of the acyl enzyme complex of a deacylation-deficient TEM β-lactamase in complex with benzylpenicillin (PDB code 1TEM). The enzyme is rendered in ribbon representation, and residues Glu-166 (9 o'clock), Lys-73 (12 o'clock), Ser-130 (1 o'clock), and Lys-234 (3 o'clock) are shown in capped sticks. An arrow is used to point to the C-6 side chain of the bound benzylpenicillin. The hydrolytic water is shown as a sphere.

in boldface refer to structures given in the figures), whose primary target is the penicillin-binding proteins (PBPs). The notion was put to practice by the development of clavulanate (**2**) in the 1970s. A mixture of clavulanate and a penicillin, a product known as Augmentin, was introduced to the clinic in the United States in 1985. This product is a success story, as the combination is quite effective, and had sales of $2.1 billion in 2002. Class A β-lactamases are to date the only antibiotic resistance enzymes whose inhibitors have found clinical use.

Besides clavulanate, the search of natural products and exploration of synthetic β-lactams in the 1970s to 1980s led to several important discoveries. These include tazobactam (**3**), a highly effective sulfone penam inhibitor, penicillanic acid sulfone sulbactam (**4**) (15), 6-β-bromopenicillanic acid (**5**) (53), and thienamycin (**6**) (21) (Fig. 8.2). Some of these inhibitors are now in use clinically in combination with other antibiotics; examples are amoxicillin-clavulanate (known as Augmentin), ampicillin-sulbactam (Unasyn), and piperacillin-tazobactam (Zocyn). At least three structural variants of thienamycin, i.e., imipenem, meropenem, and ertapenem, have also become clinical.

Shortly after the introduction of these inhibitors, bacteria began to develop resistance against them. The resistance came in the form of decreased susceptibility to the effects of the inhibitor, usually as a result of point mutations in the resistance enzymes at locations that affected effective binding or the covalent inhibition pathway. The term inhibitor-resistant TEMs (IRTs) was coined because the first resistant variants emerged in the TEM family of class A β-lactamases; in light of the emergence of inhibitor-resistant SHV variants as well, the name "inhibitor-resistant β-lactamases" would appear to be more appropriate. The identified variants of these enzymes are listed at a website maintained by George Jacoby and Karen Bush (http://www.lahey.org/studies/temtable.htm). To date, there are more than 20 reported cases of inhibitor-resistant β-lactamases.

CLAVULANIC ACID

As discussed above, the search for inhibitors of β-lactamases in the 1970s was rewarded with the discovery of clavulanate (**2**), a natural product isolated from *Streptomyces clavuligerus* (4). Clavulanate is a potent inhibitor of class A β-lactamases, which incidentally exhibits weak antimicrobial activity as well (4). Therefore, clavulanate is most effective when coadministered with a β-lactam antibiotic. The chemistry of inhibition of this inhibitor is complex, involving several intermediates along the reaction pathway (11). The process begins with the formation of an acyl enzyme species, **7**, similar to the case of a typical substrate (Fig. 8.3). This is followed by a series of rearrangements to **8** and **9**, resulting in the formation of an iminium intermediate, which is concomitant with the opening of the five-membered oxazole ring resulting in **10**. This process is facilitated by protonation of **8**. The source of proton for this process is believed to be a crystallographically seen water molecule that is sequestered by the guanidinium group of Arg-244 and the backbone carbonyl of Val-216 (19). From **10**, the reaction branches, and one of the routes leads to the irreversible inactivation of the enzyme by entrapment of Ser-130, eventually resulting in the formation of **12** (5, 19). The other leads to the formation of an enamine species that readily undergoes decarboxylation, culminating in species **14**, which could only slowly undergo ester bond hydrolysis. Evidence for the formation of this species came from X-ray crystallography (12).

PENICILLANIC ACID DERIVATIVES

Sulbactam (**4**) was the first penam sulfone to be synthesized in the late 1970s (15). Since then, a series of potent penam sulfone derivatives have emerged, among which tazobactam (**3**) has found clinical use. Recently mass and Raman spectroscopy and X-ray crystallography have contributed to the understanding of the complex inhibition chemistry of this important inhibitor. The initial parts of the chemistry of penam sulfones follow closely those of clavulanate

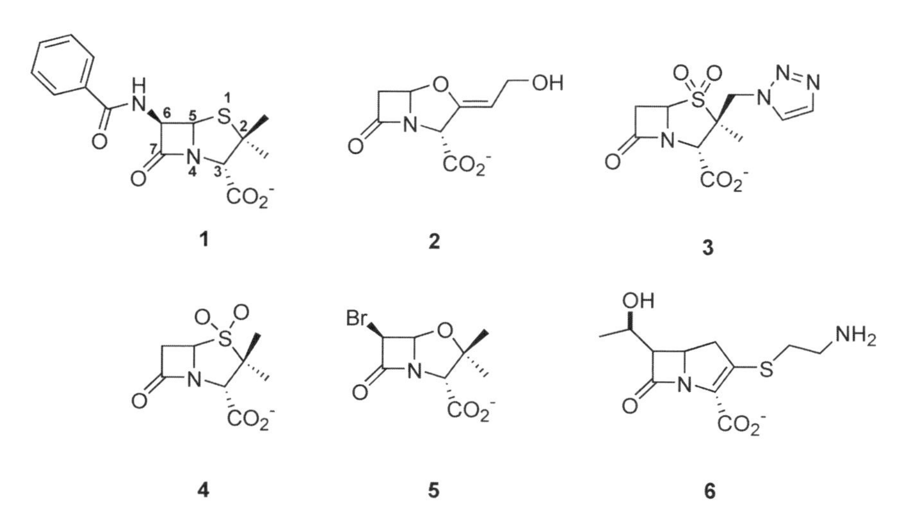

Figure 8.2 Chemical structures of some inhibitors of class A β-lactamases.

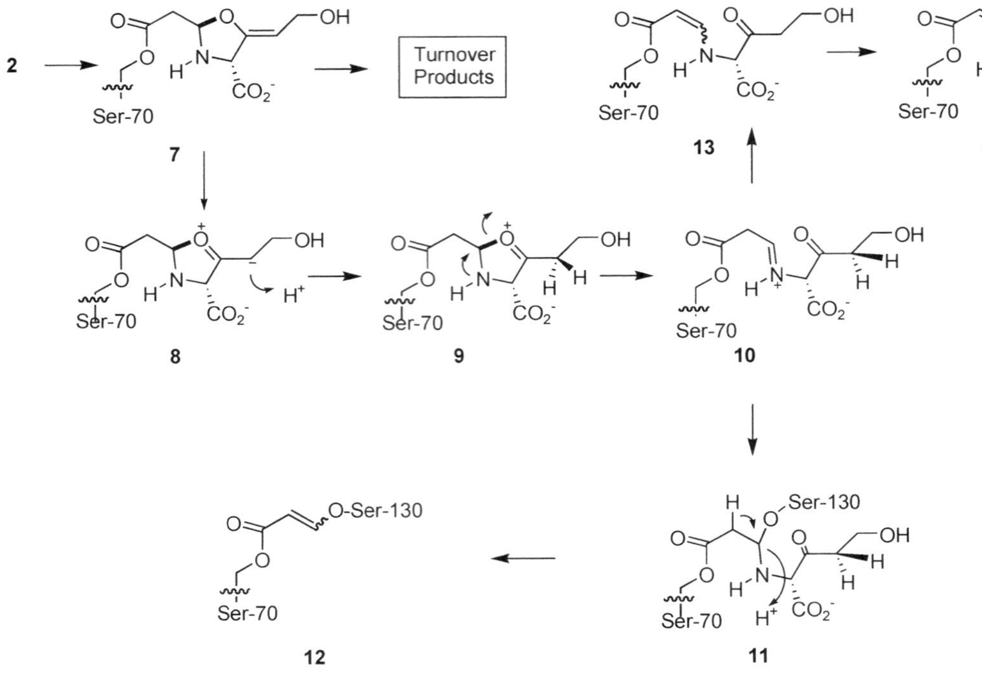

Figure 8.3 Chemistry of inhibition of class A β-lactamases by clavulanate.

with the formation of an imine species (**16**) (54) (Fig. 8.4). The structure of this intermediate bound to the active site of SHV-1 was solved by X-ray crystallography and is shown in Fig. 8.5A (24). The structure reveals that the sulfinic acid moiety and the triazolyl ring make contacts with the enzyme, while the carboxylate is oriented towards the solvent and does not interact with the protein. The imine intermediate then undergoes conversion to the enamines **17** and **18**. The formation of the enamine was monitored by Raman spectroscopy in SHV-1 (18). By monitoring a band that is specific to the enamine species, as the reaction progressed, the authors noted that tazobactam appeared to result in more of the enamine species (at least twice as much as with clavulanate and sulbactam). Quantum mechanical calculations showed that the band occurring at 1,595 cm^{-1} actually corresponded to the *trans*-enamine species. The long lifetime of this intermediate made it possible to trap it and to solve an X-ray structure with the species bound to the active site of the Glu-166-Ala SHV mutant variant at 1.67-Å resolution, as shown in Fig. 8.5B (37). It was found that one of the triazolyl nitrogens makes a 2.7-Å hydrogen bond with a buried water molecule, which in turn is hydrogen bonded to Ser-130 Oδ and Lys-234 Nδ, and another water molecule. This interaction is likely the reason for the stability of the enamine species of tazobactam being greater than that of sulbactam, which lacks the triazolyl ring. The enamine intermediate then branches into several possible routes, among them a species that involves the cross-linking of Ser-130 and Ser-70 via a portion of the inhibitor structure, leading to an irreversibly inhibited enzyme. Another intermediate (**22**), which was characterized from the X-ray diffraction data and is shown in Fig. 8.5A, forms as a result of the hydrolysis of the ester bond of acylated Ser-70 in **21**.

A compound that could simultaneously inhibit more than one β-lactamase class would be highly desirable, since more than one class of β-lactamases are often present in bacteria. A recent study (1) has reported the synthesis of a penam sulfone (**23**) (Fig. 8.6), which is active against both class A and class C β-lactamases, by systematically varying the stereochemistry of various hydroxylalkyl groups at C-6. The compound exhibited nanomolar (6 nM) and low-micromolar (~1 μM) inhibition against the class A and C β-lactamases, respectively. It was also found that the sulfone moiety was necessary for the compound to remain active against both classes. It is worth stating that the penicillanate version of compound **23** (sulfide, rather than sulfone at position 1) was synthesized earlier and was shown to be an effective inhibitor of class A TEM-1 β-lactamase (30), and the X-ray structure of the complex indicated that inhibition of this enzyme took place by acylation of the active site in a manner that the hydroxymethyl

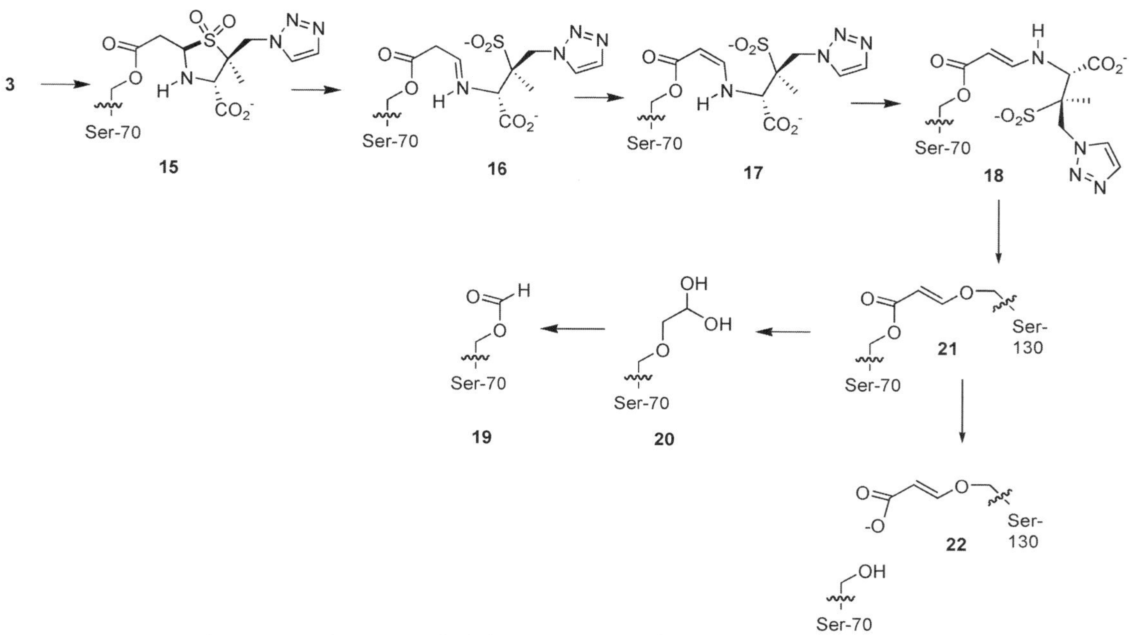

Figure 8.4 Chemistry of inhibition of class A β-lactamases by tazobactam.

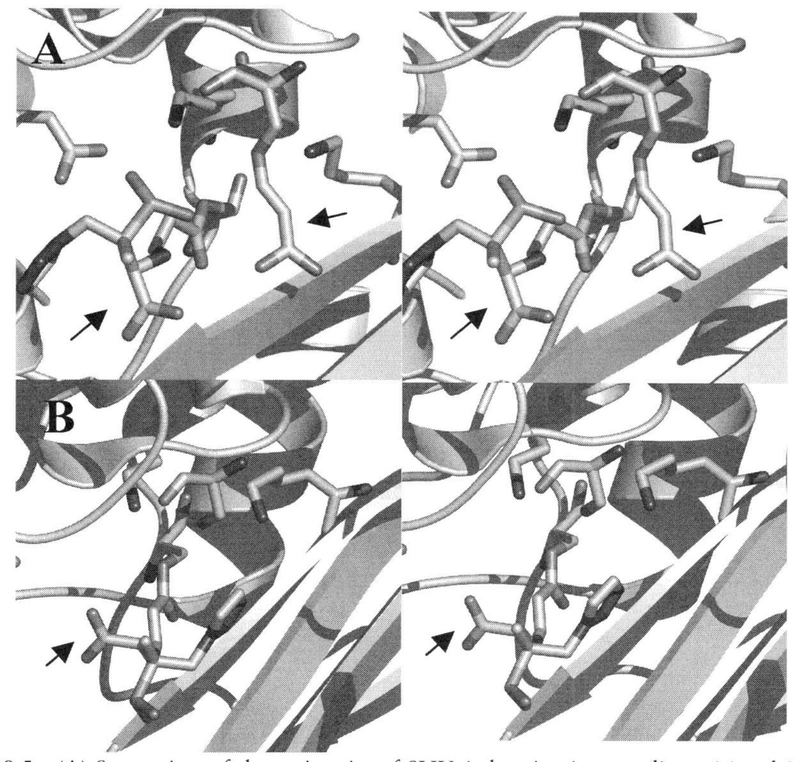

Figure 8.5 (A) Stereoview of the active site of SHV-1 showing intermediates **16** and **22** (Protein Data Bank [PDB] code 1G56). The bottom arrows point to **16**, while the top arrow points to **22**. The protein is rendered in ribbon, while Ser-70 (covalently bound to inhibitor; center), Glu-166 (10 o'clock), Lys-73 (12 o'clock), Ser-130 (covalently bound to inhibitor; 1 o'clock), and Lys-234 (3 o'clock) are shown in capped-sticks representation. (B) Intermediate **18** shown in the active site of SHV-2. The arrows point to the bound inhibitor that is attached to Ser-70 (center). Lys-73 (12 o'clock), Ser-130 (1 o'clock), and Lys-234 (2 o'clock) are shown in capped-sticks representation (PDB code 1RCJ).

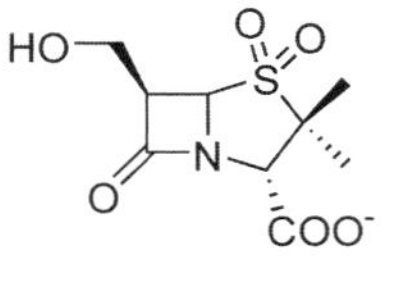

23 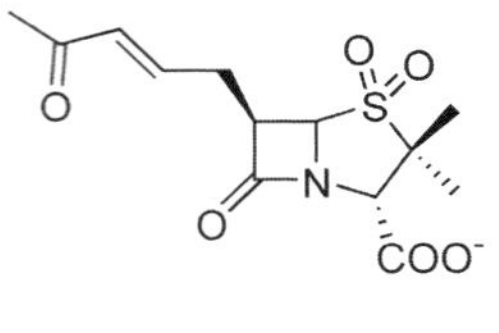 24 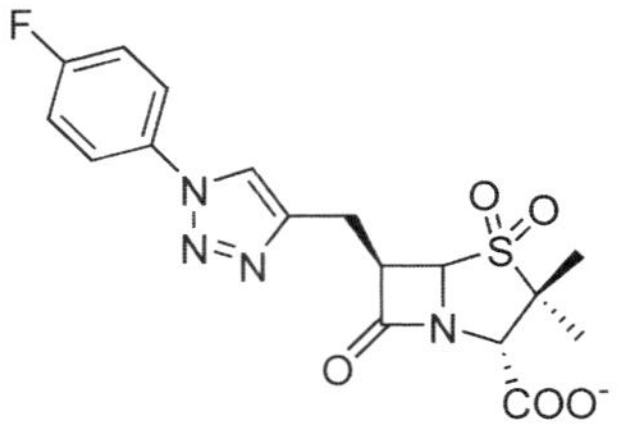 25

Figure 8.6 Chemical structures of penam sulfone (23) and derivatives.

group prevented the approach of the hydrolytic water molecule, giving the species longevity (28). This very same mechanism probably facilitates inhibition of the class A enzymes with 23. It is likely that in the case of class C enzymes, the chemistry is similar to that of sulbactam and tazobactam.

Scientists at Wyeth-Ayerst have successfully used the penam sulfone as a scaffold to arrive at a series of molecules that show good activity towards classes A, B, and C of β-lactamases (44). Among the compounds that were discovered were **24** and **25**, which showed low-micromolar inhibition for all three classes. The study found that for the alkenyl derivatives, the β isomer significantly improved activity over the α isomer for the class B β-lactamases, resulting in a decrease of more than two orders of magnitude in the 50% inhibitory concentration (IC_{50}). The triazole series, represented by **25**, also showed potency against classes A, B, and C. One interesting aspect of the triazole derivatives is that elimination of the fluorine group from the C-6 substituent of **25** resulted in a significant decrease in the activity of the compound against CcrA, a class B β-lactamase. However, an amino or hydroxyl group at that position restored activity to that enzyme. Other compounds, based on the structure of 6-(nitrileoxidomethyl)-penam sulfone, were also recently synthesized (**26–30**) (Fig. 8.7), and they were found to be active against classes A, B, and C. However, none were broad-spectrum inhibitors (46), but **26**, **27**, and **29** showed nanomolar inhibition of the TEM-1 class A β-lactamase, while **28** and **30** showed nanomolar activity against the class B CcrA enzyme.

Buynak et al. (8) have synthesized a series of molecules—using sulfoxide and sulfone penams as starting points—with sulfhydryl and sulfide moieties at C-6; the goal of this exercise was to arrive at molecules that would simultaneously inhibit classes A and B of β-lactamases. Compound **31** (Fig. 8.8) was a potent inhibitor of enzymes of classes A, B, and C with IC_{50} values in the low-nanomolar range. By comparison, the 6-mercapto-penicillanates were relatively inactive against the β-lactamases. The study also confirmed that the sulfone oxidation state of the penam thiazolidine resulted in greater inhibition.

The Buynak group has also derived powerful inhibitors of both class A and class C β-lactamases by introducing modifications at the C-6 and 2β′ positions of the penam sulfone ring (9). Two of these compounds, **32** and **33**, showed nanomolar inhibition of the class A TEM-1 and the class C P99 and PC1 β-lactamases.

CARBAPENEMS

Isolation of thienamycin (6), a potent inhibitor of β-lactamases from *Streptomyces cattleya*, prompted the synthesis of a large number of derivatives, resulting in the discovery of over 60 carbapenems (2). These compounds were found to be potent inhibitors of serine-dependent β-lactamases (17), possessed broad-spectrum activity, and

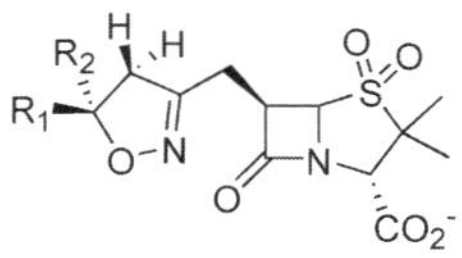

26 $R_1 = SO_2Ph$, $R_2 = H$
27 $R_1 = H$, $R_2 = SO_2Ph$

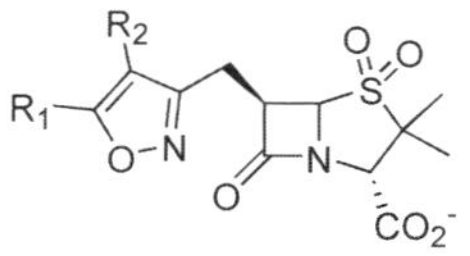

28 $R_1 = H$, $R_2 = SO_2Tol$
29 $R_1 = COPh$, $R_2 = H$
30 $R_1 = CO_2Me$, $R_2 = CO_2Me$

Figure 8.7 Synthetic inhibitors based on the structure of 6-(nitrileoxidomethyl)penam sulfone active in inhibition of classes A, B, and C of β-lactamases.

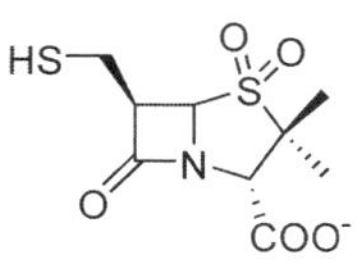

31

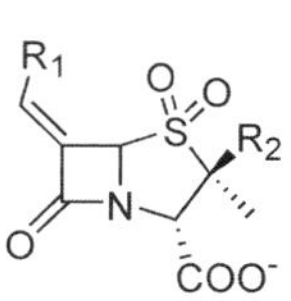

32 $R_1 = \alpha$ pyr. $R_2 = CH_2O_2CCH_3$
33 $R_1 = \alpha$ pyr $R_2 = CH_2O_2CCH_2Ph$

Figure 8.8 Chemical structures of additional inhibitors based on the penam sulfone nucleus.

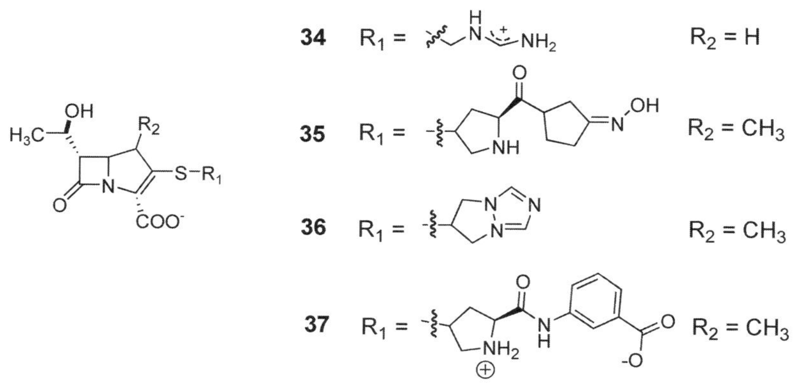

Figure 8.9 Chemical structures of representative carbapenems.

were efficacious in cases where the organism has developed resistance against other antibiotics (3); examples include imipenem (**34**), meropenem (**35**), biapenem (**36**), and ertapenem (**37**) (Fig. 8.9).

Unlike clavulanic acid and the penam sulfones, inhibition of carbapenems (**38**) is not irreversible but was early on proposed to occur through a tautomerization process of Δ^2 (**39**) to a more stable Δ^1 (**40**) in the ring of the acyl-enzyme species (**39**) (22) (Fig. 8.10). Subsequently, hydrolysis of the ester bond may take place for species **39** and **40** to **41** and **42**, respectively, to regenerate enzyme activity (11, 29). However, recent studies have shown that the inhibition process is more involved. One carbapenem in particular, imepenem (**34**), has been widely used in the clinic. Its discovery dates back to the 1980s, but its mechanism of inhibition, such as the biphasic nature of its kinetics, has only recently been elucidated by a combination of enzymology, X-ray diffraction, and computational studies (29, 50). Insight into the mode of action of imipenem was provided by the X-ray structure of the TEM-1/imipenem acyl enzyme, which reveals an unprecedented conformational change. It was shown that the ester carbonyl was not ensconced into the oxyanion hole formed by the backbone nitrogen atoms of Ala-237 and Ser-70 (the canonical conformation, with the carbonyl oxygen housed into the oxyanion hole, would be responsible for the fast phase of hydrolysis). A hydrogen bond between the hydroxyl group of the 6α-1*R*-hydroxyethyl substituent and the hydrolytic water was found, an interaction that would likely decrease its nucleophilicity. Molecular dynamics simulation—starting with the canonical TEM-1-imipenem acyl enzyme species—suggested that the steric encumbrance introduced by the hydroxymethyl moiety of imipenem facilitates transition into another conformation that is responsible for the slow rate of hydrolysis (29). Hence, the mere tautomerization of the double bond in the ring at the acyl enzyme

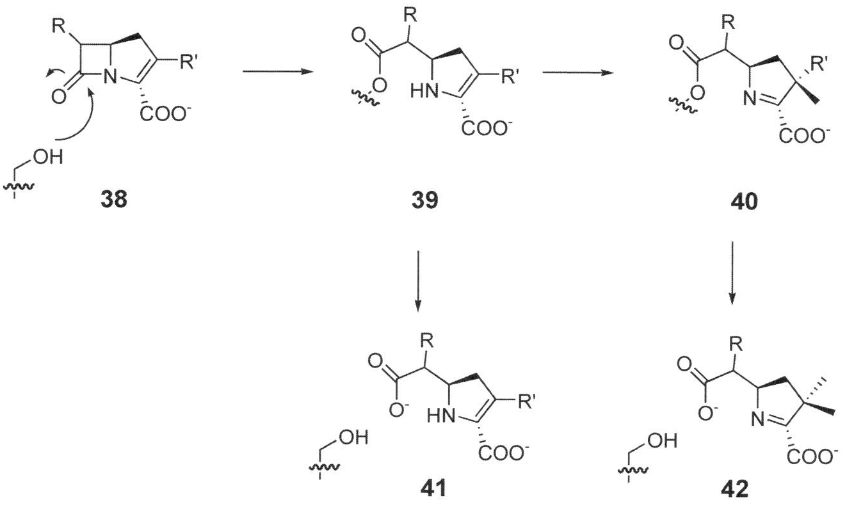

Figure 8.10 Inhibitory mechanism of β-lactamase by carbapenems.

species stage is not sufficient to explain the kinetic behavior, but rather it is a set of specific interactions with the enzyme that accounts for a critical conformational change that is at the root of inhibition.

Meropenem (**35**) is another example of a carbapenem derivative with good antimicrobial activity against gram-negative bacteria and moderate activity against gram-positive organisms. This compound, in contrast to imipenem, does not have to be coadministered with a dehydropeptidase inhibitor. Another carbapenem, ertapenem, a compound conceived at Merck laboratories, is the most recent addition that has been approved for clinical use in the United States and the European Union. (47). Like its predecessors—imipenem and meropenem—ertapenem has good antimicrobial activity against gram-positive and gram-negative organisms.

Carbapenems are effective inhibitors against class A β-lactamases, but they are less effective against the class C β-lactamases, which are a growing clinical problem (43). Recently, Copar and cowokers used computational methods to design a series of tricyclic derivatives of carbapenems (**43–46**) (Fig. 8.11) that were found to be active against class A TEM-1 β-lactamase from *E. coli* and class C β-lactamase from *Enterobacter cloacae* 908R at the micromolar level.

There have been few new carbapenems that have been reported in the past five years, and the emergence of carbapenemases such NMC-A (non metallo-carbapenemase A), a class A β-lactamase that is able to efficiently hydrolyze carbapenem-based inhibitors, has added new urgency for the development of more effective carbapenems (32).

METHYLIDENE PENEM

The introduction of a double bond at C-6 with a sulfide at position 1 of a penem resulted in compounds that were potent and irreversible inhibitors of β-lactamases. The mechanism of inhibition of one of these compounds, penem BRL 42715 (**47**), was recently elucidated (see Fig. 8.12). It was shown that the acyl enzyme species **48** would first lead to the formation of an imine intermediate, **49**, which would be followed by the opening of the five-membered

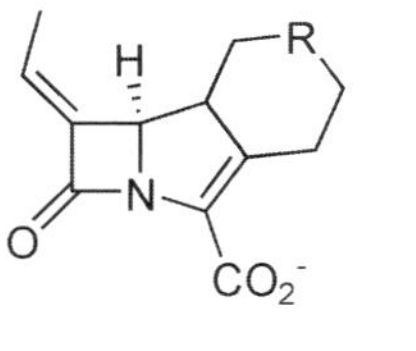

43 (8S) R=CH_2
44 (8R) R=CH_2
45 (8S) R=S
46 (8R) X=S

Figure 8.11 Synthetic carbapenem derivatives that serve as inhibitors of classes A and C of β-lactamases.

ring. The subsequent rearrangement of **49**, which involves a nucleophilic attack of the double bond initially at C-6 by the newly formed thiolate, leads to the formation of a seven-membered dihydrothiazepine **50** (7, 16, 27).

The success of BRL 42715 prompted additional efforts into compounds with a double bond at C-6, leading to the discovery of SYN-1012 (**39**)—with a methyl triazolyl moiety at C_6 instead–and another more recent methylidene penem (**50**)—with a bicyclic and heterocyclic moiety at C-6 (35); both of these compounds show good activity against class A and C β-lactamases (35, 39). The IC_{50} values of **51** were in the low-nanomolar range for TEM-1, SHV-1, and GC1 β-lactamases (35). The inactivation mechanism of **50** leads to the formation of a thiazepine product with a seven-membered ring (**51**). The formation of **52** was confirmed by electrospray ionization mass spectrometry (49) and X-ray crystallography (35). An X-ray structure of an inhibited species with **51** was solved recently (Fig. 8.13) (35). It appears that the conformation adopted by species **52** in SHV-1 and GC-1 (a class C β-lactamase) is similar, except for a 180-degree rotation of the dihydrothiazepine ring. In both cases, the carboxylate is found exposed to solvent and the carbonyl of the ester bond is located in the oxyanion hole (35).

MONOBACTAMS

The early success with inhibition of class C β-lactamases with monobactams (26, 34) was replicated with the class A β-lactamases—namely TEM-1 from *E. coli*—where derivatives similar to **53** rapidly inactivated the enzyme with a second-order rate constant k_{inact}/K_i of 3.2×10^2 $M^{-1}s^{-1}$ (6). The inhibited species was stable given the exceedingly slow recovery of activity from inhibition (<10% after 6 days). The acylated species **54** is shown to undergo elimination of the tosylate group, potentially giving rise to inhibited species **55** to **58** (Fig. 8.14). In light of the fact that a chromophore for the α,β-unsaturated ester was not seen, it was concluded that structures **55** and **58** were inconceivable and **56** or **57** was the most likely product. X-ray diffraction studies of the inhibition of the class A NMC-A β-lactamase by **53** provided insight into the identity of the species after the loss of the tosylate moiety. The X-ray structure revealed two possible binding modes of the same molecular species, and the electron density ruled out **55** and **58**. The keto species **56** was favored over **57**, as the iminium species is fully exposed to the solvent and is likely to undergo hydrolysis. For example, a water molecule was found at the oxyanion hole (2.9 and 2.4 Å of the main chain nitrogen atoms), which draws parallels to inhibition by imipenem **34**, which showed the ester carbonyl out of the oxyanion hole. The lack of water molecule in the case of

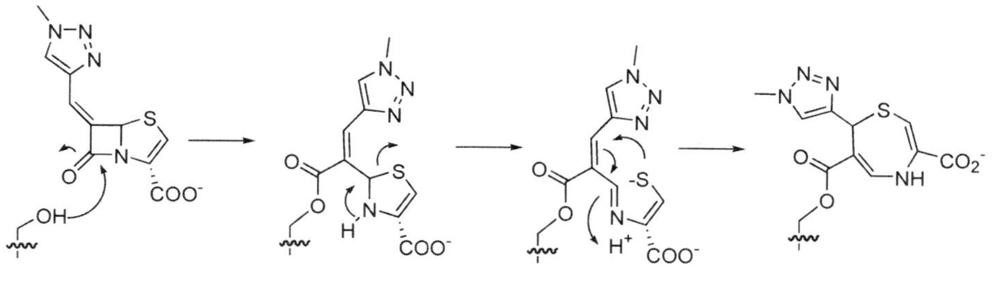

Figure 8.12 Inhibitory mechanism of penem BRL 42715 (47).

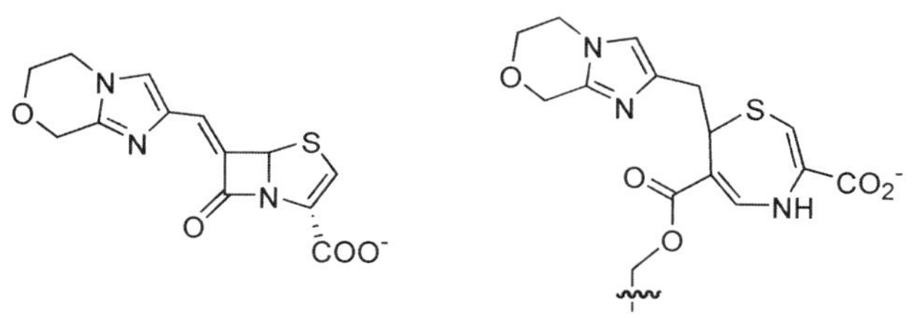

imipenem is due to the highly dynamic nature of the moiety at C-6. The presence of water for inhibition of 53 suggests added stability for the intermediate, which would explain the exceedingly slow recovery of enzymic activity from the inhibited species.

TRANSITION-STATE ANALOGS

Nucleophilic attack of the β-lactam carbonyl by the activated hydroxyl of the active-site serine subsequent to the formation of the Michaelis complex leads to the formation of a tetrahedral species. In the process, the carbonyl carbon is converted from sp^2 to sp^3 hybridization, while the anionic oxygen is sequestered within the oxyanion hole, which is formed by the main chain nitrogen atoms of Ala-237 and Ser-70 in the case of the class A β-lactamases. The unstable tetrahedral species would ultimately convert to the acyl enzyme species. This species then experiences addition of an activated water molecule and goes through a second tetrahedral species, which leads to enzyme deacylation. It would thus be desirable to design compounds that would mimic the initial stages of the binding of β-lactam antibiotic—since a compound that is similar to the enzyme's substrate is more likely to engender favorable

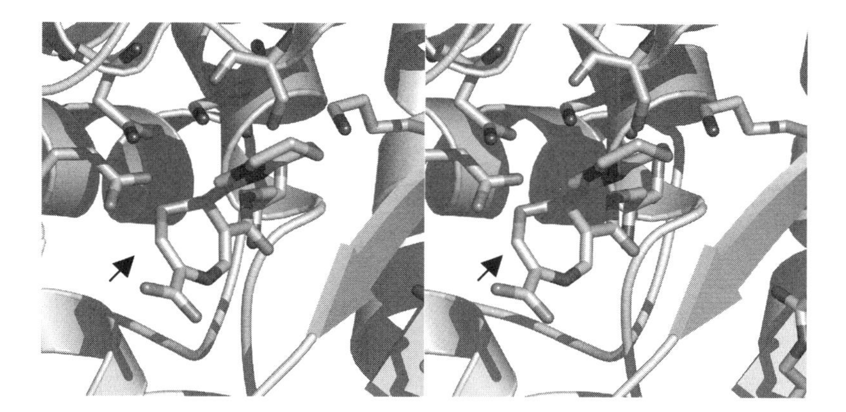

Figure 8.13 Stereoview of intermediate (49) bound to the active site of SHV-1 (Protein Data Bank [PDB] code 1ONH). The enzyme is shown in ribbon representation, while the bound ligand and active-site residues Glu-166 (9 o'clock), Lys-73 (11 o'clock), Ser-130 (12 o'clock), Lys-234 (2 o'clock), and Arg-244 (5 o'clock) are shown in capped-sticks representation. The arrows point to the thiazepine intermediate that is bound to Ser-70.

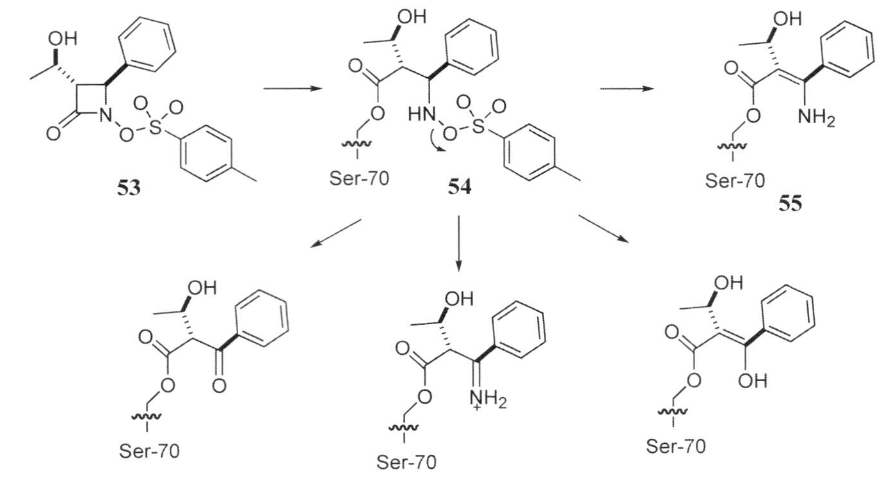

Figure 8.14 Inhibitory mechanism of monobactam derivative 53 against class A β-lactamases.

binding—but resist the hydrolytic step that ultimately destroys β-lactams. These compounds are known as transition-state analogs. Two different strategies have been adopted for the design of these inhibitors. The first strategy makes use of a phosphonate moiety, and these compounds were first designed in the late 1980s by the Pratt group (40, 42). Another strategy involves the use of a boronate moiety, which mimics the β-lactam motif (14). While the phosphonates and boronates have been shown to effectively inhibit class C β-lactamases, more recent studies have shown that some of these derivatives can be highly effective against class A enzymes (25, 33, 41, 52).

The reaction for the phosphonate compounds would proceed by nucleophilic attack at the phosphoryl moiety (**59**), leading to the formation of a pentacoordinated intermediate with bipyramidal geometry (**60**) and culminating to the formation of an inert tetrahedral species **61**, which is reminiscent of the tetrahedral species formed by β-lactam compounds (hence the name transition-state analogs) (38) (Fig. 8.15).

Recently, a series of phosphonate-based transition-state analogs have been reported. These compounds (**62** to **64**) (Fig. 8.16) were shown to inhibit the class C β-lactamase from *Enterobacter cloacae* P99 and the TEM-1 class A β-lactamase. The novelty behind these compounds is that they are mixed anhydrides, making it possible for an acyl- or phosphoryl-enzyme complex to form. Kinetic characterization of these compounds with the

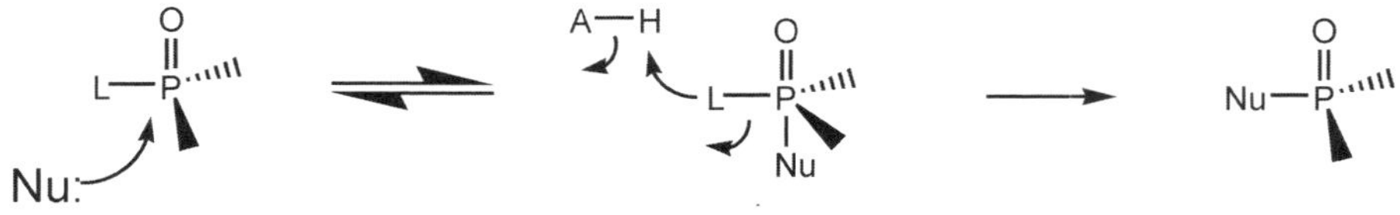

Figure 8.15 Schematic of phosphonate interactions with nucleophiles.

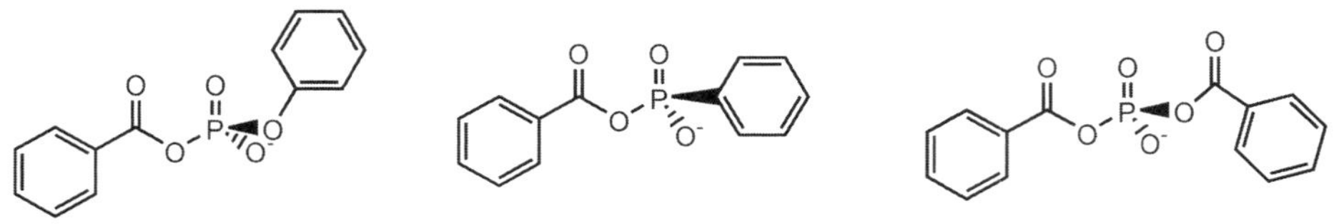

Figure 8.16 Chemical structure of synthetic phosphonate-based transition-state analog inhibitors.

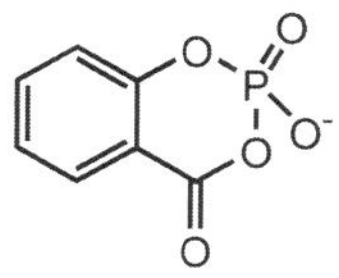

65 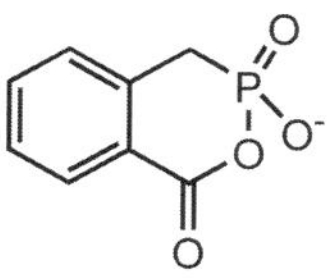 66

Figure 8.17 Chemical structure of cyclic phosphonate-based transition-state analog inhibitors.

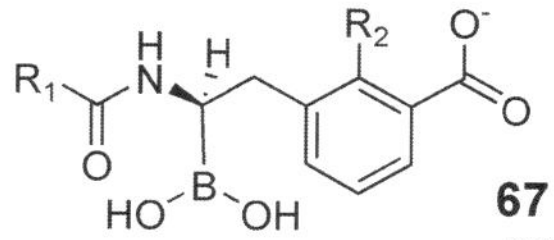

67 R_1=$PhCH_2$ R_2=H
68 R_1=CH_3 R_2=OH

Figure 8.18 Chemical structure of boronate transition-state analog inhibitors.

aforementioned enzymes showed that they were poor substrates but good inhibitors leading to complexes that were essentially irreversible.

Other phosphonate inhibitors have been designed, which include cyclic phosphates, such as compounds **65** and **66** (41) (Fig. 8.17). The strategy for the design of these compounds was that the leaving group would remain attached to the enzyme after the nucleophilic attack and result in a hydrolytically stable complex (41), reminiscent of the slow hydrolysis of acyl enzyme species in PBPs. The strategy was successful, as evidenced by the fact that the inhibitors transiently inhibited P99, TEM-1, and a DD-peptidase from *Streptomyces* R61.

Unlike the phosphonates, the boron-containing moiety in some transition-state analogs adopts a trigonal planar configuration, which closely mimics the carbonyl of a β-lactam ring (14). Recently Ness et al. (33) have used molecular modeling to design several variants of boronic acids, an exercise that resulted in the selection and synthesis of compounds **67** and **68** (Fig. 8.18). Kinetic characterization

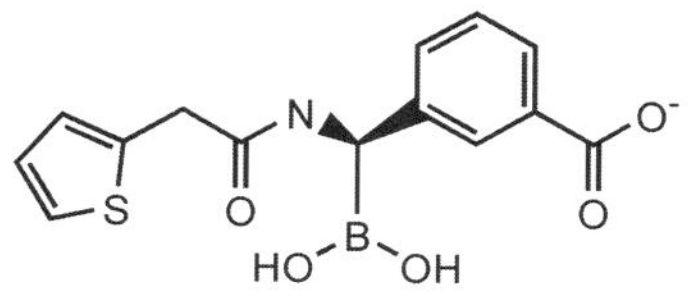

69

Figure 8.19 Chemical structure of a potent boronate inhibitor.

revealed that these compounds inhibit the TEM-1 β-lactamase with K_i values of 5.9 and 13 nM, respectively. The three-dimensional structure of the above compounds was determined by X-ray crystallography, clearly showing the formation of the tetrahedral species and providing insight into the inhibition mechanism of these compounds (33). The structure with compound **68** showed one of the oxygen atoms of the tetrahedral adduct ensconced into the oxyanion hole. A second oxygen atom of the boronate is found to form a hydrogen bond with Glu-166 and Asn-170, displacing a conserved structural water molecule that is found in most X-ray structures of TEM-1 β-lactamases, prompting the authors to conclude that this structure mimics the deacylation transition state.

More recently, the Shoichet group has reported a series of new boronate inhibitors, which were used to probe the structural bases for recognition and resistance (52). Compound **69** (Fig. 8.19) was a potent inhibitor, with an inhibition constant in the low-nanomolar region. Interestingly, this compound—which includes features that make it more substrate-like—is found to inhibit TEM-30 β-lactamase well, which is an inhibitor-resistant TEM (IRT).

CEPHALOSPORIN-BASED INHIBITORS

Cephalosporin-derived inhibitors usually show preference for the class C β-lactamases. A recent study has reported a

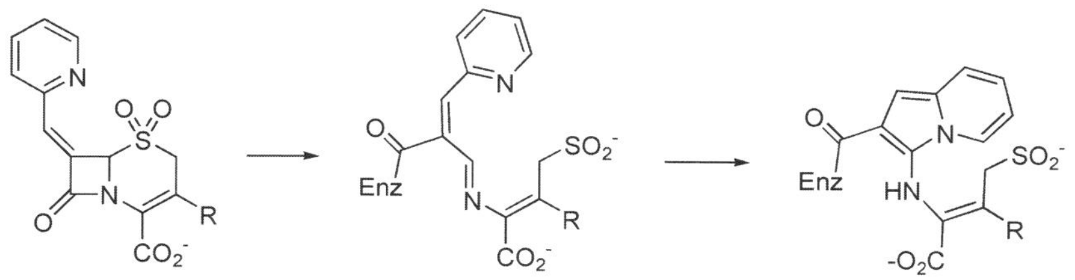

70 R= C=CH-CONH-$CH_2CH_2(CN_3H_4)$
71 R=C=CH-CONH2
72 R=SO-Me
73 R=SO_2Ph
74 R= SPh

75 76

Figure 8.20 Cephalosporin-based inhibitors and the inhibitory mechanism against class C β-lactamases.

series of cephalosporin derivatives that target both the class A and class C β-lactamases (**70** to **74**) (Fig. 8.20); the authors also reported, for the first time, cephalosporin derivatives that are selective for inhibition of class A β-lactamases (**72** and **73**) (10). X-ray crystallography of compound **71** with a class C β-lactamase was recently carried out (13). An imine species, **75**, was proposed to form en route to the final inhibited species **76**, which was the only species that was seen in the X-ray structure. The authors also noted that addition of an electronegative group at C-3 likely played a role in increasing the activity of the compounds towards the class A β-lactamases.

CONCLUDING REMARKS

The efficacy of clavulanic acid, one of the first discovered inhibitors for β-lactamases, has been challenged in the clinic with the identification of inhibitor-resistant β-lactamases. In addition, the increased incidence of resistance mediated by other classes of β-lactamases emphasizes the need for broad-spectrum inhibitors that target more than one class of β-lactamases. Several routes have been taken towards the development of more effective inhibitors including the syntheses of variants of penam sulfones, penems, alkylidenes, monobactams, transition-state analogs, and the boronates, as described in this report.

The continued interest in β-lactam antibiotics, coupled with the fact that we do not have any replacements for them in hand or in the pipeline, necessitates a vigilant search for inhibitors of β-lactamases. The discovery of the next generation of these inhibitors should prolong the utility of β-lactam antibiotics for the foreseeable future.

References

1. **Bitha, P., Z. Li, G. D. Francisco, Y. Yang, P. J. Petersen, E. Lenoy, and Y. I. Lin.** 1999. 6-(1-Hydroxyalkyl)penam sulfone derivatives as inhibitors of class A and class C beta-lactamases II. *Bioorg. Med. Chem. Lett.* **9**:997–1002.
2. **Bonfiglio, G., G. Russo, and G. Nicoletti.** 2002. Recent developments in carbapenems. *Expert Opin. Investig. Drugs* **11**:529–544.
3. **Bradley, J. S., J. Garau, H. Lode, K. V. Rolston, S. E. Wilson, and J. P. Quinn.** 1999. Carbapenems in clinical practice: a guide to their use in serious infection. *Int. J. Antimicrob. Agents* **11**:93–100.
4. **Brown, A. G., D. Butterworth, M. Cole, G. Hanscomb, J. D. Hood, C. Reading, and G. N. Rolinson.** 1976. Naturally-occurring beta-lactamase inhibitors with antibacterial activity. *J. Antibiot.* (Tokyo) **29**:668–669.
5. **Brown, R. P., R. T. Aplin, and C. J. Schofield.** 1996. Inhibition of TEM-2 beta-lactamase from *Escherichia coli* by clavulanic acid: observation of intermediates by electrospray ionization mass spectrometry. *Biochemistry* **35**:12421–12432.
6. **Bulychev, A., J. R. Bellettini, M. O'Brien, P. J. Crocker, J. P. Samama, M. J. Miller, and S. Mobashery.** 2000. N-sulfonyloxy-beta-lactam inhibitors for beta-lactamases. *Tetrahedron* **56**:5719–5728.
7. **Bulychev, A., I. Massova, S. A. Lerner, and S. Mobashery.** 1995. Penem Brl-42715—an effective inactivator for beta-lactamases. *J. Am. Chem. Soc.* **117**:4797–4801.
8. **Buynak, J. D., H. Chen, L. Vogeti, V. R. Gadhachanda, C. A. Buchanan, T. Palzkill, R. W. Shaw, J. Spencer, and T. R. Walsh.** 2004. Penicillin-derived inhibitors that simultaneously target both metallo- and serine-beta-lactamases. *Bioorg. Med. Chem. Lett.* **14**:1299–1304.
9. **Buynak, J. D., A. S. Rao, V. R. Doppalapudi, G. Adam, P. J. Petersen, and S. D. Nidamarthy.** 1999. The synthesis and evaluation of 6-alkylidene-2'beta-substituted penam sulfones as beta-lactamase inhibitors. *Bioorg. Med. Chem. Lett.* **9**: 1997–2002.
10. **Buynak, J. D., K. Wu, B. Bachmann, D. Khasnis, L. Hua, H. K. Nguyen, and C. L. Carver.** 1995. Synthesis and biological activity of 7-alkylidenecephems. *J. Med. Chem.* **38**:1022–1034.
11. **Charnas, R. L., J. Fisher, and J. R. Knowles.** 1978. Chemical studies on inactivation of *Escherichia coli* RTEM beta-lactamase by clavulanic acid. *Biochemistry* **17**:2185-2189.
12. **Chen, C. C. H., and O. Herzberg.** 1992. Inhibition of beta-lactamase by clavulanate. Trapped intermediates in cryocrystallographic studies. *J. Mol. Biol.* **224**:1103–1113.
13. **Crichlow, G. V., M. Nukaga, V. R. Doppalapudi, J. D. Buynak, and J. R. Knox.** 2001. Inhibition of class C beta-lactamases: structure of a reaction intermediate with a cephem sulfone. *Biochemistry* **40**:6233–6239.
14. **Crompton, I. E., B. K. Cuthbert, G. Lowe, and S. G. Waley.** 1988. Beta-lactamase inhibitors. The inhibition of serine beta-lactamases by specific boronic acids. *Biochem. J.* **251**: 453–459.
15. **English, A. R., J. A. Retsema, A. E. Girard, J. E. Lynch, and W. E. Barth.** 1978. CP-45,899, a beta-lactamase inhibitor that extends the antibacterial spectrum of beta-lactams: initial bacteriological characterization. *Antimicrob. Agents Chemother.* **14**:414–419.
16. **Farmer, T. H., J. W. J. Page, D. J. Payne, and D. J. C. Knowles.** 1994. Kinetic and physical studies of beta-lactamase inhibition by a novel penem, Brl-42715. *Biochem. J.* **303**: 825–830.
17. **Fisher, J.** 1984. β-Lactams resistant to hydrolysis by the β-lactamases, p. 33–79. *In* L. E. Bryan (ed.), *Antimicrobial Drug Resistance*. Academic Press, Orlando, Fla.
18. **Helfand, M. S., M. A. Totir, M. P. Carey, A. M. Hujer, R. A. Bonomo, and P. R. Carey.** 2003. Following the reactions of mechanism-based inhibitors with beta-lactamase by Raman crystallography. *Biochemistry* **42**:13386–13392.
19. **Imtiaz, U., E. Billings, J. R. Knox, E. K. Manavathu, S. A. Lerner, and S. Mobashery.** 1993. Inactivation of class-A beta-lactamases by clavulanic acid—the role of arginine-244 in a proposed nonconcerted sequence of events. *J. Am. Chem. Soc.* **115**:4435–4442.
20. **Jelsch, C., F. Lenfant, J. M. Masson, and J. P. Samama.** 1992. Beta-lactamase TEM1 of *E. coli*. Crystal structure determination at 2.5 A resolution. *FEBS Lett.* **299**:135–142.

21. **Kahan, J. S., F. M. Kahan, R. Goegelman, S. A. Currie, M. Jackson, E. O. Stapley, T. W. Miller, A. K. Miller, D. Hendlin, S. Mochales, S. Hernandez, H. B. Woodruff, and J. Birnbaum.** 1979. Thienamycin, a new beta-lactam antibiotic. I. Discovery, taxonomy, isolation and physical properties. *J. Antibiot.* (Tokyo) **32:**1–12.

22. **Knowles, J. R.** 1985. Penicillin resistance. The chemistry of beta-lactamase inhibition. *Acc. Chem. Res.* **18:**97–104.

23. **Knox, J. R., P. C. Moews, W. A. Escobar, and A. L. Fink.** 1993. A catalytically-impaired class-a beta-lactamase: 2 A crystal structure and kinetics of the *Bacillus licheniformis* E166a mutant. *Protein Eng.* **6:**11–18.

24. **Kuzin, A. P., M. Nukaga, Y. Nukaga, A. Hujer, R. A. Bonomo, and J. R. Knox.** 2001. Inhibition of the SHV-1 beta-lactamase by sulfones: crystallographic observation of two reaction intermediates with tazobactam. *Biochemistry* **40:**1861–1866.

25. **Li, N. X., and R. F. Pratt.** 1998. Inhibition of serine beta-lactamases by acyl phosph(on)ates: new source of inert acyl [and phosphyl] enzymes. *J. Am. Chem. Soc.* **120:**4264–4268.

26. **Livermore, D. M., and H. Y. Chen.** 1997. Potentiation of beta-lactams against *Pseudomonas aeruginosa* strains by Ro 48-1256, a bridged monobactam inhibitor of AmpC beta-lactamases. *J. Antimicrob. Chemother.* **40:**335–343.

27. **Matagne, A., P. Ledent, D. Monnaie, A. Felici, M. Jamin, X. Raquet, M. Galleni, D. Klein, I. Francois, and J. M. Frere.** 1995. Kinetic study of interaction between BRL 42715, beta-lactamases, and D-alanyl-D-alanine peptidases. *Antimicrob. Agents Chemother.* **39:**227–231.

28. **Maveyraud, L., I. Massova, C. Birck, K. Miyashita, J. P. Samama, and S. Mobashery.** 1996. Crystal structure of 6 alpha-(hydroxymethyl)penicillanate complexed to the TEM-1 beta-lactamase from *Escherichia coli*: evidence on the mechanism of action of a novel inhibitor designed by a computer-aided process. *J. Am. Chem. Soc.* **118:**7435–7440.

29. **Maveyraud, L., L. Mourey, L. P. Kotra, J. D. Pedelacq, V. Guillet, S. Mobashery, and J. P. Samama.** 1998. Structural basis for clinical longevity of carbapenem antibiotics in the face of challenge by the common class A beta-lactamases from the antibiotic-resistant bacteria. *J. Am. Chem. Soc.* **120:**9748–9752.

30. **Miyashita, K., I. Massova, P. Taibi, and S. Mobashery.** 1995. Design, synthesis, and evaluation of a potent mechanism-based inhibitor for the TEM beta-lactamase with implications for the enzyme mechanism. *J. Am. Chem. Soc.* **117:**11055–11059.

31. **Moews, P. C., J. R. Knox, O. Dideberg, P. Charlier, and J. M. Frere.** 1990. Beta-lactamase of *Bacillus licheniformis* 749/C at 2 A resolution. *Proteins* **7:**156–171.

32. **Mourey, L., K. Miyashita, P. Swaren, A. Bulychev, J. P. Samama, and S. Mobashery.** 1998. Inhibition of the NMC-A beta-lactamase by a penicillanic acid derivative and the structural bases for the increase in substrate profile of this antibiotic resistance enzyme. *J. Am. Chem. Soc.* **120:**9382–9383.

33. **Ness, S., R. Martin, A. M. Kindler, M. Paetzel, M. Gold, S. E. Jensen, J. B. Jones, and N. C. J. Strynadka.** 2000. Structure-based design guides the improved efficacy of deacylation transition state analogue inhibitors of TEM-1 beta-lactamase. *Biochemistry* **39:**5312–5321.

34. **Nishida, K., C. Kunugita, T. Uji, F. Higashitani, A. Hyodo, N. Unemi, S. N. Maiti, O. A. Phillips, P. Spevak, K. P. Atchison, S. M. Salama, H. Atwal, and R. G. Micetich.** 1999. In vitro and in vivo activities of Syn2190, a novel beta-lactamase inhibitor. *Antimicrob. Agents Chemother.* **43:**1895–1900.

35. **Nukaga, M., T. Abe, A. M. Venkatesan, T. S. Mansour, R. A. Bonomo, and J. R. Knox.** 2003. Inhibition of class A and class C beta-lactamases by penems: crystallographic structures of novel 1,4-thiazepine intermediate. *Biochemistry* **42:**13152–13159.

36. **Padayatti, P. S., M. S. Helfand, M. A. Totir, M. P. Carey, A. M. Hujer, P. R. Carey, R. A. Bonomo, and F. van den Akker.** 2004. Tazobactam forms a stoichiometric trans-enamine intermediate in the E166A variant of SHV-1 beta-lactamase: 1.63 angstrom crystal structure. *Biochemistry* **43:**843–848.

37. **Page, M. G.** 2000. β-Lactamase inhibitors. *Drug. Resist. Updat.* **3:**109–125.

38. **Page, M. I., and A. P. Laws.** 1998. The mechanism of catalysis and the inhibition of beta-lactamases. *Chem. Commun.* **1998:**1609–1617.

39. **Phillips, O. A., D. P. Czajkowski, P. Spevak, M. P. Singh, C. Hanehara-Kunugita, A. Hyodo, R. G. Micetich, and S. N. Maiti.** 1997. SYN-1012: a new beta-lactamase inhibitor of penem skeleton. *J. Antibiot.* (Tokyo) **50:**350–356.

40. **Pratt, R. F.** 1989. Inhibition of a class C beta-lactamase by a specific phosphonate monoester. *Science* **246:**917–919.

41. **Pratt, R. F., and N. J. Hammar.** 1998. Salicyloyl cyclic phosphate, a "penicillin-like" inhibitor of beta-lactamases. *J. Am. Chem. Soc.* **120:**3004–3006.

42. **Rahil, J., and R. F. Pratt.** 1992. Mechanism of inhibition of the class C beta-lactamase of Enterobacter cloacae P99 by phosphonate monoesters. *Biochemistry* **31:**5869–5878.

43. **Rice, L. B., and R. A. Bonomo.** 2000. β-Lactamases: which ones are clinically important? *Drug Resist. Updat.* **3:**178–189.

44. **Sandanayaka, V. P., G. B. Feigelson, A. S. Prashad, Y. J. Yang, and P. J. Petersen.** 2001. Allyl and propargyl substituted penam sulfones as versatile intermediates toward the syntheses of new beta-lactamase inhibitors. *Bioorg. Med. Chem. Lett.* **11:**997–1000.

45. **Sandanayaka, V. P., and A. S. Prashad.** 2002. Resistance to beta-lactam antibiotics: structure and mechanism based design of beta-lactamase inhibitors. *Curr. Med. Chem.* **9:**1145–1165.

46. **Sandanayaka, V. P., and Y. J. Yang.** 2000. Dipolar cycloaddition of novel 6-(nitrileoxidomethyl) penam sulfone: an efficient route to a new class of beta-lactamase inhibitors. *Org. Lett.* **2:**3087–3090.

47. **Shah, P. M., and R. D. Isaacs.** 2003. Ertapenem, the first of a new group of carbapenems. *J. Antimicrob. Chemother.* **52:**538–542.

48. **Strynadka, N. C., H. Adachi, S. E. Jensen, K. Johns, A. Sielecki, C. Betzel, K. Sutoh, and M. N. James.** 1992. Molecular structure of the acyl-enzyme intermediate in beta-lactam hydrolysis at 1.7 A resolution. *Nature* **359:**700–705.

49. **Tabei, K., X. Feng, A. M. Venkatesan, T. Abe, U. Hideki, T. S. Mansour, and M. M. Siegel.** 2004. Mechanism of inactivation of beta-lactamases by novel 6-methylidene

penems elucidated using electrospray ionization mass spectrometry. *J. Med. Chem.* **47**:3674–3688.

50. **Taibi, P., and S. Mobashery.** 1995. Mechanism of turnover of imipenem by the TEM beta-lactamase revisited. *J. Am. Chem. Soc.* **117**:7600–7605.
51. **Therrien, C., and R. C. Levesque.** 2000. Molecular basis of antibiotic resistance and beta-lactamase inhibition by mechanism-based inactivators: perspectives and future directions. *FEMS Microbiol. Rev.* **24**:251–262.
52. **Wang, X. J., G. Minasov, and B. K. Shoichet.** 2002. Noncovalent interaction energies in covalent complexes: TEM-1 beta-lactamase and beta-lactams. *Protein. Struct. Funct. Genet.* **47**:86–96.
53. **Wise, R., J. M. Andrews, and N. Patel.** 1981. 6-Beta-bromo- and 6-beta-iodo penicillanic acid, two novel beta-lactamase inhibitors. *J. Antimicrob. Chemother.* **7**:531–536.
54. **Yang, Y., K. Janota, K. Tabei, N. Huang, M. M. Siegel, Y. I. Lin, B. A. Rasmussen, and D. M. Shlaes.** 2000. Mechanism of inhibition of the class A beta-lactamases PC1 and TEM-1 by tazobactam. Observation of reaction products by electrospray ionization mass spectrometry. *J. Biol. Chem.* **275**: 26674–26682.

Enzyme-Mediated Resistance to Antibiotics: Mechanisms, Dissemination, and Prospects for Inhibition
Edited by Robert A. Bonomo and Marcelo E. Tolmasky

Gian Maria Rossolini
Jean-Denis Docquier

9 Class B β-Lactamases

Class B β-lactamases (CBBLs) are bacterial enzymes that degrade β-lactam antibiotics with the help of a metal cofactor (zinc in the natural form), and for this reason, they are also referred to as metallo-β-lactamases (MBLs). Although catalyzing the same reaction as serine-β-lactamases, CBBLs are structurally and mechanistically unrelated to the former enzymes, and their common function apparently represents an example of convergent evolution of different protein lineages within the bacterial domain.

The first CBBL was discovered in the mid-1960s (i.e., some 25 years later than serine-β-lactamases) in the gram-positive spore-forming rod *Bacillus cereus* (148). For a relatively long time these enzymes were regarded more as biochemical curiosities than as resistance determinants of any clinical importance. The lack of structural and mechanistic relatedness with serine-β-lactamases was indeed interesting for enzymologists and protein scientists, but the occurrence in a virtually nonpathogenic species did not draw much attention from clinicians and clinical microbiologists. During the 1980s, the discovery of CBBLs in species of clinical relevance (*Stenotrophomonas maltophilia* [150], some strains of *Bacteroides fragilis* [36], and *Aeromonas* spp. [72]) gradually changed this view. An increasing interest for these enzymes was also prompted by their functional properties, namely, potent carbapenemase activity and unsusceptibility to conventional β-lactamase inhibitors (clavulanate and penicillanic acid sulfones). The carbapenemase activity, in particular, had fearsome implications since it could endanger the efficacy of carbapenems, the last generation of β-lactams that showed an exceedingly broad spectrum of activity and an unprecedented stability to the vast majority of serine-β-lactamases. However, it was not until the 1990s, with the appearance of broad-spectrum CBBLs encoded by genes associated to mobile DNA elements in clinical isolates of major nosocomial pathogens (such as *Pseudomonas aeruginosa*, *Acinetobacter* spp., and members of the family *Enterobacteriaceae*), that these enzymes were included among the most worrisome resistance determinants emerging in gram-negative pathogens (19, 77).

This short historical background accounts for the increasing attention that CBBLs have recently attracted and for the several efforts that are currently under way at understanding their mechanistic aspects and at finding new inhibitors. The scope of this chapter is to provide an overview of CBBLs, considering both fundamental and clinical aspects.

DIVERSITY, EPIDEMIOLOGY, AND GENETICS OF CBBLs

Class B is one of the four classes in the structural classification of β-lactamases, which was created to accommodate the MBLs (2). Currently, more than 100 CBBLs have been

Gian Maria Rossolini, Dipartimento di Biologia Molecolare, Università di Siena, Siena, I-53100, Italy. **Jean-Denis Docquier,** Centre d'Ingénierie des Protéines & Laboratoire d'Enzymologie, Université de Liège, Liège, B-4000, Belgium.

described, including several members of quite divergent protein lineages plus a number of minor allelic variants for most of them (Table 9.1, Fig. 9.1).

According to structural relatedness, CBBLs have been grouped in three different subclasses: B1, B2, and B3 (Fig. 9.1). Subclass B1 is the most numerous and includes (i) the β-lactamase II from *Bacillus cereus* (Bc-II) and the closely related enzymes from *Bacillus anthracis* and other *Bacillus* spp.; (ii) the CcrA/CfiA enzyme and variants, present in a cluster of *Bacteroides fragilis*; (iii) the large group of related enzymes from members of the order *Flavobacteriales*, including the BlaB enzyme and variants from *Elizabethkingia meningoseptica* (formerly *Chryseobacterium meningosepticum*), the IND-1 enzyme and variants from *Chryseobacterium indologenes*, the CGB-1 enzyme from *Chryseobacterium gleum*, the EBR-1 enzyme from *Empedobacter brevis*, the JOHN-1 enzyme from *Flavobacterium johnsoniae*, and the TUS-1 and MUS-1 enzymes from *Myroides odoratus* and *Myroides odoratimimus*, respectively; (iv) the IMP, VIM, SPM, GIM, and SIM enzymes, encoded by genes associated with mobile DNA elements, that have been found in isolates of *Enterobacteriaceae*, *Pseudomonas* spp., *Acinetobacter* spp., and other nonfastidious gram-negative nonfermenters (GNNFs) (Table 9.1). Subclass B2 is the less numerous and includes the CphA enzyme and related variants from some *Aeromonas* spp., the Sfh-I enzyme found in an isolate of *Serratia fonticola*, and the CVI-1 enzyme from *Chromobacterium violaceum* (Table 9.1). Subclass B3 is quite divergent from the others and includes (i) the L1 enzyme and variants from *Stenotrophomonas maltophilia*; (ii) the FEZ-1 enzyme from *Fluoribacter* (formerly *Legionella*) *gormanii*; (iii) the GOB-1 enzyme and variants from *E. meningoseptica*; (iv) the THIN-B enzyme from *Janthinobacterium lividum*; (v) the CAU-1 enzyme and variants from *Caulobacter vibrioides* (formerly *Caulobacter crescentus*); (vi) the NOV-1 enzyme from *Novosphingobium aromaticivorans*; and (vii) the BJP-1 enzyme from *Bradyrhizobium japonicum* (Table 9.1).

Amino acid divergence (considering the entire protein and pairwise alignments) can be as high as 79% within subclass B1, 51% within subclass B2, and 78% within subclass B3, while between members of subclasses B1 and B3 divergence can be as high as 89%. The degree of diversity among CBBLs is striking in view of the conservation of the β-lactamase function and underscores the remarkable plasticity of metalloprotein structures that can catalyze degradation of β-lactam compounds.

Amino acid sequence alignments reveal that there are 17 highly conserved residues among members of subclass B1 (His116, His118, Asp120, Thr142, Gly193, His196, Asp199, Asn200, Val202, Leu217, Gly219, Gly220, Cys221, Gly232, Trp244, His263, and Thr303; the numbering is according to the BBL scheme [52, 54]) and 16 highly conserved residues among members of subclass B3 (Pro45, Gly56, Thr57, Gly79, Gly103, Asp108, His118, Asp120, His121, Ala134, Gly149, Gly183, Gly195, His196, Gly199, His 263). However, only four of these residues (His118, Asp120, His196, and His263) are strictly conserved among all CBBLs (Fig. 9.2), all of which are known to be involved in metal ion coordination (see below). Interestingly, the enzymes of subclass B1 show an overall higher degree of conservation in the C-terminal domain, while the enzymes of subclass B3 show higher conservation in the amino-terminal domain (Fig. 9.2).

Most CBBLs are encoded by genes that appear to be resident in the genome of some bacterial species (Table 9.1). The resident CBBLs are encoded by chromosomal genes, with the exception of that encoding the L1 enzyme in *S. maltophilia*, for which location on a large plasmid has been claimed (5). However, also the latter gene behaves as a typically resident determinant, and transferability to other species has never been reported. In *B. fragilis* the *cfiA* gene is present in only one monophyletic cluster of that species, suggesting a history of recent acquisition followed by genomic fixation (146).

Resident CBBL genes are usually organized as monogenic units. An interesting exception to this rule is represented by the gene encoding the CAU-1 enzyme in *C. vibrioides*, which is part of an operon-like structure consisting of four different genes: the first encoding a putative bifunctional protein whose amino-terminal and carboxy-terminal moieties exhibit similarity to transcriptional regulators of the ArsR family and to *S*-adenosylmethionine-dependent methyltransferases, respectively; the second encoding a homolog to the MetF family of bacterial methylenetetraidrolofale reductases; the third encoding CAU-1; and the fourth encoding a homolog to the amino-terminal methyltransferase domain of the MetH family of bacterial cobalamin-dependent methionine synthases (45). A similar arrangement might suggest that, in this case, the MBL could also be involved in some other function, but this issue remains to be clarified.

In some cases the expression of resident CBBLs is regulated, being low under basal conditions and inducible upon exposure to β-lactams. A similar regulated expression has been shown for the CBBLs of *Aeromonas* spp. (72), *S. maltophilia* (145, 150), and *J. lividum* (139). In *Aeromonas hydrophila*, where the regulatory system has been investigated in more detail, expression of the CBBL is coordinated with that of two other resident β-lactamases (a class C and a class D enzyme) and is apparently under the control of a two-component regulatory system (6, 116). In other instances the expression of resident CBBLs is

Table 9.1 Class B β-lactamases

Enzyme[a]	No. of allelic variants[b]	Host(s)	Accession[c] no.	Relevant reference(s)[d]	Notes[e]
Subclass B1					
Bc-II/Bla2	3	*Bacillus cereus*	P04190	71	Resident in some *Bacillus* species (*B. cereus*, *B. anthracis*)
	2	*Bacillus anthracis*		25	
	1	Alkalophilic *Bacillus* sp.		81	
CcrA/CfiA	8	*Bacteroides fragilis*	P25910	159	Resident in a monophyletic cluster of *B. fragilis*
BlaB	11	*Elizabethkingia meningoseptica*	CAA65601	141	Resident
IND	5	*Chryseobacterium indologenes*	AAD20273	10	Resident
CGB	1	*Chryseobacterium gleum*	AAL55263	11	Likely resident; according to sequence similarity, the CGB enzyme could actually be considered an allelic variant of IND
EBR	1	*Empedobacter brevis*	AAN32638	9	Likely resident
JOHN	1	*Flavobacterium johnsoniae*	AAK38324	115	Likely resident
TUS	1	*Myroides odoratus*	AAN63648	102	Likely resident
MUS	1	*Myroides odoratimimus*	AAN63647	102	Likely resident
IMP	17	GNNFs, *Enterobacteriaceae*	P52699	120	Present in some isolates; encoded by genes associated with mobile DNA (acquired by horizontal transfer)
VIM	11	GNNFs, *Enterobacteriaceae*	CAB46686	91	Present in some isolates; encoded by genes associated with mobile DNA (acquired by horizontal transfer)
SPM	1	*Pseudomonas aeruginosa*	CAD37801	163	Present in some isolates; encoded by a gene associated with mobile DNA (acquired by horizontal transfer)
GIM	1	*Pseudomonas aeruginosa*	CAF05908	24	Present in some isolates; encoded by genes associated with mobile DNA (acquired by horizontal transfer)
SIM	1	*Acinetobacter baumannii*	AAX76774	96a	Present in some isolates; encoded by genes associated with mobile DNA (acquired by horizontal transfer)
Subclass B2					
CphA/Imi	14	*Aeromonas* spp.	P26918	103	Resident in *A. hydrophila*, *A. veronii* (bv. sobria and bv. veronii), and *A. jandaei*; likely resident in *A. salmonicida* and *A. allosaccharophila*
Sfh	1	*Serratia fonticola*	AAF09244	147	Not resident; only detected in one environmental isolate
CVI	1	*Chromobacterium violaceum*	AAQ60817	—[f]	Likely resident
Subclass B3					
L1	11	*Stenotrophomonas maltophilia*	P52700	171	Resident
FEZ	1	*Fluoribacter* (*Legionella*) *gormanii*	CAB96921	16	Likely resident
GOB	8	*Elizabethkingia meningoseptica*	AAF04458	8	Resident
THIN-B	1	*Janthinobacterium lividum*	CAC33832	139	Resident
CAU/MblB	3	*Caulobacter vibrioides* (*crescentus*)	CAC87665	45, 157	Resident
NOV	1	*Novosphingobium aromaticivorans*	ZP_00303740	—[g]	Likely resident
BJP	1	*Bradyrhizobium japonicum*	NP_772870	157a	Likely resident

[a] When one enzyme has been reported by two names, both names are reported.
[b] Allelic variants can differ by single or several amino acid residues; the degree of divergence between allelic variants is lower than that with other enzymes.
[c] The accession number for the protein sequence of the first identified variant is reported; the sequences of the other variants can easily be retrieved using the Blink utility of the Protein database (http://www.ncbi.nlm.nih.gov/), and corresponding references can be retrieved in the database entries.
[d] The reference in which the first variant was described.
[e] The enzyme is reported as resident when the gene has been detected in two or more strains of the species but not outside the species (or outside the genus in the cases of Bc-II and of CphA/Imi) and there is no evidence that it is associated with mobile DNA elements; or it is reported as likely resident when the gene has been detected in a single strain of the species but not outside the species and there is no evidence that it is associated to mobile DNA elements.
[f] J. D. Docquier et al., *Abstr. 44th Intersci. Conf. Antimicrob. Agents Chemother.*, abstr. C1-289, 2004.
[g] J. D. Docquier et al., *Abstr. 14th Eur. Congr. Clin. Microbiol. Infect. Dis.*, abstr. P1460, 2004.

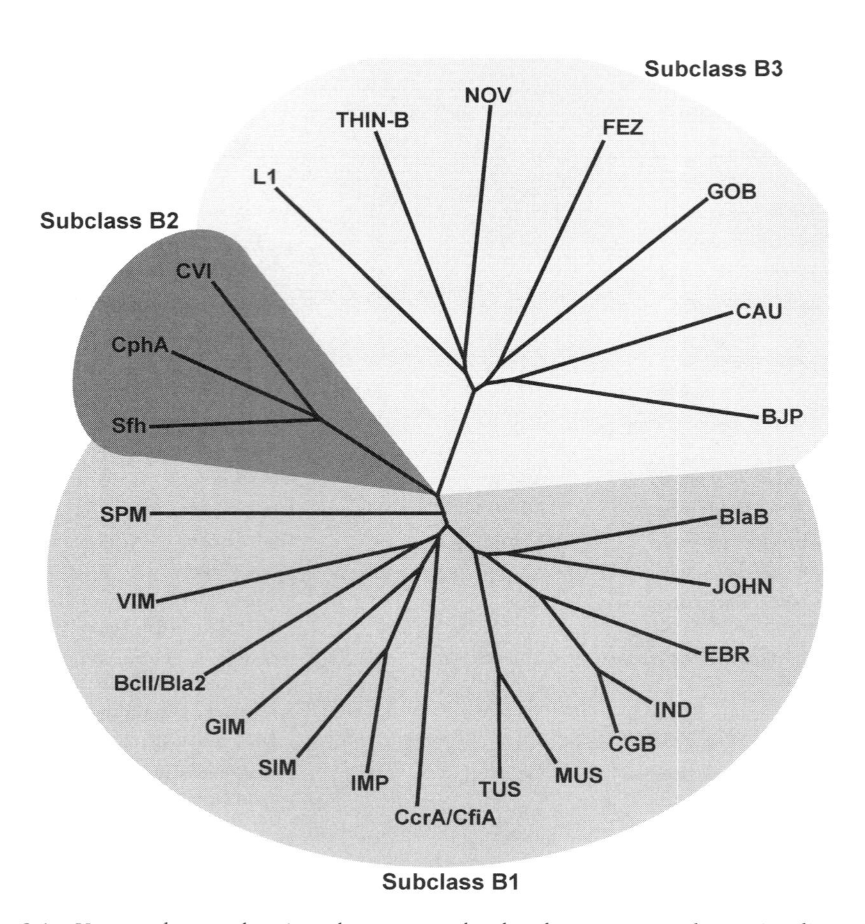

Figure 9.1 Unrooted tree showing the structural relatedness among the major lineages of CBBLs. The names of the enzymes are the same as in Table 9.1. The tree was constructed using the TREEVIEW program (121) on the basis of a sequence alignment constructed with the CLUSTAL X program (158) using the sequences whose accession numbers are reported in the fourth column of Table 9.1.

apparently constitutive, as it is the case for the BlaB enzyme of *E. meningoseptica* (140) and for other flavobacterial enzymes (CGB-1 from *C. gleum* and EBR-1 from *E. brevis*) (9, 11). In *B. fragilis,* the *cfiA/ccrA* gene is normally expressed at very low levels and expression is not inducible upon exposure to β-lactams (135). However, high-level expression of the gene can be achieved by the transposition of insertion sequence elements upstream of the ribosome-binding site, which can provide efficient signals for transcription initiation (82, 127). Regulation has not been specifically investigated for other resident CBBL genes.

Five enzymes, namely IMP, VIM, SPM, GIM, and SIM, are encoded by genes associated with mobile DNA elements, and are currently spreading by horizontal transfer among *P. aeruginosa* and other GNNFs, as well as in members of the family *Enterobacteriaceae* (Table 9.1).

The IMP-1 enzyme was originally detected in Japan, but IMP-type enzymes were subsequently reported also in other areas of the Far East, as well as in Europe, the Americas, and Australia (Table 9.2). Several different allelic variants have been described, belonging in various sublineages, that can be divergent as much as 22% from

Figure 9.2 Amino acid sequence alignment of CBBLs of subclass B1 (section A), subclass B2 (section B), and subclass B3 (section C). The alignments were constructed considering a single representative for each major lineage, using the sequences whose accession numbers are reported in the fourth column of Table 9.1. Structural elements are shown above the alignment for enzymes of each subclass, based on the three-dimensional structures of the Bc-II (23), CphA (53), and L1 (167) enzymes, respectively. The conserved residues shared by enzymes of subclass B1 and those shared by enzymes of subclass B3 are boxshaded in gray (note that the Trp244 is not conserved in IMP-18). The four residues conserved among all CBBLs are boxshaded in black. Numbering is according to the BBL scheme (52, 54).

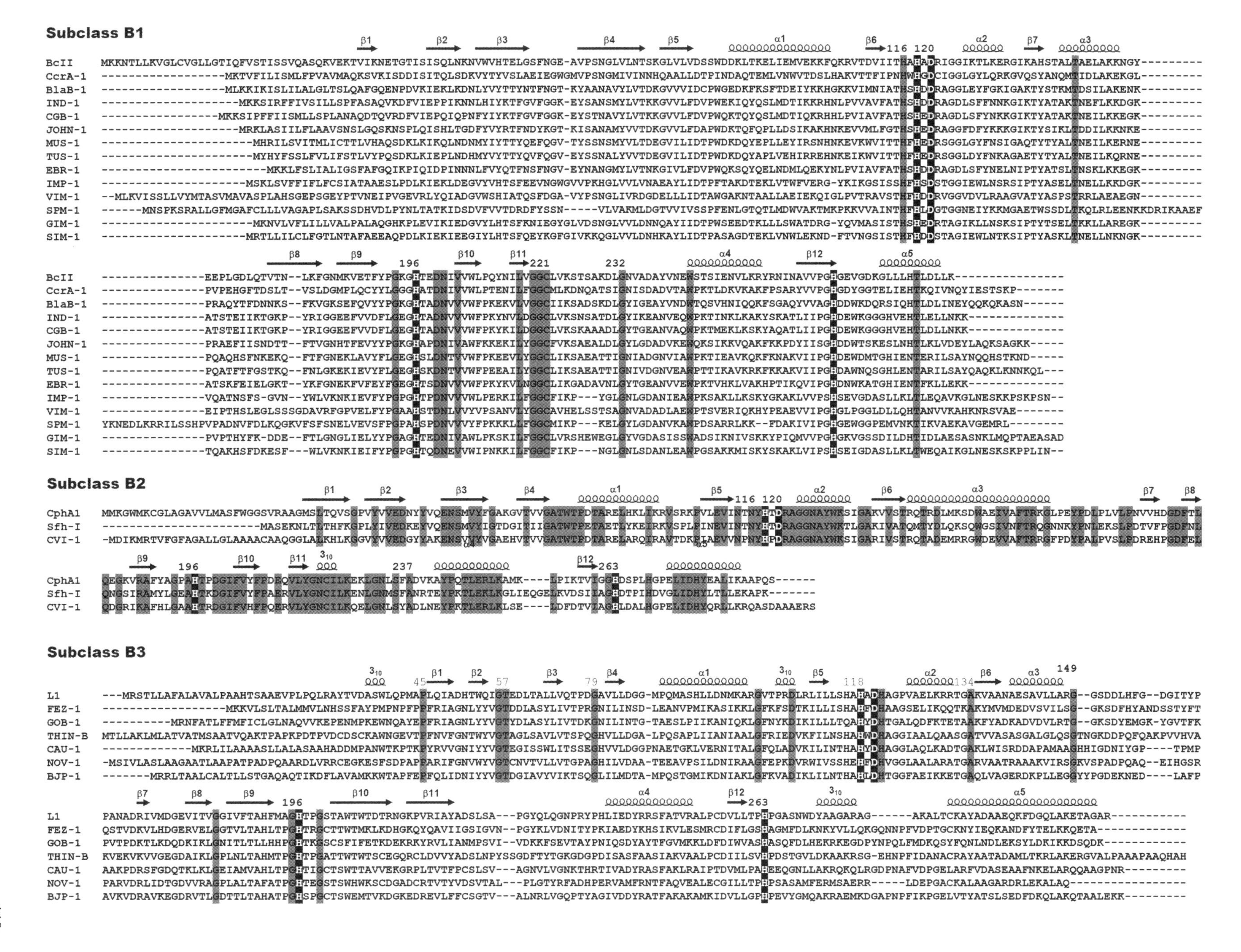
Subclass B1
Subclass B2
Subclass B3

Table 9.2 IMP-type enzymes: sublineages, allelic variants, and distribution

Sublineage[a]	Allelic variant[b]	Accession no.[c]	Geographic area	Host(s)	Reference(s)
IMP-1	IMP-1	S71932	Japan, Singapore, Korea, Taiwan, Argentina, Brazil, United Kingdom	*Pseudomonas* spp., *Acinetobacter* spp., *Burkholderia cepacia, Alcaligenes faecalis, Enterobacteriaceae*	68, 84, 94, 118, 120, 149, 154, 155, 166, 178
	IMP-3	AB010417	Japan	*Shigella flexneri*	119
	IMP-6	AB040994	Japan	*Serratia marcescens*	183
	IMP-10	AB074433	Japan	*Pseudomonas aeruginosa, Achromobacter xylosoxydans*	75
IMP-2	IMP-2	AJ243491	Italy, Japan	*Pseudomonas aeruginosa, Acinetobacter baumannii*	136, 155
	IMP-8	AF322577	Taiwan	*Klebsiella pneumoniae, Enterobacter cloacae*	179, 181
	IMP-19	AB184976	Japan	*Pseudomonas* spp., *Acinetobacter baumannii, Enterobacter cloacae, Achromobacter xylosoxydans*	155
	IMP-20	AB196988	Japan	*Pseudomonas aeruginosa*	155
IMP-4		AF288045	China, Hong Kong, Australia	*Citrobacter youngae, Acinetobacter baumannii, Pseudomonas aeruginosa*	27, 64, 70, 125
IMP-5		AF290912	Portugal	*Acinetobacter baumannii*	39
IMP-7		AF318077	Canada, Malaysia	*Pseudomonas aeruginosa*	60, 69
IMP-9		AY033653	China	*Pseudomonas aeruginosa*	175a
IMP-11		AB074437	Japan	*Pseudomonas aeruginosa, Acinetobacter baumannii*	79
IMP-12		AJ420864	Italy	*Pseudomonas putida*	46
IMP-13		AJ550807	Italy	*Pseudomonas aeruginosa*	160
IMP-16		AJ584652	Brazil	*Pseudomonas aeruginosa*	107
IMP-18		AY780674	United States	*Pseudomonas aeruginosa*	63a

[a] Sublineages are defined as allelic variants or groups of closely related variants that differ from the other sublineages by more than 2% of amino acid residues. Each sublineage was named after the first allelic variant identified.
[b] The IMP-14 and IMP-15 variants have been assigned (http://www.lahey.org/studies/webt.htm), but the sequences have not yet been released.
[c] The accession number for the first reported nucleotide sequence.

each other at the level of their primary structure (IMP-9 and IMP-18). For some sublineages, minor variants are also known (Table 9.2; Fig. 9.3). Different IMP variants tend to exhibit a geographic distribution, although some of them have been detected in very different geographic areas (Table 9.2), likely reflecting the ability for long-haul spread of these resistance determinants. To date, the IMP-type enzymes have been reported in several members of the family *Enterobacteriaceae*, in *Pseudomonas* spp., and in several other species of GNNFs (Table 9.2).

The VIM-1 enzyme was originally detected in Italy, but VIM-type enzymes were subsequently reported also in other European countries as well as in Turkey, the Far East, and the Americas (Table 9.3). Several different allelic variants have been described, belonging in three sublineages, that can be divergent as much as 27% from each other (between VIM-3/6 and VIM-7). For the VIM-1 and VIM-2 sublineages, minor variants are also known (Table 9.3; Fig. 9.4). Different VIM variants tend to exhibit a geographical distribution, except for VIM-2 that was found to be widespread in Europe and the Far East, and also present in the Americas (Table 9.3), a finding which points to a remarkable spreading attitude of this resistance determinant. To date, the VIM-type enzymes have been reported in several members of the family *Enterobacteriaceae*, in *Pseudomonas* spp., and in several other species of GNNFs (Table 9.3).

The SPM-1 enzyme was first described in a *P. aeruginosa* isolate from Brazil (163), and it was subsequently shown to be widespread in carbapenem-resistant *P. aeruginosa* isolates from that country (51). Recently, the GIM-1 and SIM-1 enzymes were reported in *P. aeruginosa*

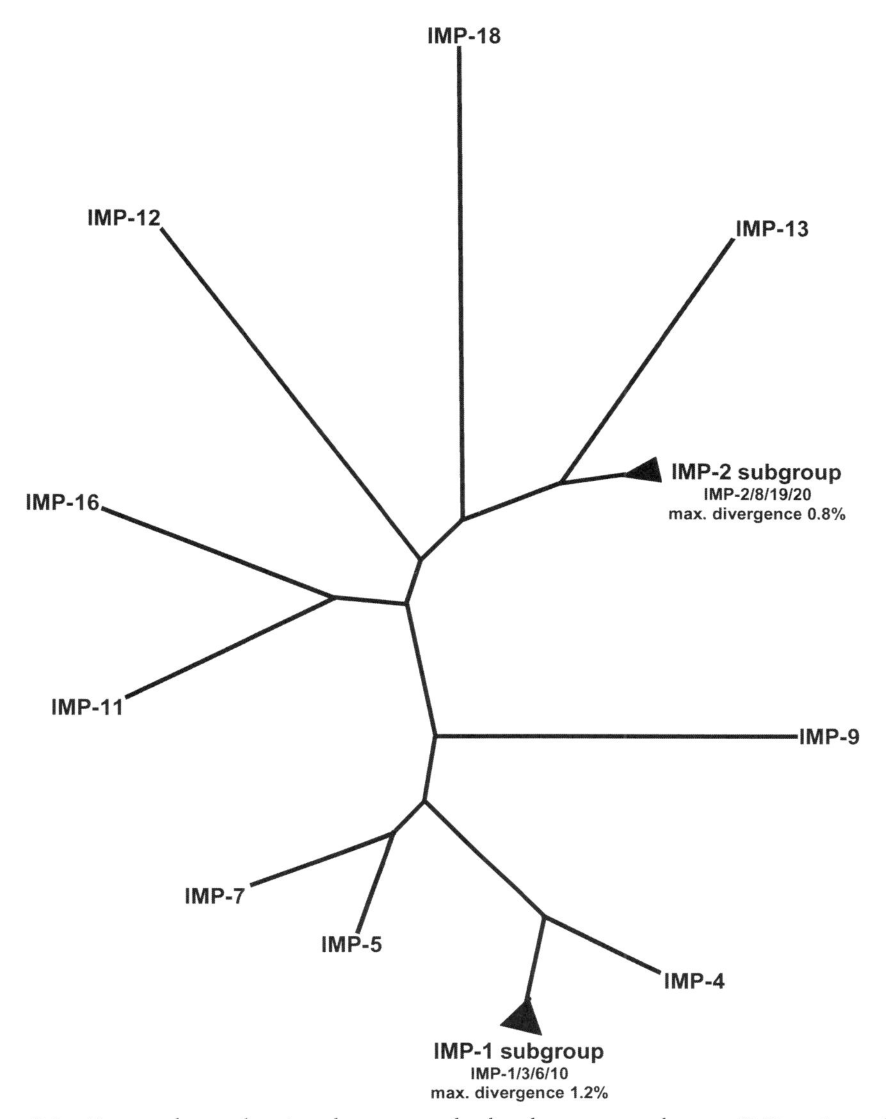

Figure 9.3 Unrooted tree showing the structural relatedness among known IMP variants. The tree was constructed using the TREEVIEW program (121) on the basis of a sequence alignment constructed with the CLUSTAL X program (158) using the sequences whose accession numbers are reported in the third column of Table 9.2.

isolates from Germany (24) and *A. baumannii* isolates from Korea (69a), respectively.

The bla_{IMP}, bla_{VIM}, bla_{GIM}, and bla_{SIM} genes are carried on mobile gene cassettes inserted into integrons (24, 96a, 117, and references therein), which accounts for their remarkable potential for horizontal spreading. The $bla_{\mathrm{SPM\text{-}1}}$ gene was found to be carried on large plasmids, but it is not associated with an integron system (128, 163). Instead, it was found to be preceded by a putative recombinase gene (*orf495*) related to a family of putative recombinase genes that also includes *orf513*, which is associated with several types of antibiotic resistance determinants (128). CBBL gene cassettes were usually found in class 1 integrons (24, 96a, 117, and references therein). Occasionally, the $bla_{\mathrm{IMP\text{-}1}}$ gene cassette has also been found inserted in class 3 integrons (3, 155). The structure of several integrons containing CBBL gene cassettes has been determined, revealing a remarkable diversity of the variable regions (Fig. 9.5). These findings indicate that the integron recombination system has largely been exploited for the dissemination of these resistance determinants. Moreover, the fact that CBBL gene cassettes are usually found in the first position of the integron cassette array (Fig. 9.5) suggests that the capture of these cassettes in integrons

Table 9.3 VIM-type enzymes: sublineages, allelic variants, and distribution

Sublineage[a]	Allelic variant[b]	Accession no.	Geographic area	Host(s)	Reference(s)
VIM-1	VIM-1	Y18050	Italy, Greece	*Pseudomonas* spp., *Achromobacter xylosoxydans*, *Escherichia coli*, *Klebsiella pneumoniae*	59, 91, 99, 110, 137
	VIM-4	AF531419	Italy, Greece, Sweden, Poland, Hungary	*Pseudomonas aeruginosa*, *Klebsiella pneumoniae*, *Enterobacter cloacae*	62, 97a, 100, 124, 131
	VIM-5	AY144612	Turkey	*Klebsiella pneumoniae*, *Pseudomonas aeruginosa*	7
VIM-2	VIM-2		Europe, Far East, South America, United States	*Pseudomonas* spp., *Acinetobacter* spp., *Enterobacteriaceae*	22, 29, 78, 95, 105, 122, 129, 132, 149, 151, 155, 174, 178, 179, 184, 186, 98a
	VIM-3	AF300454	Taiwan	*Pseudomonas aeruginosa*	178
	VIM-6	AY165025	Singapore	*Pseudomonas putida*	85
	VIM-8	AY524987	Colombia	*Pseudomonas aeruginosa*	34
	VIM-9	AY524988	United Kingdom	*Pseudomonas aeruginosa*	Woodford et al., ICAAC
	VIM-10	AY524989	United Kingdom	*Pseudomonas aeruginosa*	Woodford et al., ICAAC
	VIM-11	AY605049	Argentina	*Pseudomonas aeruginosa*	123
VIM-7		AJ536835	United States	*Pseudomonas aeruginosa*	162

[a] Sublineages are defined as allelic variants or groups of closely related variants that differ from the other sublineages by more than 2% of amino acid residues. Each sublineage was named after the first allelic variant identified.
[b] The accession number for the first reported nucleotide sequence.

circulating in the clinical setting has overall been a recent event.

The gene cassettes carrying closely related allelic variants of the bla_{IMP} or bla_{VIM} genes are usually equipped with identical or closely related *attC* recombination sites (e.g., bla_{IMP-1}, bla_{IMP-6} and bla_{IMP-10}, or bla_{VIM-1} and bla_{VIM-4}), suggesting a likely derivation from the same ancestral cassette. Gene cassettes carrying more divergent variants are often equipped with unrelated *attC* sites (e.g., bla_{IMP-1} and bla_{IMP-2} or bla_{VIM-1} and bla_{VIM-2}), a condition which likely reflects an independent phylogeny of the respective cassettes.

The integrons containing CBBL gene cassettes can be inserted in the chromosome or carried on plasmids (Fig. 9.5). The presence of bla_{IMP} and bla_{VIM} genes on conjugative plasmids (59, 73, 90, 100, 110; reference 117 and references therein; 179) is particularly worrisome, since it might greatly facilitate the dissemination of similar resistance determinants in the clinical setting. The transfer of a conjugative plasmid carrying the bla_{VIM-4} gene between a *K. pneumoniae* and an *Enterobacter cloacae* strain infecting the same patient has recently been described (100).

The original sources of the CBBL genes associated with mobile DNA remain thus far undetermined but are likely to be represented by some environmental species from which the genes, assembled on gene cassettes, have eventually escaped to the gram-negative opportunistic pathogens circulating in the nosocomial settings. In fact, a gene encoding an IMP-like enzyme (59% identical to IMP-11) has recently been detected in the metagenome from soil microbiota (169). The fact that, in most cases, these genes have initially been detected in *P. aeruginosa* and in other GNNFs and then also in *Enterobacteriaceae* would suggest that these species represent the major entry port for these genes in the clinical setting.

Expression of the CBBL genes carried on gene cassettes is normally under the control of the integron promoters (P_c and, possibly, P2) located in the 5′-conserved segment of the integron (28). This ensures a basal level of expression of variable intensity depending on the promoter variant (strong, weak, or hybrid) (97). Expression can also be influenced by the cassette position, being higher when the cassette is in the first position and reduced to a variable extent when the cassette is inserted in a more downstream position (28).

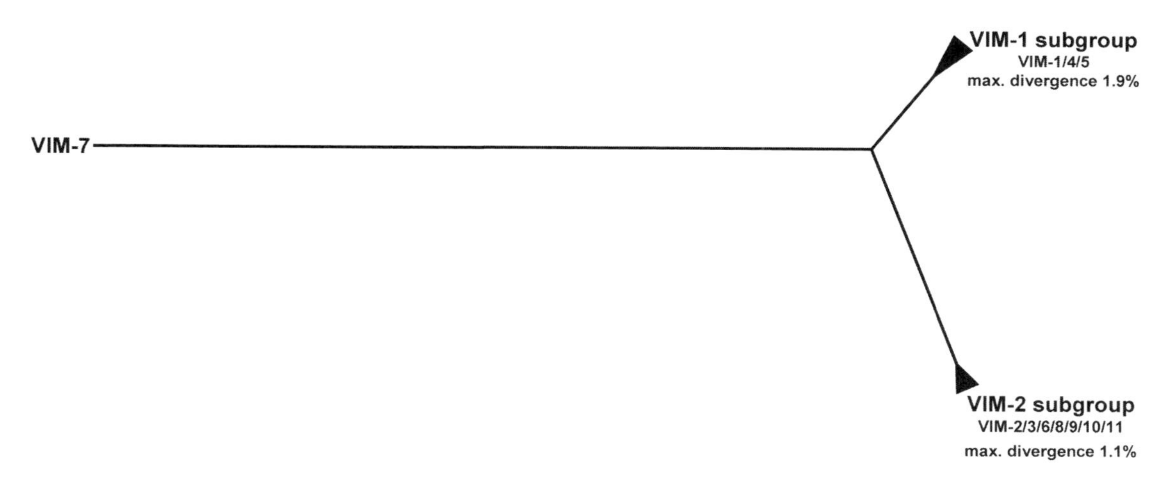

Figure 9.4 Unrooted tree showing the structural relatedness among known VIM variants. The tree was constructed using the TREEVIEW program (121) on the basis of a sequence alignment constructed with the CLUSTAL X program (158) using the sequences whose accession numbers are reported in the third column of Table 9.3.

EVOLUTION OF CBBLs

Unlike serine-β-lactamases, which share a common ancestry with penicillin-binding proteins and retain the same modality of interaction with β-lactams (56, 104), the evolutionary history of CBBLs remains unclear. CBBLs are part of a large protein family that are characterized by the presence of a conserved domain, named the MBL superfamily (Pfam accession number PF00753 [http://pfam.wustl.edu/]; InterPro accession number IPR001279 [http://www.ebi.ac.uk/interpro/]) or the zinc metallohydrolase family of the β-lactamase fold (37). In addition to CBBLs, this protein superfamily includes several enzymes of diverse functions (such as thiolesterases, glyoxalases II, arylsulfatases, enzymes involved in mRNA processing or in DNA repair, enzymes able to degrade acyl-homoserine-lactones involved in quorum sensing, proteins involved in DNA uptake that are essential for natural transformation, etc.) (21, 37, 138), and a large number of hypothetical proteins of unknown function encoded by bacterial, archaeal, and eukaryotic genomes. As of this writing, almost 4,000 members of this protein family have been detected in *Bacteria*, more than 400 in *Archaea*, and more than 500 in *Eukarya* (http://www.sanger.ac.uk/Software/Pfam/). In the bacterial domain, members of the MBL superfamily are found in representatives of most phyla (http://www.sanger.ac.uk/Software/Pfam/), while CBBLs have been detected in only some members of three phyla: BXII (*Proteobacteria*), BXIII (*Firmicutes*), and BXX (*Bacteroidetes*). Overall, these findings suggest that (i) the MBL fold is a very versatile protein scaffold of ancient origin, which has been exploited to generate proteins of very diverse functions; and (ii) apparently, evolution of these proteins toward a β-lactamase function has been relatively uncommon during the process of bacterial evolution.

Concerning the distribution of the various CBBLs in the bacterial domain, enzymes of subclasses B2 and B3 are found among *Proteobacteria* (phylum BXII), with the only exception of GOB, which is present in one member of *Bacteroidetes* (phylum BXX) (*E. meningoseptica*). On the other hand, enzymes of subclass B1 are found among *Firmicutes* (phylum BXIII) and *Bacteroidetes*, with the exception of those encoded by genes associated with mobile DNA (IMP, VIM, SPM, GIM, and SIM) that are found in strains of some members of the β- and γ-*Proteobacteria* classes (e.g., *Enterobacteriaceae*, *Pseudomonas*, *Acinetobacter*, and *Achromobacter*). According to current knowledge on bacterial evolution, which postulates a more ancient origin for *Firmicutes* and *Bacteroidetes* and a more recent origin for *Proteobacteria* (63), the observed distribution of CBBLs would suggest that (i) enzymes of subclass B1 have likely evolved earlier than those of the other subclasses; (ii) the genes encoding acquired CBBLs of subclass B1 (bla_{IMP}, bla_{VIM}, bla_{SPM}, bla_{GIM} and bla_{SIM}) that are currently spreading in pathogenic species of the β- and γ-*Proteobacteria* classes have likely a transphyletic origin; and (iii) the gene encoding the GOB enzyme of *E. meningoseptica* is probably derived from a recent acquisition by horizontal transfer followed by genomic fixation and likely has a transphyletic origin.

FUNCTIONAL FEATURES OF CBBLs

All CBBLs share some constant functional features that differentiate these enzymes from active-site serine-

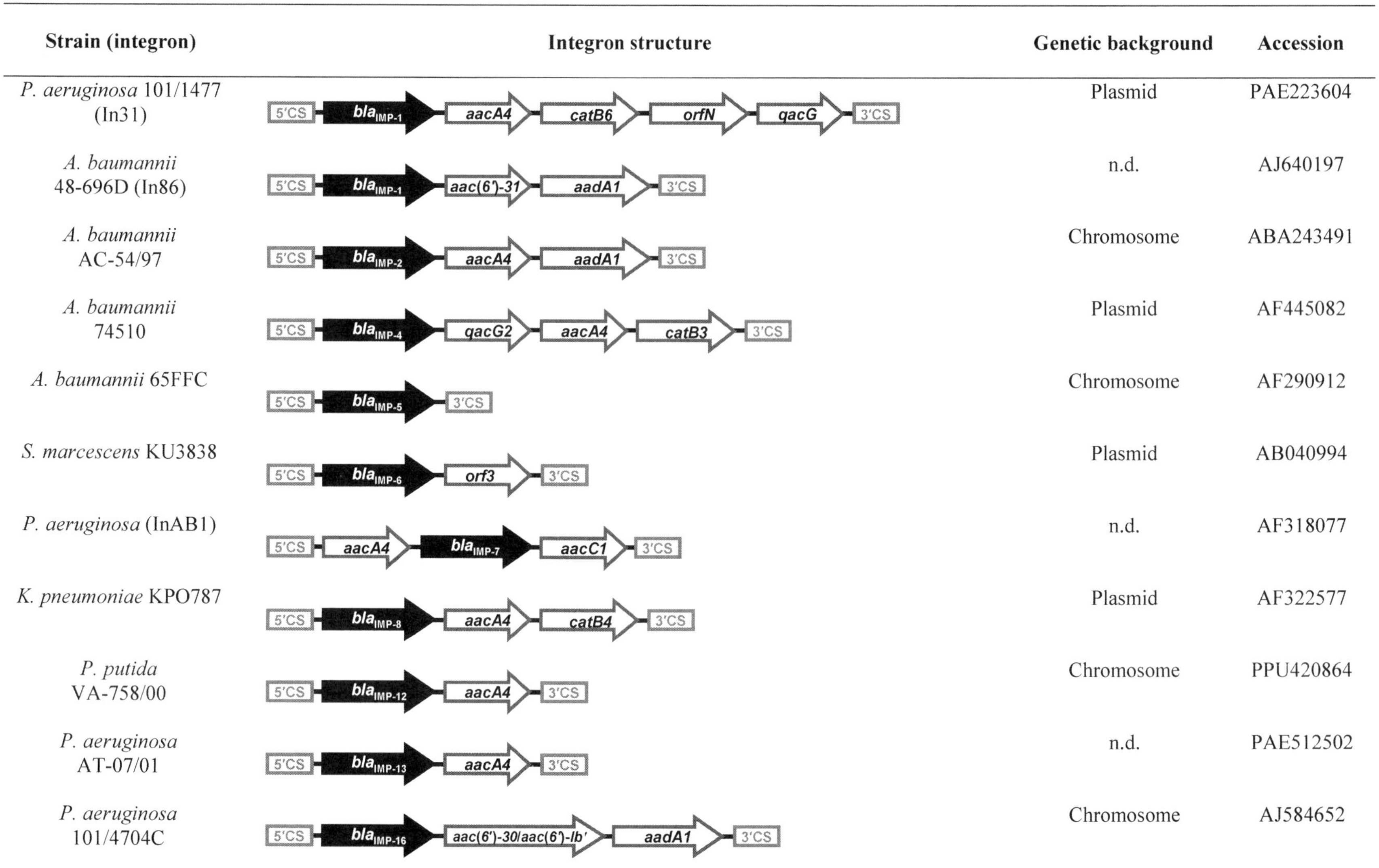

Strain (integron)	Integron structure	Genetic background	Accession
P. aeruginosa 101/1477 (In31)	5'CS – bla_{IMP-1} – *aacA4* – *catB6* – *orfN* – *qacG* – 3'CS	Plasmid	PAE223604
A. baumannii 48-696D (In86)	5'CS – bla_{IMP-1} – *aac(6')-31* – *aadA1* – 3'CS	n.d.	AJ640197
A. baumannii AC-54/97	5'CS – bla_{IMP-2} – *aacA4* – *aadA1* – 3'CS	Chromosome	ABA243491
A. baumannii 74510	5'CS – bla_{IMP-4} – *qacG2* – *aacA4* – *catB3* – 3'CS	Plasmid	AF445082
A. baumannii 65FFC	5'CS – bla_{IMP-5} – 3'CS	Chromosome	AF290912
S. marcescens KU3838	5'CS – bla_{IMP-6} – *orf3* – 3'CS	Plasmid	AB040994
P. aeruginosa (InAB1)	5'CS – *aacA4* – bla_{IMP-7} – *aacC1* – 3'CS	n.d.	AF318077
K. pneumoniae KPO787	5'CS – bla_{IMP-8} – *aacA4* – *catB4* – 3'CS	Plasmid	AF322577
P. putida VA-758/00	5'CS – bla_{IMP-12} – *aacA4* – 3'CS	Chromosome	PPU420864
P. aeruginosa AT-07/01	5'CS – bla_{IMP-13} – *aacA4* – 3'CS	n.d.	PAE512502
P. aeruginosa 101/4704C	5'CS – bla_{IMP-16} – *aac(6')-30/aac(6')-Ib'* – *aadA1* – 3'CS	Chromosome	AJ584652

Figure 9.5 (See P. 14)..

Strain (integron)	Integron structure	Genetic background	Accession
P. aeruginosa VR-143/97 (In70)	5′CS – *bla*$_{VIM-1}$ – *aacA4* – *aphA15* – *aadA1*(Δ*attC*) – 3′CS	Chromosome	Y18050
P. aeruginosa PPV-97 (In80)	5′CS – *bla*$_{VIM-1}$ – *aacA4* – *aacA4* – *bla*$_{OXA-46}$ – 3′CS	Chromosome	AF317511
P. putida VA-304/99 (In110)	5′CS – *bla*$_{VIM-1}$ – *aacA4* – *aadA1*(Δ*attC*) – 3′CS	Plasmid	AJ439689
E. coli EC23	5′CS – *aacA4* – *bla*$_{VIM-1}$ – 3′CS	n.d.	AY152821
E. coli V541	5′CS – *bla*$_{VIM-1}$ – *aacA7* – *dhfr1* – *aadA1* – 3′CS	Plasmid	AY339625
P. aeruginosa COL-1 (In56)	5′CS – *bla*$_{VIM-2}$ – 3′CS	Plasmid	AF191564
P. aeruginosa RON-1 (In58)	5′CS – *aacA7* – *bla*$_{VIM-2}$ – *aacC1* – *aacA4* – 3′CS	Chromosome	AF263520
P. aeruginosa RON-2 (In59)	5′CS – *aac29a* – *bla*$_{VIM-2}$ – *aac29b* – 3′CS	Chromosome	AF263519
P. aeruginosa VR-193/98 (In72)	5′CS – *aacA4* – *bla*$_{VIM-2}$(Δ*attC*) – 3′CS	Plasmid	AF302086
P. aeruginosa VA-182/00 (In182)	5′CS – *bla*$_{VIM-2}$ – *aacC-A5* – *cmlA7* – 3′CS	n.d.	AJ511268
A. baumannii YMC 98/7/363 (In105)	5′CS – *bla*$_{VIM-2}$ – *aacA7* – *aadA1* – 3′CS	n.d.	AF324464

Figure 9.5 (*Continued*).

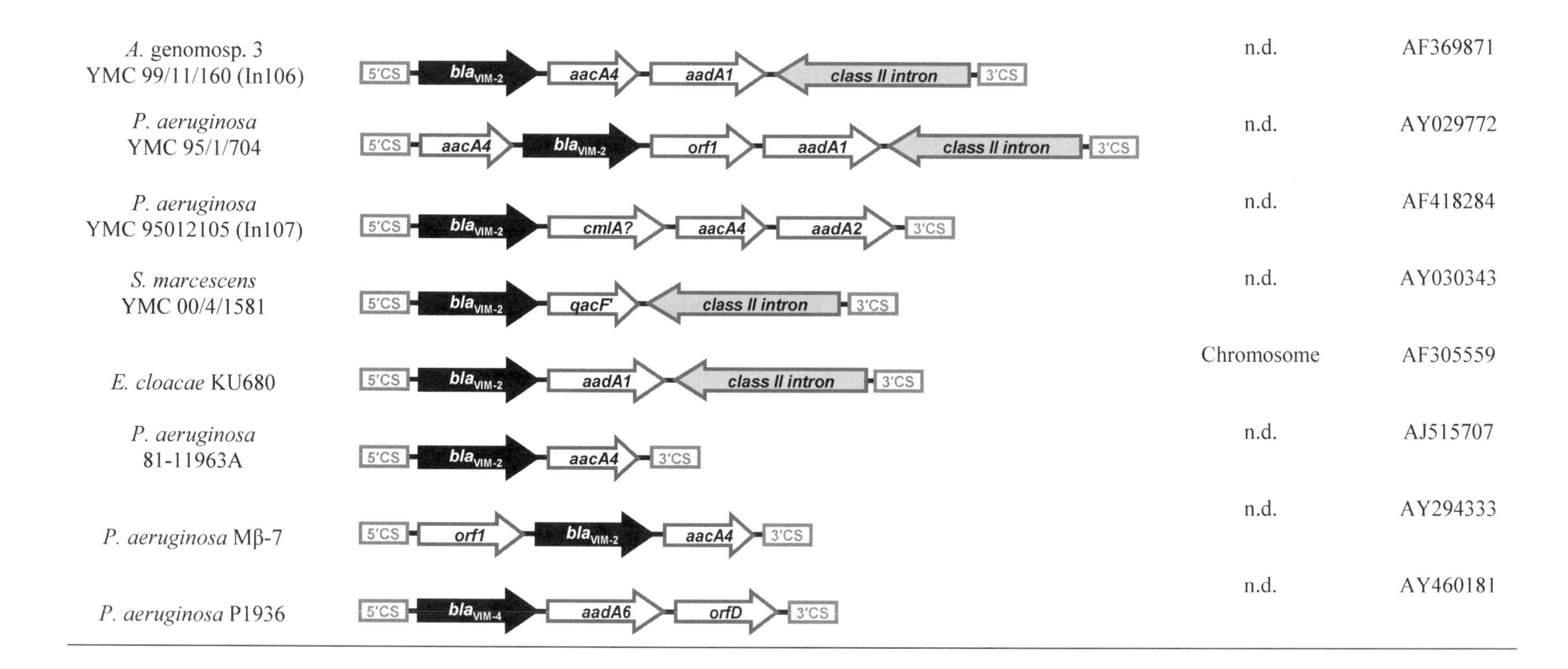

Figure 9.5 Schematic representation of the structure of the variable region of class 1 integrons containing *bla*$_{IMP}$ (section A) or *bla*$_{VIM}$ (section B) gene cassettes. The gene cassettes are indicated by arrows (those carrying CBBL genes are filled). The presence of the integron 5′-conserved segment (5′-CS) and 3′-conserved segment (3′-CS), flanking the cassette array, is also indicated. Only integrons whose variable region has been completely sequenced are shown.

β-lactamases and account for their classification in a separate group (group 3) of the functional classification of β-lactamases (20). The constant features of CBBLs include (i) good to excellent carbapenemase activity; (ii) lack of activity on monobactams, which apparently do not interact with these enzymes; and (iii) inhibition by EDTA and other metal ion chelators, and lack of inhibition by the conventional serine-β-lactamase inhibitors. Penicillanic acid sulfones and β-iodopenicillanate can actually be degraded by CBBLs (18) (Table 9.4 and references therein). Apart from these conserved features, a remarkable functional heterogeneity is observed among CBBLs in terms of substrate specificity. Based on functional features, three different subgroups have been proposed within group 3 in the functional classification of β-lactamases: subgroup 3a, including enzymes that exhibit a broad substrate specificity, with hydrolysis rates for penicillins that are faster or comparable to those for carbapenems and hydrolysis rates for cephalosporins that are overall lower that those for carbapenems (e.g., Bc-II/Bla2, CfiA/CcrA, IMP-1, and L1); subgroup 3b, including enzymes that behave as narrow-spectrum carbapenemases (e.g., CphA/Imi); and subgroup 3c, including enzymes with hydrolysis rates for cephalosporins that tend to be faster than those for the other substrates (e.g., FEZ-1) (18). Although this classification scheme was proposed when the number of known CBBLs was still relatively small and detailed kinetic characterization was available only for a few enzymes, it retains a substantial validity from the practical standpoint and most of the enzymes that have been subsequently discovered can be assigned to one of the above subgroups.

Kinetic parameters of several enzymes for various β-lactam substrates have been determined. Considering catalytic efficiency, expressed by the k_{cat}/K_m ratio, some enzymes exhibit an overall good to excellent efficiency (k_{cat}/K_m ratios, $>10^5$ $M^{-1}{\cdot}s^{-1}$) with most substrates (e.g., CcrA, IMP-1, VIM-2, SPM-1, L1, and GOB-1), while others show a good efficiency with a more limited number of substrates (e.g., Bc-II, FEZ-1, CAU-1, GIM-1, and the subclass B1 flavobacterial enzymes). An extreme situation is that of the CphA1 enzyme, which can efficiently hydrolyze only carbapenems (Table 9.4). With some enzymes, the catalytic efficiency can be greatly variable for substrates of the same β-lactam nucleus. For instance, the k_{cat}/K_m ratios of IMP-1 can vary more than 1,000-fold for different penicillin substrates (89), while the k_{cat}/K_m ratios of BlaB1 can vary more than 1,000-fold for different cephalosporin substrates (168).

As already mentioned, carbapenems are good to excellent substrates for all CBBLs (k_{cat}/K_m ratios are equal to or higher than 10^5 $M^{-1}{\cdot}s^{-1}$ for all enzymes except JOHN-1, which exhibits a somewhat lower catalytic efficiency). Among cephems, some compounds tend to behave as relatively poor substrates (k_{cat}/K_m ratios $<10^4$ $M^{-1}{\cdot}s^{-1}$) with several enzymes or groups thereof. For instance, ceftazidime is an overall poor substrate for the flavobacterial enzymes of subclass B1 and for some enzymes of subclass B3, while cefoxitin (a 7α-methoxy cephalosporin) and moxalactam tend to be relatively poor substrates for most enzymes of subclass B1 (except for IMP, VIM, and SPM). Cefepime is a relatively poor substrate for several CBBLs, and in one case (the CAU-1 enzyme), it behaves as an inhibitor (Table 9.4).

The apparent affinity of CBBLs for different substrates can be highly variable (K_m values can vary from the micromolar to the millimolar range) depending on the enzyme-substrate combination. Some CBBLs show overall high affinities for a broad repertoire of substrates (e.g., SPM-1 and L1), while others exhibit a remarkable variability in the affinity for different classes of β-lactams (e.g., IMP-1 and FEZ-1 overall exhibit a higher affinity for cephalosporins and meropenem than for penicillins) (Table 9.4). A peculiar situation is that of the CAU-1 enzyme from *C. vibrioides*, which exhibits an overall low affinity for most β-lactam substrates. For this reason, along with the unique location of that gene within an operon-like structure containing genes involved in methionine metabolism, it was hypothesized that CAU-1 might represent a sort of evolutionary intermediate between β-lactamases and members of the MBL superfamily endowed with another function (45).

Considering individual kinetic parameters, it is evident that different CBBLs have evolved very different strategies to achieve comparable catalytic efficiencies against the same substrate. For instance, efficient degradation of carbapenems can be dependent on high turnover rates associated with relatively low affinities (e.g., Bc-II, BlaB1, and FEZ-1), on high affinities associated to lower turnover rates (e.g., VIM-2), or even on conditions that are intermediate between these two extreme situations (Table 9.4).

Functional diversity may exist not only between enzymes of different subclasses or lineages but also for allelic variants of the same enzyme, as it was shown to be the case for IMP and VIM variants. The IMP enzymes overall exhibit a broad substrate specificity, with higher affinity for cephalosporins and carbapenems. However, significant differences in the kinetic behavior with several substrates (including carbapenems, cephalosporins, and penicillins) were found between different IMP variants that have been subjected to kinetic characterization (Table 9.5). Significant functional heterogeneity with several substrates was also found between VIM-1 and VIM-2 (42). In some cases the differences in individual kinetic parameters (K_m and

Table 9.4 Kinetic parameters of CBBLs against relevant β-lactam substrates

Enzyme	IPM		MEM		CEF		CXM		CTX		CAZ		FEP		FOX		MOX		PEN		AMP		PIP		Reference(s)
	K_m	k_{cat}/K_m	K_m	k_{cat}/K_m	K_m	k_{cat}/K_m	K_m	k_{cat}/K_m	K_m	k_{cat}/K_m	K_m	k_{cat}/K_m	K_m	k_{cat}/K_m	K_m	k_{cat}/K_m	K_m	k_{cat}/K_m	K_m	k_{cat}/K_m	K_m	k_{cat}/K_m	K_m	k_{cat}/K_m	
Subclass B1																									
Bc-II	>1,000	1.3×10^{6}	400	1.5×10^{6}	—[a]	—	45	7.8×10^{5}	90	6.7×10^{5}	—	—	>500	7.4×10^{2}	2,100	9.5×10^{1}	650	1.9×10^{3}	1,550	4.6×10^{5}	1,530	7.2×10^{5}	450	1.7×10^{5}	48
CcrA1	270	7.4×10^{5}	—	—	—	—	4	7.3×10^{5}	27	3.6×10^{6}	—	—	—	—	110	9.0×10^{4}	160	1.9×10^{5}	40	4.8×10^{6}	120	1.6×10^{6}	—	—	48
BlaB1	360	2.0×10^{6}	>1,650	7.6×10^{5}	6	1.6×10^{6}	6	1.0×10^{6}	17	7.9×10^{5}	875	6.6×10^{3}	>1,000	2.0×10^{2}	110	5.0×10^{4}	>2,000	8.7×10^{3}	33	2.7×10^{7}	235	7.2×10^{6}	275	8.7×10^{6}	168
EBR-1	780	2.5×10^{5}	>2,000	1.0×10^{5}	31	2.0×10^{5}	1	3.5×10^{5}	50	2.0×10^{4}	>1,000	2.0×10^{3}	n.h.[b]	n.h.	140	2.0×10^{3}	620	2.0×10^{3}	47	2.5×10^{6}	—	—	490	8.5×10^{5}	9
IND-2	170	6.0×10^{5}	920	2.0×10^{5}	16	1.2×10^{6}	—	—	10	9.0×10^{5}	440	5.0×10^{3}	2,200	$<10^{3}$	40	5.0×10^{4}	—	—	70	4.6×10^{6}	95	7.0×10^{5}	210	2.0×10^{6}	12
TUS-1	250	9.0×10^{5}	1,060	1.6×10^{5}	125	3.0×10^{5}	24	1.2×10^{6}	110	4.7×10^{5}	1,500	1.0×10^{4}	>2,000	n.d.[c]	820	2.0×10^{4}	>2,000	n.d.	80	4.0×10^{6}	—	—	75	1.4×10^{7}	102
MUS-1	470	5.0×10^{5}	1,400	1.8×10^{5}	100	9.0×10^{5}	15	8.7×10^{5}	180	7.0×10^{5}	1,700	2×10^{3}	>2,000	n.d.	1,300	5.0×10^{3}	>2,000	n.d.	40	4.2×10^{6}	—	—	160	2.0×10^{6}	102
JOHN-1	90	4.4×10^{4}	235	2.1×10^{4}	9	1.1×10^{5}	10	1.0×10^{5}	29	7.0×10^{4}	210	2.0×10^{3}	1,200	1.0×10^{2}	1	1.5×10^{4}	7	7.0×10^{3}	110	7.3×10^{4}	—	—	110	7.0×10^{4}	115
IMP-1	39	1.2×10^{6}	10	5.0×10^{5}	21	2.4×10^{6}	37	2.2×10^{5}	4	3.5×10^{5}	44	1.8×10^{5}	11	6.6×10^{5}	8	2.0×10^{6}	10	8.8×10^{6}	520	6.2×10^{5}	200	4.8×10^{6}	n.d.	7.2×10^{5}	89
VIM-2	9	3.8×10^{6}	2	2.5×10^{6}	11	1.2×10^{7}	20	4.0×10^{5}	12	5.8×10^{6}	72	5.0×10^{4}	>400	1.0×10^{5}	13	1.2×10^{6}	55	1.6×10^{6}	70	4.0×10^{6}	90	1.4×10^{6}	125	2.4×10^{6}	42
SPM-1	37	1.0×10^{6}	280	2.2×10^{5}	4	1.2×10^{7}	4	8.8×10^{6}	9	1.9×10^{6}	46	6.0×10^{5}	18	1.0×10^{6}	2	4.0×10^{6}	100	1.3×10^{5}	38	2.8×10^{6}	72	1.6×10^{6}	59	2.0×10^{6}	113
GIM-1	290	9.0×10^{4}	25	1.1×10^{5}	22	7.2×10^{5}	7	8.0×10^{5}	4	2.4×10^{5}	31	5.8×10^{5}	430	4.0×10^{4}	210	4.0×10^{4}	1,040	1.0×10^{4}	46	1.4×10^{5}	20	1.6×10^{5}	69	1.0×10^{5}	24
Subclass B2																									
CphA1	90	1.9×10^{6}	250	1.8×10^{6}	—	—	—	—	26	2.7×10^{3}	—	—	—	—	Inact.[d]	Inact.	Inact.	Inact.	680	7.4×10^{3}	—	—	55	7.8×10^{3}	47, 48
Subclass B3																									
L1	90	7.3×10^{5}	10	4.5×10^{6}	—	—	30	2.7×10^{6}	26	2.6×10^{6}	145	1.8×10^{5}	>1,000	n.d.	2	5.5×10^{5}	1	2.9×10^{5}	50	2.2×10^{7}	40	4.4×10^{6}	20	7.0×10^{6}	47, 48, 156
FEZ-1	>1,000	2.0×10^{5}	85	5.0×10^{5}	120	2.5×10^{6}	50	6.6×10^{6}	70	2.4×10^{6}	>1,000	4.0×10^{3}	>1,000	6.0×10^{3}	11	2.7×10^{5}	18	1.7×10^{5}	590	1.1×10^{5}	>5,000	1.1×10^{4}	4,200	1.2×10^{4}	108
GOB-1	60	6.6×10^{5}	5	5.3×10^{6}	24	6.7×10^{5}	27	9.8×10^{5}	51	8.5×10^{5}	71	7.6×10^{5}	50	2.0×10^{5}	47	2.5×10^{5}	80	1.3×10^{5}	110	1.9×10^{6}	—	—	170	1.7×10^{6}	8
CAU-1	1,100	2.0×10^{5}	700	2.6×10^{5}	330	4.3×10^{5}	>1000	1.4×10^{4}	—	—	>400	2.0×10^{3}	120[d]	Inhib.[e]	—	—	>500	1.2×10^{5}	400	4.5×10^{5}	440	5.0×10^{5}	115	5.7×10^{5}	45
THIN-B	80	1.5×10^{6}	40	5.0×10^{6}	—	—	50	2.8×10^{6}	40	2.0×10^{6}	140	1.4×10^{5}	>300	7.9×10^{3}	—	—	—	—	—	—	1,300	3.7×10^{5}	500	2.0×10^{5}	43

[a] —, data not available.
[b] n.h., no hydrolysis detected.
[c] n.d., not determined.
[d] Inact., the CphA1 enzyme is inactivated by cefoxitin and moxalactam (see reference 47 for further details).
[e] Inhib., the CAU-1 enzyme is inhibited by cefepime, the value reported as K_m corresponds to K_i (see reference 45 for further details).

k_{cat}) tend to balance out, eventually resulting in similar hydrolytic efficiencies, while in other cases the k_{cat}/K_m ratios are also significantly different (Table 9.5) (42). These findings imply that at least some of the differences existing between variants could bear functional significance and suggest that comparative analysis of allelic variants of CBBLs could provide interesting insights on the structure-function relationships of these enzymes. For instance the large difference in affinity for imipenem observed between IMP-1 and IMP-3, which differ from each other by only two amino acid residues (at positions 126 and 262), was shown to be related to the residue present at position 262 by site-directed mutagenesis experiments (74), suggesting a crucial role for this position in the interaction with carbapenems. This view is further supported by the fact that the Ser262→Gly substitution, which is not present in IMP variants with higher imipenem affinities, is present in IMP-12, a variant that also exhibits a remarkably low affinity for imipenem (46). The functional heterogeneity observed between VIM-1 and VIM-2 has been hypothesized to reflect the different nature of side chains of residues close to the active site. In particular, based on molecular modeling, it has been speculated that the differences existing at positions 223, 224, and 228, all of which are located in the neighborhood of the active site, could influence the enzyme interaction with β-lactam substrates (42). In other CBBLs of subclass B1, the residue at position 224 is a conserved lysine whose positively charged side chain is supposed to stabilize the enzyme-substrate complex by interacting with the β-lactam carboxylate (134). The variability encountered at this position, which is a unique feature of the VIM-type enzymes, could influence interaction with the substrate. Moreover, stabilization of the substrate in the active site could be influenced in different ways by the different residues present at position 228 (42).

All CBBLs are inhibited by EDTA and by other zinc ion chelators such as pyridine-2,6-dicarboxylic acid (dipicolinic acid) and 1,10-phenanthroline. Analysis of inactivation parameters upon exposure to chelators has been carried out with several enzymes, and differences have been found with enzymes of different subclasses. With CBBLs of subclasses B1 and B2 the inactivation mechanism apparently proceeds via the formation of a ternary enzyme-metal-chelator complex that precedes the removal of the metal from the enzyme active site, and 1,10-phenanthroline and dipicolinic acid are more efficient than EDTA as inactivators. With CBBLs of subclass B3, instead, inactivation is apparently dependent on scavenging of the free metal by the chelators, except for the THIN-B enzyme whose inactivation mechanism seems to occur via the formation of a ternary complex (Table 9.6).

Table 9.5 Kinetic parameters of different IMP variants for selected β-lactam substrates

Enzyme	IPM		MEM		CEF		CXM		CTX		CAZ		PEN		AMP		PIP		References
	K_m	k_{cat}/K_m	K_m	k_{cat}/K_m	K_m	k_{cat}/K_m	K_m	k_{cat}/K_m	K_m	k_{cat}/K_m	K_m	k_{cat}/K_m	K_m	k_{cat}/K_m	K_m	k_{cat}/K_m	K_m	k_{cat}/K_m	
IMP-1	39	1.2×10^6	10	5.0×10^5	21	2.4×10^6	37	2.2×10^5	4	3.5×10^5	44	1.8×10^5	520	6.2×10^5	200	4.8×10^6	—[a]	7.2×10^5	89
IMP-2	24	9.2×10^5	0.3	3.3×10^6	—	—	—	—	—	—	110	1.9×10^5	—	—	110	2.1×10^5	—	—	136
IMP-3	1,140	8.0×10^4	—	—	9.9	2.3×10^7	—	—	3.1	1.3×10^7	130	3.5×10^4	370	3.9×10^4	460	1.6×10^4	—	—	74
IMP-6	250	7.1×10^5	7.6	4.2×10^6	4.7	8.0×10^7	—	—	3.8	1.5×10^7	—	—	220	2.3×10^5	—	—	230	9.0×10^4	183
IMP-10	60	3.7×10^6	47	1.4×10^6	4.9	4.7×10^7	—	—	5.7	1.3×10^7	51	1.0×10^6	—	7.0×10^4	—	0.6×10^4	-	4.0×10^4	75
IMP-12	920	2.6×10^5	7.2	1.3×10^6	16	7.4×10^6	7	8.7×10^6	22	2.5×10^6	15	4.5×10^5	—	—	1,500	1.2×10^4	-	2.3×10^4	46
IMP-16	370	3.6×10^5	72	3.2×10^5	42	1.8×10^6	49	1.1×10^6	36	9.7×10^5	87	1.5×10^5	7,800	1.0×10^5	1,070	1.3×10^5	2,800	9.0×10^4	107

[a] —, data not available.

Table 9.6 Inactivation parameters of different CBBLs for selected chelating agents

	EDTA		1,10-*o*-phenanthroline		Dipicolinic acid			
Enzyme	k_{+2} or k_i (s^{-1})[a]	k_{+2}/K ($M^{-1} \cdot s^{-1}$)	k_{+2} or k_i (s^{-1})	k_{+2}/K ($M^{-1} \cdot s^{-1}$)	k_{+2} or k_i (s^{-1})	k_{+2}/K ($M^{-1} \cdot s^{-1}$)	Inactivation mechanism	Reference
IMP-1	n.d.[b]	n.d.	n.d.	120	2.4×10^{-2}	380	Via formation of a ternary complex	89
VIM-1	n.d.	n.d.	2.8×10^{-3}	215	5.4×10^{-3}	23	Via formation of a ternary complex	50
VIM-2		2.5		860		460	Via formation of a ternary complex	42
CphA1	6.0×10^{-3}	3.4	4.4×10^{-2}	440	1.9×10^{-2}	390	Via formation of a ternary complex	66
L1	1.7×10^{-2}	n.d.	—[c]	—	—	—	Chelator scavenges the free metal	13
FEZ-1	2.5×10^{-2}	n.d.	2.5×10^{-2}	n.d.	3.8×10^{-2}	n.d.	Chelator scavenges the free metal	108
CAU-1	1.9×10^{-3}	n.d.	2.5×10^{-3}	n.d.	—	—	Chelator scavenges the free metal	45
THIN-B	2.2×10^{-2}	7.0×10^{3}	1.2×10^{-2}	>240	3.9×10^{-2}	180	Via formation of a ternary complex	43

[a] The reported values are k_{+2} for IMP-1, VIM-1, VIM-2, CphA1, and THIN-B enzymes and k_i for the other enzymes.
[b] n.d., not determined (K >10 mM).
[c] —, data not available.

STRUCTURE AND MECHANISM OF CBBLs

Solving the first structure of a CBBL at high resolution (23) revealed an original protein fold, consisting of two central β-sheets flanked by two α-helix-containing modules on each face, which is completely unrelated to that of serine-β-lactamases. The structure exhibits an internal twofold symmetry contributed by two independent domains connected by a loop. The active site, containing the metal binding site, is located on one edge of the β-barrel and is formed by residues belonging to both domains (Fig. 9.6).

Currently, the structures of nine different CBBLs have been solved, including six enzymes of subclass B1 (BcII, CcrA, IMP-1, BlaB1, VIM-2, and SPM-1), one of subclass B2 (CphA1), and two of subclass B3 (L1 and FEZ-1) (Protein Data Bank [PDB], http://www.rcsb.org/pdb/index.html, and references therein). All CBBL structures present an overall similar fold with strands and helices in the following order: β1β2β3β4(β5)α1β5(β6)α2β6(β7)α3 for the N-terminal domain and (β7)β8β9β10β11α4β12α5 for the C-terminal domain. However, some remarkable structural differences are evident between the enzymes of different subclasses concerning the molecular architecture and the structure of the zinc center.

The enzymes of subclass B3 exhibit a larger molecular size than those of the other subclasses, due to a larger size of their polypeptide component (~30 kDa versus ~25 kDa). In the molecular structure, the additional amino acid residues are mainly accommodated in some strands and helices (β10, β11, and α5) of the C-terminal domain, in the loop joining the N-terminal and C-terminal domains, and at the N terminus (in the L1 enzyme) (55, 167) (Fig. 9.2 and 9.6). The enzyme of subclass B2 differs from members of the other subclasses for (i) the presence of an elongated α3 helix, which is located just above the active-site pocket and is a key element of the hydrophobic wall that defines the pocket, and (ii) the absence of mobile loops (between β3 and β4 in members of subclass B1, or between α3 and β7 in those of subclass B3) near the active site (Fig 9.2 and 9.6). This determines the presence of a very well defined active site, which accounts for the narrow substrate specificity of CphA1 (see above) (53).

Concerning the structure of the zinc center, which is located at the bottom of a shallow groove between the two β-sheets, in subclass B1 and B3 CBBLs it can accommodate two zinc ions (dinuclear zinc center): a tetrahedrally coordinated zinc ion (Zn1) and a trigonal bipyramidally coordinated zinc ion (Zn2), bridged by a water molecule/hydroxide ion (Wat1). Zn1 is coordinated by three His residues (His 116, His118, and His196) and the bridging Wat1 in all subclass B1 and B3 enzymes, while the strategy of Zn2 coordination is partially different in members of the two subclasses. In enzymes of subclass B1, Zn2 is coordinated by the bridging Wat1 and the conserved His263 and Cys221 residues in equatorial positions and by an additional water molecule (Wat2) and the conserved Asp120 residue in apical positions. In enzymes of subclass B3, Zn2 is coordinated by the bridging Wat1 and the conserved His263 and His121 residues in equatorial positions and by the conserved Asp120 residue and an additional water molecule (Wat2, which in turn is hydrogen bonded to Ser221) in apical positions (Fig. 9.6). A strikingly different strategy is adopted in the enzyme of subclass B2, in which the zinc center accommodates a single metal ion (mononuclear zinc center) in the Zn2 site (Fig. 9.6). In the native enzyme, the zinc ion was found to be coordinated by Asp120, Cys221, His263, and a carbonate ion in a tetrahedral geometry. The His118 and His196 residues (which in enzymes of the other subclasses are ligands of the Zn1 site), together with Asn116 and Lys224, are involved in forming a hydrogen bond network in the active site that involves several water molecules and apparently contributes to positioning of the zinc-bound carbonate ion (53).

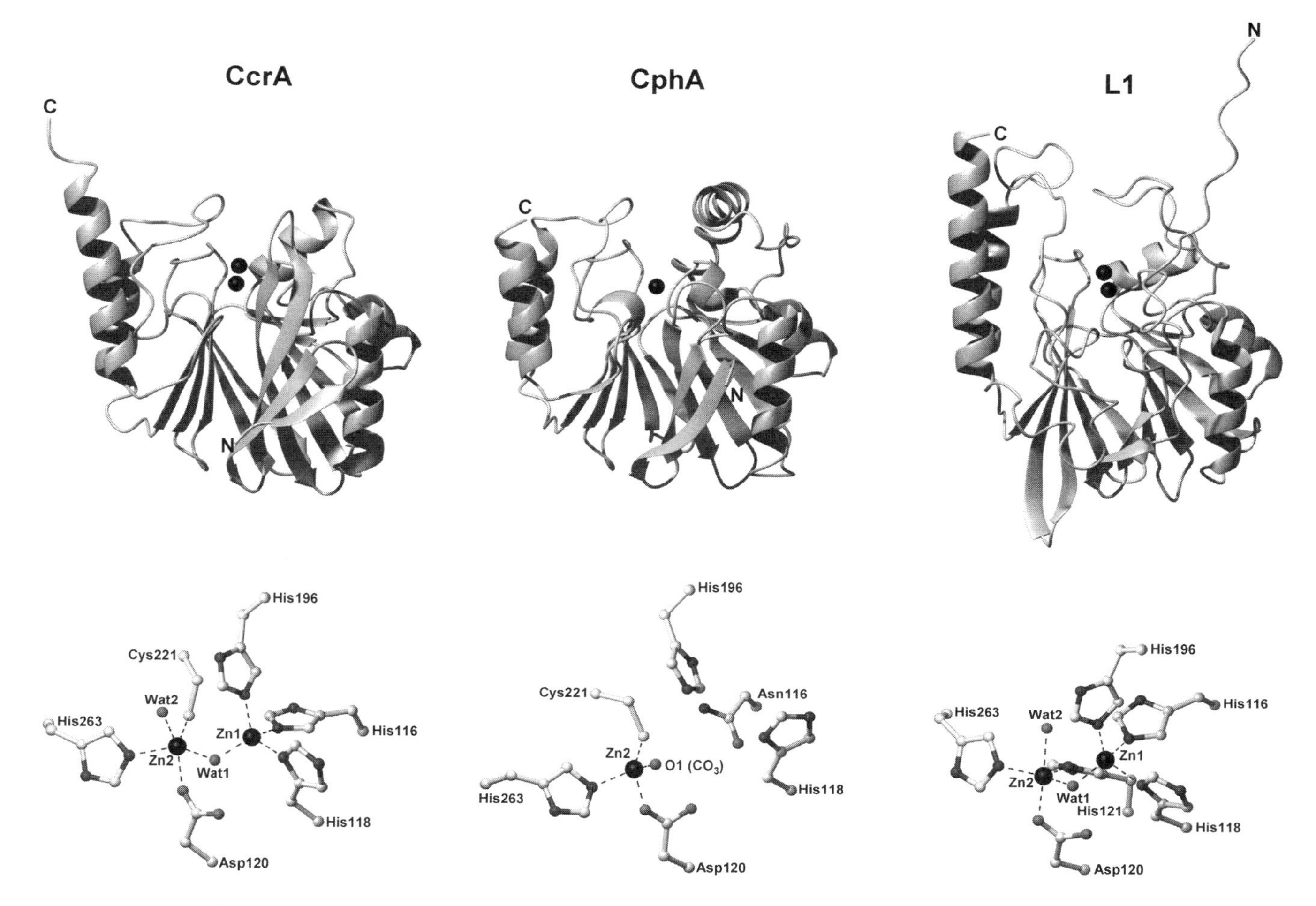

Figure 9.6 (Upper panels) Ribbon diagram of the three-dimensional structure of the CfiA/CcrA enzyme from *B. fragilis* strain QMCN3 (PDB accession no., 1ZNB), of the CphA enzyme from *A. hydrophila* AE036 (PDB accession no., 1X8G), and of the L1 enzyme from *S. maltophilia* strain IID 1275 (PDB accession, 1SML). The diagrams were constructed using the MOLMOL program (86). (Lower panels) Structure of the zinc centers of the corresponding enzymes (the water molecule is replaced by a carbonate in the CphA structure, and a single oxygen atom, involved in zinc coordination, is shown for clarity).

The quaternary structure of most CBBLs is monomeric (8, 42, 45, 102, 115; PDB [see above]). Thus far, the only exception to this rule is represented by the L1 enzyme, which is found as a homotetramer (167). In that protein, two structural elements are essential to create the intermonomeric interactions that lead to tetramerization: (i) the side chain of Met175, which is accommodated in a hydrophobic pocket formed by residues Leu154, Pro198, and Tyr236 in the other monomer, and (ii) hydrophobic residues in the long N-terminal tail, which are engaged in dimer-dimer interaction (167). The essential role of the Met175 residue in L1 oligomerization was demonstrated by the fact that an engineered mutant in which Met175 was replaced by Asp remained monomeric, while deletion of the N-terminal extension did not prevent oligomerization (156). Interestingly, enzyme functionality was strongly affected in both types of mutants (156).

CBBLs are metal-dependent enzymes in which the metal cofactor plays an essential role in catalysis. In dinuclear zinc enzymes containing two zinc ions it has been proposed that the hydroxide ion bridging the two zincs (Wat1 in Fig. 9.6) can serve as the attacking nucleophile on the carbonyl oxygen atom of the β-lactam ring (31, 167, 175). In enzymes of subclass B1, the presence and activation of Wat1 would not necessarily require the presence of both metal ions at the active site, as observed in the monozinc structure of BcII (23), explaining why monometal enzymes can still perform catalysis (14, 23, 41). However, in this case, the extent of occupancy of the zinc center positively modulates the enzyme activity (the BcII and CcrA enzymes exhibit maximal activity when two metal ions are bound [14, 35]). On the other hand, an increasing zinc concentration results in a strong noncompetitive inhibition of the mononuclear zinc enzyme of subclass B2 (CphA1) due to the occupancy of a low-affinity site by an additional zinc ion (53, 66, 67).

The available structural data for dinuclear zinc enzymes, combined with molecular modeling studies, suggest that the correct orientation of the substrate molecule could be promoted by at least two factors. First, the β-lactam carbonyl oxygen would fit in an oxyanion hole formed by Zn1 and the side chain of a variable residue (Asn233 in some enzymes of subclass B1, Tyr228 in L1, or Asn225 in FEZ-1) (31, 55, 133, 134, 167). Second, the β-lactam carboxyl group would be stabilized by interacting with the side chain of a variable residue (Lys224, which is conserved in most subclass B1 enzymes, or Ser223 in L1) (133; reference 134 and references therein; 167). Other factors might also contribute to the stabilization of the substrate-enzyme complex with some substrates, like, for instance, hydrophobic interactions involving the hydrophobic cleft formed by residues at positions 61 and 87 (in some subclass B1 enzymes) and the presence of a mobile tryptophane residue (Trp64) in a flap-like structure for which conformational changes have been detected upon inhibitor binding in the IMP-1 enzyme (30, 133, 134). The factors affecting substrate binding and catalysis could be different in enzymes of subclass B2, as revealed by the structure of the CphA-biapenem complex (53).

According to the enzyme state (mono- or dizinc form), alternative mechanisms for β-lactam hydrolysis have been proposed (17, 38, 53, 65, 111, 175), although this subject is still in open discussion. This underlines a possible important mechanistic heterogeneity encountered with CBBLs. A more detailed discussion on these aspects is beyond the scope of this chapter.

CLINICAL RELEVANCE OF CBBLs

The clinical relevance of resident CBBLs essentially reflects that of the host species, where these enzymes can variably contribute to intrinsic β-lactam resistance depending on their expression pattern and substrate specificity. Some CBBL-producing species, such as *J. lividum*, *C. vibrioides*, *N. aromaticivorans*, and *B. japonicum*, are not pathogenic for humans, and therefore, their enzymes are not of direct clinical relevance. However, those CBBLs could be interesting as investigational models for enzymes of subclass B3. The other species carrying resident CBBL genes (Table 9.1) can behave as opportunistic pathogens of variable clinical importance. The most important ones (based on the frequency of isolation from clinical specimens and of association with human infections) are *S. maltophilia*, *B. fragilis*, *Aeromonas* spp., *E. meningoseptica*, and *C. indologenes* (114).

S. maltophilia is an important opportunistic pathogen that can be responsible for a broad variety of infections in compromised hosts (bacteremia, endocarditis, skin and soft tissue infections, respiratory tract infections, etc.), which primarily occur in the hospital setting (40). It could also represent an emerging pathogen for lower respiratory tract infections in cystic fibrosis patients (61). In *S. maltophilia*, production of the L1 enzyme and that of L2 (a class A serine-β-lactamase of extended-spectrum activity [172]) contribute to the intrinsic broad-spectrum β-lactam resistance typical of this species (40). The L1 enzyme hydrolyzes most β-lactams except monobactams (see above), and, being the only one capable of degrading carbapenems, it apparently plays a crucial role in determining intrinsic resistance of *S. maltophilia* to those drugs (15).

B. fragilis is a strict anaerobe that inhabits the human intestine and is the most common anaerobe recovered in clinical specimens, especially from intra-abdominal infections but also from other sites (80). In *B. fragilis*, production of the CfiA/CcrA enzyme confers resistance to most β-lactams including carbapenems and cephamycins, which are among the drugs of choice for *Bacteroides* infections. Enzyme production causing resistance, however, is limited to a minority of clinical isolates, since the gene is present only in a cluster of *B. fragilis* which accounts for less than 5% of isolates (126, 177), while expression of the *cfiA/ccrA* gene to achieve significant levels of resistance is not ubiquitous but depends on transcriptional activation by transposition of insertion sequences in the promoter region (82, 127). In a surveillance study conducted in two different hospitals in Japan, among a total of 286 clinical isolates of *B. fragilis*, the prevalence of *cfiA/ccrA* positivity was 1.9% in one hospital and 4.1% in the other, and only 23% of the *cfiA/ccrA*-positive isolates were resistant or intermediate to imipenem (177). In a large surveillance study conducted in France, among a total of 500 isolates of *B. fragilis* randomly collected from 35 hospitals, the prevalence of *cfiA/ccrA* positivity was 2.4%, and only one-third of the *cfiA/ccrA*-positive isolates were resistant to imipenem (126).

Aeromonas spp. are associated with either intestinal or extraintestinal infections (mostly wound infections after trauma, and bacteremias in compromised hosts) (1). The CphA/Imi metalloenzyme is present in only some *Aeromonas* spp. including *A. hydrophila*, *A. salmonicida*, *A. veronii*, and *A. jandaei* (49, 143, 144, 173). Unlike most other CBBLs, the *Aeromonas* enzyme behaves as a quite specific carbapenemase (see above) and its impact essentially concerns resistance to those agents. It should be noted that most isolates of *cphA/imi*-positive *Aeromonas* species do not exhibit a carbapenem-resistant phenotype in conventional susceptibility testing. However, with a higher bacterial inoculum, carbapenem MICs usually exceed the breakpoint for resistance so that carbapenems are not recommended for treatment of infections caused by CBBL-producing *Aeromonas* spp. (143).

C. indologenes is the flavobacterial species most frequently isolated from clinical specimens, but it rarely has clinical significance and virulence is overall low (152). *E. meningoseptica* is the flavobacterial species of highest clinical importance, causing neonatal meningitis and opportunistic nosocomial infections (152). Production of the resident CBBLs (BlaB and GOB in *E. meningoseptica* and IND in *C. indologenes*) contributes to the intrinsic broad-spectrum β-lactam resistance typical of those flavobacterial species (152) and would play a crucial role in determining the intrinsic resistance to carbapenems exhibited by these species.

B. anthracis, the highly pathogenic species causing anthrax, is usually susceptible to penicillin G and other β-lactams since expression of the two chromosomal β-lactamase genes (including the *bla2* CBBL gene) is weak and not inducible (25, 26). However, the occurrence of β-lactam-resistant isolates of *B. anthracis* which produce β-lactamase activity has been reported (26, 112). In a penicillin-resistant clinical isolate in which both genes were expressed at high level, the contribution to β-lactam resistance of *bla2* expression was found to be overall marginal (26).

The CBBLs of greatest clinical importance are the broad-spectrum enzymes of the IMP and VIM types, which are encoded by genes associated with mobile DNA elements and have started spreading in major nosocomial pathogens since the 1990s (19, 77, 117). To date, diffusion of these genes has involved *P. aeruginosa*, *Acinetobacter* spp., other GNNFs, and several members of the family *Enterobacteriaceae*, and it appears to be a phenomenon of a worldwide scale (Tables 9.2 and 9.3). The IMP enzymes appear to be the most prevalent type of acquired CBBLs in Japan (79, 83, 155), while the VIM enzymes would seem to be the most common type in Europe (57, 88, 101; reference 117 and references therein; 161, 165) and in Korea (93, 94). However, current knowledge on the epidemiology of these resistance determinants is still rather limited due to the limited number of large-scale specific surveillance studies and to the fact that routine screening for acquired CBBLs is carried out in only a minority of clinical microbiology labs.

Strains producing IMP-type or VIM-type enzymes have been reported as causing sporadic cases or nosocomial outbreaks of variable sizes. In Japan, where the IMP enzymes were first detected, surveillance studies carried out in the 1990s have overall shown a low-level endemicity of IMP producers. In a large surveillance study, carried out in 1992–94 with a total of 3,700 *P. aeruginosa* isolates from 17 Japanese hospitals, the prevalence of bla_{IMP} carriage (screened among isolates resistant to ceftazidime and imipenem or showing high-level resistance to ceftazidime) was 0.4%, and most IMP-positive isolates were actually from two small outbreaks (154). In another survey, carried out in Japan in 1996–97 with almost 6,000 isolates of *P. aeruginosa* and *Serratia marcescens*, the prevalence of IMP production was found to be 1.3% among *P. aeruginosa* isolates and 4.4% among *S. marcescens* isolates, respectively (87). Low prevalences of MBL producers were also reported in a collection of 800 *Enterobacteriaceae* and 200 *P. aeruginosa* isolates from a region of Japan in 1998 and 2000 (0.7% of IMP producers) (176), in a collection of 801 gram negatives isolated from three medical centers in the period 1998–2002 (1.1% of IMP producers) (79), and in a collection of 594 *P. aeruginosa* isolates from 60 hospitals thoughout Japan in 2002 (1.9% of MBL producers, mostly of the IMP type). In Europe, where IMP- and VIM-type enzymes emerged in the late 1990s, most reports concern sporadic cases or small nosocomial outbreaks and are from the Mediterranean area (22, 32, 33, 39, 46, 59, 62, 90, 99, 100, 110; reference 117 and references therein; 124, 130, 151, 160, 161, 166, 174; N. Woodford, S. Salso, L. Tysall, J. Coelho, V. Kretchikov, A. Sinclair, M. Kaufmann, H. Schuster, G. Duckworth, and D. Livermore, *Abstr. 43rd Intersci. Conf. Antimicrob. Agents Chemother.*, abstr. C2-2019, 2003). Two recent longitudinal studies conducted in southern European hospitals revealed low prevalences (0.8 to 0.9%) of MBL producers among *Pseudomonas* spp. (101, 165). However, a major outbreak caused by a *P. aeruginosa* strain producing the VIM-2 enzyme (more than 200 isolates over a 3-year period, involving several departments) has been reported in a Greek hospital (105, 164), and recent data suggest that VIM producers could provide a remarkable contribution to carbapenem resistance in *P. aeruginosa* in various Greek hospitals (57). A worrying scenario has also been observed in a large hospital of northern Italy, where sporadic isolates of *P. aeruginosa* producing VIM type CBBLs were first detected in 1999 (142). In that hospital, multiple *P. aeruginosa* strains producing VIM-type enzymes have rapidly established a high-level endemicity and, two years after the first detection, they accounted for approximately 20% of all *P. aeruginosa* isolates and 70% of carbapenem-resistant isolates, involving several different hospital departments as well as a number of long-term care facilities which frequently exchange patients with the hospital. The phenomenon was also associated with a net increase of carbapenem-resistance rates in *P. aeruginosa* isolates (88). IMP and/or VIM producers have also been reported among clinical isolates from North America (60, 63a, 98a, 162), Latin America (34, 106, 107, 123, 149), and Australia (125). However, the epidemiological impact of the IMP- and VIM-type MBLs appears to be overall lower in those continents than in Europe and the Far East.

The impact of CBBL production among β-lactam-resistant isolates of gram-negative bacteria can be variable in different epidemiological settings. In surveys carried out in Japan in the 1990s, the rate of IMP producers among ceftazidime-resistant isolates of gram-negative bacteria was found to be around 10% (68, 153). However, in a recent study carried out in Japan (2001–2002), the prevalence of CBBL producers (including both IMP and VIM types) among gram-negative isolates resistant to both ceftazidime and cefoperazone-sulbactam was found to be 73% (155). In Korea, the prevalence of CBBL production (including both IMP and VIM types) among imipenem-resistant *P. aeruginosa* and *Acinetobacter* spp. isolated during a nationwide survey (involving 28 hospitals) in 2000–2001 was found to be 11 and 14%, respectively (94), and similar rates were observed also in 2003 (93). In Greece, the prevalence of VIM-type producers among imipenem-nonsusceptible *P. aeruginosa* strains isolated from 15 Greek hospitals in 2001 was found to be 62% (57).

Altogether, all these data strongly support the notion that IMP-type and VIM-type enzymes are emerging resistance determinants of increasing clinical importance.

More limited at this stage seems to be the epidemiological impact of the SPM-1 enzyme, which has been detected only in multidrug-resistant *P. aeruginosa* isolates from Brazilian hospitals (51, 128, 149), and that of the newly discovered GIM-1 enzyme, which has been detected in a multidrug-resistant *P. aeruginosa* strain causing a small outbreak that occurred in a German hospital (24). This more limited distribution, however, might only reflect the later discovery of these genes, which are expected to exhibit propensity for horizontal spreading due to their association with mobile DNA elements.

In *P. aeruginosa* and in other GNNFs, production of an acquired CBBL usually results in a broad-spectrum β-lactam resistance phenotype that includes penicillins, expanded-spectrum β-lactams and carbapenems, with carbapenem MICs that often are ≥64 μg/ml (reference 98 and references therein; 155). However, *Pseudomonas* and *Acinetobacter* isolates carrying acquired CBBL genes have occasionally been reported showing relatively low carbapenem MICs (83, 93). In *Enterobacteriaceae*, the impact of acquired CBBL production can be different depending on the species. In *S. marcescens* it is usually associated with carbapenem MICs that exceed the breakpoint for resistance (73, 153, 155, 187), while in *Escherichia coli*, *K. pneumoniae*, *Shigella flexneri*, *E. cloacae*, and *Citrobacter* spp., carbapenem MICs often remain lower than the breakpoint for susceptibility, although they are usually higher than the modal MIC values for the respective species (59, 64, 76, 100, 110, 119, 180). In the latter case, carbapenem MICs can readily exceed the breakpoints for resistance when larger bacterial inocula are tested and failures of carbapenem-based chemotherapy have been reported, notwithstanding an apparent in vitro susceptibility (100). Therefore, while waiting for a critical revision of breakpoints, it would seem advisable to consider all CBBL producers to be biologically resistant to carbapenems, even though they appear to be susceptible in conventional in vitro susceptibility testing. It should also be noted that, with some MBL-producing *Enterobacteriaceae*, significant discrepancies can be observed in carbapenem MICs determined with different testing systems (58).

LABORATORY DETECTION

Detection of MBL production is not of particular interest in pathogenic species with resident enzymes, such as *S. maltophilia*, *A. hydrophila*, or *E. meningoseptica*, but became an increasingly relevant issue in the clinical microbiology laboratory once CBBLs encoded by genes associated with mobile DNA started spreading among major gram-negative pathogens. In fact, surveillance of these emerging resistance determinants is very important to monitor the extent and the evolutions of the phenomenon, as well as to evaluate the efficacy of control measures.

Since carbapenemase activity is a constant feature of CBBLs and an uncommon feature among serine-β-lactamases, a carbapenem-resistant phenotype in an isolate of a gram-negative species that is naturally susceptible to carbapenems should always be considered suspicious for the production of similar enzymes. However, carbapenem resistance per se is neither a specific nor a sensitive marker for production of acquired CBBLs. In fact, in *P. aeruginosa* and in *Acinetobacter* spp. (where production of an acquired CBBL usually results in a carbapenem-resistant phenotype) acquired carbapenem resistance is often due to mechanisms other than CBBL production, such as decreased outer membrane permeability, upregulation of efflux systems, and, in *Acinetobacter*, production of OXA-type carbapenemases (117). On the other hand, in *Enterobacteriaceae*, production of an acquired CBBL may not result in a carbapenem-resistant phenotype in conventional susceptibility testing, especially in some species (see above).

The reference method for detection of MBL activity in a clinical isolate is based on testing, by a spectrophotometric assay, for the presence of EDTA-inhibitable carbapenemase activity in a sonic extract of the bacterial culture. Imipenem ($\Delta\varepsilon^{299} = -9{,}000\ M^{-1}\cdot cm^{-1}$) is generally used as a substrate (at a concentration of 100 to 150 μM), and buffer systems compatible with zinc solubility (such as HEPES/NaOH) are preferred (at an ionic strength of 10 to 100 mM and at a pH value of around 7). Upon extract addition to the imipenem-containing buffer, the A_{299} is monitored in a time course assay (usually carried out at room temperature or at 30 to 37°C). A regular decrease of the absorbance (significantly faster than that resulting from spontaneous imipenem degradation in the buffer system, under the same assay conditions) indicates the presence of carbapenemase activity. The assay is then repeated after preincubation of the cell extract (usually for 10 to 15 min, at room temperature) in the presence of EDTA (usually at a final

Table 9.7 Tests for phenotypic detection of acquired MBL production in gram-negative pathogens

Test	Format	Drug/inhibitor[a]	Reference(s)	Principle and notes
Hodge, modified	Diffusion	IMI	96	IMI and zinc sulfate on the same disk, placed on medium seeded with an *Escherichia coli* susceptible strain; the test strain is streaked from the edge of the disk outward; positivity is indicated by distortion of the inhibitory zone in presence of the test strain. Detects carbapenemase activity in general.
CAZ/MPA double-disk CAZ/SMA double-disk IMI/EDTA double-disk IMI/EDTA-SMA double-disk CAZ-CLA/MPA double-disk FEP/MPA double-disk FEP-CLA/MPA double-disk	Diffusion	CAZ/2-MPA CAZ/SMA IMI/EDTA IMI/EDTA+SMA CAZ+CLA/2-MPA FEP/2-MPA FEP+CLA/2-MPA	4, 96, 182	Antibiotics and inhibitors on separate disks; positivity is indicated by enhanced inhibition in the area between the two disks (double-disk-related synergy). Good results obtained with *Pseudomonas aeruginosa* and other GNNFs, while variable results reported with *Enterobacteriaceae*. The addition of clavulanate to the β-lactam and the use of cefepime were recently proposed to improve the performance of the test with CBBL-producing *Enterobacteriaceae*.
Etest MBL	Diffusion	IMI/EDTA	170	Etest strip with IMI on one side and IMI+EDTA on the other side; positivity is indicated by an MIC reduction in the presence of EDTA. Good results obtained with various gram-negatives producing different CBBLs (except with *Chryseobacterium* spp.). Reading can be difficult and performance reduced with isolates showing low imipenem MICs.
IMI-EDTA disk	Diffusion	IMI/EDTA	185	IMI and EDTA on the same disk; positivity is indicated by an increase of the inhibitory zone in comparison with that of IMI alone. Good results with *Pseudomonas* spp. and, to a somewhat lower extent, with *Acinetobacter* spp.
EPI microdilution	Microdilution	IMI/EDTA + PHE	109	IMI MIC is determined either in presence or in absence of the chelating mix; positivity is indicated by a reduction of the IMI MIC in presence of the chelating mix. Good results obtained with *Pseudomonas aeruginosa* (not tested with other species). Uses microtiter plates. Amenable to automated testing.

[a] IMI, imipenem; CAZ, ceftazidime; 2-MPA, 2-mercaptopropionic acid; SMA, mercaptoacetic acid; CLA, clavulanic acid; FEP, cefepime; PHE, 1,10-phenanthroline.

Table 9.8 Primers for detection of bla_{IMP} and bla_{VIM} genes by multiplex PCR assay, and strategy for presumptive identification of the sublineage by means of RFLP analysis of PCR products[a]

PCR primers	Amplicon size (bp)	Genes[b]	Enzyme[c]	Restriction profile
IMP-DIA				
/f 5′-ggAATAgAgTggCTTAATTCTC	36	$bla_{IMP-1/3/6/10}$	AluI	271 + 90 bp
/r 5′-gTgATgCgTCYCCAAYTTCACT		$bla_{IMP-2/8/12/18/19/20}$		183 + 90 + 88 bp
		$bla_{IMP-5/7}$		158 + 90 + 88 + 25 bp
		bla_{IMP-4}		183 + 75 + 60 + 28 + 15 bp
		bla_{IMP-13}		183 + 88 + 50 + 40 bp
VIM-DIA				
/f 5′-CAgATTgCCgATggTgTTTgg	52	$bla_{VIM-1/4/5}$	RsaI	259 + 264 bp
/r 5′-AggTgggCCATTCAgCCAgA		$bla_{VIM-2/3/6/8/9/10/11}$		259 + 159 + 105 bp

[a] A description of the reaction conditions can be found in reference 46.
[b] These primers are not suitable to amplify bla_{IMP-9}, bla_{IMP-11}, bla_{IMP-16}, or bla_{VIM-7} due to the presence of multiple mismatches in one or both primers.
[c] Restriction enzymes used for RFLP analysis of the amplicons.

concentration of 2 to 5 mM), to determine the extent of inhibition caused by the chelating agent.

The above method, however, is not suitable for routine use in the clinical microbiology lab. Recently, several phenotypic tests for detection of MBL activity in clinical isolates have been developed that are amenable to use in the clinical laboratory setting. Of these tests, one is a revisitation of the Hodge test for detection of β-lactamase production which actually detects carbapenemase production (92, 96), while the others are all based on the same general principle, namely the detection of an increased susceptibility to a β-lactam which is hydrolyzed by MBLs in the presence of an enzyme inhibitor. The tests based on inhibitors differ from each other in the format of the assay (diffusion versus dilution) and in the nature of the combinations of β-lactams and inhibitors. The relevant features of the various phenotypic tests that have been developed for detection of MBL production in clinical isolates are summarized in Table 9.7. Currently, these tests have been validated mostly with *P. aeruginosa* and other GNNFs, while the experience with *Enterobacteriaceae* is more limited.

Once production of an MBL activity has been detected by a phenotypic test, molecular characterization of the MBL gene is warranted to confirm its presence and to identify its nature. Filter hybridization assays (in the format of colony, dot, or Southern blots) with type-specific probes are a robust and relatively inexpensive method that can be useful to detect CBBL genes of known types, although they usually cannot distinguish between different allelic variants of the same type. Currently, PCR analysis is most commonly used for detection and identification of CBBL genes; the degree of analytical resolution of PCR analysis depends on the design of primers used for the amplification step and on the approach adopted for characterization of the amplification product. A multiplex PCR test using "broad-spectrum" primers for bla_{IMP} and bla_{VIM} genes, designed on conserved regions, in combination with analysis of the restriction fragment length polymorphisms (RFLP) of the amplification products (Table 9.8), was shown to be effective for detection of acquired CBBL genes of the above types in clinical isolates, and for rapid characterization of the gene at the sublineage level (44, 46, 99). Precise identification of the MBL determinant requires sequencing of the entire gene. For genes of known type, this is most easily done by PCR amplification of the coding sequence with external primers followed by direct sequencing of the amplification product. Since the bla_{IMP} and bla_{VIM} genes are known to be carried on gene cassettes inserted into integrons, an alternative strategy for their detection and characterization consists of PCR amplification of the integron cassette array followed by direct sequencing of the amplification products (136). A similar strategy can also allow the identification of new MBL genes carried on gene cassettes, as was the case for bla_{GIM-1} (24). If molecular probes for known CBBL genes yield negative results with an isolate for which MBL production has been confirmed, then the presence of a novel MBL determinant should be suspected and shotgun cloning and sequencing can be adopted for identification and characterization of the new gene. Protocols for shotgun cloning of MBL determinants are described in references 91 and 141.

The work on CBBLs in our lab has mostly been funded by the European Commission (since 1993, contracts ERBCHRX-CT93-026, FMRX-CT98-0232, HPRN-CT-2002-00264, and 6 PCRD LSHM-CT-2003-503335), by the Italian Ministry of University and Research (since 1999, contracts 9906404271, 2001068755_003), and by the Belgian Fonds National de la Recherche Scientifique (2003–2007).

References

1. **Abbott, S. L.** 2003. *Aeromonas*, p. 701–705. *In* P. R. Murray, E. J. Baron, J. H. Jorgensen, M. A. Pfaller, and R. H. Yolken (ed.), *Manual of Clinical Microbiology*, 8th ed. ASM Press, Washington, D.C.
2. **Ambler, R. P.** 1980. The structure of β-lactamases. *Philos. Trans. R. Soc. Lond. B Biol. Sci.* **289:**321–331.
3. **Arakawa, Y., M. Murakami, K. Suzuki, H. Ito, R. Wacharotayankun, S. Ohsuka, N. Kato, and M. Ohta.** 1995. A novel integron-like element carrying the metallo-β-lactamase gene bla_{IMP}. *Antimicrob. Agents Chemother.* **39:** 1612–1615.
4. **Arakawa, Y., N. Shibata, K. Shibayama, H. Kurokawa, T. Yagi, H. Fujiwara, and M. Goto.** 2000. Convenient test for screening metallo-β-lactamase-producing gram-negative bacteria by using thiol compounds. *J. Clin. Microbiol.* **38:** 40–43.
5. **Avison, M. B., C. S. Higgins, C. J. von Heldreich, P. M. Bennett, and T. R. Walsh.** 2001. Plasmid location and molecular heterogeneity of the L1 and L2 β-lactamase genes of *Stenotrophomonas maltophilia*. *Antimicrob. Agents Chemother.* **45:**413–419.
6. **Avison, M. B., P. Niumsup, K. Nurmahomed, T. R. Walsh, and P. M. Bennett.** 2004. Role of the 'cre/blr-tag' DNA sequence in regulation of gene expression by the *Aeromonas hydrophila* β-lactamase regulator, BlrA. *J. Antimicrob. Chemother.* **53:**197–202.
7. **Bahar, G., A. Mazzariol, R. Koncan, A. Mert, R. Fontana, G. M. Rossolini, and G. Cornaglia.** 2004. Detection of VIM-5 metallo-β-lactamase in a *Pseudomonas aeruginosa* clinical isolate from Turkey. *J. Antimicrob. Chemother.* **54:** 282–283.
8. **Bellais, S., D. Aubert, T. Naas, and P. Nordmann.** 2000. Molecular and biochemical heterogeneity of class B carbapenem-hydrolyzing β-lactamases in *Chryseobacterium meningosepticum*. *Antimicrob. Agents Chemother.* **44:** 1878–1886.
9. **Bellais, S., D. Girlich, A. Karim, and P. Nordmann.** 2002. EBR-1, a novel Ambler subclass B1 β-lactamase from *Empedobacter brevis*. *Antimicrob. Agents Chemother.* **46:** 3223–3227.
10. **Bellais, S., S. Leotard, L. Poirel, T. Naas, and P. Nordmann.** 1999. Molecular characterization of a carbapenem-hydrolyzing β-lactamase from *Chryseobacterium* (*Flavobacterium*) *indologenes*. *FEMS Microbiol. Lett.* **171:** 127–132.
11. **Bellais, S., T. Naas, and P. Nordmann.** 2002. Genetic and biochemical characterization of CGB-1, an Ambler class B carbapenem-hydrolyzing β-lactamase from *Chryseobacterium gleum*. *Antimicrob. Agents Chemother.* **46:** 2791–2796.
12. **Bellais, S., L. Poirel, S. Leotard, T. Naas, and P. Nordmann.** 2000. Genetic diversity of carbapenem-hydrolyzing metallo-β-lactamases from *Chryseobacterium* (*Flavobacterium*) *indologenes*. *Antimicrob. Agents Chemother.* **44:** 3028–3034.
13. **Bicknell, R., E. L. Emanuel, J. Gagnon, and S. G. Waley.** 1985. The production and molecular properties of the zinc β-lactamase of *Pseudomonas maltophilia IID 1275*. *Biochem. J.* **229:**791–797.
14. **Bicknell, R., and S. G. Waley.** 1985. Cryoenzymology of *Bacillus cereus* β-lactamase II. *Biochemistry* **24:**6876–6887.
15. **Bonfiglio, G., S. Stefani, and G. Nicoletti.** 1995. Clinical isolate of a *Xanthomonas maltophilia* strain producing L-1-deficient and L-2-inducible β-lactamases. *Chemotherapy* **41:**121–124.
16. **Boschi, L., P. S. Mercuri, M. L. Riccio, G. Amicosante, M. Galleni, J. M. Frere, and G. M. Rossolini.** 2000. The *Legionella* (*Fluoribacter*) *gormanii* metallo-β-lactamase: a new member of the highly divergent lineage of molecular-subclass B3 β-lactamases. *Antimicrob. Agents Chemother.* **44:**1538–1543.
17. **Bounaga, S., A. P. Laws, M. Galleni, and M. I. Page.** 1998. The mechanism of catalysis and the inhibition of the *Bacillus cereus* zinc-dependent β-lactamase. *Biochem. J.* **331**(Pt. 3):703–711.
18. **Bush, K.** 1998. Metallo-β-lactamases: a class apart. *Clin. Infect. Dis.* **27**(Suppl. 1):S48–S53.
19. **Bush, K.** 2001. New β-lactamases in gram-negative bacteria: diversity and impact on the selection of antimicrobial therapy. *Clin. Infect. Dis.* **32:**1085–1089.
20. **Bush, K., G. A. Jacoby, and A. A. Medeiros.** 1995. A functional classification scheme for β-lactamases and its correlation with molecular structure. *Antimicrob. Agents Chemother.* **39:**1211–1233.
21. **Callebaut, I., D. Moshous, J. P. Mornon, and J. P. De Villartay.** 2002. Metallo-β-lactamase fold within nucleic acids processing enzymes: the β-CASP family. *Nucleic Acids Res.* **30:**3592–3601.
22. **Cardoso, O., R. Leitao, A. Figueiredo, J. C. Sousa, A. Duarte, and L. V. Peixe.** 2002. Metallo-β-lactamase VIM-2 in clinical isolates of *Pseudomonas aeruginosa* from Portugal. *Microb. Drug Resist.* **8:**93–97.
23. **Carfi, A., S. Pares, E. Duee, M. Galleni, C. Duez, J. M. Frere, and O. Dideberg.** 1995. The 3-D structure of a zinc metallo-β-lactamase from *Bacillus cereus* reveals a new type of protein fold. *EMBO J.* **14:**4914–4921.
24. **Castanheira, M., M. A. Toleman, R. N. Jones, F. J. Schmidt, and T. R. Walsh.** 2004. Molecular characterization of a beta-lactamase gene, bla_{GIM-1}, encoding a new subclass of metallo-β-lactamase. *Antimicrob. Agents Chemother.* **48:** 4654–4661.
25. **Chen, Y., J. Succi, F. C. Tenover, and T. M. Koehler.** 2003. β-lactamase genes of the penicillin-susceptible *Bacillus anthracis* Sterne strain. *J. Bacteriol.* **185:**823–830.
26. **Chen, Y., F. C. Tenover, and T. M. Koehler.** 2004. β-Lactamase gene expression in a penicillin-resistant *Bacillus anthracis* strain. *Antimicrob. Agents Chemother.* **48:**4873–4877.
27. **Chu, Y. W., M. Afzal-Shah, E. T. Houang, M. I. Palepou, D. J. Lyon, N. Woodford, and D. M. Livermore.** 2001. IMP-4, a novel metallo-β-lactamase from nosocomial *Acinetobacter* spp. collected in Hong Kong between 1994 and 1998. *Antimicrob. Agents Chemother.* **45:**710–714.
28. **Collis, C. M., and R. M. Hall.** 1995. Expression of antibiotic resistance genes in the integrated cassettes of integrons. *Antimicrob. Agents Chemother.* **39:**155–162.
29. **Conceicao, T., A. Brizio, A. Duarte, and R. Barros.** 2005. First isolation of bla_{VIM-2} in *Klebsiella oxytoca* clinical

isolates from Portugal. *Antimicrob. Agents Chemother.* **49**:476.

30. **Concha, N. O., C. A. Janson, P. Rowling, S. Pearson, C. A. Cheever, B. P. Clarke, C. Lewis, M. Galleni, J. M. Frere, D. J. Payne, J. H. Bateson, and S. S. Abdel-Meguid.** 2000. Crystal structure of the IMP-1 metallo β-lactamase from *Pseudomonas aeruginosa* and its complex with a mercaptocarboxylate inhibitor: binding determinants of a potent, broad-spectrum inhibitor. *Biochemistry* **39**:4288–4298.
31. **Concha, N. O., B. A. Rasmussen, K. Bush, and O. Herzberg.** 1996. Crystal structure of the wide-spectrum binuclear zinc β-lactamase from *Bacteroides fragilis. Structure* **4**:823–836.
32. **Cornaglia, G., A. Mazzariol, L. Lauretti, G. M. Rossolini, and R. Fontana.** 2000. Hospital outbreak of carbapenem-resistant *Pseudomonas aeruginosa* producing VIM-1, a novel transferable metallo-β-lactamase. *Clin. Infect. Dis.* **31**:1119–1125.
33. **Cornaglia, G., M. L. Riccio, A. Mazzariol, L. Lauretti, R. Fontana, and G. M. Rossolini.** 1999. Appearance of IMP-1 metallo-β-lactamase in Europe. *Lancet* **353**:899–900.
34. **Crespo, M. P., N. Woodford, A. Sinclair, M. E. Kaufmann, J. Turton, J. Glover, J. D. Velez, C. R. Castaneda, M. Recalde, and D. M. Livermore.** 2004. Outbreak of carbapenem-resistant *Pseudomonas aeruginosa* producing VIM-8, a novel metallo-β-lactamase, in a tertiary care center in Cali, Colombia. *J. Clin. Microbiol.* **42**:5094–5101.
35. **Crowder, M. W., Z. Wang, S. L. Franklin, E. P. Zovinka, and S. J. Benkovic.** 1996. Characterization of the metal-binding sites of the β-lactamase from *Bacteroides fragilis. Biochemistry* **35**:12126–12132.
36. **Cuchural, G. J., Jr., M. H. Malamy, and F. P. Tally.** 1986. β-Lactamase-mediated imipenem resistance in *Bacteroides fragilis. Antimicrob. Agents Chemother.* **30**:645–648.
37. **Daiyasu, H., K. Osaka, Y. Ishino, and H. Toh.** 2001. Expansion of the zinc metallo-hydrolase family of the β-lactamase fold. *FEBS Lett.* **503**:1–6.
38. **Dal Peraro, M., A. J. Vila, and P. Carloni.** 2004. Substrate binding to mononuclear metallo-β-lactamase from *Bacillus cereus. Proteins* **54**:412–423.
39. **Da Silva, G. J., M. Correia, C. Vital, G. Ribeiro, J. C. Sousa, R. Leitao, L. Peixe, and A. Duarte.** 2002. Molecular characterization of bla_{IMP-5}, a new integron-borne metallo-β-lactamase gene from an *Acinetobacter baumannii* nosocomial isolate in Portugal. *FEMS Microbiol. Lett.* **215**:33–39.
40. **Denton, M., and K. G. Kerr.** 1998. Microbiological and clinical aspects of infection associated with *Stenotrophomonas maltophilia. Clin. Microbiol. Rev.* **11**:57–80.
41. **de Seny, D., C. Prosperi-Meys, C. Bebrone, G. M. Rossolini, M. I. Page, P. Noel, J. M. Frere, and M. Galleni.** 2002. Mutational analysis of the two zinc-binding sites of the *Bacillus cereus* 569/H/9 metallo-β-lactamase. *Biochem. J.* **363**:687–696.
42. **Docquier, J. D., J. Lamotte-Brasseur, M. Galleni, G. Amicosante, J. M. Frere, and G. M. Rossolini.** 2003. On functional and structural heterogeneity of VIM-type metallo-β-lactamases. *J. Antimicrob. Chemother.* **51**:257–266.
43. **Docquier, J. D., T. Lopizzo, S. Liberatori, M. Prenna, M. C. Thaller, J. M. Frere, and G. M. Rossolini.** 2004. Biochemical characterization of the THIN-B metallo-β-lactamase of *Janthinobacterium lividum. Antimicrob. Agents Chemother.* **48**:4778–4783.
44. **Docquier, J. D., F. Luzzaro, G. Amicosante, A. Toniolo, and G. M. Rossolini.** 2001. Multidrug-resistant *Pseudomonas aeruginosa* producing PER-1 extended-spectrum serine-β-lactamase and VIM-2 metallo-β-lactamase. *Emerg. Infect. Dis.* 7:910–911.
45. **Docquier, J. D., F. Pantanella, F. Giuliani, M. C. Thaller, G. Amicosante, M. Galleni, J. M. Frere, K. Bush, and G. M. Rossolini.** 2002. CAU-1, a subclass B3 metallo-β-lactamase of low substrate affinity encoded by an ortholog present in the *Caulobacter crescentus* chromosome. *Antimicrob. Agents Chemother.* **46**:1823–1830.
46. **Docquier, J. D., M. L. Riccio, C. Mugnaioli, F. Luzzaro, A. Endimiani, A. Toniolo, G. Amicosante, and G. M. Rossolini.** 2003. IMP-12, a new plasmid-encoded metallo-β-lactamase from a *Pseudomonas putida* clinical isolate. *Antimicrob. Agents Chemother.* **47**:1522–1528.
47. **Felici, A., and G. Amicosante.** 1995. Kinetic analysis of extension of substrate specificity with *Xanthomonas maltophilia, Aeromonas hydrophila,* and *Bacillus cereus* metallo-β-lactamases. *Antimicrob. Agents Chemother.* **39**: 192–199.
48. **Felici, A., G. Amicosante, A. Oratore, R. Strom, P. Ledent, B. Joris, L. Fanuel, and J. M. Frere.** 1993. An overview of the kinetic parameters of class B β-lactamases. *Biochem. J.* **291**(Pt.1):151–155.
49. **Fosse, T., C. Giraud-Morin, I. Madinier, and R. Labia.** 2003. Sequence analysis and biochemical characterisation of chromosomal CAV-1 (*Aeromonas caviae*), the parental cephalosporinase of plasmid-mediated AmpC 'FOX' cluster. *FEMS Microbiol. Lett.* **222**:93–98.
50. **Franceschini, N., B. Caravelli, J. D. Docquier, M. Galleni, J. M. Frere, G. Amicosante, and G. M. Rossolini.** 2000. Purification and biochemical characterization of the VIM-1 metallo-β-lactamase. *Antimicrob. Agents Chemother.* **44**: 3003–3007.
51. **Gales, A. C., L. C. Menezes, S. Silbert, and H. S. Sader.** 2003. Dissemination in distinct Brazilian regions of an epidemic carbapenem-resistant *Pseudomonas aeruginosa* producing SPM metallo-β-lactamase. *J. Antimicrob. Chemother.* **52**:699–702.
52. **Galleni, M., J. Lamotte-Brasseur, G. M. Rossolini, J. Spencer, O. Dideberg, and J. M. Frere.** 2001. Standard numbering scheme for class B β-lactamases. *Antimicrob. Agents Chemother.* **45**:660–663.
53. **Garau, G., C. Bebrone, C. Anne, M. Galleni, J. M. Frere, and O. Dideberg.** 2005. A metallo-β-lactamase enzyme in action: crystal structures of the monozinc carbapenemase CphA and its complex with biapenem. *J. Mol. Biol.* **345**: 785–795.
54. **Garau, G., I. Garcia-Saez, C. Bebrone, C. Anne, P. Mercuri, M. Galleni, J. M. Frere, and O. Dideberg.** 2004. Update of the standard numbering scheme for class B β-lactamases. *Antimicrob. Agents Chemother.* **48**:2347–2349.
55. **Garcia-Saez, I., P. S. Mercuri, C. Papamicael, R. Kahn, J. M. Frere, M. Galleni, G. M. Rossolini, and O. Dideberg.** 2003. Three-dimensional structure of FEZ-1, a monomeric subclass B3 metallo-β-lactamase from *Fluoribacter gormanii,* in native form and in complex with D-captopril. *J. Mol. Biol.* **325**:651–660.

56. **Ghuysen, J. M.** 1991. Serine β-lactamases and penicillin-binding proteins. *Annu. Rev. Microbiol.* **45:**37–67.

57. **Giakkoupi, P., G. Petrikkos, L. S. Tzouvelekis, S. Tsonas, N. J. Legakis, and A. C. Vatopoulos.** 2003. Spread of integron-associated VIM-type metallo-β-lactamase genes among imipenem-nonsusceptible *Pseudomonas aeruginosa* strains in Greek hospitals. *J. Clin. Microbiol.* **41:**822–825.

58. **Giakkoupi, P., L. S. Tzouvelekis, G. L. Daikos, V. Miriagou, G. Petrikkos, N. J. Legakis, and A. C. Vatopoulos.** 2005. Discrepancies and interpretation problems in susceptibility testing of VIM-1-producing *Klebsiella pneumoniae* isolates. *J. Clin. Microbiol.* **43:**494–496.

59. **Giakkoupi, P., A. Xanthaki, M. Kanelopoulou, A. Vlahaki, V. Miriagou, S. Kontou, E. Papafraggas, H. Malamou-Lada, L. S. Tzouvelekis, N. J. Legakis, and A. C. Vatopoulos.** 2003. VIM-1 metallo-β-lactamase-producing *Klebsiella pneumoniae* strains in Greek hospitals. *J. Clin. Microbiol.* **41:**3893–3896.

60. **Gibb, A. P., C. Tribuddharat, R. A. Moore, T. J. Louie, W. Krulicki, D. M. Livermore, M. F. Palepou, and N. Woodford.** 2002. Nosocomial outbreak of carbapenem-resistant *Pseudomonas aeruginosa* with a new bla_{IMP} allele, bla_{IMP-7}. *Antimicrob. Agents Chemother.* **46:**255–258.

61. **Gilligan, P. H., G. Lum, P. A. R. Vandamme, and S. Whittier.** 2003. *Burkholderia, Stenotrophomonas, Ralstonia, Brevundimonas, Comamonas, Delftia, Pandorae*, and *Acidovorax*, p. 729-748. *In* P. R. Murray, E. J. Baron, J. H. Jorgensen, M. A. Pfaller, and R. H. Yolken (ed.), Manual of Clinical Microbiology, 8th ed. ASM Press, Washington, D.C.

62. **Giske, C. G., M. Rylander, and G. Kronvall.** 2003. VIM-4 in a carbapenem-resistant strain of *Pseudomonas aeruginosa* isolated in Sweden. *Antimicrob. Agents Chemother.* **47:** 3034-3035.

63. **Gupta, R. S., and E. Griffiths.** 2002. Critical issues in bacterial phylogeny. *Theor. Popul. Biol.* **61:**423–434.

63a. **Hanson, N. D., A. Hossain, L. Buck, E. S. Moland, and K. S. Thomson.** 2006. First occurrence of a *Pseudomonas aeruginosa* isolate in the United States producing an IMP metallo-β-lactamase, IMP-18. *Antimicrob. Agents Chemother.* **50:**2272–2273.

64. **Hawkey, P. M., J. Xiong, H. Ye, H. Li, and F. H. M'Zali.** 2001. Occurrence of a new metallo-β-lactamase IMP-4 carried on a conjugative plasmid in *Citrobacter youngae* from the People's Republic of China. *FEMS Microbiol. Lett.* **194:**53–57.

65. **Heinz, U., and H. W. Adolph.** 2004. Metallo-β-lactamases: two binding sites for one catalytic metal ion? *Cell Mol. Life Sci.* **61:**2827–2839.

66. **Hernandez, V. M., A. Felici, G. Weber, H. W. Adolph, M. Zeppezauer, G. M. Rossolini, G. Amicosante, J. M. Frere, and M. Galleni.** 1997. Zn(II) dependence of the *Aeromonas hydrophila* AE036 metallo-β-lactamase activity and stability. *Biochemistry* **36:**11534–11541.

67. **Hernandez, V. M., M. Kiefer, U. Heinz, R. P. Soto, W. Meyer-Klaucke, H. F. Nolting, M. Zeppezauer, M. Galleni, J. M. Frere, G. M. Rossolini, G. Amicosante, and H. W. Adolph.** 2000. Kinetic and spectroscopic characterization of native and metal-substituted β-lactamase from *Aeromonas hydrophila* AE036. *FEBS Lett.* **467:**221–225.

68. **Hirakata, Y., K. Izumikawa, T. Yamaguchi, H. Takemura, H. Tanaka, R. Yoshida, J. Matsuda, M. Nakano, K. Tomono, S. Maesaki, M. Kaku, Y. Yamada, S. Kamihira, and S. Kohno.** 1998. Rapid detection and evaluation of clinical characteristics of emerging multiple-drug-resistant gram-negative rods carrying the metallo-β-lactamase gene bla_{IMP}. *Antimicrob. Agents Chemother.* **42:**2006–2011.

69. **Ho, S. E., G. Subramaniam, S. Palasubramaniam, and P. Navaratnam.** 2002. Carbapenem-resistant *Pseudomonas aeruginosa* in malaysia producing IMP-7 β-lactamase. *Antimicrob. Agents Chemother.* **46:**3286–3287.

70. **Houang, E. T., Y. W. Chu, W. S. Lo, K. Y. Chu, and A. F. Cheng.** 2003. Epidemiology of rifampin ADP-ribosyltransferase (*arr-2*) and metallo-β-lactamase (bla_{IMP-4}) gene cassettes in class 1 integrons in *Acinetobacter* strains isolated from blood cultures in 1997 to 2000. *Antimicrob. Agents Chemother.* **47:**1382–1390.

71. **Hussain, M., A. Carlino, M. J. Madonna, and J. O. Lampen.** 1985. Cloning and sequencing of the metallothioprotein β-lactamase II gene of *Bacillus cereus* 569/H in *Escherichia coli. J. Bacteriol.* **164:**223–229.

72. **Iaconis, J. P., and C. C. Sanders.** 1990. Purification and characterization of inducible β-lactamases in *Aeromonas* spp. *Antimicrob. Agents Chemother.* **34:**44–51.

73. **Ito, H., Y. Arakawa, S. Ohsuka, R. Wacharotayankun, N. Kato, and M. Ohta.** 1995. Plasmid-mediated dissemination of the metallo-β-lactamase gene bla_{IMP} among clinically isolated strains of *Serratia marcescens*. *Antimicrob. Agents Chemother.* **39:**824–829.

74. **Iyobe, S., H. Kusadokoro, J. Ozaki, N. Matsumura, S. Minami, S. Haruta, T. Sawai, and K. O'Hara.** 2000. Amino acid substitutions in a variant of IMP-1 metallo-β-lactamase. *Antimicrob. Agents Chemother.* **44:**2023–2027.

75. **Iyobe, S., H. Kusadokoro, A. Takahashi, S. Yomoda, T. Okubo, A. Nakamura, and K. O'Hara.** 2002. Detection of a variant metallo-β-lactamase, IMP-10, from two unrelated strains of *Pseudomonas aeruginosa* and an *Alcaligenes xylosoxidans* strain. *Antimicrob. Agents Chemother.* **46:** 2014–2016.

76. **Iyobe, S., M. Tsunoda, and S. Mitsuhashi.** 1994. Cloning and expression in *Enterobacteriaceae* of the extended-spectrum β-lactamase gene from a *Pseudomonas aeruginosa* plasmid. *FEMS Microbiol. Lett.* **121:**175–180.

77. **Jacoby, G. A., and L. S. Munoz-Price.** 2005. The new β-lactamases. *N. Engl. J. Med.* **352:**380–391.

78. **Jeong, S. H., K. Lee, Y. Chong, J. H. Yum, S. H. Lee, H. J. Choi, J. M. Kim, K. H. Park, B. H. Han, S. W. Lee, and T. S. Jeong.** 2003. Characterization of a new integron containing VIM-2, a metallo-β-lactamase gene cassette, in a clinical isolate of *Enterobacter cloacae*. *J. Antimicrob. Chemother.* **51:**397–400.

79. **Jones, R. N., L. M. Deshpande, J. M. Bell, J. D. Turnidge, S. Kohno, Y. Hirakata, Y. Ono, Y. Miyazawa, S. Kawakama, M. Inoue, Y. Hirata, and M. A. Toleman.** 2004. Evaluation of the contemporary occurrence rates of metallo-β-lactamases in multidrug-resistant Gram-negative bacilli in Japan: report from the SENTRY Antimicrobial Surveillance Program (1998–2002). *Diagn. Microbiol. Infect. Dis.* **49:** 289–294.

80. **Jousimies-Somer, H., P. H. Summanen, H. Wexler, S. M. Finegold, S. E. Gharbia, and H. N. Shah.** 2003. *Bacteroides, Porphyromonas, Prevotella, Fusobacterium,* and other anaerobic gram negative bacteria, p. 880–901. *In* P. R. Murray, E. J. Baron, J. H. Jorgensen, M. A. Pfaller, and R. H. Yolken (ed.), *Manual of Clinical Microbiology,* 8th ed. ASM Press, Washington, D.C.

81. **Kato, C., T. Kudo, K. Watanabe, and K. Horikoshi.** 1985. Nucleotide sequence of the β-lactamase gene of alkalophilic *Bacillus* sp. strain 170. *J. Gen. Microbiol.* **131**(Pt. 12):3317–3324.

82. **Kato, N., K. Yamazoe, C. G. Han, and E. Ohtsubo.** 2003. New insertion sequence elements in the upstream region of *cfiA* in imipenem-resistant *Bacteroides fragilis* strains. *Antimicrob. Agents Chemother.* **47:**979–985.

83. **Kimura, S., J. Alba, K. Shiroto, R. Sano, Y. Niki, S. Maesaki, K. Akizawa, M. Kaku, Y. Watanuki, Y. Ishii, and K. Yamaguchi.** 2005. Clonal diversity of metallo-β-lactamase-possessing *Pseudomonas aeruginosa* in geographically diverse regions of Japan. *J. Clin. Microbiol.* **43:**458–461.

84. **Koh, T. H., L. H. Sng, G. S. Babini, N. Woodford, D. M. Livermore, and L. M. Hall.** 2001. Carbapenem-resistant *Klebsiella pneumoniae* in Singapore producing IMP-1 β-lactamase and lacking an outer membrane protein. *Antimicrob. Agents Chemother.* **45:**1939–1940.

85. **Koh, T. H., G. C. Wang, and L. H. Sng.** 2004. IMP-1 and a novel metallo-β-lactamase, VIM-6, in fluorescent pseudomonads isolated in Singapore. *Antimicrob. Agents Chemother.* **48:**2334–2336.

86. **Koradi, R., M. Billeter, and K. Wuthrich.** 1996. MOLMOL: a program for display and analysis of macromolecular structures. *J. Mol. Graph.* **14:**51–32.

87. **Kurokawa, H., T. Yagi, N. Shibata, K. Shibayama, and Y. Arakawa.** 1999. Worldwide proliferation of carbapenem-resistant gram-negative bacteria. *Lancet* **354:**955.

88. **Lagatolla, C., E. A. Tonin, C. Monti-Bragadin, L. Dolzani, F. Gombac, C. Bearzi, E. Edalucci, F. Gionechetti, and G. M. Rossolini.** 2004. Endemic carbapenem-resistant *Pseudomonas aeruginosa* with acquired metallo-β-lactamase determinants in European hospitals. *Emerg. Infect. Dis.* **10:** 535–538.

89. **Laraki, N., N. Franceschini, G. M. Rossolini, P. Santucci, C. Meunier, E. de Pauw, G. Amicosante, J. M. Frere, and M. Galleni.** 1999. Biochemical characterization of the *Pseudomonas aeruginosa* 101/1477 metallo-β-lactamase IMP-1 produced by *Escherichia coli. Antimicrob. Agents Chemother.* **43:**902–906.

90. **Lartigue, M. F., L. Poirel, and P. Nordmann.** 2004. First detection of a carbapenem-hydrolyzing metalloenzyme in an *Enterobacteriaceae* isolate in France. *Antimicrob. Agents Chemother.* **48:**4929–4930.

91. **Lauretti, L., M. L. Riccio, A. Mazzariol, G. Cornaglia, G. Amicosante, R. Fontana, and G. M. Rossolini.** 1999. Cloning and characterization of bla_{VIM}, a new integron-borne metallo-β-lactamase gene from a *Pseudomonas aeruginosa* clinical isolate. *Antimicrob. Agents Chemother.* **43:**1584–1590.

92. **Lee, K., Y. Chong, H. B. Shin, Y. A. Kim, D. Yong, and J. H. Yum.** 2001. Modified Hodge and EDTA-disk synergy tests to screen metallo-β-lactamase-producing strains of *Pseudomonas* and *Acinetobacter* species. *Clin. Microbiol. Infect.* **7:**88–91.

93. **Lee, K., G. Y. Ha, B. M. Shin, J. J. Kim, J. O. Kang, S. J. Jang, D. Yong, and Y. Chong.** 2004. Metallo-β-lactamase-producing Gram-negative bacilli in Korean Nationwide Surveillance of Antimicrobial Resistance group hospitals in 2003: continued prevalence of VIM-producing *Pseudomonas* spp. and increase of IMP-producing *Acinetobacter* spp. *Diagn. Microbiol. Infect. Dis.* **50:**51–58.

94. **Lee, K., W. G. Lee, Y. Uh, G. Y. Ha, J. Cho, and Y. Chong.** 2003. VIM- and IMP-type metallo-β-lactamase-producing *Pseudomonas* spp. and *Acinetobacter* spp. in Korean hospitals. *Emerg. Infect. Dis.* **9:**868–871.

95. **Lee, K., J. B. Lim, J. H. Yum, D. Yong, Y. Chong, J. M. Kim, and D. M. Livermore.** 2002. bla_{VIM-2} cassette-containing novel integrons in metallo-β-lactamase-producing *Pseudomonas aeruginosa* and *Pseudomonas putida* isolates disseminated in a Korean hospital. *Antimicrob. Agents Chemother.* **46:**1053–1058.

96. **Lee, K., Y. S. Lim, D. Yong, J. H. Yum, and Y. Chong.** 2003. Evaluation of the Hodge test and the imipenem-EDTA double-disk synergy test for differentiating metallo-β-lactamase-producing isolates of *Pseudomonas* spp. and *Acinetobacter* spp. *J. Clin. Microbiol.* **41:**4623–4629.

96a. **Lee, K., J. H. Yum, D. Yong, H. M. Lee, H. D. Kim, J. D. Docquier, G. M. Rossolini, and Y. Chong.** 2005. Novel acquired metallo-b-lactamase gene, bla_{SIM-1}, in a class 1 integron from *Acinetobacter baumannii clinical* isolates from Korea. *Antimicrob. Agents Chemother.* **49:**4485–4491.

97. **Levesque, C., S. Brassard, J. Lapointe, and P. H. Roy.** 1994. Diversity and relative strength of tandem promoters for the antibiotic-resistance genes of several integrons. *Gene* **142:**49–54.

97a. **Libisch, B., M. Gacs, K. Csiszar, M. Muzslay, L. Rokusz, and M. Fuzi.** 2004. Isolation of an integron-borne bla_{VIM-4} type metallo-β-lactamase gene from a carbapenem-resistant *Pseudomonas aeruginosa* clinical isolate in Hungary. *Antimicrob. Agents Chemother.* **48:**3576–3578.

98. **Livermore, D. M., and N. Woodford.** 2000. Carbapenemases: a problem in waiting? *Curr. Opin. Microbiol.* **3:** 489–495.

98a. **Lolans, K., A. M. Queenan, K. Bush, A. Sahud, and J. P. Quinn.** 2005. First nosocomial outbreak of *Pseudomonas aeruginosa* producing an integron-borne metallo-β-lactamase (VIM-2) in the United States. *Antimicrob. Agents Chemother.* **49:**3538–3540.

99. **Lombardi, G., F. Luzzaro, J. D. Docquier, M. L. Riccio, M. Perilli, A. Coli, G. Amicosante, G. M. Rossolini, and A. Toniolo.** 2002. Nosocomial infections caused by multidrug-resistant isolates of *Pseudomonas putida* producing VIM-1 metallo-β-lactamase. *J. Clin. Microbiol.* **40:**4051–4055.

100. **Luzzaro, F., J. D. Docquier, C. Colinon, A. Endimiani, G. Lombardi, G. Amicosante, G. M. Rossolini, and A. Toniolo.** 2004. Emergence in *Klebsiella pneumoniae* and *Enterobacter cloacae* clinical isolates of the VIM-4 metallo-β-lactamase encoded by a conjugative plasmid. *Antimicrob. Agents Chemother.* **48:**648–650.

101. **Luzzaro, F., A. Endimiani, J. D. Docquier, C. Mugnaioli, M. Bonsignori, G. Amicosante, G. M. Rossolini, and A.**

Toniolo. 2004. Prevalence and characterization of metallo-β-lactamases in clinical isolates of *Pseudomonas aeruginosa*. *Diagn. Microbiol. Infect. Dis.* **48**:131–135.

102. **Mammeri, H., S. Bellais, and P. Nordmann**. 2002. Chromosome-encoded β-lactamases TUS-1 and MUS-1 from *Myroides odoratus* and *Myroides odoratimimus* (formerly *Flavobacterium odoratum*), new members of the lineage of molecular subclass B1 metalloenzymes. *Antimicrob. Agents Chemother.* **46**:3561–3567.

103. **Massidda, O., G. M. Rossolini, and G. Satta**. 1991. The *Aeromonas hydrophila cphA* gene: molecular heterogeneity among class B metallo-β-lactamases. *J. Bacteriol.* **173**:4611–4617.

104. **Massova, I., and S. Mobashery**. 1998. Kinship and diversification of bacterial penicillin-binding proteins and β-lactamases. *Antimicrob. Agents Chemother.* **42**:1–17.

105. **Mavroidi, A., A. Tsakris, E. Tzelepi, S. Pournaras, V. Loukova, and L. S. Tzouvelekis**. 2000. Carbapenem-hydrolysing VIM-2 metallo-β-lactamase in *Pseudomonas aeruginosa* from Greece. *J. Antimicrob. Chemother.* **46**: 1041–1042.

106. **Mendes, R. E., M. Castanheira, P. Garcia, M. Guzman, M. A. Toleman, T. R. Walsh, and R. N. Jones**. 2004. First isolation of bla_{VIM-2} in Latin America: report from the SENTRY Antimicrobial Surveillance Program. *Antimicrob. Agents Chemother.* **48**:1433–1434.

107. **Mendes, R. E., M. A. Toleman, J. Ribeiro, H. S. Sader, R. N. Jones, and T. R. Walsh**. 2004. Integron carrying a novel metallo-β-lactamase gene, bla_{IMP-16}, and a fused form of aminoglycoside-resistant gene *aac*(6')-30/*aac*(6')-Ib': report from the SENTRY Antimicrobial Surveillance Program. *Antimicrob. Agents Chemother.* **48**:4693–4702.

108. **Mercuri, P. S., F. Bouillenne, L. Boschi, J. Lamotte-Brasseur, G. Amicosante, B. Devreese, J. Van Beeumen, J. M. Frere, G. M. Rossolini, and M. Galleni**. 2001. Biochemical characterization of the FEZ-1 metallo-β-lactamase of *Legionella gormanii* ATCC 33297T produced in *Escherichia coli*. *Antimicrob. Agents Chemother.* **45**:1254–1262.

109. **Migliavacca, R., J. D. Docquier, C. Mugnaioli, G. Amicosante, R. Daturi, K. Lee, G. M. Rossolini, and L. Pagani**. 2002. Simple microdilution test for detection of metallo-β-lactamase production in *Pseudomonas aeruginosa*. *J. Clin. Microbiol.* **40**:4388–4390.

110. **Miriagou, V., E. Tzelepi, D. Gianneli, and L. S. Tzouvelekis**. 2003. *Escherichia coli* with a self-transferable, multiresistant plasmid coding for metallo-β-lactamase VIM-1. *Antimicrob. Agents Chemother.* **47**:395–397.

111. **Moali, C., C. Anne, J. Lamotte-Brasseur, S. Groslambert, B. Devreese, J. Van Beeumen, M. Galleni, and J. M. Frere**. 2003. Analysis of the importance of the metallo-β-lactamase active site loop in substrate binding and catalysis. *Chem. Biol.* **10**:319–329.

112. **Mohammed, M. J., C. K. Marston, T. Popovic, R. S. Weyant, and F. C. Tenover**. 2002. Antimicrobial susceptibility testing of *Bacillus anthracis*: comparison of results obtained by using the National Committee for Clinical Laboratory Standards broth microdilution reference and Etest agar gradient diffusion methods. *J. Clin. Microbiol.* **40**:1902–1907.

113. **Murphy, T. A., A. M. Simm, M. A. Toleman, R. N. Jones, and T. R. Walsh**. 2003. Biochemical characterization of the acquired metallo-β-lactamase SPM-1 from *Pseudomonas aeruginosa*. *Antimicrob. Agents Chemother.* **47**:582–587.

114. **Manrrray, P. R., E. J. Boron, J. H. Jorgansen, M. A. Pfaller, and R. H. Yolken (ed.)**. 2003. *Manual of Clinical Microbiology*, 8th ed. ASM Press, Washington, D.C.

115. **Naas, T., S. Bellais, and P. Nordmann**. 2003. Molecular and biochemical characterization of a carbapenem-hydrolysing β-lactamase from *Flavobacterium johnsoniae*. *J. Antimicrob. Chemother.* **51**:267–273.

116. **Niumsup, P., A. M. Simm, K. Nurmahomed, T. R. Walsh, P. M. Bennett, and M. B. Avison**. 2003. Genetic linkage of the penicillinase gene, *amp*, and *blrAB*, encoding the regulator of β-lactamase expression in *Aeromonas* spp. *J. Antimicrob. Chemother.* **51**:1351–1358.

117. **Nordmann, P., and L. Poirel**. 2002. Emerging carbapenemases in Gram-negative aerobes. *Clin. Microbiol. Infect.* **8**:321–331.

118. **Oh, E. J., S. Lee, Y. J. Park, J. J. Park, K. Park, S. I. Kim, M. W. Kang, and B. K. Kim**. 2003. Prevalence of metallo-β-lactamase among *Pseudomonas aeruginosa* and *Acinetobacter baumannii* in a Korean university hospital and comparison of screening methods for detecting metallo-β-lactamase. *J. Microbiol. Methods* **54**:411–418.

119. **O'Hara, K., S. Haruta, T. Sawai, M. Tsunoda, and S. Iyobe**. 1998. Novel metallo β-lactamase mediated by a *Shigella flexneri* plasmid. *FEMS Microbiol. Lett.* **162**:201–206.

120. **Osano, E., Y. Arakawa, R. Wacharotayankun, M. Ohta, T. Horii, H. Ito, F. Yoshimura, and N. Kato**. 1994. Molecular characterization of an enterobacterial metallo-β-lactamase found in a clinical isolate of *Serratia marcescens* that shows imipenem resistance. *Antimicrob. Agents Chemother.* **38**: 71–78.

121. **Page, R. D**. 1996. TreeView: an application to display phylogenetic trees on personal computers. *Comput. Appl. Biosci.* **12**:357–358.

122. **Pallecchi, L., M. L. Riccio, J. D. Docquier, R. Fontana, and G. M. Rossolini**. 2001. Molecular heterogeneity of bla_{VIM-2}-containing integrons from *Pseudomonas aeruginosa* plasmids encoding the VIM-2 metallo-β-lactamase. *FEMS Microbiol. Lett.* **195**:145–150.

123. **Pasteran, F., D. Faccone, A. Petroni, M. Rapoport, M. Galas, M. Vazquez, and A. Procopio**. 2005. Novel variant (bla_{VIM-11}) of the metallo-β-lactamase bla_{VIM} family in a GES-1 extended-spectrum-β-lactamase-producing *Pseudomonas aeruginosa* clinical isolate in Argentina. *Antimicrob. Agents Chemother.* **49**:474–475.

124. **Patzer, J., M. A. Toleman, L. M. Deshpande, W. Kaminska, D. Dzierzanowska, P. M. Bennett, R. N. Jones, and T. R. Walsh**. 2004. *Pseudomonas aeruginosa* strains harbouring an unusual bla_{VIM-4} gene cassette isolated from hospitalized children in Poland (1998–2001). *J. Antimicrob. Chemother.*

125. **Peleg, A. Y., C. Franklin, J. Bell, and D. W. Spelman**. 2004. Emergence of IMP-4 metallo-β-lactamase in a clinical isolate from Australia. *J. Antimicrob. Chemother.* **54**:699–700.

126. **Podglajen, I., J. Breuil, I. Casin, and E. Collatz**. 1995. Genotypic identification of two groups within the species

Bacteroides fragilis by ribotyping and by analysis of PCR-generated fragment patterns and insertion sequence content. *J. Bacteriol.* **177**:5270–5275.

127. **Podglajen, I., J. Breuil, and E. Collatz.** 1994. Insertion of a novel DNA sequence, 1S1186, upstream of the silent carbapenemase gene *cfiA*, promotes expression of carbapenem resistance in clinical isolates of *Bacteroides fragilis*. *Mol. Microbiol.* **12**:105–114.

128. **Poirel, L., M. Magalhaes, M. Lopes, and P. Nordmann.** 2004. Molecular analysis of metallo-β-lactamase gene bla_{SPM-1}-surrounding sequences from disseminated *Pseudomonas aeruginosa* isolates in Recife, Brazil. *Antimicrob. Agents Chemother.* **48**:1406–1409.

129. **Poirel, L., T. Naas, D. Nicolas, L. Collet, S. Bellais, J. D. Cavallo, and P. Nordmann.** 2000. Characterization of VIM-2, a carbapenem-hydrolyzing metallo-β-lactamase and its plasmid- and integron-borne gene from a *Pseudomonas aeruginosa* clinical isolate in France. *Antimicrob. Agents Chemother.* **44**:891–897.

130. **Pournaras, S., M. Maniati, E. Petinaki, L. S. Tzouvelekis, A. Tsakris, N. J. Legakis, and A. N. Maniatis.** 2003. Hospital outbreak of multiple clones of *Pseudomonas aeruginosa* carrying the unrelated metallo-β-lactamase gene variants bla_{VIM-2} and bla_{VIM-4}. *J. Antimicrob. Chemother.* **51**:1409–1414.

131. **Pournaras, S., A. Tsakris, M. Maniati, L. S. Tzouvelekis, and A. N. Maniatis.** 2002. Novel variant (bla_{VIM-4}) of the metallo-β-lactamase gene bla_{VIM-1} in a clinical strain of *Pseudomonas aeruginosa*. *Antimicrob. Agents Chemother.* **46**:4026–4028.

132. **Prats, G., E. Miro, B. Mirelis, L. Poirel, S. Bellais, and P. Nordmann.** 2002. First isolation of a carbapenem-hydrolyzing β-lactamase in *Pseudomonas aeruginosa* in Spain. *Antimicrob. Agents Chemother.* **46**:932–933.

133. **Prosperi-Meys, C., D. de Seny, G. Llabres, M. Galleni, and J. Lamotte-Brasseur.** 2002. Active-site mutants of class B β-lactamases: substrate binding and mechanistic study. *Cell Mol. Life Sci.* **59**:2136–2143.

134. **Prosperi-Meys, C., J. Wouters, M. Galleni, and J. Lamotte-Brasseur.** 2001. Substrate binding and catalytic mechanism of class B β-lactamases: a molecular modelling study. *Cell Mol. Life Sci.* **58**:2136–2143.

135. **Rasmussen, B. A., and K. Bush.** 1997. Carbapenem-hydrolyzing β-lactamases. *Antimicrob. Agents Chemother.* **41**:223–232.

136. **Riccio, M. L., N. Franceschini, L. Boschi, B. Caravelli, G. Cornaglia, R. Fontana, G. Amicosante, and G. M. Rossolini.** 2000. Characterization of the metallo-β-lactamase determinant of *Acinetobacter baumannii* AC-54/97 reveals the existence of bla_{IMP} allelic variants carried by gene cassettes of different phylogeny. *Antimicrob. Agents Chemother.* **44**:1229–1235.

137. **Riccio, M. L., L. Pallecchi, R. Fontana, and G. M. Rossolini.** 2001. In70 of plasmid pAX22, a bla_{VIM-1}-containing integron carrying a new aminoglycoside phosphotransferase gene cassette. *Antimicrob. Agents Chemother.* **45**:1249–1253.

138. **Roche, D. M., J. T. Byers, D. S. Smith, F. G. Glansdorp, D. R. Spring, and M. Welch.** 2004. Communications blackout? Do N-acylhomoserine-lactone-degrading enzymes have any role in quorum sensing? *Microbiology* **150**:2023–2028.

139. **Rossolini, G. M., M. A. Condemi, F. Pantanella, J. D. Docquier, G. Amicosante, and M. C. Thaller.** 2001. Metallo-β-lactamase producers in environmental microbiota: new molecular class B enzyme in *Janthinobacterium lividum*. *Antimicrob. Agents Chemother.* **45**:837–844.

140. **Rossolini, G. M., N. Franceschini, L. Lauretti, B. Caravelli, M. L. Riccio, M. Galleni, J. M. Frere, and G. Amicosante.** 1999. Cloning of a *Chryseobacterium* (*Flavobacterium*) *meningosepticum* chromosomal gene $blaA_{CME}$ encoding an extended-spectrum class A β-lactamase related to the *Bacteroides* cephalosporinases and the VEB-1 and PER β-lactamases. *Antimicrob. Agents Chemother.* **43**:2193–2199.

141. **Rossolini, G. M., N. Franceschini, M. L. Riccio, P. S. Mercuri, M. Perilli, M. Galleni, J. M. Frere, and G. Amicosante.** 1998. Characterization and sequence of the *Chryseobacterium* (*Flavobacterium*) *meningosepticum* carbapenemase: a new molecular class B β-lactamase showing a broad substrate profile. *Biochem. J.* **332**(Pt. 1):145–152.

142. **Rossolini, G. M., M. L. Riccio, G. Cornaglia, L. Pagani, C. Lagatolla, L. Selan, and R. Fontana.** 2000. Carbapenem-resistant *Pseudomonas aeruginosa* with acquired bla_{VIM} metallo-β-lactamase determinants, Italy. *Emerg. Infect. Dis.* **6**:312–313.

143. **Rossolini, G. M., T. Walsh, and G. Amicosante.** 1996. The *Aeromonas* metallo-β-lactamases: genetics, enzymology, and contribution to drug resistance. *Microb. Drug Resist.* **2**:245–252.

144. **Rossolini, G. M., A. Zanchi, A. Chiesurin, G. Amicosante, G. Satta, and P. Guglielmetti.** 1995. Distribution of *cphA* or related carbapenemase-encoding genes and production of carbapenemase activity in members of the genus *Aeromonas*. *Antimicrob. Agents Chemother.* **39**:346–349.

145. **Rosta, S., and H. Mett.** 1989. Physiological studies of the regulation of β-lactamase expression in *Pseudomonas maltophilia*. *J. Bacteriol.* **171**:483–487.

146. **Ruimy, R., I. Podglajen, J. Breuil, R. Christen, and E. Collatz.** 1996. A recent fixation of *cfiA* genes in a monophyletic cluster of *Bacteroides fragilis* is correlated with the presence of multiple insertion elements. *J. Bacteriol.* **178**:1914–1918.

147. **Saavedra, M. J., L. Peixe, J. C. Sousa, I. Henriques, A. Alves, and A. Correia.** 2003. Sfh-I, a subclass B2 metallo-β-lactamase from a *Serratia fonticola* environmental isolate. *Antimicrob. Agents Chemother.* **47**:2330–2333.

148. **Sabath, L. D., and E. P. Abraham.** 1966. Zinc as a cofactor for cephalosporinase from *Bacillus cereus* 569. *Biochem. J.* **98**:11C–13C.

149. **Sader, H. S., M. Castanheira, R. E. Mendes, M. Toleman, T. R. Walsh, and R. N. Jones.** 2005. Dissemination and diversity of metallo-beta-lactamases in Latin America: report from the SENTRY Antimicrobial Surveillance Program. *Int. J. Antimicrob. Agents* **25**:57–61.

150. **Saino, Y., F. Kobayashi, M. Inoue, and S. Mitsuhashi.** 1982. Purification and properties of inducible penicillin β-lactamase isolated from *Pseudomonas maltophilia*. *Antimicrob. Agents Chemother.* **22**:564–570.

151. **Sardelic, S., L. Pallecchi, V. Punda-Polic, and G. M. Rossolini.** 2003. Carbapenem-resistant *Pseudomonas aeruginosa*-carrying VIM-2 metallo-β-lactamase determinants, Croatia. *Emerg. Infect. Dis.* **9:**1022–1023.

152. **Schreckenberger, P. C., M. I. Daneshvar, R. S. Weyant, and D. G. Hollis.** 2003. *Acinetobacter*, *Achromobacter*, *Chryseobacterium*, *Moraxella*, and other nonfermentative gram-negative rods., p. 749–779. *In* P. R. Murray, E. J. Baron, J. H. Jorgensen, M. A. Pfaller, and R. H. Yolken (ed.), *Manual of Clinical Microbiology*, 8th ed. ASM Press, Washington, D.C.

153. **Senda, K., Y. Arakawa, S. Ichiyama, K. Nakashima, H. Ito, S. Ohsuka, K. Shimokata, N. Kato, and M. Ohta.** 1996. PCR detection of metallo-β-lactamase gene bla_{IMP} in gram-negative rods resistant to broad-spectrum β-lactams. *J. Clin. Microbiol.* **34:**2909–2913.

154. **Senda, K., Y. Arakawa, K. Nakashima, H. Ito, S. Ichiyama, K. Shimokata, N. Kato, and M. Ohta.** 1996. Multifocal outbreaks of metallo-β-lactamase-producing *Pseudomonas aeruginosa* resistant to broad-spectrum β-lactams, including carbapenems. *Antimicrob. Agents Chemother.* **40:** 349–353.

155. **Shibata, N., Y. Doi, K. Yamane, T. Yagi, H. Kurokawa, K. Shibayama, H. Kato, K. Kai, and Y. Arakawa.** 2003. PCR typing of genetic determinants for metallo-β-lactamases and integrases carried by gram-negative bacteria isolated in Japan, with focus on the class 3 integron. *J. Clin. Microbiol.* **41:**5407–5413.

156. **Simm, A. M., C. S. Higgins, A. L. Carenbauer, M. W. Crowder, J. H. Bateson, P. M. Bennett, A. R. Clarke, S. E. Halford, and T. R. Walsh.** 2002. Characterization of monomeric L1 metallo-β-lactamase and the role of the N-terminal extension in negative cooperativity and antibiotic hydrolysis. *J. Biol. Chem.* **277:**24744–24752.

157. **Simm, A. M., C. S. Higgins, S. T. Pullan, M. B. Avison, P. Niumsup, O. Erdozain, P. M. Bennett, and T. R. Walsh.** 2001. A novel metallo-β-lactamase, Mbl1b, produced by the environmental bacterium *Caulobacter crescentus*. *FEBS Lett.* **509:**350–354.

157a. **Stoczko, M., J. M. Frère, G. M. Rossolini, and J. D. Docquier.** 2006. Postgenomic scan of metallo-β-lactamase homologues in Rhizobacteria: identification and characterization of BJP-1, a subclass B3 ortholog from *Bradyrhizobium japonicum*. *Antimicrob. Agents Chetmother.* **50:**1973–1981.

158. **Thompson, J. D., T. J. Gibson, F. Plewniak, F. Jeanmougin, and D. G. Higgins.** 1997. The CLUSTAL_X windows interface: flexible strategies for multiple sequence alignment aided by quality analysis tools. *Nucleic Acids Res.* **25:** 4876–4882.

159. **Thompson, J. S., and M. H. Malamy.** 1990. Sequencing the gene for an imipenem-cefoxitin-hydrolyzing enzyme (CfiA) from *Bacteroides fragilis* TAL2480 reveals strong similarity between CfiA and *Bacillus cereus* β-lactamase II. *J. Bacteriol.* **172:**2584–2593.

160. **Toleman, M. A., D. Biedenbach, D. Bennett, R. N. Jones, and T. R. Walsh.** 2003. Genetic characterization of a novel metallo-β-lactamase gene, bla_{IMP-13}, harboured by a novel Tn*5051*-type transposon disseminating carbapenemase genes in Europe: report from the SENTRY worldwide antimicrobial surveillance programme. *J. Antimicrob. Chemother.* **52:**583–590.

161. **Toleman, M. A., D. Biedenbach, D. M. Bennett, R. N. Jones, and T. R. Walsh.** 2005. Italian metallo-β-lactamases: a national problem? Report from the SENTRY Antimicrobial Surveillance Programme. *J. Antimicrob. Chemother.* **55:**61–70.

162. **Toleman, M. A., K. Rolston, R. N. Jones, and T. R. Walsh.** 2004. bla_{VIM-7}, an evolutionarily distinct metallo-β-lactamase gene in a *Pseudomonas aeruginosa* isolate from the United States. *Antimicrob. Agents Chemother.* **48:**329–332.

163. **Toleman, M. A., A. M. Simm, T. A. Murphy, A. C. Gales, D. J. Biedenbach, R. N. Jones, and T. R. Walsh.** 2002. Molecular characterization of SPM-1, a novel metallo-β-lactamase isolated in Latin America: report from the SENTRY antimicrobial surveillance programme. *J. Antimicrob. Chemother.* **50:**673–679.

164. **Tsakris, A., S. Pournaras, N. Woodford, M. F. Palepou, G. S. Babini, J. Douboyas, and D. M. Livermore.** 2000. Outbreak of infections caused by *Pseudomonas aeruginosa* producing VIM-1 carbapenemase in Greece. *J. Clin. Microbiol.* **38:**1290–1292.

165. **Tsakris, A., P. T. Tassios, F. Polydorou, A. Papa, E. Malaka, A. Antoniadis, and N. J. Legakis.** 2003. Infrequent detection of acquired metallo-β-lactamases among carbapenem-resistant *Pseudomonas* isolates in a Greek hospital. *Clin. Microbiol. Infect.* **9:**846–851.

166. **Tysall, L., M. W. Stockdale, P. R. Chadwick, M. F. Palepou, K. J. Towner, D. M. Livermore, and N. Woodford.** 2002. IMP-1 carbapenemase detected in an *Acinetobacter* clinical isolate from the UK. *J. Antimicrob. Chemother.* **49:**217–218.

167. **Ullah, J. H., T. R. Walsh, I. A. Taylor, D. C. Emery, C. S. Verma, S. J. Gamblin, and J. Spencer.** 1998. The crystal structure of the L1 metallo-β-lactamase from *Stenotrophomonas maltophilia* at 1.7 A resolution. *J. Mol. Biol.* **284:** 125–136.

168. **Vessillier, S., J. D. Docquier, S. Rival, J. M. Frere, M. Galleni, G. Amicosante, G. M. Rossolini, and N. Franceschini.** 2002. Overproduction and biochemical characterization of the *Chryseobacterium meningosepticum* BlaB metallo-β-lactamase. *Antimicrob. Agents Chemother.* **46:**1921–1927.

169. **Voget, S., C. Leggewie, A. Uesbeck, C. Raasch, K. E. Jaeger, and W. R. Streit.** 2003. Prospecting for novel biocatalysts in a soil metagenome. *Appl. Environ. Microbiol.* **69:**6235–6242.

170. **Walsh, T. R., A. Bolmstrom, A. Qwarnstrom, and A. Gales.** 2002. Evaluation of a new Etest for detecting metallo-β-lactamases in routine clinical testing. *J. Clin. Microbiol.* **40:**2755–2759.

171. **Walsh, T. R., L. Hall, S. J. Assinder, W. W. Nichols, S. J. Cartwright, A. P. MacGowan, and P. M. Bennett.** 1994. Sequence analysis of the L1 metallo-β-lactamase from *Xanthomonas maltophilia*. *Biochim. Biophys. Acta* **1218:** 199–201.

172. **Walsh, T. R., A. P. MacGowan, and P. M. Bennett.** 1997. Sequence analysis and enzyme kinetics of the L2 serine β-lactamase from *Stenotrophomonas maltophilia*. *Antimicrob. Agents Chemother.* **41:**1460–1464.

173. **Walsh, T. R., R. A. Stunt, J. A. Nabi, A. P. MacGowan, and P. M. Bennett.** 1997. Distribution and expression of β-lactamase genes among *Aeromonas* spp. *J. Antimicrob. Chemother.* **40:**171–178.

174. **Walsh, T. R., M. A. Toleman, W. Hryniewicz, P. M. Bennett, and R. N. Jones.** 2003. Evolution of an integron carrying bla_{VIM-2} in Eastern Europe: report from the SENTRY Antimicrobial Surveillance Program. *J. Antimicrob. Chemother.* **52:**116–119.

175. **Wang, Z., W. Fast, A. M. Valentine, and S. J. Benkovic.** 1999. Metallo-β-lactamase: structure and mechanism. *Curr. Opin. Chem. Biol.* **3:**614–622.

175a. **Xiong, J., M. F. Hynes, H. Ye, H. Chen, Y. Yang, F. M'zali, and P. M. Hawkey.** 2006. bla_{IMP-9} and its association with large plasmids carried by *Pseudomonas aeruginosa* isolates from the People's Republic of China. *Antimicrob. Agents Chemother.* **50:**355–358.

176. **Yamasaki, K., M. Komatsu, T. Yamashita, K. Shimakawa, T. Ura, H. Nishio, K. Satoh, R. Washidu, S. Kinoshita, and M. Aihara.** 2003. Production of CTX-M-3 extended-spectrum β-lactamase and IMP-1 metallo β-lactamase by five Gram-negative bacilli: survey of clinical isolates from seven laboratories collected in 1998 and 2000, in the Kinki region of Japan. *J. Antimicrob. Chemother.* **51:**631–638.

177. **Yamazoe, K., N. Kato, H. Kato, K. Tanaka, Y. Katagiri, and K. Watanabe.** 1999. Distribution of the *cfiA* gene among *Bacteroides fragilis* strains in Japan and relatedness of cfiA to imipenem resistance. *Antimicrob. Agents Chemother.* **43:**2808–2810.

178. **Yan, J. J., P. R. Hsueh, W. C. Ko, K. T. Luh, S. H. Tsai, H. M. Wu, and J. J. Wu.** 2001. Metallo-β-lactamases in clinical *Pseudomonas* isolates in Taiwan and identification of VIM-3, a novel variant of the VIM-2 enzyme. *Antimicrob. Agents Chemother.* **45:**2224–2228.

179. **Yan, J. J., W. C. Ko, C. L. Chuang, and J. J. Wu.** 2002. Metallo-β-lactamase-producing *Enterobacteriaceae* isolates in a university hospital in Taiwan: prevalence of IMP-8 in *Enterobacter cloacae* and first identification of VIM-2 in *Citrobacter freundii. J. Antimicrob. Chemother.* **50:**503–511.

180. **Yan, J. J., W. C. Ko, S. H. Tsai, H. M. Wu, and J. J. Wu.** 2001. Outbreak of infection with multidrug-resistant *Klebsiella pneumoniae* carrying bla_{IMP-8} in a university medical center in Taiwan. *J. Clin. Microbiol.* **39:**4433–4439.

181. **Yan, J. J., W. C. Ko, and J. J. Wu.** 2001. Identification of a plasmid encoding SHV-12, TEM-1, and a variant of IMP-2 metallo-β-lactamase, IMP-8, from a clinical isolate of *Klebsiella pneumoniae. Antimicrob. Agents Chemother.* **45:**2368–2371.

182. **Yan, J. J., J. J. Wu, S. H. Tsai, and C. L. Chuang.** 2004. Comparison of the double-disk, combined disk, and Etest methods for detecting metallo-β-lactamases in gram-negative bacilli. *Diagn. Microbiol. Infect. Dis.* **49:**5–11.

183. **Yano, H., A. Kuga, R. Okamoto, H. Kitasato, T. Kobayashi, and M. Inoue.** 2001. Plasmid-encoded metallo-β-lactamase (IMP-6) conferring resistance to carbapenems, especially meropenem. *Antimicrob. Agents Chemother.* **45:**1343–1348.

184. **Yatsuyanagi, J., S. Saito, S. Harata, N. Suzuki, Y. Ito, K. Amano, and K. Enomoto.** 2004. Class 1 integron containing metallo-β-lactamase gene bla_{VIM-2} in *Pseudomonas aeruginosa* clinical strains isolated in Japan. *Antimicrob. Agents Chemother.* **48:**626–628.

185. **Yong, D., K. Lee, J. H. Yum, H. B. Shin, G. M. Rossolini, and Y. Chong.** 2002. Imipenem-EDTA disk method for differentiation of metallo-β-lactamase-producing clinical isolates of *Pseudomonas* spp. and *Acinetobacter* spp. *J. Clin. Microbiol.* **40:**3798–3801.

186. **Yum, J. H., K. Yi, H. Lee, D. Yong, K. Lee, J. M. Kim, G. M. Rossolini, and Y. Chong.** 2002. Molecular characterization of metallo-β-lactamase-producing *Acinetobacter baumannii* and *Acinetobacter* genomospecies 3 from Korea: identification of two new integrons carrying the bla_{VIM-2} gene cassettes. *J. Antimicrob. Chemother.* **49:**837–840.

187. **Yum, J. H., D. Yong, K. Lee, H. S. Kim, and Y. Chong.** 2002. A new integron carrying VIM-2 metallo-β-lactamase gene cassette in a *Serratia marcescens* isolate. *Diagn. Microbiol. Infect. Dis.* **42:**217–219.

Enzyme-Mediated Resistance to Antibiotics: Mechanisms, Dissemination, and Prospects for Inhibition
Edited by Robert A. Bonomo and Marcelo E. Tolmasky

C. Bauvois
J. Wouters

10 Crystal Structures of Class C β-Lactamases: Mechanistic Implications and Perspectives in Drug Design

When the novel classification proposed by Ambler was introduced in 1980 (1), there were only three classes of β-lactamases (classes A, B, and C) and only two enzymes were assigned as belonging to the class C: AmpC from *Escherichia coli* K-12 and *Pseudomonas aeruginosa* (18). Indeed, their primary sequences showed no significant homology with other class A or class B enzymes and were much longer: the number of residues per molecule was large, about 377 (15) for *E. coli* and 397 (19) for *P. aeruginosa* AmpC β-lactamases. After translocation into the periplasmic space and removal of the signal peptide, mature proteins presented a relatively high molecular mass (around 39 kDa) compared to enzymes of other classes (around 29 kDa [20] for class A and 25 kDa [38] for class B enzymes). These two enzymes were also characterized by a highly basic isoelectric point.

As for class A enzymes, some evidence indicated that the reaction mechanism involves the formation of an acyl enzyme and that this acylation occurs on serine residues (17). AmpCs were inhibited reversibly by several boronic acid derivatives (5), which was, at that time, considered a serine active β-lactamase-specific property (9). Regarding their activity profile, these enzymes hydrolyze cephalosporins at a high rate and the hosts that overproduce them become β-lactam resistant. Some substrates, for example, cloxacillin, considered a good substrate for other classes of enzymes, behave as inactivator for AmpC (17, 20). For those reasons, AmpCs were commonly referred to as cephalosporinases (6). Although this denomination was rapidly considered a misnomer (18), it remains in use.

Knott-Hunziker et al. (18) showed that, unlike class A β-lactamases, those enzymes were able to replace water by alcohols and carry out alcoholysis instead of hydrolysis. Moreover, these authors demonstrated that the *Pseudomonas* β-lactamase can function as an esterase. A more worrying kinetic property bore on the inefficiency of clavulanic acid to inactivate class C β-lactamases (18). Indeed, this compound was increasingly used in combination with another β-lactam antibiotic (like Augmentin, which combines clavulanic acid and amoxicillin) to counter the emergence of bacteria producing β-lactamases. This practice probably has contributed to the subsequent propagation of strains able to produce an AmpC-type enzyme. Finally, another divergence has to do with the gene encoding AmpC, which is located on the bacterial chromosome and not on a transferable plasmid.

C. Bauvois, Institut de Recherches Microbiologiques Wiame, Campus Ceria, 1 Av E. Gryzon, B-1070 Brussels, Belgium. **J. Wouters**, Institut de Recherches Microbiologiques Wiame, Campus Ceria, 1 Av E. Gryzon, B-1070 Brussels, and Department of Chemistry, University of Namur, 61 Rue de Bruxelles, B-5000 Namur, Belgium.

In the case of the *E. coli* enzyme, no expression regulation was observed; β-lactamase is produced constitutively at a low level. In contrast, AmpC from *P. aeruginosa* is induced when β-lactam antibiotics are added to the culture medium.

From that time, numerous AmpC-type variants have been identified in many bacterial species, especially in the *Enterobacteriaceae* family including pathogens such as *Citrobacter*, *Salmonella*, and *Shigella*. Extensive studies of several of those enzymes concerning their biochemical and kinetic as well as structural properties have allowed to further define this group. For example, the isoelectric point should no longer be considered to be discriminating since on one hand, different class C enzymes present acidic isoelectric points and on the other hand, some β-lactamases from other classes have basic ones (as CTX-M-12 with a pI of 9.0 [8, 16]). Moreover, at first exclusively found on the bacterial chromosome, the *ampC* genes were more often located on transferable plasmid (for a review, see Philippon et al. [28]).

However, despite their number, class C β-lactamases form a very homogeneous group compared to the other classes. They are all purified as monomeric proteins, presenting a molecular mass of around 39 kDa. They are exclusively produced by gram-negative bacteria and are located in the periplasmic space. Only one exception is known: AmpC from *Psychrobacter immobilis* A5, which is secreted into the external medium (12). Kinetic analysis (2, 13, 14) showed that AmpC-type enzymes have relatively low k_{cat} and K_m values for many substrates, which is, in all investigated cases, related to their relatively poor deacylation capacity. Consequently, class C enzymes are generally considered to be less efficient than their class A counterparts. Their principal characteristic is their remarkable ability to accommodate diverse substrates. For instance, these enzymes are able to function also as peptidases (32) as shown by their ability to accept the acyclic linear thiolesters (20) and depsipeptides (31) as a substrate.

Unlike a large majority of class A enzymes (21), they are able to hydrolyze efficiently cephamycins characterized by a methoxy group on the α-face of the β-lactam ring (as cefoxitin). As is discussed below, this is a direct consequence of the geometry of their active site.

Finally, they are not very sensitive to β-lactamase inhibitors such as clavulanic acid, sulbactam, or tazobactam. On the contrary, they are inactivated by isoxazoyl penicillin derivatives (such as oxacillin or cloxacillin) and monobactams (such as azthreonam) and reversibly inhibited by boronic acid. The last property was exploited during purification by affinity chromatography and for the mechanistic studies. More significantly, it has allowed the design of novel inhibitors, as is discussed in the next sections of this work (23, 39).

The class C enzymes are considered to be the most homogeneous and least effective among the different groups of β-lactamases. Their genes were first exclusively located on the bacterial chromosome, which does not enhance horizontal interspecies transfer. In our sense, several of those characteristics are not devoid of physiological implications. Indeed, in many cases AmpC-related bacterial resistance is principally due to expression deregulation leading to overproduction of AmpC. This is frequently associated with other resistance factors like membrane permeability alteration and/or coproduction of several different β-lactamase types. There are not a lot of genuine extended-spectrum (ES) AmpCs characterized, the lone exception being the GC1 mutant, which presents an unusually high hydrolytic activity towards ceftazidime. Moreover, AmpC-associated resistance emergence is frequently encountered in clinical cases absolutely requiring the use of specific inhibitors (like clavulanic acid) or substrates (as observed in the case of the host producing CMY-2 and ceftriaxone). Nevertheless, class C β-lactamases should not be underestimated, because first, as proved by the GC1 mutant, these enzymes have not ceased to evolve and, like the other classes of enzymes, they are still able to change their substrate profiles against clinically used compounds; and secondly, the recent transfer of the *ampC* gene on a plasmid will undoubtedly provoke, as already observed, the appearance of new variants among different bacterial species. Inevitably, this will increase the group diversity and enhance the coming out of novel ES AmpC. These arguments justify the interest for this class of enzymes, both from a biochemical point of view and as important drug targets for the design of antibiotics.

STUCTURAL FEATURES OF CLASS C β-LACTAMASES

Table 10.1 contains a selection of three-dimensional structures of class C β-lactamases that are available at the Protein Data Bank (PDB) (7). Crystallographic structures from distinct AmpC-type enzymes have been resolved, among which three are chromosomal, i.e., AmpC from *Enterobacter cloacae* P99, AmpC from *E. coli*, and AmpC from *Citrobacter freundii*, and two are plasmid-mediated enzymes, i.e., ACT-1 and CMY-2. The last two enzymes are so close to the equivalent chromosomal proteins (AmpC from *E. coli* and *C. freundii*, respectively) that their structures present no significant discrepancies.

The overall structure of AmpCs, like for the other classes of β-lactamases, is formed from two distinct domains: one contains only α-helices (α domain), and the

Table 10.1 Selected structures of class C β-lactamases that were obtained by crystallography[a]

Organism and code (resolution [Å])	Details
Enterobacter cloacae	
1XXT[b] (1.88)	P99
1BLS (2.3)	P99 complex with phosphonate analogue
1S6R (2.24)	908R complex with boronic acid
1Y54 (2.07)	908R complex with BRL-42715 (penem inhibitor)
1GCE (1.80)	GC1
1GA0 (1.60)	GC1 with a cephalosporin sulfone inhibitor
1ONH (1.38)	GC1 with a penem inhibitor
1RGZ (1.37)	GC1 complexed with transition state analogue of cefotaxime
Escherichia coli AMPC	
1KE4[c] (1.72)	AMPC
1L0D (1.53)	S64D mutant
1L0E (1.90)	K67Q mutant
1L0F (1.66)	N152H mutant
1L0G (1.50)	S64G mutant
1FCO (2.2)	Covalently acylated with moxalactam
1C3B (2.25)	Complex with benzo(b)thiophene-2-boronic acid
1FSW (1.90)	Complex with inhibitor cephalothinboronic acid
1FSY (1.75)	Complex with inhibitor cloxacillinboronic acid
1GA9 (2.10)	Complex with ETP
1IEL (2.00)	Complex with ceftazidime
1IEM (2.30)	Complex with a boronic acid inhibitor (CEFB4)
1KDS (2.15)	Complex with 3-nitrophenylboronic acid
1KDW (2.28)	Complex with 4-carboxyphenylboronic acid
1KE0 (2.30)	Complex with 4-(carboxyvin-2-yl) phenylboronic acid
1KE3 (2.15)	Complex with 4,4-biphenyldiboronic acid
1KVM (2.06)	Complex with covalently bound cephalothin
1L2S (1.94)	Complex with a dock-predicted noncovalent inhibitor STC
1LL5 (1.80)	Complex with covalently bound imipenem
1LL9 (1.87)	Complex with amoxicillin
1LLB (1.72)	Complex with atmo-penicillin
1MXO (1.83)	Complex with SM3
1MY8 (1.72)	Complex with SM2
1XGI (196)	Complex with 3-{[(3-nitroaniline]sulfonyl} thiophene-2-carboxylic acid
1XGJ (197)	Complex with 3-{[(4-carboxy-2 hydroxy) aniline]sulfonyl} thiophene-2-carboxylic acid
1I5Q (1.83)	Mutant N152A covalently acylated with moxalactam
1KVL (1.53)	Mutant S64G in complex with cephalothin
1PI4 (1.39)	Mutant N289A mutant in complex with SM3
1PI5 (1.49)	Mutant N289A mutant in complex with SM2
1FCM (2.46)	Mutant Q120L/Y150E covalently acylated with cloxacillin
1FCN (2.35)	Mutant Q120L/Y150E covalently acylated with loracarbef
1O07 (1.71)	Mutant Q120L/Y150E with a β-lactam inhibitor (MXG)
Citrobacter freundii	
1FR1 (2.00)	
1FR6 (2.50)	Complex with aztreonam
1RGY (1.52)	Complex with phosphonic acid PTX
Staphylococcus aureus	
1GHI (2.30)	Mutant GLU166ASP/ASN170GLN

[a] The structures are available at the PDB and can be accessed at http://www.rcsb.org/pdb/ via their four-character code. Abbreviations: ETP, 2, 3-(4-benzenesulfonyl-thiophene-2-sulfonylamino)-phenylboronic acid; MXG, 2-(1-{2-[4-(2-acetyl-amino-propionylamino)-4-carboxy-butyrylamino]-6-amino-hexanoylamino}-2-oxo-ethyl)-5-methylene-5,6-dihydro-2h-[1,3]thiazine-4-carboxylic acid; STC, 3-[(4-chloroanilino)sulfonyl]thiophene-2-carboxylic acid; ATMO, 2-{1-[2-(2-amino-thiazol-4-yl)-2-methoxyimino-acetylamino]-2-oxo-ethyl}-5,5-dimethyl-thiazolidine-4-carboxylic acid; SM2, carboxyphenylglycylboronic acid bearing the cephalothin R1 side chain; SM3, phenylglycylboronic acid bearing the cephalothin R1 side chain; PTX, (2E)-{2-[(2-amino-1,3-thiazol-4-yl)-2-(methoxyimino)ethanoyl]amino}methylphosphonic acid.
[b] PDB entry 2BLT has been replaced by this entry.
[c] Entries 2BLS and 3BLS have been replaced by this entry.

other contains a β-sheet and α-helices (α/β domain) (Fig. 10.1 [top]). All class C β-lactamases share in common several sequence motifs separated in the primary structure at constant intervals: S_{64}XXK, Y_{150}(A/S)N, and K_{315}TG (Fig. 10.1 [middle]). Actually, these motifs are located in close vicinity in tertiary structures and form the so-called active site. They are all implicated in catalytic mechanism, which explains why those residues are well conserved in sequence. Ser64 corresponds to the so-called active serine, which is the serine that realizes nucleophilic attack onto β-lactam carbonyl and where the acylation takes place. Tyr150, as shown by directed mutagenesis, is an essential residue and is generally considered to be the general basis increasing serine and water nucleophilicity during the acylation step and the deacylation step, respectively, as is

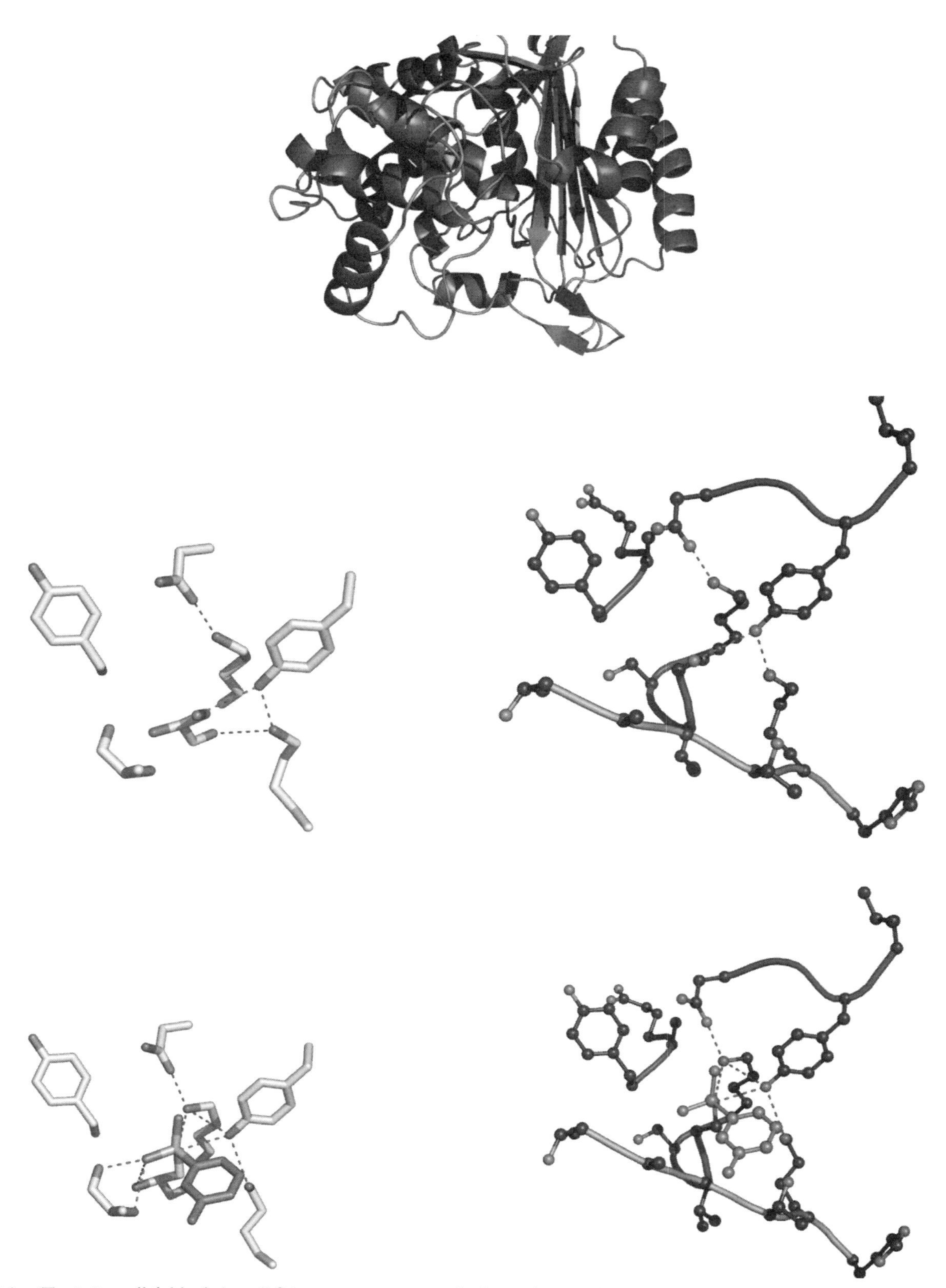

Figure 10.1 (Top) Overall fold of class C β-lactamases composed of two distinct domains: an all-α domain (left) and an α/β domain (right). (Middle) Views of the active site showing the conserved sequence motifs: S_{64}XXK, Y_{150}(A/S)N, and K_{315}TG. (Bottom) Views of the active site of Ampc C in complex with *m*-aminophenylboronic acid. Coordinates are taken from PDB entries 2BLS and 3BLS.

detailed in "Structural Features of the Active Sites of Class C β-Lactamases and Mechanistic Considerations" below.

Besides the "classical" AmpC class C β-lactamases, two related enzymes can be mentioned. The first is an ES class C β-lactamase (*E. cloacae* GC1) that deserves special attention from a structural point of view, to understand both its activity and inhibition potential. The second is a cold-adapted enzyme from a psychrophilic microorganism (*Psychrobacter immobilis* A5) that could lead to a better understanding of bacterial adaptation to temperature.

In 1995, a class C chromosomal ES β-lactamase was discovered in *E. cloacae* GC1 (25). Kinetic and mutagenesis experiments with the ES GC1 enzyme revealed an important deacylation activity that is somehow due to a unique three-residue repeat insertion (27). The crystal structure of the ES GC1 enzyme (PDB code, 1GCE) shows that the insertion makes a loop (called the omega loop) at the bottom of the binding site that is more flexible than the parental wild-type *E. cloacae* P99 enzyme.

It was suggested that a conformational change in ES GC1 permits the enzyme to accommodate cephalosporins with large side chains and allows more conformational freedom for the acyl intermediate. Indeed, first crystallographic analysis of the unliganded GC1 β-lactamase (PDB code, 1GCE) revealed disorder below the binding site, in the so-called Ω-loop (residues 189 to 226). The disordered segment included part of an unusual tandem tripeptide insertion, with residues 213 to 215 invisible in the electron density map. The insertion moved the Ω loop 1 to 2 Å away from the reactive Ser64 and appeared to enlarge the binding cavity for β-lactam entry. These findings supported the hypothesis that the three-residue insertion permits the enzyme to accommodate β-lactams with larger substituents. In addition, the larger cavity allows the acyl intermediate more conformational freedom, with increased hydrolysis. This was further studied by structure determination of complexes (Table 10.1, 1GA0, 1ONH, and 1RGZ), as is presented later in this chapter.

A heat-labile β-lactamase has been purified from culture supernatants of *Psychrobacter immobilis* A5 grown at 4°C, and the corresponding chromosomal *ampC* gene has been cloned and sequenced (12). All structural and kinetic properties clearly relate this enzyme to class C β-lactamases. The kinetic parameters of *P. immobilis* β-lactamase for the hydrolysis of some β-lactam antibiotics are in the same range as the values recorded for the highly specialized cephalosporinases from pathogenic mesophilic bacteria. By contrast, the enzyme displays a low apparent optimum temperature of activity and a reduced thermal stability. Structural factors responsible for the latter property were analyzed from the three-dimensional structure built by homology modeling. The deletion of proline residues in loops, the low number of arginine-mediated H-bonds and aromatic-aromatic interactions, the lower global hydrophobicity, and the improved solvent interactions through additional surface acidic residues appear to be the main determinants of the enzyme's flexibility.

STUCTURAL FEATURES OF THE ACTIVE SITES OF CLASS C β-LACTAMASES AND MECHANISTIC CONSIDERATIONS

Structures of complexes with class C β-lactamases were used to approach the mechanism of these enzymes. First studies of *C. freundii* AmpC suggested the importance of Ser64 and Tyr150 in the catalytical mechanism of class C β-lactamases. This was confirmed by the crystal structure of the *E. cloacae* enzyme. Among those early works, one also finds the structures of AmpC β-lactamase from *E. coli*, alone and in complex with a transition state analogue, *m*-aminophenylboronic acid (Fig. 10.2, **1**) (Fig. 10.1 [bottom]) (36).

The structure of AmpC from *E. coli* resembles those previously determined for the class C enzymes from *E. cloacae* and *C. freundii*. The transition state boronic acid analogue makes several interactions with AmpC that were unexpected. In particular, the putative "oxyanion" of the boronic acid forms a hydrogen bond with the backbone carbonyl oxygen of Ala318, suggesting that the high-energy intermediate for amide hydrolysis by β-lactamases and related enzymes could involve a hydroxyl and not an oxyanion. The involvement of the main-chain carbonyl in ligand and transition state stabilization would thus be a distinguishing feature between serine β-lactamases and serine proteases, to which they are often compared. The structure of the *m*-aminophenylboronic acid adduct also suggested several ways to improve the affinity of this class of inhibitor as is discussed in "Complexes with Class C β-Lactamases and Drug Design" below.

The mechanism of inhibition of the class C β-lactamase AmpC by imipenem (**2**) has also been approached by crystallography (Fig. 10.3). The X-ray crystal structure of the acyl enzyme complex was determined to a resolution of 1.80 Å (code 1LL5) (3). In the complex, the lactam carbonyl oxygen of imipenem has flipped by approximately 180 degrees compared to its expected position; the electrophilic acyl center is thus displaced from the point of hydrolytic attack. This conformation resembles that of imipenem bound to the class A enzyme TEM-1 but is different from that of moxalactam (**3**) bound to AmpC (Fig. 10.3).

The structure of the S64G mutant AmpC in complex with the β-lactam cephalothin (**4**) in its substrate and product forms was also determined by X-ray crystallography to 1.53 Å resolution (code 1KVL) (Fig. 10.4) (4). The acyl-enzyme intermediate between AmpC and cephalothin

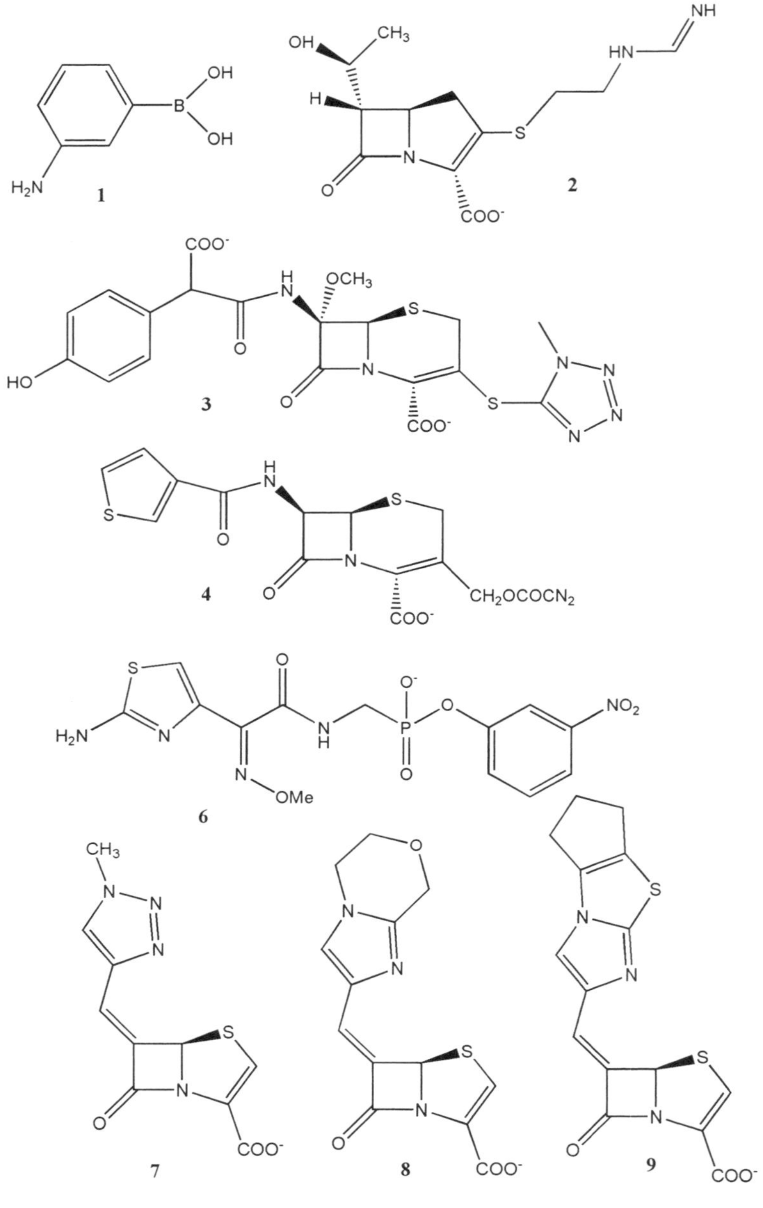

Figure 10.2 Chemical diagrams of compounds **1** to **4** and **6** to **9**.

was determined to 2.06 Å resolution (1KVM). The ligand undergoes a dramatic conformational change as the reaction progresses, with the characteristic six-membered dihydrothiazine ring of cephalothin rotating by 109 degrees. These structures correspond to all three intermediates along the reaction path and provide insights into substrate recognition, catalysis, and product expulsion.

ES Class C β-Lactamase

As already mentioned, bacterial resistance to the ES cephalosporins is an issue of great concern in current antibiotic therapeutics. An important source of this resistance comes from production of ES β-lactamases by bacteria. The *E. cloacae* GC1 enzyme is an example of a class C ES β-lactamase. Unlike wild-type (WT) forms, such as the *E. cloacae* P99 and *C. freundii* enzymes, the ES GC1 β-lactamase is able to rapidly hydrolyze ES cephalosporins such as cefotaxime and ceftazidime.

The crystallographic structure of the *E. cloacae* GC1 ES class C β-lactamase, inhibited by **5**, an alkylidenecephalosporin sulfone (Fig. 10.5), has been determined by X-ray diffraction at 100 K to a resolution of 1.6 Å (code 1GA0

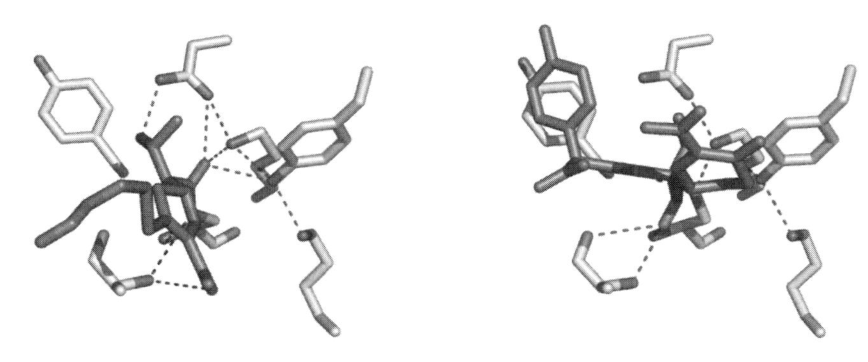

Figure 10.3 Comparison of the active site of *E. coli* AmpC in complex with imipenem 2 (left) and moxalactam 3 (right). Coordinates are taken from crystallographic structures 1LL5 and 1FCO, respectively.

[11]), revealing the structure of a reaction intermediate (5*) with a cephem sulfone. The crystal structure was solved by molecular replacement using the unliganded structure (code 1GCE) (10). Cryoquenching of the reaction of the sulfone with the enzyme produced an intermediate that is covalently bound via Ser64 (Fig. 10.6).

After acylation of the β-lactam ring, the dihydrothiazine dioxide ring opened with departure of the sulfinate. Nucleophilic attack of a side chain pyridine nitrogen atom on the C-6 atom of the resultant imine yielded a bicyclic aromatic system which helps to stabilize the acyl enzyme to hydrolysis. Positioning of the anionic sulfinate group close to Tyr150 (the catalytic base) and the acyl ester bond could block the approach of a potentially deacylating water molecule and explain the observation of the reaction intermediate 5*.

Comparison of the liganded and unliganded protein structures showed that a major movement (up to 7 Å) and refolding of part of the Ω loop (residues 215 to 224) accompany the binding of the inhibitor. This conformational flexibility in the Ω loop may form the basis of an ES activity of class C β-lactamases against modern cephalosporins.

This mechanistic study was followed by a similar one using *m*-nitrophenyl 2-(2-aminothiazol-4-yl)-2-[(Z)-metho xyimino]acetylaminomethyl phosphonate (6) (26). This phosphonate was designed to generate a transition state analogue for turnover of cefotaxime.

The crystal structures of complexes of the phosphonate with both ES GC1 and WT *C. freundii* GN346 β-lactamases have been determined to high resolution (1.4 to 1.5 Å; codes 1RGZ and 1RGY). In both WT GN346 and ES GC1 complexes, the catalytic Ser64 has been phosphorylated by 6 with the departure of the *m*-nitrophenol group. The terminal P-O_1 group displaces a water molecule from the oxyanion hole formed by the backbone amide groups of Ser64 and Ser318/321 (using numbering for the WT GN346 and ES GC1 β-lactamases, respectively). The oxygen atom is asymmetrically hydrogen-bonded to the amides, with the stronger and more linear hydrogen bond to the NH on the B3 β-strand. The remaining P-O_2 group is exposed. It forms hydrogen bonds to the hydroxyl group of Tyr150 and to a water molecule bridging to Thr316/319 on the B3 strand. Each complex has a similar conformation in the Ser64-Oγ-P linkage.

However, orientation and conformation of the cefotaxime-like side chain of the phosphonate are markedly different in the two complexes. In the WT GN346 binding site, the oxyimino branch is solvent exposed, but in the ES GC1 complex the oxyimino branch is quite buried. The serine-bound analogue of the tetrahedral transition state for deacylation thus exhibits a very different binding geometry in each enzyme. In the WT β-lactamase the cefotaxime-like side chain is crowded against the Ω loop and must protrude from the binding site with its methyloxime branch exposed. In the ES enzyme, a mutated Ω loop adopts an alternate conformation, allowing the side chain to be much more buried. During the binding and turnover of the cefotaxime substrate by this ES enzyme, it is proposed that ligand-protein contacts and intraligand contacts are considerably relieved relative to WT, facilitating positioning and activation of the hydrolytic water molecule. The ES β-lactamase is thus able to efficiently inactivate ES cephalosporins.

Mechanism of Inhibition by Methylidene Penems

Novel penem molecules with heterocycle substitutions at the 7 position via a methylidene linkage (e.g., 7, 8, and 9) have been investigated at the end of the 1990s for their activities and efficacy as β-lactamase inhibitors. The concentrations of these molecules that result in 50% inhibition of enzyme activity (IC_{50}) are in the low nanomolar ranges both for the TEM-1 enzyme and for AmpC. Some of them

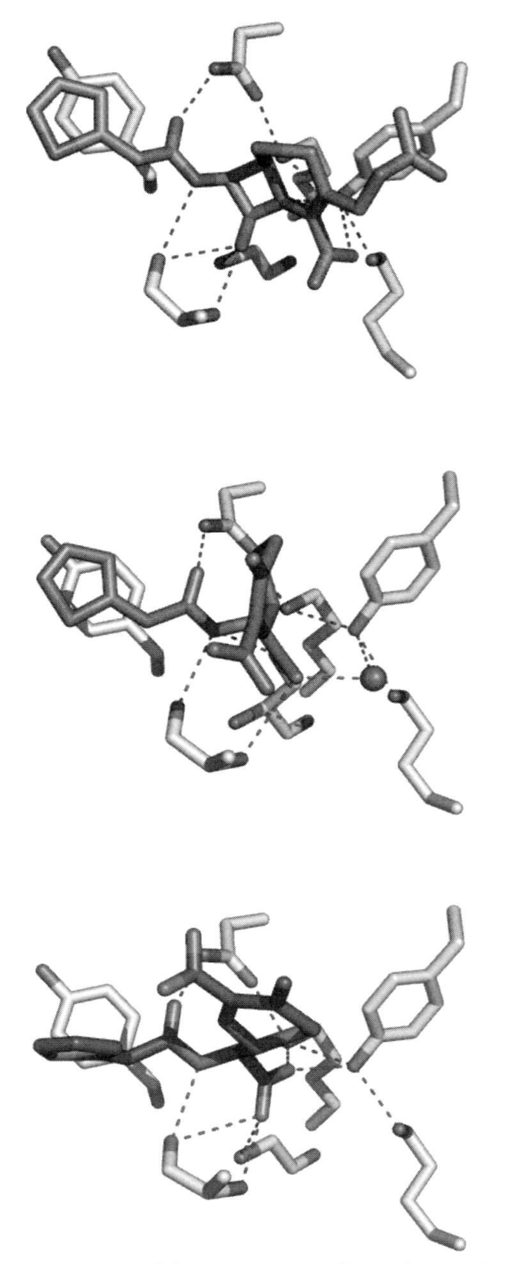

Figure 10.4 Views of the active sites of *E. coli* AmpC inactive mutant (S64G) in complex with cephalothin (**4**) showing snapshots along the catalytic pathway of hydrolysis by the β-lactamase: ES (enzyme-substrate complex), EI (acyl-enzyme intermediate), and EP (enzyme product complex). Coordinates are taken from crystallographic structures 1KVL and 1KVM.

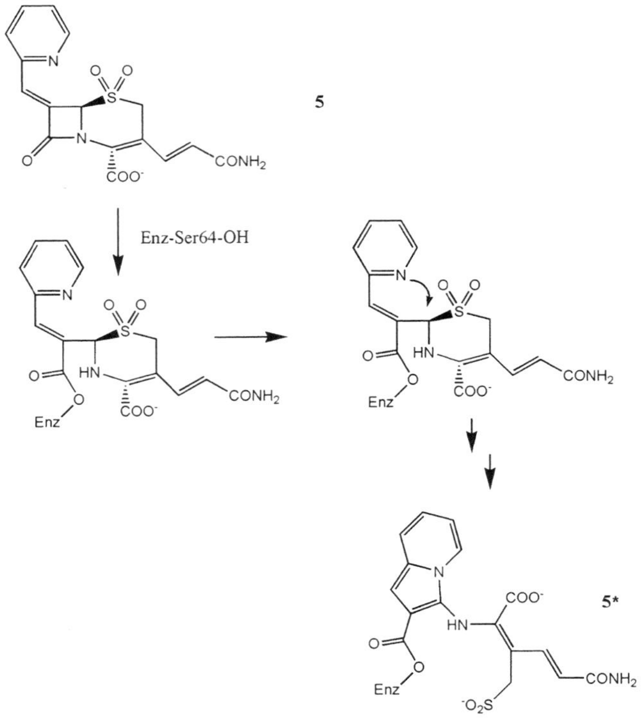

also inhibit the metalloenzyme class B β-lactamases. These inhibitors are usually more stable than imipenem against hydrolysis by dehydropeptidases. Their interest and development as potential new antibiotics are briefly discussed in the next section. Those compounds also present interest from a mechanistic point of view as illustrated by a

Figure 10.5 Chemical diagrams of compounds **5** and **5***.

structural study of BRL 42715, C7-(N1-methyl-1,2,3-triazolylmethylene)penem, **7**, an active-site-directed inactivator of a broad range of bacterial β-lactamases, including the class C enzymes. This compound is 10 to 100 times more active than other clinically used β-lactamase inactivators like clavulanic acid, sulbactam, and tazobactam. These potent inactivating activities are reflected in the low concentrations of **7** needed to potentiate the antibacterial activity of β-lactamase-susceptible β-lactams. Although **7** has not been developed because of its chemical instability and short half-life in humans, it is still used as a reference inactivator and thus remains a useful pharmacologic tool for mechanistic investigations.

In particular, the mechanism of action of **7** towards class C β-lactamases was investigated and the crystal structure of its complex with *E. cloacae* 908R β-lactamase (1Y54) was reported (22). A stable covalent adduct, a cyclic β-aminoacrylate-enzyme complex, resulting from acylation of the active-site serine by the penem followed by intramolecular rearrangement leads to the corresponding dihydrothiazepine (Fig. 10.7).

This crystal structure confirms a mechanism that implies opening of the five-membered thiazole ring system at the C5-S bond upon alcoholysis and rearrangement via a Michael addition to form a seven-membered

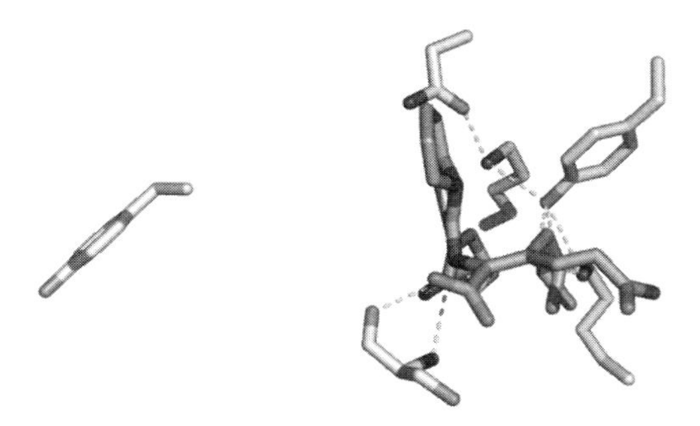

Figure 10.6 Active site of *E. cloacae* GC1 ES β-lactamase inhibited by cephem **5**, leading to intermediate **5*** after acylation by reactive serine Ser64 and intramolecular rearrangement. Coordinates are taken from crystallographic structure 1GA0.

dihydrothiazepine ring system (**7***, Fig. 10.8). This is in good agreement with spectral properties of the product that are identical to those of the dihydrothiazepine obtained after sodium hydroxide hydrolysis and with mass spectrometry results.

The binding mode and geometry of the covalent adduct of **7*** in class C 908R β-lactamase are quite different from that of penem analogue **8** in complex with class C ES GC1 β-lactamase (1ONH). It is, however, consistent with differences observed for a transition state analogue of cefotaxime in the parental and ES enzymes. In particular, the absolute configuration of the cyclic β-aminoacrylate-enzyme complex **7*** is *S*, while the other enantiomer of **8** forms upon complexation with GC1 β-lactamase. As already mentioned, the overall structure of the GC1 enzyme is equivalent to that of the 908R enzyme. In the latter structure of complex with **8**, the so-called Ω loop (residues 189 to 226) is well defined. The three additional residues (213 to 215) after position 210 in this Ω loop in GC1 lead to a more flexible binding site and allow an alternative fold ("open folding") in which Tyr224 (equivalent to Tyr221 in 908R) is displaced by ~6 Å relative to its position in the free enzyme. As a result, the main difference between the two enzyme complexes is the position of this tyrosine residue (Fig. 10.7). This has consequences for the binding mode of the ligand. Indeed, in 908R, the triazolyl cycle of **7** is able to stack with Tyr221, unlike the heterocyclic double ring of **8** in GC1. Moreover, the carboxylate moiety of **7** does not interact with Gln120 and Asn152 while it does so for **8** in CG1. This could explain why the absolute configuration is reversed.

Docking and energy minimization studies were performed to further understand and quantify differences in conformation and stereochemistry between the complexes observed with the 908R and GC1 class C β-lactamases. In the case of **7**, the triazolyl moiety is close to Tyr221 and the thiolate group lies along the B3 β-strand. The *S* isomer of the compound is therefore strongly favored.

Indeed, the thiolate group is able to react with only one side of the double bond. In contrast, the heterocyclic ring of **8** binds along the B3 β-strand. The thiolate moiety is near Leu119 and Gln120 and is positioned in such a way that only the product with *R* configuration can be produced (22).

COMPLEXES WITH CLASS C β-LACTAMASES AND DRUG DESIGN

In addition to their interest for mechanistic studies, structures of complexes between β-lactamases and synthetic compounds can also provide a rational basis for the design of original antibiotics.

Together with class A enzymes, class C β-lactamases are clinically the most commonly encountered. As already mentioned, class C enzymes such as AmpC are mainly chromosomal and are typically synthesized by gram-negative organisms. Through the 1990s, approximately one or two new plasmid-encoded AmpC β-lactamases were

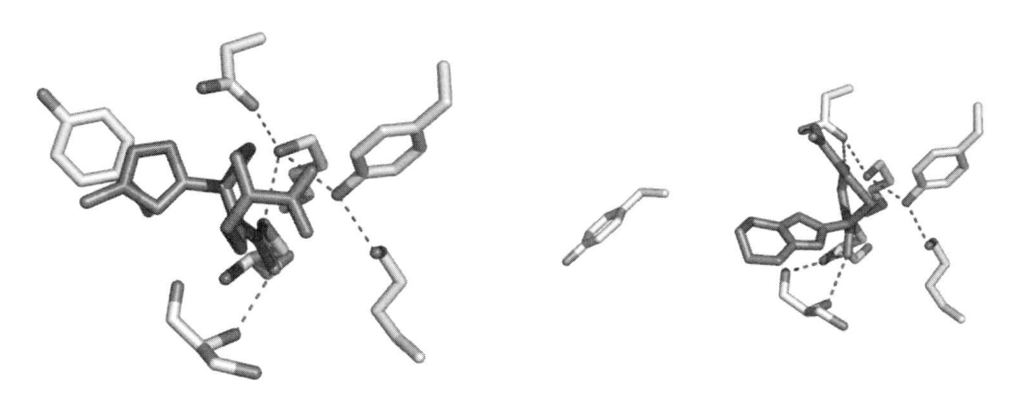

Figure 10.7 Comparison of the active sites of *E. cloacae* 908R (left) and ES CG1 (right) class C β-lactamases after reaction with methylidene penems **7** and **8**, respectively. Coordinates were taken from PDB files 1Y54 and 1ONH.

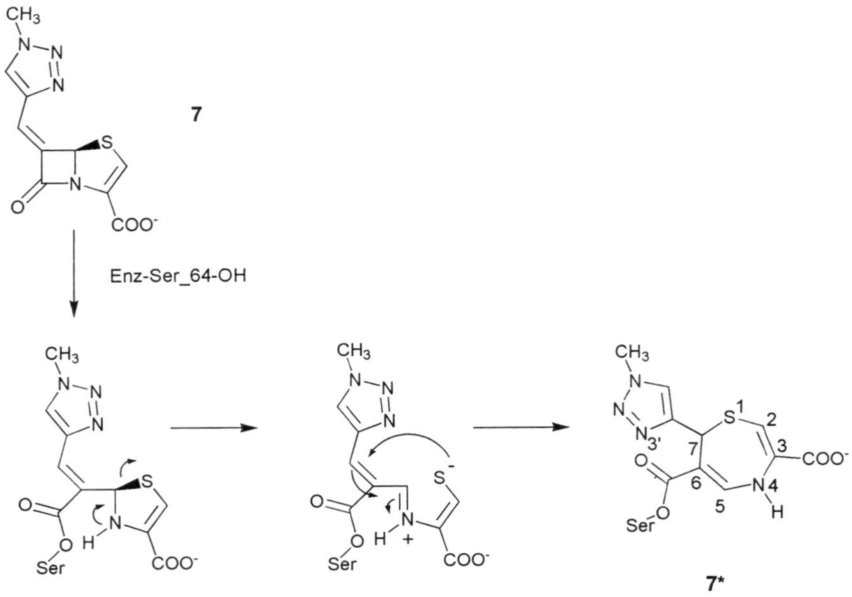

Figure 10.8 Possible mechanism of inhibition of class C β-lactamases by methylidene penems leading to a seven-membered dihydrothiazepine ring system after acylation and intramolecular rearrangement.

discovered annually, so that this class of enzymes, both chromosomal and plasmide borne, are now present in 10 to 50% of the patients infected with *C. freundii, E. cloacae, Serratia marcescens*, and *P. aeruginosa*.

Complexes of class C β-lactamases with compounds belonging to different families (β-lactams, boronic acids, phosphonates, etc.) have been obtained (Table 10.1). Those structures not only led to a better understanding of the mechanism of this class of hydrolases as discussed in "Structural Features of the Active Sites of Class C β-Lactamases and Mechanistic Considerations" above, but also provided valuable starting points for the design of new selective inhibitors as illustrated in the following examples.

AmpC Inhibition by Boronic Acids

In an effort to identify non-β-lactam-based β-lactamase inhibitors, the crystallographic structure of *E. coli* AmpC β-lactamase in complex with *m*-aminophenylboronic acid (**1**) was used to design modifications that might enhance the affinity of boronic acid-based inhibitors for class C β-lactamases. Several types of compounds were modeled into the AmpC binding site, and a total of 37 boronic acids were ultimately tested for β-lactamase inhibition (39). The most potent of these compounds, benzo[b]thiophene-2-boronic acid (Fig. 10.9, **10**), has an affinity for *E. coli* AmpC of 27 nM (Table 10.1; PDB code 1C3B). The wide range of functionality represented by these compounds allowed for the steric and chemical "mapping" of the AmpC active site in the region of the catalytic Ser64 residue, which may be useful in subsequent inhibitor discovery efforts. Also, the new boronic acid-based inhibitors were found to potentiate the activity of β-lactam antibiotics, such as amoxicillin and ceftazidime, against bacteria expressing class C β-lactamases. This suggested that boronic acid-based compounds may serve as leads for the development of therapeutic agents for the treatment of β-lactam-resistant infections.

To investigate if larger boronic compounds would better interact with AmpC, structure-guided in-parallel synthesis was used to explore new inhibitors of AmpC (35). Twenty-eight derivatives of the lead compound, 3-aminophenylboronic acid, led to an inhibitor with 80-fold better binding. Based on molecular docking results, 12 derivatives were synthesized, leading to inhibitors with K_i values of 60 nM and with improved solubility. Several of these inhibitors reversed the resistance of nosocomial gram-positive bacteria, though they showed little activity against gram-negative bacteria. The X-ray crystal structure of compound **11** in complex with AmpC was subsequently determined to 2.1 Å resolution (1GA9). In the experimental structure, the inhibitor interacts with conserved residues in the active site, whose role in recognition had not been previously explored. Similar recognition sites were observed in the structure of the class C β-lactamase from *E. cloacae* 908R in complex with iodo-acetamido-phenyl boronic acid (**12**) (PDB code 1S6R) (40).

The crystal structures of complexes between AmpC and other boronic acid inhibitors were further studied by computational methods, and consensus binding sites were identified. Along with determinations of AmpC structures in complexes with boronic acid inhibitors and β-lactams, the programs GRID, MCSS, and X-SITE were used to predict potential binding site hot spots on AmpC. Several consensus binding sites were identified from the crystal structures, and predictions by the computational programs showed some correlation with the experimentally observed binding sites (30). Several sites were not predicted, but novel binding sites were suggested. Taken together, a map of binding site hot spots found on AmpC, along with information on the functionality recognized at each site, was constructed. As a result of those structure-based drug design approaches, new carboxyphenyl-glycylboronic acid transition state analogue inhibitors of AmpC were designed and evaluated. The new compounds improve inhibition by over 2 orders of magnitude compared to analogous glycylboronic acids, with K_i values as low as 1 nM (23, 33). In this context, Morandi et al. designed compound **13** based on the β-lactam antibiotic cephalothin (23). The benzoic acid group of **13** was intended to mimic, in the covalent adduct **13***, the dihydrothiazine ring system of the high-energy intermediate of the substrate cephalothin (Fig. 10.10, **14***).

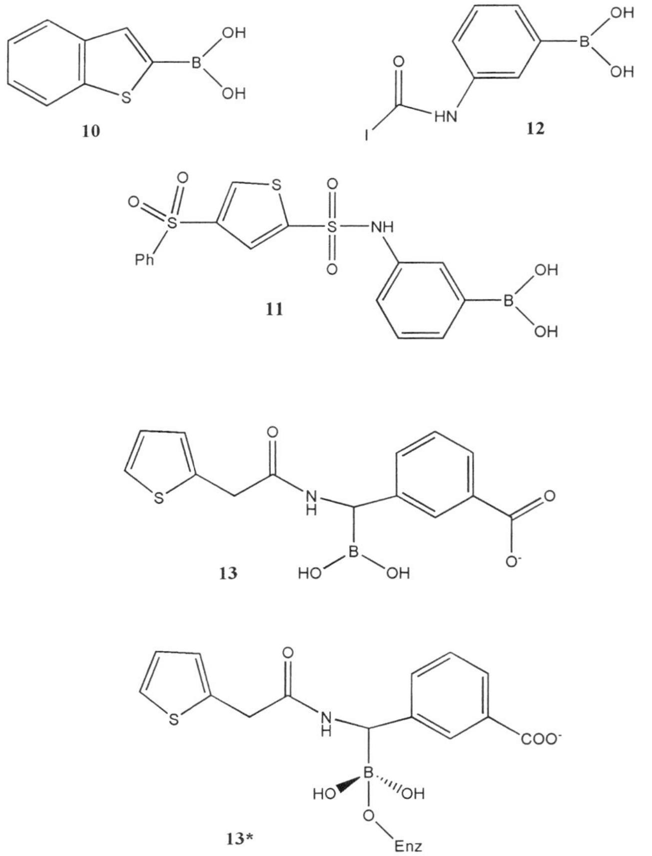

Indeed, β-lactams bear negatively charged groups at this position, and structural analysis of the binding determinants of AmpC suggests that there is a complementary carboxylate binding site on the enzyme (30). Among known structures of AmpC with bound substrate, the analogous carboxylate hydrogen bonds with either Asn343 or Asn346 (4). Thus, it was expected that the carboxylate of **13** would form a hydrogen bond with at least one of those residues. Surprisingly, the crystal structure of **13** in complex with AmpC revealed a hydrogen bond between the *m*-carboxylate of the benzoic acid and Asn289 (23) and not Asn343 or Asn346, as was expected and designed.

To understand the origins of this affinity, and to guide the design of future inhibitors, further structural investigations were undertaken. An unexpected hydrogen bond between the nonconserved Asn289 and a key inhibitor carboxylate was observed in the X-ray crystal structure of **13** in complex with AmpC β-lactamase. To investigate the energy of this hydrogen bond, the mutant enzyme N289A was made, as was an analogue of **13** that lacked the carboxylate (compound **14**).

The differential affinity of the four different protein and analogue complexes indicates that the carboxylate-amide hydrogen bond contributes 1.7 kcal/mol to overall binding affinity. The synthesis of an analogue of **13** in which the carboxylate was replaced with an aldehyde led to an inhibitor that lost all this hydrogen bond energy, consistent with the importance of the ionic nature of this hydrogen bond. To investigate the structural bases of these energies, X-ray crystal structures of N289A/**13** and N289A/**14** were determined to 1.49 and 1.39 Å, respectively (PDB codes 1PI4 and 1PI5). These structures suggest that no significant rearrangement occurs in the mutant versus the WT complexes with both compounds. The mutant enzymes L119A and L293A were made to investigate the interaction between

Figure 10.9 Chemical diagrams of compounds **10** to **13***.

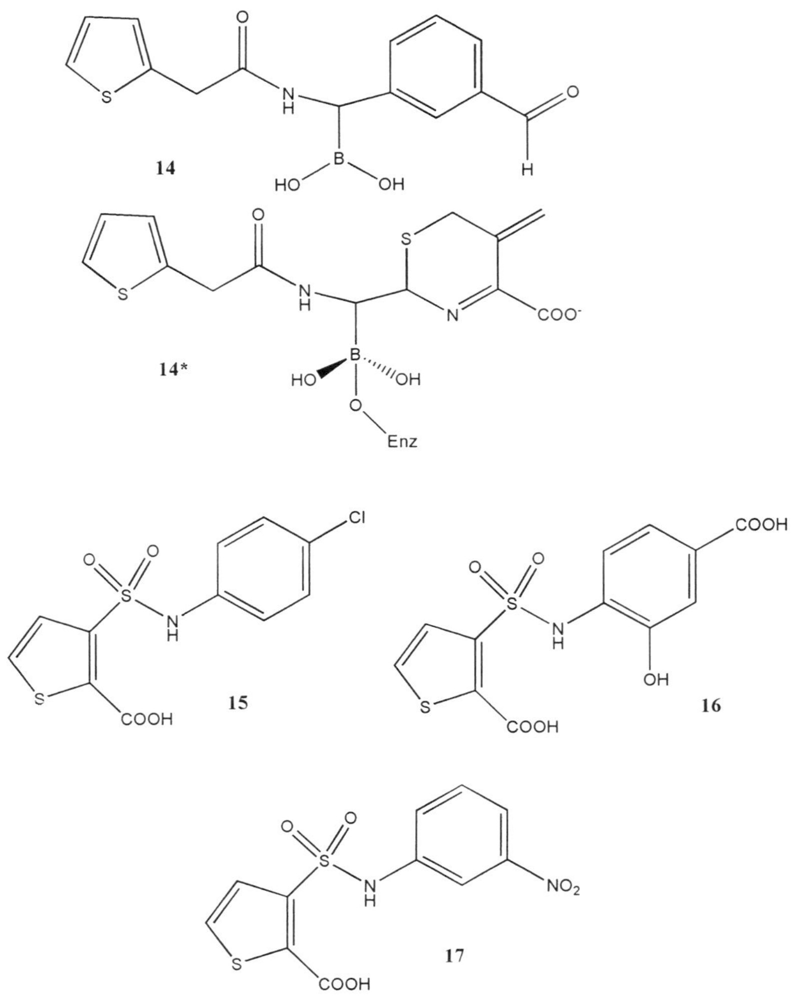

Figure 10.10 Chemical diagrams of compounds **14** to **17**.

a phenyl ring in **13** and these residues. Whereas deletion of the phenyl itself diminishes affinity fivefold, the double-mutant cycles suggest that this energy does not come through interaction with the leucines, despite the close contact in the structure. The energies of these interactions provide key information for the design of improved inhibitors against β-lactamases. The high magnitude of the ion-dipole interaction between Asn289 and the carboxylate of **13** is consistent with the idea that ionic interactions can provide significant net affinity in inhibitor complexes in this series of promising compounds.

Structure-Based Discovery of Novel Noncovalent Inhibitors of AmpC β-Lactamase

Not only boronic acid inhibitors were designed on the basis of X-ray structures. A database of over 200,000 compounds was docked to the active site of AmpC β-lactamase to identify potential inhibitors (29). Fifty-six compounds were tested, and three had K_i values of 650 μM or better. The best of these, 3-[(4-chloroanilino)sulfonyl]thiophene-2-carboxylic acid, **15**, was a competitive noncovalent inhibitor (K_i = 26 μM), which also reversed resistance to β-lactams in bacteria expressing AmpC. The structure of AmpC in complex with this compound was determined by X-ray crystallography (1L2S) and revealed that the inhibitor interacts with key active-site residues in sites targeted in the docking calculation. Indeed, the experimentally determined conformation of the inhibitor closely resembles the predicted one.

Lead compound **15** is dissimilar to penicillins and cephalosporins and binds to the enzymes noncovalently and reversibly, in contrast to the β-lactam substrates and boronic inhibitors. Despite these differences, the X-ray crystal structure of the AmpC/**15** complex (1L2S) revealed that **15** complements the core of the active site, interacting with key residues involved in β-lactam recognition and hydrolysis such as Ser64, Lys67, Asn152, and Tyr221 (Fig. 10.11) (34). Morandi et al. concluded that the ligand recognition encoded by the AmpC structure was plastic enough to accommodate inhibitors genuinely dissimilar to β-lactams, allowing a new departure in the medicinal chemistry of their inhibitors.

Efforts to improve this series of inhibitors focused on a structure-based approach. In the structure of the AmpC/**15** complex, the inhibitor complements the core of the active site but leaves a distal region open. Morandi et al. sought derivatives to take advantage of this region that were relatively easy to synthesize and would not diminish the solubility and drug-like properties of the inhibitors. They therefore focused on derivates that made new interactions with polar residues in the distal part of the AmpC site, including Arg204, or that tested features of the ligands that appeared important for binding. Fourteen analogues were synthesized, the best of which bound to AmpC with 26-fold improved affinity over **15**. To understand their improved affinity, X-ray crystal structures were determined for two of the new analogues in complex with AmpC. To explore their biological potential, several of the new inhibitors were tested in bacterial cell culture against clinical pathogens. Intriguing differences in their biological effects on resistant bacteria compared to classic, β-lactam-based inhibitors such as clavulanate and cefoxitin were observed.

To investigate the structural basis for the higher affinities found for the new series, the crystal structures of AmpC in complex with compound **16** (1XGJ) and with compound **17** (1XGI) were determined at 1.97 and 1.96 Å resolution, respectively.

Overall, the complexes with **16** and **17** resemble that of the AmpC/**15** complex, with a few interesting differences (Fig. 10.11). Key hydrogen bond interactions are observed in the conserved parts of all three structures. The thiophene carboxylate in both **16** and **17** hydrogen bonds with both the Oγ and the main-chain NH of the hydrolytic Ser64.

As with the lead compound **15**, the thiophene ring is within van der Waals distance of residues Leu119 and Leu293, which form a hydrophobic patch on AmpC. A sulfonamide oxygen interacts with the Oγ of Ser64 and the Nζ of Lys67; the other hydrogen bonds with the Nδ2 of Asn152. As in the AmpC/**15** complex, the nitrogen atom of the sulfonamide interacts with the main-chain oxygen of Ala318. Differences emerge in the 4′-substituted benzene ring, which is the point of substitution for compound **15**. The ring itself appears to form edge-to-face quadrupole interactions with Tyr221 via the introduced 2′-hydroxyl of **16**. This interaction is also present in the structure of the lead **15**, but in this case it appears to be a ring-ring or quadrupolar interaction, whereas with **16** the intercession of the hydroxyl makes it a dipole-quadrupolar interaction. The presence of the 2′-hydroxyl also displaces **16** slightly "up" in the site, moving the ring away from Tyr221 and toward polar residues such as Arg204 and Thr319. The 4′-carboxylate points toward the solvent-accessible entry of the binding site and hydrogen bonds with an ordered water molecule, which in turn hydrogen bonds with the Nε of Arg204.

In the structure of the AmpC/**17** complex, the inhibitor is moved slightly down in the site, keeping the quadrupole edge-to-face stacking with Tyr221. The absence of the 2′-hydroxyl in this analogue excludes interaction of the phenyl ring with an ordered water and with Arg204. The loss of these polar interactions may be consistent with the diminished affinity for this inhibitor. From a molecular recognition standpoint, and as we noted, it is surprising that a water-mediated hydrogen bond between the carboxylate and Arg204, in a relatively solvent-exposed region, can significantly improve affinity. Detailed conclusions as to the importance of these interactions must remain tentative pending more rigorous analyses. What can be said, and this was one of the conclusions of this work, is that the X-ray structures of these complexes provide templates for designing and understanding such experiments.

Structure-Based Drug Design among Methylidene Penems

As mentioned in "Structural Features of the Active Sites of Class C β-Lactamases and Mechanistic Considerations" above, novel penem molecules with heterocycle substitutions at the 6 position via a methylidene linkage (**7**, **8**, and **9**) were investigated at the end of the 1990s for their activities and efficacy as β-lactamase inhibitors. A new β-lactamase inhibitor, a methylidene penem having a 5,6-dihydro-8H-imidazo[2,1-c][1,4]oxazine heterocyclic substituent at the C-6 position with a Z configuration (**8**), irreversibly inhibits both class A and class C serine β-lactamases with IC_{50} values of 0.4 and 9.0 nM for TEM-1 and SHV-1 (class A), respectively, and 4.8 nM in AmpC (class C) β-lactamases. The compound also inhibits irreversibly the class C ES GC1 β-lactamase (IC_{50}=6.2 nM). High-resolution crystallographic structures of a reaction intermediate of **8** with the SHV-1 β-lactamase and with the GC1 β-lactamase have been determined by X-ray

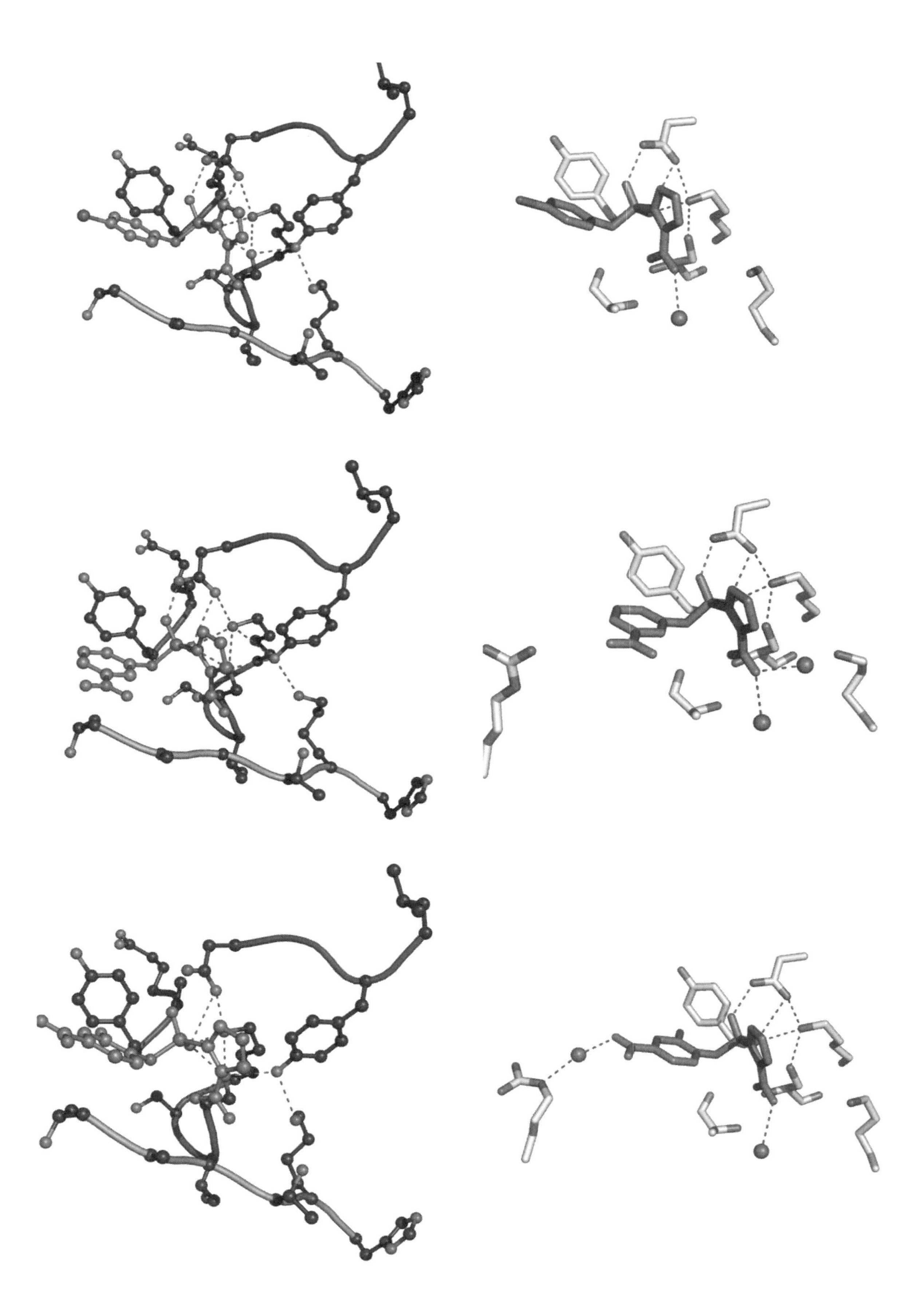

diffraction to resolutions of 1.10 and 1.38 Å, respectively (24). Cryoquenching of the reaction of **8** with each β-lactamase crystal produced a common, covalently bound intermediate. After acylation of the serine, a nucleophilic attack by the departing thiolate on the C-6′ atom yielded a novel seven-membered 1,4-thiazepine ring having *R* stereochemistry at the new C-7 moiety (Fig. 10.8). The orientation of this ring in each complex differs by a 180 degree rotation about the bond to the acylated serine. As already discussed (see "Structural Features of the Active Sites ..." above), the stereochemistry of the product of cyclization of **8** in CG1 (and SHV-1) was different in the complex of **7** with the parent 908R *E. cloacae* enzyme (22).

On the basis of those first structures of complexes with methylidene penem **8**, the design and synthesis of a series of seven tricyclic 6-methylidene penems as novel class A and C serine β-lactamase inhibitors have been recently described (37).

Indeed, an unexpected feature reported in the study of complexes with **8** indicated that the 1,4-thiazepine intermediate is oriented differently in each complex following the acylation reaction between **8** and the active serine residue. This finding is significant in the context of designing new analogues with the desired potency against both class A and C enzymes. Accordingly, on the basis of these findings, Knox et al. undertook the synthesis and biological evaluation of novel tricyclic methylidene penems. Seven compounds were tested against TEM-1 (class A), CcrA (class B), and AmpC (class C) β-lactamases. In comparison with tazobactam, the new inhibitors exhibited excellent activity against those enzymes. These compounds were very potent inhibitors of TEM-1 and AmpC enzymes with IC_{50} values in the ranges of 1 to 18 and 1 to 3 nM, respectively. Furthermore, in comparison to tazobactam, they exhibited almost 5- to 100-fold greater potency against TEM-1 enzyme and were about 28,000- to 84,000-fold more potent against AmpC enzymes. Within this series of new inhibitors, comparison of their potencies reveals that any substitution on the aromatic ring or any modification did not alter their potency against the AmpC enzyme significantly. In vitro evaluation in a cell-based assay (MIC) reaffirmed the effectiveness of those compounds as potent broad-spectrum β-lactamase inhibitors.

In this context, Knox et al. also determined the crystallographic structure of reaction intermediates of **9** with the SHV-1 and GC1 β-lactamases. They found that the 1,4-thiazepine intermediate has distinctive hydrophobic contacts within the binding site of each β-lactamase. This time, the absolute configuration at carbon C-6′ is variable in the two complexes. A single conformer with the *R* configuration is observed in the SHV complex, whereas in the GC1 complex the conformer with the *R* configuration has only 30% occupancy and coexists with the predominant *S* conformer (70%). It is thus possible that both stereoisomers are formed initially in each β-lactamase, but the form more stable to hydrolysis predominates in the crystallographic maps. Alternatively, stereospecific reaction pathways may differ in each active site, producing one chiral form in preference to the other as discussed by Michaux et al. (22). The inhibitory reaction of **9** with the GC1 class C β-lactamases as monitored by X-ray crystallography also revealed the formation of a 1,4-thiazepine intermediate with unexpected modes of binding. The significance of hydrophobic stacking of the C-6′ heterocycle with Tyr105 and Tyr224 in SHV-1 and GC1, respectively, and its possible role in protecting the intermediate from hydrolytic attack were presented. At this stage, it would be very interesting to complement this study by getting a complex of **9** with 908R or P99 *E. cloacae* β-lactamase and compare the results with the ES enzyme. In any case, even if pending questions remain, the reported crystallographic complexes with bicyclic and tricyclic penems disclose new insights and considerations into β-lactamase inhibition and further stress the benefit of those structure-based approaches in the design of new inhibitors and understanding of their mechanism(s) of inhibition.

CONCLUSION

The increasing number and variety of β-lactamases produced by bacteria represent a serious threat to the clinical utility of β-lactams. The introduction of β-lactamase inhibitors was thought to have alleviated this problem; however, alterations and mutations within β-lactamases have allowed bacteria to overcome the effects of the β-lactamase inhibitors. The currently marketed inhibitors (tazobactam, clavulanate, or sulbactam) are not active against all β-lactamases. Among the most problematic β-lactamases are the molecular class C enzymes. Bacterial genera known to produce inducible or stably derepressed

Figure 10.11 Comparison of the crystal structures of complexes between *E. coli* AmpC and thiophene-sulfonamide derivatives designed as original non-β-lactam inhibitors. Coordinates of those complexes between AmpC and compounds **15** (top), **16** (middle), and **17** (bottom) were obtained from PDB files 1L2S, 1XGI, and 1XGJ.

AmpC β-lactamases are *Citrobacter*, *Enterobacter*, *Hafnia*, *Pseudomonas*, *Morganella*, *Providencia*, and *Serratia*. Expression of the class C enzymes by these organisms has been shown to confer resistance to the current commercial inhibitor combinations as well as ES cephalosporins. Crystallographic structures of a large number of class C β-lactamases (alone or in complex with substrate analogues or inhibitors) have been reported and studied (Table 10.1). As was presented in "Structural Features of the Active Sites of Class C β-Lactamases and Mechanistic Considerations" above, those structures produced a clear picture of different intermediates along the catalytic route of class C β-lactamases. The section entitled "Complexes with Class C β-Lactamases and Drug Design" selected a number of examples where structure-based drug design approaches were successful in identifying and optimizing original inhibitors.

References

1. **Ambler, R. P.** 1980. The structure of beta-lactamases. *Philos. Trans. R. Soc. Lond. B Biol. Sci.* **289:**321–331.
2. **Bauvois, C., A. S. Ibuka, A. Celso, J. Alba, Y. Ishii, J. M. Frere, and M. Galleni.** 2005. Kinetic properties of four plasmid-mediated AmpC beta-lactamases. *Antimicrob. Agents Chemother.* **49:**4240–4246.
3. **Beadle, B. M., and B. K. Shoichet.** 2002. Structural basis for imipenem inhibition of class C beta-lactamases. *Antimicrob. Agents Chemother.* **46:**3978–3980.
4. **Beadle, B. M., I. Trehan, P. J. Focia, and B. K. Shoichet.** 2002. Structural milestones in the reaction pathway of an amide hydrolase: substrate, acyl, and product complexes of cephalothin with AmpC beta-lactamase. *Structure* **10:** 413–424.
5. **Beesley, T., N. Gascoyne, V. Knott-Hunziker, S. Petursson, S. G. Waley, B. Jaurin, and T. Grundstrom.** 1983. The inhibition of class C beta-lactamases by boronic acids. *Biochem. J.* **209:**229–233.
6. **Bergstrom, S., and S. Normark.** 1979. Beta-lactam resistance in clinical isolates of *Escherichia coli* caused by elevated production of the ampC-mediated chromosomal beta-lactamase. *Antimicrob. Agents Chemother.* **16:**427–433.
7. **Berman, H. M., J. Westbrook, Z. Feng, G. Gilliland, T. N. Bhat, H. Weissig, I. N. Shindyalov, and P. E. Bourne.** 2000. The Protein Data Bank. *Nucleic Acids Res.* **28:**235–242.
8. **Bonnet, R.** 2004. Growing group of extended-spectrum beta-lactamases: the CTX-M enzymes. *Antimicrob. Agents Chemother.* **48:**1–14.
9. **Cartwright, S. J., and S. G. Waley.** 1984. Purification of beta-lactamases by affinity chromatography on phenylboronic acid-agarose. *Biochem. J.* **221:**505–512.
10. **Crichlow, G. V., A. P. Kuzin, M. Nukaga, K. Mayama, T. Sawai, and J. R. Knox.** 1999. Structure of the extended-spectrum class C beta-lactamase of Enterobacter cloacae GC1, a natural mutant with a tandem tripeptide insertion. *Biochemistry* **38:**10256–10261.
11. **Crichlow, G. V., M. Nukaga, V. R. Doppalapudi, J. D. Buynak, and J. R. Knox.** 2001. Inhibition of class C beta-lactamases: structure of a reaction intermediate with a cephem sulfone. *Biochemistry* **40:**6233–6239.
12. **Feller, G., Z. Zekhnini, J. Lamotte-Brasseur, and C. Gerday.** 1997. Enzymes from cold-adapted microorganisms. The class C beta-lactamase from the antarctic psychrophile *Psychrobacter immobilis* A5. *Eur. J. Biochem.* **244:**186–191.
13. **Galleni, M., G. Amicosante, and J. M. Frere.** 1988. A survey of the kinetic parameters of class C beta-lactamases. Cephalosporins and other beta-lactam compounds. *Biochem. J.* **255:**123–129.
14. **Galleni, M., and J. M. Frere.** 1988. A survey of the kinetic parameters of class C beta-lactamases. Penicillins. *Biochem. J.* **255:**119–122.
15. **Jaurin, B., and T. Grundstrom.** 1981. ampC cephalosporinase of Escherichia coli K-12 has a different evolutionary origin from that of beta-lactamases of the penicillinase type. *Proc. Natl. Acad. Sci. USA* **78:**4897–4901.
16. **Kariuki, S., J. E. Corkill, G. Revathi, R. Musoke, and C. A. Hart.** 2001. Molecular characterization of a novel plasmid-encoded cefotaximase (CTX-M-12) found in clinical *Klebsiella pneumoniae* isolates from Kenya. *Antimicrob. Agents Chemother.* **45:**2141–2143.
17. **Knott-Hunziker, V., S. Petursson, G. S. Jayatilake, S. G. Waley, B. Jaurin, and T. Grundstrom.** 1982. Active sites of beta-lactamases. The chromosomal beta-lactamases of Pseudomonas aeruginosa and Escherichia coli. *Biochem. J.* **201:**621–627.
18. **Knott-Hunziker, V., S. Petursson, S. G. Waley, B. Jaurin, and T. Grundstrom.** 1982. The acyl-enzyme mechanism of beta-lactamase action. The evidence for class C beta-lactamases. *Biochem. J.* **207:**315–322.
19. **Lodge, J. M., S. D. Minchin, L. J. Piddock, and J. W. Busby.** 1990. Cloning, sequencing and analysis of the structural gene and regulatory region of the Pseudomonas aeruginosa chromosomal ampC beta-lactamase. *Biochem. J.* **272:**627–631.
20. **Matagne, A., A. Dubus, M. Galleni, and J. M. Frere.** 1999. The beta-lactamase cycle: a tale of selective pressure and bacterial ingenuity. *Nat. Prod. Rep.* **16:**1–19.
21. **Matagne, A., J. Lamotte-Brasseur, G. Dive, J. R. Knox, and J. M. Frere.** 1993. Interactions between active-site-serine beta-lactamases and compounds bearing a methoxy side chain on the alpha-face of the beta-lactam ring: kinetic and molecular modelling studies. *Biochem. J.* **293**(Pt. 3)**:**607–611.
22. **Michaux, C., P. Charlier, J. M. Frere, and J. Wouters.** 2005. Crystal structure of BRL 42715, C6-(N1-methyl-1,2,3-triazolylmethylene)penem, in complex with Enterobacter cloacae 908R beta-lactamase: evidence for a stereoselective mechanism from docking studies. *J. Am. Chem. Soc.* **127:**3262–3263.
23. **Morandi, F., E. Caselli, S. Morandi, P. J. Focia, J. Blazquez, B. K. Shoichet, and F. Prati.** 2003. Nanomolar inhibitors of AmpC beta-lactamase. *J. Am. Chem. Soc.* **125:**685–695.
24. **Nukaga, M., T. Abe, A. M. Venkatesan, T. S. Mansour, R. A. Bonomo, and J. R. Knox.** 2003. Inhibition of class A and class C beta-lactamases by penems: crystallographic

structures of a novel 1,4-thiazepine intermediate. *Biochemistry* **42:**13152–13159.

25. **Nukaga, M., S. Haruta, K. Tanimoto, K. Kogure, K. Taniguchi, M. Tamaki, and T. Sawai.** 1995. Molecular evolution of a class C beta-lactamase extending its substrate specificity. *J. Biol. Chem.* **270:**5729–5735.
26. **Nukaga, M., S. Kumar, K. Nukaga, R. F. Pratt, and J. R. Knox.** 2004. Hydrolysis of third-generation cephalosporins by class C beta-lactamases. Structures of a transition state analog of cefotoxamine in wild-type and extended spectrum enzymes. *J. Biol. Chem.* **279:**9344–9352.
27. **Nukaga, M., K. Taniguchi, Y. Washio, and T. Sawai.** 1998. Effect of an amino acid insertion into the omega loop region of a class C beta-lactamase on its substrate specificity. *Biochemistry* **37:**10461–10468.
28. **Philippon, A., G. Arlet, and G. A. Jacoby.** 2002. Plasmid-determined AmpC-type beta-lactamases. *Antimicrob. Agents Chemother.* **46:**1–11.
29. **Powers, R. A., F. Morandi, and B. K. Shoichet.** 2002. Structure-based discovery of a novel, noncovalent inhibitor of AmpC beta-lactamase. *Structure* **10:**1013–1023.
30. **Powers, R. A., and B. K. Shoichet.** 2002. Structure-based approach for binding site identification on AmpC beta-lactamase. *J. Med. Chem.* **45:**3222–3234.
31. **Pratt, R. F., and C. P. Govardhan.** 1984. beta-Lactamase-catalyzed hydrolysis of acyclic depsipeptides and acyl transfer to specific amino acid acceptors. *Proc. Natl. Acad. Sci. USA* **81:**1302–1306.
32. **Rhazi, N., M. Galleni, M. I. Page, and J. M. Frere.** 1999. Peptidase activity of beta-lactamases. *Biochem. J.* **341**(Pt. 2)**:** 409–413.
33. **Roth, T. A., G. Minasov, S. Morandi, F. Prati, and B. K. Shoichet.** 2003. Thermodynamic cycle analysis and inhibitor design against beta-lactamase. *Biochemistry* **42:** 14483–14491.
34. **Tondi, D., F. Morandi, R. Bonnet, M. P. Costi, and B. K. Shoichet.** 2005. Structure-based optimization of a non-beta-lactam lead results in inhibitors that do not up-regulate beta-lactamase expression in cell culture. *J. Am. Chem. Soc.* **127:**4632–4639.
35. **Tondi, D., R. A. Powers, E. Caselli, M. C. Negri, J. Blazquez, M. P. Costi, and B. K. Shoichet.** 2001. Structure-based design and in-parallel synthesis of inhibitors of AmpC beta-lactamase. *Chem. Biol.* **8:**593–611.
36. **Usher, K. C., L. C. Blaszczak, G. S. Weston, B. K. Shoichet, and S. J. Remington.** 1998. Three-dimensional structure of AmpC beta-lactamase from Escherichia coli bound to a transition-state analogue: possible implications for the oxyanion hypothesis and for inhibitor design. *Biochemistry* **37:** 16082–16092.
37. **Venkatesan, A. M., Y. Gu, O. Dos Santos, T. Abe, A. Agarwal, Y. Yang, P. J. Petersen, W. J. Weiss, T. S. Mansour, M. Nukaga, A. M. Hujer, R. A. Bonomo, and J. R. Knox.** 2004. Structure-activity relationship of 6-methylidene penems bearing tricyclic heterocycles as broad-spectrum beta-lactamase inhibitors: crystallographic structures show unexpected binding of 1,4-thiazepine intermediates. *J. Med. Chem.* **47:**6556–6568.
38. **Wang, Z., W. Fast, A. M. Valentine, and S. J. Benkovic.** 1999. Metallo-beta-lactamase: structure and mechanism. *Curr. Opin. Chem. Biol.* **3:**614–622.
39. **Weston, G. S., J. Blazquez, F. Baquero, and B. K. Shoichet.** 1998. Structure-based enhancement of boronic acid-based inhibitors of AmpC beta-lactamase. *J. Med. Chem.* **41:**4577–4586.
40. **Wouters, J., E. Fonze, M. Vermeire, J. M. Frere, and P. Charlier.** 2003. Crystal structure of Enterobacter cloacae 908R class C beta-lactamase bound to iodo-acetamido-phenyl boronic acid, a transition-state analogue. *Cell. Mol. Life Sci.* **60:**1764–1773.

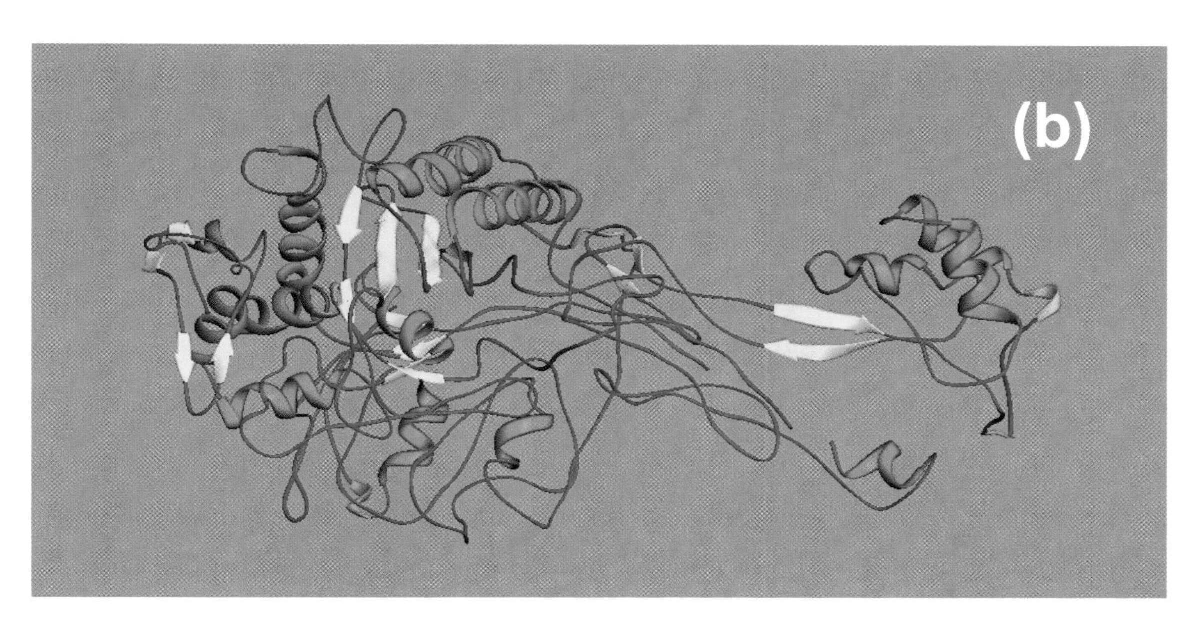

Color Plate 1 (chapter 7) Structures of PBPs. Ribbon diagrams showing the structures of *E. faecium* PBP5 (type B1) (a) and *S. pneumoniae* PBP2x (type B4) (b).

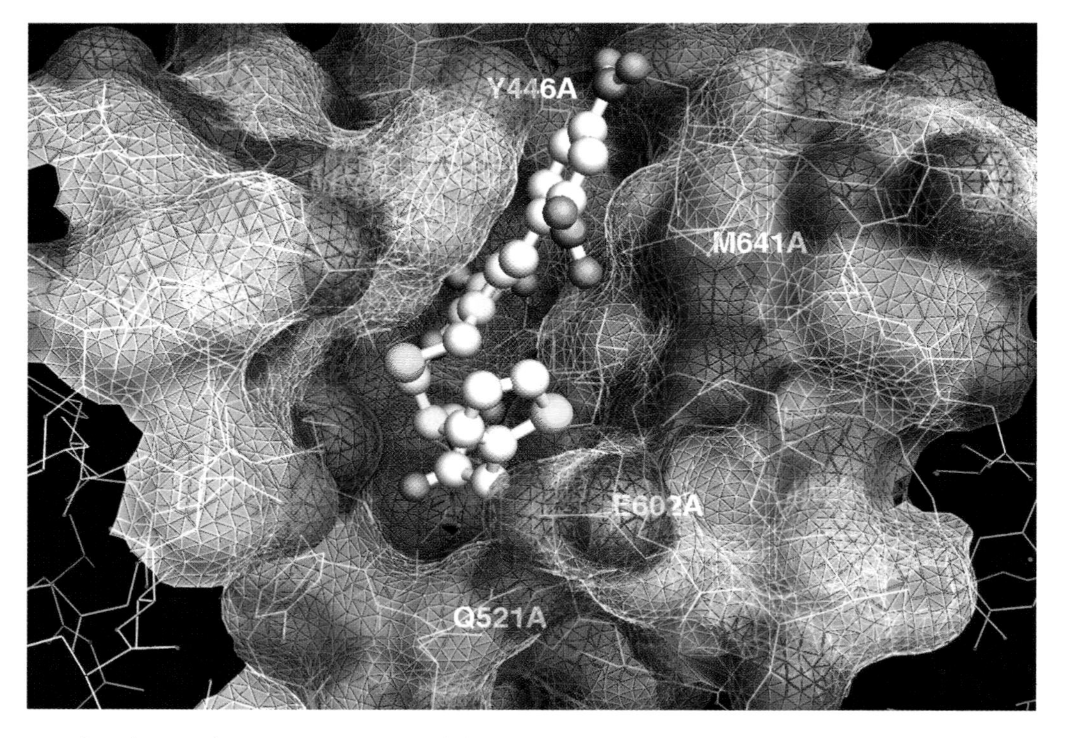

Color Plate 2 (chapter 7) Structure of the active site of *S. aureus* PBP2′ acyl enzyme complex with nitrocefin. The structure (78) is shown as ball and stick model with the protein overlaid by an electrostatic surface calculated using MOLOC.

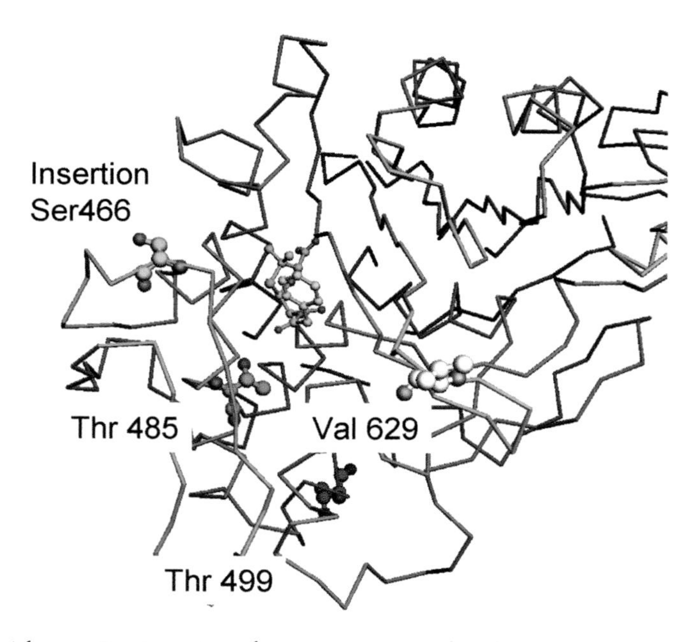

Color Plate 3 (chapter 7) Location of point mutations of *E. faecium* PBP5 conferring β-lactam resistance. The alpha-carbon chain of the structure of the transpeptidase domain of the acyl enzyme complex (120) formed with benzylpenicillin is shown with the inhibitor and the side chains of the residues implicated in β-lactam resistance shown as ball and stick models.

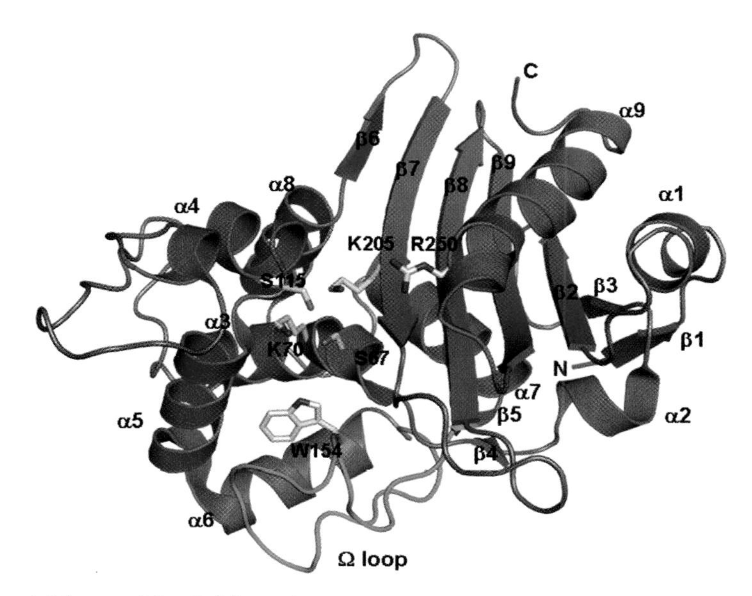

Color Plate 4 (chapter 11) Folding of OXA-10 protein (1FOF) as a ribbon presentation, with OXA-10 numbering. Key residues in the active site and the omega loop are indicated.

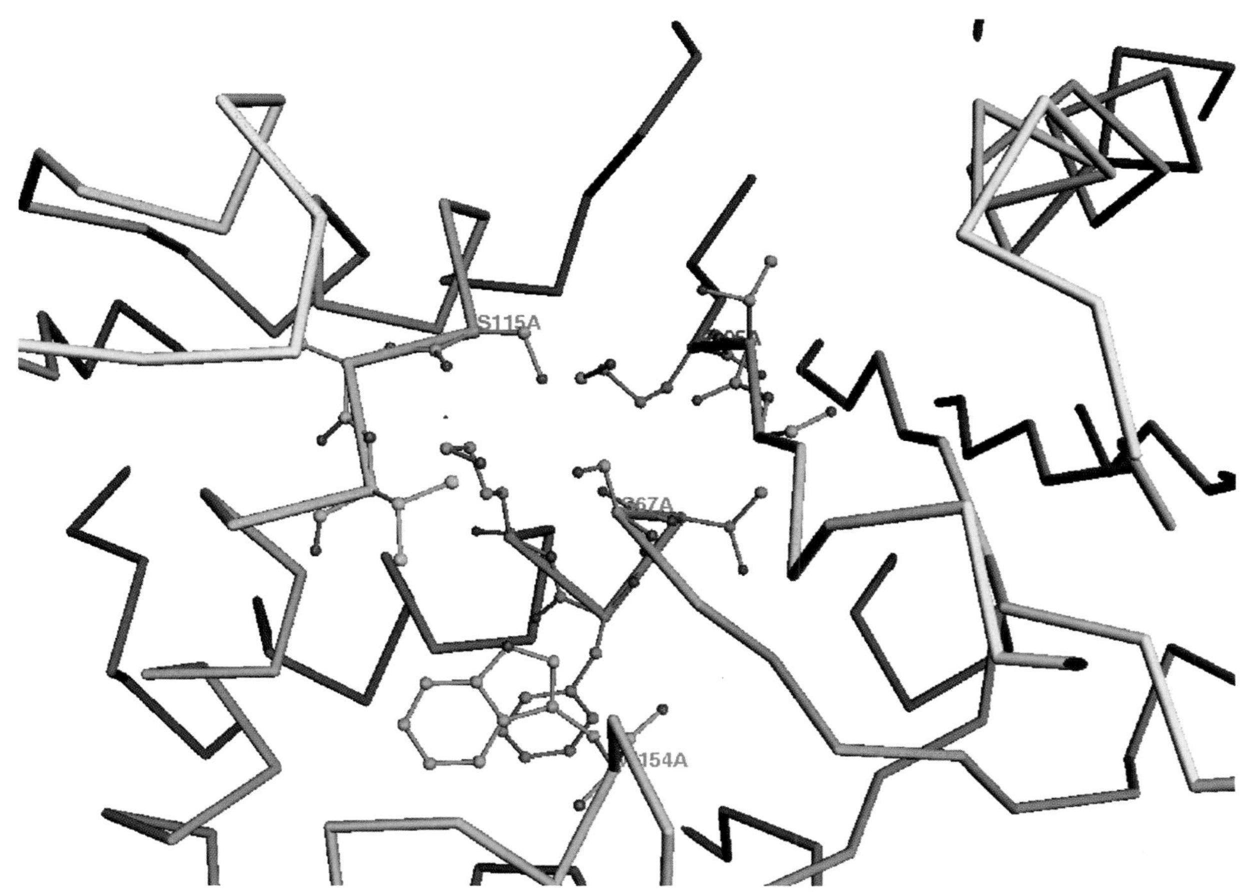

Color Plate 5 (chapter 11) Conserved residues in OXA-10 (1FOF) ($_{67}$Ser-Xxx-Xxx-Lys, $_{115}$Ser-Xxx-Val, and $_{205}$Lys-Xxx-Gly [OXA-10 numbering]).

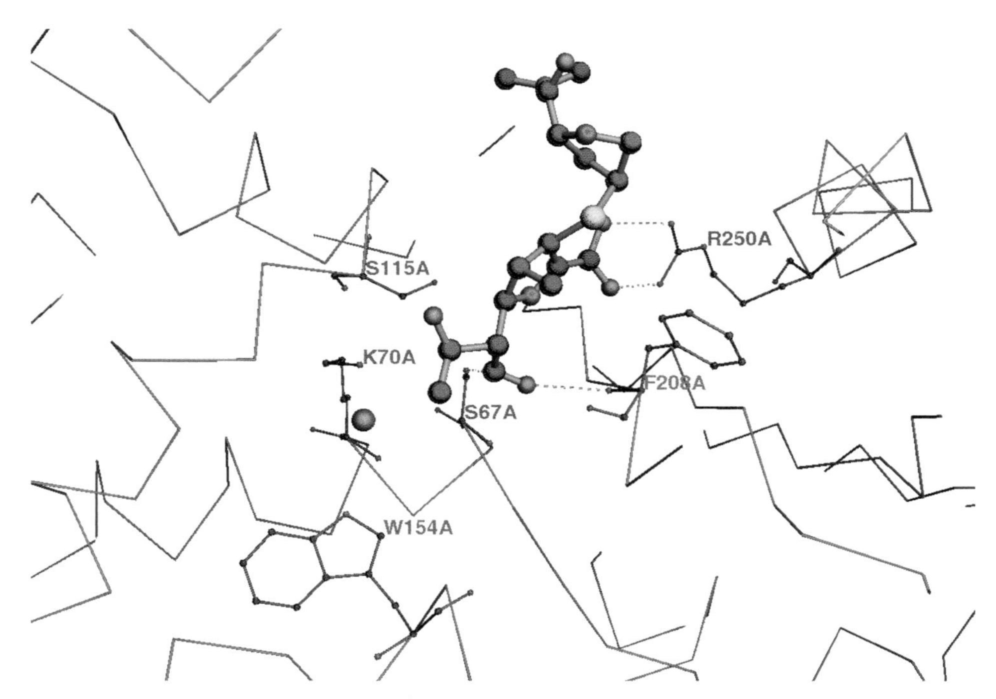

Color Plate 6 (chapter 11) OXA-13 β-lactamase, acylated by meropenem. The structure is a closed form with a noncarboxylated lysine. A water molecule (the big ball) is positioned at the end of Lys73 (70 in OXA-13) at a position similar to that of the carboxylate, and is correctly positioned to be catalytic. Strong hydrogen bonds are represented as dashed lines. Residues are numbered using OXA-10 numbering.

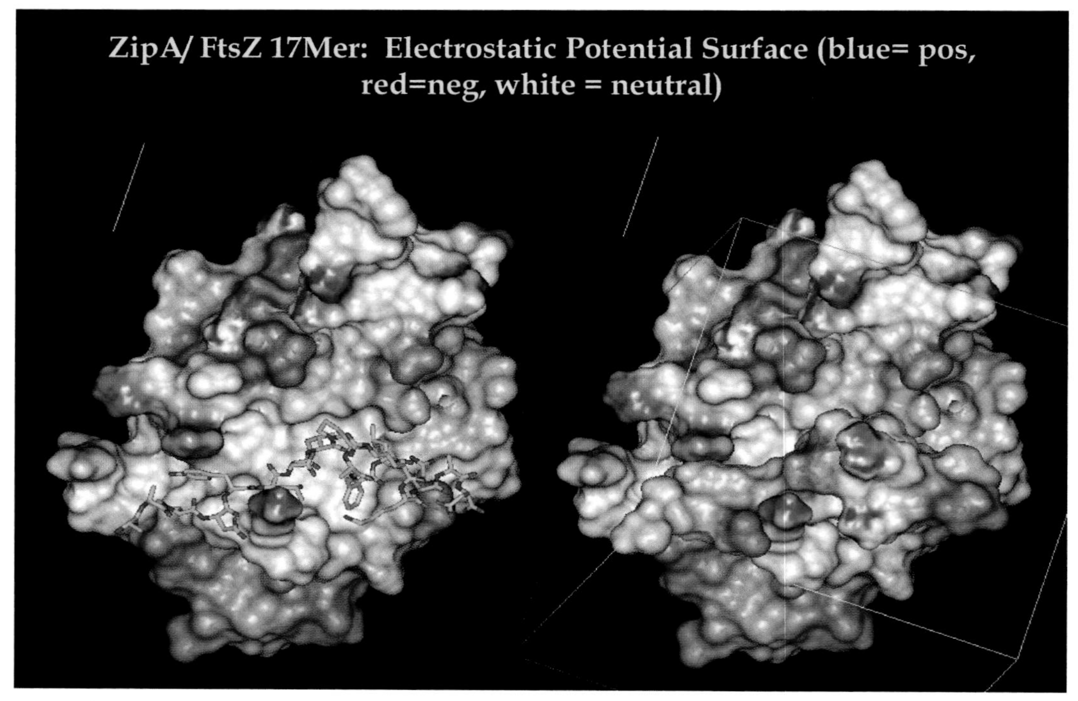

Color Plate 7 (chapter 13) X-ray crystal structure of ZipA (space-filling depiction in both panels) interacting with the carboxy-terminal peptide of FtsZ (ball and stick on left and space filling on right).

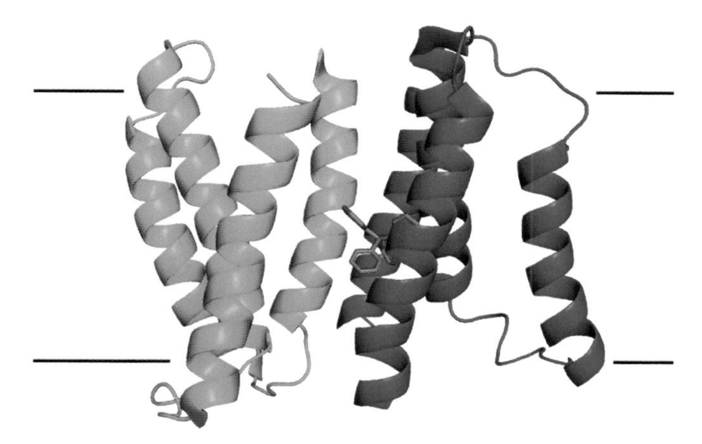

Color Plate 8 (chapter 15) Three-dimensional structure of EmrE from *E. coli* (Protein Data Bank [PDB] code 2f2m). Monomers are represented as green and blue ribbons. A bound TPP molecule is shown in red, and black lines indicate a proposed position of the lipid bilayer.

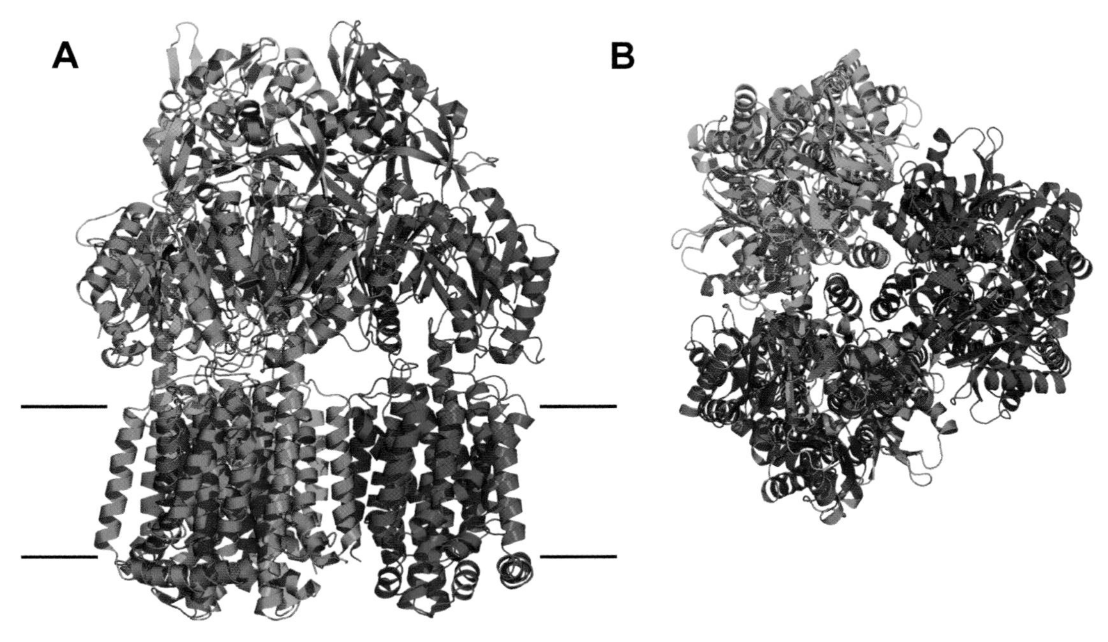

Color Plate 9 (chapter 15) (A) Three-dimensional structure of AcrB from *E. coli* (Protein Data Bank [PDB] code 1iwg). Monomers are shown as red, blue, and green ribbons. Black lines indicate the approximate position of the inner membrane. (B) Top view of an AcrB trimer as seen from the periplasmic side.

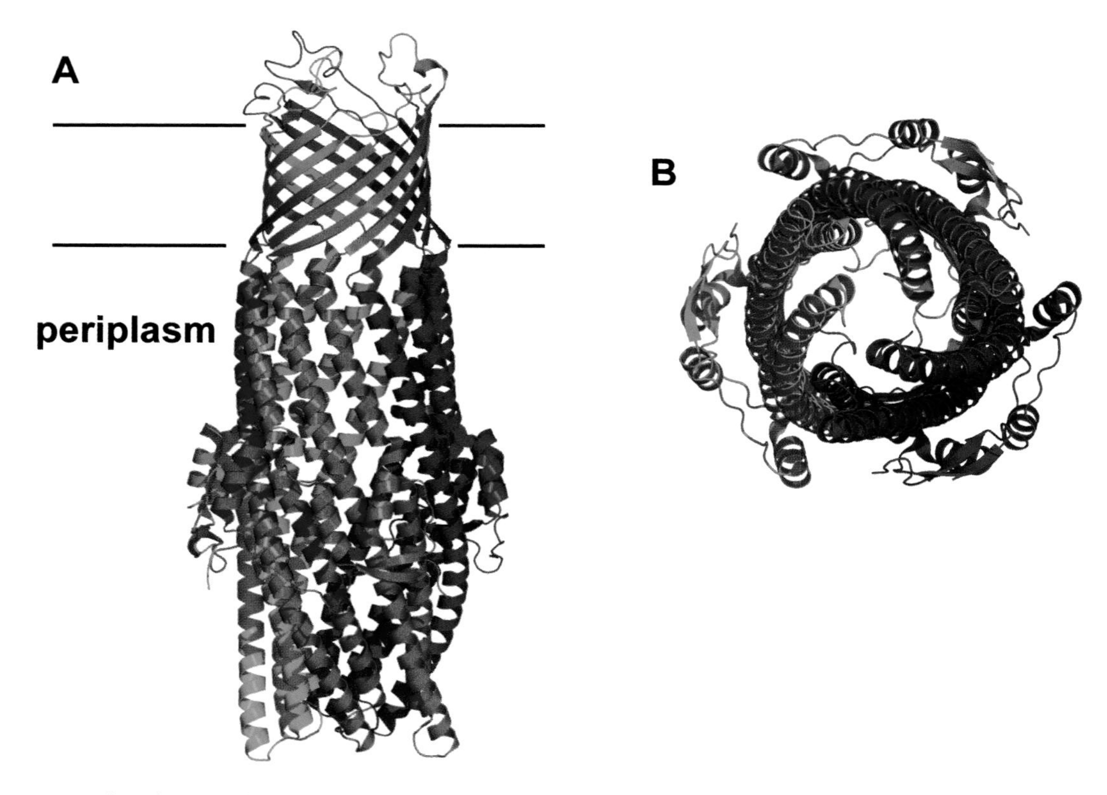

Color Plate 10 (chapter 15) (A) Three-dimensional structure of TolC from *E. coli* (Protein Data Base [PDB] code 1ek9). Monomers are shown as red, blue, and green ribbons. Black lines indicate the approximate position of the outer membrane. (B) Bottom view of a TolC trimer as seen from the periplasmic side.

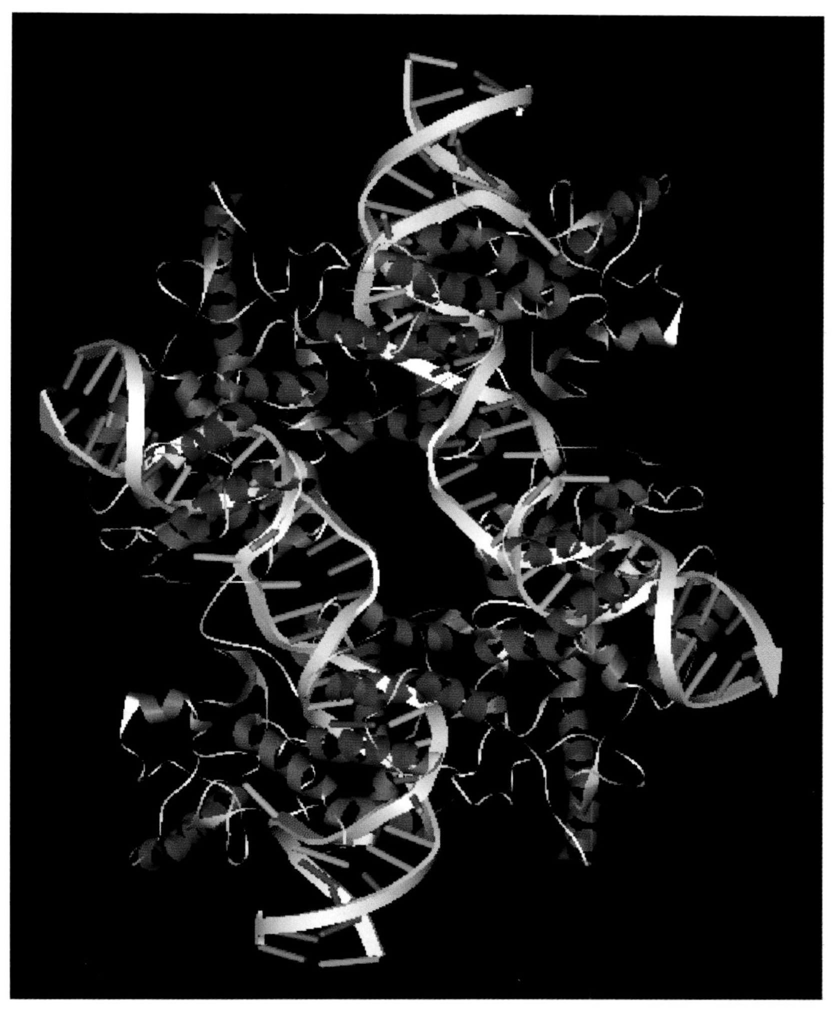

Color Plate 11 (chapter 20) Structure of the class 1 integrase complex with DNA (Protein Data Bank).

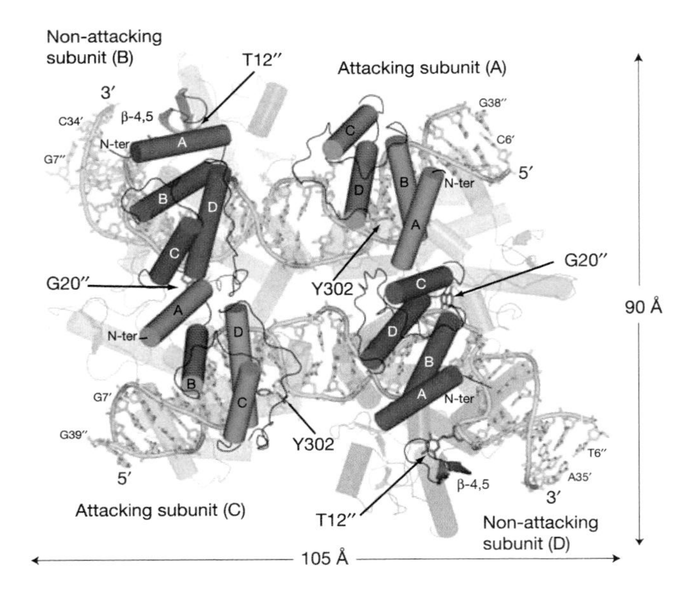

Color Plate 12 (chapter 20) Binding of the integrase to DNA. From MacDonald et al. (8).

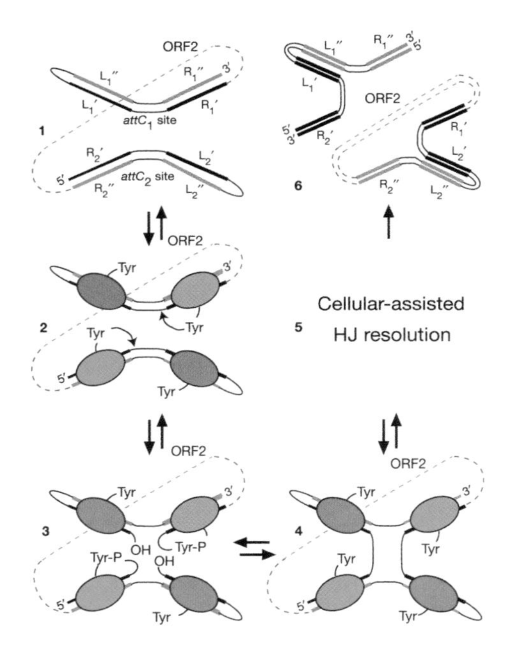

Color Plate 13 (chapter 20) Proposed IntI excision through a single-stranded DNA substrate pathway based upon a previously proposed model by Mazen et al. The bottom strand of the integron element, produced by conjugation or transformation, folds upon itself to yield an active stem-loop substrate (step 1). Two IntI molecules bind each folded *attC* site to form an antiparallel recombination synapse (step 2). Here, the HJ intermediate requires cellular components in order to be resolved (steps 5 and 6). IntI molecules colored green or magenta are potentially active or nonactive for cleavage, respectively.

Enzyme-Mediated Resistance to Antibiotics: Mechanisms, Dissemination, and Prospects for Inhibition
Edited by Robert A. Bonomo and Marcelo E. Tolmasky

Franck Danel
Malcolm G. P. Page
David M. Livermore

11

Class D β-Lactamases

β-Lactams are the most widely used antimicrobial agents. This success reflects their broad spectrum, low toxicity, limited side effects, bactericidal action, good pharmacokinetic properties, and the diversity of useful analogs that can be synthesized, including penams, cephems, oxapenems, carbapenems, and monobactams.

However, resistant bacteria began to accumulate as soon as the first β-lactam, benzylpenicillin, entered the clinic, more than 60 years ago. The predominant mechanism of resistance among gram-negative bacteria is hydrolysis of the β-lactam ring by β-lactamases. Pharmaceutical companies have sought to overcome these enzymes by adding side chains that sterically hinder hydrolysis or by combining β-lactam antibiotics with β-lactamase inhibitors. Nevertheless, none of the 70 or thereabouts β-lactams now commercially available worldwide wholly evades β-lactamase-mediated resistance, and the diversity of these enzymes presents a great challenge to the further development of β-lactams for clinical use.

A classification for β-lactamases based on amino acid sequence was first proposed by Ambler (4) and has been subsequently expanded to define four classes (A to D) (3). Class A, C, and D enzymes have a serine in the active site, whereas the class B enzymes have either one or two zinc ions as part of the catalytic center. The serine β-lactamases have common structural elements (Table 11.1) (62) and are acylated by β-lactams. The acylation rate can be so fast that the catalytic efficiency (k_{cat}/K_m) of good substrates approaches the diffusion limit of 10^7 to 10^8 $M^{-1}s^{-1}$ (48); deacylation, regenerating the active enzyme, occurs more slowly and may require rearrangement in the catalytic center (33, 60).

Class A β-lactamases form the largest and most diverse class, while the class C enzymes are the most homogeneous. The class D β-lactamases form a highly diverse group, comprising more than 100 enzymes. The first class D enzyme (OXA-2) was identified in 1965 (25), and the moniker "OXA," applied to most class D β-lactamases, is short for oxacillinase and denotes the ability of most of these enzymes to hydrolyze oxacillin at a rate of at least 50% of that for benzylpenicillin. This trait is in contrast to most β-lactamases belonging to classes A and C, for which oxacillin is a poor substrate. Class D corresponds to group 2d of the Bush functional classification (12), which is, in large part, defined by oxacillinase activity.

EPIDEMIOLOGY OF CLASS D β-LACTAMASES

Most of the class D β-lactamases found in important pathogens are encoded by acquired genes, and the dissemination

Franck Danel and Malcolm G. P. Page, Basilea Pharmaceutica Ltd., Grenzacherstrasse 487, CH-4058 Basel, Switzerland. **David M. Livermore,** Antibiotic Resistance Monitoring & Reference Laboratory, Centre for Infections, Health Protection Agency, 61 Colindale Avenue, London NW9 5EQ, United Kingdom.

Table 11.1 Conserved elements of transpeptidases and serine β-lactamases[a]

Enzyme	Element 1	Element 2	Element 3
Transpeptidase	Ser*-Xxx-Xxx-Lys	Ser-Xxx-Asn	Lys-Thr/Ser-Gly
β-Lactamase			
Class A	$_{70}$Ser*-Xxx-Xxx-Lys	$_{130}$Ser-Xxx-Asn	$_{234}$Lys/Arg-Ser/Thr-Gly
Class C	$_{64}$Ser*-Xxx-Xxx-Lys	$_{150}$Tyr-Xxx-Asn	$_{314}$Lys-Thr-Gly
Class D	$_{70}$Ser*-Xxx-Xxx-Lys	$_{118}$Ser-Xxx-Val	$_{216}$ Lys-Thr/Ser-Gly

[a]Xxx, nonconserved amino acid; *, active serine.

of their host plasmids defines their distribution. Only a few acquired OXA enzymes, specifically OXA-1, -2, -3, and -10, are widespread and frequently encountered. Many others, e.g., OXA-5, -6, -7, -8, and -9, are scarce and may be known only from single reports (44). A few opportunistic pathogens produce chromosomally mediated OXA enzymes: examples include OXA-12 from *Aeromonas jandaei*, OXA-22 and -60 from *Ralstonia* (*Pseudomonas*) *pickettii*, and an enzyme from *Burkholderia pseudomallei* (Table 11.2). Very recently, and surprisingly, it has been revealed that *Pseudomonas aeruginosa* has a long-missed chromosomal class D β-lactamase, OXA-50, though this is only weakly expressed and makes only a minimal contribution to resistance (29). Chromosomal class D β-lactamases (LoxA and OXA-29, respectively) have also been described from single strains of *Legionella pneumophila* and *Legionella gormanii*, but their distribution in these species remains to be established (7, 28). Interestingly, the *A. jandaei*, *B. pseudomallei*, and *L. gormanii* hosts all have additional class A and/or B β-lactamases whereas *P. aeruginosa* has an AmpC enzyme. The reasons for producing such multiple enzymes are unclear. The gene for a carbapenem-hydrolyzing class D enzyme, $bla_{\text{OXA-51-like}}$, is intrinsic to *Acinetobacter baumannii*, but expression, and contingent resistance, requires upstream migration of IS*Aba1* or maybe other insertions (78b).

OXA-1, -2, and -3 are the predominant acquired class D types in *Enterobacteriaceae* (44), whereas OXA-10 and its extended-spectrum relatives have been recorded mostly in *P. aeruginosa*, a species notable not only for the diversity of acquired β-lactamases that it can host, but also for the low frequency at which they occur (43). Most acquired OXA enzymes, including OXA-1, -2, -3, and -10 and their immediate relatives, are determined by type 1 integrons (Table 11.2), which often also carry the sulfonamide resistance determinant *sul1* and the streptomycin nucleotidyltransferase gene, *aad3*, together with a variable spectrum of further gene cassettes. These integrons can be located within transposons and within more or less transmissible plasmids or may be inserted into the chromosome (56).

In large-scale surveys it has been found that OXA-1 accounts for 1 to 10% of all acquired β-lactamases in *Escherichia coli* (44), thus vying with SHV-1 and TEM-2 as the second-most-prevalent acquired β-lactamase after TEM-1. OXA-1 is less frequent in other *Enterobacteriaceae* species besides *E. coli* and has been found on only a few occasions in *P. aeruginosa*. OXA-2 and -3, which share 94.5% amino acid identity with each other, occur sporadically in *E. coli* and *P. aeruginosa* but only ever account for a small proportion (<1%) of all β-lactamases found in large surveys. OXA-10 is uncommon too but has been found on a dozen or more occasions since its original description in 1974, mostly from *P. aeruginosa* but occasionally in *Enterobacteriaceae* (11, 45). Because the first OXA-10-encoding plasmid, R151, was transferable only among pseudomonads, the enzyme was initially called PSE (for *Pseudomonas*-specific enzyme)-2 but was renamed after its subsequent discovery in *Enterobacteriaceae* and the realization that it was an OXA type (32). The other PSE-type enzymes belong to molecular class A and are also known as CARB types.

Derivatives of OXA-2 and OXA-10 with increased ability to confer resistance to oxyimino-cephalosporins include OXA-11, -14, -15, -16, and -17. These have been found mostly in Turkey, although slightly more distant relatives of OXA-10 have been found in a few *P. aeruginosa* isolates from France (56). These latter include the OXA-19 and OXA-28 extended-spectrum β-lactamases (ESBLs) and the similar, but narrower-spectrum, enzymes, OXA-13 and OXA-35. It remains unclear whether these associations with Turkey and France reflect the true distribution of extended-spectrum OXA β-lactamases or whether they simply have been sought more intensively in these countries; in particular, it is uncertain whether Turkey is the epicenter or the western edge of a Middle-Eastern distribution.

The OXA carbapenemases, which form five clusters (OXA-23, -24, -48, -51, and -58 subgroups), have been found mostly in *Acinetobacter* spp. (12a, 13), the exceptions being single reports of OXA-23 in *Proteus mirabilis* and OXA-48 in *Klebsiella pneumoniae*. *Acinetobacter* spp. isolates with OXA-23-related enzymes have been found sporadically in the United Kingdom, Brazil, and the Far East since 1985; those with OXA-24/40-related enzymes

Table 11.2 Class D β-lactamases: pI, original host, and genetic location

β-Lactamase[a]	pI	Nucleotide GenBank no.	Original clinical isolates	Location	Reference(s)
OXA-1	7.4	J02967	*E. coli*	Integron	56, 76
OXA-2	7.7	X07260	*S. enterica* serovar Typhimurium	Integron	56
OXA-3	7.1	L07945	*K. pneumoniae*	Integron	56
OXA-4	7.5	AY162283	*E. coli*	Integron	56
OXA-5	7.6	X58272	*P. aeruginosa*	Integron	56
OXA-6	7.7		*P. aeruginosa*		56
OXA-7	7.7	X75562	*E. coli*	Integron	56
OXA-9	6.9	M55547	*K. pneumoniae*	Integron	56
OXA-10	6.1	U37105	*P. aeruginosa*	Integron	56
OXA-11	6.4	Z22590	*P. aeruginosa*	Plasmid	56
OXA-12	8.6	U10251	*A. jandaei*		71
OXA-13	8.0	U59183	*P. aeruginosa*	Integron	56
OXA-14	6.2	L38523	*P. aeruginosa*	Plasmid	56
OXA-15	8.7	U63835	*P. aeruginosa*	Integron	56
OXA-16	6.2	AF043100	*P. aeruginosa*	Transposon	56
OXA-17	6.1	AF060206	*P. aeruginosa*		56
OXA-18	5.5	U85514	*P. aeruginosa*	Integron	56
OXA-19	7.6	AF043381	*P. aeruginosa*	Integron	56
OXA-20	7.4	AF024602	*P. aeruginosa*	Integron	56
OXA-21	7.0	Y10693	*A. baumannii*	Integron	56
OXA-22	7.0	AF064820	*R. pickettii*	Chromosomal	59
OXA-23	6.7	AJ132105	*A. baumannii*		26
OXA-24	9.0	AJ239129	*A. baumannii*		9
OXA-25	8.0	AF201826	*A. baumannii*		1
OXA-26	7.9	AF201827	*A. baumannii*		1
OXA-27	6.8	AF201828	*A. baumannii*		1
OXA-28	8.1	AF231133	*P. aeruginosa*	Integron	67
OXA-29	≥9.0	AJ400619	*Legionella gormanii*		28
OXA-31	7.5	AF294653	*P. aeruginosa*	Integron	6
OXA-32	7.7	AF315351	*P. aeruginosa*		68
OXA-33		AY008291	*P. aeruginosa*		Assigned
OXA-34		AF350424	*P. aeruginosa*		Assigned
OXA-35	8.0	AF315786	*P. aeruginosa*	Integron	5
OXA-36		AF300985			Assigned
OXA-37	7.4	AY007784	*A. baumannii*	Integron	57
OXA-40	8.6	AF509241	*A. baumannii*		47
OXA-41			*P. aeruginosa*		Assigned
OXA-42		AJ488302	*B. pseudomallei*	Chromosomal	58
OXA-43		AJ488303	*B. pseudomallei*	Chromosomal	58
OXA-45	8.8	AJ519683	*P. aeruginosa*		78
OXA-46	7.8	AF317511	*P. aeruginosa*	Integron	30a
OXA-47	7.4	AY237830	*K. pneumoniae*	Plasmid	66
OXA-48	7.2	AY236073	*Shewanella oneidensis*	Chromosomal	65
OXA-49		AY288523	*A. baumannii*		Assigned
OXA-50	8.6	AY306130	*P. aeruginosa*	Chromosomal	29

Table 11.2 *(Continued)*

β-Lactamase[a]	pI	Nucleotide GenBank no.	Original clinical isolates	Location	Reference(s)
OXA-51	7.0	AJ309734	*A. baumannii*	Chromosomal	10
OXA-53	6.9	AY289608	*S. enterica*		55
OXA-54	6.8	AY500137	*S. oneidensis*	Chromosomal	65
OXA-55	8.6	AY343493	*Shewanella algae*	Chromosomal	34
OXA-56	6.5	AY445080	*P. aeruginosa*	Integron	70
OXA-57		AJ631966	*B. pseudomallei*		37a
OXA-58	7.2	AY570763	*A. baumannii*	Plasmid	69
OXA-59		AJ632249	*B. pseudomallei*		37a
OXA-60	5.1	AF525303	*R. pickettii*	Chromosomal	30
OXA-61		AY587956	*Campylobacter jejuni*		Assigned
OXA-69	8.4	AY750911	*A. baumannii*	Chromosomal	35a
OXA-71		AY750913	*A. baumannii*	Chromosomal	9a
OXA-72		AY739646	*A. baumannii*		Assigned
OXA-75		AY859529	*A. baumannii*	Chromosomal	35a
AmpS	7.9	X80276	*A. sobria*	Chromosomal	56
LCR-1	6.5	X56809	*P. aeruginosa*	Integron	56
LoxA	>8.0	YP095645	*L. pneumophila*	Chromosomal	7
OXA-62	>9	AY423074	*Pandoraea pnomenusa*	Chromosomal	73
OXA-64		AY750907	*A. baumannii*	Chromosomal	35a
OXA-65		AY750908	*A. baumannii*	Chromosomal	35a
OXA-66		AY750909	*A. baumannii*	Chromosomal	35a
OXA-68		AY750910	*A. baumannii*	Chromosomal	35a
OXA-70		AY750912	*A. baumannii*	Chromosomal	35a
OXA-76		AY949203	*A. baumannii*	Chromosomal	35a
OXA-77		AY949202	*A. baumannii*	Chromosomal	35a
OXA-83		DQ309277	*A. baumannii*	Chromosomal	78b
OXA-84		DQ309276	*A. baumannii*	Chromosomal	78b
OXA-85	5.3	AY227054	*Fusobacterium nucleatum*	Chromosomal	80
OXA-89		DQ445683	*A. baumannii*	Chromosomal	52a
OXA-92		DQ335566	*A. baumannii*	Chromosomal	78b
AmpG		AJ276031	*A. hydrophila*	Chromosomal	6a

[a]Unpublished enzymes are detailed in the Lahey website (http://www.lahey.org/Studies/webt.asp).

(which have only 60% sequence homology with the OXA-23 family) are linked mostly to Spain and Portugal, where strains with OXA-40 enzyme have spread widely (24). Two *A. baumannii* strains with OXA-23 have spread among hospitals in and around London (12a, 13), with little evidence that these outbreaks are yet under control. Among the more-recently described OXA carbapenemases, OXA-58 was first described from an *Acinetobacter* isolate collected in France (69) but has since been found in earlier isolates from Argentina, Austria, and Kuwait, suggesting that it has a wide distribution (12b). It has only 47 to 48% homology with the OXA-23 and OXA-24 clusters. It now appears (10) that OXA-51 is a chromosomal β-lactamase ubiquitous in *A. baumannii*. Whether or not it confers carbapenem resistance depends on whether its expression is activated by upstream insertion of IS*Aba1* or similar elements, as has occurred in one clone prevalent in southeast England (the SE clone) (78b). It is also present, though unexpressed, in prevalent clones that also have OXA-23 carbapenemase (12a). These OXA carbapenemases are the most frequent cause of carbapenem resistance in *A. baumannii* isolates in Europe, whereas metallo-β-lactamases dominate among carbapenem-resistant *Acinetobacter* from the Far East. Carbapenem resistance is inexorably rising among *Acinetobacter* spp. in North America too, from 2% in 1996 to 9% in 2003 according to data obtained from the 250 hospitals participating in the TSN Databases surveillance (42), but there is a surprising lack of analysis of the underlying resistance mechanisms responsible, though it is now apparent that OXA-40-producing clones have spread around Chicago.

The OXA carbapenemases are exceptional among OXA enzymes in never having been associated with integrons or transposons. In the original producer strain, collected in Scotland in 1985, OXA-23 was encoded by a plasmid that was transferable to another *Acinetobacter* sp. (73); most subsequent producers of OXA-carbapenemases, by contrast, fail to transfer their resistance under laboratory conditions.

LABORATORY DETECTION AND INVESTIGATION OF CLASS D β-LACTAMASES

No general test has been proposed for the detection of class D β-lactamases. This is partly because the class is so diverse and partly because its members have been much less important in terms of clinical resistance than the hugely prevalent TEM (class A) and AmpC (class C) enzymes.

It might be possible to devise a screen for class D enzymes by using clover-leaf (Hodge) tests, with the test isolate streaked towards an oxacillin disk at the center of a plate seeded with an oxacillin-susceptible *Staphylococcus* or *Micrococcus*. Indentation of the oxacillin zone along the streaks of growth would then imply oxacillinase activity, suggesting a class D enzyme. In practice, such a test would be bedeviled by false-positive results from other β-lactamases with weak oxacillinase activity. More usefully, a few phenotypic traits do give clues that an isolate may have a class D β-lactamase:

- *E. coli* isolates with moderate-level resistance to ampicillin (MIC, 32 to 512 mg/liter), not reversed by clavulanate, are likely to have OXA-1 (mostly), OXA-2, or OXA-3 enzymes, though this phenotype may also be caused by inhibitor-resistant TEM (class A) mutants (46).
- *P. aeruginosa* isolates with high-level resistance to ceftazidime (MIC, >128 mg/liter), not reversed by clavulanate or penicillanic acid sulfones, may have extended-spectrum OXA-2 or -10 variants. A screen for the OXA-10 variants has been proposed, based on colony hybridization tests followed by PvuII and HaeIII digestion of PCR products, but is not widely used (80).
- *Acinetobacter* spp. isolates with carbapenem MICs of 2 to 64 mg/liter (compared with modal values of 0.12 to 0.5 mg/liter) may have class D carbapenemases. Suspect isolates can be screened by PCR with primers to bla_{OXA-23}, bla_{OXA-24}, and bla_{OXA-58}, and for the bla_{OXA-51}-IS*Aba1* link (78b, 81a), but other carbapenem-inactivating oxacillinases may exist. Detection based on carbapenem hydrolysis is difficult, as extractable activity is often weak. In some cases isolates with class D carbapenemases may give false-positive results in EDTA-based screening tests for metallo-β-lactamases (e.g., Etest) (D. Livermore, unpublished data).

Precise identification of class D β-lactamases demands PCR, and often, sequencing. Kinetic investigation of these enzymes is frequently complicated by nonlinear kinetics (see "Substrate Specificity and Inhibition Profiles" below).

PHYLOGENY: PRIMARY STRUCTURE ANALYSIS AND CLASSIFICATION

Bayesian phylogeny analysis shows that OXA genes are very old, with ancestral types present before the divergence of the *Cyanobacteria* about 2.5 billion years ago. Moreover, most of the diversity within the OXA family appears to be due to ancient events (8). It seems likely, therefore, that OXA enzymes play an important role in bacterial fitness, though the precise nature of this role is uncertain. Such a view accords with the growing evidence that many nonfermenters, including *Pseudomonas*, *Acinetobacter*, and *Ralstonia* spp., have chromosomally mediated class D β-lactamases (see above).

The transfer of the OXA genes to plasmids must have occurred several times during the last 500 million years in order to generate the diversity of transferable types that now occur. Transfer of OXA-1 appears to have been the most recent event, whereas transfer of the OXA-2 and OXA-10 genes was estimated to have occurred 42 ± 9 and 116 ± 25 million years ago, respectively (8).

A sequence-numbering scheme for class D β-lactamases, named DBL, that parallels the Ambler (4) (ABL) numbering system for class A enzymes has been proposed (14), with the active-site serine taken as amino acid 70. In this article, we have adopted the DBL numbering system as a general scheme and with the actual position in the sequence of a specific enzyme indicated afterwards in brackets. Alignment of 87-amino-acid sequences for class D β-lactamase suggested that only 9 amino acids are strictly conserved throughout the whole class, including Ser70 in the active site. Several of these conserved residues lie in conserved elements also present in other classes of serine β-lactamases and transpeptidases (Table 11.1) (62), and three elements in particular need highlighting. The first element comprises Ser70 and Lys73, separated by two amino acids. Generally this tetrad is Ser-Thr-Phe-Lys in class D enzymes, rarely Ser-Ser-Phe-Lys (OXA-9 and -85) or Ser-Thr-Tyr-Lys (OXA-62 and OXA-50). The second element generally comprises Ser118 and Val120, separated by one amino acid; rarely, Ser118 is replaced by Pro as in OXA-43 and Val120 by Ile in the carbapenemase of group VI subgroup 3 (see below) and Leu in the case of OXA-62 and -83. Replacement of Val120 by Asn, which is the only residue to be found in this position in the other classes of serine β-lactamase, has not been reported in class D. The last conserved element (residues 216 to 218) is usually Lys-Thr-Gly, although the Thr217 may be replaced by Ser, e.g., in OXA-9 and in many of the carbapenemases belonging to group VI.

The first OXA β-lactamase sequence determined was OXA-2, and this allowed identification of the first element, containing the active Ser, which was confirmed unambiguously by labeling with 6-β-iodo-[^{3}H]-penicillanic acid (40). The third element also became apparent very rapidly, as more enzymes were sequenced (83), but for the second element, two sections of sequence were contenders (Tyr_{144}-Gly-Asn_{146} and Ser_{118}-Xxx-Val_{120}), with the correct candidate becoming clear only once the first OXA structure was elucidated.

Overall, across the whole family, class D β-lactamases share only 12.6% amino acid identity and 25.2% similarity (Table 11.3 and Fig. 11.1), and, based on the homology between primary sequences, Sanschagrin et al. (72) proposed a five-branched relationship tree, defining the homology groups I to V (Fig. 11.1). An additional branch is now required to include the OXA carbapenemases:

Group I: OXA-5, -7, -10, -11, -13, -14, -16, -17, -19, -28, -35, -48, -54, -55, -56, and -101.
Group II: OXA-2, -3, -15, -20, -21, -32, -34, -36, -37, -46, and -53.
Group III: OXA-1 (-30), -4 (-35), -31, -33, and -47.
Group IV: OXA-9, -12, -18, -22, -29, -42, -43, -45, -57, -59, AmpS, and LoxA.
Group V: LCR-1.
Group VI: OXA-23, -24, -25, -26, -27, -40, -49, -50, -51, -58, -60, -69 to -73, -75 to -78, -83 to -89, -91 to -95, and -99.

Generally, enzymes belonging to one group share 50% amino acid similarity or more but exceptions exist (Table 11.3). Some of the recently described enzymes are difficult to group because they are relatively close to more than one cluster. For example, OXA-50 and OXA-60, which share 44.9% identity and 58.8% similarity, both have around 50% similarity to individual enzymes of groups I and VI (Table 11.3). Overall, these enzymes seem best classified in group VI. OXA-60 too is classified in this group, as it shares greatest similarity with OXA-58, whereas the OXA-48, -54, and -55 carbapenemases are closest to group I, but also have a strong homology to groups II and IV enzymes (Table 11.3). OXA-55 is the most divergent enzyme, but this divergence largely reflects its unusual N and C termini. It might be proposed that this enzyme should form a different group like OXA-62 and OXA-63.

Besides this classification, based on primary sequence homology, other schemes have been proposed, based on structure in relation to that of OXA-10. On this basis, Maveyraud et al. (50) suggested classification of class D β-lactamases into three groups, A, B, and C. Group A includes homology group I and part of group VI; group B comprises homology groups II and V, but group C is very diverse, comprising groups III and IV enzymes. Compared with structural group A, the group C enzymes have insertions in the omega loop (amino acids 145 to 160) and in the connection between β-sheets 5 and 6; group B enzymes have a shorter connection.

Pernot et al. (63) proposed another classification, based on structural features and particular sequence elements. Here, the first cluster contains homology groups I and II, which have the following sequence features: an aromatic residue (Phe, Tyr, or Trp) at position 219 just after the third conserved element (Lys-Thr-Gly); Ala at position 69, just before conserved element 1; a polar residue (Ser, Asn, or His) at position 76 and Gly (or in the extended-spectrum variants of OXA-10, Asp) at position 167. The second cluster comprises OXA-1, -18, and -22 (groups I and IV), which have the following sequence features: Asp, Met, or Ala at position 69; Leu at position 76; Ser at position 167; and Ala or Thr at position 219. These amino acid differences do reflect underlying structural differences, but this classification has the problem that a single amino acid substitution can compromise the clustering, even if most amino acids are conserved; nevertheless, it is a useful adjunct to homology clustering for those enzymes that are difficult to group.

MOLECULAR MECHANISM AND SUBSTRATE SPECIFICITY PROFILES

General Molecular Mechanism of Class D β-Lactamases

Three-Dimensional Structures

Tertiary structures have been solved for several class D β-lactamases, including OXA-1 (Protein Data Bank [PDB] entry 1M6K [76]), OXA-2 (PDB entry 1K38; P. Charlier, unpublished data), OXA-10 (PDB entry 1EWZ [50, 51], and PDB entry 1FOF [62]), and OXA-13 (PDB entries 1H8Z [63]). All these proteins are composed of two domains, one with five α-helices (α 3, 4, 5, 6, and 8) and the other mixed, containing a seven-stranded antiparallel β-sheet with α-helices at its N and C termini. The active site lies between these two domains (Color Plate 4). No structure is yet available from any member of homology groups IV to VI.

Table 11.3 Amino acid similarity among class D β-lactamases and homology group[a]

Group	β-lactamase	OXA -2	OXA -34	OXA -32	OXA -53	OXA -3	OXA -20	LCR -1	OXA -10	OXA -35	OXA -19	OXA -13	OXA -7	OXA -56	OXA -5	OXA -48	OXA -54	XA -55	OXA -23	OXA -27	OXA -40	OXA -72	OXA -51	OXA -71	OXA -69	OXA -75	OXA -58	OXA -50	OXA -60	OXA -33	OXA -31	OXA -1	OXA -47	OXA -43	OXA -59	XA -42	OXA -22	OXA -18	OXA -45	OXA -9	OXA -12	OXA -29
II	OXA-34	93.1																																								
II	OXA-36	93.1	99.2																																							
II	OXA-53	92.7	86.5	86.5																																						
II	OXA-3	94.5	87.6	87.6	92.7																																					
II	OXA-20	80.4	80.1	80.1	81.1	80.4																																				
V	LCR-1	45.9	45.2	45.6	46.6	45.2	48.9																																			
I	OXA-10	45.4	46	46.3	45.7	43.6	45.8	43.5																																		
I	OXA-35	45	45.6	46	45.4	43.3	45.4	43.5	97.7																																	
I	OXA-19	44.7	45.2	45.6	45	42.9	45.1	43.1	97.4	99.6																																
I	OXA-13	45	45.6	46	45.4	43.3	45.4	43.1	97.4	99.6	99.2																															
I	OXA-7	44.3	44.9	45.2	44.7	42.6	45.1	43.1	96.3	98.5	98.1	98.1																														
I	OXA-56	44.7	45.2	45.6	45	42.9	45.1	43.1	97	99.2	98.9	98.9	99.3																													
I	OXA-5	45	45.4	45.8	46.8	44.7	47.6	45.7	87.6	88	87.6	87.6	86.9	87.3																												
I	OXA-48	52	52.6	52.6	50.5	50.2	53	47.6	59.5	59.1	58.7	58.7	58.5	58.7	57.4																											
I	OXA-54	52.7	53.4	53.4	51.3	50.9	52.6	46.5	57.6	57.6	57.2	57.2	57	57.2	55.9	95.8																										
I	OXA-55	43.2	43.5	43.5	42.9	43.6	45.2	42.7	51.5	51.2	50.9	51.5	50.7	51.2	50	59.5	59.5																									
VI	OXA-23	39	39.3	39.3	38.7	39.4	39.2	38.7	46.1	45.4	45	45	45.2	45.7	43.8	48.9	49.3	47.2																								
VI	OXA-27	39.4	39.6	39.6	39	39.7	39.6	39.4	46.5	45.7	45.4	45.4	45.6	46.1	44.2	49.3	49.6	47.6	99.3																							
VI	OXA-40	36.4	36.9	37.2	36.4	36.4	37.5	38.4	45.8	45.4	45.1	45.1	45.3	45.4	40.7	45.7	45	45.9	68.8	69.6																						
VI	OXA-72	36.4	36.9	37.2	36.1	36.4	37.5	38.4	46.1	45.8	45.4	45.4	45.6	45.8	41.1	45.7	45	45.9	69.2	69.9	99.6																					
VI	OXA-51	34.6	35	35	35.6	35.6	36.4	36.2	41.8	40.8	40.4	40.4	41	41.1	40.3	48.2	48.6	43.1	70.1	70.8	70.7	71																				
VI	OXA-71	34.6	35	35	35.6	35.6	36.4	36.2	41.8	40.8	40.4	40.4	41	41.1	40.3	48.2	48.6	43.4	70.4	70.4	70.7	71	98.9																			
VI	OXA-69	34.2	34.6	34.6	35.3	35.3	36	36.2	41.8	40.8	40.4	40.4	41	41.1	39.9	47.1	47.5	43.4	69.7	70.4	71.4	71.7	97.4	97.8																		
VI	OXA-75	33.9	34.3	34.3	34.9	34.9	36	36.5	41.5	40.4	40.1	40.1	40.6	40.8	39.6	47.5	47.8	43.4	69	69.7	71.4	71.7	96.7	97.1	97.8																	
VI	OXA-58	35.8	36.3	36.6	36.9	35.2	39.1	39.6	45.8	44.7	44.4	44.4	44.9	45.1	44	46.3	46.6	44.2	57.5	58.2	62.3	61.9	62.1	61.4	60.7	60.4																
VI	OXA-50	49.5	49.3	49.3	49.1	48.4	48.9	45.6	49.6	49.3	48.9	48.9	48.3	48.9	50.6	50	51.1	47.9	50.2	50.5	45.2	45.5	50.9	51.3	51.3	51.3	50.7															
VI	OXA-60	41.5	40.6	41	40.1	39.7	40.3	43.7	47.1	47.1	46.8	46.8	47.3	47.5	49.3	44.4	44	41.2	48.2	48.6	47.5	47.9	47.8	47.8	47.1	47.5	49.5	58.8														
III	OXA-33	30.5	30.5	30.5	31.9	30.5	33.6	35	32.9	32.5	32.2	32.5	32.4	32.5	33.8	35.6	35.2	33.1	30.4	30.4	29.5	29.5	30.7	30.7	31.4	31.8	31.5	34.9	31.4													
III	OXA-31	30.2	30.2	30.2	31.5	30.2	33.2	35.3	32.9	32.5	32.2	32.5	32.4	32.5	33.8	35.6	35.2	33.1	30.4	30.4	29.5	29.5	30.7	30.7	31.4	31.8	31.5	34.9	31.4	99.6												
III	OXA-1	30.5	30.5	30.5	31.9	30.5	33.6	34.6	33.2	32.9	32.5	32.9	32.8	32.9	34.1	35.6	35.2	33.1	30.4	30.4	29.9	29.9	31.1	31.1	31.8	32.1	31.9	34.5	31.1	99.6	99.3											
III	OXA-47	30.2	30.2	30.2	31.5	30.2	32.9	34.3	32.9	32.5	32.2	32.5	32.4	32.5	33.4	34.9	34.5	33.4	30.4	30.4	29.9	29.9	30.7	30.7	31.4	31.8	31.9	34.5	31.1	98.2	97.8	98.6										
IV	OXA-43	31.9	31.4	31.4	31.5	31.2	31.4	32.3	35.1	35.8	35.4	35.8	35.3	35.8	35.4	33.3	33.7	30.4	29.4	29.4	26.2	26.5	29	29	29	29.3	28.9	33.3	31.8	44.4	44.1	44.8	44.8									
IV	OXA-59	32.2	31.7	31.7	31.9	31.5	31.7	32.6	35.4	36.1	35.8	36.1	35.6	36.1	35.8	33.7	34	30.8	29.8	29.8	26.5	26.8	29.3	29.3	29.3	29.7	29.2	33.7	32.1	44.8	44.4	45.1	45.1	99.6								
IV	OXA-42	32.6	32.1	32.1	32.2	31.9	32.1	32.6	35.8	36.5	36.1	36.5	35.6	36.1	36.1	33.7	34	30.8	29.8	29.8	26.5	26.8	29.3	29.3	29.3	29.7	29.2	33.7	32.1	44.8	44.4	45.1	45.1	99.3	99.6							
IV	OXA-22	31.5	30.7	30.7	30.8	30.5	32.4	31.6	33	33.7	33.3	33.7	33.6	33.7	34.4	33	33.7	28.9	26.2	26.2	25.3	25.6	26.8	26.8	27.1	26.5	27.4	32	29.8	40.8	40.8	41.2	41.2	63.4	63.8	63.8						
IV	OXA-18	32.7	31.9	31.9	32.7	32	32.2	31.1	32	31.3	31	31.3	30.8	31.3	31.3	34.1	34.1	31.2	29.1	29.1	25.2	25.2	26.4	26.4	26.4	26.4	30.3	31.4	29.5	43.6	43.3	44	44.3	51.2	51.6	51.6	48.4					
IV	OXA-45	31.6	30.8	30.8	31.6	31.3	31.5	32.8	36.5	35.8	35.4	35.8	35.3	35.8	36.8	34.1	34.1	30.2	29.2	29.2	24.9	24.9	24.4	24.4	24.7	24.7	29	34.1	29.5	40.4	40	40.7	41.1	55.4	55.7	55.7	51.4	76				
IV	OXA-9	30.5	29.9	29.9	29.5	29.5	28.6	30.5	32.9	32.5	32.2	32.5	32.4	32.9	32.2	30.8	30.8	30.8	26.7	26.3	26.2	26.2	27.3	27.7	27.3	27.7	26.6	31.2	32.1	44.5	44.2	44.9	44.9	50.5	50.9	50.9	47.4	56.9	54.8			
IV	OXA-12	30.5	30	30	29.5	29.2	31.7	31	31	30.7	30.3	30.7	30.6	30.7	31.4	31.3	31.6	28.2	26.4	26.4	24.2	24.2	25.3	25.7	26.3	26.3	27.6	30.6	32.8	41.7	42	42	42.4	53.7	54.1	54.1	47.4	49.6	51.3	50.2		
IV	OXA-29	30.2	29.8	29.8	30.5	29.9	30.8	32.5	31.4	30.7	30.3	30.7	30.2	30.7	30.7	33.3	34.4	27.9	28.1	28.1	25.6	25.6	24.8	24.8	25.4	25.1	27	27.8	27.1	46.6	46.2	46.9	47.2	52	52.4	52.4	47.3	49.3	48.4	45.4	53.5	
IV	LoxA	30.9	30.5	30.5	30.9	30.5	32.5	34.9	32.4	32.1	31.7	32.1	31.6	31.7	32.8	35.7	36.1	30.8	31.5	31.5	26.9	26.9	27.1	27.1	27.7	27.4	28.3	32	30.4	45.5	45.2	45.9	45.9	50.9	51.3	51.3	49.1	49.3	49.8	45.4	56.1	85

[a]This similarity table is based on CLUSTAL alignment (http://services.bioasp.nl/blast/cgi-bin/clustal.cgi) and similarity is calculated using Ident and Sim program (http://www.ualberta.ca/~stothard/javascript/index.html), counting the following residues as similar: Ile, Leu, and Val; Phe, Trp, and Tyr; Lys, Arg, and His; Asp and Glu; Gly, Ala, and Ser; and Thr, Asn, Gln, and Met. Similarity values of ≥50% are underlined. The shaded boxes represent the similarity for enzymes belonging to the same homology group.

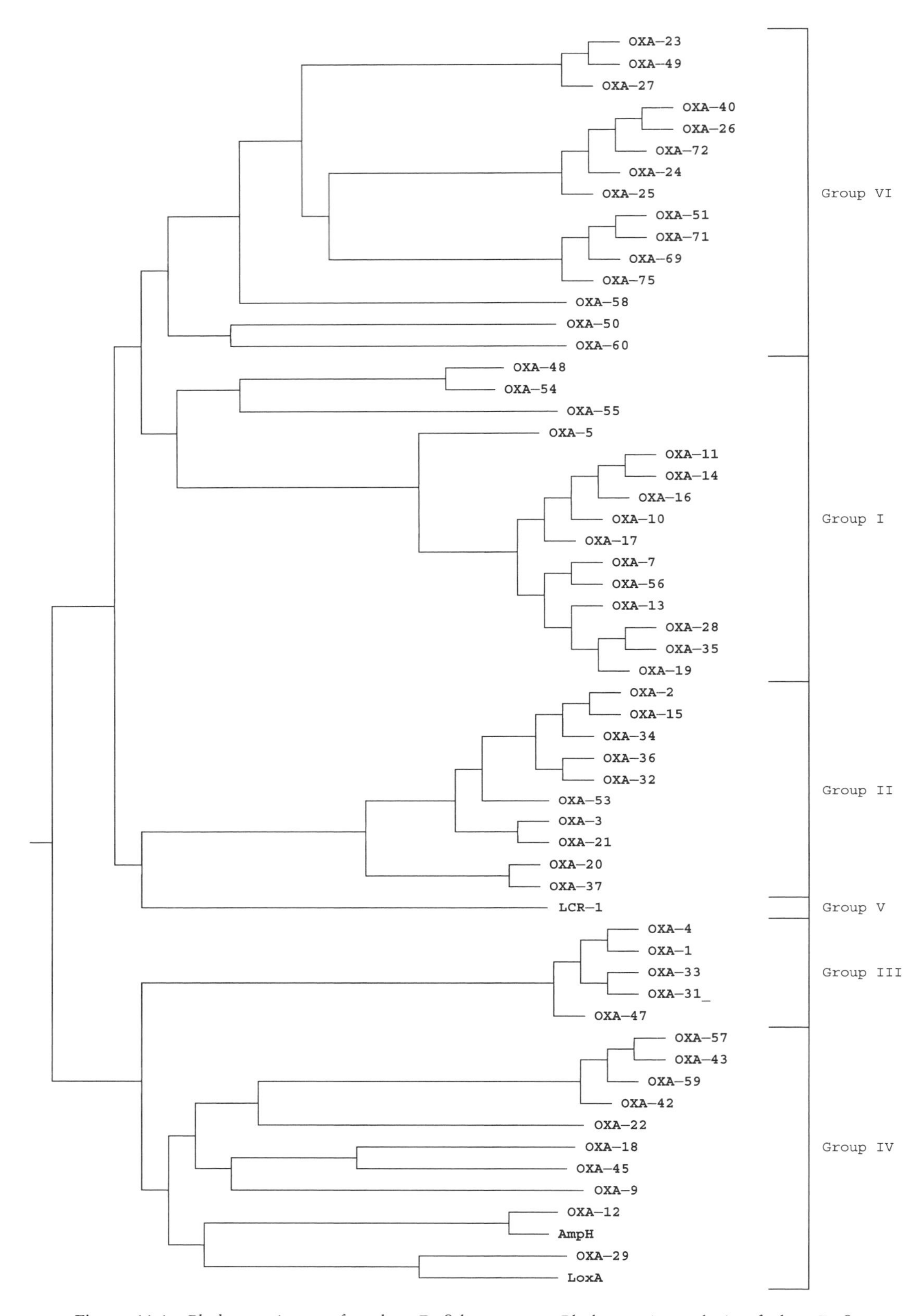

Figure 11.1 Phylogenetic tree for class D β-lactamases. Phylogenetic analysis of class D β-lactamases was performed with the ClustalW program. The amino acid sequences are listed in Table 11.2, with the Lahey website (http://www.lahey.org/studies/webt.asp) used to cross-reference the OXA numbering system and the GenBank number.

The general folding of class D β-lactamase resembles that of other serine β-lactamase classes. Thus, the root mean square deviations for the α-carbon alignment of OXA-10 to TEM-1 (class A) and *Citrobacter freundii* AmpC (class C) enzymes are only 1.9 and 2.1 Å, respectively (62). Some regions of the class D enzymes are structurally very similar to class A enzymes, especially β-sheets 2 to 5 and 7 to 9 and α-helices 1, 3 and 5 to 9; nevertheless, the binding sites of the known class D β-lactamases are more extended than those of class A enzymes, with the C-terminal α-helix (lying on the front of the β-sheet) situated 3 Å closer to the substrate-binding cleft. This binding site is also more hydrophobic than in class A and C enzymes, possibly explaining why OXA-2, a member of homology group II, is inhibited by hydrophobic dyes and can be purified by Sepharose-blue chromatography (53).

Tertiary Structure of Class D β-Lactamases

Homology group I β-lactamases. During 2000, the structure of the OXA-10 apoenzyme was determined at 2.4 and 2.0 Å by two independent groups (Maveyraud et al. [50] and Paetzel et al. [62]); one year later, the OXA-13 structure was determined at 1.8 Å by Pernot et al. (63). OXA-13 differs from OXA-10 by only nine amino acids (Table 11.4), and two crystal forms were obtained. Both forms were structurally very similar to OXA-10 but, for OXA-13 form 1 (unlike form 2), the loop between α-helices 3 and 4 showed no electron density and revealed a flexibility. Form 1 was used to resolve the structure.

Both OXA-10 and OXA-13 have a single disulfide bridge between Cys 45 (44 OXA-10) and Cys 54 (51 OXA-10), linking β-sheets 2 and 3, and this is conserved in all other group I enzymes except the carbapenemases OXA-48, -54, and -55. In both OXA-10 and -13 enzymes, the β-lactam-binding pocket is a long groove delineated on one side by the residues Gln116 (113 OXA-10) to Val120 (117 OXA-10), which lie between α-helices 4 and 5 and contain conserved element 2 (Table 11.1). The other side of the groove is delineated by β-strand 7, including residues Lys216 (205 OXA-10) to Ser220 (209 OXA-10), which contains conserved element 3 (Table 11.1 and Color Plates 4 and 5). The bottom of the groove is formed by the side chain of Leu155 (165 OXA-10) and the N terminus of α-helix 3, with the conserved element 1. The only real difference between the active sites of OXA-10 and OXA-13 is at residue 76 (73 OXA-10), where OXA-13 has Ser and OXA-10 has Asn. The active site of OXA-13 in crystal form 1 is slightly more open than for OXA-10, owing to a different organization in the loop connecting α-helices 5 and 6 (residue 94 to 107). A consequence is that Ser118 (115 OXA-10) of conserved element 2 cannot form hydrogen bonds with Lys216 (205 OXA-10) in OXA-13 as it does in OXA-10. In form 2, the

Table 11.4 Amino acid differences between OXA-10 β-lactamase and its close relatives (homology group I)[a]

DBL	10	20	28	52–53	58	70–73	76	92	110	118–120	127	144–146	164	167	184	208	216–218	240–241	258	272	ESBL
OXA-10 numbering	10	20	27	—50	55	67–70	73	89	107	115–117	124	141–143	154	157	174	197	205–207	229–230	245	259	
OXA-10	Ile	Gly	Ser	—Ser	Asp	**SerThrPheLys**	Asn	Val	Thr	**SerXxxVal**	Ala	**TyrGlyAsn**	Trp	Gly	Tyr	Ala	**LysThrGly**	GluThr	Ser	Glu	
OXA-11												TyrGlySer		Asp							X
OXA-14														Asp							X
OXA-16											Thr			Asp							X
OXA-17							Ser														X
OXA-M101[b]													Leu								X
OXA-M102[b]											Ser			Asp							X
OXA-M103[b]												TyrGlyLys									X
OXA-13	Thr	Ser			Asn		Ser		Ser						Phe			GlyThr	Asn	Ala	X
OXA-19	Thr	Ser			Asn				Ser					Asp	Phe			GlyThr	Asn	Ala	X
OXA-28	Thr	Ser			Asn				Ser				Gly		Phe			GlyThr	Asn	Ala	X
OXA-35	Thr	Ser			Asn				Ser						Phe			GlyThr	Asn	Ala	
OXA-13-1	Thr	Ser			Asn		Ser		Ser					Asp	Phe			GlyThr	Asn	Ala	X
OXA-56	Thr	Ser	Phe		Asn			Ile	Ser						Phe			GlyAla	Asn	Ala	
OXA-7	Thr	Ser	Phe	LeuAla	Asn			Ile	Ser						Phe	—		GlyAla	Asn	Ala	

[a]Amino acids are numbered according to the DBL scheme (14). The amino acids up to DBL position 20 are part of the signal peptide and are eliminated from the mature protein. —, amino acid gap in the alignment.

[b]Laboratory-selected mutants.

OXA-13 structure is less open and the active-site conformation is closer to that of OXA-10 (Fig. 11.2). Comparison of OXA-13 and -10 structures shows that the only difference in the binding site was due to the smaller size of the Ser76 residue, which allows an additional water molecule in OXA-10. The additional water forms a hydrogen bond with a water molecule proposed to be involved in catalysis and could help to position this molecule correctly, in association with Lys73 and Trp164. Such a precisely positioned water in OXA-10 favors penicillinase activity more efficient than that of OXA-13.

The active-site structures for OXA-10 proposed by Golemi et al. (31) and Paetzel et al. (62) differ in that the former group reported carbamylation of Lys73 (70 OXA-10) whereas the latter group did not observe the electron density consistent with this modification; neither of the two crystal forms of OXA-13 had a carbamylated Lys. The noncarbamylated and the carbamylated OXA-10 structures were obtained at pH 6.5 and 7.8, respectively, while the noncarbamylated OXA-13 structure was obtained at pH 5.5, and it may be that the degree of carbamylation reflects the pH. If so, pH and the presence of bicarbonate might be expected to affect activity, as the carbamylate group would form a hydrogen bond with the side chain amide of Trp164 (154 OXA-10), a widely conserved residue in the omega loop. At pH 7.5, OXA-10 was more active in the presence of 25 mM $NaHCO_3$ than in its absence, while removal of the bicarbonate by gel filtration led to a 50% loss of activity, which was reversible by readdition of bicarbonate (31). After incubation at pH 4.5 under vacuum, to facilitate full decarbamylation, only 12% of the activity remained after readjusting to pH 7.5. The in vivo physiological concentration of carbon dioxide in the blood has been estimated at 1.3 mM, a level at which the OXA-10 enzyme should be saturated with CO_2.

The structure of OXA-10 was also obtained at pH 8.5, 7.5, 6.5, and 6.0 by Maveyraud et al. (51). The structure at pH 7.5 was carbamylated, unlike that at pH 6.0, while the structures at pH 6.5 and 7.0 were mixtures of both forms, further supporting the view that the pH affects carbamylation. Interestingly, the absence of the carbamylate residue at pH 6.8 altered the binding site, with the Lys73 (70 OXA-10) side chain repositioned 4.6 Å away from the active Ser70 (67 OXA-10) and with neither side chain able to form a strong hydrogen bond such as that observed in the Paetzel et al. (62) structure at pH 6.5 (Fig. 11.2). Rather, at lower pH, the Lys side chain interacts with main chain oxygen atom of Val117 (114 OXA-10) (2.8 Å) while the

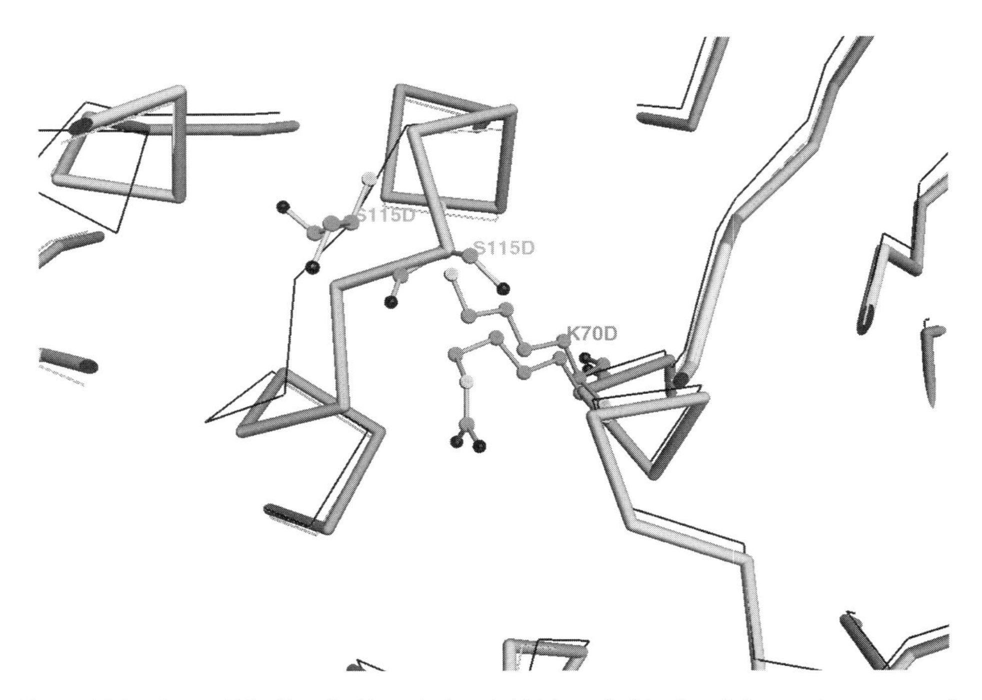

Figure 11.2 Open (thin line dark) and closed (thick and thin line light gray) structures for OXA-10 β-lactamase. The two closed structures (1FOF, 1EWZ) represent carbamylated and noncarbamylated forms (both structures are superimposable), with Lys73 (70 in OXA-10) bent. In the open structure, Lys73 is straight. In the noncarbamylated structure (1FOF), the structure is identical to the carbamylated structure (1EWZ) but with a water replacing the carbamylate. Residues are indicated using OXA-10 numbering.

main chain atoms of residues 117 to 122 (115-119 OXA-10) are moved out of the active site and there is a 180-degree rotation of the peptide bond between Val117 (114 OXA-10) and Ser118 (115 OXA-10). Owing to this rotation, Ser118 in conserved element 2 is moved away from the active Ser, from 3.2 to 7.1 Å. These observations predict that full carbamylation of the Lys73 (70 OXA-10) will only arise above pH 7.5, with partial carbamylation between pH 6.5 and 7.0, and full decarbamylation below pH 6.0, supporting Maveyraud's experimental data and also the structural effect of the carbamylated lysine.

The structures of acyl enzyme complexes were investigated by soaking OXA-10 crystals with α- and β-hydroxyisopropylpenicillanate (PDB entry 1K54) (51) and moxalactam (PDB entry 1K6R) and by soaking OXA-13 crystals with imipenem (PDB entry 1H5X) and meropenem (PDB entry 1H8Y) (63). These structures were resolved at 1.7, 2.3, 1.9, and 2.0 Å, respectively. Only the OXA-10 structures were carbamylated, while OXA-13 complexes, which were soaked at pH 5.5, were not carbamylated. It should be added that α-hydroxyisopropylpenicillanate (I) and β-hydroxyisopropylpenicillanate (II) are useful β-lactams in studies to determine the direction of approach of the water to attack the acyl enzyme intermediate (Fig. 11.3). With class A enzymes, water attacks from the α-side, so these enzymes are inhibited by α-hydroxyisopropylpenicillanate (I). With class C β-lactamases the water attacks from the other side, and the enzymes are inhibited by compound II only. Surprisingly, both compounds I and II inhibited OXA-10 β-lactamase similarly, with K_i values of 300 ± 100 μM and 240 ± 10 μM respectively. In one set of studies, both compounds were soaked in OXA-10 crystals. Compound I, but not II, led to the destruction of the crystals, probably by inducing a conformational change in the protein molecules. ^{13}C nuclear magnetic resonance data showed that compound I, unlike II, had a very weak interaction with the carbamylate. In the structure with compound II, the main part of the β-lactam, including the carboxylate group of the thiazoline ring, was directed toward the outside of the enzyme and the open ring form of compound II

OH S O CO_2H I

OH S O CO_2H II

Figure 11.3 Chemical structures of α- and β-hydroxyisopropylpenicillanates (I and II).

was rotated after the formation of the acyl enzyme, highlighting a less favorable interaction of the initial acyl enzyme complex.

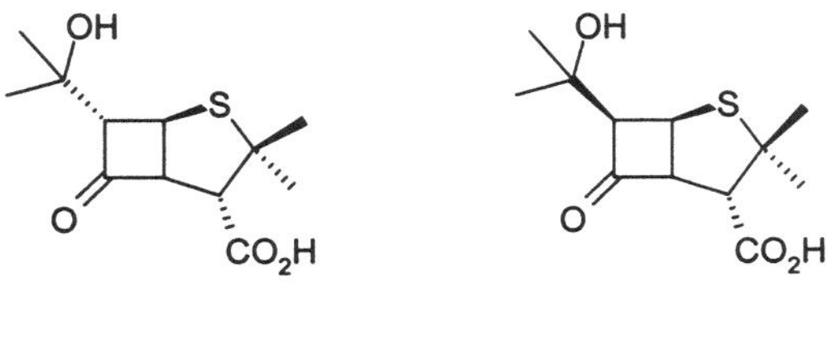

In both the OXA-10 and -13 structures, the carbonyl oxygen atom of the acyl enzyme ester is situated in an oxyanion hole at hydrogen bond distance from Phe219 (208 OXA-10) and the backbone nitrogen of Ser70 (67 OXA-10) (Color Plate 6). These interactions are very similar to those generally observed with class A and C β-lactamases. For meropenem and imipenem (which are not substrates for OXA-10 and -13), the positioning of the β-lactam seems to depend very much on the salt bridge between the carboxylate of the carbapenem and the guanidine group of Arg263 (250 OXA-10). A similar interaction is observed with Arg244 among class A β-lactamases.

Different structures were observed for the complexes formed with moxalactam and each of the two monomers in the dimeric OXA-10 protein. In the first monomer unit, the structure of the active site was similar to 1EWZ and 1K54 and acylation by the moxalactam was observed. In the other monomer, the carboxyl group of the oxacephem ring did not interact with Arg263 (250 OXA-10); rather, the carboxyl group of the C7 side chain interacted with this residue. In the latter monomer, the α-carbon chain from Thr110 (107 OXA-10) to Pro121 (118 OXA-10) was moved, and the structure of the active site was more open than in the first monomer; nevertheless, acylation by moxalactam was observed. It is interesting that, even with noncarbamylated Lys, an acyl enzyme was obtained in the closed conformation of the OXA-13 carbapenem complex.

The available structures for OXA-10 and OXA-13 represent two enzyme conformations and, between pH 6.5 and 7.5, both conformations exist in crystals. The structure of the more open form suggests that it is less active than the closed form. Nevertheless, the open enzyme retains some activity because a structure of an open form with acylated β-lactam has been obtained (PDB entry 1K6R).

Numerous uncertainties remain, not least of which is whether the monomeric enzyme is more likely to adopt the open conformation. The OXA-10 structure obtained by Paetzel et al. (62) in the presence of divalent metal cations at pH 6.5 was only in the closed form, while a mixture of open and closed forms might have been expected. This is the only structure obtained with metal ions present, and it may be that these prevent the enzyme from adopting an open conformation. Moreover, no open structure has been obtained with a carbamylated Lys; therefore, does a carbamylated Lys prevent the enzyme from adopting the open form? If this were so, the carbamylate would have a structural role rather than catalytic. In this context, it is interesting that, from the structural information, 100% conversion to the closed structure, and with the carbamylated Lys, can be obtained only at pH 8.5, one pH unit above the physiological level.

Homology group II β-lactamases. The first structure of a group II OXA β-lactamase was obtained for OXA-2 at pH 9.0, with 1.5 Å resolution (PDB entry 1k38). The overall structure was very similar to that of OXA-10, although the loop connecting the β-strands 9 and 10 was shorter. Lys73 (the numbering of amino acids for OXA-2 is shown in Table 11.5) was carbamylated, and Trp219 (position 213 in OXA-2) and Ile165 (160 in OXA-2) were proposed to play roles similar to those of Phe219 (208 in OXA-10) and Leu165 (155 in OXA-10) in OXA-10 (38).

Homology group III β-lactamases. The structure of OXA-1, also called OXA-30, was solved at 1.5 Å and was obtained at pH 7.5. It has a carbamylated Lys73 (75, 76). Its general folding is very similar to class D β-lactamases belonging to groups I and II. One major difference, however, is the absence of Arg263, the guanidine group of which points into the active site in group I β-lactamases. OXA-1, by contrast, has Ser263 (270 in OXA-1) at a spatially equivalent position and also differs from group I and II class D β-lactamase structures in having an omega loop that is six amino acids longer. The position of β-strand 8 is also slightly different, and the loop connecting the β-strands 9 and 10 is closer to the substrate binding site than in the OXA-10 structure. Lastly, OXA-1 and other members of group III do not have an aromatic residue such as Trp, Tyr, or Phe after the conserved element 3 (Lys-Thr-Gly), as do enzymes belonging to groups I or II; instead, they have Ala219 (218 in OXA-1). Leu260 (267 OXA-1) is directed toward Ala219, possibly compensating for the missing aromatic residue, at least in OXA-1.

Quaternary Structures of Class D β-Lactamases

The class D enzymes of homology groups I (closely related to OXA-10), II, and V are believed to form dimers. Dimers have also been reported among members of homology group IV (a diverse group), whereas homology group III enzymes appear to be monomers and no conclusion is yet possible for the homology group VI enzymes (Table 11.6). OXA-1 (homology group III) was confirmed by analytical centrifugation to be a monomer (76); other OXA types, including OXA-2, -3, -6, -14, and -29 and LCR-1, are inferred to be dimers based on comparing the molecular weights estimated by gel filtration or analytical centrifugation with those predicted from sequence data (Table 11.6).

Homology group I β-lactamases. OXA-10 has been shown to be a dimer by analytical centrifugation by two different groups (50, 51, 60). The monomers interact symmetrically via salt bridges, hydrophobic interactions, 10 direct hydrogen bonds, and many water-mediated

Table 11.5 Amino acid differences among enzymes closely related to OXA-2 β-lactamase (homology group II)[a]

DBL	21	30	35	52	54	59–60	63	70–73	99	101	107	118–120	133	141	143	148	143	164	169	185	187	216–218	228	243	267	272	ESBL
OXA-2 numbering	23	32	37	54	56	61–62	65	72–75	101	103	109	120–122	135	143	145	150	155	159	164	180	182	210–212	216	230	253	258	
OXA-2	Glu	Arg	Glu	Arg	Met	**ProVal**	Lys	**SerThrPheLys**	Asn	Gly	Gln	**SerThrVal**	Asp	Lys	Asp	Asp	Asn	Trp	Leu	Arg	Glu	**LysThrGly**	Arg	Ser	Val	Arg	
OXA-32																			Ile								X
OXA-15																Gly											X
OXA-34																		Cys									
OXA-36																Tyr											
OXA-53	Asp	Gly	Asp	His	Ile	GlnAla	Met		Lys	Ser	Lys		Gly	Gln	Gly		His			Gln	Asp		Ser	Pro	Ala	Leu	X

[a] Amino acids are numbered according to the DBL scheme for class D β-lactamases. The amino acids up to DBL position 17 are part of the signal peptide and are omitted.

Table 11.6 Molecular masses of class D β-lactamases: evidence for dimerization

β-Lactamase	Molecular mass (kDa)		Quaternary structure[a]	Reference(s)
	Determined by gel filtration	Predicted from sequence		
Group I				
OXA-5	27	27.5	Dimer*	52
OXA-6	40	27.5	Dimer	52
OXA-7	25.3	27.5	Dimer*	52
OXA-10	12.4	27.5	Dimer	49
	35.3			41
	55			17
OXA -14	41–47	27.5	Dimer	17
Group II				
OXA-2	44.1	29.6	Dimer	15, 16
	44.6			49
	43.6			52
	50.6			41
OXA-3	41.2	29.7	Dimer	49
OXA-46	50.0	28.5	Dimer	30a
Group III				
OXA-1	23.3	28.8	Monomer	49
	24.6			41
	28.4			75
OXA-4	23	28.8	Monomer	52
Group IV				
OXA-29	55	28.5	Dimer	28
Group V				
LCR-1	44	27.5	Unknown	74

[a]*, enzyme predicted to be a dimer by homology.

hydrogen bonds through about 11% of their solvent-accessible surfaces, mainly via residues from α-helix 9 and β-strand 4. The β-sheets are organized in antiparallel directions and interact via hydrogen bonds and hydrophobic interactions (50, 62). A large symmetrical cluster of residues involved in hydrophobic interactions is observed between the α-helices 9 of each monomer. Residues involved in these interactions include Ala207 (196 in OXA-10), Leu198 (186 in OXA-10), Ile197 (187 in OXA-10), and Val204 (193 in OXA-10). Amino acids involved in the formation of hydrogen bonds or salt bridges include Glu89 (86 in OXA-10) with Lys192 (182 in OXA-10), Lys200 (189 in OXA-10) with Asn185 (176 in OXA-10), His90 (87 in OXA-10) with Tyr183 (174 in OXA-10), Arg112 (109 in OXA-10) with Ala208 (197 in OXA-10), and Glu201 (190 in OXA-10) with His214 (203 OXA-10).

Paetzel et al. (62) identified two cobalt ions at the interface between the monomer units in the OXA-10 structure. These were bound by the carboxylate oxygen of Glu201 (190 in OXA-10) of one monomer and the carboxylate oxygen of Glu238 (227 in OXA-10) and the imidazole group of His214 (203 in OXA-10) of the second monomer. Two water molecules also were present, allowing the octahedral coordination typical of zinc(II) or cobalt(II) ions. Cobalt was found because it was present at 10 mM in the crystallization conditions, but among divalent metals ions, the potential to stabilize the dimer was in the rank order $Cd^{2+} > Cu^{2+} > Zn^{2+} > Co^{2+} > Ni^{2+} > Mn^{2+} > Ca^{2+} > Mg^{2+}$ (17). Zn^{2+} is most likely to be bound to OXA-10 in nature, given its relative affinity and abundance. The K_d values of Cd^{2+} and Zn^{2+} were 7.8 and 5.7 μM, compared with a monomer/dimer K_d of 45 μM for OXA-10, meaning that these metal cations strongly influence the dimer-monomer equilibrium. The presence of a low concentration of divalent metal ions also increased the enzyme activity of OXA-10, with an optimum Zn^{2+} concentration of 0.2 mM. The pH optimum for dimerization in the presence of 0.1 mM Zn^{2+} was pH 7.5.

The sequences of other group I enzymes, including OXA-5, OXA-7, and OXA-13, are very similar to that of OXA-10, and it is likely that they, as well as the OXA-10 point mutants (OXA-11, -14, -16, and -17), all form dimers stabilized by divalent cations. Many of the residues involved in hydrogen bonds and hydrophobic interactions between monomers also are conserved in OXA-48, -54, and -55 (which form a divergent cluster within group I [Fig. 11.1]), but the metal binding site is absent because His214 is replaced by an Arg (as in the homology group II β-lactamase). It is likely, but not certain, that these carbapenemase enzymes form dimers.

Homology group II β-lactamases. OXA-2 forms a dimer, with contact between the monomers mediated by residues from β-sheet 4 and α-helix 6. In the dimer interface, a salt bridge is formed between Asp201 (195 in OXA-2) and Arg214 (208 in OXA-2). These two residues replace Glu201 (190 in OXA-10) and His214 (203 OXA-10) of OXA-10, which are part of the cation-binding site that mediates the dimer formation in group I enzymes. In consequence, OXA-2 and other group II enzymes are very unlikely to bind metals in the manner of OXA-10. Otherwise, interactions between the OXA-2 monomers were at positions very similar to those in OXA-10, involving the same α-helix and β-sheet. Salt bridges were observed between Arg209 (203 OXA-2) and Glu264 (250 OXA-2), Glu89 (91 OXA-2) and Arg197 (191 OXA-2), as well as a hydrophobic interaction involving Ile212 (206 OXA-2), Ile204 (198 OXA-2), and Ala207 (201 OXA-2). Other enzymes of this group are also expected to form dimers.

Enzymes from other homology groups. OXA-1, the only group III enzyme to be crystallized, was monomeric in the crystal structure (75); it was also found to be monomeric in solution by gel filtration and analytical centrifugation (Table 11.6).

OXA-29, belonging to homology group IV, was found to be a dimer by gel filtration, with an apparent molecular mass of 55 kDa, compared with an expected molecular mass of 28 kDa from the monomer sequence (Table 11.6). Among this group, OXA-57 and -59 were also reported to form dimers. The presence of EDTA or zinc did not interfere with the apparent molecular mass of OXA-29, implying a different mode of association from that seen with OXA-10 β-lactamase.

LCR-1, the only known member of group V, is likely to be a dimer based on the apparent molecular mass values found by different methods, but prediction of the amino acids involved in the dimerization is not yet possible because this enzyme has little homology to those for which there are good structural models.

Catalytic Mechanism of Class D β-Lactamases

Class D β-lactamases are similar to class A and C types in their three-dimensional structure, with the three same conserved elements, positioned similarly (37). The general mechanism of catalysis is also similar, with an activated Ser attacking the β-lactam ring, and with a water molecule hydrolyzing the resultant acyl enzyme intermediate. However, the details of the catalytic mechanism seem quite different in each class. For those class D enzymes studied in detail (OXA-1, -2, and -10), several hypotheses have been proposed, based on mainly structural data and contingent on the presence or absence of a carbamylate function on Lys73.

Acylation

If Lys73 is carbamylated, this carbamylate may activate Ser70 by deprotonation before it attacks the carbonyl of the β-lactam (76). The conserved residues Ser118 and Lys73 are then suggested to participate in the proton transfer to the nitrogen atom of the β-lactam ring (Fig. 11.4), with the carbamyl group acting as a proton acceptor (51).

A different mechanism was proposed based on the noncarbamylated Lys73 structure. Here, the amine side chain of the Lys was postulated to activate Ser70, which then attacks the β-lactam carbonyl group, as is believed to occur also in class A and class C enzymes. Ser118 and Lys73 are suggested to play key roles in proton transfer to the nitrogen group of the β-lactam ring (62, 63). The last transfer is facilitated by Lys73 and Lys216, which are positioned closely to Ser118 (Fig. 11.4). The pKa values of the ε-amino group of these Lys residues may be influenced by the generally hydrophobic environment, facilitating the reaction.

Deacylation

In the case of class D β-lactamases, there is no water molecule positioned correctly in the carbamylated enzyme structures to attack the acyl enzyme intermediate. This situation differs from that observed among class A and class C β-lactamases, where the hydrolytic water is present in all apo-structures, and from the noncarbamylated structures of OXA-10 and OXA-13, where a water molecule is at the position of the carbamylate (Fig. 11.2 and Color Plate 4), hydrogen bonded to the side chain nitrogen of Lys73 and to Trp164. This water molecule is correctly positioned for nucleophilic attack on the carbonyl carbon of the acyl enzyme intermediate, with Lys70 acting as a general base (62).

The possibility that acylation occurs with a carbamylated Lys73 whereas deacylation occurs with a noncarbamylated Lys cannot be excluded. In this hypothesis, Lys73 is decarbamylated during the formation of the acyl

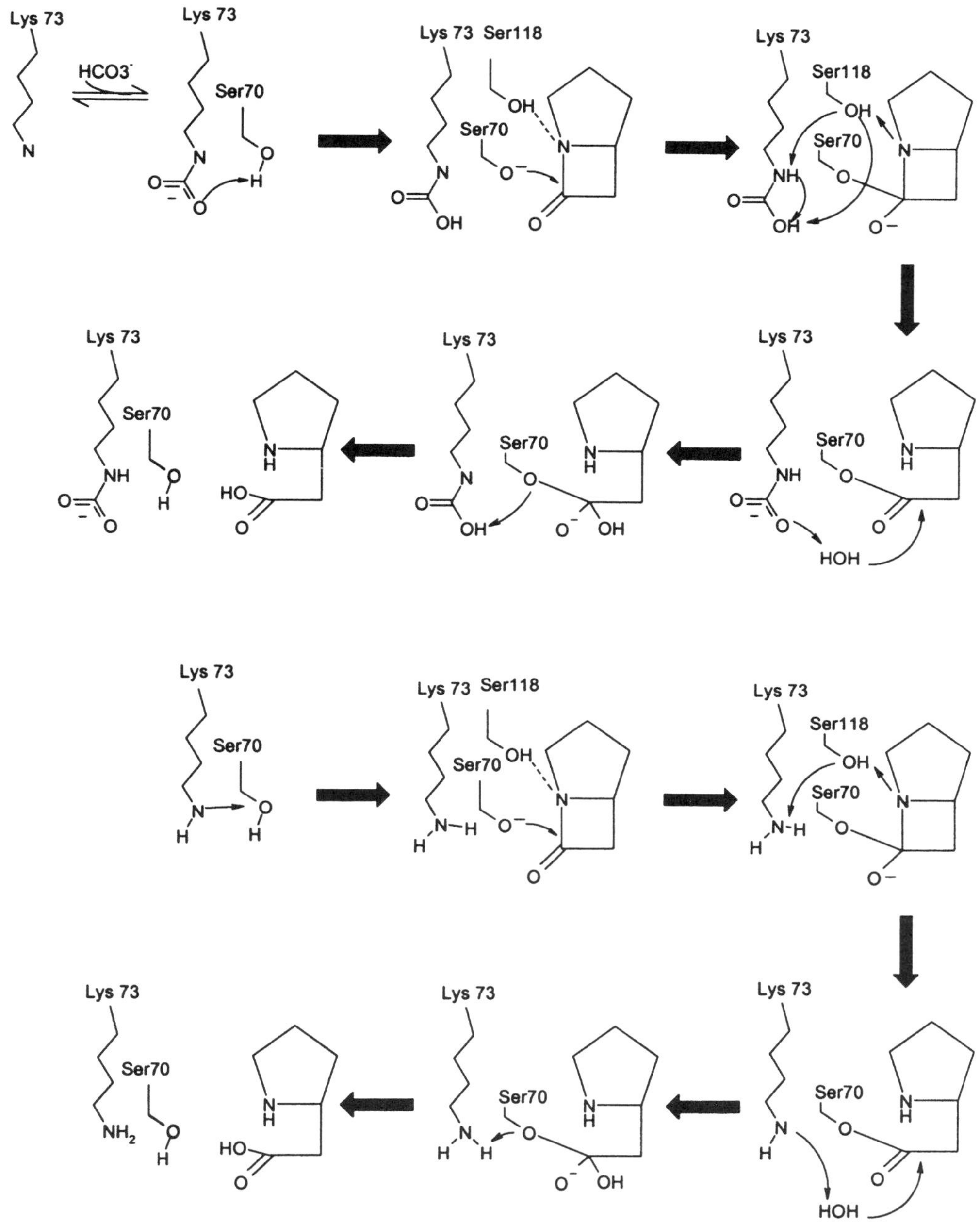

Figure 11.4 Hydrolytic mechanism for class D β-lactamase as proposed by Sun et al. (75) with carbamylated Lys73 (upper diagram). The other possible mechanism is with a noncarbamylated lysine (lower diagram).

enzyme, perhaps because the affinity of the lysine for the carbonate is modified as a result of the formation of the acyl enzyme intermediate. In this context, it is interesting that the constant of dissociation (K_d) for CO_2 from the OXA-10 monomer (12 μM) estimated in the presence of benzylpenicillin was 52-fold higher than that estimated for the dimer in the absence of a substrate (0.23 μM) (31). This might be explained by a difference in affinity for CO_2 of the Lys between monomeric and dimeric enzyme environments, but the possibility that benzylpenicillin might affect the affinity of the enzyme for carbon dioxide cannot be excluded. In any event, the enzyme should be fully carbamylated at the CO_2 concentration present in the blood.

Roles of the Three Conserved Elements of Class D Ser β-Lactamases

The first element, Ser-Xxx-Xxx-Lys. The first element, Ser-Xxx-Xxx-Lys, is positioned on an α-helix, with Lys73 just one turn after the active Ser70, so that its side chain is oriented toward the active site and hydrogen-bonded to Ser70. As with other Ser β-lactamases, the generally agreed reaction mechanism is that the activated Ser residue attacks the β-lactam carbonyl group to form an acyl enzyme intermediate. Lys73 is believed to be involved in the activation of Ser70 either directly (62, 63) or indirectly (51, 76).

The second element, Ser-Xxx-Val. The second conserved element, Ser-Xxx-Val, is situated in a loop that delineates one side of the binding pocket. It comprises Ser-Xxx-Val for class D β-lactamases as opposed to Ser-Xxx-Asn for class A β-lactamases and most D-Ala-D-Ala transpeptidases and D-D carboxypeptidases, or Tyr-Xxx-Asn for class C β-lactamase and a few D-D-transpeptidases or carboxypeptidases. All these conserved residues have their side chains directed toward the active site. The Ser residue has been proposed to act as a proton donor to the nitrogen atom of the β-lactam ring via the Lys73 of the first element (76). The Val in this element must have a unique role in the class D enzymes; in class A and class C β-lactamases, it is replaced by Asn and the polar side chain of this residue is hydrogen-bonded to Lys73 of the first element and the carbonyl group of the substrate side chain (60, 61). This Asn residue is also involved in the control of the replenishment of the catalytic water in the class A enzymes (39) and has an essential role in deacylation in class C β-lactamases (27). The Val in class D enzymes cannot fulfill similar interactions, and its role must be very different.

The third element, Lys-Xxx-Gly. The three residues Lys, Xxx, and Gly are situated on the strand of the core β-sheet that defines one edge of the active site, lying opposite element 1. The Lys residue is involved in hydrogen bonding with Ser of element 2. The Gly residue is strictly conserved because a residue with a side chain at this position would interfere with the formation of the oxyanion hole that binds and polarizes the carbonyl oxygen of the β-lactam during approach and the acyl intermediate during deacylation (48, 60).

Substrate Specificity and Inhibition Profiles

Sykes and Matthew (77) defined the OXA enzymes then known (OXA-1, -2, and -3) as enzymes that hydrolyzed methicillin and isoxazoyl β-lactam and were resistant to inhibition by cloxacillin. Further enzymes were added to the class as they were also found to fulfill these criteria. In 1995, nearly all the Ambler class D β-lactamases were placed into group 2d of the functional classification proposed by Bush et al. (12) based on the criteria of (i) not being inhibited by EDTA and (ii) being penicillinases able to hydrolyze cloxacillin or oxacillin at a rate ≥50% of that for benzylpenicillin.

Beyond their penicillinase and oxacillinase activity, the kinetics of class D β-lactamases are complex and often nonlinear. As a result, the data reported for the same enzyme can differ considerably between groups. Many enzymes show burst kinetics, whereby the initial hydrolysis rate declines more rapidly than would be expected as a result of substrate depletion (23). This behavior may account for the variability between published data sets, as its ramifications are not always reported or taken into account. In many cases, the behavior itself cannot be explained by a slow hydrolysis of the acyl enzyme intermediate, by β-lactam-induced inactivation, or by product inhibition (81). Rather, at least for group I enzymes, it seems to reflect dilution-dependent interconversion of the enzyme between monomer and dimer (23) and (possibly) between carbamylated and noncarbamylated forms (Table 11.7) (31).

Taking all the factors into account, published substrate and inhibition profile data for class D enzymes should be treated with caution. Moreover, these complexities confound attempts to relate kinetic efficiency (k_{cat}/K_m) to ability to confer resistance, in the way that can be done for class A and C enzymes.

Homology Group I β-Lactamases

OXA-10 β-lactamase, the most prevalent member of group I, hydrolyzes a wide range of β-lactams, including aztreonam, cefotaxime, and ceftriaxone (Table 11.8), but not ceftazidime. Under conditions where the enzyme is largely monomeric (i.e., at low concentration in assay) it is a penicillinase in terms of catalytic efficiency (k_{cat}/K_m) (Table 11.7), largely owing to higher turnover numbers (k_{cat}) for penicillins than for cephalosporins. Classical β-lactamase inhibitors like clavulanate, tazobactam, and sulbactam only weakly inhibit OXA-10 and other group I β-lactamases (Table 11.8), and imipenem is the only potent inhibitor (the 50% inhibitory concentration [IC_{50}] is in the low nanomolar range, at least for OXA-13).

Addition of 50 mM $NaHCO_3$ far above the physiological concentration to the phosphate assay buffer eliminated the biphasic kinetics of OXA-10 for ampicillin, carbenicillin, cephalothin, and cephaloridine, but not those for oxacillin and cloxacillin (31). Moreover, kinetic data for the initial phase showed that the main effect of HCO_3^-

Table 11.7 Kinetic parameters for homology group I class D β-lactamases (part 1)[a]

β-Lactams	OXA-10 (100 mM PB pH 7.0)						OXA-10 (100 mM PB pH 7.0, 50 mM $NaHCO_3$)						OXA-11 (100 mM PB pH 7.0)		
	V_o			V_{ss}			V_o			V_{ss}					
	k_{cat} [s^{-1} (%)]	K_m (μM)	k_{cat}/K_m [$mM^{-1}s^{-1}$ (%)]	k_{cat} [s^{-1} (%)]	K_m (μM)	k_{cat}/K_m [$mM^{-1}s^{-1}$ (%)]	k_{cat} [s^{-1} (%)]	K_m (μM)	k_{cat}/K_m [$mM^{-1}s^{-1}$ (%)]	k_{cat} [s^{-1} (%)]	K_m (μM)	k_{cat}/K_m [$mM^{-1}s^{-1}$ (%)]	k_{cat} %	K_m (μM)	k_{cat}/K_M (%)
Benzylpenicillin	89 (100)	63	1,412 (100)	—	—	—	109 (100)	23	5,000 (100)	—	—	—	100	42	100
Ampicillin	531 (597)	77	6,896 (488)	587 (660)	235	2,500 (177)	143 (131)	34	4,200 (84)	—	—	—	72	35	90
Carbenicillin	126 (142)	370	340 (24)	31 (35)	195	159 (11)	112 (103)	92	1,200 (24)	—	—	—	3.8	—	—
Piperacillin	—	—	—	—	—	—	—	—	—	—	—	—	—	—	—
Oxacillin	660 (742)	96	6,870 (487)	608 (683)	222	2,739 (194)	1,261 (1,157)	29	43,000 (860)	331 (304)	101	3,300 (66)	526	Nonlinear	—
Cloxacillin	2,682 (3,013)	1,110	2,416 (171)	520 (584)	2,640	196 (14)	1,533 (1,406)	114	13,000 (260)	129 (118)	110	1,200 (24)	—	—	—
Cephaloridine	39 (44)	395	98 (7)	79 (89)	2,340	33 (2)	57 (52)	374	150 (3)	—	—	—	0.6	3.6	7
Cephalothin	6 (7)	38	158 (11)	—	—	—	8.3 (8)	32	260 (5)	—	—	—	1.7	12	6
Cefuroxime	—	—	—	—	—	—	—	—	—	—	—	—	—	—	—
Cefotaxime	9 (10)	346	26 (2)	—	—	—	—	—	—	—	—	—	1	8.5	5
Ceftriaxone	3 (3)	55	54 (4)	—	—	—	—	—	—	—	—	—	—	—	—
Ceftazidime	ND	ND	—	—	—	—	—	—	—	—	—	—	0.6	1.1	23
Cefoxitin	—	—	—	—	—	—	—	—	—	—	—	—	<0.1	—	—
Aztreonam	—	—	—	—	—	—	—	—	—	—	—	—	—	—	—
Imipenem	—	—	—	—	—	—	—	—	—	—	—	—	<0.1	—	—
Meropenem	—	—	—	—	—	—	—	—	—	—	—	—	<0.1	—	—

[a]Data from references 17, 20, 22, 31, 32, 34, 54, and 65. The k_{cat} (%) (also reported in literature as V_{max}) and k_{cat}/K_m (%) values are relative to benzylpenicillin, which was set at 100%. ND, hydrolysis not detected; —, not determined or not applicable.

Table 11.7 Kinetic parameters for homology group I class D β-lactamases (part 2)[a]

β-Lactams	OXA-13 (50 mM PB pH 7.0)			OXA-14 (100 mM PB pH 7.0)						OXA-16 (100 mM PB pH 7.0)						OXA-17 (100 mM PB pH 7.0)		
				V_o			V_{ss}			V_o			V_{ss}					
	k_{cat} [s^{-1} (%)]	K_m (μM)	k_{cat}/K_m [$mM^{-1}s^{-1}$ (%)]	k_{cat} [s^{-1} (%)]	K_m (μM)	k_{cat}/K_m [$mM^{-1}s^{-1}$ (%)]	k_{cat} [s^{-1} (%)]	K_m (μM)	k_{cat}/K_m [$mM^{-1}s^{-1}$ (%)]	k_{cat} [s^{-1} (%)]	K_m (μM)	k_{cat}/K_m [$mM^{-1}s^{-1}$ (%)]	k_{cat} [s^{-1} (%)]	K_m (μM)	k_{cat}/K_m [$mM^{-1}s^{-1}$ (%)]	k_{cat} [s^{-1} (%)]	K_m (μM)	k_{cat}/K_m [$mM^{-1}s^{-1}$ (%)]
Benzylpenicillin	2.2 (100)	7	314 (100)	160 (100)	53	2,943 (100)	40 (25)	190	210 (7)	>60	<20	3,000	48 (80)	65	738 (25)	5 (100)	34	147 (100)
Ampicillin	3.4 (170)	74	46 (15)	450 (281)	230	1,957 (66)	40 (25)	170	235 (8)	163 (272)	142	1,150 (38)	97 (162)	205	473 (16)	26 (520)	245	106 (72)
Carbenicillin	—	—	—	—	>1,000	—	—	>1,000	—	69 (115)	150	460 (15)	17 (28)	129	132 (4)	2 (40)	296	21 (14)
Piperacillin	0.8 (40)	176	4.5 (1)	—	—	—	—	—	—	—	—	—	—	—	—	—	—	—
Oxacillin	8.5 (425)	105	81 (26)	510 (319)	370	1,367 (46)	57 (36)	700	81 (3)	455 (758)	660	758 (25)	411 (685)	960	428 (14)	120 (2,400)	153	784 (533)
Cloxacillin	—	—	—	1,110 (694)	1,600	708 (24)	36 (23)	1,300	27 (1)	1,200 (2,000)	1,120	1,070 (36)	264 (440)	2,080	127 (4)	20 (400)	573	35 (24)
Cephaloridine	7.2 (360)	1,625	4.5 (1)	43 (27)	1,000	42 (1)	30 (19)	3,700	8 (0)	30 (50)	379	79 (3)	21 (35)	424	50 (2)	23 (460)	2,940	8 (5)
Cephalothin	1.3 (65)	43	30 (10)	5 (3)	14	357 (12)	3 (2)	150	20 (1)	3 (5)	32	94 (3)	—	—	—	5 (100)	286	17 (12)
Cefuroxime	7.6 (380)	278	27 (9)	—	—	—	—	—	—	—	—	—	—	—	—	—	—	—
Cefotaxime	3.6 (180)	660	5.5 (2)	<1	<0.2	—	—	—	—	10 (17)	157	64 (2)	6 (10)	346	17 (1)	22 (440)	2,240	10 (7)
Ceftriaxone	—	—	—	<1	<0.2	—	—	—	—	1 (2)	36	39 (1)	—	—	—	1 (29)	544	20 (14)
Ceftazidime	ND	ND	—	—	—	—	—	—	—	ND	ND	—	ND	ND	ND	ND	ND	ND
Cefoxitin	0.03 (1.4)	1,200	0.02 (0.01)	—	—	—	—	—	—	—	—	—	—	—	—	—	—	—
Aztreonam	ND	ND	—	—	—	—	—	—	—	—	—	—	—	—	—	—	—	—
Imipenem	ND	ND	—	—	—	—	—	—	—	—	—	—	—	—	—	—	—	—
Meropenem	ND	ND	—	—	—	—	—	—	—	—	—	—	—	—	—	—	—	—

Table 11.7 Kinetic parameters for homology group I class D β-lactamases (part 3)[a]

β-Lactams	OXA-48			OXA-54			OXA-55		
	k_{cat} [s^{-1} (%)]	K_m (μM)	k_{cat}/K_m [mM^{-1} s^{-1} (%)]	k_{cat} [s^{-1} (%)]	K_m (μM)	k_{cat}/K_m [mM^{-1} s^{-1} (%)]	k_{cat} [s^{-1} (%)]	K_m (μM)	k_{cat}/K_m [mM^{-1} s^{-1} (%)]
Benzylpenicillin	245 (100)	40	6,100 (100)	120 (100)	60	2 (100)	4 (100)	25	160 (100)
Ampicillin	340 (138.8)	5,200	65 (1.1)	540 (450)	4300	0.125 (6.25)	8 (200)	500	15 (9.4)
Piperacillin	75 (30.6)	410	180 (3.0)	20 (16.7)	240	0.085 (4.25)	3 (75)	110	30 (18.8)
Oxacillin	25 (10.2)	30	850 (13.9)	35 (29.2)	75	0.5 (25)	5 (125)	390	10 (6.3)
Cloxacillin	— (—)	—	— (—)	30 (25)	240	0.125 (6.25)	0.5 (12.5)	190	3 (1.9)
Cephaloridine	2 (0.8)	27	75 (1.23)	1 (0.8)	45	0.02 (1)	10 (250)	750	15 (9.4)
Cephalothin	3 (1.2)	20	150 (2.5)	3 (2.5)	230	0.015 (0.75)	0.6 (15)	70	10 (6.2)
Cefotaxime	10 (4.1)	190	60 (1)	15 (12.5)	1,600	0.01 (0.5)	<0.01 (<0.2)	—	—
Ceftazidime	4 (1.6)	5,100	1 (0.02)	<0.01 (<0.01)	—	—	2 (50)	4,700	0.5 (0.31)
Aztreonam	ND	—	—	<0.01 (<0.01)	—	—	<0.01 (0.25)	—	—
Imipenem	2 (0.8)	14	145 (2.4)	1 (0.8)	4	0.25 (12.5)	0.1 (2.5)	20	5 (3.1)
Meropenem	0.1 (0.04)	200	0.5 (0.01	0.1 (0.08)	125	0.001 (0.05)	0.05 (1.25)	500	5 (0.06)

[a]—, not determined or not applicable; ND, hydrolysis not detected.

was to lower the K_m for penicillins, whereas k_{cat} was little affected for benzylpenicillin or carbenicillin, raised for isoxazoyl β-lactams, and lowered for ampicillin (Table 11.7). The kinetic parameters for the initial phase of hydrolysis of cephalothin and cephaloridine were little affected by the presence of HCO_3^-. It is interesting that the k_{cat} (the parameter that is expected to be the most affected if only part of the enzyme is active) is actually less clearly affected by the presence of HCO_3^- than is the K_m (Table 11.7), which is not expected to be so affected by the proportion of enzyme in its more active form. The β-lactams where the effect on k_{cat} was least were poor substrates ($k_{cat} < 130$ s^{-1} in the initial phase) and, consequently, were those for which the most concentrated enzyme was used, perhaps thereby creating an artifact, as the high protein concentration would disfavor conversion to the less active monomeric form. If the limiting factor for the hydrolysis of the β-lactam at saturated substrate concentration is deacylation of the enzyme, the k_{cat} will be equal to the rate of deacylation of the enzyme. Based on this hypothesis and the observation that the carbamylation affects the k_{cat} less than the K_m, it might be that the carbamylation of Lys73 is not involved in the deacylation mechanism. Several hypotheses might explain the effect of carbamylation on the K_m. For example, it could be a structural effect as reported for other enzymes (the closed form of the protein obtained in presence of the carbamylated Lys might have a higher affinity for β-lactams) or it might be a mechanistic effect (an increase in one or more of the microscopic rate constants for the steps involved in formation of the acyl enzyme complex would appear in the K_m term of the steady-state rate equation). It might also be argued that the concentration of $NaHCO_3$ used in the assay was far above the physiological concentration, thus interfering with the enzyme mechanism.

The enzymes most closely related to OXA-10 include 12 sequenced variants (OXA-5, -7, -11, -13, -14, -16, -17, -19, -28, -31, -35, and -101) (Fig. 11.1). OXA-6 β-lactamase probably also belongs to this group as it cross-hybridizes with DNA probes for bla_{OXA-10} (36) and has an apparent molecular weight close to that expected for a dimer. OXA-5 is rather more remote from the other members of the group (Table 11.3). Several members of the group (OXA-11, -14, -16, and -17) are extended-spectrum OXA-10 derivatives with one or more amino acid substitutions that increase the ability to confer resistance to oxyimino-cephalosporins, particularly ceftazidime (Table 11.8). OXA-14 differs from OXA-10 only by a Gly167Asp substitution (OXA-10 numbering of the mutations is shown in Table 11.4) but confers much greater resistance to all extended-spectrum cephalosporins in *P. aeruginosa*. Moreover, OXA-14 showed more-marked burst kinetics than OXA-10 for all of the 17 β-lactams tested (23). The fact that such a seemingly minor change

Table 11.8 Relative hydrolysis rates for class D β-lactamases

β-Lactamase	Relative rate of hydrolysis[a]													IC_{50} (μM)				Reference(s)
	PEN	AMP	CAR	CLX	OXA	MET	CLR	LOT	FOX	CTX	CAZ	ATM	IPM	CLA	SUL	TZB	IPM	
Homology group I																		
OXA-10	100	660	35	584	683	230	89	7	<0.1	10	0.12	6.1	0.05	0.81	37	0.94	—	56
OXA-5	100	190	40	260	210	110	89	180	10	49	—	—	—	3.1	18	0.25	—	56
OXA-6	100	600	46	300	1,000	590	150	24	<0.2	28	—	—	—	1.6	51	1.7	—	56
OXA-7	100	540	48	490	700	420	140	51	4	31	—	—	—	0.36	40	0.61	—	56
OXA-11	100	72	3.8	—	530	—	0.6	1.7	<0.1	1	0.6	—	<0.1	4.5	—	0.5	—	56
OXA-13	100	281	—	—	425	—	360	65	1.4	180	—	6	—	54	—	—	<0.01	63
OXA-14	100	103	—	90	142	—	75	8	—	—	—	—	—	—	—	—	—	17, 18
	400[c]	*1,125*	—	*2,775*	*1,275*	—	*67*	*125*	—	—	—	—	—	—	—	—	—	
OXA-16	100	202	35	520	856	—	44	6	—	13	3	—	—	—	—	—	—	
	>125[c]	*339*	*144*	*2,500*	*948*	—	*62*	—	—	*21*	—	—	—	—	—	—	—	22
OXA-17	100	520	40	400	2,400	—	460	100	—	440	—	—	—	—	—	—	—	20
OXA-48	100	139	—	—	10.2	—	0.8	1.2	—	4.1	1.6	<0.001	0.8	16	50	1.7	—	66
OXA-54	100	450	—	25	29.2	—	0.8	2.5	—	12.5	<0.001	<0.001	0.8	500	>2,000	350	—	65
OXA-55	100	200	—	12.5	125	—	250	15	—	<0.02	50	<0.001	2.5	700	300	12	—	34
Homology group II																		
OXA-2	100	2.2	2.1	50	1,000	17	42	3.8	2	0.4	0.02	3.6	—	1.4	0.14	0.01	—	56
OXA-3	100	180	10	350	340	—	44	10	—	—	—	—	—	—	—	—	—	56
OXA-15	100	44	—	—	681	—	400	148	—	5	≤1	—	≤1	1.5	—	—	—	21
	—	—	—	—	—	—	—	*297*[c]	—	*65*	—	—	—	—	—	—	—	
OXA-20	100	307	—	269	446	—	77	50	—	—	—	23	—	2.2	—	—	—	56
OXA-32	100	—	—	71	—	—	—	86	—	—	—	—	—	8	—	0.05	0.01	68

(*Continued next page*)

Table 11.8 (*Continued*)

β-Lactamase	Relative rate of hydrolysis[a]													IC$_{50}$ (μM)				Reference(s)
	PEN	AMP	CAR	CLX	OXA	MET	CLR	LOT	FOX	CTX	CAZ	ATM	IPM	CLA	SUL	TZB	IPM	
Homology group III																		
OXA-1	100	360	63	360	180	390	30	100	—	110	—	—	—	1.8	4.7	1.4	—	56
OXA-4	100	440	39	64	220	710	190	83	<2	64	—	—	—	8.4	16	5.6	—	56
OXA-31	100	—	—	400	—	—	—	80	—	75	—	—	—	—	—	—	—	7
Homology group IV																		
OXA-9	100	110	200	—	81	—	—	—	—	—	—	—	—	<10	—	—	—	56
OXA-12	100	—	160	190	210	—	14	—	—	—	≤2	—	≤1	0.01	0.24	0.03	—	56
OXA-18	100	63	—	150	36	—	80	81	<0.1	470	808	63	<0.1	0.08	0.56	0.13	0.01	56
OXA-22	100	—	—	700	<1	—	5500	900	50	<1	<1	<1	—	1.2	—	6.5	—	59
OXA-29	100	164	20	—	52	41	—	—	—	0.26	<0.4	0.3	<0.1	>1,000	48	3.2	—	28
OXA-45	100	98	—	—	34	—	—	—	<1	192	1,150	1	<0.1	0.04	—	—	0.6	78
AmpS	100	64	20	2	78	—	2.1	3.2	—	—	—	—	—	—	—	—	—	56
LoxA[b]	—	100	11	—	36	—	—	—	—	—	—	—	—	—	—	—	—	7
Homology group V																		
LCR-1	100	150	4	<8	63	20	55	24	—	—	—	9	—	100	—	—	—	56
Homology group VI																		
OXA-24	100	—	—	<1	<1	<1	64	—	—	—	—	—	4	50	40	0.5	—	9
OXA-25	100	21	76	—	440	—	33	3	—	0.2	0.01	—	3	0.1	—	2	—	1
OXA-26	100	55	25	—	500	—	27	8	—	<0.003	0.1	<0.003	2.4	0.04	—	1.6	—	1
OXA-27	100	6	<0.0005	—	3	—	6	0.3	—	0.2	0.0005	<0.0005	—	0.65	—	2.5	—	1
OXA-40	100	100	—	<0.02	40	—	100	60	—	<0.02	400	—	2	300	190	180	—	35
OXA-50	100	<2.7	—	—	0.2	—	1.8	0.18	—	—	—	—	0.09	500	>2,000	350	—	30
OXA-51	—	100	—	2.1	32.3	—	10	—	—	<0.001	<0.001	—	3.1	100	50	—	2.5	11
OXA-58	100	18.2	—	—	27.3	—	—	1.8	—	<0.2	<0.2	<0.2	1.8	310	60	2500	—	69
OXA-60	100	—	—	17	31	—	0.12	<0.01	<0.01	<0.01	0.02	<0.01	2.29	450	320	73	—	30

[a]—, not determined or not applicable. AMP, ampicillin; ATM, aztreonam; PEN, benzylpenicillin; CLA, clavulanate; CAR, carbenicillin; CLR, cephaloridine; CLX, cloxacillin; CTX, cefotaxime; CTZ, ceftazidime; FOX, cefoxitin; IPM, imipenem; CEF, cephalothin; MET, methicillin; OXA, oxacillin; TZB, tazobactam.
[b]Relative hydrolysis rates to benzylpenicillin hydrolysis, which was set at 100%.
[c]Rapid initial phase of biphasic kinetics (data in italics), relative to second, steady-state phase hydrolysis rate for benzylpenicillin.

has such a major effect may be because, in the basic OXA-10 structure, the amino acid side chain at position 167 interacts with the backbone of the loop containing the Trp164 (Fig. 11.5). The only way to accommodate aspartate at position 167 is to reposition the loop, and this may have a profound effect. In any event, the particularly strong burst kinetics observed with OXA-14 are explained by the conversion of a highly active dimer to a less active monomer following dilution into the assay cuvette. The dissociation constant for the dimer to monomer equilibrium determined by gel filtration was close to 45 μM for both OXA-14 and its parent enzyme, OXA-10. From the calculated concentration of enzyme in the bacterial periplasmic space, the enzyme is a dimer in the cell, whereas the monomer predominates under most assay conditions.

Besides OXA-14, the other OXA-10 variants have further changes besides the Gly167Asp: OXA-11 has the additional mutation Asn146Ser, and OXA-16 has Ala127Thr (Table 11.4). The quantitative resistance phenotypes conferred in *P. aeruginosa* transconjugants OXA-11, OXA-14, and OXA-16 enzymes were identical, but unlike OXA-14, the double mutants (OXA-11 and OXA-16) did not show biphasic kinetics for all substrates; moreover, OXA-16 achieved a faster hydrolysis rate of cefotaxime than did any of the other mutants and, while all were able to confer ceftazidime resistance, hydrolysis was readily detectable only with OXA-11 (Table 11.8). Several OXA-10 laboratory mutants were obtained in *P. aeruginosa* by selecting in vitro with ceftazidime (Table 11.4) (19). These included one identical to wild-type OXA-14 and OXA-M102, an enzyme similar to OXA-16 but with Ala127Ser rather than Ala127Thr. Other ESBL mutants of OXA-10 selected in vitro, e.g., OXA-M101 with Trp164Leu and OXA-M103 with Asn146Lys, had sequence changes that have not been reported yet from clinical isolates; nevertheless, mutations affecting the same positions occur in OXA-28 and OXA-11, respectively.

Many of the mutations observed in OXA-10 derivatives are not in the active site but lie close to it. The effect of Gly167Asp has been discussed already. The additional Asn146Ser mutation in OXA-11 cannot be accommodated in the structure described for OXA-10 and very probably forces repositioning of Trp164 and Leu165 (155 in OXA-10). This is because Asn at position 146 has unfavorable interactions, via its side chain nitrogen, with Gln168 (158 in OXA-10) and, via its carbonyl, with Arg170 (160 in OXA-10), both residues that lie at the beginning of the loop containing the Trp164 (Fig. 11.5). A similar situation arises with OXA-16: Ala127 of OXA-10 is normally in van der Waals contact with Trp164, but, when it is replaced by Thr in OXA-16 (or Ser in OXA-M102), this interaction is likely to be affected. The Asn146Lys substitution of OXA-M103 also may change the positioning of the omega loop containing Trp164, especially if the Lys residue forms a salt

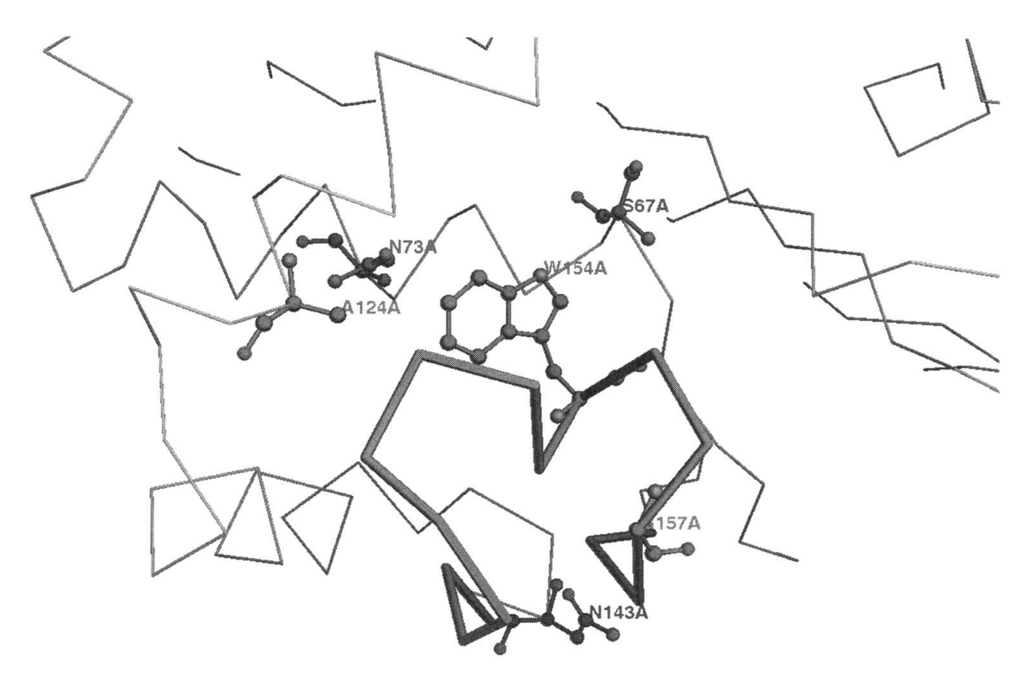

Figure 11.5 Amino acid replacements in the ESBL derivatives of OXA-10. The active serine is represented, with the omega loop containing Trp164 (154 in OXA-10) shown by a thicker line (OXA-10 numbering).

bridge with Glu65 (62 in OXA-10). In OXA-M101, Leu164 cannot engage in the hydrogen bonding that is performed by the Trp normally present at this position, precluding interaction with the carbamylate of Lys73.

Collectively, these data suggest that repositioning or removal of Trp164 is necessary to confer the ability to cause resistance to ceftazidime. Compared with OXA-10 in the absence of added carbonate, these substitutions reduce efficiency against ampicillin. OXA-14 and OXA-16 were also less efficient than OXA-10 against oxacillin, owing to higher K_m values. Despite the increased resistance to ceftazidime associated with OXA-11, -14, and -16, hydrolysis of ceftazidime was not detected or, in the case of OXA-11, remained very slow in vitro, perhaps because of differences in kinetic behavior between a high periplasmic enzyme concentration and a low enzyme concentration in the assay cuvette.

OXA-17, another derivative of OXA-10, was associated with increased resistance to cefotaxime, but not ceftazidime, once it was cloned into *E. coli* (investigation in the original *P. aeruginosa* host was precluded by the simultaneous presence of OXA-2 and PER-1 enzymes) (20). This cefotaxime resistance correlated with a higher k_{cat} for OXA-17, which, compared to OXA-10, had an Asn76Ser substitution. Interestingly, OXA-13, another group I enzyme, also has this mutation, along with eight other substitutions (Table 11.4) and was also relatively more active than OXA-10 against cefotaxime and cephaloridine. *E. coli* strains with cloned OXA-13, like those with OXA-17, were less susceptible to cefotaxime and aztreonam. OXA-17 had up to 20-fold lower k_{cat} values for ampicillin and benzylpenicillin than OXA-10 (Table 11.7) and although the k_{cat} for cefotaxime was raised 2.4-fold, the K_m was increased 7.4-fold, decreasing catalytic efficiency. More generally, OXA-13 and OXA-17 were stronger cephalosporinases than OXA-10 in terms of relative k_{cat}. As already noted, however, the relationship of kinetic data to resistance data is difficult to determine with OXA enzymes, owing to nonlinear behavior.

OXA-13-1, a spontaneous laboratory mutant selected with ceftazidime from a *P. aeruginosa* strain producing OXA-13 (54), had the Gly167Asp substitution found in OXA-14. This mutant conferred high-level ceftazidime resistance, though resistance to ticarcillin was reduced, both in the original *P. aeruginosa* host and after transfer to *E. coli*. OXA-19, another derivative of OXA-13, was found in a *P. aeruginosa* isolate highly resistant to ceftazidime and corresponded to OXA-13-1, except that Asn was present at position 76, as in native OXA-10 (Table 11.4). A further OXA-13 derivative, OXA-35, found in a *P. aeruginosa* isolate resistant to ureidopenicillins and intermediately resistant to ceftazidime, also had Asn76. No increase in resistance to cefotaxime was observed when this enzyme was transferred into *E. coli*, confirming the role of the Asn76Ser substitution in raised cefotaxime resistance. OXA-101 is also a close analogue and only differs from OXA-35 by one mutation— Ser27Phe. A similar mutation is also found in OXA-7 and OXA-56. A further OXA-10-related enzyme is OXA-28, described from clinical isolates of *P. aeruginosa* resistant to ceftazidime and intermediately resistant to ureidopenicillins (67). Once cloned into *E. coli*, OXA-28 was again associated with resistance to ceftazidime, and reduced susceptibility or resistance to cefotaxime. This enzyme hydrolyzed cloxacillin, benzylpenicillin, cefotaxime, and ceftazidime, but not piperacillin; clavulanic acid had an IC_{50} of 10 μM. OXA-28 had a Trp164Gly substitution, reinforcing the view that the Trp164 side chain ordinarily limits the spectrum of OXA-10-related enzymes.

Aside from the OXA-10 mutants, some other enzymes belonging to group I also have higher relative hydrolysis rates for cefotaxime than does OXA-10 but do not confer ceftazidime resistance. Cefotaxime hydrolysis is especially rapid with OXA-5, an enzyme also strongly active against cephalothin but less so against ampicillin and oxacillin (Table 11.8). OXA-7, another strong cefotaximase, is a unique member of the group, having an insertion and a deletion compared with all the other group I enzymes (Table 11.4). OXA-56 is identical to OXA-7, except that it does not have this insertion and deletion and is closely related to OXA-19, with differences for only three amino acids.

The OXA-48, -54, and -55 enzymes have some resemblance in sequence to group VI enzymes but nevertheless more strongly resemble group I types (65, 66). They are penicillinases rather than cephalosporinases but, as with group VI enzymes, can hydrolyze imipenem at 2 to 12% of the rate for benzylpenicillin; meropenem is a poorer substrate (Table 11.7). The single *K. pneumoniae* clinical isolate producing OXA-48 was highly resistant to all β-lactams tested including imipenem, meropenem, and cefoxitin, but, besides OXA-48, it also had the narrow-spectrum enzymes OXA-47 and TEM-1 and the ESBL SHV-2a and had lost a porin. In these complex circumstances it is not possible to attribute resistance to single factors.

Homology Group II β-Lactamases

Fourteen group II OXA β-lactamases have been reported (Fig. 11.1). These include OXA-2 and -3, which are among the more frequent OXA type and share 90.5% amino acid identity. They have 73% identity to OXA-20, another group II enzyme. All three of these β-lactamases are penicillinases but, compared with group I oxacillinases, have slow relative hydrolysis rates for carbenicillin (Table 11.8). OXA-2 hydrolyzes narrow-spectrum cephalosporins, but not the oxyimino broad-spectrum and "fourth-generation" agents. Its mutants, OXA-15 and -32, with Asp148Gly and Leu169Ile substitutions, respectively (Table 11.5), are able to confer resistance to ceftazidime but not cefotaxime in *E. coli* and *P. aeruginosa*. As with the OXA-10 mutants, this ability to confer ceftazidime resistance exists despite low k_{cat} values and a very high K_m for OXA-32 (Table 11.9), both of which would be predicted to impair efficiency. Hydrolysis is, however, biphasic, and once again the natural activity in the periplasm may be greater than that observed in the assay cuvette. Residue 148 in OXA-2 is at the beginning of the loop containing Trp164, which is involved in the hydrogen bonding with the Lys73 carbamylate, and mutation of this residue (as in OXA-15) may affect the positioning of Trp164 or the flexibility of the omega loop. The Leu169Ile substitution in OXA-32 seems more trivial, and it is surprising that it was associated with resistance to ceftazidime. However, this Leu residue is conserved among all other class D β-lactamases except OXA-48 and -54, suggesting that it may be of some importance, perhaps because it is buried. We postulate that the methyl group of the Ile will point towards the α carbon of Trp164 and, very probably, be displaced by it. Other derivatives of OXA-2 include OXA-34 and -36 (Table 11.5), but their activity has not yet been reported. OXA-36 has a mutation at the same position as in OXA-15, but with an Asp148Tyr substitution. OXA-34 has Trp169 (which is involved in hydrogen bonding, with the carbamylated Lys73) mutated to cysteine (Table 11.5). From the mutations observed in OXA-15, -24, and -36, it seems that the repositioning or removal of Trp164 is critical to obtaining ESBL activity in the OXA-2 group.

OXA-21 is a close relative of OXA-3, but has Ile225 (217 in OXA-3) replaced by Met, which is the residue found at the corresponding site in OXA-2. Since this mutation lies in an amino acid sequence otherwise conserved between OXA-2 and OXA-3, it seems unlikely to affect activity strongly. OXA-37 differs from OXA-20 in having the conservative substitution Glu21Asp (position 23 in OXA-3). None of these enzymes is known to have extended-spectrum activity. OXA-53, reported from *Salmonella enterica*, shares only 90% amino acid identity with OXA-2 and was associated, after transfer into E. coli, with reduced susceptibility to ceftazidime (55) (Table 11.5).

From the data available, tazobactam is a relatively potent inhibitor of OXA-2 and OXA-32, whereas clavulanate and sulbactam are weaker inhibitors; OXA-53 seems unusual in being readily inhibited by clavulanate (55). Imipenem is a potent inhibitor of OXA-32 and, very probably, also of the other members of the group (Table 11.8) (68).

Homology Group III β-Lactamases

The most important member of this group, OXA-1, is the classical and most prevalent class D penicillinase; other members include OXA-4 (also named OXA-35) and -31. OXA-1 has a relatively high k_{cat} for cloxacillin, with a very low K_m (unlike, say, OXA-10), meaning that this compound is the most efficient substrate (also for OXA-31) (Table 11.9). OXA-1 and OXA-31 differ by five amino acids, but the two enzymes had very similar kinetic parameters (Table 11.9). Curiously, both enzymes have k_{cat} values for cefepime and cefpirome 15- to 18-fold higher than for benzylpenicillin but, owing to high K_m values, the catalytic efficiency (k_{cat}/K_m) is only half that for benzylpenicillin. Resistance to these fourth-generation cephalosporins was observed in *P. aeruginosa* strains with OXA-1 and -31 enzymes, whereas only decreased susceptibility was seen in *E. coli*, presumably reflecting the lower permeability and greater efflux function in the former species. The effects of these enzymes on cefotaxime susceptibility were only marginal, and the greater activity against cefepime and cefpirome, along with the particularly strong oxacillinase activity, may be explained by the large omega loop of group III enzymes (75).

OXA-4 has two residues replaced compared with OXA-1. The first replacement is conservative, with Glu208Asp (207 OXA-1), leaving an acidic side chain directed toward the solvent. The other replacement is Ala48Val (49 OXA-1). This mutation is far from the active site and is conservative but is situated at the beginning of a β-turn and may change the positioning of some secondary element, allowing accommodation of larger residues. OXA-1 and OXA-31 differ by three amino acids, with the following substitutions: Ala48Val (49 OXA-1), Ala67Pro (68 OXA-1), and Asp208Glu (207 OXA-1). Two of these mutations are rather conservative, and one gives a sequence corresponding to that present in OXA-4. The last substitution, Ala-67Pro (68 OXA-1) in OXA-31, probably has very little effect in the structure. OXA-47, another group III enzyme with a narrow-spectrum activity, differs from OXA-1 by six residues.

None of the marketed β-lactamase inhibitors has good activity against group III enzymes (Table 11.8), although carbapenems may be stronger inhibitors.

Homology Group IV β-Lactamases

Group IV comprises diverse enzymes, not as closely related as those in other homology groups (Fig. 11.1). OXA-12, for example, shares only 31 to 55% similarity to OXA-9, -18, -22, -42, and -43 (Table 11.3). This diversity is reflected in their activity. Several of these enzymes are clearly penicillinases, for example, the OXA-9, OXA-29, LoxA,

Table 11.9 Kinetic parameters for homology group II, III and V class D β-lactamases[a]

	Group II						Group III						Group V		
	OXA-2			OXA-32			OXA-1			OXA-31			LCR-1		
β-Lactam	k_{cat} [s^{-1} (%)]	K_m (μM)	k_{cat}/K_m [mM^{-1} s^{-1} (%)]	k_{cat} [s^{-1} (%)]	K_m (μM)	k_{cat}/K_m [mM^{-1} s^{-1} (%)]	k_{cat} (%)	K_m (μM)	k_{cat}/K_m (%)	k_{cat} (%)	K_m (μM)	k_{cat}/K_m (%)	k_{cat} (%)	K_m (μM)	k_{cat}/K_m (%)
Benzylpenicillin	90 (100)	4	22,500 (100)	3.5 (100)	45	80 (100)	100	5	100	100	5	100	100	7	100
Amoxicillin	—	—	—	1.5 (43)	220	70 (87)	360	53	30	350	60	30	—	—	—
Ampicillin	110 (122)	25	4,400 (19.6)	—	—	—	—	—	—	—	—	—	83	21	28
Carbenicillin	2 (2)	150	13 (0.1)	—	—	—	—	—	—	—	—	—	3.5	15	1.6
Cloxacillin	>45 (50)	>650	69 (3)	25 (71)	110	20 (2.5)	360	2	300	400	2	1,000	<8	ND	—
Methicillin	15 (17)	200	75 (0.3)	—	—	—	—	—	—	—	—	—	<50	ND	—
Oxacillin	900 (1,000)	360	2,500 (11)	—	—	—	—	—	—	—	—	—	63	120	3.7
Piperacillin	—	—	—	3 (86)	155	20 (25)	—	—	—	—	—	—	—	—	—
Ticarcillin	—	—	—	1 (29)	60	15 (19)	—	—	—	—	—	—	—	—	—
Cephaloridine	38 (42)	52	730 (3.2)	2 (57)	360	6 (7.5)	—	—	—	—	—	—	21	1,400	0.1
Cephalothin	3.4 (3.8)	7	485 (2.2)	3 (86)	60	50 (62)	100	40	10	80	45	10	—	—	—
Cefepime	—	—	—	—	—	—	1,630	215	40	1,790	200	75	—	—	—
Cefepirome	—	—	—	—	—	—	1,410	110	65	1,810	120	75	—	—	—
Cefotaxime	—	—	—	—	—	—	110	35	20	75	35	10	<0.2	ND	
Ceftazidime	0.02 (0.02)	12.0	1.7 (0.01)	H[b]	>3,000	—	—	—	—	—	—	—	—	—	—
Aztreonam	—	—	—	—	—	—	—	—	—	—	—	—	—	—	—
Imipenem	—	—	—	—	—	—	—	—	—	—	—	—	<0.3	ND	—

[a]Data from references 6, 41, 68, and 82. —, not determined or not applicable; ND, hydrolysis not detected.
[b]H, hydrolysis detected but k_{cat} could not be determined due to high K_m value.

and the OXA-12 and AmpS enzymes of *Aeromonas* spp., all of which have particularly weak activity against cephalosporins (Table 11.8). By contrast, other members, for example, OXA-18 and -45, have 8- to 10-fold-faster relative hydrolysis rates for oxyimino-cephalosporins, such as cefotaxime and ceftazidime, than for benzylpenicillin (Table 11.10). OXA-18 is surprising in hydrolyzing cefepime and aztreonam more rapidly than cephaloridine (64). OXA-42 and OXA-43, which differ by only two amino acids, were associated with ceftazidime resistance in *Burkholderia pseudomallei*, where they are chromosomal (58). Several members of this group (OXA-12, -18, and -45) are inhibited strongly by clavulanic acid, tazobactam, and sulbactam, whereas OXA-29 is not inhibited even by clavulanic acid concentrations even in the millimolar range (Table 11.8).

Homology Group V β-Lactamases

Group V presently comprises a single enzyme, LCR-1. This is a penicillinase and oxacillinase in terms of relative k_{cat} and K_m (Table 11.9) (82). It is not strongly inhibited by clavulanate (Table 11.8).

Homology Group VI β-Lactamases

Group VI comprises class D enzymes that have been associated with carbapenem resistance or with situations in which imipenem hydrolysis was detected. Most of its members have been found predominantly or exclusively in *Acinetobacter baumannii*. The first producer was found in 1985 in the United Kingdom; its enzyme was originally named ARI-1 but, after sequencing, was renamed OXA-23 (26).

Seven subgroups exist, based on sequence (91). The first comprises OXA-23, -27, and -49; the second comprises OXA-24, -25, -26, -40, and -72 (Fig. 11.1); the third contains OXA-51, -69, -71, and -75; the fourth consists of OXA-58; and the last two subgroups consist of OXA-50 and -60, respectively. The first three subgroups are more closely related, and the diversity is associated with the N- and C-terminal parts of the protein. Curiously, OXA-24 (which appears, on sequencing, to be identical to OXA-40) was reported to hydrolyze cloxacillin, oxacillin, or methicillin (9) very slowly, although these compounds are substrates for its close relatives OXA-25 and OXA-26.

OXA-27 (1), belonging to the other major cluster, hydrolyzes oxacillin rather slowly; there are no data available for OXA-23, as the more typical member of this subgroup.

For all these enzymes hydrolysis of imipenem is slow in vitro, with relative k_{cat} values between 2 and 4% of those observed for benzylpenicillin; hydrolysis of meropenem is even slower. In the case of OXA-26 and OXA-27, however, hydrolysis of imipenem was associated with a low K_m, increasing the catalytic efficiency (Table 11.11) and, once again, it is plausible that many of these enzymes are more efficient under periplasmic conditions or are produced in large quantities to compensate for the low turnover. More generally, OXA-25 and OXA-26 have greater relative activity against penicillins than cephalosporins, whereas OXA-27 has greater relative activity against cephalosporins, owing to very low K_m values (Table 11.11).

The OXA-23 and -24 related carbapenemases have Tyr-Gly-Asn at positions 144 to 146 (154-156 OXA-23), replacing the Phe-Gly-Asn sequence, which is conserved in other OXA β-lactamases including carbapenemase of group VI subgroup 4 (i.e., OXA-50 and -60) and group I (i.e., OXA-48, -54, and -55). Although it is tempting to speculate that Phe154 may have a role in the carbapenemase activity, there are two lines of evidence that point against this view. First, in the OXA-10 structure, Tyr144 (154 in OXA-10) is buried in a hydrophobic environment, with its hydroxyl directed away from the active site. Consequently, it is unlikely to greatly affect the hydrolytic spectrum. Second, in one study on OXA-40, Phe154 (144 OXA-40) was replaced by a Tyr by site-directed mutagenesis (35) and the mutant enzyme, expressed in *E. coli*, conferred more resistance to piperacillin and less to ceftazidime, while the MICs of imipenem and meropenem were not affected. Similarly, when Tyr154 was replaced by Phe in OXA-1, the main effect was a four- to eightfold reduction of the MICs of piperacillin, cefepime, and cefpirome but, again, the MICs of imipenem or meropenem were not affected. Otherwise, OXA-40-Tyr had higher k_{cat} rates for all substrates tested except ceftazidime than did wild-type OXA-40, while effects on K_m values generally were not significant (Table 11.11). It does seem that enzymes with a Phe144 instead of Tyr are unusually resistant to inhibition by NaCl, differentiating them from other class D carbapenemases.

The third subgroup of homology group VI comprises 25 enzymes which differ from each other by not more than 45 amino acids. OXA-57 and -66 seem the most widespread forms of the subfamily. All these enzymes were isolated from *A. baumannii* and have changes in the second and third conserved elements. In particular, Ser-Xxx-Val is modified to Ser-Xxx-Ile Leu and the standard Lys-Thr-Gly motif of the second element is replaced by Lys-Ser-Gly. The last motif is also found in the OXA-23 and -40 clusters and in OXA-9 and -58. OXA-51 and -69 are the only enzymes from this subgroup with published kinetic information (10). OXA-51 is a penicillinase with ampicillin as its best substrate; cloxacillin and oxacillin were weaker substrates. Cefuroxime and broad-spectrum cephalosporins are stable; indeed, the only cephalosporin readily hydrolyzed by OXA-51 was cephaloridine, and that with a

Table 11.10 Kinetic parameters for class D enzymes belonging to homology group IV[a]

	OXA-12			OXA-18			OXA-22			OXA-29		
β-Lactam	k_{cat} (%)	K_m (μM)	k_{cat}/K_m (%)	k_{cat} (%)	K_m (μM)	k_{cat}/K_m (%)	k_{cat} [s^{-1} (%)]	K_m (μM)	k_{cat}/K_m [mM^{-1} s^{-1} (%)]	k_{cat} [(s^{-1}) k_{cat} (%)]	K_m (μM)	k_{cat}/K_m [mM^{-1} s^{-1} (%)]
Benzylpenicillin	100	13	100	100	2.50	100	0.1 (100)	<2	>0.04 (>250)	65 (100)	10	6,500 (100)
Amoxicillin	—	—	—	—	—	—	0.1 (100)	6	0.016 (100)	—	—	—
Ampicillin	—	—	—	63	1.30	121	—	—	—	107 (165)	16	6,700 (103)
Carbenicillin	160	43	48	—	—	—	—	—	—	13 (20)	16	810 (12.5)
Cloxacillin	190	480	5.10	150	9	40	0.7 (700)	5	0.14 (875)	—	—	—
Methicillin	—	—	—	—	—	—	—	—	—	27 (41.54)	41	660 (10)
Oxacillin	210	7	400	36	12	7.5	<0.01 (<0.1)	—	—	34 (52.31)	4.40	7,700 (119)
Piperacillin	—	—	—	—	—	—	<0.01 (<0.1)	—	—	31 (47)	39	790 (12)
Ticarcillin	—	—	—	229	12	48	0.1 (100)	9	0.01 (62.5)	—	—	—
Cephaloridine	14	ND	—	80	132	1.5	5.5 (5,500)	10	0.55 (3,437)	—	—	—
Cephalothin	—	—	—	81	26.5	7.5	0.9 (900)	3	0.33 (2,062)	—	—	—
Cefazolin	—	—	—	—	—	—	—	—	—	11 (17)	30	370 (6)
Cefepime	—	—	—	160	64	6.3	—	—	—	—	—	—
Cefotaxime	—	—	—	470	23	51.3	<0.01 (<0.1)	—	—	0.17 (0.3)	128	1.3 (0.02)
Cefsulodine	—	—	—	261	87	7.5	—	—	—	—	—	—
Ceftazidime	<2	ND	—	808	499	4	<0.01 (<0.1)	—	—	>0.2	>500	0.49 (0.01)
Cefuroxime	—	—	—	—	—	—	0.05 (50)	5	0.01 (62.5)	>0.2	>250	0.79 (0.01)
Nitrocefin	—	—	—	—	—	—	—	—	—	1,800 (2,769)	96	19,000 (288)
Aztreonam	—	—	—	63	3	52.5	<0.01 (<0.1)	—	—	0.2 (0.3)	210	1 (0.02)
Imipenem	<1	ND	ND	<0.1	ND	ND	—	—	—	ND	ND	ND

[a]Data from references 28, 29, 30, 59, 64, and 71. —, not determined or not applicable; ND, hydrolysis not detected.

Table 11.11 Kinetic parameters for homology group VI class D β-lactamases

β-Lactam	OXA-25 k_{cat} (%)	OXA-25 K_m (μM)	OXA-25 k_{cat}/K_m (%)	OXA-26 k_{cat} (%)	OXA-26 K_m (μM)	OXA-26 k_{cat}/K_m (%)	OXA-27 k_{cat} (%)	OXA-27 K_m (μM)	OXA-27 k_{cat}/K_m (%)
Benzylpenicillin	100	100	1,000	100	25	100	100	88	100
Ampicillin	21	21	100	55	15	91.67	6	3	176
Piperacillin	22	55	40	53	10	132.5	4	10	35.2
Carbenicillin	76	300	25.33	25	210	2.97	<0.0005	—	—
Oxacillin	440	840	52.38	500	580	21.5	3	402	40
Cloxacillin	—	—	—	—	—	—	—	—	—
Cephaloridine	33	590	5.59	27	640	1.05	6	3	176
Cephalothin	3	80	3.75	8	90	2.22	0.3	260	0.1
Cefotaxime	0.2	35	0.57	<0.003	—	—	0.2	0.1	176
Cefepime	—	—	—	—	—	—	—	—	—
Cefpirome	—	—	—	—	—	—	—	—	—
Cefuroxime	0.4	105	0.38	0.04	7	0.14	1	2	44
Ceftazidime	0.01	—	—	0.1	—	—	0.0005	—	—
Imipenem	3	11	27.30	2.4	3	20	0.1	20	0.44
Meropenem	0.4	12	3.33	0.4	3	3.33	0.04	15	0.23
Aztreonam	—	—	—	<0.003	—	—	<0.0005	—	—

β-Lactam	OXA-40 k_{cat} [s^{-1} (%)]	OXA-40 K_m (μM)	OXA-40 k_{cat}/K_m [mM^{-1} s^{-1} (%)]	OXA-40-Tyr k_{cat} [s^{-1} (%)]	OXA-40-Tyr K_m (μM)	OXA-40-Tyr k_{cat}/K_m [mM^{-1} s^{-1} (%)]	OXA-50 k_{cat} [s^{-1} (%)]	OXA-50 K_m (μM)	OXA-50 k_{cat}/K_m [mM^{-1} s^{-1} (%)]
Benzylpenicillin	5 (100)	23	220 (100)	30 (100)	9	3,000 (100)	110 (100)	800	140 (100)
Ampicillin	5 (100)	220	20 (9.09)	40 (133.33)	180	220 (7.33)	<3 (2.7)	>1,000	<3 (<2.14)
Piperacillin	1 (20)	23	50 (22.73)	15 (50)	40	350 (11.66)	<2 (<1.8)	>1,000	<2 (<1.43)
Carbenicillin	—	—	—	—	—	—	—	—	—
Oxacillin	2 (40)	876	5 (2.2)	6 (20)	155	40 (1.3)	<0.2 (<0.18)	>1,000	<0.2 (<0.14)
Cloxacillin	ND	—	—	35 (116.67)	780	50 (1.66)	—	—	—
Cephaloridine	5 (100)	1000	5 (2.27)	15 (50)	950	15 (0.5)	<2 (<1.8)	>1,000	<2 (<1.43)
Cephalothin	—	—	—	5 (1.6)	120	30 (1)	0.2 (0.18)	900	<1 (<0.71)
Cefotaxime	—	—	—	ND	—	—	—	—	—
Cefepime	ND	—	—	2 (6.67)	3000	1 (0.03)	—	—	—
Cefpirome	ND	—	—	1 (3.33)	620	2 (0.07)	—	—	—
Cefuroxime	—	—	—	—	—	—	—	—	—
Ceftazidime	20 (400)	2500	10 (4.55)	ND	—	—	—	—	—
Imipenem	0.1 (2)	6.5	15 (6.82)	0.5 (1.67)	1	360 (12)	0.1 (0.09)	20	5 (3.57)
Meropenem	ND	—	—	—	—	—	—	—	—
Aztreonam	—	—	—	—	—	—	—	—	—

β-Lactam	OXA-51 k_{cat} (%)	OXA-51 K_m (μM)	OXA-51 k_{cat}/K_m (%)	OXA-58 k_{cat} [(s^{-1}) (%)]	OXA-58 K_m (μM)	OXA-58 k_{cat}/K_m [mM^{-1} s^{-1} (%)]	OXA-60 k_{cat} [s^{-1} (%)]	OXA-60 K_m (μM)	OXA-60 k_{cat}/K_m [mM^{-1} s^{-1} (%)]
Benzylpenicillin	—	—	—	5.5 (100)	50	110 (100)	420 (100)	40	11,350 (100)
Ampicillin	100	24	100	1 (18.2)	130	8 (7)	—	—	—
Piperacillin	—	—	—	—	50	50 (48)	<300 (<71)	>2,000	<150 (<1.3)
Carbenicillin	—	—	—	—	—	—	—	—	—
Oxacillin	32.3	531	1.46	1.5 (27.3)	70	2 (2)	130 (<31)	>2,000	<65 (<0.6)
Cloxacillin	2.1	129	0.4	—	—	—	—	—	—
Cephaloridine	10	—	3.1	—	—	—	<0.5(<0.12)	>1,000	<0.5 (<0.004)
Cephalothin	—	—	—	0.1 (1.8)	150	1 (1)	<0.01 (<0.002)	—	—
Cefotaxime	ND	—	—	<0.01 (<0.001)	—	—	<0.01 (<0.002)	—	—
Cefepime	—	—	—	<0.01 (<0.001)	—	—	—	—	—
Cefpirome	—	—	—	0.1 (1.8)	200	0.5 (0.5)	—	—	—
Cefuroxime	ND	—	—		—	—	—	—	—
Ceftazidime	ND	—	—	<0.01 (<0.001)	—	—	2 (0.48)	1,000	2 (0.02)
Imipenem	3.1	11	6.8	0.1 (1.8)	7.5	13.5 (13)	0.5 (0.12)	2.00	260 (2.3)
Meropenem	ND	—	—	<0.01 (<0.001)	0.075	<0.15 (<0.7)	—	—	—
Aztreonam	—	—	—	<0.01 (<0.001)	—	—	<0.01 (<0.002)	—	—

[a]Data from references 1 and 35. —, not determined or not applicable; ND, hydrolysis not detected.

relative efficiency poorer than for imipenem; meropenem had a high affinity (IC_{50}, 4 nM) but slow turnover. The enzyme was only inhibited by NaCl concentration in the millimolar range. OXA-69, a chromosomally encoded β-lactamase in *A. baumannii*, hydrolyzes β-lactams with a very weak efficiency, and penicillins are the best substrates for this enzyme.

The fourth subgroup is composed (at the moment) of only OXA-58, reported to be active mainly against penicillins and, more surprisingly, against imipenem. In terms of k_{cat}, OXA-58 had values 200- to 300-fold lower than did TEM-1 and OXA-10. For imipenem, k_{cat} was actually lower than for TEM-1 (48) and an *Acinetobacter* transconjugant with OXA-58 required carbapenem MICs at least 16-fold lower than for the original isolate, underlining the role of other mechanism of resistance (e.g., porin loss) to confer a high-level resistance. Nevertheless (and unlike TEM-1), OXA-58 alone could confer some small rise in carbapenem MICs in *Acinetobacter* transconjugants, just as OXA-23 and -40 can.

The last subgroup of homology group VI enzymes comprises OXA-50 and OXA-60 β-lactamases: chromosomal class D types recently described from *P. aeruginosa* and *R. pickettii*, respectively (29, 30). These enzymes are difficult to group as they are intermediate between homology groups I and VI (Table 11.3), but if only the core sequence regions are compared, they seem closer to the latter group (for example, OXA-50 has 65% similarity to OXA-23 and 57.7% similarity to OXA-10 in this region). OXA-50 is expressed constitutively and has a spectrum encompassing penicillins, nitrocefin, cephaloridine, cephalothin, and—more surprisingly—imipenem (Table 11.11); however, expression is weak in the *P. aeruginosa* cell; the enzyme had little effect on susceptibility. OXA-60 is also a narrow-spectrum β-lactamase but, like OXA-50, hydrolyzes imipenem efficiently (Table 11.11). Clavulanate, tazobactam, and sulbactam are weak inhibitors of both enzymes.

CLINICAL IMPORTANCE OF CLASS D β-LACTAMASES

Class D β-lactamases are of modest clinical importance compared with either the class A enzymes such the TEM-1, SHV, and CTX-M families or the AmpC (class C) chromosomal β-lactamases.

The classical, and still the commonest, OXA enzymes of *Enterobacteriaceae*—namely, OXA-1, -2, and -3—are narrow-spectrum enzymes, conferring resistance only to anti-gram-negative penicillins and weakly, if at all, to narrow-spectrum cephalosporins. It might be anticipated that heavy use of amoxicillin-clavulanic acid and ampicillin-sulbactam would select for these enzymes, which are rather resistant to inhibition, but there is no evidence for this, and resistance to inhibitor combinations is more often associated with hyperproduction of TEM-related enzymes or with the production of multiple TEM and SHV enzymes. Classical OXA-10 enzyme does have a wider spectrum of activity, raising also the MICs of aztreonam and ceftriaxone, but is mostly (though still rarely) seen in *P. aeruginosa*, where the latter drug, at least, is of little relevance.

Extended-spectrum OXA-10 and OXA-2 mutants cause serious therapeutic problems in infections due to those few *P. aeruginosa* isolates where they occur, conferring resistance to all β-lactams except carbapenems (43). These problems are compounded when, as often applies, the host strain is also multiresistant to fluoroquinolones and aminoglycosides. Nevertheless, the problem should not be exaggerated: *P. aeruginosa* strains with extended-spectrum OXA enzymes are widespread only in Turkey, sporadic in France, and virtually unknown elsewhere. Broad-spectrum resistance to β-lactams in *P. aeruginosa* is much more often associated with multiple chromosomal mutations or, increasingly, with acquired metallo-β-lactamases belonging to class B (43). In Turkey itself, PER-1 β-lactamase, a class A ESBL remote from the TEM, SHV, and CTX-M families, is a more widespread agent of resistance to oxyimino-cephalosporins in *P. aeruginosa* than is any class D ESBL (2, 79).

The OXA carbapenemases of *Acinetobacter* spp. are a greater emerging problem and are the one group of OXA enzymes that is beginning to pose a major clinical problem. There are three reasons for this: (i) they are the most prevalent source of carbapenem resistance in the genus, at least in Europe (13); (ii) some producers cause large outbreaks; and (iii) therapeutic options against *Acinetobacter* already are very narrow (42). Many strains producing these enzymes are resistant to all β-lactams, aminoglycosides, and fluoroquinolones. Consequently, the available therapeutic choices lie between polymyxins, which have significant toxicity and questionable efficacy in the treatment of pneumonia (one of the commonest diseases caused by infection by *Acinetobacter*); sulbactam, which is not active against all producers; and tigecycline, whose efficiency remains to be proven in severe infections among heavily compromised patients.

We are deeply grateful to Hans-Peter Weber for interesting and helpful discussion about the structure of OXA β-lactamases.

References

1. **Afzal-Shah, M., N. Woodford, and D. M. Livermore.** 2001. Characterization of OXA-25, OXA-26, and OXA-27, molecular class D β-lactamases associated with carbapenem resistance in clinical isolates of *Acinetobacter baumannii*. *Antimicrob. Agents Chemother.* **45:**583–588.

2. **Aktas, Z., L. Poirel, M. Salcioglu, P. E. Ozcan, K. Midilli, C. Bal, O. Ang, and P. Nordmann.** 2005. PER-1- and OXA-10-like β-lactamases in ceftazidime-resistant *Pseudomonas aeruginosa* isolates from intensive care unit patients in Istanbul, Turkey. *Clin. Microbiol. Infect.* **11:**193–198.

2a. **Alfredson, D. A., and V. Korolik.** 2005. Isolation and expression of a novel molecular class D beta-lactamase, OXA-61, from *Campylobacter jejuni. Antimicrob. Agents Chemother.* **49:**2515–2518.

3. **Ambler, R. P., A. F. Coulson, J. M. Frere, J. M. Ghuysen, B. Joris, M. Forsman, R. C. Levesque, G. Tiraby, and S. G. Waley.** 1991. A standard numbering scheme for the class A β-lactamases. *Biochem. J.* **276:**269–270.

4. **Ambler, R. P.** 1980. The structure of β-lactamases. *Philos. Trans. R. Soc. Lond. B* **289:**321–331.

5. **Aubert, D., L. Poirel, A. B. Ali, F. W. Goldstein, and P. Nordmann.** 2001. OXA-35 is an OXA-10-related β-lactamase from *Pseudomonas aeruginosa. J. Antimicrob. Chemother.* **48:**717–721.

6. **Aubert, D., L. Poirel, J. Chevalier, S. Leotard, J. M. Pages, and P. Nordmann.** 2001. Oxacillinase-mediated resistance to cefepime and susceptibility to ceftazidime in *Pseudomonas aeruginosa. Antimicrob. Agents Chemother.* **45:**1615–1620.

6a. **Avison, M. B., P. Niumsup, T. R. Walsh, and P. M. Bennett.** 2000. *Aeromonas hydrophila* AmpH and CepH beta-lactamases: derepressed expression in mutants of *Escherichia coli* lacking *creB. J. Antimicrob. Chemother.* **46:**695–702.

7. **Avison, M. B., and A. M. Simm.** 2002. Sequence and genome context analysis of a new molecular class D β-lactamase gene from *Legionella pneumophila. J. Antimicrob. Chemother.* **50:**331–338.

8. **Barlow, M., and B. G. Hall.** 2002. Phylogenetic analysis shows that the OXA β-lactamase genes have been on plasmids for millions of years. *J. Mol. Evol.* **55:**314–321.

9. **Bou, G., A. Oliver, and J. Martinez-Beltran.** 2000. OXA-24, a novel class D β-lactamase with carbapenemase activity in an *Acinetobacter baumannii* clinical strain. *Antimicrob. Agents Chemother.* **44:**1556–1561.

9a. **Brown, S., and S. G. B. Amyes.** 2005. The sequences of seven class D β-lactamases isolated from carbapenem-resistant *Acinetobacter baumannii* from four continents. *Clin. Microb. Infect.* **11:**326–329.

10. **Brown, S., H. K. Young, and S. G. Amyes.** 2005. Characterisation of OXA-51, a novel class D carbapenemase found in genetically unrelated clinical strains of *Acinetobacter baumannii* from Argentina. *Clin. Microbiol. Infect.* **11:** 15–23.

11. **Bryan, L. E., M. S. Shahrabadi, and H. M. van den Elzen.** 1974. Gentamicin resistance in *Pseudomonas aeruginosa*: R-factor-mediated resistance. *Antimicrob. Agents Chemother.* **6:**191–199.

12. **Bush, K., G. A. Jacoby, and A. A. Medeiros.** 1995. A functional classification scheme for β-lactamases and its correlation with molecular structure. *Antimicrob. Agents Chemother.* **39:**1211–1233.

12a. **Coelho, J. M., J. F. Turton, M. E. Kaufmann, J. Glover, N. Woodford, M. Warner, M. F. Palepou, R. Pike, T. L. Pitt, B. C. Patel, and D. M. Livermore.** 2006. Occurrence of carbapenem-resistant *Acinetobacter baumannii* clones at multiple hospitals in London and Southeast England. *J. Clin. Microbiol.* **44:**3623–3627.

12b. **Coelho, J., N. Woodford, M. Afzal-Shah, and D. M. Livermore.** 2006. Occurrence of OXA-58-like carbapenemases in *Acinetobacter* spp. collected over 10 years in three continents. *Antimicrob. Agents Chemother.* **50:**756–758.

13. **Coelho, J., N. Woodford, J. Turton, and D. M. Livermore.** 2004 Multiresistant acinetobacter in the UK: how big a threat? *J. Hosp. Infect.* **58:**167–169.

14. **Couture, F., J. Lachapelle, and R. C. Lévesque.** 1992. Phylogeny of LCR-1 and OXA-5 with class A and class D β-lactamases. *Mol. Microbiol.* **6:**1693–1705.

15. **Dale, J. W., and J. T. Smith.** 1971. The purification and properties of the β-lactamase specified by the resistance factor R-1818 in *Escherichia coli* and *Proteus mirabilis. Biochem. J.* **123:**493–500.

16. **Dale, J. W., and J. T. Smith.** 1976. The dimeric nature of an R-factor mediated β-lactamase. *Biochem. Biophys. Res. Commun.* **68:** 1000–1005.

17. **Danel, F., J. M. Frere, and D. M. Livermore.** 2001. Evidence of dimerisation among class D β-lactamases: kinetics of OXA-14 β-lactamase. *Biochim. Biophys. Acta* **1546:** 132–142.

18. **Danel, F., L. M. Hall, D. Gur, and D. M. Livermore.** 1995. OXA-14, another extended-spectrum variant of OXA-10 (PSE-2) β-lactamase from *Pseudomonas aeruginosa. Antimicrob. Agents Chemother.* **39:**1881–1884.

19. **Danel, F., L. M. Hall, and D. M. Livermore.** 1999. Laboratory mutants of OXA-10 β-lactamase giving ceftazidime resistance in *Pseudomonas aeruginosa. J. Antimicrob. Chemother.* **43:**339–344.

20. **Danel, F., L. M. Hall, B. Duke, D. Gur, and D. M. Livermore.** 1999. OXA-17, a further extended-spectrum variant of OXA-10 β-lactamase, isolated from *Pseudomonas aeruginosa. Antimicrob. Agents Chemother.* **43:**1362–1366.

21. **Danel, F., L. M. Hall, D. Gur, and D. M. Livermore.** 1997. OXA-15, an extended-spectrum variant of OXA-2 β-lactamase, isolated from a *Pseudomonas aeruginosa* strain. *Antimicrob. Agents Chemother.* **41:**785–790.

22. **Danel, F., L. M. Hall, D. Gur, and D. M. Livermore.** 1998. OXA-16, a further extended-spectrum variant of OXA-10 β-lactamase, from two *Pseudomonas aeruginosa* isolates. *Antimicrob. Agents Chemother.* **42:**3117–3122.

23. **Danel, F., M. Paetzel, N. C. J. Strynadka, and M. G. P. Page.** 2001. Effect of divalent metal cations on the dimerization of OXA-10 and -14 class D β-lactamases from *Pseudomonas aeruginosa. Biochemistry* **40:**9412–9420.

24. **Da Silva, G. J., S. Quinteira, E. Bertolo, J. C. Sousa, L. Gallego, A. Duarte, and L. Peixe.** 2004. Long-term dissemination of an OXA-40 carbapenemase-producing *Acinetobacter baumannii* clone in the Iberian Peninsula. *J. Antimicrob. Chemother.* **54:**255–258.

25. **Datta, N., and P. Kontomichalou.** 1965. Penicillinase synthesis controlled by infectious R factors in Enterobacteriaceae. *Nature* **208:**239–241.

26. **Donald, H. M., W. Scaife, S. G. B. Amyes, and H. K. Young.** 2000. Sequence analysis of ARI-1, a novel OXA β-lactamase, responsible for imipenem resistance in *Acinetobacter baumannii* 6B92. *Antimicrob. Agents Chemother.* **44:**196–199.

27. **Dubus, A., S. Normark, M. Kania, and M. G. Page.** 1995. Role of asparagine 152 in catalysis of β-lactam hydrolysis by *Escherichia coli* AmpC β-lactamase studied by site-directed mutagenesis. *Biochemistry* **34:**7757–7764.

28. **Franceschini, N., L. Boschi, S. Pollini, R. Herman, M. Perilli, M. Galleni, J. M. Frere, G. Amicosante, and G. M. Rossolini.** 2001. Characterization of OXA-29 from *Legionella (Fluoribacter) gormanii*: molecular class D β-lactamase with unusual properties. *Antimicrob. Agents Chemother.* **45**:3509–3516.

29. **Girlich, D., T. Naas, and P. Nordmann.** 2004. Biochemical characterization of the naturally occurring oxacillinase OXA-50 of *Pseudomonas aeruginosa. Antimicrob. Agents Chemother.* **48**:2043–2048.

30. **Girlich, D., T. Naas, and P. Nordmann.** 2004. OXA-60, a chromosomal, inducible, and imipenem-hydrolyzing class D β-lactamase from *Ralstonia pickettii. Antimicrob. Agents Chemother.* **48**:4217–4225.

30a. **Giuliani, F., J. D. Docquier, M. L. Riccio, L. Pagani, and G. M. Rossolini.** 2005. OXA-46, a new class D beta-lactamase of narrow substrate specificity encoded by a blaVIM-1-containing integron from a *Pseudomonas aeruginosa* clinical isolate. *Antimicrob. Agents Chemother.* **49**:1973–1980.

31. **Golemi, D., L. Maveyraud, S. Vakulenko, J.-P. Samama, and S. Mobashery.** 2001. Critical involvement of a carbamylated Lys in catalytic function of class D β-lactamases. *Proc. Natl. Acad. Sci. USA* **98**:14280–14285.

32. **Hall, L. M., D. M. Livermore, D. Gur, M. Akova, and H. E. Akalin.** 1993. OXA-11, an extended-spectrum variant of OXA-10 (PSE-2) β-lactamase from *Pseudomonas aeruginosa. Antimicrob. Agents Chemother.* **37**:1637–1644.

33. **Heinze-Krauss, I., P. Angehrn, R. L. Charnas, K. Gubernator, E. M. Gutknecht, C. Hubschwerlen, M. Kania, C. Oefner, M. G. Page, S. Sogabe, J. L. Specklin, and F. Winkler.** 1998. Structure-based design of β-lactamase inhibitors. 1. Synthesis and evaluation of bridged monobactams. *J. Med. Chem.* **41**:3961–3971.

34. **Heritier, C., L. Poirel, and P. Nordmann.** 2004. Genetic and biochemical characterization of a chromosome-encoded carbapenem-hydrolyzing ambler class D β-lactamase from *Shewanella algae. Antimicrob. Agents Chemother.* **48**:1670–1675.

35. **Heritier, C., L. Poirel, D. Aubert, and P. Nordmann.** 2003. Genetic and functional analysis of the chromosome-encoded carbapenem-hydrolyzing oxacillinase OXA-40 of *Acinetobacter baumannii. Antimicrob. Agents Chemother.* **47**:268–273.

35a. **Heritier, C., L. Poirel, P. E. Fournier, J. M. Claverie, D. Raoult, and P. Nordmann.** 2005. Characterization of the naturally occurring oxacillinase of *Acinetobacter baumannii. Antimicrob. Agents Chemother.* **49**:4174–4179.

35b. **Heritier, C., L. Poirel, T. Lambert, and P. Nordmann.** 2005. Contribution of acquired carbapenem-hydrolyzing oxacillinases to carbapenem resistance in *Acinetobacter baumannii. Antimicrob. Agents Chemother.* **49**:3198–3202.

36. **Huovinen, S., P. Huovinen, and G. A. Jacoby.** 1988. Detection of plasmid-mediated β-lactamases with DNA probes. *Antimicrob. Agents Chemother.* **32**:175–179.

37. **Joris, B., P. Ledent, O. Dideberg, E. Fonze, J. Lamotte-Brasseur, J. A. Kelly, J. M. Ghuysen, and J. M. Frere.** 1991. Comparison of the sequences of class A β-lactamases and of the secondary structure elements of penicillin-recognizing proteins. *Antimicrob. Agents Chemother.* **35**:2294–2301.

37a. **Keith, K. E., P. C. Oyston, B. Crossett, N. F. Fairweather, R. W. Titball, T. R. Walsh, and K. A. Brown.** 2005. Functional characterization of OXA-57, a class D beta-lactamase from *Burkholderia pseudomallei. Antimicrob. Agents Chemother.* **49**:1639–1641.

38. **Kerff, F., E. Fonze, and P. Charlier.** 2002. *Structure of the OXA 2 β-Lactamase.* Eighth β-Lactamase Workshop, Holy Island, Northumberland, United Kingdom.

39. **Knox, J. R., and P. C. Moews.** 2004. β-Lactamase of *Bacillus licheniformis* 749/C. Refinement at 2 A resolution and analysis of hydration. *Antimicrob. Agents Chemother.* **48**:2043–2048.

40. **Ledent, P., X. Raquet, B. Joris, J. Van Beeumen, and J.-M. Frère.** 1993. A comparative study of class-D β-lactamases. *Biochem. J.* **292**:555–562.

41. **Ledent, P., and J.-M. Frère.** 1993. Substrate-induced inactivation of the OXA2 β-lactamase. *Biochem. J.* **295**:871–878.

42. **Livermore, D. M.** 2003. The threat from the pink corner. *Ann. Med.* **35**:226–234.

43. **Livermore, D. M.** 2002. Multiple mechanisms of antimicrobial resistance in *Pseudomonas aeruginosa*: our worst nightmare? *Clin. Infect. Dis.* **34**:634–640.

44. **Livermore, D. M.** 1995. β-Lactamases in laboratory and clinical resistance. *Clin. Microbiol. Rev.* **8**:557–584.

45. **Livermore, D. M., J. P. Maskell, and J. D. Williams.** 1984. Detection of PSE-2 β-lactamase in enterobacteria. *Antimicrob. Agents Chemother.* **25**:268–272.

46. **Livermore, D. M., T. G. Winstanley, and K. P. Shannon.** 2001. Interpretative reading: recognizing the unusual and inferring resistance mechanisms from resistance phenotypes. *J. Antimicrob. Chemother.* **48**(Suppl. 1):87–102.

47. **Lopez-Otsoa, F., L. Gallego, K. J. Towner, L. Tysall, N. Woodford, and D. M. Livermore.** 2002. Endemic carbapenem resistance associated with OXA-40 carbapenemase among *Acinetobacter baumannii* isolates from a hospital in northern Spain. *J. Clin. Microbiol.* **40**:4741–4743.

48. **Matagne, A., J. Lamotte-Brasseur, and J. M. Frere.** 1998. Catalytic properties of class A β-lactamases: efficiency and diversity. *Biochem. J.* **330**:581–598.

49. **Matthew, M.** 1978. Properties of the β-lactamase specified by the Pseudomonas plasmid R151. *FEMS Microbiol. Lett.* **4**:241–244.

50. **Maveyraud, L., D. Golemi, L. P. Kotra, S. Tranier, S. Vakulenko, S. Mobashery, and J.-P. Samama.** 2000. Insights into class D β-lactamases are revealed by the crystal structure of the OXA10 enzyme from *Pseudomonas aeruginosa. Structure* **8**:1289–1298.

51. **Maveyraud, L., D. Golemi-Kotra, A. Ishiwata, O. Meroueh, S. Mobashery, and J.-P. Samama.** 2002. High-resolution x-ray structure of an acyl-enzyme species for the class D OXA-10 β-lactamase. *J. Am. Chem. Soc.* **124**:2461–2465.

52. **Medeiros, A. A., M. Cohenford, and G. A. Jacoby.** 1985. Five novel plasmid-determined β-lactamases. *Antimicrob. Agents Chemother.* **27**:715–719.

52a. **Merkier, A. K., and D. Centron.** 2006. bla(OXA-51)-type beta-lactamase genes are ubiquitous and vary within a strain in *Acinetobacter baumannii. Int. J. Antimicrob. Agents* **28**:110–113.

53. **Monaghan, C., S. Holland, and J. W. Dale.** 1982. The interaction of anthraquinone dyes with the plasmid-mediated OXA-2 β-lactamase. *Biochem. J.* **205**:413–417.

54. **Mugnier, P., I. Podglajen, F. W. Goldstein, and E. Collatz.** 1998. Carbapenems as inhibitors of OXA-13, a novel, integron-encoded β-lactamase in *Pseudomonas aeruginosa*. *Microbiology* **144**(Pt 4):1021–1031.

55. **Mulvey, M. R., D. A. Boyd, L. Baker, O. Mykytczuk, E. M. Reis, M. D. Asensi, D. P. Rodrigues, and L. K. Ng.** 2004. Characterization of a *Salmonella enterica* serovar Agona strain harbouring a class 1 integron containing novel OXA-type β-lactamase (bla_{OXA-53}) and 6′-N-aminoglycoside acetyltransferase genes [aac(6′)-I30]. *J. Antimicrob. Chemother.* **54**:354–359.

56. **Naas, T., and P. Nordmann.** 1999. OXA-type β-lactamases. *Curr. Pharm. Des.* **5**:865–879.

57. **Navia, M. M., J. Ruiz, and J. Vila.** 2002. Characterization of an integron carrying a new class D β-lactamase (OXA-37) in *Acinetobacter baumannii*. *Microb. Drug Resist.* **8**:261–265.

58. **Niumsup, P., and V. Wuthiekanun.** 2002. Cloning of the class D β-lactamase gene from *Burkholderia pseudomallei* and studies on its expression in ceftazidime-susceptible and -resistant strains. *J. Antimicrob. Chemother.* **50**:445–455.

59. **Nordmann, P., L. Poirel, M. Kubina, A. Casetta, and T. Naas.** 2000. Biochemical-genetic characterization and distribution of OXA-22, a chromosomal and inducible class D β-lactamase from *Ralstonia (Pseudomonas) pickettii*. *Antimicrob. Agents Chemother.* **44**:2201–2204.

60. **Oefner, C., A. D'Arcy, J. J. Daly, K. Gubernator, R. L. Charnas, I. Heinze, C. Hubschwerlen, and F. K. Winkler.** 1990. Refined crystal structure of β-lactamase from *Citrobacter freundii* indicates a mechanism for β-lactam hydrolysis. *Nature* **343**:284–288.

61. **Osuna, J., H. Viadiu, A. L. Fink, and X. Soberon.** 1995. Substitution of Asp for Asn at position 132 in the active site of TEM β-lactamase. Activity toward different substrates and effects of neighboring residues. *J. Biol. Chem.* **270**:775–780.

62. **Paetzel, M., F. Danel, L. Castro, S. C. Mosimann, M. G. P. Page, and N. C. J. Strynadka.** 2000. Crystal structure of the class D β-lactamase OXA-10. *Nat. Struct. Biol.* **7**:918–925.

63. **Pernot, L., F. Frenois, T. Rybkine, G. L'Hermite, S. Petrella, J. Delettre, V. Jarlier, E. Collatz, and W. Sougakoff.** 2001. Crystal structures of the class D β-lactamase OXA-13 in the native form and in complex with meropenem. *J. Mol. Biol.* **310**:859–874.

64. **Philippon, L. N., T. Naas, A.-T. Bouthors, V. Barakett, and P. Nordmann.** 1997. OXA-18, a class D clavulanic acid-inhibited extended-spectrum β-lactamase from *Pseudomonas aeruginosa*. *Antimicrob. Agents Chemother.* **41**:2188–2195.

65. **Poirel, L., C. Heritier, and P. Nordmann.** 2004. Chromosome-encoded ambler class D β-lactamase of *Shewanella oneidensis* as a progenitor of carbapenem-hydrolyzing oxacillinase. *Antimicrob. Agents Chemother.* **48**:348–351.

66. **Poirel, L., C. Heritier, V. Tolun, and P. Nordmann.** 2004. Emergence of oxacillinase-mediated resistance to imipenem in *Klebsiella pneumoniae*. *Antimicrob. Agents Chemother.* **48**:15–22.

67. **Poirel, L., D. Girlich, T. Naas, and P. Nordmann.** 2001. OXA-28, an extended-spectrum variant of OXA-10 β-lactamase from *Pseudomonas aeruginosa* and its plasmid- and integron-located gene. *Antimicrob. Agents Chemother.* **45**:447–453.

68. **Poirel, L., P. Gerome, C. De Champs, J. Stephanazzi, T. Naas, and P. Nordmann.** 2002. Integron-located oxa-32 gene cassette encoding an extended-spectrum variant of OXA-2 β-lactamase from *Pseudomonas aeruginosa*. *Antimicrob. Agents Chemother.* **46**:566–569.

69. **Poirel, L., S. Marque, C. Heritier, C. Segonds, G. Chabanon, and P. Nordmann.** 2005. OXA-58, a novel class D β-lactamase involved in resistance to carbapenems in *Acinetobacter baumannii*. *Antimicrob. Agents Chemother.* **49**:202–208.

70. **Poirel, L., M. Magalhaes, M. Lopes, and P. Nordmann.** 2004. Molecular analysis of metallo-β-lactamase gene bla(SPM-1)-surrounding sequences from disseminated *Pseudomonas aeruginosa* isolates in Recife, Brazil. *Antimicrob. Agents Chemother.* **48**:1406–1409.

71. **Rasmussen, B. A., D. Keeney, Y. Yang, and K. Bush.** 1994. Cloning and expression of a cloxacillin-hydrolyzing enzyme and a cephalosporinase from *Aeromonas sobria* AER 14M in *Escherichia coli*: requirement for an *E. coli* chromosomal mutation for efficient expression of the class D enzyme. *Antimicrob. Agents Chemother.* **38**:2078–2085.

72. **Sanschagrin, F., F. Couture, and R. C. Levesque.** 1995. Primary structure of OXA-3 and phylogeny of oxacillin-hydrolyzing class D β-lactamases. *Antimicrob. Agents Chemother.* **39**:887–893.

73. **Scaife, W., H. K. Young, R. H. Paton, and S. G. Amyes.** 1995. Transferable imipenem-resistance in Acinetobacter species from a clinical source. *J. Antimicrob. Chemother.* **36**:585–586.

73a. **Schneider, I., A. M. Queenan, and A. Bauernfeind.** 2006. Novel carbapenem-hydrolyzing oxacillinase OXA-62 from *Pandoraea pnomenusa*. *Antimicrob. Agents Chemother.* **50**:1330–1335.

74. **Simpson, I. N., S. J. Plested, M. J. Budin-Jones, J. Lees, R. W. Hedges, and G. A. Jacoby.** 1983. Characterisation of a novel plasmid-mediated β-lactamase and its contribution to β-lactam resistance in *Pseudomonas aeruginosa*. *FEMS Microbiol. Lett.* **19**:23–27.

75. **Sun, T., M. Nukaga, K. Mayama, E. H. Braswell, and J. R. Knox.** 2003. Comparison of β-lactamases of classes A and D: 1.5-A crystallographic structure of the class D OXA-1 oxacillinase. *Protein Sci.* **12**:82–91.

76. **Sun, T., M. Nukaga, K. Mayama, G. V. Crichlow, A. P. Kuzin, and J. R. Knox.** 2001. Crystallization and preliminary X-ray study of OXA-1, a class D β-lactamase. *Acta Crystallogr. D Biol. Crystallogr.* **57**:1912–1914.

77. **Sykes, R. B., and M. Matthew.** 1976. The β-lactamases of gram-negative bacteria and their role in resistance to β-lactam antibiotics. *J. Antimicrob. Chemother.* **2**:115–157.

78. **Toleman, M. A., K. Rolston, R. N. Jones, and T. R. Walsh.** 2003. Molecular and biochemical characterization of OXA-45, an extended-spectrum class 2d' β-lactamase in *Pseudomonas aeruginosa*. *Antimicrob. Agents Chemother.* **47**:2859–2863.

78a. **Tsakris, A., A. Ikonomidis, N. Spanakis, S. Pournaras, and K. Bethimouti.** 2007. Identification of a novel blaOXA-51 variant, blaOXA-92, from a clinical isolate of *Acinetobacter baumannii*. *Clin. Microb. Infect.* **13**:348–349.

78b. **Turton, J. F., M. E. Ward, N. Woodford, M. E. Kaufmann, R. Pike, D. M. Livermore, and T. L. Pitt.** 2006. The role of IS*Aba1* in expression of OXA carbapenemase genes in

Acinetobacter baumannii. *FEMS Microbiol. Lett.* **258:** 72–77.

79. **Vahaboglu, H., F. Coskunkan, O. Tansel, R. Ozturk, N. Sahin, I. Koksal, B. Kocazeybek, M. Tatman-Otkun, H. Leblebicioglu, M. A. Ozinel, H. Akalin, S. Kocagoz, and V. Korten.** 2001. Clinical importance of extended-spectrum β-lactamase (PER-1-type)-producing *Acinetobacter* spp. and *Pseudomonas aeruginosa* strains. *J. Med. Microbiol.* **50:** 642–645.

80. **Vahaboglu, H., R. Ozturk, H. Akbal, S. Saribas, O. Tansel, and F. Coskunkan.** 1998. Practical approach for detection and identification of OXA-10-derived ceftazidime-hydrolyzing extended-spectrum β-lactamases. *J. Clin. Microbiol.* **36:**827–829.

80a. **Voha, C., J. D. Docquier, G. M. Rossolini, and T. Fosse.** 2006. Genetic and biochemical characterization of FUS-1 (OXA-85), a narrow-spectrum class D beta-lactamase from *Fusobacterium nucleatum* subsp. *polymorphum*. *Antimicrob. Agents Chemother.* **50:**2673–2679.

81. **Waley, S. G.** 1991. The kinetics of substrate-induced inactivation. *Biochem. J.* **279:**87–94.

81a. **Walther-Rasmussen, J., and N. Høiby.** 2006. OXA-type carbapenemases. *J. Antimicrob. Chemother.* **57:**373–383.

81b. **Woodford, N., M. J. Ellington, J. M. Coelho, J. F. Turton, M. E. Ward, S. Brown, S. G. Amyes, and D. M. Livermore.** 2006. Multiplex PCR for genes encoding prevalent OXA carbapenemases in *Acinetobacter* spp. *Int. J. Antimicrob. Agents* **27:**351–353.

82. **Yang, Y., and K. Bush.** 1995. Oxacillin hydrolysis by the LCR-1 β-lactamase. *Antimicrob. Agents Chemother.* **39:** 1209.

83. **Zhu, Y. F., I. H. Curran, B. Joris, J. M. Ghuysen, and J. O. Lampen.** 1990. Identification of BlaR, the signal transducer for β-lactamase production in *Bacillus licheniformis*, as a penicillin-binding protein with strong homology to the OXA-2 β-lactamase (class D) of *Salmonella typhimurium*. *J. Bacteriol.* **172:**1137–1141.

Enzyme-Mediated Resistance to Antibiotics: Mechanisms, Dissemination, and Prospects for Inhibition
Edited by Robert A. Bonomo and Marcelo E. Tolmasky

Moreno Galleni
Jean-Marie Frère

12 Kinetics of β-Lactamases and Penicillin-Binding Proteins

A BRIEF SURVEY OF ENZYME KINETICS

The discovery of novel β-lactamases and penicillin-binding proteins (PBPs) often requires kinetic characterization. As such, the rate at which a β-lactamase hydrolyzes a β-lactam is influenced by several factors. The first is concentration of β-lactam, which is designated [S] and is expressed in units of molarity. The second is temperature. As the temperature rises, molecular motion, and hence collisions between β-lactamase and β-lactam, and the rates of interconversion of intermediates increase. Usually, temperature is held constant in studies examining the hydrolytic properties of β-lactamases. As β-lactamases are proteins, there is an upper limit beyond which they become denatured and inactive. The third factor is the presence of inhibitors. β-Lactamase inhibitors are clinically used to hinder the activity of the β-lactamase. The last is pH: the charge of active-site groups and the conformation of a protein are influenced by pH, and enzyme activity is crucially dependent on both these factors.

The study of the rate at which an enzyme works is called enzyme kinetics. The equations of enzyme kinetics are conceptual tools that allow us to interpret quantitative measurements of enzyme activity.

Let us examine enzyme kinetics as a function of the concentration of substrate [S] available to the enzyme. A simple enzyme-catalyzed reaction can be described by the following scheme:

$$E + S \underset{k_{-1}}{\overset{k_{+1}}{\Leftrightarrow}} ES \xrightarrow{k_{+2}} ES' \xrightarrow{k_{+3}} ES'' \rightarrow \rightarrow \rightarrow \xrightarrow{k_{+n}} E + P$$

Clearly, all steps can be reversible and involve kinetic constants k_{-2}, k_{-3}, ... and k_{-n}, but, for our purpose, it is sufficient to assume that the only significantly reversible step is the first one. More complex, branched pathways can also occur and they are briefly discussed below.

Steady-State Conditions

When the initial substrate concentration (S_0) is much larger than the initial enzyme concentration (E_0), a steady state is rapidly established. During this steady state, the concentrations of the intermediates ES, ES′, and ES″ remain constant and the rate of substrate disappearance is equal to that of product formation. The steady state is characterized by a linear accumulation of product (or substrate disappearance) versus time. It is also generally assumed that the initial substrate concentration does not significantly decrease during the time course of the experiment. In consequence, the steady-state rate is often referred to as the initial rate.

Moreno Galleni and Jean-Marie Frère, Center for Protein Engineering, University of Liege, Institut de Chimie B6, Sart Tilman, B-4000 Liège, Belgium.

The hyperbolic Henri-Michaelis equation characterizes the initial rate v (sometimes referred to as v_0):

$$v = VS_0/(K_m + S_0) \tag{I-1}$$

where V is the maximum rate, observed when $S_0 >> K_m$ (substrate saturation). It depends on the values of all the constants in the pathway (with the exception of k_{+1} and k_{-1}) which precede an eventual strongly rate-limiting step.

The catalytic constant, k_{cat}, is defined as V/E_0. It is a first-order rate constant (expressed in $second^{-1}$) and represents the number of substrate molecules that one enzyme molecule transforms per second at substrate saturation (see the Appendix about the order of reactions). K_m is the Michaelis constant. It represents the substrate concentration for which $v = V/2$.

Under these conditions, the sum of the concentrations of all the intermediates is equal to 50% of E_0. K_m and V can be determined by directly fitting the experimental results to equation I-1 (Fig. 12.1A). Alternatively, linearized equations can be utilized. Of these, the Hanes-Woolf equation is probably the best:

$$S_0/v = (K_m + S_0)/V \tag{I-2}$$

A plot of S_0/v versus S_0 (Fig. 12.1B) yields 1/V (slope) and K_m/V (intercept with the ordinate).

The often-used Lineweaver-Burk or double reciprocal plot ($1/v$ versus $1/S_0$) is considered to be strongly biased and should be avoided.

When $S_0 << K_m$, equation I-1 becomes $v = VS_0/K_m$ and the reaction is first order versus S. V/K_m is thus a pseudo-first-order rate constant because it is proportional to E_0. In turn, k_{cat}/K_m is a second-order rate constant expressed in per molar concentration per second. Both V and V/K_m are first-order versus E_0.

In practice, the different S_0 values should range from $K_m/4$ to $4\ K_m$. If all the S_0 values are well below K_m, only k_{cat}/K_m can be accurately determined. Conversely, if the S_0 values are too high, only k_{cat} can be derived.

It is sometimes erroneously assumed that if an enzymatic system obeys the Henri-Michaelis equation, it is adequately represented by the simplest scheme involving one single intermediate ES. This is clearly not true since the equation is valid whatever the number of intermediates. The number of steps in the reaction pathway determines the values of k_{cat} and K_m as functions of those of the individual constants k_{+i} and k_{-i}. An example with two intermediates is developed below.

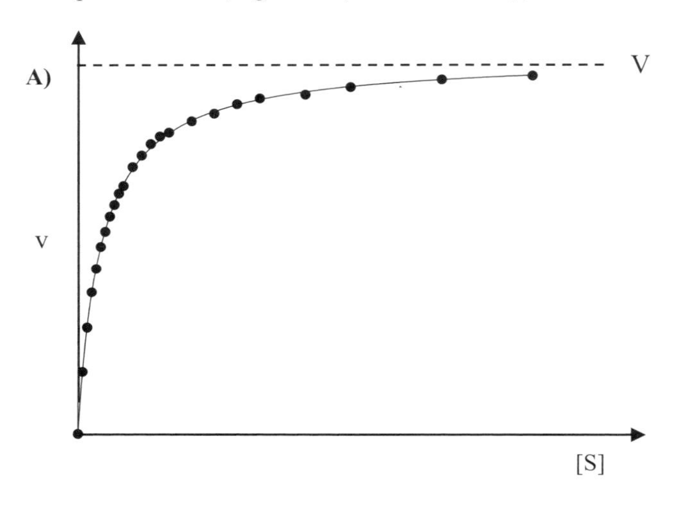

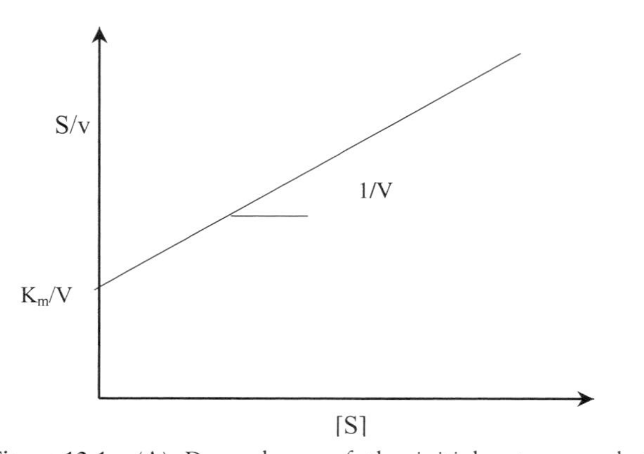

Figure 12.1 (A) Dependence of the initial rate on substrate concentration in the Henri-Michaelis-Menten model. (B) Hanes-Woolf plot for the determination of the kinetic parameters K_m and V.

Integrated Equation

When the complete time course of the reaction is monitored, and if the enzyme remains perfectly stable throughout the experiment and no inhibition by-product or excess substrate occurs, the Henri-Michaelis equation can be integrated, yielding equation I-3:

$$(1/t)\ \ln[S_0/(S_0 - P_t)] = (1/K_m)[V - P_t/t] \tag{I-3}$$

where P_t is the concentration of product at time t.

A plot of $(1/t)\ \ln[S_0/(S_0 - P_t)]$ versus P_t/t yields a line of slope $-1/K_m$ which intercepts the ordinate at V/K_m (Fig. 12.2). Both V and K_m can thus be obtained from a single experiment, but it is always necessary to perform several experiments with different S_0 values which should all be above the K_m value (but not too much, a range of two to five times the K_m value being advisable).

Pre-Steady-State or Transient-State Conditions

Steady-state measurements do not allow the determination of the individual values of all the kinetic constants. For example, in the simplest possible mechanism where there is only one intermediate ES, $k_{cat} = k_2$ but $K_m = (k_{-1} + k_2)/k_{+1}$ and k_{+1} and k_{-1} cannot be derived. Additional information is thus needed.

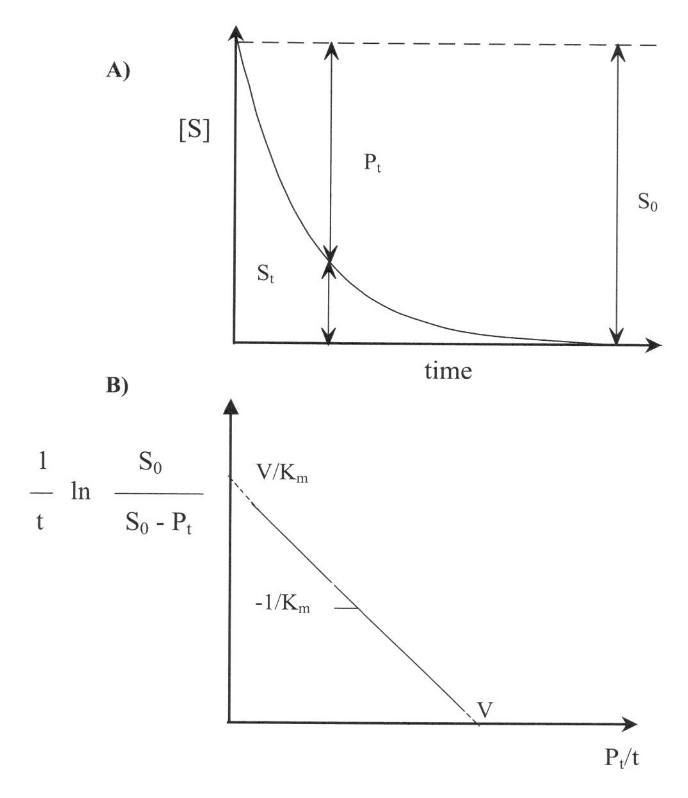

Figure 12.2 Complete time course of the hydrolysis of a substrate. (A) Decrease of the substrate concentration versus time. (B) Plot of $1/t$ ($\ln S_0/S_0 - P$) versus P/t.

Before the steady state is established, the concentrations of the intermediates progressively increase and the rate of substrate disappearance is larger than that of product formation. Measuring the rates of intermediate formation, of substrate disappearance, or of product formation often requires the utilization of rapid kinetic techniques such as stopped-flow and can supply additional information concerning the values of the kinetic constants.

In the simplest case again, and when $S_0 \gg E_0$ (the S_0/E_0 ratio can be as low as 5, but it is then advisable to utilize only the first 30 to 40% of the curve), the rate of ES formation is characterized by a pseudo-first-order rate constant:

$$k_a = k_{-1} + k_2 + k_{+1}S_0 \tag{I-4}$$

If one measures k_a for various S_0 values, k_{+1} can be determined from the slope of the linear plot of k_a versus S_0 and $(k_{-1} + k_2)$ from the intercept. By using the k_2 value determined under steady-state conditions, one can thus obtain the three individual rate constants.

Experiments can also be performed under conditions where $E_0 = S_0$ or $E_0 > S_0$. These are called single turnover experiments but are not discussed here.

β-LACTAMASES AND PBPs

Linear Models and General Equations

PBPs and β-lactamases catalyze two types of reactions: (i) the hydrolysis of β-lactams and (ii) the transfer of the acyl moiety of linear substrates of general structure R-CO-X-R′ (where X can be NH, O, or S) on an acyl group acceptor HY where Y can be OH (hydrolysis reaction), or OR″ or HN-R″ (transfer reactions). In all cases, the general model 1 applies:

$$E + S \underset{k_{-1}}{\overset{k_{+1}}{\leftrightarrow}} ES \overset{k_2}{\rightarrow} \underset{(+P_2)}{ES^*} \overset{k_3\,(HY)}{\rightarrow} E + P_1 \qquad \text{(model 1)}$$

where E is the enzyme, S is the substrate, ES is the noncovalent Henri-Michaelis complex, ES* is the acyl enzyme, P_2 is the leaving group R′-XH (in the case of linear substrates), and P_1 is the hydrolysis or transfer product (16, 25). Note that in some cases, hydrolysis of the acyl enzyme can occur concomitantly to the transfer onto an adequate acceptor (R″OH or R″-NH_2), resulting in the formation of two different P_1 products, R-COOH and R-CO-OR″ or R-CO-NH-R″. This complicates the analysis and is discussed below in the section devoted to branched pathways.

Steady-State Equations

The steady-state equations are as follows:

$$k_{cat} = k_2k_3/(k_2 + k_3) \tag{1}$$

and

$$K_m = (k_{-1} + k_2)\, k_3/k_{+1}\,(k_2 + k_3) \tag{2}$$

or

$$K_m = k_3K'/(k_2 + k_3) \tag{3}$$

where

$$K' = (k_{-1} + k_2)/k_{+1} \tag{4}$$

The k_{cat}/K_m ratio is equal to k_2/K'. It is easy to deduce that if $k_{-1} \gg k_2$, $K' = k_{-1}/k_{+1} = K$, the dissociation constant of ES.

Transient-State Equations

If the initial concentration of S (S_0) is much larger than that of the enzyme (E_0), the formation of the acyl enzyme usually follows pseudo-first-order kinetics and is characterized by equation 5:

$$(ES^*)/(ES^*)_{SS} = 1 - e^{-k_a t} \tag{5}$$

where $(ES^*)_{SS}$ is the concentration of acyl enzyme at the steady state, k_a is the pseudo-first-order rate constant, and t is the time of contact.

The value of k_a depends upon the S_0 value according to equation 6:

$$k_a = k_3 + k_2S_0/(K' + S_0) \qquad (6)$$

The curve k_a versus S_0 is a rectangular hyperbola which intercepts the ordinate at $k_a = k_3$ (for $S_0 = 0$) and with a horizontal asymptote $k_a = k_3 + k_2$ (for $S_0 \gg K'$).

If S_0 is $\ll K'$, $k_a = k_3 + k_2S_0/K'$ (Fig. 12.3A), from which k_3 can be obtained as the intercept of the line k_a versus S_0 with the ordinate. For larger S_0 values, equation 6 applies (Fig. 12.3B). It can be linearized as $S_0/(k_a - k_3) = (K' + S_0)/k_2$ (Fig. 12.3C), from which the individual values of k_2 and K' can be deduced if k_3 has been determined in a preliminary experiment.

If the excess of substrate is eliminated, the recovery of free enzyme also follows first-order kinetics:

$$(ES^*)/(ES^*)_0 = e^{-k_3t} \qquad (7)$$

where $(ES^*)_0$ is the concentration of ES^* when the excess of S is eliminated. The half-life of ES^* is equal to $0.69/k_3$.

Low k_3 Values: PBPs with β-Lactams, β-Lactamases with Some β-Lactam Substrates

When k_3 is low, the acylation and deacylation reactions can be studied separately. This situation occurs in the interactions between nearly all PBPs and β-lactams and between some β-lactamases and a few substrates, the best examples being the reaction of class C β-lactamases with aztreonam (28).

In these cases, the acyl enzyme can usually be easily isolated, for instance, by gel filtration at low temperature. Several methods are available to measure k_3. If the PBP exhibits a measurable enzymatic activity, the recovery of this activity is monitored after isolation of the acyl enzyme by gel filtration or, more simply, after destruction of the excess of β-lactam by a β-lactamase (24). If a radioactive or fluorescent β-lactam is studied, the disappearance of the enzyme-linked radioactivity or fluorescence is followed, usually by gel electrophoresis (1). In the other cases the free enzyme resulting from the k_3 step is back-titrated by the addition of an excess of radioactively labeled or fluorescent

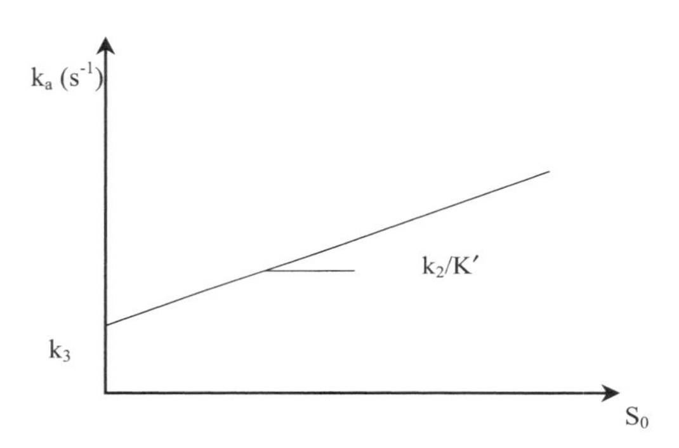

B) General case

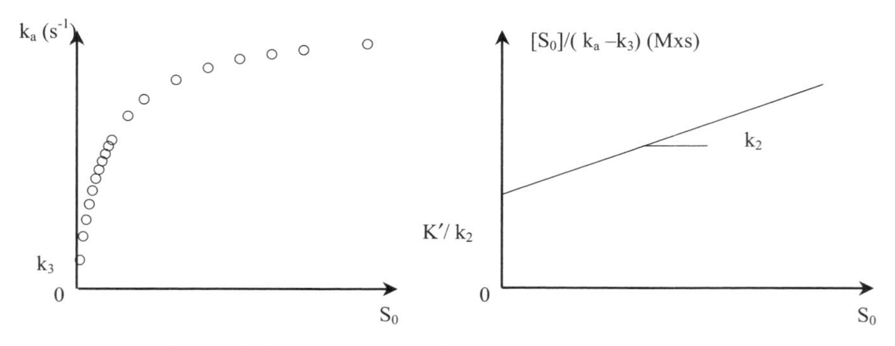

Figure 12.3 Dependence of the rate constant for acyl enzyme accumulation (k_a) on substrate concentration. (A) $K' >> S_0$. (B) General case.

β-lactam and quantification of the labeled protein after gel electrophoresis (41).

High k_3 Values

High k_3 values occur with β-lactamases and a large number of β-lactams and linear substrates and with PBPs and their linear peptide (X = NH), ester (X = O), and thiolester (X = S) substrates. In these cases, the determination of the steady-state parameters does not allow to obtain the individual k_2 and k_3 values and rapid kinetic methods must be utilized. Note that when k_3 is much greater than k_2, its value can never be accurately determined unless similar substrates (same R-CO-group) with much better leaving groups are available.

Acylation of PBPs

Acylation by β-Lactams

The k_3 *Value Is Much Smaller than* k_a

In most cases, the k_3 value is so low (10^{-4} s^{-1} or less) that the acyl enzyme represents close to 100% of the enzyme at the steady state. In equation 5 $(ES^*)_{SS}$ can then be replaced by E_0 and equation 6 then simplifies to

$$k_a = k_2 S_0/(K' + S_0) \quad (8)$$

Now the hyperbola intercepts the ordinate at $k_a = 0$, and the horizontal asymptote is $k_a = k_2$.

Accumulation of acyl enzyme can be easily monitored with a radioactive, chromogenic, or fluorescent β-lactam. Chromogenic compounds such as nitrocefin or CENTA allow a direct and continuous evaluation of the proportion of acyl enzyme (27). The visible spectrum of nitrocefin is significantly modified when the β-lactam ring is opened. With CENTA, the C-3′ leaving group is rapidly expelled upon opening of the β-lactam ring, yielding a characteristic yellow color (Fig. 12.4). Opening of the β-lactam ring of all other cephalosporins and of that of imipenem also results in decreased absorbances at 260 and 300 nm, respectively (27). The drawback of these direct methods is that a sufficiently high enzyme concentration (5 to 10 μM) must be used to obtain a measurable absorbance variation, since the variation of the extinction coefficient is in the 5,000 to 15,000 $M^{-1}cm^{-1}$ range. If one remembers that the S_0 value must be at least five times that of the enzyme, it can be seen that one will be able to follow only rather slow reactions with these methods ($k_2/K' < 100$ $M^{-1}s^{-1}$) if rapid kinetic equipment is not to be utilized. A sufficient quantity of enzyme must also be available to perform these experiments. With a good stopped-flow apparatus, 1,000-fold-higher k_2/K' values can be determined (27).

With radioactively labeled or fluorescent β-lactams, much lower concentrations of enzyme can be used, but the acyl enzyme must be separated from the excess of β-lactam, most often by gel electrophoresis (38). This method is, however, much more time-consuming because each time point requires a separate assay.

In a few cases, the fluorescence emission of the enzyme is quenched upon acylation (22). This also allows the utilization of rather low enzyme concentrations, around 10 to 100 nM. The reaction can then be directly monitored as was done in the first determination of the K' and k_2 constants for a PBP, the *Streptomyces* R61 DD-carboxypeptidase-transpeptidase. The quenching was related to the presence of a tryptophan residue near the enzyme active site. It also occurred upon acylation by a thiolester linear substrate.

Finally, when the enzymatic activity of the PBP can be easily measured, the time course of its disappearance can be monitored by taking samples after various periods of time and determining the residual activity. Alternatively, the reporter substrate method (34) can be used if an adequate substrate is available (see below).

More recently, mass spectroscopic methods have been used to quantify the formation of the acyl enzyme (2, 42).

In many cases, the K' value is too high to be determined. This is, however, not a major problem, since the relevant parameter at physiologically significant β-lactam concentrations is then k_2/K'. Indeed, if S_0 is much smaller than K',

$$k_a = k_2 S_0/K' \quad (9)$$

and k_a versus S_0 is a line extrapolating to the origin.

When the k_2/K' value has been determined for a β-lactam, a simple direct competition method can be used to determine those of all other compounds, as long as no significant degradation of the two acyl enzymes occurs during the time course of the experiment. If one adds the enzyme with a mixture of the reference compound R and the one to be assayed (A), one can show that, when the enzyme is completely acylated,

$$(ER^*)/(EA^*) = (k_2/K')_R\, R_0/(k_2/K')_A A_0 \quad (10)$$

and the decrease in the amount of ER* is thus related to the value of $(k_2/K')_A$, which is the only unknown in the equation (26). The advantage of this method is that by increasing both the R_0 and A_0 values, the reaction can be terminated in a very short time, so that the degradation of both ER* and EA* remains negligible. It is, however, more careful to first determine the k_3 values for both compounds.

If the values of k_2 and K' are known for the reference compound, the direct competition method can also be utilized to obtain the corresponding individual constants for the competing β-lactam, but this is a little more delicate (25a). Here, for the sake of simplicity, we assume that $K' = K$, i.e., that k_{-1} is much greater than k_2, a situation which appears to prevail with PBPs. These experiments can be

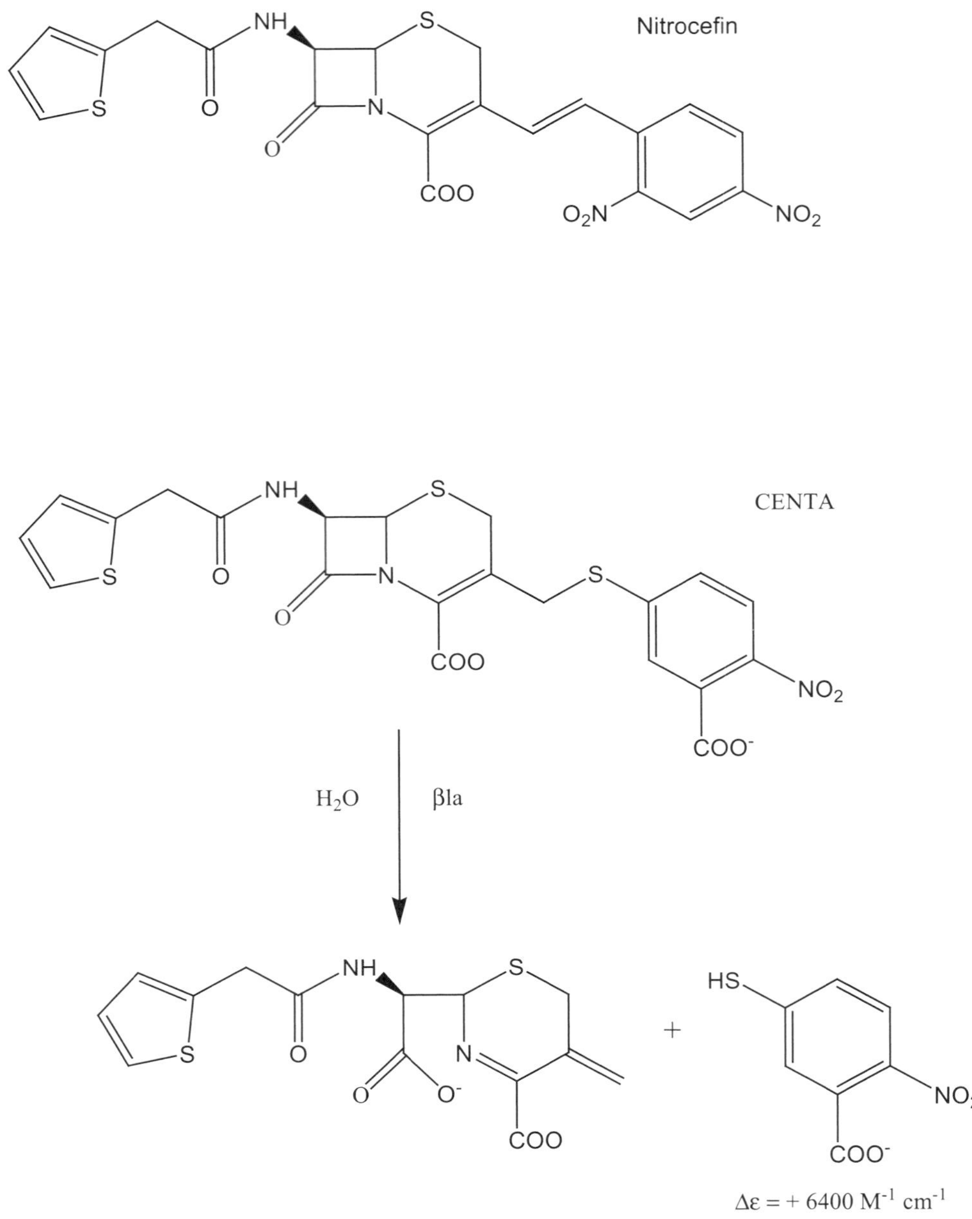

Nitrocefin

CENTA

$\Delta\varepsilon = +\,6400\ M^{-1}\ cm^{-1}$

Figure 12.4 Structures of two chromogenic substrates.

performed if the formation of ER* can be continuously monitored. Nitrocefin is the most practical reference compound, since the accumulation of ER* can be monitored at 480 to 490 nm and no interference is expected with most other β-lactams which do not yield acyl enzymes absorbing in this wavelength range. Interestingly, the disappearance of the free enzyme and the appearance of EA* and ER* occur with the same rate constant, k_a, although at different absolute rates, resulting in the (ER*)/(EA*) ratio defined by equation 10. The value of k_a is given by the following formula:

$$\begin{aligned} k_a &= (k_{2,R}R_0)/(K_R + R_0 + K_R A_0/K_A) \\ &\quad + (k_{2,A}\,A_0)/(K_A + A_0 + K_A R_0/K_R) \\ &= (k_{2,R}K_A R_0 + k_{2,A}K_R A_0)/(K_A R_0 + K_R K_A + K_R A_0) \qquad (11) \end{aligned}$$

and K_A and $k_{2,A}$ can be derived by fitting the k_a values obtained for different A_0 concentrations to the equation.

Active-Site Titration

If conditions can be found under which the acylation is completed before significant degradation of the acyl enzyme occurs ($k_a \gg k_3$), PBPs (and, with very poor substrates, some β-lactamases) can be titrated with a β-lactam. Increasing quantities of β-lactam are added to protein samples and when the quantity of β-lactam is equal to that of the PBP, the enzymatic activity of the latter drops to zero (24). Alternatively, if the PBP has no measurable activity, one can show that it is no longer able to bind a second, labeled β-lactam. Precautions must be taken to make sure that the acylation reaction is completed before the residual activity or binding capacity of the PBP is measured and that no significant deacylation occurs over the time course of the incubation. This sometimes requires that a sufficient concentration of the PBP is utilized since, at equal concentrations of PBP and β-lactam, the reaction is second-order (see the Appendix). Generally, the k_3 values are so low that the method can be easily applied, even with relatively low k_2/K' ratios. For instance, if $k_2/K' = 2{,}000\ M^{-1}s^{-1}$ and the concentrations of PBP and β-lactam are 10 μM, it takes 16 min for the acylation reaction to be 95% complete. The method thus allows a fair estimation of the PBP concentration (with an error by default of <10%) if the half-life of ES* is larger than 320 min, i.e., if k_3 is greater than $3 \times 10^{-5} s^{-1}$. If the k_2/K' value is only 200 $M^{-1}s^{-1}$, a 100 μM concentration of the PBP is needed to perform a similar experiment yielding a good approximation of the PBP concentration. This method is interesting because it allows the estimation of the active PBP concentration in an apparently homogeneous preparation. It is surprising that it is not utilized more often. Note that it is not applicable to "penicillin-resistant" variants or proteins because the concentrations which should be used become impractically high. By contrast, it can also be very useful with β-lactamases if a substrate exhibiting a low k_3 value can be found (Fig. 12.5 and 12.9).

The k_3 *Value Is Not Much Smaller than* k_3

Instances in which the k_3 value is not much smaller than k_a are rather exceptional when interactions between PBPs and β-lactams are studied. This situation, however, can prevail when low k_a values are measured. The general equation 6 then applies, and the concentration of ES* at the steady state is not equal to E_0. This is the reason why it is always much more prudent to determine the k_3 value before starting experiments to measure k_2/K'. No major difficulties are, however, expected if one of the continuous methods is used to estimate the concentration of the acyl enzyme or if a rapid denaturation of the protein is performed before the gel electrophoresis (usually by heating the samples at 100°C in the presence of sodium dodecyl sulfate). In short, the following precautions allow to avoid poor interpretations of the results: first determine k_3; then, make sure to use $(ES^*)_{SS}$ in equation 5; and finally, utilize the general equation 6. In this situation, the reporter substrate method (9, 10) is particularly useful.

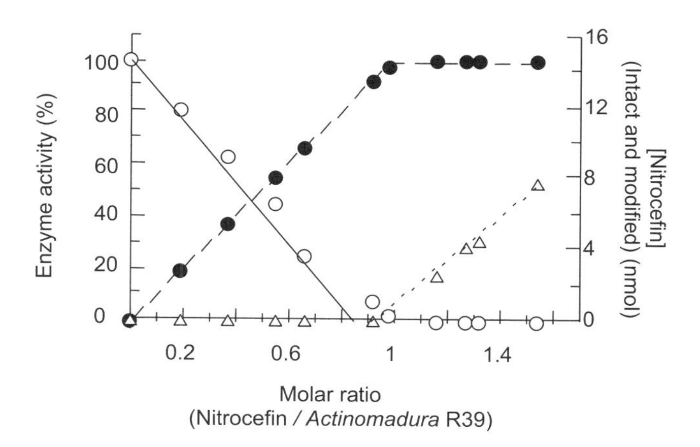

Figure 12.5 Titration of the *Actinomadura* R39 DD-carboxypeptidase with nitrocefin, based on the inhibition of the enzyme activity and on alteration of the antibiotic molecule. Symbols: ○, residual activity; ●, concentration of hydrolyzed nitrocefin; and △, concentration of intact nitrocefin. Adapted from reference 4.

The Reporter Substrate Method

If a substrate whose utilization by the enzyme can be easily monitored is available, the values of the transient-state kinetic constants can be more easily obtained by mixing the enzyme with this substrate and the β-lactam and directly recording the slow "death" of the enzyme by watching the progressive disappearance of its activity (Fig. 12.6). Ideally, this should be done under conditions

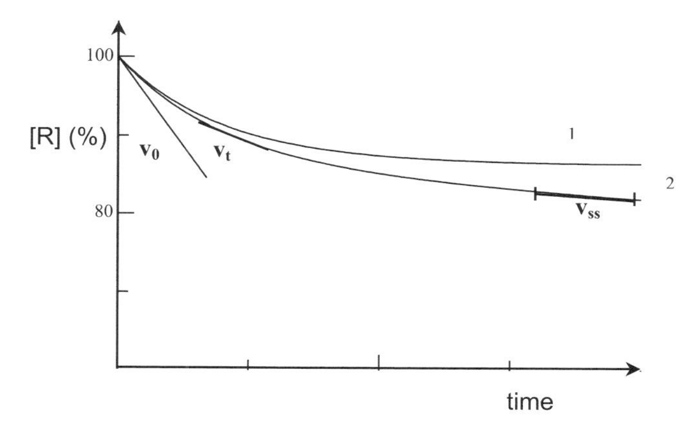

Figure 12.6 The reporter substrate method. Disappearance of the reporter substrate (R) in the absence (v_0) and in the presence (curves 1 and 2) of an inactivator or an alternative substrate for which k_3 is 0 or negligible compared to $k_2S_0/(K'+S_0)$ (curve 1, $v_{ss}=0$) or not (curve 2, $v_{ss}>0$).

where, in the absence of β-lactam, utilization of the substrate remains linear throughout the duration of the experiment. When this is not the case, a correction can be applied (9), but this is outside the scope of this chapter. Let us thus assume that the rate of substrate disappearance remains constant in the absence of β-lactam. In the presence of the latter, this rate, which is proportional to the free enzyme concentration, progressively decreases until a steady state is reached where substrate utilization becomes linear again, but at a reduced rate. It is easy to understand that the decrease of the rate of reporter sustrate utilization (v) obeys an equation similar to equation 6.

$$v - v_{SS} = (v_0 - v_{SS})e^{-k_a t} \quad (12)$$

where v_0 and v_{SS} are the rates at time 0 and after the steady state is established, respectively. The value of k_a can thus be easily deduced from the time course of reporter substrate disappearance (or product appearance). Moreover, the v_{SS}/v_0 ratio supplies additional information. Indeed

$$v_{SS}/v_0 = k_3/k_a \quad (13)$$

Clearly, if k_3 is much smaller than k_a, the activity at the steady state is negligible, and equation 8 can be used rather than equation 6.

The value of k_a is, however, decreased by the presence of the reporter substrate which protects the enzyme from inactivation. The true value of k_a, $(k_a)_0$ is easily derived from equation 14

$$k_a = (k_a)_0(K_{mR})/[K_{mR} + (R)] \quad (14)$$

where (R) and K_{mR} are, respectively, the concentration of the reporter substrate and its K_m value. Note that when k_3 is not negligible, one should make sure that the concentration of β-lactam does not significantly decrease over the time course of the experiment.

This method has been successfully utilized to determine the k_2/K' values for the interaction between various β-lactams and several PBPs, using thiolesters as reporter substrates.

Similarly, the k_3 value can be determined by mixing the completely inactivated enzyme isolated, e.g., by gel filtration, and the reporter substrate (28). The recovery of activity is then characterized by equation 15:

$$v/v_f = 1 - e^{-k_3 t} \quad (15)$$

where v_f is now the rate of reporter substrate utilization after the enzyme has completely "recovered."

Acylation by Substrates

Peptide substrates of PBPs usually exhibit rate-limiting acylation. This means that the k_3 value cannot be determined, that k_{cat} is equal to k_2, and that K_m is equal to K'.

Conversely, with thiolesters deacylation is rate limiting in some cases, so that the values of K', k_2, and k_3 can be obtained by combining transient- and steady-state experiments (33). The R61 enzyme hydrolyzes the thiolester benzoyl-Gly-S-CH_2-COO^- with a k_{cat} of 5 s^{-1}. That k_3 was rate limiting was shown by adding D-alanine, which increased the k_{cat} value up to 100 s^{-1} (see "In the Presence of an Acceptor" under "Branched Pathways" below). The k_2 and K' values were derived by measuring the rate of acyl enzyme formation thanks to the quenching of the enzyme fluorescence under conditions such that k_a was much greater than k_3.

Activity of β-Lactamases

Steady-State Parameters

When the hydrolysis of a β-lactam is studied, the general equations 1 to 4 apply. Determination of the steady-state parameters k_{cat} and K_m alone does not allow one either to derive the individual values of the rate constants or to conclude which of the two intermediates accumulates at the steady state. In practice, k_{cat} and K_m for a set of representative substrates supply a full characterization of the enzyme activity spectrum and sometimes allow one to rationalize the MIC values for a β-lactamase-producing gram-negative bacterial strain (19, 48). In the literature, hydrolysis rates below 1% of that of the best substrate are sometimes considered "insignificant." This is a dangerous assumption since in overproducing strains, very poor activities towards some substrates can nonetheless result in significantly increased MIC values for these compounds (39).

The k_{cat} and K_m values can be derived from initial rate measurements or from complete time courses. With the latter, care must be taken to make sure that no enzyme inactivation and no product inhibition occur. Although product inhibition is rather exceptional, substrate-induced inactivation is often observed (see below). Product inhibition can easily be detected by estimating the effect of product addition on the initial rate (Fig. 12.7) or by adding a new aliquot of substrate to the sample after a first hydrolysis reaction is completed. Moreover, initial rates and complete time courses yield different values for the kinetic parameters (Table 12.1).

The analysis of complete time courses rests on a careful determination of the final value of the measured parameter (most often, absorbance). Any side reaction which might modify this value can thus interfere with the analysis. For example, when the hydrolysis of cephalexin is monitored by UV absorbance measurements, the product slowly decays, resulting in a further decrease of the absorbance (44). It is thus important to record a completely stable final value of the measured parameter. In all cases, it is advisable to confirm the k_{cat} and K_m values derived

Table 12.1 Kinetic parameters for the hydrolysis of biapenem by the metallo-β-lactamase CphA

Derivation method	k_{cat} (s^{-1})	K_m (μM)	k_{cat}/K_m (μM^{-1} s^{-1})
Initial rate measurements	290	170	1.75
Complete time courses	790	860	0.91

[a]Results are typical of product inhibition (reference 29a and Bebrone and Galleni, unpublished).

from complete time course analyses by a few initial rate measurements.

When the K_m values are very low, it becomes impossible to determine them directly because, for instance, the absorbance variation becomes too small. A low K_m value can then be estimated as a K_i by using the tested compound as a competitive inhibitor for a good substrate for which the kinetic parameters are known (28).

In a few cases, inhibition by excess substrate is observed (Fig. 12.8).

The k_2/k_3 Ratio

Accumulation of Acyl Enzyme

When the k_2/k_3 ratio is high, the acyl enzyme accumulates and k_{cat} is equal to k_3. If accumulation of the acyl enzyme can be monitored, for instance, by recording the opening of the β-lactam ring spectrophotometrically (which yields a curve similar to curve 2 in Fig. 12.6), a k_a value can be determined which obeys equation 6 (47). In the best cases, k_a varies with S_0 according to a hyperbola and both k_2 and K' can be derived (Fig. 12.3). However, very large

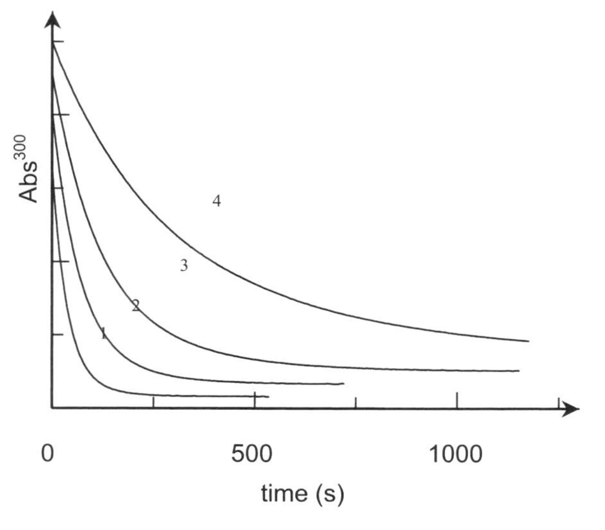

Figure 12.7 Product inhibition of CphA by hydrolyzed biapenem. Curve 1 represents the complete hydrolysis time course of 200 μM biapenem. Curves 2, 3, and 4 are the complete hydrolysis time courses of 200 μM biapenem recorded in the presence of 200, 400, and 600 μM hydrolyzed biapenem, respectively (reference 29a and Bebrone and Galleni, unpublished data).

(>500 s^{-1}) k_a values cannot be determined with standard stopped-flow equipment and only the k_2/K' ratio can be obtained if k_a remains linear versus S_0 up to 300 to 500 s^{-1}.

Acyl Enzyme at the Steady State

In theory, measuring the acyl enzyme concentration at the steady state allows one to derive the k_2/k_3 ratio. Indeed, at the steady state,

$$(ES^*)/E_0 = k_f/(k_3 + k_f) \quad (16)$$

where $k_f = k_2S_0/(K' + S_0)$.

At substrate concentrations much larger than K', the acyl enzyme concentration is thus $k_2E_0/(k_2 + k_3)$.

Equation 16 can also be written as follows: $E_0/(ES^*) = (k_2 + k_3)/k_2 + k_3K'/k_2S_0$.

Measuring the acyl enzyme concentration can most easily be performed with a radioactive substrate by quenching the reaction, for instance, with strong acid (43). But even in this case, if a good substrate is studied, the reaction proceeds so rapidly that the substrate concentration significantly decreases over the time of manual mixing followed by manual quenching and that a quenched-flow apparatus must be used. With penicillins, Martin and Waley also devised a quenched-flow method whereby detection of the acyl enzyme was based on the penamaldate reaction (43). This allowed them to measure the k_2 and k_3 values of some compounds with the *Bacillus cereus* I enzyme. Whatever the method, it is quite difficult, on this basis, to determine k_2/k_3 ratios higher than 4 because (ES*) is then close to E_0 and experimental errors strongly influence the result. It is possible that mass spectrometric methods will in the future allow a more accurate determination of the $(ES^*)/E_0$ ratio since they should visualize both ES* and E (and maybe possibly ES), but to our knowledge, these methods remain to be applied to β-lactamases at least with good substrates (50).

When the k_2/k_3 ratio is very low, k_{cat} is equal to k_2, the acyl enzyme does not accumulate, and the k_3 value becomes impossible to estimate.

Note that when K' is very large, k_f can be much smaller than k_3 with practically attainable S_0 values even if k_2 is $>k_3$. In consequence, the fact that no acyl enzyme can be detected does not unequivocally demonstrate that k_2 is $<k_3$ if the maximum possible substrate concentration is still below K_m.

k_{+1} *and* k_{-1}

When the value of k_2 is relatively low (e.g., <100 s^{-1}), it can usually be safely assumed that it is smaller than k_{-1} (although exceptions can be encountered) and that K' is equal to K. It is then not possible to obtain the values k_{+1} and k_{-1}, at least with the presently available methods and

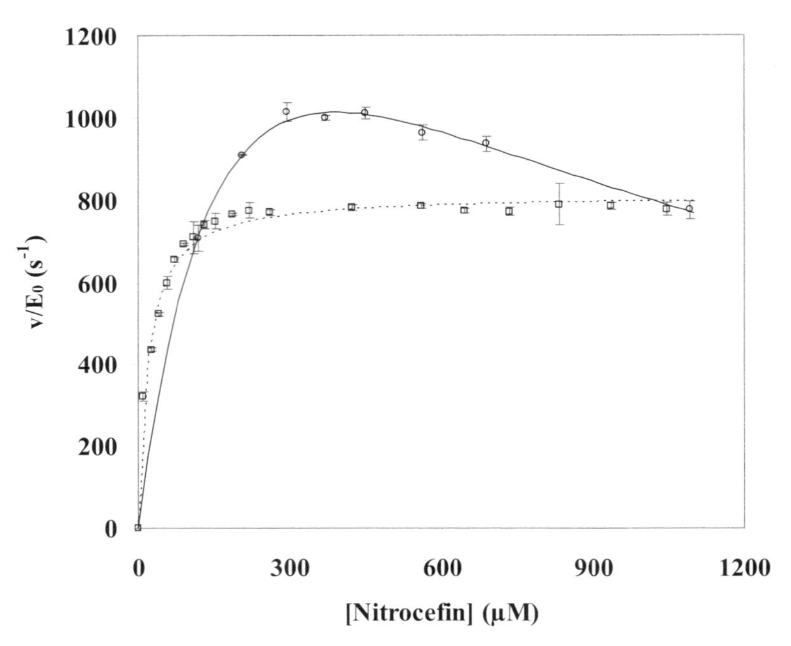

Figure 12.8 Inhibition by excess substrate. Inhibition of the CMY-1 β-lactamase by excess nitrocefin. The activities of CMY-1 (○) and of the chromosomal *Enterobacter cloacae* P99 β-lactamase (□) were determined as the initial rates of hydrolysis of nitrocefin solutions prepared in 50 mM MOPS (morpholinepropanesulfonic acid) buffer pH 7.0, containing 50 mM NaCl. The dotted and continuous lines were obtained by fitting the data to the Henri-Michaelis (for P99) and to the substrate inhibition equations (for CMY-1), respectively. Both enzymes are class C β-lactamases. The equation for substrate inhibition is : $v = V\ S_0/(S_0 + Km + S_0^2/K')$, where K' is the dissociation constant of the inactive ES_2 complex (2a).

unless the ES noncovalent complex were found to exhibit specific measurable properties (which has not been the case up to now).

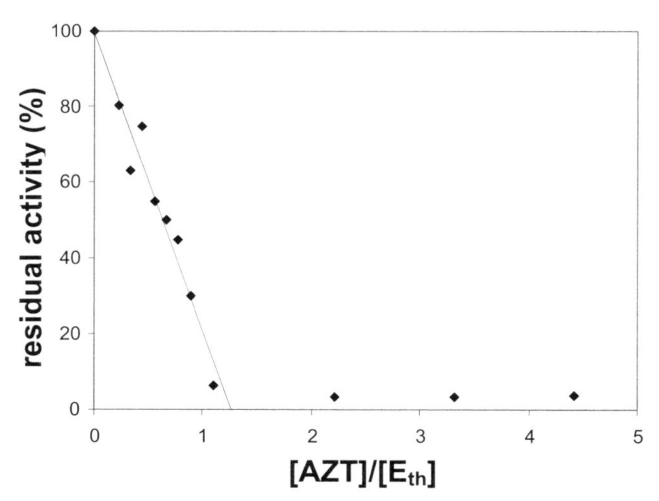

Figure 12.9 Titration curve of the ACT-1 β-lactamase by aztreonam. The residual activity of ACT-1 versus the [aztreonam]/[ACT-1] ratio is shown. The linear regression allows one to determine the actual concentrations of active enzyme (2a).

If the k_2 and k_{-1} values are similar or if k_2 is larger than k_{-1}, the reaction becomes diffusion limited. In this case, the viscosity of the solvent influences K_m because k_{+1} and k_{-1} decrease with increasing solvent viscosity, but not k_2 and k_3. The first thing to verify is thus that the viscosity does not modify k_{cat}. If this condition is fulfilled, k_{+1} and k_{-1} can be derived. Indeed,

$$K_m/k_{cat} = \eta_r/k_{+1o} + k_{-1o}/k_{+1o}\ k_2$$

where η_r is the relative viscosity, k_{+1o} and k_{-1o} are the values of the parameters in the absence of viscogen, and k_{+1o} can be obtained from the slope of the line K_m/k_{cat} versus η_r.

This method, together with the estimation of the concentration of acyl enzyme at the steady state, has successfully been applied by Waley and his coworkers to determine all the constants involved in the interaction between benzylpenicillin and several enzymes. With the TEM β-lactamase, the following values were found at 20°C (5), indicating that the enzyme closely approached "catalytic perfection":

$$k_{+1} = 120\ \mu M^{-1}s^{-1},\ k_{-1} = 11{,}800\ s^{-1},\ k_2 = 2{,}800\ s^{-1},\ k_3 = 1{,}500\ s^{-1}$$

Measuring K′ Directly?

It has been suggested that the K' value of a tested substrate could be determined by utilizing it as a competitive inhibitor of a good (reporter) substrate over short periods of time (59). If manual mixing methods are used, this is possible only if k_2 is quite low, less than 0.01 s^{-1}. In most of the examples proposed by the authors, the measured K' value was in fact the K_m of the tested substrate. In theory, the method could be used with higher k_2 values, but this requires the utilization of stopped-flow equipment under conditions where the steady state is established with the reporter substrate within the dead-time of the equipment. By combining the K' value with those derived from steady-state measurements, the individual k_2 and k_3 values can indeed be derived.

"Slow" Substrates

Poor substrates are characterized by high K' values, low k_2 or k_3 values, or a combination thereof.

Class C Enzymes: Rate-Limiting Deacylation

Some interactions between class C enzymes and poor substrates supply very good examples of low k_3 values (28, 29). The *Enterobacter cloacae* P99 and 908R enzymes exhibit k_3 values in the range of 0.2×10^{-3} to 10×10^{-3} s^{-1} with substrates such as aztreonam, imipenem, oxacillin, and carbenicillin. The k_2 values are all above 150 s^{-1} with the exception of carbenicillin (3.3 s^{-1}). In all cases, the k_2/K' values are quite high (15 to 10,000 $mM^{-1}s^{-1}$) and with all compounds but carbenicillin, the acylation reaction is so fast at high concentrations that the individual k_2 and K' values cannot be obtained even by stopped-flow methods since the k_a value is linear versus S_0 throughout the range of tested concentrations (47). Since the accumulating acyl enzyme is expectedly inactive versus good substrates, the reporter substrate method has been successfully utilized to measure the k_2/K' values of several poor substrates of class C β-lactamases by using nitrocefin or cephaloridine as reporter substrates. An interesting additional information resulting from these measurements is due to the fact that $k_2/K' = k_{cat}/K_m$. For the poor substrates, k_{cat} was measured as V/E_0 directly at substrate saturation (which is easy, since the K_m values are in the range of 1 nM to 1 μM), and K_m was measured as a K_i in a simple competition experiment with a good substrate. The independent measurement of k_2/K' thus allows one to verify the k_{cat}/K_m value and supplies a rough estimate of the purity of the enzyme since the enzyme concentration does not intervene in the determination of k_2/K' while it directly reflects on k_{cat}/K_m since k_{cat} is obtained as V/E_0 (28, 29).

When the k_3 value is low enough, the enzyme can be titrated with the antibiotic as explained above (Fig. 12.9).

Rate-Limiting Acylation

Conversely, acylation remains rate limiting in the hydrolysis of ceftazidime by the class A TEM-1 enzyme. The k_{cat} and K_m values are 0.3 s^{-1} and 4 mM, respectively, and the second-order acylation rate constant, k_2/K', is thus 70 $M^{-1}s^{-1}$. However, K' is higher than the largest substrate concentrations which can be practically achieved and it is not possible to tell whether k_{cat} represents k_2, k_3, or a combination of both (54). As explained above, the fact that no acyl enzyme can be detected in such a case does not mean that k_3 is larger than k_2. But it can be safely concluded that within the range of physiologically significant ceftazidime concentrations, acylation remains rate limiting and no acyl enzyme accumulates. In turn, the values of the three constants have been determined for the interaction between the same enzyme and cefoxitin (14). Both k_2 (10×10^{-3} s^{-1}) and k_3 (2×10^{-3} s^{-1}) are low, and K' is high (10 mM), so that the acyl enzyme can be visualized, but only at high cefoxitin concentrations (>5 mM). At cefoxitin concentrations below 1 mM, acylation thus becomes rate limiting since $k_2(S)/K'$ is $< 10^{-3}$ s^{-1} and k_3 is equal to 2×10^{-3} s^{-1}.

One can be tempted to conclude that, again within the range of physiologically meaningful concentrations, poor substrates are characterized by a slow deacylation rate with class C and a slow acylation rate with class A, but this generalization should be considered with a lot of care.

Activity versus Linear Substrates

β-Lactamases hydrolyze linear esters, thiolesters, and even amides (Fig. 12.10) with highly variable efficiencies (53, 55).

Some esters and thiolesters are hydrolyzed relatively well, mainly by the class C enzymes. With these, transfer reactions onto a suitable acceptor have been observed (51). Peptides are very poor substrates, but in a few cases the enzyme rate enhancement factor determined by Rhazi et al. is far from negligible, due to the high intrinsic stability of peptides (55). With the P99 class C enzyme, accumulation of acyl enzyme has been observed with some linear substrates, and the reaction rate increases in the presence of an acceptor (61).

Branched Pathways

In the Presence of an Acceptor

As mentioned above, PBPs can concomitantly catalyze the simple hydrolysis of a substrate and the transfer of the acyl R-CO moiety onto a suitable acceptor (23, 35). Similarly, the class C β-lactamases can transfer the penicilloyl moiety of penicillins onto methanol (37). When the acyl enzyme accumulates, addition of the acceptor increases both the k_{cat} and the K_m values by identical factors so that the k_{cat}/K_m ratio remains unchanged (33, 37). When no such accumulation occurs, no increase of the reaction rate is observed.

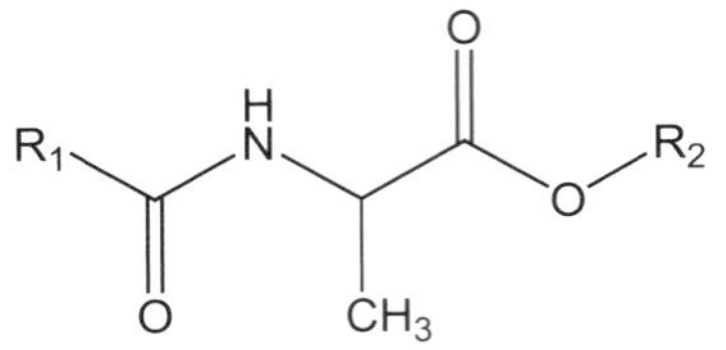

Ester

$R_1 = PhCH_2$- , R_2 = m-carboxyphenyl-

Thioester

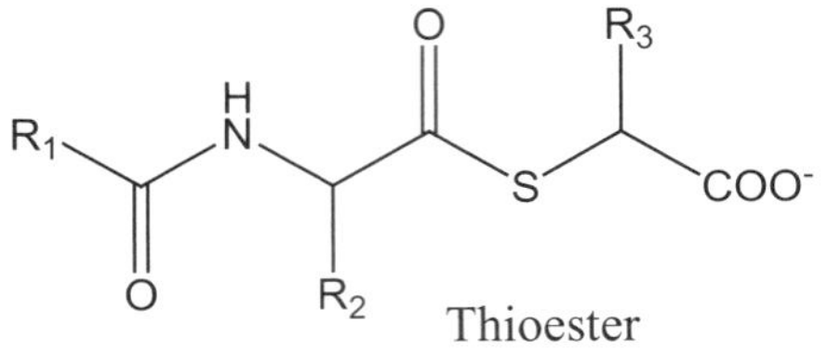

$R_1 = C_6H_5$- or C_6H_5-CH_2-
$R_2 = CH_3$(D) or H
R_3 = H or CH_3(D)

Figure 12.10 Structure of good linear substrates of β-lactamases. The peptides C_6H_5-CO- and C_6H_5-CO-O-D-Ala-D-Ala are very poor substrates.

In both cases, the acyl enzyme partitions between the hydrolysis and transfer pathways and the ratio between the transfer and hydrolysis products increases with the acceptor concentration. In a simple branched pathway where the two competing reactions are as follows:

$$ES^* \xrightarrow{k_3\,(H_2O)} E + P_1$$

$$ES^* \xrightarrow{k_4\,(HY)} E + P_3$$

the P_3/P_1 ratio should be proportional to the acceptor concentration (17). This does not always occur, and more complex pathways must be assumed (35). Similarly, complex pathways are also needed to account for the interactions mentioned above where the acceptor accelerates the reaction despite the fact that no acyl enzyme accumulates in its absence (52).

Cephalosporins: Expulsion of the C-3′ Leaving Group

Many cephalosporins contain a good leaving group (for instance, O-CO-CH_3 or O-CO-NH_2) on C-3′, i.e., the C atom on C-3 of the dihydrothiazine cycle. When the fused β-lactam ring is opened, this group is expelled at a rate depending on the exact structure of the cephalosporin molecule (11, 12). This expulsion occurs at the levels of both the product and the acyl enzyme leading to the formation of an exomethylene group (Fig. 12.11). The rearranged acyl enzyme ES** undergoes hydrolysis more slowly than ES*, but an analysis of the mechanism shows that its formation decreases k_{cat} and K_m by the same factor and does not affect k_{cat}/K_m. The reaction pathway is decribed by the following scheme:

$$\begin{array}{ccccccc} E+S & \rightleftarrows & ES & \rightarrow & ES^* & \rightarrow & E+P_1 \\ & & & & \uparrow\downarrow & & \uparrow\downarrow \\ & & & & ES^{**} & \rightarrow & E+P_2 \\ & & & & +LG & & +LG \end{array}$$

where LG is the leaving group.

With PBPs, the rearrangement simply stabilizes the acyl enzyme. Since k_2/K' is not modified, the sensitivity of the PBP is not affected because the rate of ES* hydrolysis is physiologically irrelevant anyway. Conversely, with β-lactamases exhibiting very high k_3 values, the cephalosporin can be completely hydrolyzed before ES** significantly accumulates and the kinetic parameters determined both under initial rate or complete time course conditions are identical. However, when complete time courses are recorded, a fraction of acyl enzyme equal to $k_4/(k_3+k_4)$ undergoes rearrangement at each turnover and this induces a progressive decrease of the reaction rate characterized as "substrate-induced inactivation." If the reaction is not completed before a significant amount of ES** has accumulated, the k_{cat} and K_m values thus obtained will be lower than those derived from initial rate experiments, while the k_{cat}/K_m ratio remains the same. Using a constant initial substrate concentration, k_{cat} and K_m determined on the basis of complete time courses decrease progressively with decreasing total enzyme concentrations. When possible, the occurrence of this phenomenon can also be demonstrated by using an initial substrate concentration well above the K_m value (determined on the basis of initial rates) and showing that the time course is not zero order but that the rate decreases with time faster than predicted on the basis of substrate depletion. The interactions between β-lactamases

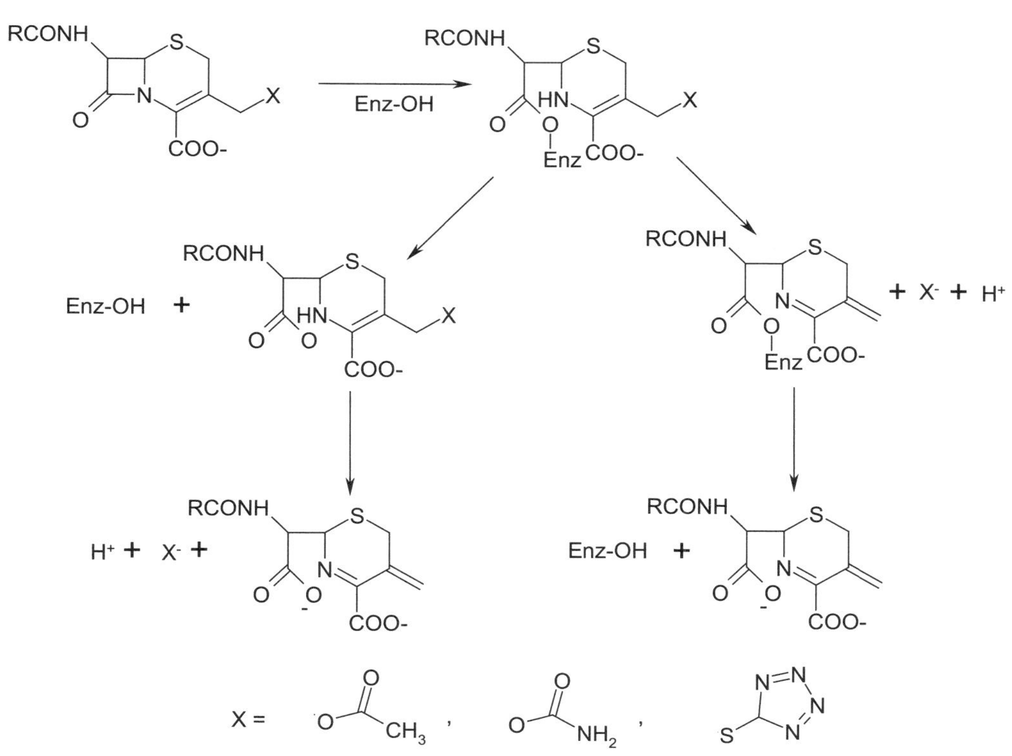

Figure 12.11 Expulsion of the C-3′ leaving group during (right) and after (left) hydrolysis of cephalosporins. The reaction is described by the scheme shown in the text (adapted from reference 38).

and cephalosporins have been analyzed in depth by Faraci and Pratt (11, 12).

Simple Suicide Substrates of β-Lactamases

Other β-lactam compounds (β-halogenopenicillanates, penicillanic acid sulfones, and clavulanic acid) also rearrange at the acyl enzyme level (7, 15). Clavulanic acid and β-lactamases often react according to very complex pathways which are briefly examined below. The situation with the other compounds is more simple. The acyl enzyme irreversibly rearranges into a stable ES** adduct:

$$\begin{array}{ccccccccc} & & & & & k_3 & \\ E+S & \rightleftarrows & ES & \rightarrow & ES^* & \rightarrow & E+P \\ & & & k_4 & \downarrow & & \\ & & & & ES^{**} & & \end{array}$$

The final fate of the enzyme depends on the S_0/E_0 ratio. If this ratio is greater than or equal to $(k_3+k_4)/k_4$, the enzyme is completely inactivated. When S_0/E_0 is smaller than $(k_3+k_4)/k_4$, the suicide substrate is completely hydrolyzed before the enzyme is completely inactivated and, at the end of the reaction, a residual activity is observed:

$$(E)/E_0 = 1 - k_4S_0/(k_3+k_4)E_0$$

The enzyme can thus be "titrated" with the substrate, and the k_3/k_4 ratio can be derived from such a titration. Clearly, the simplest situation prevails when $k_3 < k_4$, where the enzyme is completely inactivated for $S_0 \approx E_0$. Detailed studies of various examples have been performed by Knowles, Waley, Frère, and their coworkers (15, 20, 36).

Clavulanic Acid

The simplest scheme explaining the interaction between clavulanic acid and the TEM β-lactamase has been proposed by Knowles and his coworkers (4, 15). It involves three different adducts: the acyl enzyme ES*, a second, reversibly rearranged adduct ES**, and a third adduct, ES_i, which is irreversibly modified.

$$\begin{array}{ccccccc} & & & & ES^{**} & & \\ & & & & \upharpoonleft\downharpoonright & & \\ E+S & \rightleftarrows & ES & \rightarrow & ES^* & \rightarrow & E+P \\ & & & & \downarrow & & \\ & & & & ES_i & & \end{array}$$

The characteristic of this 3-branch pathway is that, at relatively low (<200 in the case of the TEM-clavulanic acid interaction) S_0/E_0 ratios, a transient inactivation is observed (Fig. 12.12). When all the substrate is hydrolyzed, a part of the activity is recovered. If the S_0/E_0 ratio is sufficient, inactivation is complete and no recovery occurs. Brown et al. (3) have performed a very elegant analysis of this interaction by mass spectrometry and proposed an even more complex pathway.

Not all β-lactamases react with clavulanic acid according to such a complex pathway. Although the three-branch pathway appears to be valid for the *Actinomadura* R39 class A enzyme, that from *Streptomyces albus* G, another class A enzyme, appears to be inactivated according to the more simple scheme described in "Simple Suicide Substrates of β-Lactamases" above (21). Class C enzymes also appear to follow this more simple scheme, but the efficiency of the inactivation is very low, due to a low k_2/K' value (46). In consequence, clavulanic acid is of no clinical utility against strains producing the class C enzymes.

Inactivation Due to an Enzyme Conformation Change

Citri and his coworkers (6) showed that the hydrolysis of a few penicillin substrates (methicillin, oxacillin, and cloxacillin) by the *B. cereus* I (BcI) class A β-lactamase was accompanied by a reversible substrate-induced inactivation. It was later noticed that k_{cat} and K_m decreased by similar factors so that the k_{cat}/K_m ratio was not affected (18). This could again be explained by a conformation change at the level of the acyl enzyme. In contrast to what has been described above for cephalosporins, a chemical rearrangement of the penicilloyl moiety can be safely excluded and many results suggested a conformation change of the enzyme itself, induced by the sterically hindered side chain of the substrates. However, from a kinetic point of view, the equations are the same as those derived for the situation described in "Cephalosporins: Expulsion of the C-3′ Leaving Group" above, but it was never determined if the modified acyl enzyme retained some catalytic competence or not.

Class D β-Lactamases

Several class D β-lactamases also exhibit substrate-induced inactivation (or biphasic kinetics) with a significant number of substrates. In the case of the OXA-2 enzyme, substrates exhibiting a normal Henri-Michaelis behavior are even the exceptions (40). But in contrast to all the cases discussed above, the k_{cat}/K_m ratio is decreased in the steady-state phase when compared to the early phase of the reaction. This shows that the phenomenon cannot be explained by an event taking place at the acyl enzyme level. Class D enzymes exhibit two specific properties. They have a tendency to dimerize (8), and the Lys side chain that is found three residues after the active serine can be reversibly carboxylated (30, 49), the carboxylated form being more active than the unmodified one. Both properties can explain the peculiar substrate-induced inactivation if one assumes that binding of the substrate can modify the monomer-dimer equibrium or the equililibrium between the free and

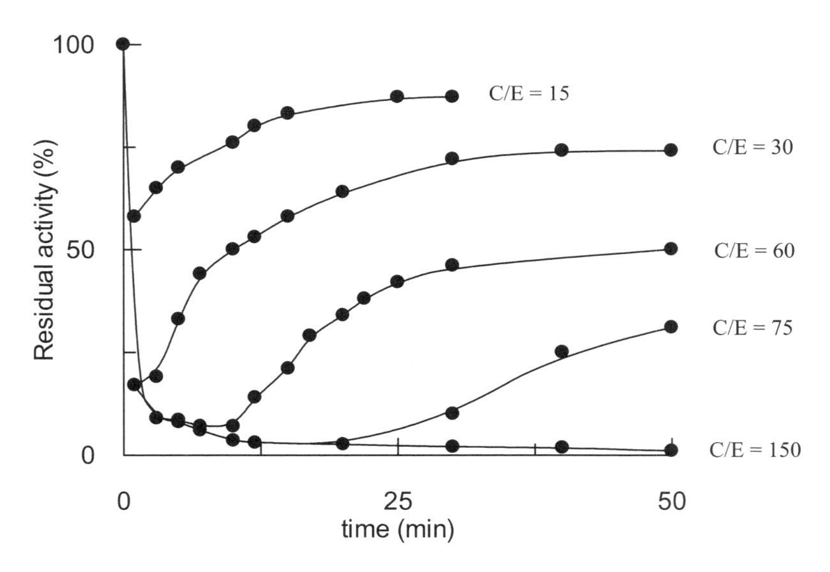

Figure 12.12 The time dependence of TEM-1 β-lactamase activity in the presence of clavulanic acid. C/E represents the molar ratio of the concentration of clavulanic acid (C) and the enzyme (E) (adapted from reference 41).

the carboxlyated Lys side chain. Indeed, in some but not all cases, the substrate-induced inactivation disappeared in the presence of a saturating concentration of bicarbonate which was assumed to completely maintain the Lys in the carboxylated form (30). This suggests that the interaction with the substrate somewhat destabilizes the carbamyl group and displaces the equilibrium in favor of the free Lys amino group, but to explain the variation of the k_{cat}/K_m ratio, one must assume that this destabilization already occurs at the level of the noncovalent ES complex. Similarly, a monomer-dimer equilibrium should also be perturbed at the level of the ES complex. With the data presently available, such a perturbation can be expected to take place in the cases where the biphasic behavior does not disappear in the presence of saturating concentrations of bicarbonate.

The Class B Metallo-β-Lactamases

The class B metallo-β-lactamases hydrolyze nearly all the families of β-lactams (Fig. 12.13), with the exception of monobactams. Their high activity against carbapenems is particularly remakable (and worrying). They contain one or two Zn^{2+} ions in their active site. They do not form covalent adducts with their substrates, at least not during a "normal" turnover. An intermediate exhibiting very special properties has been characterized only with the chromogenic substrate nitrocefin (32).

Nitrocefin

With several enzymes, the hydrolysis of nitrocefin appears to follow the mechanism described by the linear model 1. Here, however, ES* is a noncovalent complex where the β-lactam amide bond has been opened, but the nitrogen

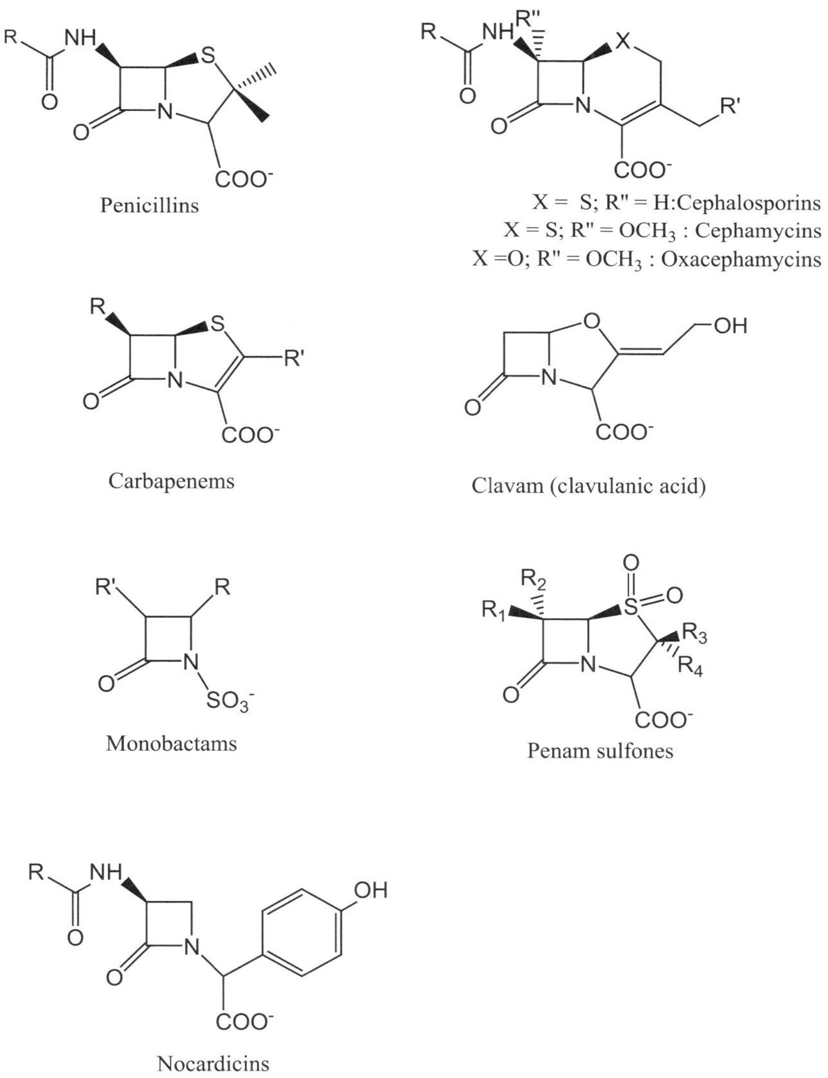

Penicillins

X = S; R" = H:Cephalosporins
X = S; R" = OCH_3 : Cephamycins
X =O; R" = OCH_3 : Oxacephamycins

Carbapenems

Clavam (clavulanic acid)

Monobactams

Penam sulfones

Nocardicins

Figure 12.13 Structures of the different β-lactam families.

atom remains to be protonated so that it bears a negative charge (60). This intermediate has been detected thanks to its spectral properties: it exhibits an absorption maximum at 665 nm. At the steady state and at substrate saturation, it accumulates in very different proportions according to the enzyme (13, 45, 60): from nearly 100% with the *Stenotrophomonas maltophilia* enzyme to less than 3% with that from *B. cereus* (BcII). It is probably stabilized by the possibility of electron delocalization over the unique dinitrostyryl side chain of nitrocefin (Fig. 12.14).

Other Substrates

The other substrates do not contain a side chain which might stabilize a negative charge on the leaving nitrogen atom upon opening of the β-lactam amide bond. In consequence, it seems likely that protonation of the nitrogen is concomitant with the opening of the C-N bond. No intermediate but the classical ES complex has thus been detected with all other substrates, including the most clinically relevant carbapenems.

One or Two Zn^{2+} Ions?

Some class B1 enzymes are active with one Zn^{2+} per monomer and are slightly activated by a second one (Fig. 12.15A), while others bind the two ions very tightly. It has not been possible to prepare the mono-Zn^{2+} form of IMP-1 without first denaturing the protein (56). Interestingly, after removal of the two Zn^{2+} ions from the active site of the enzyme, the IMP-1 apoenzyme becomes unable to bind more than one Zn^{2+} and the kinetic properties of the new "mono" form are quite different from those of the native "di-Zn^{2+}" enzyme (56). The structural basis of this phenomenon remains to be explained.

The class B2 enzyme from *Aeromonas hydrophila* is optimally active with one Zn^{2+} (31) and is even inhibited when binding a second Zn^{2+} ion (Fig. 12.15 B). Finally, the class B3 enzymes tightly bind two Zn^{2+} ions per monomer. The *Stenotrophomonas maltophilia* enzyme is a tetramer (58), which exhibits weak negative cooperativity with nitrocefin (57). To our knowledge, no mono-Zn^{2+} form of a class B3 enzyme has ever been obtained.

APPENDIX

The Order of Reactions

The rate of a reaction involving several reagents, A, B, ... N, can be characterized by the equation

$$v = k(\mathrm{A})^a(\mathrm{B})^b....(\mathrm{N})^n$$

where a, b, and n represent the orders of the reaction versus A, B, and N, respectively. The total order of the reaction is thus $a+b+....+n$. Note that a, b, and n do not need to be integers. They can even change during the time course of a given reaction (see below).

Zero-Order Reaction

The rate of the reaction does not depend on the reagent concentration (A): $v=k$, and $(\mathrm{A})=\mathrm{A}_0-kt$ (valid as long as kt is greater than A_0).

k is the zero-order rate constant (expressed in M s^{-1}), and the half-reaction time, $t_{1/2}$, is $t_{1/2}=\mathrm{A}_0/2k$.

First-Order Reaction

The rate is directly proportional to the reagent concentration: $v=k(\mathrm{A})$ and $(\mathrm{A})=\mathrm{A}_0\,e^{-kt}$, which can also be written

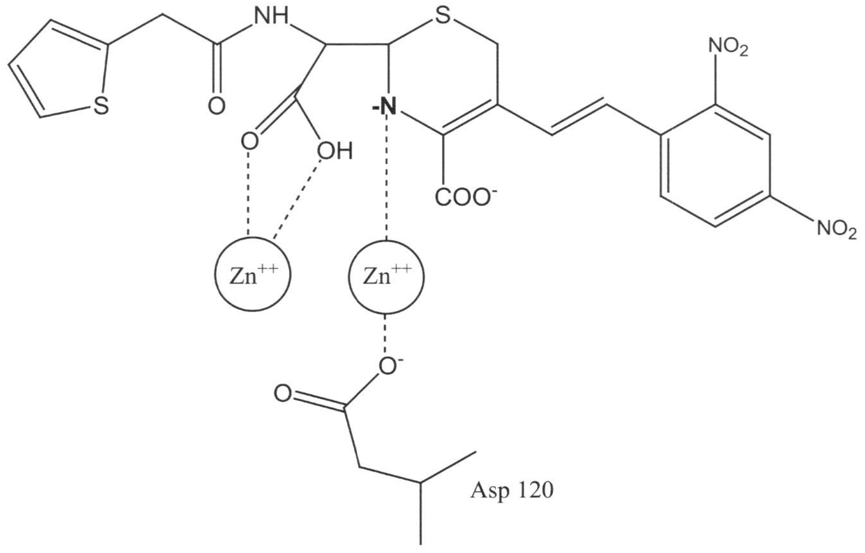

Figure 12.14 Structure of the intermediate postulated on the hydrolysis pathway of nitrocefin by metallo-β-lactamases.

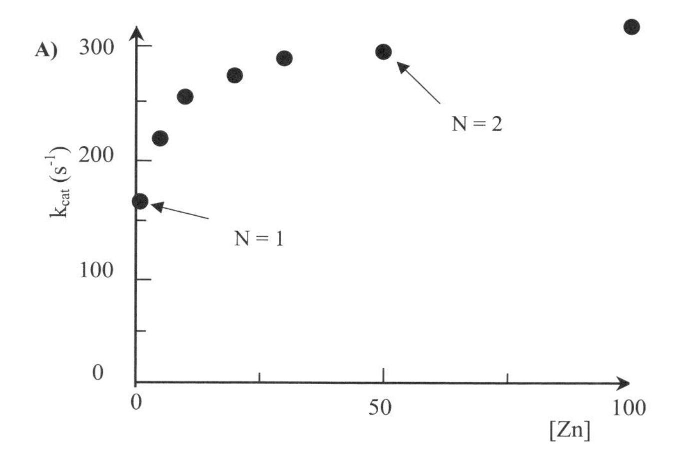

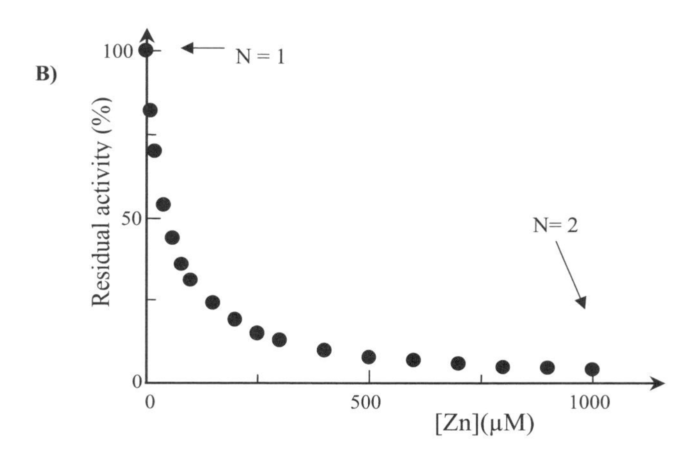

Figure 12.15 (A) Dependence of k_{cat} values versus [Zn^{2+}] for the hydrolysis of benzylpenicillin by the subclass B1 *B. cereus* 569H β-lactamase BcII. (B) Dependence of the residual activity on [Zn^{2+}] for the subclass B2 CphA zinc β-lactamase. In both cases, N represents the number of Zn ions bound per molecule of enzyme.

$$\ln(A) = \ln A_0 - kt$$

where A_0 is the initial concentration of A; k is the first-order rate constant (expressed in s^{-1}), and

$$t_{1/2} = 0.69/k$$

Second-Order Reactions
If there is only one reagent,

$$v = k(A)^2$$

and $$1/(A) = 1/A_0 + kt$$

k is the second-order rate constant (expressed in $M^{-1}s^{-1}$), and

$$t_{1/2} = 1/kA_0$$

If there are two reagents, $v = k(A)(B)$, the reaction is first order versus each reagent, and the total order of the reaction is 2.

If the concentration of one of the reagents is much larger than that of the other one, it can be considered as constant, and $v = kA_0(B) = k'(B)$; k' is a pseudo-first-order rate constant (expressed in $seconds^{-1}$) which is directly proportional to A_0.

Application to Enzyme Kinetics
According to the Henri-Michaelis equation,

$$v = k_{cat}E_0\,(S)/([K_m + (S)]$$

The reaction is first order versus E_0, but versus (S), the order is not an integer: it lies between 0 and 1 and increases over the time course of the reaction.

If (S) is $\gg K_m$,

$$v = k_{cat}E_0$$

and the reaction is zero-order versus (S). Conversely, if (S) is $\ll Km$,

$$v = k_{cat}E_0/Km$$

and the reaction is first order versus (S).

In consequence, if a reaction starts at (S) $\gg K_m$, it is first zero order versus (S) and this order subsequently increases until it reaches a value of 1.

References

1. **Adam, M., C. Fraipont, N. Rhazi, M. Nguyen-Disteche, B. Lakaye, J. M. Frère, B. Devreese, J. Van Beeumen, Y. van Heijenoort, J. van Heijenoort, and J. M. Ghuysen.** 1997. The bimodular G57-V577 polypeptide chain of the class B penicillin-binding protein 3 of *Escherichia coli* catalyzes peptide bond formation from thiolesters and does not catalyze glycan chain polymerization from the lipid II intermediate. *J. Bacteriol.* **179:**6005–6009.
2. **Aplin, R. T., J. E. Baldwin, C. J. Schofield, and S. G. Waley.** 1990. Use of electrospray mass spectrometry to directly observe an acyl enzyme intermediate in beta-lactamase catalysis. *FEBS Lett.* **277:**212–214.

2a. **Bauvois, C., A. S. Ibuka, A. Celso, J. Alba, Y. Ishii, J. M. Frère, and M. Galleni.** 2005. Kinetic properties of four plasmid-mediated AmpC β-lactamases. *Antimicrob. Agents Chemother.* **49:**4240–4246.

3. **Brown, R. P., R. T. Aplin, and C. J. Schofield.** 1996. Inhibition of TEM-2 beta-lactamase from *Escherichia coli* by clavulanic acid: observation of intermediates by electrospray ionization mass spectrometry. *Biochemistry* **35:**12421–12432.
4. **Charnas, R. L., J. Fisher, and J. R. Knowles.** 1978. Chemical studies on the inactivation of *Escherichia coli* RTEM beta-lactamase by clavulanic acid. *Biochemistry* **17:**2185–2189.
5. **Christensen, H., M. T. Martin, and S. G. Waley.** 1990. Beta-lactamases as fully efficient enzymes. Determination of all the rate constants in the acyl-enzyme mechanism. *Biochem. J.* **266:**853–861.
6. **Citri, N., A. Samuni, and N. Zyk.** 1976. Acquisition of substrate-specific parameters during the catalytic reaction of penicillinase. *Proc. Natl. Acad. Sci. USA* **73:**1048–1052.

7. **Cohen, S. A., and R. F. Pratt.** 1980. Inactivation of *Bacillus cereus* beta-lactamase I by 6 beta-bromopencillanic acid: mechanism. *Biochemistry* **19:**3996–4003.

8. **Danel, F., M. Paetzel, N. C. Strynadka, and M. G. Page.** 2001. Effect of divalent metal cations on the dimerization of OXA-10 and -14 class D beta-lactamases from *Pseudomonas aeruginosa. Biochemistry* **40:**9412–9420.

9. **De Meester, F., B. Joris, G. Reckinger, C. Bellefroid-Bourguignon, J. M. Frère, and S. G. Waley.** 1987. Automated analysis of enzyme inactivation phenomena. Application to beta-lactamases and DD-peptidases. *Biochem. Pharmacol.* **36:** 2393–2403.

10. **Duggleby, R. G., P. V. Attwood, J. C. Wallace, and D. B. Keech.** 1982. Avidin is a slow-binding inhibitor of pyruvate carboxylase. *Biochemistry* **21:**3364–3370.

11. **Faraci, W. S., and R. F. Pratt.** 1986. Mechanism of inhibition of RTEM-2 beta-lactamase by cephamycins: relative importance of the 7 alpha-methoxy group and the 3′ leaving group. *Biochemistry* **25:**2934–2941.

12. **Faraci, W. S., and R. F. Pratt.** 1985. Mechanism of inhibition of the PC1 beta-lactamase of *Staphylococcus aureus* by cephalosporins: importance of the 3′-leaving group. *Biochemistry* **24:**903–910.

13. **Fast, W., Z. Wang, and S. J. Benkovic.** 2001. Familial mutations and zinc stoichiometry determine the rate-limiting step of nitrocefin hydrolysis by metallo-beta-lactamase from *Bacteroides fragilis. Biochemistry* **40:**1640–1650.

14. **Fisher, J., J. G. Belasco, S. Khosla, and J. R. Knowles.** 1980. beta-Lactamase proceeds via an acyl-enzyme intermediate. Interaction of the *Escherichia coli* RTEM enzyme with cefoxitin. *Biochemistry* **19:**2895–2901.

15. **Fisher, J., R. L. Charnas, and J. R. Knowles.** 1978. Kinetic studies on the inactivation of *Escherichia coli* RTEM beta-lactamase by clavulanic acid. *Biochemistry* **17:**2180–2184.

16. **Frère, J. M.** 1995. Beta-lactamases and bacterial resistance to antibiotics. *Mol. Microbiol.* **16:**385–395.

17. **Frère, J. M.** 1973. Enzymic mechanisms involving concomitant transfer and hydrolysis reactions. *Biochem. J.* **135:** 469–481.

18. **Frère, J. M.** 1981. Interaction between serine beta-lactamases and class A substrates: a kinetic analysis and a reaction pathway hypothesis. *Biochem. Pharmacol.* **30:**549–552.

19. **Frère, J. M.** 1989. Quantitative relationship between sensitivity to beta-lactam antibiotics and beta-lactamase production in gram-negative bacteria–I. Steady-state treatment. *Biochem. Pharmacol.* **38:**1415–1426.

20. **Frère, J. M., C. Dormans, C. Duyckaerts, and J. De Graeve.** 1982. Interaction of beta-iodopenicillanate with the beta-lactamases of *Streptomyces albus* G and *Actinomadura* R39. *Biochem. J.* **207:**437–444.

21. **Frère, J. M., C. Dormans, V. M. Lenzini, and C. Duyckaerts.** 1982. Interaction of clavulanate with the beta-lactamases of *Streptomyces albus* G and *Actinomadura* R39. *Biochem. J.* **207:**429–436.

22. **Frère, J. M., J. M. Ghuysen, and M. Iwatsubo.** 1975. Kinetics of interaction between the exocellular DD-carboxypeptidase-transpeptidase from *Streptomyces* R61 and beta-lactam antibiotics. A choice of models. *Eur. J. Biochem.* **57:** 343–351.

23. **Frère, J. M., J. M. Ghuysen, H. R. Perkins, and M. Nieto.** 1973. Kinetics of concomitant transfer and hydrolysis reactions catalysed by the exocellular DD-carboxypeptidase-transpeptidase of *Streptomyces* R61. *Biochem. J.* **135:**483–492.

24. **Frère, J. M., J. M. Ghuysen, P. E. Reynolds, and R. Moreno.** 1974. Binding of beta-lactam antibiotics to the exocellular DD-carboxypeptidase-transpeptidase of *Streptomyces* R39. *Biochem. J.* **143:**241–249.

25. **Frère, J. M., and B. Joris.** 1985. Penicillin-sensitive enzymes in peptidoglycan biosynthesis. *Crit. Rev. Microbiol.* **11:**299–396.

25a. **Frère, J. M., and P. Marchot.** 2005. Inactivators in competiton. How to deal with them ... and not! *Biochem. Pharmacol.* **70:**1417–1423.

26. **Frère, J. M., M. Nguyen-Distèche, J. Coyette, and B. Joris.** 1992. Mode of action: interactions with the penicillin-binding proteins, p. 148–197. *In* M. Page (ed.), *The Chemistry of Beta-Lactams.* Chapman & Hall Ltd., Andover, England.

27. **Fuad, N., J. M. Frère, J. M. Ghuysen, C. Duez, and M. Iwatsubo.** 1976. Mode of interaction between beta-lactam antibiotics and the exocellular DD-carboxypeptidase-transpeptidase from *Streptomyces* R39. *Biochem. J.* **155:**623–629.

28. **Galleni, M., G. Amicosante, and J. M. Frère.** 1988. A survey of the kinetic parameters of class C beta-lactamases. Cephalosporins and other beta-lactam compounds. *Biochem. J.* **255:**123–129.

29. **Galleni, M., and J. M. Frère.** 1988. A survey of the kinetic parameters of class C beta-lactamases. Penicillins. *Biochem. J.* **255:**119–122.

29a. **Garau, G., C. Bebrone, C. Anne, M. Galleni, J. M. Frère, and O. Dideberg.** 2005. A metallo-β-lactamase enzyme in action: crystal structure of the monozinc carbapenemase CphA and its complex with biapenem. *J. Mol. Biol.* **345:**785–795.

30. **Golemi, D., L. Maveyraud, S. Vakulenko, J. P. Samama, and S. Mobashery.** 2001. Critical involvement of a carbamylated lysine in catalytic function of class D beta-lactamases. *Proc. Natl. Acad. Sci. USA* **98:**14280–14285.

31. **Hernandez Valladares, M., A. Felici, G. Weber, H. W. Adolph, M. Zeppezauer, G. M. Rossolini, G. Amicosante, J. M. Frère, and M. Galleni.** 1997. Zn(II) dependence of the *Aeromonas hydrophila* AE036 metallo-beta-lactamase activity and stability. *Biochemistry* **36:**11534–11541.

32. **Herzberg, O., and P. M. D. Fitzgerald.** 2004. *Metallo Beta-Lactamases*, vol. 3. John Wiley & Sons, Chichester, United Kingdom.

33. **Jamin, M., M. Adam, C. Damblon, L. Christiaens, and J. M. Frère.** 1991. Accumulation of acyl-enzyme in DD-peptidase-catalysed reactions with analogues of peptide substrates. *Biochem. J.* **280**(Pt. 2):499–506.

34. **Jamin, M., C. Damblon, S. Millier, R. Hakenbeck, and J. M. Frère.** 1993. Penicillin-binding protein 2x of *Streptococcus pneumoniae*: enzymic activities and interactions with beta-lactams. *Biochem. J.* **292**(Pt. 3):735–741.

35. **Jamin, M., J. M. Wilkin, and J. M. Frère.** 1993. A new kinetic mechanism for the concomitant hydrolysis and transfer reactions catalyzed by bacterial DD-peptidases. *Biochemistry* **32:**7278–7285.

36. **Knott-Hunziker, V., B. S. Orlek, P. G. Sammes, and S. G. Waley.** 1979. 6 beta-Bromopenicillanic acid inactivates beta-lactamase I. *Biochem. J.* **177:**365–367.

37. **Knott-Hunziker, V., S. Petursson, S. G. Waley, B. Jaurin, and T. Grundstrom.** 1982. The acyl-enzyme mechanism of beta-lactamase action. The evidence for class C beta-lactamases. *Biochem. J.* **207:**315–322.

38. **Lakaye, B., C. Damblon, M. Jamin, M. Galleni, S. Lepage, B. Joris, J. Marchand-Brynaert, C. Frydrych, and J. M. Frère.** 1994. Synthesis, purification and kinetic properties of fluorescein-labelled penicillins. *Biochem. J.* **300**(Pt. 1):141–145.

39. **Lakaye, B., A. Dubus, S. Lepage, S. Groslambert, and J. M. Frère.** 1999. When drug inactivation renders the target irrelevant to antibiotic resistance: a case story with beta-lactams. *Mol. Microbiol.* **31:**89–101.

40. **Ledent, P., X. Raquet, B. Joris, J. Van Beeumen, and J. M. Frère.** 1993. A comparative study of class-D beta-lactamases. *Biochem. J.* **292**(Pt. 2):555–562.

41. **Leyh-Bouille, M., M. Nguyen-Disteche, S. Pirlot, A. Veithen, C. Bourguignon, and J. M. Ghuysen.** 1986. *Streptomyces* K15 DD-peptidase-catalysed reactions with suicide beta-lactam carbonyl donors. *Biochem. J.* **235:**177–182.

42. **Lu, W. P., E. Kincaid, Y. Sun, and M. D. Bauer.** 2001. Kinetics of beta-lactam interactions with penicillin-susceptible and -resistant penicillin-binding protein 2x proteins from *Streptococcus pneumoniae*. Involvement of acylation and deacylation in beta-lactam resistance. *J. Biol. Chem.* **276:** 31494–31501.

43. **Martin, M. T., and S. G. Waley.** 1988. Kinetic characterization of the acyl-enzyme mechanism for beta-lactamase I. *Biochem. J.* **254:**923–925.

44. **Matagne, A., A. M. Misselyn-Bauduin, B. Joris, T. Erpicum, B. Granier, and J. M. Frère.** 1990. The diversity of the catalytic properties of class A beta-lactamases. *Biochem. J.* **265:** 131–146.

45. **Moali, C., C. Anne, J. Lamotte-Brasseur, S. Groslambert, B. Devreese, J. Van Beeumen, M. Galleni, and J. M. Frère.** 2003. Analysis of the importance of the metallo-beta-lactamase active site loop in substrate binding and catalysis. *Chem. Biol.* **10:**319–329.

46. **Monnaie, D., and J. M. Frère.** 1993. Interaction of clavulanate with class C beta-lactamases. *FEBS Lett.* **334:**269–271.

47. **Monnaie, D., R. Virden, and J. M. Frère.** 1992. A rapid-kinetic study of the class C beta-lactamase of *Enterobacter cloacae* 908R. *FEBS Lett.* **306:**108–112.

48. **Nikaido, H., and S. Normark.** 1987. Sensitivity of Escherichia coli to various beta-lactams is determined by the interplay of outer membrane permeability and degradation by periplasmic beta-lactamases: a quantitative predictive treatment. *Mol. Microbiol.* **1:**29–36.

49. **Paetzel, M., F. Danel, L. de Castro, S. C. Mosimann, M. G. Page, and N. C. Strynadka.** 2000. Crystal structure of the class D beta-lactamase OXA-10. *Nat. Struct. Biol.* **7:** 918–925.

50. **Pagan-Rodriguez, D., X. Zhou, R. Simmons, C. R. Bethel, A. M. Hujer, M. S. Helfand, Z. Jin, B. Guo, V. E. Anderson, L. M. Ng, and R. A. Bonomo.** 2004. Tazobactam inactivation of SHV-1 and the inhibitor-resistant Ser130 -->Gly SHV-1 beta-lactamase: insights into the mechanism of inhibition. *J. Biol. Chem.* **279:**19494–19501.

51. **Pazhanisamy, S., C. P. Govardhan, and R. F. Pratt.** 1989. Beta-lactamase-catalyzed aminolysis of depsipeptides: amine specificity and steady-state kinetics. *Biochemistry* **28:**6863–6870.

52. **Pazhanisamy, S., and R. F. Pratt.** 1989. Beta-lactamase-catalyzed aminolysis of depsipeptides: proof of the nonexistence of a specific D-phenylalanine/enzyme complex by double-label isotope trapping. *Biochemistry* **28:**6870–6875.

53. **Pratt, R. F., and C. P. Govardhan.** 1984. beta-Lactamase-catalyzed hydrolysis of acyclic depsipeptides and acyl transfer to specific amino acid acceptors. *Proc. Natl. Acad. Sci. USA* **81:**1302–1306.

54. **Raquet, X., J. Lamotte-Brasseur, E. Fonze, S. Goussard, P. Courvalin, and J. M. Frère.** 1994. TEM beta-lactamase mutants hydrolysing third-generation cephalosporins. A kinetic and molecular modelling analysis. *J. Mol. Biol.* **244:** 625–639.

55. **Rhazi, N., M. Galleni, M. I. Page, and J. M. Frère.** 1999. Peptidase activity of beta-lactamases. *Biochem. J.* **341**(Pt. 2): 409–413.

56. **Siemann, S., D. Brewer, A. J. Clarke, G. I. Dmitrienko, G. Lajoie, and T. Viswanatha.** 2002. IMP-1 metallo-beta-lactamase: effect of chelators and assessment of metal requirement by electrospray mass spectrometry. *Biochim. Biophys. Acta* **1571:**190–200.

57. **Simm, A. M., C. S. Higgins, A. L. Carenbauer, M. W. Crowder, J. H. Bateson, P. M. Bennett, A. R. Clarke, S. E. Halford, and T. R. Walsh.** 2002. Characterization of monomeric L1 metallo-beta-lactamase and the role of the N-terminal extension in negative cooperativity and antibiotic hydrolysis. *J. Biol. Chem.* **277:**24744–24752.

58. **Spencer, J., A. R. Clarke, and T. R. Walsh.** 2001. Novel mechanism of hydrolysis of therapeutic beta-lactams by *Stenotrophomonas maltophilia* L1 metallo-beta-lactamase. *J Biol Chem.* **276:**33638–33644.

59. **Vakulenko, S. B., P. Taibi-Tronche, M. Toth, I. Massova, S. A. Lerner, and S. Mobashery.** 1999. Effects on substrate profile by mutational substitutions at positions 164 and 179 of the class A TEM(pUC19) beta-lactamase from *Escherichia coli*. *J. Biol. Chem.* **274:**23052–23060.

60. **Wang, Z., W. Fast, and S. J. Benkovic.** 1999. On the mechanism of the metallo-beta-lactamase from *Bacteroides fragilis*. *Biochemistry* **38:**10013–10023.

61. **Xu, Y., G. Soto, K. R. Hirsch, and R. F. Pratt.** 1996. Kinetics and mechanism of the hydrolysis of depsipeptides catalyzed by the beta-lactamase of *Enterobacter cloacae* P99. *Biochemistry* **35:**3595–3603.

Novel Approaches and Future Prospects

C

Enzyme-Mediated Resistance to Antibiotics: Mechanisms, Dissemination, and Prospects for Inhibition
Edited by Robert A. Bonomo and Marcelo E. Tolmasky

David M. Shlaes
Lefa Alksne
Steven J. Projan

13

The Pharmaceutical Industry and Inhibitors of Bacterial Enzymes: Implications for Drug Development

Within the large, general topic of enzyme inhibitors as antimicrobial agents, there are a smaller number of subtopics that we highlight in this chapter. There are a few key principles which go beyond our discussion but which are essential to understanding the leap from the concepts elucidated here to subsequent drug development. First, inhibitors of enzyme activity or even of protein-protein interactions are not difficult to identify in the usual screening assays utilized within the pharmaceutical industry. Second, inhibitors so identified are most often nonspecific or, frequently for physicochemical reasons, are unsuitable as starting points for chemical optimization. Third, any inhibitors that pass these very early hurdles must subsequently pass hurdles related to practical scale-up synthesis, bioavailability in animals suggesting applicability to humans, and toxicological testing. Therefore, while inhibitors are not difficult to identify, drugs or drug candidates are extremely difficult to find and develop.

There are basically three non-mutually-exclusive approaches utilized within the industry to identify drug candidates: (i) random screening of libraries of chemical compounds against some "target" in vitro; (ii) optimization of a lead discovered by someone else and for which additional novelty is achievable; and (iii) structure-based screening "in silico," where computational algorithms are used to identify ligands which can subsequently be assayed. All three approaches have their challenges.

One key basic difficulty has been the nature of chemical libraries available for screening either in in vitro assays or in silico. Two key problems (among others) have been identified. First, libraries do not cover all theoretical chemical space (42, 48). In fact, some estimate that even large libraries of over a million compounds may cover only about 20 to 30% of potential chemical space. Therefore, the random screen is biased by the library screened. Second, the compound libraries which exist in industry tend to be biased, for historical reasons, towards pharmaceutically unfavorable (hydrophobic) compounds, making the actives identified in a screen more likely to be nonspecific or poor starting points for chemical optimization based on their physicochemical properties. The industry has attempted, in some cases, to ameliorate this by using computational approaches to prioritize compounds within their libraries. An example of this is the wide use of the Lipinsky rules (or variants thereof) to weed out pharmaceutically disadvantageous compounds (26). The Lipinsky rules classify compounds according to molecular weight, logP, and the relative numbers of carbon, nitrogen, and oxygen atoms in the molecule. The classification attempts to predict solubility and oral bioavailability based on empirically identified

David M. Shlaes, Anti-infectives Consulting, Stonington, CT 06378. **Lefa Alksne,** Wyeth Research, Pearl River, NY 10965. **Steven J. Projan,** Biological Technologies, Wyeth Research, Cambridge, MA 02140.

properties of orally bioavailable marketed drugs. While these rules are not without controversy, they attempt to deal with a problem that most people agree is a real one.

TARGETS FOR SCREENING

One might argue that the existence of the modern pharmaceutical industry relies on the discovery of compounds first identified by their ability to inhibit the growth of bacteria on agar (e.g., sulfonamides and penicillin). This screening method led to the subsequent identification of essentially all the antibacterial compounds currently marketed. This includes the first new class of compounds to hit the market since rifampicin, the oxazolidinones (e.g., linezolid [Zyvox]). If one defines target validation as the demonstration of clinical efficacy in human trials, then it follows that screening for antimicrobial activity using whole bacterial cells as the target is the most validated target in the history of pharmaceutics. Unfortunately, although many companies, large and small, have continued this practice, only one novel class of antibacterials has made it to the marketplace based on this screening paradigm in the last 40 years.

Given these difficulties and the emergence of genomics, proteomics, structure-based design, sophisticated computational methods, and high-throughput screening technology, a totally different set of approaches to the discovery of new antibiotics are now available to the pharmaceutical industry. So far, none of these technologies has delivered a drug to the antibacterial marketplace. Nevertheless, the lack of progress in the traditional approaches forces the industry to continue to explore new paradigms. Some, including us, argue that the new technologies are not yet mature enough to deliver antibacterials, especially given the safety hurdles for antibacterials (currently marketed products are among the safest ever produced) and the desire for broad-spectrum activity expressed by the marketers and many physicians.

The choice of genes or cellular functions to target is also a difficult one. A general choice exists between targeting genes required for cellular growth in vitro (in vitro essential), or those required either for growth in vivo (in vivo essential) or for initiation and maintenance of the infection (virulence). While most in vitro essential genes are also essential for growth in vivo, this may not be true in the other sense. Most agree that targeting in vitro essential genes would certainly be the most well-established and reliable approach, but the need for novel therapies and approaches to avoid resistance makes the other approaches attractive as well.

In the early days of genomics, given the level of homology between *Escherichia coli* and *Pseudomonas aeruginosa* and even *Staphylococcus aureus*, many argued that the genome sequence of one bacterial species should be adequate for antibacterial discovery. This argument was misguided since, for an individual target, there might be considerable variation among bacterial genera. An example of this is the fatty acid biosynthesis pathway, an essential pathway, in *E. coli*. Unlike *E. coli*, gram-positive organisms do not contain homologs of *fabA* or *fabB* (13, 38). The *fabI* gene in *E. coli* is replaced by *fabK* in *Streptococcus pneumoniae*, while some pathogens like *Enterococcus faecalis* and *P. aeruginosa* have both *fabI* and *fabK* (13, 38). There are also examples of this sort of difficulty in the cell wall synthesis pathway. So, the sequences of many genera are required in order to make intelligent choices among targets.

On a practical level, targeting in vitro essential genes is the most straightforward approach. In a typical bacterial genome, about 30% of the open reading frames encode essential genes, but about 30% of these are genes of unknown function. Since the majority encode known enzymatic functions, simple in vitro enzyme assays can be formatted for high-throughput screening. In vitro essential genes can also be targeted using under- and overexpressing systems in bacterial cells and carrying out screenings for differential inhibition of growth. In such a format, many essential genes can be screened simultaneously (so-called multiplex screening) to help weed out nonspecific inhibitors of growth. The over- and underexpressing systems have also been used in an attempt to target essential genes of unknown function (about 30% of essential genes). To our knowledge, although active molecules have been identified in such screens, none have progressed into clinical trials.

An extension of the concept of essential gene targeting is the idea of attacking an essential metabolic pathway. The pathway approach is attractive for a number of reasons. First, substrates of various enzymes within a pathway, by necessity, resemble each other closely. Therefore, it is not too far a stretch to imagine an inhibitor that might be active against neighboring enzymes in the pathway. In some pathways, such as cell wall synthesis, the enzymes have a similar activity (in this example, amino acid transfer). The potential advantages of such a strategy for avoidance of the emergence of resistance are obvious. A number of approaches to this have been elucidated and are discussed below.

Essential protein-protein interaction targets have tempted drug hunters since the emergence of the yeast two-hybrid technology in the 1990s, which allowed the development of assays for the first time. Unfortunately, experience has shown that these targets are exceedingly challenging to validate and, ultimately, to inhibit with a pharmaceutically acceptable compound. We discuss one such example below.

Although it is beyond the scope of this chapter to delve into the controversies surrounding the identification of so-called in vivo essential genes, we also examine the issue of virulence functions as targets below.

ESSENTIAL PATHWAYS AS TARGETS

One approach to avoid the rapid emergence of resistance is to develop compounds which target more than one cellular enzyme. The problem is that given the diversity of enzymatic active sites, this is usually impractical. One potential application of this approach is the use of enzymatic pathways. Given that there are extensive similarities among the substrates being handled within a pathway, it seems more likely to be able to identify single inhibitors that might target more than one of these enzymes. The challenge has been to first have a sensitive screening assay that would allow for the identification of inhibitors of any enzyme within the pathway and then have secondary assays that would discern which of the enzymes were being inhibited.

Two pathways that have been studied are cell wall synthesis and fatty acid biosynthesis.

One general approach to the development of pathway screens has been to use reporter constructs. In the case of fatty acid biosynthesis, in one study, transcriptional profiling revealed that inhibition of any step of the pathway in *Bacillus subtilis* resulted in large increases in expression of *fabx* (Fig. 13.1) (10, 33). A reporter-based screen was constructed based on these results (Fig. 13.2). For cell wall synthesis, one such screen took advantage of the induction of β-lactamase expression after inhibition of the cross-linking step of peptidoglycan synthesis (9). The researchers went on to show that inhibition of any step along the pathway could lead to this induction, and they demonstrated that novel inhibitors of the late stages of cell wall synthesis were identified by the assay.

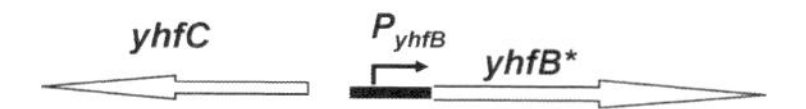

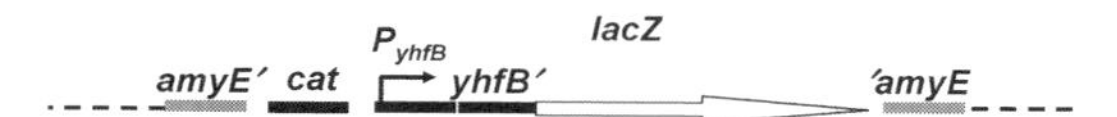

Figure 13.2 Construction of a reporter assay for inhibitors of the fatty acid biosynthesis pathway (33).

Another approach has been to either study the pathway in whole cells by using an assay dependent on some late step or to use a mixture of enzymes in vitro (4). Bona fide inhibitors have been discovered by all methods, although we are still awaiting the development of a new antibacterial drug from these approaches.

ESSENTIAL PROTEIN-PROTEIN INTERACTIONS AS TARGETS

The concept of protein-protein interactions as drug targets has been a controversial one in the pharmaceutical industry. Historically, many of the interactions that have been described occur over a large surface area. That is, the energy of interaction is widely dispersed. Therefore, it seems unlikely that one would be able to identify a small molecule to inhibit such an interaction. The recent approval of enfuvirtide (Fuzeon), a peptide inhibitor of human immunodeficiency virus fusion that acts by inhibiting a critical intraprotein interaction, has encouraged

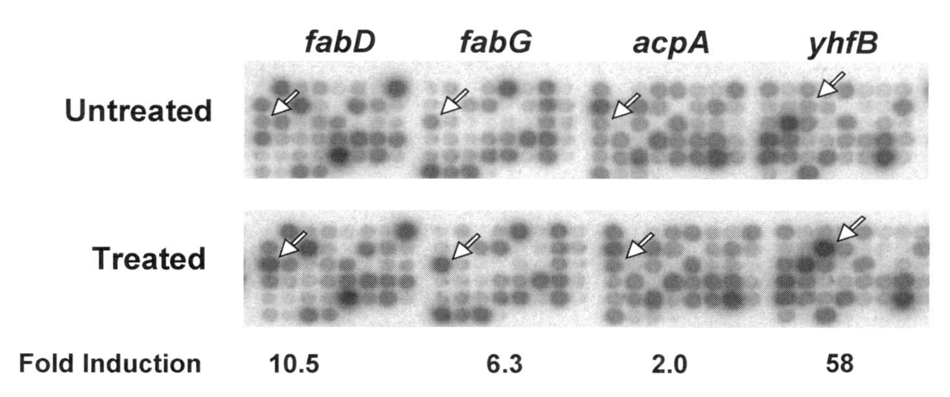

Figure 13.1 Transcriptional profiling of the fatty acid biosynthesis pathway in *B. subtilis* (33), using cells exposed to the inhibitor cerulenin (0.5 μg/ml [0.06 × MIC] for 30 min) or untreated.

researchers who had been discouraged after many years of work in this difficult area.

Protein-protein interactions have been known and have been identified as potential targets for drug discovery for many years (3) (Table 13.1). Two technological advances have made high- or medium-throughput screening for inhibitors of protein-protein interactions practical. The first was the development of the yeast 2-hybrid system, and the second was the advent of surface plasma resonance (BiaCore). The yeast 2-hybrid system uses a DNA binding domain fusion and an activating domain fusion where the fused components include interacting proteins. In this construct, inhibitors reduce the expression of a signal, in the most common method, resulting in decreased growth of the yeast on minimal media. Appropriate controls are, obviously, essential. In the BiaCore approach, a protein is bound to a chip. An interacting protein is passed in solution over the chip, and binding is measured by a resulting increase in plasma resonance. Inhibitors included in the solution with the interacting protein would prevent this increased resonance. Again, appropriate controls are critical. Each approach has its advantages and disadvantages, and both can be complementary to each other.

One example of an approach to protein-protein interactions is that of ZipA, an essential protein involved in cell division in *E. coli* and other bacteria. Cell division in bacteria is mediated by a membrane-associated organelle called the septal ring or the FtsZ ring, where FtsZ is a tubulin-like GTPase which can form protofilaments. The septal ring is composed of a number of interacting proteins, including ZipA (Fig. 13.3). Experiments using the yeast 2-hybrid system demonstrated that the interaction between ZipA and FtsZ occurred with the carboxy terminus of FtsZ and that a single amino acid substitution could abolish the interaction in the yeast 2-hybrid system (12). Mutants which abolish the interaction were shown to be lethal in *E. coli* (12). ZipA was purified and crystallized, and along with its FtsZ interacting peptide, its structure was determined by both nuclear magnetic resonance and crystallography (Color Plate 4) (32). Based on this structure and on alanine-scanning mutagenesis along with measurement of

Table 13.1 Validation of protein-protein interactions as an antibacterial target

1. Demonstrate that the interaction per se is essential.
2. Show that the interaction is susceptible to inhibition by a putative small molecule.
3. Show that compensatory mutations are rare or detrimental.
4. Develop assays to support identification of hits and a chemical optimization program — this usually requires structural information.

K_d for wild-type and mutant peptides using the BiaCore, a 400-Å^2 interaction area was defined, which accounted for over 90% of the binding energy between the FtsZ carboxy-terminal peptide and ZipA. These efforts provided scientific validation for the FtsZ-ZipA interaction as a potential drug target.

The effort involved in providing scientific validation for this target was substantially greater than for most other targets that one would consider for antibacterial discovery. To our knowledge, no inhibitor of this interaction active against bacteria has been discovered in spite of many years of effort in an academic-industrial collaboration.

VIRULENCE AS A TARGET FOR ANTIBACTERIALS

As resistance to currently prescribed antibiotic classes continues to rise among pathogenic bacteria (35, 40), researchers have set their sights on novel targets for drug discovery, including the pathogenic process itself (1, 25). Inhibitors of virulence processes may have a unique mode of action in that they may not result in direct antibiotic activity but rather can act to inhibit the establishment of and/or maintenance of an infection. This would, in effect, disable the invading bacterium and result in an enhanced ability of the host to eliminate the pathogen. One of the most intriguing targets within this class is sortase, a ubiquitous enzyme found in gram-positive bacteria (27, 28, 44) (Fig. 13.4). Sortase is a transpeptidase that is required for the covalent anchoring of many cell wall-associated virulence factors into the peptidoglycan. These factors, commonly called "MSCRAMMs" (acronym for microbial surface components recognizing adhesive matrix molecules), are key to bacterial attachment to host tissue surfaces and to the evasion of host defenses (37). It has been hypothesized that inhibition of sortase should result in the loss of cell surface anchoring of these proteins, therefore weakening the bacterium's ability to establish foci of infection. Because sortase inhibitors would have a downstream effect on multiple substrates, it has been suggested that this method of antibacterial chemotherapy might be less prone to the development of resistance than conventional methods (2). Sortase inhibitors may be efficacious prophylactically in cases where the risk of gram-positive infection is high, for example in some types of surgery, in dialysis patients, and even in dental procedures for patients at risk for endocarditis.

The role of sortase in virulence has been validated in a number of animal models. Knockout mutations of sortases in *S. aureus*, *Streptococcus gordonii*, and *Listeria monocytogenes*, among others, result in decreased lethality, septic arthritis, colonization, adherence, and

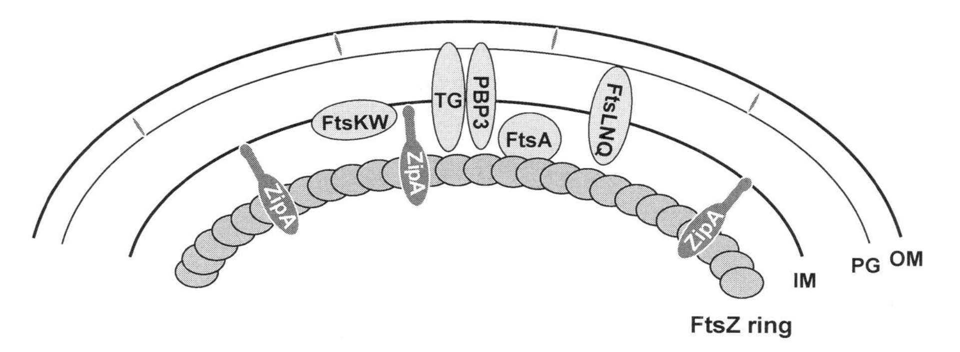

Figure 13.3 Schematic representation of interacting proteins in the septal ring of *E. coli*.

invasiveness (6, 7, 17, 29, 49). In vitro, many studies have shown decreased cell surface expression and function of major virulence factors upon inactivation or mutation of sortase.

The function and enzymological properties of SrtA, the sortase from *S. aureus*, have been intensively studied (11, 14, 16, 21, 45–47). SrtA is a 206-amino-acid protease that recognizes a unique consensus sequence at the carboxy terminus of its protein substrates. The sequence consists of the amino acid residues "LPXTG" followed by a hydrophobic region and a positively charged tail. An active-site cysteine 184 is entirely conserved among the sortases identified thus far and is essential for enzyme function in vitro and in vivo. This residue attacks the scissile bond between the threonine and glycine and, in vivo, catalyzes a transpeptidation of the protein substrate onto a growing peptidoglycan cross-bridge (in the case of *Staphylococcus*, pentaglycine). There is evidence to suggest that the actual acceptor in the cell wall is lipid II (39, 41). In vitro, a nucleophilic attack of the acyl enzyme, either by water or by an added cosubstrate mimicking the cross-bridge, releases the enzyme from the substrate (14).

The structure of SrtA has been elucidated by nuclear magnetic resonance and crystallization and contains a unique beta-barrel structure with eight strands in parallel and antiparallel alignment (15, 16, 51). The active site contains a groove that has been speculated to be involved in scanning precursor proteins for the consensus cleavage site. Recent studies have suggested that the distal binding of a single calcium ion significantly alters the flexibility and geometry of the binding pocket and enhances substrate binding (34). The crystal structure for SrtB has also been

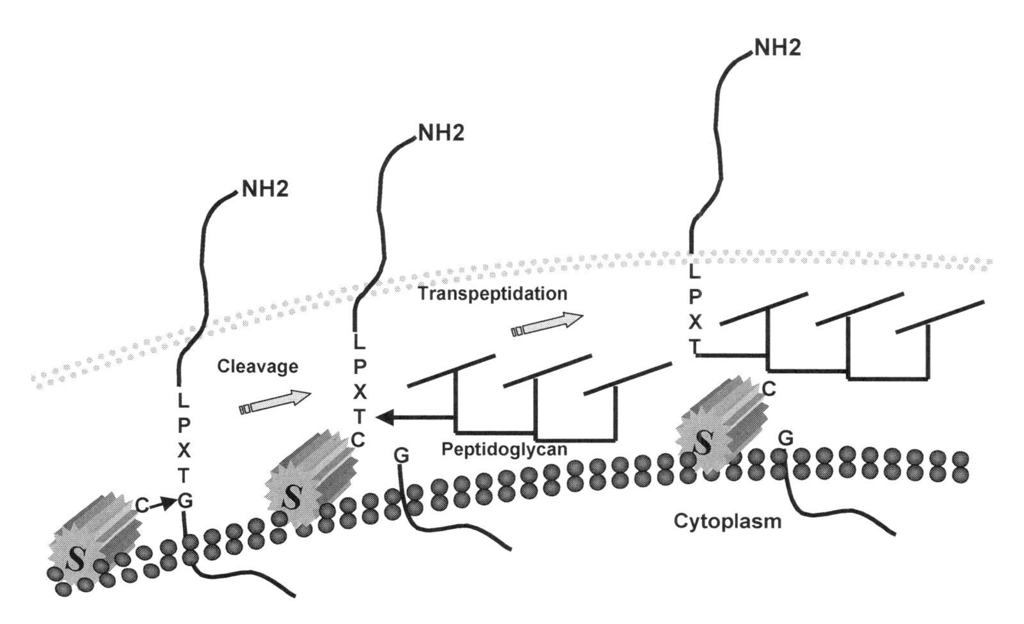

Figure 13.4 Sortase activity results in the transpeptidation of an MSCRAMM protein with its LPXTG motif to the cross-bridge of the gram-positive bacterial peptidoglycan.

deduced, showing considerable structural homology within the active site to SrtA, regardless of the fact that the sequence homology between the two enzymes is only 40% (52).

The sorting reaction of sortase can be divided into two steps in vitro — a cleavage step and a subsequent transpeptidation step. Multiple studies have attempted to elucidate the structural interaction of sortase with its substrate and the kinetic mechanism of transpeptidation. Due to the essential nature of the histidine 120 residue in the active site, it was originally hypothesized that catalysis occurs via a thiolate-imidazolium ion interaction with His120, comparable to the reaction of the enzyme papain (16). However, the cysteine and histidine residues point away from each other and are likely too far apart to form a pair (50). Other data showed that the pK_as of the cysteine and histidine pair are not consistent with this mechanism, and a general base catalysis mechanism was considered to be more likely (8). A triad thiol activation model was suggested by active-site structures in the SrtB isoform (50). Careful study of the kinetics of the sorting reaction in vitro concluded that both hydrolysis and transpeptidation occur by a ping-pong mechanism, in which a common acyl intermediate is formed (14). However, the two reactions have different rate-limiting steps: in the transpeptidation reaction, the formation of the acyl intermediate is rate limiting, and transpeptidation is quite rapid. In the absence of a transpeptidation partner, the hydrolysis is slow. Most recently, a ping-pong bi-bi hydrolytic shunt mechanism has been proposed (11), which may satisfy all of the structural and kinetic observations made to date.

Studies of SrtA activity and kinetic mechanism have been hindered by the weak activity of the enzyme displayed in vitro. In most studies, fluorescence resonance energy transfer (FRET) assays have made use of fluor-quenching pairs flanking the "LPXTG" consensus site (11, 14, 16, 22, 28, 45–47). These assays have demonstrated low turnover rates and require large amounts of enzyme, substrate, and transpeptidation substrate (typically triglycine has been used). Recent studies have made use of high-performance liquid chromatography to follow the reaction, as there is some evidence that internal quenching in the FRET assays results in artifactually low activity measures (21). According to the proposed kinetic mechanism of Frankel et al., low activity is actually due to only a small portion of the enzyme being in an active form. In the meantime, functional assays have focused on the cell surface display of specific MSCRAMMs by examining the difference in the ability of sortase mutants and isogenic wild-type bacteria to adhere to host matrix proteins, either on coated plates or in solution (see references 7, 29, and 49).

While the universal presence of sortase genes among gram-positive organisms is not surprising, the number of alleles present in many species is certainly unusual (36). Each species appears to contain at least two alleles of the gene, with at least one species containing as many as 10 alleles. The similarity between the alleles can be low, but in each case the catalytic consensus of TLXTC is conserved. SrtA from *S. aureus*, in fact, is more of an anomaly than most. In *S. aureus*, a consensus recognition sequence differing from LPXTG has been identified for SrtB, i.e., NPQTN (30). SrtB appears to specifically anchor IsdC, the product of a gene in the same operon. IsdC expression is regulated by iron and is involved in the transfer of heme to the cell wall (31). Studies in vivo in which animals are infected by bacteria containing knockout mutations in either SrtA or SrtB suggest unique roles for the two genes in infection, with SrtA perhaps playing a critical role in the establishment of infection, and SrtB being more critical to its maintenance (17, 49). In *Streptococcus pyogenes*, a different allele apparently identifies a unique subset of substrates and acts to anchor them to the surface instead of SrtA (5). These and other studies are beginning to suggest a common theme in which the SrtA homolog plays a central role in the cell surface attachment of a large number of proteins, while other homologs may play more specified roles in anchoring specific substrates, perhaps under specific conditions (36).

The existence of so many disparate homologs, substrates, and apparent functions represents a challenge to understanding the nature of the sortase active site and its specificity. Recent studies have made use of synthetic substrate libraries to begin to probe the substrate preferences of the different sortases (23). They have confirmed the recognition requirements of SrtA and that there is apparently a complete lack of cross-recognition of substrates by SrtA and SrtB from *S. aureus*.

Several groups have been successful in identifying inhibitors of SrtA that may provide information useful in drug discovery efforts. Substrate-derived uncleavable affinity labels were found to irreversibly inhibit SrtA and can be used to probe the active site of sortase, as well as to detect the enzyme in cell lysates (43). Inhibitors of the enzyme have now been purified from plant extracts, namely, *Coptis chinensis* and *Fritillaria verticillata* (18–20) among others. It is unlikely that these inhibitors are specific inhibitors of the sortase enzyme, especially as they appear to have antimicrobial activity, a feature not expected for a sortase inhibitor in vitro.

The distinctive nature of virulence inhibition as a method to eradicate bacterial infection will likely require special tools in the drug development process. If inhibition of virulence does not result in antibacterial activity in vitro per se, the standard use of antimicrobial susceptibility testing will be ineffective as a diagnostic tool. Consequently, the use of animal models may become more critical earlier on in the process. Furthermore, regulatory requirements for antimicrobial agents may not be applicable to an antivirulence agent. In particular, sortase offers several significant and unique hurdles as a target for antimicrobial chemotherapy. The in vitro studies used to analyze its activity inherently cannot reproduce the unique microenvironment of the peptidoglycan, in which both enzyme and substrate are tethered to the wall and may be spatially constrained. To our knowledge the recombinant enzymes used thus far all have had their amino termini removed, including the membrane anchor, and this may alter the structure and flexibility of the protein. Additionally, it may be that the enzyme requires a cofactor, or at the very least interaction with the secretion machinery, to function optimally. Intriguingly, a recent report described a unique, putative, nonribosomally produced, heavily glycosylated "LPXTGase," the function of which in vivo is unknown (24). Future research will need to develop better in vitro tools to study the anchoring process. In the long run, drug discovery efforts in this arena will need to address the variability of the sortase enzyme itself both in form and in function. Should a preference be made between targeting transpeptidation and targeting hydrolysis? Ultimately it remains to be determined if the optimal sortase inhibitor will be SrtA specific or have broader coverage of the overabundance of sortases identified to date.

CONCLUSIONS

The topic of enzyme inhibitors as antibacterial therapeutics must be taken in the context of an overall decrease in the effort to discover such therapies in the pharmaceutical industry. Given the difficulties outlined in this chapter, the challenge for researchers to meet the demands of physicians and their patients for therapies for infections increasingly caused by resistant pathogens becomes all that much greater. Nevertheless, it is clear to us that the combination of progress in science and the pressure of resistance will lead to novel therapies. The question is when, and at what price.

Note Added in Proof

Since this manuscript was submitted, Cubicin (daptomycin), a second novel class of antibiotic originally identified by whole-cell screening methods, was approved by the FDA.

References

1. **Alksne, L. E., and S. J. Projan.** 2000. Bacterial virulence as a target for antimicrobial chemotherapy. *Curr. Opin. Biotechnol.* **11:**625–636.
2. **Alksne, L. E. 2002.** Virulence as a target for antimicrobial chemotherapy. *Expert Opin. Investig. Drugs* **11:**1149–1159.
3. **Archakov, A. I., V. M. Govorun, A. V. Dubanov, Y. D. Ivanov, A. V. Veselovsky, P. Lewi, and P. Janssen.** 2003. Protein-protein interactions as a target for drugs in proteomics. *Proteomics* **3:**380–391.
4. **Barbosa, M. D. F. S., G. Yang, M. G. Kurilla, and D. L. Pompliano.** 2002. Development of a whole cell assay for peptidoglycan biosynthesis inhibitors. *Antimicrob. Agents Chemother.* **46:**943–946.
5. **Barnett, T. C., and J. R. Scott.** 2002. Differential recognition of surface proteins in *Streptococcus pyogenes* by two sortase gene homologs. *J. Bacteriol.* **184:**2181–2191.
6. **Bierne, H., S. K. Mazmanian, M. Trost, M. G. Pucciarelli, G. Liu, P. Dehoux, L. Jansch, F. Garcia-del Portillo, O. Schneewind, P. Cossart, and the European Listeria Genome Consortium.** 2002. Inactivation of the srtA gene in Listeria monocytogenes inhibits anchoring of surface proteins and affects virulence. *Mol. Microbiol.* **43:**869–881.
7. **Bolken, T. C., C. A. Franke, K. F. Jones, Z. O. Zeller, C. H. Jones, E. K. Dutton, and D. E. Hruby.** 2001. Inactivation of the srtA gene in *Streptococcus gordonii* inhibits cell wall anchoring of surface proteins and decreases in vitro and in vivo adhesion. *Infect. Immun.* **69:**75–80.
8. **Connolly, K. M., B. T. Smith, R. Pilpa, U. Ilanogovan, M. E. Jung, and R. T. Clubb.** 2003. Sortase from *Staphylococcus aureus* does not contain a thiolate-imidazolium ion pair in its active site. *J. Biol. Chem.* **278:**34061–34065.
9. **DeCenzo, M., M. Kuranda, S. Cohen, J. Babiak, Z. Jiang, D. Sun, M. Hickey, P. Sanchetti, P. A. Bradford, P. Youngman, S. Projan, and D. M. Rothstein.** 2002. Identification of compounds that inhibit late steps of peptidoglycan synthesis in bacteria. *J. Antibiot.* **55:**288–295.
10. **Fischer, H. P., N. A. Brunner, B. Wieland, J. Paquette, L. Macko, K. Ziegelbauer, and C. Freiberg.** 2004. Identification of antibiotic stress-inducible promoters: a systematic approach to novel pathway-specific reporter assays for antibacterial drug discovery. *Genome Res.* **14:**90–98.
11. **Frankel, B. A., R. G. Kruger, D. E. Robinson, N. L. Kelleher, and D. G. McCafferty.** 2005. *Staphylococcus aureus* sortase transpeptidatse SrtA: insight into the kinetic mechanism and evidence for a reverse protonation catalytic mechanism. *Biochemistry* **44:**11188–11200.
12. **Haney, S. A., E. Glasfeld, C. Hale, D. Keeney, Z. He, and P. DeBoer.** 2001. Genetic analysis of the *Escherichia coli* FtsZ-ZipA interaction in the yeast two-hybrid system; characterization of the FtsZ residues essential for the interaction with ZipA and with FtsA. *J. Biol. Chem.* **276:**11980–11987.
13. **Heath, R. J., S. W. White, and C. O. Rock.** 2001. Lipid biosynthesis as a target for antibacterial agents. *Prog. Lipid Res.* **40:**467–497.
14. **Huang, X., A. Aulabaugh, W. Ding, B. Kapoor, L. Alksne, K. Tabei, G. Ellestad.** 2003. Kinetic mechanism of *Staphylococcus aureus* sortase SrtA. *Biochemistry* **42:**11307–11315.

15. **Ilangovan, U., J. Iwahara, H. Ton-That, O. Schneewind, and R. T. Clubb.** 2001. Assignment of the 1H, 13C and 15N signals of sortase. *J. Biomol. NMR* **19:**379–380.
16. **Ilangovan, U., H. Ton-That, J. Iwahara, O. Schneewind, and R. T. Clubb.** 2001. Structure of sortase, the transpeptidase that anchors proteins to the cell wall of *Staphylococcus aureus*. *Proc. Natl. Acad. Sci. USA* **98:**6056–6061.
17. **Jonsson, I. M., S. K. Mazmanian, O. Schneewind, T. Bremell, and A. Tarkowski.** 2003. The role of *Staphylococcus aureus* sortase A and sortase B in murine arthritis. *Microbes Infect.* **5:**775–780.
18. **Kim, S. W., I. M. Chang, and K. B. Oh.** 2002. Inhibition of the bacterial surface protein anchoring transpeptidase sortase by medicinal plants. *Biosci. Biotechnol. Biochem.* **66:**2751–2754.
19. **Kim, S. H., D. S. Shin, M. N. Oh, S. C. Chung, J. S. Lee, I. M. Chang, and K. B. Oh.** 2003. Inhibition of sortase, a bacterial surface protein anchoring transpeptidase by beta-sitosterol-3-O-glucopyranoside from *Fritillaria verticillata*. *Biosci. Biotechnol. Biochem.* **67:**2477–2479.
20. **Kim, S. H., D. S. Shin, M. N. Oh, S. C. Chung, J. S. Lee, and K. B. Oh.** 2004. Inhibition of the bacterial surface protein anchoring transpeptidase sortase by isoquinoline alkaloids. *Biosci. Biotechnol. Biochem.* **68:**421–424.
21. **Kruger, R. G., P. Dostal, and D. G. McCafferty.** 2004. Development of a high-performance liquid chromatography assay and revision of kinetic parameters for the *Staphylococcus aureus* sortase transpeptidase SrtA. *Anal. Biochem.* **326:** 42–48.
22. **Kruger, R. G., P. Dostal, and D. G. McCafferty.** 2002. An economical and preparative orthogonal solid phase synthesis of fluorescein and rhodamine derivatized peptides: FRET substrates for the *Staphylococcus aureus* sortase SrtA transpeptidase reaction. *Chem. Commun.* **18:**2092–2093.
23. **Kruger, R. G., B. Otvos, B. A. Frankel, M. Bentley, P. Dostal, and D. G. McCafferty.** 2004. Analysis of the substrate specificity of the *Staphylococcus aureus* sortase transpeptidase SrtA. *Biochemistry* **43:**1541–1551.
24. **Lee, S. G., V. Pancholi, and V. A. Fischetti.** 2002. Characterization of a unique glycosylated anchor endopeptidase that cleaves the LPXTG sequence motif of cell surface proteins. *J. Biol. Chem.* **277:**46912–46922.
25. **Lee, Y. M., F. Almqvuist, and S. J. Hultgren.** 2003. Targeting virulence for antimicrobial chemotherapy. *Curr. Opin. Pharmacal.* **3:**513–519
26. **Lipinski, C. A.** 2000. Drug-like properties and the causes of poor solubility and poor permeability. *J. Pharmacol. Toxicol. Methods* **44:**235–249.
27. **Mazmanian, S. K., G. Liu, H. Ton-That, and O. Schneewind.** 1999. *Staphylococcus aureus* sortase, an enzyme that anchors surface proteins to the cell wall. *Science* **285:**760–763.
28. **Mazmanian, S. K., H. Ton-That, and O. Schneewind.** 2001. Sortase-catalysed anchoring of surface proteins to the cell wall of *Staphylococcus aureus*. *Mol. Microbial.* **40:**1049–1057.
29. **Mazmanian, S. K., G. Liu, E. R. Jensen, E. Lenoy, and O. Schneewind.** 2000. *Staphylococcus aureus* sortase mutants defective in the display of surface proteins and in the pathogenesis of animal infections. *Proc. Natl. Acad. Sci. USA* **97:**5510–5515.
30. **Mazmanian, S. K., H. Ton-That, K. Su, and O. Schneewind.** 2002. An iron-regulated sortase anchors a class of surface protein during *Staphylococcus aureus* pathogenesis. *Proc. Natl. Acad. Sci. USA* **99:**2293–2298.
31. **Mazmanian, S. K., E. P. Skaar, A. H. Gaspar, M. Humayun, P. Gornicki, J. Jelenska, A. Joachmiak, D. M. Missiakas, and O. Schneewind.** 2003. Passage of heme-iron across the envelope of *Staphylococcus aureus*. *Science* **299:**906–909.
32. **Moysak, L., Y. Zhang, E. Glasfeld, S. Haney, M. Stahl, J. Seehra, and W. S. Somers.** 2000. The bacterial cell-division protein ZipA and its interaction with an FtsZ fragment revealed by X-ray crystallography. *EMBO J.* **19:**3179–3191.
33. **Murphy, C., and P. Youngman.** 2003. High throughput screen for inhibitors of fatty acid biosynthesis in bacteria. U.S. patent 6,656,703 B1.
34. **Naik, M. T., N. Suree, U. Ilangovan, C. K. Liew, W. Thieu, D. O. Campbell, J. J. Clemens, M. E. Jung, and R. T. Clubb.** 2006. *Staphylococcus aureus* sortase A transpeptidase: calcium promotes sorting signal binding by altering the mobility and structure of an active site loop. *J. Biol. Chem.* **281:**1817–1826.
35. **Normark, B. H., and S. Normark** 2002. Evolution and spread of antibiotic resistance. *J. Intern. Med.* **252:**91–106.
36. **Paterson, G. K., and T. J. Mitchell.** 2004. The biology of Gram-positive sortase enzymes. *Trends Microbiol.* **12:**89–95.
37. **Patti, J. M., B. L. Allen, M. J. McGavin, and M. Hook.** 1994. MSCRAMM-mediated adherence of microorganisms to host tissues. *Annu. Rev. Microbiol.* **48:**585–561.
38. **Payne, D. J., W. H. Miller, V. Berry, J. Brosky, W. J. Burgess, E. Chen, W. D. DeWolf Jr., W. Fosberry, R. Greenwood, M. S. Head, D. A. Heerding, C. A. Janson, D. D. Jaworski, P. M. Keller, P. J. Manley, T. D. Moore, K. A. Newlander, S. Pearson, B. J. Polizzi, X. Qiu, S. F. Rittenhouse, C. Slater-Radosti, K. L. Salyers, M. A. Seefeld, M. G. Smyth, D. T. Takata, I. N. Uzinskas, K. Vaidya, N. G. Wallis, S. B. Winram, C. C. K. Yuan, and W. F. Huffman.** 2002. Discovery of a novel and potent class of FabI-directed antibacterial agents. *Antimicrob. Agents Chemother.* **46:**3118–3124.
39. **Perry, A. M., H. Ton-That, S. K. Mazmanian, and O. Schneewind.** 2002. Anchoring of surface proteins to the cell wall of *Staphylococcus aureus*. III. Lipid II is an in vivo peptidoglycan substrate for sortase-catalyzed surface protein anchoring. *J. Biol. Chem.* **277:**16241–16248.
40. **Rice, L. B.** 2003. Do we really need new anti-infective drugs? *Curr. Opin. Pharmacol.* **3:**459.
41. **Ruzin, A., A. Severin, F. Ritacco, K. Tabei, G. Singh, P. A. Bradford, M. M. Siegel, S. J. Projan, and D. M. Shlaes.** 2002. Further evidence that a cell wall precursor [C(55)-MurNAc-(peptide)-GlcNAc] serves as an acceptor in a sorting reaction. *J. Bacteriol.* **184:**2141–2147.
42. **Schreiber, S. L.** 2000. Target-oriented and diversity-oriented organic synthesis in drug discovery. *Science* **287:**1964–1969.
43. **Scott, C. J., A. McDowell, S. L. Martin, J. F. Lynas, K. Vandenbroeck, and B. Walker.** 2002. Irreversible inhibition of the bacterial cysteine protease-transpeptidase sortase (SrtA) by substrate-derived affinity labels. *Biochem. J.* **366:**953–958.
44. **Ton-That, H., G. Liu, S. K. Mazmanian, K. F. Faull, and O. Schneewind.** 1999. Purification and characterization of

sortase, the transpeptidase that cleaves surface proteins of *Staphylococcus aureus* at the LPXTG motif. *Proc. Natl. Acad. Sci. USA* **96:**12424–12429.

45. **Ton-That, H., and O. Schneewind.** 1999. Anchor structure of staphylococcal surface proteins. IV. Inhibitors of the cell wall sorting reaction. *J. Biol. Chem.* **274:**24316–24320.

46. **Ton-That, H., S. K. Mazmanian, K. F. Faull, and O. Schneewind.** 2000. Anchoring of surface proteins to the cell wall of *Staphylococcus aureus*. Sortase catalyzed in vitro transpeptidation reaction using LPXTG peptide and NH(2)-Gly(3) substrates. *J. Biol. Chem.* **275:**9876–9881.

47. **Ton-That, H., S. K. Mazmanian, L. Alksne, and O. Schneewind.** 2002. Anchoring of surface proteins to the cell wall of *Staphylococcus aureus*. Cysteine 184 and histidine 120 of sortase form a thiolate-imidazolium ion pair for catalysis. *J. Biol. Chem.* **277:**7447–7452.

48. **Weber, L.** 2002. The application of multicomponent reactions in drug discovery. *Curr. Med. Chem.* **9:**2085–2093.

49. **Weiss, W. J., E. Lenoy, T. Murphy, L. Tardio, P. Burgio, S. J. Projan, O. Schneewind, and L. Alksne.** 2004. Effect of srtA and srtB gene expression on the virulence of *Staphylococcus aureus* in animal models of infection. *J. Antimicrob. Chemother.* **53:**480–486.

50. **Zhang, R., R. Wu, G. Joachimiak, S. K. Mazmanian, D. M. Missiakas, P. Gornicki, O. Schneewind, and A. Joachimiak.** 2004. Structures of sortase B from *Staphylococcus aureus* and *Bacillus anthracis* reveal catalytic amino acid triad in the active site. *Structure* **12:**1147–1156.

51. **Zong, Y., T. W. Bice, H. Ton-That, O. Schneewind, and S. V. L. Narayana.** 2004. Crystal structures of *Staphylococcus aureus* sortase A and its substrate complex. *J. Biol. Chem.* **279:**31383–31389.

52. **Zong, Y., S. K. Mazmanian, O. Schneewind, and S. V. L. Narayana.** 2004. The structure of Sortase B, a cysteine transpeptidase that tethers surface protein to the *Staphyloccus aureus* cell wall. *Structure* **12:**105.

Enzyme-Mediated Resistance to Antibiotics: Mechanisms, Dissemination, and Prospects for Inhibition
Edited by Robert A. Bonomo and Marcelo E. Tolmasky

Zhen Zhang
Timothy Palzkill

14

β-Lactamase Inhibitory Proteins

The discovery and development of β-lactam antibiotics, such as penicillins and cephalosporins, were significant achievements for the pharmaceutical industry in the 20th century. β-Lactam antibiotics are able to kill bacteria by inhibiting the function of penicillin-binding proteins (PBPs), which are essential enzymes for the synthesis of bacterial cell wall. The widespread use of β-lactam drugs has applied selection pressure to accelerate the evolution and spread of resistance to these drugs.

The production of β-lactamases by bacteria is the most prevalent cause of resistance to β-lactam antibiotics. β-Lactamases are grouped into four classes, A, B, C, and D, based on primary amino acid sequence alignments. Members of classes A, C, and D are active-site serine enzymes and share a similar fold. Class B β-lactamases are metalloenzymes and are believed to have evolved independently of the serine enzymes. Class A β-lactamases such as TEM-1 and SHV-1 are the most common plasmid-encoded enzymes in gram-negative bacteria. To overcome the antibiotic resistance mediated by β-lactamases, mechanism-based small-molecule inhibitors such as clavulanic acid, sulbactam, and tazobactam have been used. Mutants of the TEM and SHV β-lactamases have evolved, however, that are not susceptible to the action of these inhibitors. Therefore, there is still a need for novel inhibitors of these enzymes.

β-Lactamase inhibitory proteins (BLIPs) are proteinaceous inhibitors of β-lactamases that have been discovered in *Streptomyces* species. To date, three β-lactamase inhibitory proteins (BLIP, BLIP-I, and BLIP-II) have been discovered and shown to be potent inhibitors of several class A β-lactamases (5, 9, 10). The crystal structures of complexes of TEM-1 β-lactamase/BLIP and TEM-1/BLIP-II have been solved (11, 20). The TEM-1/BLIP interaction has become a popular model system for studying protein-protein interactions and is also a potential starting point for the design of inhibitors of β-lactamases. This chapter discusses structure and function studies of BLIP with a focus on the interactions between TEM-1 β-lactamase and BLIP.

STRUCTURE AND FUNCTION OF ACTIVE-SITE SERINE β-LACTAMASES

The structure of TEM-1 β-lactamase is shown in Fig. 14.1 (8). All serine β-lactamases share a similar fold consisting of two structural domains, one all-α-helix and one α-helix/β-sheet, with the active site located in the crevice between

Zhen Zhang, Structural & Computational Biology and Molecular Biophysics, Baylor College of Medicine, One Baylor Plaza, Houston, TX 77030.
Timothy Palzkill, Department of Molecular Virology and Microbiology, Baylor College of Medicine, One Baylor Plaza, Houston, TX 77030.

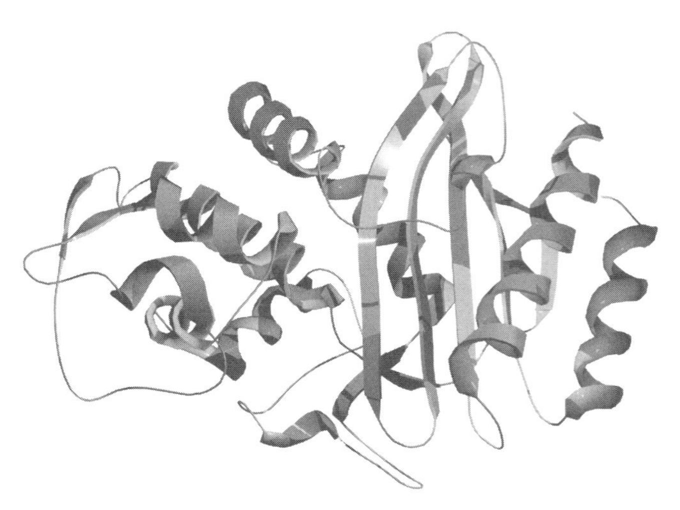

Figure 14.1 Structure of TEM-1 β-lactamase. The image was generated using Swiss-PdbViewer and rendered by POV-Ray using Protein Data Bank coordinates 1BTL (8).

the two domains (8). Serine β-lactamases also share three conserved sequence motifs: SXXK, Y(S)XN, and K(R)T(S)G. The SXXK motif contains the active-site serine residue, which is located in the center of the active-site pocket. The Y(S)XN motif resides on one wall of the active site, and K(R)T(S)G is on the opposite wall. One additional motif, Glu166 in the omega loop near the base of the active-site pocket, is also conserved among class A β-lactamases (12). The residues collectively create a net positive charge in the active site, which is common through the various serine β-lactamases.

The hydrolysis of β-lactam antibiotics by serine β-lactamases is generally believed to consist of two steps: acylation and deacylation (Fig. 14.2) (12). During acylation, the active-site serine functions as a nucleophile to attack the amine bond of the β-lactam ring to form an intermediate acyl enzyme. A water molecule then attacks the intermediate to hydrolyze the antibiotic into an inactive acid and release the free enzyme. Glu166 in class A β-lactamases activates the water molecule for deacylation. It is not clear which residue plays the same role in the β-lactamases from other classes.

Mechanism-based inhibitors, such as clavulanic acid or sulbactam, also contain a β-lactam ring structure. These molecules are poor antibiotics but are good inhibitors of some β-lactamases. Poor substrates of β-lactamases can also function as inhibitors. Non-β-lactam-based inorganic chemicals, including boronate and phosphonate derivatives, have also been shown to inhibit some β-lactamases.

PROTEIN INHIBITORS OF β-LACTAMASES

β-Lactamase inhibitory proteins inhibit class A β-lactamases with a wide range of affinities (20, 22). The first proteinaceous inhibitor of β-lactamases, BLIP, was isolated from *Streptomyces clavuligerus* by Jenson's group in 1990 (5). *S. clavuligerus* also produces several antibiotics such as penicillin N and cephamycin C and several antifungal agents such as clavam-2-carboxylate and hydroxymethyl clavam. *S. clavuligerus* is also the source of the widely used small-molecule β-lactamase inhibitor clavulanic acid. *S. clavuligerus* contains a β-lactamase-encoding gene, which produces a class A β-lactamase that has low activity towards penicillin G and can be inhibited weakly by BLIP. *S. clavuligerus* also exhibits resistance to penicillin G, which is not due to the production of the innate

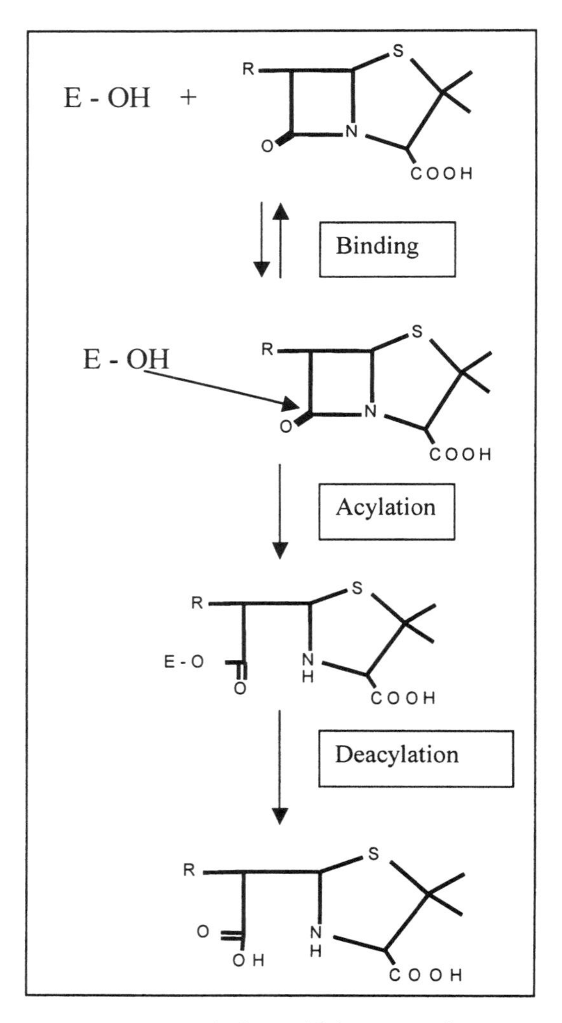

Figure 14.2 Hydrolysis of β-lactam antibiotics.

β-lactamase but rather to the low affinity of the β-lactam for its penicillin-binding proteins (21).

The biological function of BLIP is currently unknown. A BLIP nonproducer mutant and a BLIP/clavulanic acid nonproducer double mutant were constructed with the aim of elucidating BLIP's physiological function (21). Neither mutant, however, exhibited a significant difference from the wild-type strain in growth morphology, sporulation, or production of antibiotics. Two hypotheses have been proposed for the function of BLIP (5). First, BLIP may be produced in response to the production of β-lactamases by other organisms in the surrounding environment in order to inhibit these β-lactamases and prevent the hydrolysis of antibiotics produced by *S. clavuligerus*. Alternatively, BLIP may play a role in cell wall growth or morphogenesis. To date, insufficient data exist to distinguish these possibilities.

Recently, two additional inhibitory proteins, BLIP-I and BLIP-II, have been isolated from culture filtrates of *Streptomyces exfoliatus* SMF19 (10). While BLIP-I-exhibits 38% sequence identity with BLIP, BLIP-II shows no significant homology to BLIP or BLIP-I. In contrast to the BLIP knockout mutant, disruption of the BLIP-I encoding gene in *S. exfoliatus* imposes a bald phenotype in the mutant strain, which suggests a role for BLIP-I in the growth and development of the strain (9).

Strynadka and colleagues have solved the crystal structure of the complex of TEM-1 β-lactamase with BLIP-II (Fig. 14.3B) (11). BLIP-II has a funnel-shaped, seven-bladed β-propeller fold with no similarity to the BLIP fold. BLIP-II interacts with the loop-helix domain of TEM-1, as does BLIP. In contrast to BLIP, however, BLIP-II does not make direct contact with the active site of TEM-1 β-lactamase. BLIP-II renders the active site of TEM-1 inaccessible to substrate by spanning across the active site (11).

PROTEIN-PROTEIN INTERACTIONS AND HOT SPOTS

Protein-protein interactions play a significant role in most cellular processes. The importance of such interactions in biology has made protein-protein recognition an area of considerable interest. A protein displays a specific combination of 20 amino acids at an interface, which determines the affinity and specificity of a protein-protein interaction. Protein-protein interfaces are usually flat, burying a large number of residues (>10 amino acids) and a large portion of surface area (>900 Å^2) (3). Although crystallographic structures provide information on protein-protein interfaces at atomic resolution, structural data alone do not reveal the determinants of affinity or specificity for an interaction.

Alanine-scanning mutagenesis is a powerful tool for detecting and analyzing protein-protein interfaces. The technique involves systematic mutation of interface residues to alanine followed by a determination of the effect of the substitution on binding affinity (4). The alanine substitution deletes a side chain beyond β-carbon with minimal

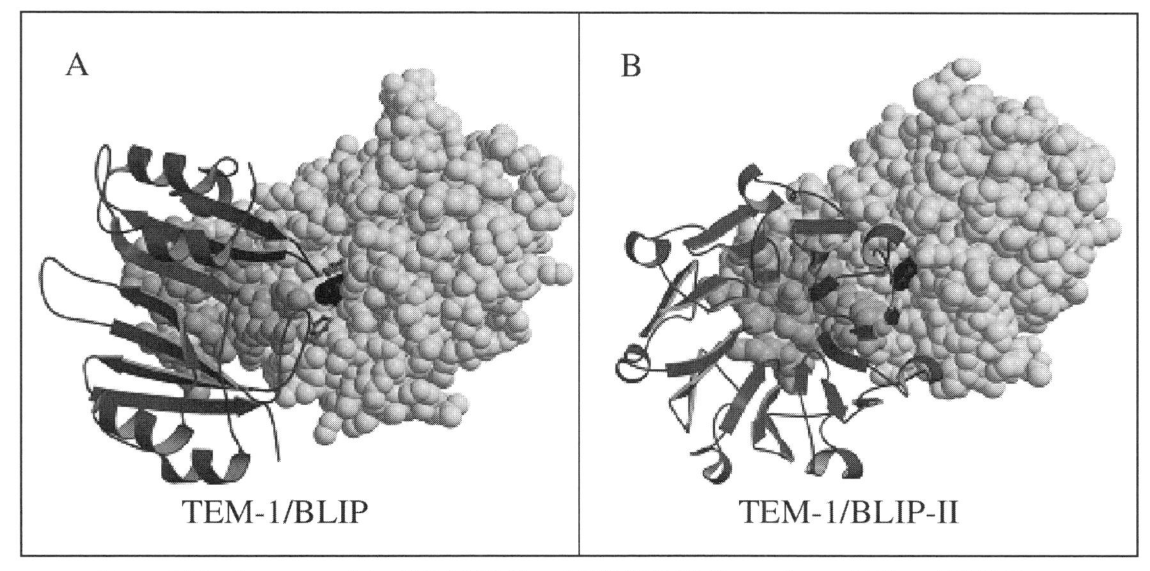

Figure 14.3 Structure of the TEM-1/BLIP and TEM1/BLIP-II complexes. BLIP and BLIP-II are shown in ribbon format. TEM-1 β-lactamase is shown in spacefill format. The darkened region is the loop-helix domain of TEM-1 that interacts with BLIP (99–114). The active-site residue (S70) of TEM-1 β-lactamase is shown in black. The images were generated with Molscript using Protein Data Bank coordinates 1JTG (TEM-1/BLIP) and 1JTD (TEM-1/BLIP-II).

structural perturbation on the protein backbone. The binding assay should therefore reveal the energetic contribution of the deleted side chain. Wells and colleagues initially utilized the technique to dissect the binding interface between human growth hormone (hGH) and its receptor (hGHR) (4). They found that the binding affinity does not evenly distribute over the interface. Instead, a small number of residues on the hGH/hGHR interface make a significant energetic contribution and these residues form a cluster, which was defined as a hot spot. Subsequent studies of many other interfaces suggest that the existence of a hot spot may be a general characteristic of protein-protein interactions (3). The existence of hot spots has implications for drug design and protein engineering. As discussed above, protein-protein interfaces consist of large numbers of interactions, which makes the rational modification of the interface by either small molecules or mutagenesis difficult. The finding that most of the binding energy exists in a small number of hot spot residues provides a focus for design.

Determination of the chemical features that uniquely define a hot spot such that one could accurately predict a hot spot based on structure remains a challenge. Hot spots are frequently observed to contain a hydrophobic core surrounded by a hydrophilic ring composed of residues that themselves do not contribute to affinity (3). Electrostatic residues are also energetically significant for some interfaces, such as the barstar/barnase interaction (18). Polar residues (Arg, Gln, His, Asp, and Asn) are generally conserved in interfaces, and it has been proposed that conserved polar residues constitute hot spots (6). More studies are required, however, to understand the molecular basis of protein-protein interactions.

STRUCTURAL STUDIES OF TEM-1/BLIP INTERACTION

BLIP is a 165-amino-acid protein consisting of two tandem repeats (20). The two domains join with each other to form an 8-strand antiparallel β-sheet. There are two small α-helixes in each domain located on one side of the β-sheet. BLIP uses the other side of the β-sheet to clamp over the loop-helix domain of TEM-1 β-lactamase (Fig. 14.3A). In addition, two loops, one from each domain of BLIP, insert into the active-site pocket of TEM-1 and mimic the interaction between TEM-1 and the substrate penicillin G. One of the loops, containing residues 46 to 51, possesses a rigid β-turn structure and makes multiple interactions with essential residues in the TEM-1 β-lactamase active site (20). For example, Asp49 on the β-turn forms two salt bridges and two hydrogen bonds with four essential TEM-1 residues: Ser130, Lys234, Arg244, and Ser235. BLIP residues Tyr50 and Tyr51 lie on one face of the β-turn and pack with a conserved proline at position 107 in TEM-1 as well as forming van der Waals interactions with several residues on the β3 strand of TEM-1. The other face of the β-turn consists of two alanine residues and a glycine. The bare nature of this face is proposed to be essential to fit the β-turn into the active site without steric clashes with the β-strand (Lys234-Glu240) of TEM-1 β-lactamase (Fig. 14.4) (20).

BIOCHEMICAL AND BIOPHYSICAL STUDIES OF TEM-1/BLIP INTERACTION

Although the physiological function of BLIP remains unknown, it has generated interest as a potential starting point for inhibitor design as well as a model system for studies of protein-protein interactions. The interface between TEM-1 β-lactamase and BLIP is one of the largest among known protein-protein interactions (2,636 Å^2) as shown by crystallography (20). The binding interaction with TEM-1 β-lactamase is strong with a K_i of 0.5 nM. In addition to TEM-1, BLIP inhibits other class A β-lactamases with a wide range of affinities (Table 14.1) (20, 22). Thus, the BLIP-β-lactamase interface represents not simply a single protein-protein interaction but rather a family of protein-protein interactions. The study of a series of homologous interfaces provides an opportunity to study specificity as well as overall affinity determinants.

Recent studies suggest that the TEM-1 β-lactamase/BLIP complex is stabilized mainly by hydrophobic interactions (1, 22). The hot spot of BLIP for binding to TEM-1 consists of Asp49, Phe142, and two patches containing mainly aromatic residues (patch 1 contains Phe36, His41,

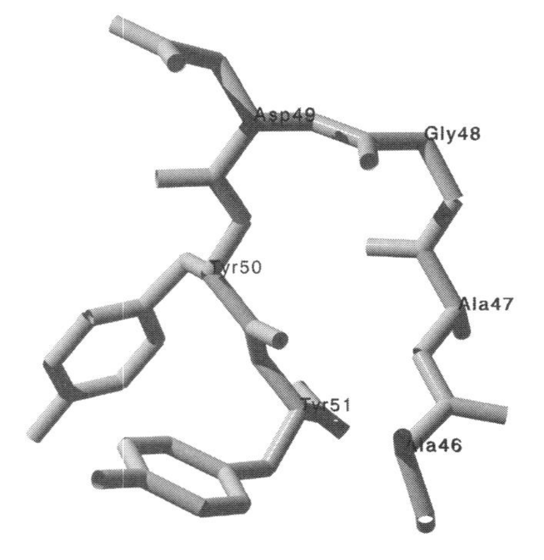

Figure 14.4 β-Hairpin structure of BLIP residues 46 to 51. The image was generated using Swiss-PdbViewer and rendered with POV-Ray using Protein Data Bank coordinates 1BTL.

Table 14.1 Inhibition of class A β-lactamases by BLIP[a]

Enzymes	K_i	Sequence identity to TEM-1 (%)
E. coli TEM-1	0.5 nM	100
Klebsiella pneumoniae SHV-1	1.1 μM	68
Serratia marcescens SME-1	2.4 nM	33
Bacillus anthracis Bla1	2.5 nM	38
Klebsiella oxytoca 1	1 nM	35
Lysobacter enzymogenes	1 nM	37
Proteus vulgaris 1028	18 pM	32
Actinomadura R39	11 μM	29
Bacillus licheniformis 749	3 μM	35
Streptomyces albus G	3 μM	32
Pseudomonas aeruginosa PSE-1	3 nM	41
Enterococcus faecalis PBP5	12 μM	11

[a]Data from references 20, 22, and 23.

and Tyr53; patch 2 contains Trp112, His148, Trp150, Arg160, and Trp162 as shown in Fig. 14.5) (22). The critical residues in TEM-1 β-lactamase for binding BLIP are mainly hydrophobic residues and are in contact with the residues within the hot spot of BLIP in the complex (15). The aliphatic portion of the side chain of Arg160 of BLIP makes close contact with other residues in patch 2 with its positively charged head group pointing into the solution. Patch 1 is surrounded by a few hydrophilic residues, which do not make a significant contribution to the binding affinity. This organization of residues has been observed in hot spots from other protein-protein interactions, consistent with the hypothesis that a ring of hydrophilic residues is required to protect the hot spot from solvent (3). However, a ring of hydrophilic residues is not evident around the patch 2 hot spot of BLIP. The residues in this patch maintain partial solvent accessibility, suggesting that a hydrophilic ring is not a required condition for a hot spot.

Electrostatic interactions play a significant role in the specificity of BLIP for binding to various β-lactamases. For example, the Glu73-Lys74 residues of BLIP have been shown to have diverse effects on binding to β-lactamases. The Glu73-Lys74 motif is buried in the interface and is in contact with Glu104-Tyr105 of TEM-1 β-lactamase. Lys74 of BLIP forms a salt bridge with Glu104 of TEM-1, and Glu73 of BLIP is stabilized by an interaction with the backbone of Glu104 and Tyr105 of TEM-1. Replacement of Lys74 with alanine decreases the binding affinity 92-fold, while replacement of Glu73 of BLIP with alanine does not impact binding affinity with TEM-1 β-lactamase (22). The effect of alanine substitutions at Glu73 and Lys74 is strongly dependent on the β-lactamase binding partner. For example, SME-1 β-lactamase is inhibited by BLIP at 2.5 nM K_i and replacement of Lys74 of BLIP with alanine has no effect on binding, while Glu73Ala BLIP is a 1,000-fold-weaker inhibitor of SME-1 than the wild type (22).

Schreiber and colleagues have studied the effects of electrostatic interactions on the TEM-1 β-lactamase/BLIP complex with respect to the association (k_{on}) and dissociation rates (k_{off}) (1, 2). The TEM-1/BLIP complex exhibits an association rate of 10^5 M^{-1} s^{-1} and a slow dissociation rate of 10^{-4} s^{-1}, which results in a dissociation constant in the low nanomolar range (1). Both long-range and short-range forces can affect the dissociation rate, k_{off}. In the case of the TEM-1/BLIP interaction, short-range forces contributed by hydrophobic interactions appear to be important based on the alanine-scanning mutagenesis studies described above. Both TEM-1 β-lactamase and BLIP are negatively charged at pH 7.0. The association rate, k_{on}, is highly sensitive to changes of charge on the two proteins because k_{on} is mainly dependent on long-range forces (1, 2). It is difficult to design protein complexes with slower k_{off} values because a physical understanding of the short-range interactions on the protein-protein binding interface and a method to calculate conformational changes caused by introducing mutations are required. In contrast, Selzer et al. demonstrated that optimizing electrostatic attractions between the proteins can increase the association rate, k_{on}

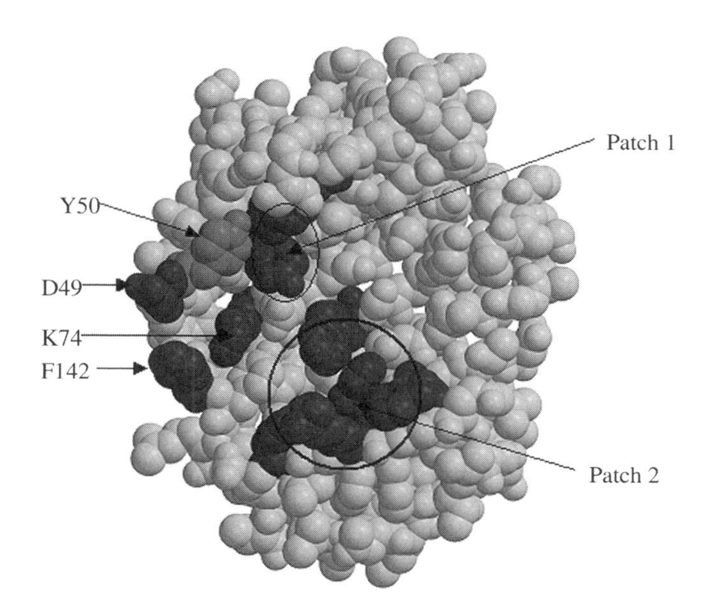

Figure 14.5 BLIP residues critical for binding to TEM-1 β-lactamase. BLIP is shown in spacefill format. The residues determined to be in the hot spot based on loss of binding affinity after substitution with alanine are shown in black. The residue Tyr50 is labeled in light gray because the Tyr50Ala mutation increases the binding affinity of BLIP for TEM-1 β-lactamase. The image was generated using Molscript with Protein Data Bank coordinates 1JTG.

(19). A computer program called Protein Association Rate Enhancement (PARE) was developed that calculates the value of k_{on} of mutant complexes based on the Debye-Huckel energy of interaction between proteins and the experimentally determined k_{on} of wild-type complexes (19). Although it is difficult to calculate the absolute value of either the electrostatic interaction energy or k_{on} of a complex, the calculation of the differences in these values between mutant and wild-type complexes can be quite accurate. The experimental value of k_{on} of wild-type complexes serves as a reference state to calibrate the calculation of all mutants. The accuracy of PARE to predict values of k_{on} of mutants was evaluated for three complexes, barnase/barstar, hirudin/thrombin, and acetylcholinesterase/fasciculin, and the predicted values closely matched with experimental data. With the help of PARE, Selzer et al. designed a BLIP variant that bound TEM-1 β-lactamase with 250-fold stronger affinity than wild-type BLIP by introducing positively charged residues onto BLIP in the vicinity of the binding interface (19). The substitutions specifically increased k_{on} without affecting k_{off} as determined by surface plasmon resonance measurements.

The interaction of the two loops of BLIP with active-site residues of TEM-1 β-lactamase has been extensively studied. Schreiber's group analyzed the role of four critical interactions (two hydrogen bonds and two salt bridges) made by Asp49 of BLIP with Ser130, Arg244, Lys234, and Ser235 of TEM-1 β-lactamase (1, 2). All possible single, double, triple, and quadruple mutants of TEM-1 involving these four residues were constructed. The binding affinities of these TEM-1 mutants with either wild-type BLIP or Asp49Ala BLIP were determined by surface plasmon resonance. The single substitution of each of the four residues of TEM-1 had a moderate effect on the binding affinity, while mutation of Asp49 of BLIP to alanine decreased the binding affinity by 1 order of magnitude (2). Petrosino et al. also identified the importance of Asp49 and Phe142 of BLIP for the TEM-1/BLIP interaction (13). In addition, Huang et al. identified critical residues in the two BLIP loops for binding to TEM-1 β-lactamase by using phage display methods (7).

Results from alanine-scanning mutagenesis of BLIP interface residues indicate that polar interactions do not contribute significantly to the total binding free energy of the TEM-1/BLIP complex. The mutation of any of four serine residues on the BLIP interface does not decrease the binding affinity significantly (22). Moreover, the binding affinities of the serine-to-alanine mutants of BLIP to three other class A β-lactamases (SME-1, SHV-1, and Bla1) did not differ significantly from that of wild-type BLIP. Analysis of an alanine-scanning mutagenesis database indicates that serine and threonine residues, on average, exhibit little effect on the binding affinity of a protein-protein interaction when replaced by alanine (3). However, polar residues such as serine and threonine are frequently found on the binding interface. Therefore, the role of polar residues on protein-protein interactions in general is unclear. It is clear, however, that the BLIP polar residues are not critical for strong binding interactions between BLIP and several class A β-lactamases.

INTERACTION OF BLIP WITH OTHER CLASS A β-LACTAMASES

In addition to TEM-1, BLIP inhibits other class A β-lactamases with a wide range of affinities (Table 14.1). Zhang and Palzkill determined the BLIP residues that are important for binding to three other enzymes including the SHV-1, SME-1, and Bla1 β-lactamases (22, 23). Although SHV-1 shares 68% amino acid sequence identity with TEM-1, it is inhibited by BLIP with an apparent binding affinity (K_i) of 1.1 μM, which is 2,200-fold weaker than the BLIP-TEM-1 β-lactamase interaction. In contrast, the SME-1 and Bla1 enzymes share approximately 30% identity with TEM-1 but are inhibited by BLIP (K_i ~ 3 nM) nearly as strongly as is TEM-1 β-lactamase (22, 23). Residues in the two hot spots (patches 1 and 2) and Asp49 of BLIP for binding TEM-1 are also important affinity determinants for binding the SME-1 and Bla1 β-lactamases. In contrast, the residues in patch 2 do not contribute as strongly to the binding of SHV-1 as do other enzymes. This lack of contribution may explain the weak binding between SHV-1 and BLIP (Table 14.2). Glu73-Lys74 residues of BLIP, which are buried in the interface, are important specificity determinants. In addition, two tyrosine residues (Tyr50 and Tyr143) located on each of the loops of BLIP that insert into the active site of TEM-1 are also significant specificity determinants (22, 23).

The information provided by alanine-scanning mutagenesis has been used to facilitate protein engineering. Based on the binding data, a double mutant (Tyr50Ala/Glu73Ala) of BLIP was predicted to have a dramatic change on specificity for binding to TEM-1 and SME-1 β-lactamase. This prediction was confirmed in that the double mutant exhibits a 220,000-fold change in binding preference for the TEM-1 versus the SME-1 enzyme (22).

MINIMIZATION OF BLIP

The design of small-molecule inhibitors that mimic functional epitopes of proteins is of considerable interest for drug design. Peptide inhibitors that mimic the functional epitopes of protease inhibitors of several proteases have been reported. In addition, small peptide interleukin-1

Table 14.2 Role of interface residues in BLIP for binding to class A β-lactamases

Enzyme	A[a]	B[a]	C[a]
TEM-1	D49, F142, K74, F36, H41, Y53, W112, H148, W150, R160, W162	Y50	E31, S35, S39, G48, Y51, S71, E73, S113, G141, Y143, R144
SHV-1	D49, F36, H41, Y53, W150	Y50, E73, Y143	E31, S35, S39, G48, Y51, S71, W112, S113, G141, F142, R144, H148, R160, W162
SME-1	D49, Y50, Y51, E73, F36, H41, Y53, W112, H148, W150, R160, W162	K74	E31, S35, S39, G48, S71, S113, G141, F142, Y143, R144
Bla1	D49, F36, H41, Y53, H148, W150, R160, W162	E73, Y143, S39, R144	E31, S35, G48, Y50, Y51, S71, K74, W112, S113, G141, F142

[a]Substitution of residues in group A results in significantly decreased binding affinity, while substitution of residues in group B results in significantly increased binding affinity. Substitution of residues in group C does not result in changes in binding affinity.

receptor antagonists that exhibit modest inhibitory activity have been designed (14). An advantage of peptide inhibitors is that it is easy to produce a large collection of peptides with different chemical compositions for screening. For example, phage display or peptide array methods can be applied to screen thousands to millions of peptides for binding to a given target.

Phage display is a powerful technique for studying protein-ligand interactions (16). It is an effective means of screening large collections of random sequence peptides and mutant libraries of proteins for variants that bind a target molecule with high affinity. The peptides or proteins to be screened are usually fused to the N terminus of the gene III phage protein. The gene III protein is a minor coat protein of a filamentous bacteriophage (three to five copies per phage), which is located at the tip of the phage and is responsible for attachment to the bacterial F pilus during normal infection process. Therefore, the peptides or proteins will be displayed on the tip of the phage. The binding affinity between the displayed peptides and the immobilized target determines if the phage is retained on the target protein. Multiple rounds of enrichment for binders from a phage library can result in the identification of tight binding peptides or proteins.

Although the cocrystal structure of TEM-1 with BLIP has been solved, the interface consists of several residues from each partner and buries 2,636 Å^2 of surface area, making it difficult to construct a small-molecule inhibitor to such a large region (20). Phage display has been used to identify a small domain of BLIP that contains a cluster of residues that are sufficient for binding and inhibition of TEM-1 β-lactamase (16). The strategy involves building a phage library by fusing random fragments of BLIP to the gVIIIp gene of M13 phage. The fragments with affinity for TEM-1 β-lactamase were identified after three rounds of binding enrichment (16). Eight in-frame BLIP-derived fragments were identified, and their capacity for binding to TEM-1 β-lactamase was confirmed by phage enzyme-linked immunosorbent assay ELISA. Six of these eight sequences contained the same 20-mer sequence, Cys30-Asp49. Two peptides, Cys30-Asp49 and Lys8-Asp49, were synthesized and tested for inhibitory activity towards TEM-1 β-lactamase. It was found that both peptides inhibit TEM-1, albeit with weak affinity. Two cysteine residues (Cys30 and Cys42) are present in both peptides, and it has been proposed that constraining the peptide at its optimal conformation may be important for binding affinity. An excellent example is the design of an inhibitor of the interaction between fibrinogen and glycoprotein IIb/IIIa (17). A β-turn region in fibrinogen containing the sequence RGD mediates the binding between fibrinogen and glycoprotein IIB/IIIa. Fixing the peptide in a rigid β-turn structure by flanking the peptide with two cysteine residues and replacing backbone hydrogen by methyl group increases the binding affinity of the peptide for glycoprotein IIB/IIIa by three orders of magnitude. Because loop 46–51 of BLIP, which inserts into the active pocket of TEM-1 β-lactamase, is also a β-turn, peptides containing residues 46 to 51 of BLIP were cyclized by oxidizing cysteine residues flanking the 46 to 51 residues. However, the oxidized peptides displayed weaker interactions with TEM-1 β-lactamases than reduced peptides, suggesting that the oxidized peptides adopt a conformation that is not favorable for binding to β-lactamases (14). More studies are required to find the optimal conformation of the peptides for binding to these targets.

SUMMARY

BLIP is a 17-kDa protein isolated from *S. clavuligerus* that binds and inhibits several class A β-lactamases (5, 20). The physiological function of this protein is still unknown. Two other BLIPs (BLIP-I and BLIP-II) were recently identified from *S. exfoliatus* (9, 10). BLIP-I shares 38% sequence identity to BLIP, while BLIP-II is not homologous to BLIP. Alanine-scanning mutagenesis has been carried out to dissect the binding interface between class A β-lactamases and BLIP (22, 23). Two patches on BLIP, which are dominated by aromatic residues, were found to contribute most of the energy for binding to the TEM-1, SME-1, and Bla1 β-lactamases. The hot spot from the second domain of BLIP, however, does not make substantial contributions to SHV-1 binding. This may explain why BLIP binds to SHV-1 β-lactamase with weaker affinity than to the other three enzymes. Three regions, including the two loops that insert into the active pocket of β-lactamase as well as a Glu73Lys74 motif, play a role in determining the specificity of binding. Certain mutants of either TEM-1 or BLIP are able to increase the binding affinity of the complex, indicating that the wild-type interaction is not optimized. Finally, the binding specificity of BLIP to class A β-lactamases can be changed dramatically with minor amino acid substitutions.

References

1. **Albeck, S., and G. Schreiber.** 1999. Biophysical characterization of the interaction of the β-lactamase TEM-1 with its protein inhibitor BLIP. *Biochemistry* **38:**11–21.
2. **Albeck, S., R. Unger, and G. Schreiber.** 2000. Evaluation of direct and cooperative contributions towards the strength of buried hydrogen bonds and salt bridges. *J. Mol. Biol.* **298:**503–520.
3. **Bogan, A. A., and K. S. Thorn.** 1998. Anatomy of hot spots in protein interfaces. *J. Mol. Biol.* **280:**1–9.
4. **Cunningham, B., and J. A. Wells.** 1989. High-resolution epitope mapping of hGH-receptor interactions by alanine-scanning mutagenesis. *Science* **244:**1081–1085.
5. **Doran, J. L., B. K. Leskiw, S. Aippersbach, and S. E. Jensen.** 1990. Isolation and characterization of a beta-lactamase-inhibitory protein from *Streptomyces clavuligerus* and cloning and analysis of the corresponding gene. *J. Bacteriol.* **172:**4909–4918.
6. **Hu, Z., B. Ma, H. Wolfson, and R. Nussinov.** 2000. Conservation of polar residues as hot spots at protein interfaces. *Proteins* **39:**331–342.
7. **Huang, W., Z. Zhang, and T. Palzkill.** 2000. Design of potent β-lactamase inhibitors by phage display of β-lactamase inhibitory protein. *J. Biol. Chem.* **275:**14964–14968.
8. **Jelsch, C., L. Mourey, J. M. Masson, and J. P. Samama.** 1993. Crystal structure of Escherichia coli TEM1 β-lactamase at 1.8 A resolution. *Proteins* **16:**364–383.
9. **Kang, S. G., H. U. Park, H. S. Lee, H. T. Kim, and K. J. Lee.** 2000. New beta -lactamase inhibitory protein (BLIP-I) from *Streptomyces exfoliatus* SMF19 and its roles on the morphological differentiation. *J. Biol. Chem.* **275:**16851–16856.
10. **Kim, M., and K. J. Lee.** 1994. Characteristics of β-lactamase-inhibiting proteins from *Streptomyces exfoliatus* SMF19. *Appl. Environ. Microbiol.* **60:**1029–1032.
11. **Lim, D., H. U. Park, L. De Castro, S. G. Kang, H. S. Lee, S. Jensen, K. J. Lee, and N. C. Strynadka.** 2001. Crystal structure and kinetic analysis of β-lactamase inhibitor protein-II in complex with TEM-1 β-lactamase. *Nat. Struct. Biol.* **8:**848–852.
12. **Matagne, A., J. Lamotte-Brasseur, and J. M. Frere.** 1998. Catalytic properties of class A β-lactamases: efficiency and diversity. *Biochem. J.* **330:**581–598.
13. **Petrosino, J., G. Rudgers, H. Gilbert, and T. Palzkill.** 1999. Contributions of aspartate 49 and phenylalanine 142 residues of a tight binding inhibitory protein of β-lactamases. *J. Biol. Chem.* **274:**2394–2400.
14. **Rudgers, G. W., W. Huang, and T. Palzkill.** 2001. Binding properties of a peptide derived from β-lactamase inhibitory protein. *Antimicrob. Agents Chemother.* **45:**3279–3286.
15. **Rudgers, G. W., and T. Palzkill.** 1999. Identification of residues in β-lactamase critical for binding β-lactamase inhibitory protein. *J. Biol. Chem.* **274:**6963–6971.
16. **Rudgers, G. W., and T. Palzkill.** 2001. Protein minimization by random fragmentation and selection. *Protein Eng.* **14:**487–492.
17. **Samanen, J., F. Ali, T. Romoff, R. Calvo, E. Sorenson, J. Vasko, B. Storer, D. Berry, D. M. Bennett, M. Strohsacker, et al.** 1991. Development of a small RGD peptide fibrinogen receptor antagonist with potent anti-aggregatory activity *in vitro*. *J. Med. Chem.* **34:**3114–3125.
18. **Schreiber, G., and A. R. Fersht.** 1995. Energetics of protein-protein interactions: analysis of the barnase-barstar interface by single mutations and double mutant cycles. *J. Mol. Biol.* **248:**478–486.
19. **Selzer, T., S. Albeck, and G. Schreiber.** 2000. Rational design of faster associating and tighter binding protein complexes. *Nat. Struct. Biol.* **7:**537–541.
20. **Strynadka, N. C., S. E. Jensen, P. M. Alzari, and M. N. James.** 1996. A potent new mode of β-lactamase inhibition revealed by the 1.7 A X-ray crystallographic structure of the TEM-1-BLIP complex. *Nat. Struct. Biol.* **3:**290–297.
21. **Thai, W., A. S. Paradkar, and S. E. Jensen.** 2001. Construction and analysis of β-lactamase-inhibitory protein (BLIP) non-producer mutants of *Streptomyces clavuligerus*. *Microbiology* **147:**325–335.
22. **Zhang, Z., and T. Palzkill.** 2003. Determinants of binding affinity and specificity for the interaction of TEM-1 and SME-1 β-lactamase with β-lactamase inhibitory protein. *J. Biol. Chem.* **278:**45706–45712.
23. **Zhang, Z., and T. Palzkill.** 2004. Dissecting the protein-protein interface between β-lactamase inhibitory protein and class A β-lactamases. *J. Biol. Chem.* **279:**42860–42866.

Enzyme-Mediated Resistance to Antibiotics: Mechanisms, Dissemination, and Prospects for Inhibition
Edited by Robert A. Bonomo and Marcelo E. Tolmasky

Jürg Dreier

15 Active Drug Efflux in Bacteria

Active efflux of drugs is today recognized as an important mechanism of antibiotic resistance. Transport of drugs out of bacterial cells can confer resistance on its own, or it may contribute significantly to high resistance levels. Drug-specific systems like the tetracycline or chloramphenicol transporters were the first to be identified, whereas multidrug export systems have emerged more recently as clinically important resistance determinants. Membrane-associated efflux pumps are now perceived as the predominant underlying mechanism of multiantibiotic resistance in bacteria (36, 147, 155, 156, 178, 209, 218–220, 223, 228, 270, 295, 306, 308, 319, 322).

The cell envelope of most gram-positive bacteria consists of the cytoplasmic membrane and a surrounding peptidoglycan layer. An additional outer membrane is present in gram-negative bacteria. Hence, gram-negative bacteria face the task of exporting molecules across two membranes and the intervening periplasmic space. Depending on the physiological state of a cell, molecules have to be transported over a total distance of 100 to 125 Å (79, 101). Such an export process requires energy and is driven by pumps located in the inner membrane. These energy-transducing proteins have to communicate with partner proteins in the periplasm and the outer membrane in order to assemble a system that can connect the cytoplasm with the extracellular environment. Multidrug efflux pumps are usually encoded by genes on the chromosome, whereas the specific systems are often also found on mobile genetic elements (48).

Active drug transporters are divided into two major classes according to their mechanism of energization (Fig. 15.1; Table 15.1). Most drug efflux systems known to date belong to the class of the secondary transporters using energy stored in the transmembrane electrochemical potential of protons or sodium. Primary transporters, on the other hand, use a primary cellular energy source, like the ABC-type transporters, which hydrolyze ATP to power drug efflux (230). The class of transporters driven by the proton motive force (PMF) can be further divided into four families based on size as well as primary and secondary structural features. These are the major facilitator superfamily (MFS), the small multidrug resistance family (SMR), the resistance nodulation division family (RND), and the multidrug and toxic compound extrusion family (MATE).

The MFS comprises drug-specific pumps like the *tet*- and *otrB*-encoded tetracycline transporters found in many gram-positive and gram-negative bacteria. This class also includes multidrug transporters exemplified by NorA from *Staphylococcus aureus* or MdfA from *Escherichia coli*.

Jürg Dreier, Basilea Pharmaceutica Ltd., Grenzacherstrasse 487, CH-4005 Basel, Switzerland.

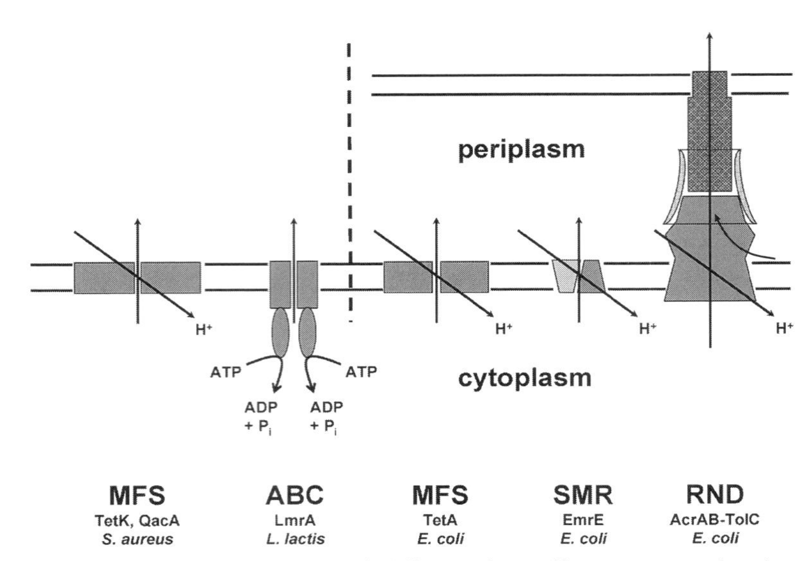

Figure 15.1 Schematic representations of different drug efflux systems. The shapes depict major characteristics where known. More-detailed information about the topology, occurrence in different organisms, and function is given in the text and in Table 15.1.

Members of the SMR family are the smallest known efflux pumps. They generally accept hydrophobic cationic substrates including antiseptics and disinfectants.

Pumps of the RND family are highly relevant in terms of multidrug efflux (289). RND systems accept a wide variety of substrates including antibiotics, hydrophobic cations, and detergents such as bile salts and sodium dodecyl sulfate (SDS). Well-known members are AcrAB-TolC in *E. coli* or MexAB-OprM in *Pseudomonas aeruginosa*. Functional RND transporters across the bacterial envelope are tripartite assemblies of an inner membrane associated pump, a periplasmic adaptor, and an outer membrane factor.

The MATE family of transporters has been identified only recently. The first member described was NorM from *Vibrio parahaemolyticus*, and only a few more have been discovered until now. Most of the MATE transporters use sodium ion gradients across the membrane rather than proton gradients to drive substrate transport.

ABC transporters form a large family of translocases, which hydrolyze ATP to energize the uptake of nutrients or to power the efflux of antibiotics and toxins. This type of transporter is found in bacteria as well as eukaryotes. Bacterial systems include multidrug efflux systems, but also more specific pumps are associated with ATP hydrolysis (see "ABC Transporters" below).

The expression of efflux systems is regulated and responsive to environmental stress factors (102). Enhanced efflux gene expression can cause clinically relevant phenotypes. For example, overexpression of MexAB-OprM can substantially contribute to dual resistance to fluoroquinolones and β-lactams in *P. aeruginosa* (150, 157). In the same bacterium, MexXY, if overexpressed, contributes to decreased amikacin susceptibility and coresistance to fluoroquinolones, carbapenems, and ceftazidime (125, 130, 274, 303).

Recent advances in the crystallization and structure elucidation of membrane-embedded transporters added a lot to the understanding of drug export mechanisms, but the natural function of multidrug transport systems is still debated. The wide variety of substrates indicates a general role in the protection from potentially harmful compounds (154, 155). Certain substrates like bile salts or quorum-signaling molecules are indicative of specific physiological roles, for example in colonization and survival in a host (216). Efflux pumps basically support the bacterial defense against antibiotics and may have evolved or are still evolving under the selective pressure of antibiotics.

EPIDEMIOLOGY

Efflux-mediated resistance has been reported from all over the world. The widespread occurrence of drug-specific systems is thought to originate from transfer of mobile genetic elements. But also the chromosomally encoded multidrug efflux systems are found in most bacteria, probably because of their supposed natural function, the extrusion of dangerous compounds.

Tetracycline transporter genes are widely distributed among bacteria isolated from humans, animals, and the environment. It appears that the situation in commensal bacteria is similar to, and sometimes even worse (as reported, for the genus *Neisseria*) than, the situation in pathogenic bacteria (64, 247). Gram-negative tetracycline

Table 15.1 Examples of drug export systems[a]

System	Organism[b]	Substrates[c]	TMS	Reference(s)
MFS			12 or 14	
TetA	*E. coli*	TET	12	36, 64
NorA	*S. aureus*	ACR, CHL; EtBr, FQ, H33342, NOR, PM, QAC, R6G, TPP	12	36, 318
MdfA	*E. coli*	AG, BA, CHL, CIP, DAR, DM, DXR, ERY, EtBr, IPTG, NEO, NOR, PM, QAC, R6G, RIF, TET, TPP	12	36, 222
Mef(A)	*S. pyogenes*	AZM, CLR, ERY	12	99
QacA	*S. aureus*	CH, EtBr, PI, QAC	14	155
TetK	*S. aureus*	TET	14	36, 64
OxlT	*O. formigenes*	OXL, FOR	12	253
SMR family			4	
EmrE	*E. coli*	ACR, EST, EtBr, MV, MPP, PRG, TET, TPP		36, 82, 222
Smr	*S. aureus*	CV, EtBr, MV, QAC, TPP		36
RND family			12	
AcrB	*E. coli*	ACR, AH, β-LAC, BS, CHL, CV, ERY, EST, EtBr, FQ, FUS, HCS; MMC, NAL, NBD-PC, NOV, OS, PRG, SDS, TET, TX-100		36, 82, 222
MexB	*P. aeruginosa*	ACR, AH, AZM, β-LAC, CHL, CER, CIP, CV, ERY, EtBr, NOV, $3O_{12}$-HSL, RIF, SDS, TCS, TET, TIG, TLM, TMP, TPP, TX-100		36, 161, 222, 270
SmeE	*S. maltophilia*	ERY, EtBr, TET, FQ		11, 36, 222
MtrD	*N. gonorrhoeae*	AZM, β-LAC, CIP, CV, ERY, NOV, PA, PEN, RIF, SPE, TET, TX-100		36, 109, 222
MATE family			12	
NorM	*V. parahaemolyticus*	CHL, EtBr, NOR		36, 222
YdhE	*E. coli*	ACR, CIP, KAN, NOR, STR, TPP		36
PmpM	*P. aeruginosa*	ACR, BA, CIP, EtBr, NOR, OFX, TPP		111
MepA	*S. aureus*	CH, DA, EtBr, NOR, QAC, TIG, TPP		137, 183
ABC transporter				
P-gp	*H. sapiens*	CC, DAR, DXR, DXM, H33342, IS, NBD-PC, NBD-PE, NCP, PM, QIN, R123G, R6G, STE, TAX, VB, VC, VP	12	36
LmrA	*L. lactis*	CC, DAR, DM, DXR, EtBr, H33342, IS, NBD-PE/VP, NCP, QIN, R123G, R6G, TPP, VB, VC	6	36
MacB	*E. coli*	AZM, CLR, ERY, OL	4	36, 142
Msr	*S. rochei*	DXR, ERY, OL, SPI, TET	—	36
MsbA	*E. coli*	LIP	6	36

[a] Additional lists can be found in references 36, 64, 147, 161, 222, 223, and 270.
[b] *E. coli, Escherichia coli; H. sapiens, Homo sapiens; L. lactis, Lactococcus lactis; N. gonorrhoeae, Neisseria gonorrhoeae; O. formigenes, Oxalobacter formigenes; P. aeruginosa, Pseudomonas aeruginosa; S. aureus, Staphylococcus aureus; S. pyogenes, Streptococcus pyogenes, S. rochei, Streptomyces rochei; S. maltophilia, Stenotrophomonas maltophilia; V. parahaemolyticus, Vibrio parahaemolyticus.*
[c] ACR, acriflavine; AG, aminoglycosides; AH, aromatic hydrocarbons; AZM, azithromycin; β-LAC, β-lactams; BA, benzalkonium; BS, bile salts; CHL, chloramphenicol; CC, colchicine; CER, cerulenin; CH, chlorhexidine; CIP, ciprofloxacin; CLR, clarithromycin; CV, crystal violet; DA, diamine; DAR, daunorubicin; DM, daunomycin; DXM, dexamethasone; DXR, doxorubicin; EST, estradiol; ERY, erythromycin; EtBr, ethidium bromide; FQ, fluoroquinolones; FUS, fusidic acid; FOR, formiate; H33342, Hoechst 33342; HCS, hydrocortisone; $3O_{12}$HSL, *N*-(3-oxodecanoyl) homoserine lactone; IPTG, isopropyl-1-thio-β-D-galactopyranoside; IS, indolizine sulfones; KAN, kanamycin; LIP, lipids; MMC, mitomycin; MPP, methyltriphenylphosphonium; MV, methyl viologen; NAL, nalidixic acid; NBD-PC, 7-nitrobenzo-2-oxa-1,3-diazolphosphatidylcholine; NBD-PE, 1-myristoyl-2-[6-(NBD)aminocaproyl]phosphatidylethanolamine; NCP, nicardipine; NEO, neomycin; NOR, norfloxacin; NOV, novobiocin; OFX, ofloxacin; OL, oleandomycin; OS, organic solvents; OXL, oxalate; PA, peptide antimicrobials; PEN, penicillin; PI, pentamidine isothionate; PM, puromycin; PRG, progesterone; QAC, quaternary amine compounds; QIN, quinine/quinidine; R123G, rhodamine-123G; R6G, rhodamine-6G; RIF, rifampicin; SDS, sodium dodecyl sulfate; SPE, spermicides; SPI, spiramycin; STE, steroids; STR, streptomycin; TAX, taxol; TCS, triclosan (irgasan); TET, tetracycline; TIG, tigecycline; TLM, thiolactomycin; TMP, trimethoprim; TPP, tetraphenyl phosphonium; TX-100, Triton X-100; VB, vinblastin; VC, vincristine; VP, verapamil.

efflux pumps are found on plasmid-integrated transposons, whereas gram-positive systems are connected to small plasmids.

In 1953, a *Shigella dysenteriae* strain was isolated as the first tetracycline-resistant bacterium. Between 1988 and 1993 already more than 60% of the *Shigella flexneri* strains isolated in Brazil were tetracycline resistant. Gene transfer by conjugation is the most likely mechanism of transfer in the case of tetracycline resistance. This was suggested by several studies (for an example, see reference 110) and corroborated by the observation that stable tetracycline resistance does not seem to occur in obligatory intracellular organisms like *Chlamydia* and *Rickettsia*. The only known example of horizontal acquisition of a *tet* gene by an obligate intracellular bacterium is a case of *Chlamydia suis* harboring *tetC* (81). The intracellular life cycle of these organisms is not beneficial for conjugational gene transfer, which requires donor and receptor bacteria to grow in very close proximity.

Macrolide resistance efflux genes (*mef*) genes are found worldwide. [In this chapter, the term *mef*(A) is used for both *mef*(A) and *mef*(E) genes as proposed earlier (248).] A lot of information was gained from several survey programs, monitoring the gobal occurrence of macrolide resistance in the gram-positive pathogens *Streptococcus pneumoniae* and *Streptococcus pyogenes*. It is interesting that in some countries *mef*(A) was more often detected than ribosomal methylases (*erm* genes). Several studies suggested that *mef*(A) is prevalent in the United States but not in Europe (for examples, see references 45, 123, 134, 175, 182, 265, and 266). As for every generalization, there are notable exceptions to the rule. Mef(A) was found in the majority of *S. pneumoniae* clinical isolates during recent studies in Germany, Finland, and Spain (217, 233, 234). Other studies indicated *mef*(A) prevalence in erythromycin-resistant *S. pyogenes* from Germany and Spain (12, 213, 235). In Germany, erythromycin resistance in *S. pyogenes* increased from 4% in 1992 to 13.7% in 2000. A nationwide survey in 2004 detected the *mef*(A) gene in 55.6% of erythromycin-resistant isolates. The majority of macrolide-resistant pneumococci in the United States, Canada, South America, Hong Kong, Singapore, Thailand, and Malaysia were cases of *mef*-mediated resistance (87, 187). On the other hand, *mef*(A) genes have been less often detected in central and east Europe, where 26% of *S. pyogenes* isolates carried the *mef*(A) gene (39). A Russian study reported that *mef*(A) was found in only 2.5 and 7.6% of erythromycin-resistant *S. pneumoniae* and *S. pyogenes*, respectively (146).

Some studies suggested that the dissemination of *mef*(A) is a reason for increasing occurrence of macrolide-resistant *S. pneumoniae*. A study in Atlanta, Ga., showed an increase of macrolide resistance from 16% in 1994 to 32% in 1999. The frequency of ribosomal methylation phenotypes remained stable, but *mef*(A) increased from 9 to 26% and was even noted in new serotypes (95). Signs are that *mef*(A) is also emerging as a macrolide resistance determinant in countries where now mostly macrolide-lincosamide-streptogramin B (MLS_B) phenotypes (caused by ribosomal methylation) are detected. This seems to be true for South Africa, for example, where the prevalence of macrolide resistance in *S. pneumoniae* is relatively low, but also for Japan, where the prevalence of macrolide-resistant pneumococci is high (128, 184, 307). Despite the survey programs mentioned above, one should not forget that *mef*(A) is not confined to streptococci but is found in a variety of gram-negative and gram-positive bacteria (41, 207).

It is tempting to take the growing number of reports on identification of efflux genes in clinical isolates as an indication for increased incidence of drug efflux. There may be some truth to this hypothesis, but it has to be interpreted with caution. The fact that all bacterial genomes studied until now contain efflux pumps indicates that they existed before the antibiotic era. Thus, the increased perception of efflux not only reflects increased occurrence of efflux pumps resulting from the use of antibacterial agents but also reflects a growing appreciation of this mechanism as an important contributor to drug resistance.

LABORATORY DETECTION

The mere presence of genes coding for efflux pumps is not sufficient to conclude a phenotypical relevance. Expression of these genes is subject to regulation and may not be high enough, unless induced, to confer a significant efflux phenotype. In addition, efflux on its own is often insufficient for a clinically relevant resistance level. Thus, PCR analysis may reveal the potential for drug efflux but should always be accompanied by complementary phenotypical tests. Although efflux as a sole cause may be too weak for high-level resistance, it can substantially increase the level of tolerability towards antibiotics in synergy with other mechanisms. Fluoroquinolone resistance in gram-negative bacteria due to target site alteration can be substantially supported by efflux (222). Multiple-antibiotic-resistant (*mar*) mutants of *E. coli* have decreased outer membrane permeability due to reduced porin expression and at the same time show increased efflux (232). Macrolide efflux on its own confers intermediate levels of resistance but contributes to high-level resistance in combination with other mechanisms.

It is certainly possible to assess the contribution of drug efflux to an observed resistance pattern, but this usually requires rather elaborate and time-consuming tests.

Although there is no standardized routine test, the following methods may help to diagnose efflux as the mechanistic basis of an observed resistance pattern.

Determination of MIC values in the presence and absence of an efflux inhibitor is a relatively simple way to test for active drug export. Lower MIC values are expected in combination with the inhibitor if efflux is contributing significantly to the resistance phenotype. Reserpine, a plant alkaloid, is often used as a nonspecific efflux inhibitor (32, 42, 177, 195). Reserpine may have a general activity, but its mode of action is not entirely understood. Another complication is that the synergistic effect depends on the drug tested and may be rather low. Newer inhibitors, such as MC-207, 110 (Phe-Arg-β-naphthylamide; see "Strategies To Overcome Drug Efflux" below), often give a much stronger response but usually increase the activity of specific drugs only (84, 100). Sensitivity towards ethidium bromide or crystal violet, both substrates of several RND pumps, has been frequently used to probe for efflux. In fact, ethidium bromide proved to be useful for the detection of the staphylococcal multidrug export systems QacA and QacB (see also "Multidrug Binding Sites" below).

For drug-specific systems, it is possible to conclude an influence of efflux pumps from the observed resistance phenotype. Macrolide resistance conferred by *mef*(A) efflux leads to the so-called M type of resistance. These cells are resistant to intermediate levels (4 to 16 μg/ml) of 14- and 15-membered macrolides like erythromycin and azithromycin, respectively, but remain susceptible to streptogramins, clindamycin, and other lincosamides. Msr efflux pumps confer inducible resistance to 14- and 15-membered macrolides and streptogramin, but not to lincosamides in staphylococci. This induced resistance can reach MIC values of 32 to 128 μg/ml. If a high level of macrolide resistance (i.e., 32 to 128 μg/ml) is accompanied by lincosamide and streptogramin B resistance (i.e., MSL_B phenotype), methylation of rRNA, alone or in combination with *mef*(A), is encountered.

The same approach can be taken with β-lactam antibiotics. Sensitivity to piperacillin in combination with an intermediate to high resistance towards ticarcillin is indicative of efflux in *P. aeruginosa*. Efflux should also confer resistance at intermediate to high levels to ticarcillin-clavulanic acid and aztreonam (29).

Resistance patterns indicating antibiotic efflux should be corroborated by other methods to confirm the presence of efflux systems and to look for other resistance determinants which could lead to a similar result.

MOLECULAR MECHANISM

Many genetic and biochemical experiments have built up a large body of knowledge on substrate transport across the bacterial membrane. More recently, crystal structures of key drug efflux proteins were solved and provided insight at the molecular level. The following paragraphs describe the structure and function of efflux systems.

MF Superfamily

MFS transporters usually fold into a structure with 12-membrane-spanning helices having both N and C termini at the cytoplasmic side of the membrane (258, 260). Some members of the MFS have 14 transmembrane segments (TMS). Thus, the family can be divided into two groups based on the number of TMS.

Tetracycline efflux pumps are prominent and well-studied members of the MFS. They are found in gram-negative and gram-positive bacteria and can be divided into six groups (64, 247). Resistance to tetracycline is conferred by active transport of a tetracycline/cation complex driven by concomitant inward proton flux. TetA is a member of proteins encoded by group 1 *tet* genes, found mainly in gram-negative bacteria. The 401 amino acids of TetA from *E. coli* represent a typical size of MFS transporters. Twelve TMS form two domains, α and β, of about equal size, separated by a putative, large cytoplasmic loop (313). It is thought that the two domains are evolutionarily related and that the 12-TMS architecture is the result of an internal duplication of a domain with six-membrane-spanning α-helices. The construction of chimeric tetracycline pumps showed that α and β domains functioned together only if they originated from closely related systems (254). Interaction between Arg70 and Asp120 in TetB was required for correct positioning of theTMS in the cytoplasmic membrane. Transmembrane helix 4 provided part of the substrate-binding site motif, and together with helices 2, 5, 8, and 11, a water-filled channel seemed to be formed. A topological model for such a channel was derived from a cysteine scan of the complete protein, where all amino acids (except the initial methionine) were mutated to cysteine and probed with *N*-ethylmaleimide (280). Genetic evidence suggested that TetA forms multimers in vivo by interaction between α-subunits only (185, 313). Mutations, which apparently affected energy coupling, were described for the cytoplasmic loops 2–3 and 10–11 (64). This model was supported by a low-resolution X-ray structure of TetA(B) two-dimensional crystals (313). The crystal lattice indicated trimers of about 100-Å side lengths and an interior depression of 25 to 30 Å.

Group 2 transporters, including TetK and TetL, are primarily found in gram-positive bacteria. TetK and TetL have a predicted 14-transmembrane helix structure and a much smaller cytoplasmic loop. These systems confer resistance to tetracycline and chlortetracycline but not to minocycline or glycylcyclines.

Multiple sequence alignments identified several conserved amino acid motifs in 12-TMS and/or 14-TMS MFS transporters (228). Some of these conserved amino acids were found in all MFS transporters and are thought to be important for the structure and function. β-Turns in the cytoplasmic loops between TMS2 and 3 are supposedly required for opening and closing of the channel. Another example is an absolutely conserved arginine in TMS4, with a possible role in proton transport.

The three-dimensional structure of the oxalate transporter OxlT from *Oxalobacter formigenes* was solved to a resolution of 6.5 Å (118). A two-dimensional projection structure of OxlT is available at 3.4-Å resolution (113). The 12 TMS of OxlT were found to be arranged into two sets of six helices related by an approximate twofold axis. Helices 2, 5, 8, and 11 were lining a central cavity across the membrane, which widened at the center of the membrane.

X-ray crystallographic structures of the lactose permease (LacY) and the P_i/glycerol-3-phosphate antiporter (GlpT) were resolved at 3.5 and 3.3 Å, respectively (2, 126). Both transporters belong to the MFS whose members seem to have preserved secondary and tertiary structural elements (259, 298). Thus, the LacY and GlpT structures are of great importance for the understanding of the function of this family of membrane transporters. Like OxlT, LacY and GlpT have two domains of six TMS each. The structures of both LacY and GlpT showed a large internal cavity, open towards the cytoplasm but closed to the periplasm. A large hydrophilic cavity with inner dimensions of 25 by 15 Å was seen at the center of the LacY structure.

The two domains of the LacY monomer were connected by a long loop between helices 6 and 7, a topology also seen in OxlT and GlpT and predicted for tetracycline transporters. Parallel to the membrane, LacY had a heart-shaped appearance of about 60 Å in diameter and length. The transporter had an oval shape of 30 by 60 Å if viewed normal to the membrane.

LacY was crystallized with and without the substrate β-D-galactopyranosyl-1-thio-β-D-galactopyranoside. This substrate bound to the central hydrophilic cavity near the pseudo-twofold axis of LacY, roughly at the same distance from both sides of the membrane. Thus, a likely model for substrate translocation predicts conformational switches, which make the substrate-binding site alternatively accessible to either side of the membrane. The required structural rearrangements could be achieved through rotation of the N- and C-terminal domains around the substrate-binding site. The LacY structure represented the conformation open to the cytoplasm, i.e., inwards. A 60° relative rotation between the two domains would result in a putative structure open to the periplasm. The latter conformation also agreed with results from thiol cross-linking experiments (1). Substrate translocation and proton flux would be linked to conformational changes by a defined series of H-bond and salt bridge formations and disruptions. The structure predicted that Arg302 is directly involved in proton transport, whereas His322 and Glu269 seemed to couple proton transport with substrate binding.

The OxlT structure provided a model for the substrate-bound, closed state of an MFS transporter (114, 119). Two positively charged residues, Arg 272 from helix 8 and Lys 355 from helix 11, were located in the central cavity and were therefore proposed to interact with the negatively charged substrates oxalate and formiate. Similarly, Arg45 and Arg269 in the central pore of GlpT were thought to represent the substrate-binding site. The florfenicol-chloramphenicol resistance marker FloR was described as a 12-TMS MFS transporter driven by the PMF (40). Mutational studies identified Asp23 in TMS1 and Arg109 in TMS4 as crucial for substrate binding. The charges at these positions were essential for the function of the transporter. A similar requirement for charged amino acids at defined positions is known from the LacY permease. Arg144 in TMS5 and Glu126 in TMS4 are in close proximity and form part of the substrate-binding site. An Arg144Lys mutation abolished substrate binding, and Glu126Asp drastically reduced affinity.

Glu26, in the periplasmic half of TMS1, is the only charged, membrane-embedded amino acid in Mdf from *E. coli* (3–5). In contrast to the systems mentioned above, Glu26 could be replaced by various amino acids. A negative charge seemed to be required to export positively charged substrates. The relaxed specificity for residues at position 26 probably reflected the broad substrate specificity of the multidrug transporter Mdf. Even a Glu26Lys mutant was active, but the specificity changed to uncharged substrates like chloramphenicol. Secondary mutations that restored the activity of inactive Glu26 mutants mapped to a region on the cytoplasmic side (TMS4–6) and to another region on the periplasmic side (TMS1+2). The two regions probably defined a part of the translocation pathway.

A general model for the mechanism of MFS transporters was derived from a comparison of the closed conformation of OxlT with the open conformations of LacY and GlpT (119). In this model, the transporter would progress through three different conformations. The transition from a conformation open towards the cytoplasm into a closed state is achieved by a movement of the two halves of the protein relative to each other. The cytoplasmic ends of the two halves are brought together as a consequence of this change. Another movement of the two domains allows the transformation into the third state of the transporter, which is open to the periplasm. As indicated earlier, tetracycline efflux may be achieved by the same way although

LacY functions by lactose/proton symport and tetracycline transporters are fueled by proton antiport. Tetracycline pumps have two domains linked by a flexible loop and seem to form a hydrophilic pore, which is indeed a reasonable feature for the transport of a cationic (i.e., [tetracycline-Mg]$^+$ complex) substrate.

This general model can describe all three basic types of proton-coupled transport. The difference between symport (coupled transport of two substrates in the same direction), antiport (coupled transport of two substrates in opposite directions), and uniport (uncoupled transport of one substrate) would originate from properties that control the conformational switches as a function of substrate binding. Uniporters could switch in the presence or absence of bound substrate. Antiporters would switch only in the presence of bound substrate, and symporters would switch only if both substrates are bound simultaneously.

SMR Family

SMR pumps consist of around 100 amino acids that putatively fold into four helices (66, 211). The multidrug export system EmrE from *E. coli* is the best characterized among these smallest known bacterial drug pumps. Homologues of EmrE were found in many gram-positive and gram-negative bacteria. They accept hydrophobic cationic substrates and, when overexpressed, confer resistance to drugs like ethidium bromide and tetracycline (Table 15.1) (312). EmrE is a small hydrophobic protein of 12 kDa (110 amino acids) with only eight charged residues. One of these residues, the indispensable Glu14 in TMS1, was shown to be crucial for substrate binding. The structure of EmrE was solved to 3.8 Å using a Cys41Ala mutant enzyme and to 3.7 Å in complex with the substrate tetraphenylphosphonium (173, 225). EmrE monomers folded into four helices with the first three forming a loose left-handed bundle. The position of the fourth helix varied between the different monomers.

EmrE was characterized to function as a homo-oligomer, and evidence for dimers, trimers, and tetramers as the molecular unit was proposed. A trimeric assembly, similar to the tertiary structure of MFS transporters, was supported by equilibrium binding studies on isolated EmrE (200, 206, 251). These studies indicated binding of TPP to EmrE in a 1:3 stoichiometry. The tetrameric form was supported by cross-linking experiments, site-directed spin-labeling studies, and an X-ray structure of unliganded EmrE. The X-ray structure of EmrE in complex with TPP indicated that the dimer is likely to be the functional unit (225). The monomers packed in an inverted orientation relative to each other, forming antiparallel assemblies with one TPP molecule bound per dimer (Color Plate 8). An asymmetric dimer structure and a 2:1 stoichiometry of monomers and substrate agreed with other studies (49, 91, 225, 282, 291, 292).

The monomer interface is likely to be the substrate translocation pathway, with Glu14 being crucial for substrate binding. Glu14 is also assumed to link proton flux with drug transport, similar to a glutamate residue near the extracellular side of lactose permease (see "MF Superfamily" above). A central structural feature is a V-shaped spatial arrangement of helices 1 from interacting monomers to form part of the TPP binding site in the middle of the dimer. Mutational analysis identified a cluster of residues that form a common substrate- and proton-binding site including a pair of the functionally important Glu14. These residues together with Glu14 mapped to the proposed translocation pathway (108).

Although the physiologically relevant structure of EmrE is still debated, the general model suggests that the transporter adopts at least two conformations with the drug binding cavity accessible either to the cytoplasm or to the periplasm. Drug and/or proton binding is thought to trigger the interconversion between the two conformations, opening cone-shaped pockets of the channel to one side of the lipid bilayer while closing it to the other. A possible sequence of events would be that a substrate binds to Glu14 of one monomer and is then passed onto Glu14 of the other monomer. Both bindings would happen in exchange for a proton. Structural rearrangements cause a switch from an inward-facing to an outward-facing conformation to release the transported substrate to the outside. Reprotonation of Glu14 would then reset the system for a new unidirectional transport.

RND Family

Gram-negative bacteria use tripartite systems to export substrates across the inner and outer membrane without release into the periplasmic space. The inner membrane transporter is energized by the proton gradient across the inner membrane. The adaptor protein forms a specific complex with the cognate transporter, and together with the outer membrane-spanning channel they constitute the functional efflux system.

Crystal structures were solved for representatives of all three components of tripartite efflux systems, which allowed sophisticated speculation about the structure of entire RND-type efflux systems.

The Inner Membrane Transporter of RND Transporters

With about 1,000 amino acids, RND pumps are distinctly larger than MFS transporters. Nevertheless, they also adopt a 12-TMS structure but, unlike members of the MFS, have two large periplasmic loops between TMS1 and 2 and between TMS7 and 8. High-resolution crystal structures of AcrB significantly contributed to the understanding of membrane-associated pumps in general and RND pumps in particular (197, 198, 226, 227, 271, 317).

AcrB is the inner membrane component of a tripartite RND multidrug pump in *E. coli*. AcrB crystallized as a homotrimer with a membrane-embedded moiety, a central cavity at the bottom of the headpiece, and a funnel-shaped upper part (Color Plate 9). The trimer comprised a 50-Å transmembrane region, with 12 α-helices from each monomer, plus a 70-Å long periplasmic part. The periplasmic headpiece was further divided into an upper TolC docking domain and a lower pore (or porter) domain with thicknesses of 30 and 40 Å, respectively. The pore domain comprised four subdomains: PN1, PN2, PC1, and PC2. PN1 and PN2 corresponded to the polypeptide segments between TMS1 and TMS2, whereas PC1 and PC2 corresponded to the segments between TMS7 and TMS8. Viewed from the top (Color Plate 9), PN1 domains were located inside the headpiece, and the other subdomains were located outside the headpiece. All four subdomains were built from characteristic repeats of β-strand-α-helix-β-strand motifs. An additional extramembrane α-helix of the membrane domain was attached to the cytoplasmic membrane surface. The central cavity at the bottom of the periplasmic part had three openings (or vestibules), one between each protomer, opening into the periplasm. The upper part of the central cavity contained many hydrophobic residues, including 12 conserved phenylalanines. Each monomer contributed four phenylalanine residues, which were involved in substrate binding (317). However, substrate binding was not exclusively achieved by hydrophobic interactions. The highly conserved Asp99, for example, was involved in binding of cationic drugs via electrostatic interaction. Hydrophobic, aromatic stacking and van der Waals interactions seemed to govern substrate binding.

Each monomer also contributed the second α-helix of PN2 to the formation of a central pore, which connected the cavity at the bottom of the headpiece with the bottom of the funnel. Cocrystals of AcrB with four different substrates showed that substrates bound to various places within the central cavity and interacted with different subsets of amino acids (317). The ligand-binding cavity close to the outer leaflet of the inner membrane was remarkably large (~ 5,000 $Å^3$) and able to accommodate several ligand molecules at the same time. Crystal structures of AcrB in complex with doxorubicin or minocycline revealed binding to one monomer only (197). Bound substrate was found at the end of a pocket, rich in aromatic Phe residues, between PN2 and PC1. No substrate was seen in the central cavity.

Each AcrB monomer formed a groove in the outer surface of the transmembrane region extending from the central cavity to the cytoplasmic surface of the inner membrane. The groove was shallow at the cytoplasmic end and deep at the periplasmic end and may be involved in substrate channeling. Further structural analysis of AcrB revealed that the monomers in the observed trimers adopted different conformations (197, 271) which led to the proposal of a rotating, peristaltic-pump-like mechanism of translocation (see "Assembly and Function of a Tripantite RND Pump" below).

Trimer formation was also proposed for MexB from *P. aeruginosa*. Mex B, like AcrB, is an inner membrane multidrug pump from the RND family. Mutants indicated that trimerization of the MexB transporter is important for the interaction with the trimeric outer membrane factor (189). Mutational analysis further suggested close packing of TMS4, 11, and 12 in the three-dimensional structure. Salt bridges between the negatively charged Asp407 and Asp408 in TMS4 and the positively charged Lys940 in TMS10 were suggested to explain results from site-directed mutagenesis (105, 314). In contrast to the translocation model for LacY, the proton and the substrate translocation pathways appeared to be separated from each other in AcrB. These findings agreed well with the crystal structure of AcrB, and a similar overall structure may be expected for MexB.

The Periplasmic Adaptor Protein of RND Transporters

High-resolution crystal structures were published for MexA, the *P. aeruginosa* homologue of AcrA (9, 115) and of a central fragment of AcrA (190). MexA and AcrA monomers consisted of a β-barrel, a lipoyl domain, and an α-helical hairpin subunit in linear succession (Fig. 15.2). The β-barrel contained a barrel of six antiparallel β-strands and one α-helix at one entrance. Such a motif is often found in ligand-binding proteins and may reflect a role of the adaptor protein in efflux substrate binding (115). In fact, direct binding of substrate molecules was reported for another periplasmic adaptor, the ethidium multidrug resistance protein EmrA from *E. coli* (35).

The lipoyl subunit was built up from eight β-sheets arranged similarly to the lipoyl domain of pyruvate dehydrogenase (28) or the biotinyl domain of acetyl coenzyme A carboxylase (20). The lipoyl β-domain structure is conserved among the adaptor proteins as two interlocking sets of four β-strands connected by an intervening sequence of variable length.

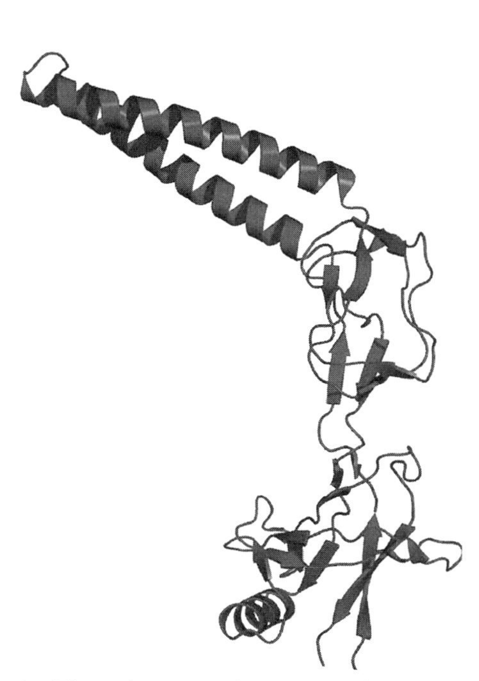

Figure 15.2 Three-dimensional structure of a MexA monomer from *P. aeruginosa* (Protein Data Base [PDB] code 1vf7).

The α-helical hairpin domains of MexA and AcrA consisted of four and five heptad repeats of 47 and 58 Å in length, respectively. The hydrophobic core of the hairpin is conserved in periplasmic adaptor subunits with hydrophobic side chains at positions a and d of the α-helical heptad repeat (133). Alanine is found in position f, whereas serine and glutamic acid are found at position c of the α-helical heptad. A fourth, unstructured region comprised the C and N termini of the protein and made up about one-third of the entire protein. A low-resolution structure of AcrA (20 Å) suggested an elongated shape reminiscent of the MexA monomer structure (22).

Two multimeric structures of six and seven monomers were found in the MexA crystals and could be modeled to form a cylindrical structure of 13 monomers in total. Although the observed multimers were most probably crystallographic artifacts, a spiral assembly of MexA monomers in the form of a sleeve-like structure to stabilize the tripartite system seemed highly attractive. MexA would be oriented such that the α-hairpin is pointing towards the outer membrane and the β-barrel domain makes contacts to the inner membrane via the N-terminal lipid modification.

The Outer Membrane Factor of RND Transporters

The outer membrane factors (OMF) of tripartite multidrug pumps in gram-negative bacteria belong to a large family of membrane proteins involved in the export of various substrates (15, 46, 272). The 2.1-Å crystal structure of TolC from *E. coli* and the 2.56-Å structure of OprM from *P. aeruginosa* showed trimers forming a tunnel-like structure (8, 145) (Color Plate 10). Two-dimensional electron crystallography of OprN and OprM also indicated trimeric complexes of high structural homology to TolC (149, 310).

Monomers were divided into a β-domain, an α-helical domain, and an equatorial domain. Each trimer formed a 12-stranded α/β barrel with a 100-Å left-twisted α-helical barrel projecting across the periplasmic space and a 40-Å outer membrane embedded, right-twisted β-barrel. Aromatic amino acid residues formed a hydrophobic ring around the periplasmic end of the β-barrel. These amino acids are common to all known TolC homologues and were also found in other crystallized outer membrane proteins. The hydrophobic ring probably has an anchor function (15). Two types of helices were found in the α-barrel: long helices, which extended the full 100 Å, and shorter ones of only about one-half that length. The shorter helices stacked end-to-end to form a structure resembling the long helices. In TolC conventional coiled coils were formed by α-helices below the equatorial domain (see below) (Color Plate 10), whereas helices in the upper part packed side to side and built up an almost uniform cylinder.

Four α-helices from each monomer formed six pairs of two-stranded coiled coils. Three of the coiled coils (i.e., one from each monomer) deviated from the cylindrical surface geometry and folded inwards to taper the periplasmic end of TolC. The tapered end of TolC had an effective diameter of about 3.9 Å (Color Plate 10B). The α-barrel of OprM was gradually constricted towards the proximal end and was totally closed at the tip. A closed state was also reflected by a low conductance of TolC in lipid bilayers (16, 27). Interdomain linkers mediated the change from β-barrel to α-barrel. Prolines in this region as well as a structural repeat of α-helices and β-strands are conserved among bacterial outer membrane efflux proteins (133). Aromatic Phe residues supported the structure of the junction in OprM. The polypeptide chains between the short helices of the α-barrel expanded outwards and formed an equatorial mixed α/β domain in TolC. This equatorial collar around the mid-region of the periplasmic barrel has the form of a mixed α/β domain. In OprM right-handed twists of α-helices formed a bulge near the equator.

Outer membrane factors formed a single pore of about 140 to 150 Å in length and a diameter of about 35 Å. The average accessible inner diameter measured 19.8 Å in TolC. The inner surface was predominantly nonpolar but became increasingly negative towards the periplasmic end. Aspartic acid residues at positions 371 and 374 of each monomer formed a ring of six aspartic acid residues in the constriction of the periplasmic entrance of TolC trimers (115). Electronegative entrances are widely conserved throughout the TolC family (133). Actually, the distinct TolC architecture may be found in all members of the TolC family because the key amino acids are conserved (272). A detailed analysis of the TolC structural features, especially of the α-barrel, led to the definition of a TolC-specific heptadic repeat (15, 50, 272). These distinct features may be used to identify other potential α-barrel structures. The periplasmic end of OprM was sealed by a triplet of hydrophobic Leu residues that were sandwiched by two layers of hydrophilic amino acids.

OprM contained covalently attached N-terminal fatty acids that were thought to support membrane anchoring. However, mutational analysis revealed that acylation of the highly conserved N-terminal cysteine is not essential for OprM function (159, 189).

The structure of VceC, the outer membrane channel of a putative tripartite drug efflux system from *Vibrio cholerae*, was solved with a resolution of 1.8 Å (88). VceC showed the same overall structure as TolC and OprM and also shared the feature of a predominantly electronegative channel interior. However, there is only a very low degree of sequence identity between VceC, TolC, and OprM. Functionally important amino acids in TolC are not conserved in VceC. The loops between strands S1 and S2 of each protomer were five amino acids longer in VceC than in TolC and formed a smaller pore at the end of the β-barrel with a diameter of about 6 Å instead of about 13 Å. Hence, the mechanism of VceC and its interaction with other proteins might differ from TolC. This agrees with the fact that VceB, the inner membrane pump, does not belong to the RND family but is a member of the MFS superfamily. An interesting aspect of the VceC structure is that the resolved positions of two octyl-β-glucoside molecules were as expected for lipopolysaccharides (LPS) of the bacterial outer membrane.

Assembly and Function of a Tripartite RND Pump

The wealth of experimental data obtained with AcrAB-TolC from *E. coli* and MexAB-OprM from *P. aeruginosa* led to the proposal of a three-dimensional model for a tripartite RND pump (Fig. 15.3). The inner diameter of about 30 Å in both the AcrB funnel and the TolC channel suggested a direct contact between these structures. The length of the periplasmic domains of AcrB and TolC together (i.e., about 170 Å) would be sufficient to span the periplasmic space. Indeed, binary AcrB-TolC complexes could be detected by cross-linking experiments in AcrA-deficient *E. coli* (281, 288). Head-to-tail complexes were formed by contacts between the ends of TolC coiled coils and periplasmic hairpins of AcrB. These results indicated close proximity of AcrB and TolC in vivo, but no interaction of the isolated proteins was observed in calorimetric analysis (288). Thus, binary complexes between AcrB-TolC may only be transient assemblies, which are later stabilized with AcrA. Binary complexes between adaptor protein and inner membrane pump were reported for AcrAB and MexAB (202, 288).

Genetic analysis (21, 96, 115) and cross-linking experiments (285, 288) pointed to a necessity of the adaptor protein for the assembly and function of the entire efflux system. Stable AcrAB-TolC and MexAB-OprM complexes could be purified via affinity tags attached to either subunit (193, 287). Analysis by isothermal calorimetry showed energetically favorable interactions of AcrA with both AcrB and TolC, and AcrA could be cross-linked to TolC and AcrB independently. Evidence from genetic analysis

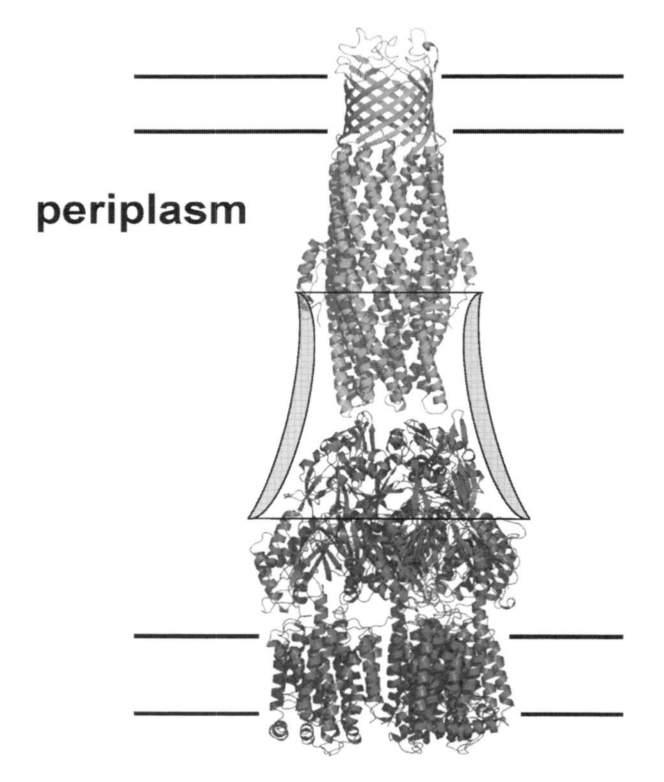

Figure 15.3 Model for a tripartite AcrAB-TolC efflux system. The hypothetical collar-like structure of an AcrA multimer is indicated by the sketch, which overlaps the AcrB and TolC trimers. The boundaries of lipid bilayers are indicated as black lines.

indicated that the β-barrel domain of AcrA made contacts with TolC. AcrA-mediated suppression of TolC mutations required AcrB, suggesting that functional interaction of AcrA with TolC depended on AcrB in vivo (96).

The architecture of the oligomeric assembly of MexA is subject to speculation, and the number of MexA monomers varies between different models from 6 (i.e., three dimers) to 9 (115) or 12 (9). Mutations of conserved amino acids in the N-terminal part (and some at the less conserved C terminus) compromised multimerization and function of MexA (202). In vivo cross-linking of the protein export system HlyB-HlyD indicated that the adaptor protein formed oligomers (285, 321). *P. aeruginosa* was found to contain MexA, MexB, and OprM in a stoichiometry of roughly 2:1:1 (201). Hence, six MexA monomers would associate in the tripartite pump if MexB and OprM formed trimers. Disulfide cross-linking experiments supported a model with two AcrA docking sites per TolC monomer leading to a hexameric adaptor-protein assembly too (276).

The length of the hairpin structure varies between different adaptor molecules and may be the basis of the specificity of transporter-adaptor complex formation (115). Adaptor proteins with short α-helices would fit to a transporter with long periplasmic loops and vice versa. A more detailed explanation for the aforementioned specificity was found by mutational analysis. The N-terminal half of AcrB, especially residues 590 to 612, seemed to be of importance for AcrA binding (286), which agreed well with the proposed model derived from the AcrB crystal structure (198). Work on MexA-MexB interaction indicated that the N terminus of MexA had only an indirect influence on MexB binding. It seemed that the N-terminal part of MexA was required for correct multimerization, a prerequisite for subsequent MexB binding. Mutations of nonconserved amino acids in the C-terminal end of MexA did not affect oligomerization but blocked MexB binding. Thus, the C-terminal part seemed to play a specific role in promoting MexA-MexB recognition (202). Domain-swapping experiments showed that the hairpin domain governed the specificity of the interaction of the adaptor protein with the inner membrane pump (276).

Tripartite efflux pumps have a remarkable although not absolute flexibility with respect to OMF recruitment. Depending on the substrate, TolC can form functional translocases with a range of periplasmic and inner membrane subunits. A hemolysin export system is formed with HlyB and HlyD. HlyBD-TolC is energized by ATP hydrolysis, whereas the multidrug efflux system AcrAB-TolC is using the proton gradient as energy source. Work on the HlyBD-TolC protein export system indicated that the translocase-adaptor complex of efflux pumps was not constantly associated with the outer membrane channel. A transient assembly was triggered when the translocase was loaded with substrate (285). The pump disassembled again after the toxin export was completed. Although essential for hemolysin export, ATP hydrolysis was not required for subunit assembly. This was consistent with the finding that the formation of AcrAB-TolC complexes did not require PMF energy coupling (281, 287, 288). In contrast to HlyBD-TolC, the AcrAB-TolC system seemed to be constitutively assembled based on in vivo cross-linking results in the presence and absence of export substrate (281, 287, 288).

The negatively charged amino acid side chains in the TolC duct may exert an important influence on substrate transport (18). An early pulse of protons could favor the entry of acidic and hydrophilic molecules, whereas a late pulse would promote the release of basic molecules.

The AcrAB system was shown to export certain β-lactams known not to cross the inner membrane (204, 205). This implied that RND pumps are able to take up substrate molecules directly from the periplasmic space or at the outer leaflet of the cytoplasmic membrane and agreed with a proposed direct role of the adaptor protein in substrate binding (see above). Domain swapping between AcrB and AcrD or AcrB and MexB as well as mutational analysis of MexD suggested a role of the periplasmic domain of the transporter subunit in substrate recognition (83, 174, 286). Reconstitution experiments indicated that the RND pump AcrD from *E. coli* could capture aminoglycosides from the periplasm (7). On the other hand, mutations within the major periplasmic loop regions of MexB did not directly change substrate-binding properties but appeared to influence pump assembly and activity (189). Experimental evidence for substrate uptake from the cell membrane was also provided for the MFS transporter QacA (44). The vestibules seen in the AcrB trimeric structure are supposed to be the openings through which substrates from the periplasm and from the outer leaflet of the cytoplasmic membrane gain access to the central cavity.

AcrB crystal structures with several, structurally diverse ligands indicated that amino acids located in the vestibule and in the central cavity region are involved in substrate binding (197, 316, 317). An additional periplasmic binding site within a deep external depression was identified with a mutant (N109A) AcrB. Generally, substrate binding appeared to be rather loose, hinting at a flexible and dynamic recognition and translocation process. The translocation mechanism is expected to allow the passage of substrates into the pump and to restrict the loss of substrate back into the periplasm. Flexible substrate binding through multiple contacts within large, flexible binding pockets was also given as a plausible explanation for the difficulty to map substrate-binding sites by mutations in pumps like MexB and AcrB (189, 316). The different substrate interaction sites found within AcrB possibly reflect the path of molecules during export. However, the exact route of transport remains largely hypothetical. Substrate translocation across the inner membrane, for example, may occur through the central pore or alternatively along a groove observed at the membrane-exposed part of AcrB monomers. The hydrophobic nature of the groove surface suggests that amphiphilic substrates associated with the inner leaflet of the cytoplasmic membrane are favored and would agree with the substrate spectrum of AcrAB-TolC.

Reconstitution experiments with AcrB and AcrD showed that these RND pumps function as proton-drug-antiporters (7, 320). Three charged amino acids (Asp407, Asp480, and Lys940) in the transmembrane region of AcrB formed functionally essential ion pairs possibly involved in proton translocation (198). Protonation of the aspartic acids would disrupt the ion pairs and possibly trigger conformational changes of the α-helices containing the three charged residues. Pore opening would eventually be coupled to the induced changes by remote structural interactions.

The central pore of AcrB was closed in the crystal structure and supposedly only opens when substrates are translocated (Color Plate 9B). A rigid-body rotation of each AcrB subunit upon substrate binding would enlarge the diameter of the periplasmic domain by about 2.5 Å (317). Such a conformational change could lead to a pore opening and subsequent drug extrusion through the central pore into the outer membrane factor. In the crystalline form, TolC and OprM were open towards the outside but closed at the periplasmic end (Color Plate 10B) (8, 9, 115). An elegant mechanism for opening of the TolC channel was proposed by Koronakis et al. (145). Induced, allosteric realignment of inner coiled coils with respect to outer coiled coils would transiently open the TolC channel by an iris-like shutter mechanism. Rotation of the inner pair of coiled coils could open the channel by 25 Å. Such a relative axial displacement of α-helices may in fact turn out to be a general mechanism for conformational switches and was also proposed as an opening mechanism for OprM (8, 50, 272). More recent biochemical experiments showed that efflux is significantly reduced if the flexibility of the coiled coils is constrained by additional disulfide bridges or by chemical cross-linking (85). A phenotype that was consistent with a permanently open TolC channel was observed when specific salt bridges and hydrogen bonds, important for the α-helical arrangement, were disrupted by mutations. Open and closed forms of TolC were confirmed by resistance phenotypes or conductance measurements in lipid bilayers (17, 21). The outer membrane channel should open only during substrate export. The substrate-loaded and energized complex of inner membrane translocase with the periplasmic adaptor possibly triggers channel opening when recruiting TolC. The functional link is most probably made via contacts to the equatorial domain of TolC (272). Such interactions of the different subunits could ensure a control mechanism for the connection of the cytosol with the external environment.

As discussed above, the architecture of the adaptor protein in the functional pump is not entirely clear. MexA oligomers could form an elongated or contracted structure, and conformational flexibility was also suggested by electron paramagnetic resonance studies (129). A hinge region between the α-hairpin and the lipoyl domain was proposed to be flexible because different AcrA conformations were found in a crystal (190). Domain motions around this hinge were also deduced from molecular dynamics simulations of MexA (293). Rather than only forming a passive structural element, AcrA may be involved in opening and closing of the channel. A transport mechanism whereby a contractible adaptor protein brings the inner and outer membranes together (129, 321) seems unlikely in the light of crystal structures showing N and C termini in close proximity.

X-ray analysis of trimeric AcrB complexes with asymmetric monomer structures directly led to the proposal of a rotating, peristaltic-pump-like mechanism of drug transport (197, 271). Only one protomer had substrates bound. One of the three helices in the central pore was tilted towards this protomer and closed a potential exit from the substrate binding pocket into the pore. The protomer that contributed this closing helix showed an open exit and reflected a state expected after substrate extrusion. The third protomer represented an intermediate conformation presumably ready for a new substrate binding event. A drug export pathway was reconstructed based on the assumption that the monomers represented consecutive steps of a transport cycle. Substrates access the binding site from the periplasm through a channel in the periplasmic domain that branches from the vestibule and reaches the central cavity. The drug-binding cavity becomes larger to accommodate the substrate. At this stage, the vestibule is still open but the path to the exit funnel is blocked in the binding protomer by the inclined central helix. Considerable conformational changes cause a closing of the uptake channel from the vestibule and an opening of the exit to the funnel-like opening at the distal end which connects to the TolC duct. Opening and closing of the channel from the vestibule are achieved by movement of the PC2 domain and partial un- and rewinding of the periplasmic end of TMS8. The bound substrate is finally released by a shrinking of the binding pocket.

The TolC docking domain did not undergo significant conformational changes, suggesting that proton translocation and drug binding are transmitted via AcrA.

MATE Family

The MATE family of transporters is the least characterized. The first member, NorM from *V. parahaemolyticus*, was discovered in 1998 only (194). Soon after this discovery, NorM and the *E. coli* homologue YdhE led to the description of the formerly unrecognized MATE efflux protein family (43). The family is represented in all three kingdoms of life, *Eukarya*, *Archaea*, and *Eubacteria*. Further bacterial homologues were later found in *Neisseria gonorrhoeae, Neisseria meningitidis, V. parahaemolyticus, Clostridium difficile, P. aeruginosa, Haemophilus influenzae, V. cholerae, S. aureus*, and *Bacteroides thetaiotaomicron* (43, 59, 78, 111, 127, 137, 183, 192, 252, 311). MepA, capable of extruding clinically relevant antimicrobial agents including tigecycline, was described as the first MATE system in the gram-positive bacterium *S. aureus*. MATE transporters have about 450 amino acids and putatively adopt a 12-TMS structure. The majority of the MATE transporters identified until now use the electrochemical potential of Na^+ across the membrane to drive multidrug export. However, this does not seem to be a general feature of the MATE family because PmpM from *P. aeruginosa* was reported to be a proton gradient-driven MATE pump (111).

ABC Transporters

ABC transporters form one of the largest protein superfamilies. They are found in prokaryotes as well as eukaryotes and transport a large variety of substrates across the membrane (Table 15.1). The observation that bacteria contain both ABC importers and exporters, whereas eukaryotes display ABC export pumps only, suggests an early evolutionary divergence (262). This family of transporters energizes substrate translocation by hydrolysis of ATP. In fact, the presence of so-called ATP binding cassettes (hence the name ABC) is a distinctive feature of ABC transporters. For LmrA from *Lactococcus lactis*, it was even shown that removal of the ABC domain caused the transporter to behave like a secondary transporter using the proton gradient rather than ATP hydrolysis to transport ethidium (299). All ABC transporters contain minimally two ABC domains, also called nucleotide-binding domain (NBD), and two transmembrane domains (TMD) each arranged into 6 TMS (56). Multidrug ABC transporters contribute to multiple antibiotic resistance in bacteria and cause multiple-cancer-drug resistance in humans. The latter problem is to a large extent linked to one of the most intensely studied multidrug ABC transporters, the human P-glycoprotein (P-gp) or MDR1/ABCB1 (14). Multidrug ABC pumps transport hydrophobic drugs and lipids from the inner to the outer leaflet of the membrane. MsbA is an essential flippase found in various bacteria that is responsible for the transport of lipid A, a major component of the bacterial outer membrane. MsbA and LmrA have similar substrate spectra (76). LmrA in fact substituted for the function of a temperature-sensitive mutant MsbA at nonpermissive temperature in *E. coli* (240). LmrA was even able to functionally substitute for P-gp in lung fibroblast cells (296). The human P-gp has all four required domains encoded by a single gene, whereas the bacterial LmrA and MsbA combine only one NBD and one TMD in one protein. Interestingly, MsbA has a high protein sequence similarity to the N-terminal half of P-gp and crystallized as a dimer (see below). LmrA is predicted to form dimers based on homology to the two halves of MDR1. It is believed that the mechanism for ATP hydrolysis is conserved among ABC transporters but that structural and mechanistic details vary. However, valuable insight into the function of ABC transporters was gained by the crystal structures of MsbA from *E. coli* and *V. cholerae* at 4.5 and 3.8 Å, respectively (57, 58, 242). The two structures represented open (*E. coli* MsbA) and closed (*V. cholerae* MsbA) conformations, suggesting large conformational changes during substrate transport. Evidence for a conformation after ATP hydrolysis was provided by the structure of MsbA from *Salmonella enterica* serovar Typhimurium in complex with ADP, Mg^{2+}, inorganic vanadate, and the substrate rough-chemotype LPS (241). The structure of the bacterial ABC transporter Sav1866, solved in complex with ADP to a resolution of 3 Å, seemed to reflect the ATP-bound state of an ABC transporter (73). Sav1866 crystallized as a dimer showing the general features of an ABC transporter but with an architecture considerably different from that of MsbA.

Low-resolution structures of human P-gp and YvcC from *Bacillus subtilis* also suggested that large conformational changes occur during the transport process (54, 250).

MsbA formed a heterodimer of approximately 140 Å in length with a 50-Å TMD and a total molecular mass of about 130 kDa. Each monomer consisted of an NBD linked to a TMD by an α-helical intracellular domain (ICD). The ICD consisted of conserved amino acids and seemed to provide the structural flexibility required to achieve an ATP-competent state by dimerization of NBDs. The 12 TMS of the TMD formed a cone-shaped cavity, which was lined with charged and polar residues. On either side of the cavity, TMS2 and TMS5 defined two substantial openings of 12 to 25 Å facing the lipid bilayer. Contacts between TMS2 of one monomer and TMS5 of the other monomer formed the main dimerization interface and served as a hinge region for chamber closure. The structures of the open and closed MsbA conformations supported the hypothesis that the TMD adopts a conserved structure. NBDs were substantially rotated and took different positions in the two structures but kept the same contact to the TMD. A similar contact was described for the structure of a vitamin B_{12} importer, BtuCD, in *E. coli* and is possibly the basis of energy transduction from the NBD to the TMD common to all ABC transporters (166).

NBDs are the most conserved part of ABC transporters. They consist of two domains, which contain the hallmark Walker ATP-binding consensus sequences. The Walker B and ABC consensus sequences are part of the helical α domain, while the β domain contains the Walker A sequence. Comparison to the structure of BtuCD indicated a movement of the β domain towards the α domain, resulting in a catalytically active conformation.

There is biochemical evidence for MsbA and P-gp that substrate binding to the TMD is triggering ATP binding to the NBD (160, 186, 304). Bound nucleotides significantly altered the structure of the ATP-binding domain of a glucose uptake system from *Sulfolobus solfataricus* compared to the nucleotide-free state (300). Rearrangement of NBD domains could probably control the affinity for ATP and contribute to the regulation of substrate recognition and ATP hydrolysis. In a general model, ATP binding to NBD elicits tight dimerization and mediates subsequent movement of the α-helices in the TMD. Such a movement would mediate a chamber closing and a concomitant reduction of its volume. The positively charged arginine and lysine residues, which line the central chamber, are most probably solvated and therefore represent an unfavorable environment for hydrophobic substrates. The substrate then spontaneously flips to its transported position as a consequence of the unfavorable situation. An alternative pathway was suggested, in which only the sugar headgroups of LPS are sequestered and moved in the chamber of the transporter. In this model, the hydrophobic tails would be dragged through the bilayer (242). Site-directed spin labeling and electron paramagnetic resonance studies showed a cycle of dramatic conformational changes that coupled ATP hydrolysis with substrate transport (77). Conformational rearrangements in the NBD could be propagated to the TMD via the ICD and cause the channel to open and to release the substrate. NBDs would finally dissociate, and the ABC transporter would adopt the resting state again. One molecule of ADP and two molecules of LPS were bound per dimer in *S. enterica* MsbA crystals. Thus, two LPS molecules seem to be flipped per molecule of ATP hydrolyzed. While the flippase mechanism adequately explains how substrates like lipid A are moved from the inner to the outer leaflet of the membrane, it might not satisfactorily explain the mechanism of other ABC transporters. Experiments with LmrA suggested that fluorescent substrates not only were flipped within the membrane but also were actively brought into the extracellular environment (297).

The macrolide-specific antiporter MacB from *E. coli* is an ABC-type membrane transporter with only four TMS (141). MacB requires the periplasmic adaptor protein MacA and the outer membrane factor TolC for efflux activity (142). MacB may function as a trimer, assuming that MacAB-TolC forms complexes similar to the RND systems AcrAB-TolC and MexAB-TolC. Hence, the functional MacB trimer would have 12 TMS like the human P-gp or the dimeric forms of the bacterial systems LmrA and MsbA (Table 15.1). On the other hand, MacB is a half-type ABC transporter with respect to its single NBD domain and may therefore form dimers. MacB has one large, hydrophilic periplasmic loop of 200 amino acids between TMS1 and TMS2. If six periplasmic loops are needed for the interaction with TolC, analogous to the AcrB-TolC complex (see "RND Family" above), then MacB may even assemble into hexamers.

Members of a subfamily of the ABC transporters have been found to confer resistance to MLS antibiotics. These antibiotic resistance (ARE) proteins are remarkable because they consist of only two NBDs on a single polypeptide chain (38). Several members of the ARE subfamily were found in MLS-producing *Streptomyces* but also in pathogens like streptococci, enterococci, and staphylococci (188, 243). The staphylococcal systems VgaA and MsrA are structurally similar but have different substrate specificities. For VgaA it was shown that the two NBDs must be located on the same polypeptide to confer streptogramin A resistance. Mutational analysis revealed a requirement of both sites for function (60). Mel, like MsrA, contains an ATP-binding domain but lacks hydrophobic segments. Mel contributes to erythromycin resistance in *S. pneumoniae*, where it is found together with *mef* on conjugative transposon-related elements (13).

Until now, all efforts to identify permeases expected to interact with ARE proteins to form functional ABC transporters have failed. VgaA was found in the membrane fraction of bacterial clinical isolates (60), but it remained unclear how VgaA was able to interact with the membrane. Interaction of the ARE component with the as yet unidentified membrane-associated protein(s) would be an attractive explanation for membrane anchoring.

Multidrug Binding Sites

The broad substrate range of multidrug export systems is based on binding sites, which are able to accommodate structurally diverse molecules. Structures of AcrB in complex with different ligands provided important clues about multisubstrate binding (197, 316, 317). It appears that different sets of residues are used to interact with different substrates. In addition, regulators of multidrug efflux pump expression often bind to similar compounds as their cognate transporters (102). QacR, a member of the TetR/CamR family of transcription regulators, controls the expression of the QacA efflux pump in *S. aureus*. QacR binds a variety of compounds including quaternary ammonium compounds, which are also substrates of QacA (103). Thus, the architecture of multidrug-binding sites was further clarified by the high-resolution structure of QacR repressor bound to each of six different substrates (268). The proton antiporter QacA exports mono- or divalent cationic, lipophilic antimicrobial compounds including antiseptics and disinfectants like cetrimide, benzalkonium chloride, and chlorhexidine (44, 191). Different binding sites for mono- and divalent cations were proposed because monovalent cations (e.g., benzalkonium) competitively inhibited ethidium efflux whereas divalent cations (e.g., chlorhexidine) were noncompetitive inhibitors. QacB, in contrast to QacA, confers only low resistance to divalent cations. This clear phenotypical difference is based on a difference of only seven amino acids between QacA and QacB (210). In this context it is remarkable that QacA has negatively charged residues, important to confer resistance to divalent cations, at positions 322 or 323, where QacB has uncharged residues.

Crystal structures of QacR showed a large (~ 5,000 $Å^3$) substrate-binding pocket with several glutamates and a large number of aromatic residues (268). QacR was crystallized with simultaneously bound ethidium bromide and proflavine (267). A multidrug-binding pocket that can accommodate a large number of chemically and structurally diverse molecules was also found in MexB as described in "The Inner Membrane Transporter of RND Transporters" above. Multiple aromatic residues seem to be a universal feature of multisubstrate binding pockets. They are found not only in the binding pockets of QacR and MexB but also, for example, in the RND pump EmhB from *Pseudomonas fluorescens* (112), the multidrug-binding protein BmrR from *B. subtilis* (324), and the human xenobiotic receptor PXR (305).

A negatively charged Glu26 in the substrate-binding site of MdfA from *E. coli* was essential for the recognition of some but not all cationic substrates (5). This result was explained by the involvement of a second multidrug recognition site located on a flexible loop (4). Contacts to these different binding determinants would likely depend on the structure of the substrate. Further analysis of the MdfA multidrug-binding site showed that mutations affecting transport activity clustered in two regions (3). One cluster was located at the cytoplasmic side of the membrane, and the other one was at the periplasmic side. It was speculated that the first cluster contributes to the initial substrate recognition and that substrates pass on to the second cluster on their way to the periplasm.

Studies with AcrB and EmhB indicated that Asp101 was involved in substrate recognition and was therefore important for antibiotic resistance (112, 199, 316).

CLINICAL IMPORTANCE

Resistance against Clinically Important Antibiotics

Apart from contributing to resistance against anticancer chemotherapy, drug efflux is highly relevant for the successful treatment of bacterial infections by tetracycline, macrolide, and fluoroquinolone antibiotics.

The spread of tetracycline resistance genes since the 1950s has significantly impaired the efficiency of tetracycline for the treatment of infections. By the mid-1970s tetracycline resistance was common among *Enterobacteriaceae*, staphylococci, and bacteroids. Nowadays, tetracycline resistance is also commonly found among the respiratory tract infection pathogens *S. pneumoniae*, nonencapsulated *H. influenzae*, and *Moraxella catarrhalis*. Generally, most of the tetracycline resistance genes code for efflux and few code for ribosomal modification (64, 246, 247). In gram-negative bacteria, efflux is even being recognized as the major mechanism for tetracycline resistance. The most widely distributed efflux genes are those coding for TetA/TetB in gram-negative bacteria and for TetK/TetL in gram-positive bacteria.

While efflux clearly has a high impact on tetracycline activity, the situation is somewhat more complex in the case of macrolide antibiotics (248). Macrolides are recommended and commonly used in the treatment of respiratory tract infections because many pneumococci become resistant to β-lactam antibiotics. Thus, the emergence of macrolide-resistant pneumococci is a problem. Macrolide efflux (*mef*) genes are found in *S. pneumoniae*, *S. pyogenes*, corynebacteria, enterococci, and micrococci. *mef* genes code for MFS pumps, which are specific for 14- and 15-membered macrolides and therefore cause the so-called M phenotype of macrolide resistance (see also "Laboratory Detection" above). Not all experts, however, share the same opinion about the clinical relevance of the intermediate M-type resistance levels of 4 to 16 μg/ml (170, 172).

A different MFS transporter, called MreA (for macrolide resistance efflux), was proposed to be responsible for resistance against 14-, 15-, and 16-membered macrolides in *Streptococcus agalactiae* (67). The *mreA* gene product was later reported to have a flavokinase activity, and no correlation to macrolide resistance in *S. agalactiae* was found (68). Nevertheless, both studies agreed that *mre*A was linked to drug efflux when the gene was cloned into *E. coli*.

MsrA and MsrB are responsible for the MS phenotype in staphylococci, which describes resistance to 14- and 15-membered macrolides and streptogramin B. The related system MsrC can increase MIC levels eightfold in *Enterococcus faecium*. Recently, *msrA*-dependent efflux was even linked to erythromycin resistance in community-associated methicillin-resistant *S. aureus* (MRSA) (47).

For the time being, even highly erythromycin-resistant pathogens are reported to be susceptible to telithromycin as observed for *S. pneumoniae* (86, 87, 123, 128, 146, 187, 217).

NorA in *S. aureus*, PmrA in *S. pneumoniae*, and (clinically less relevant) Bmr and Blt in *B. subtilis* are clear examples of the high relevance of drug efflux for fluoroquinolone resistance in gram-positive bacteria. Based on phenotypic characterization, drug efflux has also been related to fluoroquinolone resistance in viridans group streptococci (90, 104, 106). Although fluoroquinolone efflux is known to occur in other gram-positive pathogens, the contribution to resistance phenotypes is often unclear or negligible. Enhanced quinolone efflux was reported for clinical *S. pneumoniae* isolates from England (132, 214) and for about one-third of *S. aureus* isolates (MRSA and methicillin-susceptible *S. aureus*) from 23 European hospitals (264).

The MFS members CmlA and FloR specifically export chloramphenicol and the structurally related veterinary drug florfenicol (269). They confer inducible resistance in many gram-negative bacteria including *P. aeruginosa*, *H. influenzae*, and the *Enterobacteriaceae*. High resistance levels (≥128 μg/ml) against chloramphenicol and florfenicol were measured for *E. coli*.

MIC values of 128 μg/ml for chloramphenicol and florfenicol resulted from TexA-mediated efflux in the gram-positive bacterium *Staphylococcus lentus* after induction with either drug (148, 294).

Not only are drug-specific systems of concern, but also multidrug exporters seem to cause clinically relevant phenotypes (215). The emergence of multiresistant gram-negative bacteria is of growing concern especially in hospitals. The main gram-negative problem pathogens due to increasing resistance are *P. aeruginosa*, *Stenotrophomonas maltophilia*, *Acinetobacter baumannii*, and the *Enterobacteriaceae* (34, 162, 245). *P. aeruginosa* already accounts for 10% of inpatient isolates in the United Kingdom (163). Predominantly strains of *P. aeruginosa* and *A. baumannii* have become resistant to most antibiotics and cause problems in specialist units and in the treatment of compromised hospital patients. Multidrug resistance has evolved in *Enterobacteriaceae* to a point where only carbapenems remain a good choice for treatment. Resistant pathogenic bacteria are also a threat to particular groups of patients, for example, *P. aeruginosa* in the context of cystic fibrosis (162, 163). Transporters of the RND family, which is most relevant for multidrug efflux, recognize aminoglycosides, macrolides, β-lactams, chloramphenicol, fluoroquinolones, tetracyclines, antifolates, and other substrates (Table 15.1) (222). Efflux is often essential not only for intrinsic drug resistance but also for elevated levels of resistance if overproduced. Up-regulation of MexCD-OprJ and MexEF-OprN was selected in *P. aeruginosa* by ciprofloxacin therapy in cystic fibrosis patients as the predominant fluoroquinolone resistance mechanism (131). Overexpression of MexXY and MexAB-OprM was reported for multiresistant *P. aeruginosa* from Europe and the United States (53, 124, 165, 309). Such clones gain even further importance if they become the origin of nosocomial outbreaks, as reported from France (30, 80, 124).

High-level resistance is often the result of drug efflux combined with target site mutations like nalidixic acid resistance due to efflux and *gyrA* mutation (52). *A. baumannii* has a reputation for causing clonal outbreaks in hospitals (37, 117, 229) and can rapidly, i.e., within weeks, develop fluoroquinolone resistance through a combination of *gyrA* and *parC* mutations with up-regulation of AdeB efflux (117). The perception of efflux as an important factor to aggravate resistance in concert with other determinants is further corroborated by several other observations. Concomitant reduction in porin expression and increase in MexEF-OprN expression were shown for the *nfxC* mutant of *P. aeruginosa*, manifesting resistance towards quinolones, chloramphenicol, and trimethoprim (143). The underlying mutation affected the regulatory gene *mexT* and caused both phenomena at the same time. Another example of a single mutation causing multidrug resistance was found in *P. aeruginosa*. In this case, *mexR* mutations led to resistance towards β-lactams (not imipenem), fluoroquinolones, chloramphenicol, macrolides, disinfectants, detergents, and organic solvents as a result of up-regulated expression of MexAB-OprM (220).

Drug efflux systems not only cause problems now but also are likely to have an impact on the future development of antibiotic resistance. Overexpression of efflux pumps apparently promotes the development of mutations in the genes coding for the target site of an antibiotic (140, 177). *E. coli mar* mutants generated fluoroquinolone-resistant mutants much faster than did wild-type strains (70). Similar observations were also made for *S. pneumoniae* (176) and *S. aureus* (179) (see also "Antiseptics and Biocides" below). Thus, induction can cause multiresistance to literally all known classes of antibiotics in one step (156). Even the combination of a multidrug efflux pump with a drug-specific transporter has been reported. *S. enterica* serovar Typhimurium combined AcrAB-TolC with FloR and TetG to achieve high-level chloramphenicol, florfenicol, and tetracycline resistance (25).

Efflux systems can be subject to regulation by a global stress response circuit. Mar and Sox affected over 60 genes in *E. coli* including *acrAB*, *ompF*, and a chromosomally encoded tetracycline efflux gene (23, 151, 232). Similar to the case of MexR described earlier, the induced Mar phenotype caused upregulated AcrAB expression and reduced OmpF expression (70). Like this, susceptibility to norfloxacin, chloramphenicol, tetracycline, nalidixic acid, and β-lactams decreased in *Klebsiella pneumoniae* (283).

Several genes were identified in the genome of *Mycobacterium tuberculosis* that putatively code for efflux pumps of the MFS and ABC transporter families. This notion is worrying because efflux may contribute to an increased incidence of multidrug-resistant *M. tuberculosis*, leading to a serious problem for the control of tuberculosis (219, 302). Efflux-mediated resistance towards metronidazole and other antibiotics was reported for *Helicobacter pylori* (148, 294). Thus, efflux may weaken the effectiveness of the commonly used combination of metronidazole with amoxicillin and/or clarithromycin.

Antiseptics and Biocides

The finding that antiseptics and biocides like benzalkonium, cetrimide, and chlorheximide are substrates of efflux systems caused a debate about the danger of cross-resistance between biocides and antibiotics (92–94, 98, 152, 153, 221, 222, 255–257). Biocide resistance may in fact become an issue in hospitals. In addition to biocide tolerance in *P. aeruginosa* and *S. maltophilia*, clonal spread of QacA and QacB in *S. aureus* was reported. A study in Japan revealed that 32.6% of MRSA were harboring the *qac* genes (10).

Efflux systems like AcrAB in *E. coli* or MexAB-OprM, MexCD-OprJ, MexEF-OprN, and MexJK in *P. aeruginosa* recognize both biocides and antibiotics as substrates. The *P. aeruginosa nfx*B mutation, for example, leads to overexpression of MexCD-OprJ and therefore causes resistance to triclosan as well as to fluoroquinolones, chloramphenicol, β-lactams, and trimethoprim (65). Biocide-induced multidrug resistance was recently shown in vitro for *S. maltophilia* (261). Triclosan-selected mutants of *S. maltophilia* overexpressed the multidrug efflux pump SmeDEF and showed increased MIC values for chloramphenicol (2-fold), tetracycline (2- to 4-fold), and ciprofloxacin (8- to 16-fold).

Several bacterial efflux systems are known to export metal ions including cadmium, zinc, cobalt, nickel, copper, silver, arsenate, chromate, and antimonite (203, 249). Silver cations are microcidal and therefore used against bacteria, viruses, and eukaryotic microorganisms. Silver-containing formulations are widely used to treat burns, wounds, and ulcers and to coat catheters in order to prevent biofilm formation (273). Similar to triclosan, silver-containing products are used not only in the clinical environment but also in household products. Genetic analysis of a silver-resistant *Salmonella* strain from a burn ward revealed a silver-specific efflux system (107). A periplasmic silver-binding protein was associated with two pumps. One is a P-type ATPase (SilP), and the other is a membrane potential-dependent RND-type cation/proton antiporter (SilCBA). Expression of *sil* genes is regulated by a two-component membrane sensor and transcription responder system. *sil* genes were further found in *E. coli* K-12 and *E. coli* O157:H7 (273). Earlier work already indicated that *E. coli* mutants with increased active efflux could be selected by silver in vitro (158).

Concern about solvent-antibiotic cross-resistance is nourished by the fact that some RND efflux pumps have a broad enough substrate specificity to export organic solvents in addition to antibiotics (89, 231). The *P. aeruginosa* systems MexAB-OprM and, less efficiently, also MexCD-OprM and MexEF-OprN, are able to pump out *n*-hexane and *p*-xylene. Cyclohexane, *p*-xylene, and pentane are substrates of AcrAB-TolC in *E. coli*. Overexpression of the *mar*A gene increased antibiotic resistance in *E. coli* as discussed above and also led to increased cyclohexane tolerance (19).

It is currently not clear if the improper use of detergents and biocides could lead to clinically relevant resistance phenotypes. A recent study found no correlation between the use of antibacterial cleaning agents and antibiotic-resistant bacteria in the home environment (71). However, more work is required to assess the potential risk indicated by experimental evidence.

Virulence and Biofilm Formation

P. aeruginosa produces autoinducers such as *N*-butyryl-L-homoserine lactone (C_4-HSL) and *N*-(3-oxododecanoyl)-L-homoserine lactone (3-oxo-C_{12}-HSL). These molecules mediate cell-density-dependent gene expression, which is crucial for biofilm formation and toxin expression. Accumulation of C_4-HSL was observed with increasing cell density of *P. aeruginosa* cultures, and recent work showed that extracellularly added C_4-HSL enhanced the expression of MexAB-OprM in stationary-phase cells (181, 263). MexAB-OprM from *P. aeruginosa* was shown to transport 3-oxo-C_{12}-HSL but not C_4-HSL (212). These findings indicated a direct link of MexAB-OprM expression with cell density. On first sight HSL-mediated efflux pump expression seems to be an explanation for the intrinsic antibiotic resistance of *P. aeruginosa* biofilms. However, the influence of multidrug transporters on the antibiotic resistance of biofilms is probably not that simple since other studies found no increase of MexAB-OprM, MexCD-OprJ, MexEF-OprN, and MexXY expression in *P. aeruginosa* biofilms. Hence, only a minor role in antibiotic resistance of the biofilm was attributed to these efflux systems (75).

Virulence of *P. aeruginosa* is intimately linked to quorum sensing by HSL-induced virulence factor production. Multiresistant *P. aeruginosa*, which overexpressed MexEF-OprN (*nfxC*-encoded phenotype), were shown to produce less C_4-HSL and to make reduced levels of the virulence factors pyocyanin, rhamnolipid, and elastase (72, 144). It was further shown that overexpression of MexEF-OprN per se was responsible for the loss of virulence in a rat pneumonia model, but the functional link remained unclear. In contrast to MexEF-OprN, overexpression of MexAB-OprM (*nalB*- and *nalC*-encoded phenotypes) or MexCD-OprJ (*nfxB*-encoded phenotype) did not affect virulence. On the other hand, deletion of MexAB-OprM in *P. aeruginosa* PA01 led to mutant strains with significantly reduced invasiveness (120). The mutant strain was, in contrast to the wild-type cells, unable to cause fatal infections in a murine model. MexCD-OprJ deletions had no effect, and MexB or MexXY deletions only an intermediate effect, on virulence. Attenuated virulence of *P. aeruginosa* in rats and plants was also found when MexI and OpmD of the MexGHI-OpmD efflux system were mutated (6). MexI and OpmD mutants displayed impaired production of the quorum-sensing molecules 3-oxo-C_{12}-HSL and C_4-HSL. A comparable observation was made with *Burkholderia pseudomallei*, in which the RND system BpeAB-OprB was linked to virulence via the inhibition of quorum sensing (55).

Efflux pumps are probably involved in quorum sensing because they have an impact on the intracellular concentration of mediator molecules and therefore influence the timing of cell-density-dependent gene expression.

It remains to be clarified if efflux pumps really offer a possibility to block biofilm formation and/or reduce virulence.

Strategies To Overcome Drug Efflux

The development of new generations of existing antibiotics is one possible way to react to the occurrence of efflux systems (224). The goal to find molecules, which keep the target specificity but are not or only badly recognized by efflux pumps, was reached in the case of macrolides and tetracyclines. The ketolide telithromycin is active against macrolide-resistant pathogens including those with the *mef*-mediated M phenotype (see also "Laboratory Detection" and "Resistance against Clinically Important Antibiotics" above). However, the activity of telithromycin is slightly lowered by efflux, and telithromycin was shown to be a substrate of *mef*(A) in *S. pyogenes* (51, 62).

Members of the glycylcyclines, a new class of tetracyclines, overcome efflux-mediated tetracycline resistance (63, 121, 164, 275, 279, 323). Tigecycline, the first glycylcycline on the market, is active against multiresistant *S. maltophilia* and *A. baumannii* and bacteria with acquired tetracycline efflux (TetA through E and TetK) (31, 164). Unfortunately, tigecycline does not evade chromosomal efflux systems in *Proteeae* and *Pseudomonas* and the acquired tetracycline efflux pumps are likely to evolve towards efficient export of the new-generation drugs too. In fact, *tetA* and *tetB* could be mutated in vitro to confer glycylcycline resistance in *E. coli* (63). Veterinary *Salmonella* isolates were found with mutated *tetA* genes conferring reduced susceptibility to earlier, investigational glycylcyclines (290). Tigecycline was also shown to be transported by the multidrug efflux pumps MexAB-OprM and MexCD-OprJ from *P. aeruginosa* (74) and most probably by AcrAB in *Proteus mirabilis* and *E. coli* (301). In addition, MepA, a new MATE-type efflux pump, was identified to be responsible for reduced sensitivity of *S. aureus* towards tigecycline (137, 183). Thus, alternatives to the "new-generation" strategy are needed to combat efflux-mediated resistance.

Several efflux inhibitors have been developed in an effort to restore the potency of antibiotics that lost power due to efflux-mediated resistance mechanisms (136, 138, 167, 169, 171, 180, 208, 236, 315). A lot of work focused on the development of tetracycline efflux inhibitors. Various semisynthetic tetracycline derivatives were made and led to 13-CPTC (13-cyclopentylthio-5-hydroxy tetracycline), a competitive inhibitor of the TetB pump. The combination of 13-CPTC and doxycycline demonstrated synergistic activity against TetA- or TetB-expressing *E. coli*. Other inhibitors described were only weakly active (Ro 07-3149) or seemed to indirectly inhibit tetracycline efflux by membrane effects (UK-57,562) (122, 135).

Much effort was also concentrated on inhibiting fluoroquinolone efflux in *P. aeruginosa* and *S. aureus*. Interesting targets were therefore MexAB-OprM, MexCD-OprJ, MexEF-OprN, and NorA. Peptidomimetic compounds with the ability to enhance levofloxacin potency in *P. aeruginosa* were identified and optimized (168, 238). Representatives of this class are MC-02,595 (D-Orn-D-hPhe-3-aminoquinoline) and MC-207,110 (L-Phe-L-Arg-β-naphthylamine) (69, 237, 239). A concentration of 5 to 10 μg/ml of MC-207,110 caused an eightfold decrease of levofloxacin MICs against intrinsically resistant *P. aeruginosa*.

In the case of resistance due to overexpression of efflux pumps, a 32- to 64-fold reduction of levofloxacin MIC was detected. Eightfold-decreased nalidixic acid MICs were measured at 20-μg/ml MC-207,110 for some clinical isolates of *A. baumannii* and *S. maltophilia* (244). Further optimization of the MC-207,110 lead compound led to inhibitors with activity against *P. aeruginosa* in a murine neutropenic thigh model (24). Other compounds of the same class were potentiating macrolides in *H. influenzae* and chloramphenicol in *E. coli* and *Salmonella*.

Synthetic NorA inhibitors and reserpine (see "Laboratory Detection" above) were found to enhance the activity of fluoroquinolones against *S. aureus* (179). Several plant natural products, including 5′-methoxyhydnocarpin also inhibited NorA and had a synergistic effect with berberine against *S. aureus* (26, 277, 278, 284). Inhibition of NorA turned out to lower the frequency of mutations causing fluoroquinolone resistance (see also "Resistance against Clinically Important Antibiotics" above). This finding was in agreement with the observation that MC-207,110 to a large extent prevented the appearance of both efflux-mediated resistance and gyrase mutations as a cause of fluoroquinolone resistance (155). Pyridoquinoline derivatives, at the rather high dose of 128 μg/ml, increased the potency of norfloxacin and ciprofloxacin towards *Enterobacter aerogenes* with a *mar*-mediated multidrug resistance phenotype (61). Experimental evidence indicated that aryl piperazines enhanced the activity of levofloxacin, oxacillin, rifampin, chloramphenicol, and clarithromycin against *E. coli* by inhibition of the RND systems AcrAB and AcrEF (33).

Increased awareness of efflux as an important factor in antibiotic resistance triggered a general interest for inhibitors of eukaryotic drug transporters to block bacterial drug export systems. This approach was taken with phenylpiperidine-selective serotonin reuptake inhibitor or inhibitors of P-gp-mediated multidrug resistance in human tumor therapy (139). A synthetic P-gp inhibitor (GG918) moderately enhanced the in vitro activity of norfloxacin and ciprofloxacin against *S. aureus* that expressed NorA, MsrA, or TetK (97). Other inhibitors of mammalian multidrug transporters (biricodar [VX-710] and timcodar [VX-853]) inhibited growth of *S. aureus*, *Enterococcus faecalis*, and *S. pneumoniae* in combination with ciprofloxacin or norfloxacin (196).

Several hypothetical ways to block drug efflux emerged from detailed information gained by high-resolution structures of membrane-associated transporters. Crystal structures of TolC identified aspartic acid residues at the periplasmic end of TolC, which bind $Co(NH_3)_6^{3+}$ in the nanomolar range (18, 116). Cr^{3+} (which bound even irreversibly) and $Co(NH_3)_6^{3+}$ both blocked the conductance of TolC inserted into artificial membranes. This result raised hope for a general way to block drug efflux because the electronegative entrance is conserved throughout the TolC family.

Interference with gene expression may prove to be a powerful approach as well. Initial attempts would probably target transcriptional regulators like MexT (see also "Resistance against Clinically Important Antibiotics" above). However, the rather complex mechanisms make modification of gene regulation a nontrivial task. Inhibition of global regulators (*mar*) could be a promising approach to down-regulate efflux pump expression (23).

Another, certainly appealing, strategy is to interfere with substrate binding. Studies with QacR demonstrated that the binding affinity for different, simultaneously bound substrates depended on the order of binding (267). This may be a general mechanistic property of multidrug exporters and could therefore be used for the design of inhibitors. The ideal inhibitor would block the interaction of the first substrate with the transporter and concomitantly reduce the affinity for other substrates.

One could also envisage the prevention of subunit assembly or the disruption of functional systems. The specificity of such inhibitors would rely on the role of conserved amino acids in subunit contact formation. Crystal structures and mutational analysis should identify promising target sites.

Finally, molecules that specifically block the energy supply could make a new class of efflux inhibitors. Most antibiotic transporters rely on the proton gradient across the bacterial membrane. Hence, it may be advisable to look for proton flux inhibitors. A first target for this purpose could be the conserved aspartic acid residues in TMS4 and the conserved lysine in TMS10 of AcrB and corresponding amino acids in other pumps as described in "RND Family" above.

I am very grateful to Malcolm Page for critically reviewing this chapter. I also thank Stefan Reinelt for help in preparing the figures.

References

1. **Abramson, J., S. Iwata, and H. R. Kaback.** 2004. Lactose permease as a paradigm for membrane transport proteins. *Mol. Membr. Biol.* **21:**227–236.
2. **Abramson, J., I. Smirnova, V. Kasho, G. Verner, H. R. Kaback, and S. Iwata.** 2003. Structure and mechanism of the lactose permease of *Escherichia coli*. *Science* **301:**610–615.
3. **Adler, J., and E. Bibi.** 2004. Determinants of substrate recognition by the *Escherichia coli* multidrug transporter MdfA identified on both sides of the membrane. *J. Biol. Chem.* **279:**8957–8965.

4. **Adler, J., and E. Bibi.** 2005. Promiscuity in the geometry of electrostatic interactions between the *Escherichia coli* multidrug resistance transporter MdfA and cationic substrates. *J. Biol. Chem.* **280:**2721–2729.

5. **Adler, J., O. Lewinson, and E. Bibi.** 2004. Role of a conserved membrane-embedded acidic residue in the multidrug transporter MdfA. *Biochemistry* **43:**518–525.

6. **Aendekerk, S., S. P. Diggle, Z. Song, N. Hoiby, P. Cornelis, P. Williams, and M. Camara.** 2005. The MexGHI-OpmD multidrug efflux pump controls growth, antibiotic susceptibility and virulence in *Pseudomonas aeruginosa* via 4-quinolone-dependent cell-to-cell communication. *Microbiology* **151:**1113–1125.

7. **Aires, J. R., and H. Nikaido.** 2005. Aminoglycosides are captured from both periplasm and cytoplasm by the AcrD multidrug efflux transporter of *Escherichia coli*. *J. Bacteriol.* **187:**1923–1929.

8. **Akama, H., M. Kanemaki, M. Yoshimura, T. Tsukihara, T. Kashiwagi, H. Yoneyama, S. Narita, A. Nakagawa, and T. Nakae.** 2004. Crystal structure of the drug discharge outer membrane protein, OprM, of *Pseudomonas aeruginosa*: dual modes of membrane anchoring and occluded cavity end. *J. Biol. Chem.* **279:**52816–52819.

9. **Akama, H., T. Matsuura, S. Kashiwagi, H. Yoneyama, S. Narita, T. Tsukihara, A. Nakagawa, and T. Nakae.** 2004. Crystal structure of the membrane fusion protein, MexA, of the multidrug transporter in *Pseudomonas aeruginosa*. *J. Biol. Chem.* **279:**25939–25942.

10. **Alam, M. M., N. Kobayashi, N. Uehara, and N. Watanabe.** 2003. Analysis on distribution and genomic diversity of high-level antiseptic resistance genes qacA and qacB in human clinical isolates of *Staphylococcus aureus*. *Microb. Drug Resist.* **9:**109–121.

11. **Alonso, A., and J. L. Martinez.** 2000. Cloning and characterization of SmeDEF, a novel multidrug efflux pump from *Stenotrophomonas maltophilia*. *Antimicrob. Agents Chemother.* **44:**3079–3086.

12. **Alos, J. I., B. Aracil, J. Oteo, C. Torres, and J. L. Gomez-Garces.** 2000. High prevalence of erythromycin-resistant, clindamycin/miocamycin-susceptible (M phenotype) *Streptococcus pyogenes*: results of a Spanish multicentre study in 1998. *J. Antimicrob. Chemother.* **45:**605–609.

13. **Ambrose, K. D., R. Nisbet, and D. S. Stephens.** 2005. Macrolide efflux in *Streptococcus pneumoniae* is mediated by a dual efflux pump (*mel* and *mef*) and is erythromycin inducible. *Antimicrob. Agents Chemother.* **49:**4203–4209.

14. **Ambudkar, S. V., S. Dey, C. A. Hrycyna, M. Ramachandra, I. Pastan, and M. M. Gottesman.** 1999. Biochemical, cellular, and pharmacological aspects of the multidrug transporter. *Annu. Rev. Pharmacol. Toxicol.* **39:**361–398.

15. **Andersen, C., C. Hughes, and V. Koronakis.** 2000. Chunnel vision. Export and efflux through bacterial channel-tunnels. *EMBO Rep.* **1:**313–318.

16. **Andersen, C., C. Hughes, and V. Koronakis.** 2002. Electrophysiological behavior of the TolC channel-tunnel in planar lipid bilayers. *J. Membr. Biol.* **185:**83–92.

17. **Andersen, C., E. Koronakis, E. Bokma, J. Eswaran, D. Humphreys, C. Hughes, and V. Koronakis.** 2002. Transition to the open state of the TolC periplasmic tunnel entrance. *Proc. Natl. Acad. Sci. USA* **99:**11103–11108.

18. **Andersen, C., E. Koronakis, C. Hughes, and V. Koronakis.** 2002. An aspartate ring at the TolC tunnel entrance determines ion selectivity and presents a target for blocking by large cations. *Mol. Microbiol.* **44:**1131–1139.

19. **Asako, H., H. Nakajima, K. Kobayashi, M. Kobayashi, and R. Aono.** 1997. Organic solvent tolerance and antibiotic resistance increased by overexpression of *marA* in *Escherichia coli*. *Appl. Environ. Microbiol.* **63:**1428–1433.

20. **Athappilly, F. K., and W. A. Hendrickson.** 1995. Structure of the biotinyl domain of acetyl-coenzyme A carboxylase determined by MAD phasing. *Structure* **3:**1407–1419.

21. **Augustus, A. M., T. Celaya, F. Husain, M. Humbard, and R. Misra.** 2004. Antibiotic-sensitive TolC mutants and their suppressors. *J. Bacteriol.* **186:**1851–1860.

22. **Avila-Sakar, A. J., S. Misaghi, E. M. Wilson-Kubalek, K. H. Downing, H. Zgurskaya, H. Nikaido, and E. Nogales.** 2001. Lipid-layer crystallization and preliminary three-dimensional structural analysis of AcrA, the periplasmic component of a bacterial multidrug efflux pump. *J. Struct. Biol.* **136:**81–88.

23. **Barbosa, T. M., and S. B. Levy.** 2000. Differential expression of over 60 chromosomal genes in *Escherichia coli* by constitutive expression of MarA. *J. Bacteriol.* **182:**3467–3474.

24. **Barrett, J. F.** 2001. MC-207110 Daiichi Seiyaku/Microcide Pharmaceuticals. *Curr. Opin. Investig. Drugs* **2:**212–215.

25. **Baucheron, S., S. Tyler, D. Boyd, M. R. Mulvey, E. Chaslus-Dancla, and A. Cloeckaert.** 2004. AcrAB-TolC directs efflux-mediated multidrug resistance in *Salmonella enterica* serovar Typhimurium DT104. *Antimicrob. Agents Chemother.* **48:**3729–3735.

26. **Belofsky, G., D. Percivill, K. Lewis, G. P. Tegos, and J. Ekart.** 2004. Phenolic metabolites of Dalea versicolor that enhance antibiotic activity against model pathogenic bacteria. *J. Nat. Prod.* **67:**481–484.

27. **Benz, R., E. Maier, and I. Gentschev.** 1993. TolC of *Escherichia coli* functions as an outer membrane channel. *Zentbl. Bakteriol.* **278:**187–196.

28. **Berg, A., J. Vervoort, and A. de Kok.** 1997. Three-dimensional structure in solution of the N-terminal lipoyl domain of the pyruvate dehydrogenase complex from *Azotobacter vinelandii*. *Eur. J. Biochem.* **244:**352–360.

29. **Bert, F., Z. Ould-Hocine, M. Juvin, V. Dubois, V. Loncle-Provot, V. Lefranc, C. Quentin, N. Lambert, and G. Arlet.** 2003. Evaluation of the Osiris expert system for identification of beta-lactam phenotypes in isolates of *Pseudomonas aeruginosa*. *J. Clin. Microbiol.* **41:**3712–3718.

30. **Bertrand, X., P. Bailly, G. Blasco, P. Balvay, A. Boillot, and D. Talon.** 2000. Large outbreak in a surgical intensive care unit of colonization or infection with *Pseudomonas aeruginosa* that overexpressed an active efflux pump. *Clin. Infect. Dis.* **31:**E9–E14.

31. **Betriu, C., I. Rodriguez-Avial, B. A. Sanchez, M. Gomez, J. Alvarez, and J. J. Picazo.** 2002. In vitro activities of tigecycline (GAR-936) against recently isolated clinical bacteria in Spain. *Antimicrob. Agents Chemother.* **46:**892–895.

32. **Beyer, R., E. Pestova, J. J. Millichap, V. Stosor, G. A. Noskin, and L. R. Peterson.** 2000. A convenient assay for estimating the possible involvement of efflux of fluoroquinolones by *Streptococcus pneumoniae* and *Staphylococcus*

aureus: evidence for diminished moxifloxacin, sparfloxacin, and trovafloxacin efflux. *Antimicrob. Agents Chemother.* **44**:798–801.

33. **Bohnert, J. A., and W. V. Kern.** 2005. Selected arylpiperazines are capable of reversing multidrug resistance in *Escherichia coli* overexpressing RND efflux pumps. *Antimicrob. Agents Chemother.* **49**:849–852.
34. **Bonomo, R. A., and D. Szabo.** 2006. Mechanisms of multidrug resistance in *Acinetobacter* species and *Pseudomonas aeruginosa*. *Clin. Infect. Dis.* **43**(Suppl. 2):S49–S56.
35. **Borges-Walmsley, M. I., J. Beauchamp, S. M. Kelly, K. Jumel, D. Candlish, S. E. Harding, N. C. Price, and A. R. Walmsley.** 2003. Identification of oligomerization and drug-binding domains of the membrane fusion protein EmrA. *J. Biol. Chem.* **278**:12903–12912.
36. **Borges-Walmsley, M. I., K. S. McKeegan, and A. R. Walmsley.** 2003. Structure and function of efflux pumps that confer resistance to drugs. *Biochem. J.* **376**:313–338.
37. **Bou, G., G. Cervero, M. A. Dominguez, C. Quereda, and J. Martinez-Beltran.** 2000. Characterization of a nosocomial outbreak caused by a multiresistant *Acinetobacter baumannii* strain with a carbapenem-hydrolyzing enzyme: high-level carbapenem resistance in *A. baumannii* is not due solely to the presence of beta-lactamases. *J. Clin. Microbiol.* **38**:3299–3305.
38. **Bouige, P., D. Laurent, L. Piloyan, and E. Dassa.** 2002. Phylogenetic and functional classification of ATP-binding cassette (ABC) systems. *Curr. Protein Pept. Sci.* **3**:541–559.
39. **Bozdogan, B., P. C. Appelbaum, L. M. Kelly, D. B. Hoellman, A. Tambic-Andrasevic, L. Drukalska, W. Hryniewicz, H. Hupkova, M. R. Jacobs, J. Kolman, M. Konkoly-Thege, J. Miciuleviciene, M. Pana, L. Setchanova, J. Trupl, and P. Urbaskova.** 2003. Activity of telithromycin compared with seven other agents against 1039 *Streptococcus pyogenes* pediatric isolates from ten centers in central and eastern Europe. *Clin. Microbiol. Infect.* **9**:741–745.
40. **Braibant, M., J. Chevalier, E. Chaslus-Dancla, J. M. Pages, and A. Cloeckaert.** 2005. Structural and functional study of the phenicol-specific efflux pump FloR belonging to the major facilitator superfamily. *Antimicrob. Agents Chemother.* **49**:2965–2971.
41. **Brenciani, A., K. K. Ojo, A. Monachetti, S. Menzo, M. C. Roberts, P. E. Varaldo, and E. Giovanetti.** 2004. Distribution and molecular analysis of *mef*(A)-containing elements in tetracycline-susceptible and -resistant *Streptococcus pyogenes* clinical isolates with efflux-mediated erythromycin resistance. *J. Antimicrob. Chemother.* **54**:991–998.
42. **Brenwald, N. P., M. J. Gill, and R. Wise.** 1997. The effect of reserpine, an inhibitor of multi-drug efflux pumps, on the *in-vitro* susceptibilities of fluoroquinolone-resistant strains of *Streptococcus pneumoniae* to norfloxacin. *J. Antimicrob. Chemother.* **40**:458–460.
43. **Brown, M. H., I. T. Paulsen, and R. A. Skurray.** 1999. The multidrug efflux protein NorM is a prototype of a new family of transporters. *Mol. Microbiol.* **31**:394–395.
44. **Brown, M. H., and R. A. Skurray.** 2001. Staphylococcal multidrug efflux protein QacA. *J. Mol. Microbiol. Biotechnol.* **3**:163–170.
45. **Brown, S. D., D. J. Farrell, and I. Morrissey.** 2004. Prevalence and molecular analysis of macrolide and fluoroquinolone resistance among isolates of *Streptococcus pneumoniae* collected during the 2000–2001 PROTEKT US Study. *J. Clin. Microbiol.* **42**:4980–4987.
46. **Buchanan, S. K.** 2001. Type I secretion and multidrug efflux: transport through the TolC channel-tunnel. *Trends Biochem. Sci.* **26**:3–6.
47. **Buckingham, S. C., L. K. McDougal, L. D. Cathey, K. Comeaux, A. S. Craig, S. K. Fridkin, and F. C. Tenover.** 2004. Emergence of community-associated methicillin-resistant *Staphylococcus aureus* at a Memphis, Tennessee Children's Hospital. *Pediatr. Infect. Dis. J.* **23**:619–624.
48. **Butaye, P., A. Cloeckaert, and S. Schwarz.** 2003. Mobile genes coding for efflux-mediated antimicrobial resistance in Gram-positive and Gram-negative bacteria. *Int. J. Antimicrob. Agents* **22**:205–210.
49. **Butler, P. J., I. Ubarretxena-Belandia, T. Warne, and C. G. Tate.** 2004. The *Escherichia coli* multidrug transporter EmrE is a dimer in the detergent-solubilised state. *J. Mol. Biol.* **340**:797–808.
50. **Calladine, C. R., A. Sharff, and B. Luisi.** 2001. How to untwist an alpha-helix: structural principles of an alpha-helical barrel. *J. Mol. Biol.* **305**:603–618.
51. **Canton, R., A. Mazzariol, M. I. Morosini, F. Baquero, and G. Cornaglia.** 2005. Telithromycin activity is reduced by efflux in *Streptococcus pyogenes*. *J. Antimicrob. Chemother.* **55**:489–495.
52. **Capilla, S., J. Ruiz, P. Goni, J. Castillo, M. C. Rubio, M. T. Jimenez de Anta, R. Gomez-Lus, and J. Vila.** 2004. Characterization of the molecular mechanisms of quinolone resistance in *Yersinia enterocolitica* O:3 clinical isolates. *J. Antimicrob. Chemother.* **53**:1068–1071.
53. **Cavallo, J. D., P. Plesiat, G. Couetdic, F. Leblanc, and R. Fabre.** 2002. Mechanisms of beta-lactam resistance in *Pseudomonas aeruginosa*: prevalence of OprM-overproducing strains in a French multicentre study (1997). *J. Antimicrob. Chemother.* **50**:1039–1043.
54. **Chami, M., E. Steinfels, C. Orelle, J. M. Jault, A. Di Pietro, J. L. Rigaud, and S. Marco.** 2002. Three-dimensional structure by cryo-electron microscopy of YvcC, an homodimeric ATP-binding cassette transporter from *Bacillus subtilis*. *J. Mol. Biol.* **315**:1075–1085.
55. **Chan, Y. Y., and K. L. Chua.** 2005. The *Burkholderia pseudomallei* BpeAB-OprB efflux pump: expression and impact on quorum sensing and virulence. *J. Bacteriol.* **187**:4707–4719.
56. **Chang, G.** 2003. Multidrug resistance ABC transporters. *FEBS Lett.* **555**:102–105.
57. **Chang, G.** 2003. Structure of MsbA from *Vibrio cholerae*: a multidrug resistance ABC transporter homolog in a closed conformation. *J. Mol. Biol.* **330**:419–430.
58. **Chang, G., and C. B. Roth.** 2001. Structure of MsbA from *E. coli*: a homolog of the multidrug resistance ATP binding cassette (ABC) transporters. *Science* **293**:1793–1800.
59. **Chen, J., Y. Morita, M. N. Huda, T. Kuroda, T. Mizushima, and T. Tsuchiya.** 2002. VmrA, a member of a novel class of Na(+)-coupled multidrug efflux pumps from *Vibrio parahaemolyticus*. *J. Bacteriol.* **184**:572–576.
60. **Chesneau, O., H. Ligeret, N. Hosan-Aghaie, A. Morvan, and E. Dassa.** 2005. Molecular analysis of resistance to streptogramin A compounds conferred by the *vga* proteins

of staphylococci. *Antimicrob. Agents Chemother.* **49:**973–980.

61. **Chevalier, J., S. Atifi, A. Eyraud, A. Mahamoud, J. Barbe, and J. M. Pages.** 2001. New pyridoquinoline derivatives as potential inhibitors of the fluoroquinolone efflux pump in resistant *Enterobacter aerogenes* strains. *J. Med. Chem.* **44:**4023–4026.
62. **Chollet, R., J. Chevalier, A. Bryskier, and J. M. Pages.** 2004. The AcrAB-TolC pump is involved in macrolide resistance but not in telithromycin efflux in *Enterobacter aerogenes* and *Escherichia coli. Antimicrob. Agents Chemother.* **48:**3621–3624.
63. **Chopra, I.** 2002. New developments in tetracycline antibiotics: glycylcyclines and tetracycline efflux pump inhibitors. *Drug Resist. Updat.* **5:**119–125.
64. **Chopra, I., and M. Roberts.** 2001. Tetracycline antibiotics: mode of action, applications, molecular biology, and epidemiology of bacterial resistance. *Microbiol. Mol. Biol. Rev.* **65:**232–260.
65. **Chuanchuen, R., K. Beinlich, T. T. Hoang, A. Becher, R. R. Karkhoff-Schweizer, and H. P. Schweizer.** 2001. Cross-resistance between triclosan and antibiotics in *Pseudomonas aeruginosa* is mediated by multidrug efflux pumps: exposure of a susceptible mutant strain to triclosan selects *nfxB* mutants overexpressing MexCD-OprJ. *Antimicrob. Agents Chemother.* **45:**428–432.
66. **Chung, Y. J., and M. H. Saier, Jr.** 2001. SMR-type multidrug resistance pumps. *Curr. Opin. Drug Discov. Devel.* **4:**237–245.
67. **Clancy, J., F. Dib-Hajj, J. W. Petitpas, and W. Yuan.** 1997. Cloning and characterization of a novel macrolide efflux gene, *mreA*, from *Streptococcus agalactiae. Antimicrob. Agents Chemother.* **41:**2719–2723.
68. **Clarebout, G., C. Villers, and R. Leclercq.** 2001. Macrolide resistance gene *mreA* of *Streptococcus agalactiae* encodes a flavokinase. *Antimicrob. Agents Chemother.* **45:**2280–2286.
69. **Coban, A. Y., B. Ekinci, and B. Durupinar.** 2004. A multidrug efflux pump inhibitor reduces fluoroquinolone resistance in *Pseudomonas aeruginosa* isolates. *Chemotherapy* **50:**22–26.
70. **Cohen, S. P., L. M. McMurry, D. C. Hooper, J. S. Wolfson, and S. B. Levy.** 1989. Cross-resistance to fluoroquinolones in multiple-antibiotic-resistant (Mar) *Escherichia coli* selected by tetracycline or chloramphenicol: decreased drug accumulation associated with membrane changes in addition to OmpF reduction. *Antimicrob. Agents Chemother.* **33:**1318–1325.
71. **Cole, E. C., R. M. Addison, J. R. Rubino, K. E. Leese, P. D. Dulaney, M. S. Newell, J. Wilkins, D. J. Gaber, T. Wineinger, and D. A. Criger.** 2003. Investigation of antibiotic and antibacterial agent cross-resistance in target bacteria from homes of antibacterial product users and nonusers. *J. Appl. Microbiol.* **95:**664–676.
72. **Cosson, P., L. Zulianello, O. Join-Lambert, F. Faurisson, L. Gebbie, M. Benghezal, C. Van Delden, L. K. Curty, and T. Kohler.** 2002. *Pseudomonas aeruginosa* virulence analyzed in a *Dictyostelium discoideum* host system. *J. Bacteriol.* **184:**3027–3033.
73. **Dawson, R. J., and K. P. Locher.** 2006. Structure of a bacterial multidrug ABC transporter. *Nature* **443:**180–185.
74. **Dean, C. R., M. A. Visalli, S. J. Projan, P. E. Sum, and P. A. Bradford.** 2003. Efflux-mediated resistance to tigecycline (GAR-936) in *Pseudomonas aeruginosa* PAO1. *Antimicrob. Agents Chemother.* **47:**972–978.
75. **De Kievit, T. R., M. D. Parkins, R. J. Gillis, R. Srikumar, H. Ceri, K. Poole, B. H. Iglewski, and D. G. Storey.** 2001. Multidrug efflux pumps: expression patterns and contribution to antibiotic resistance in *Pseudomonas aeruginosa* biofilms. *Antimicrob. Agents Chemother.* **45:**1761–1770.
76. **Doerrler, W. T., and C. R. Raetz.** 2002. ATPase activity of the MsbA lipid flippase of *Escherichia coli. J. Biol. Chem.* **277:**36697–36705.
77. **Dong, J., G. Yang, and H. S. McHaourab.** 2005. Structural basis of energy transduction in the transport cycle of MsbA. *Science* **308:**1023–1028.
78. **Dridi, L., J. Tankovic, and J. C. Petit.** 2004. CdeA of *Clostridium difficile*, a new multidrug efflux transporter of the MATE family. *Microb. Drug Resist.* **10:**191–196.
79. **Dubochet, J., A. W. McDowall, B. Menge, E. N. Schmid, and K. G. Lickfeld.** 1983. Electron microscopy of frozen-hydrated bacteria. *J. Bacteriol.* **155:**381–390.
80. **Dubois, V., C. Arpin, M. Melon, B. Melon, C. Andre, C. Frigo, and C. Quentin.** 2001. Nosocomial outbreak due to a multiresistant strain of *Pseudomonas aeruginosa* P12: efficacy of cefepime-amikacin therapy and analysis of beta-lactam resistance. *J. Clin. Microbiol.* **39:**2072–2078.
81. **Dugan, J., D. D. Rockey, L. Jones, and A. A. Andersen.** 2004. Tetracycline resistance in *Chlamydia suis* mediated by genomic islands inserted into the chlamydial *inv*-like gene. *Antimicrob. Agents Chemother.* **48:**3989–3995.
82. **Elkins, C. A., and L. B. Mullis.** 2006. Mammalian steroid hormones are substrates for the major RND- and MFS-type tripartite multidrug efflux pumps of *Escherichia coli. J. Bacteriol.* **188:**1191–1195.
83. **Elkins, C. A., and H. Nikaido.** 2002. Substrate specificity of the RND-type multidrug efflux pumps AcrB and AcrD of *Escherichia coli* is determined predominantly by two large periplasmic loops. *J. Bacteriol.* **184:**6490–6498.
84. **Escribano, I., J. C. Rodriguez, L. Cebrian, and G. Royo.** 2004. The importance of active efflux systems in the quinolone resistance of clinical isolates of *Salmonella* spp. *Int. J. Antimicrob. Agents* **24:**428–432.
85. **Eswaran, J., C. Hughes, and V. Koronakis.** 2003. Locking TolC entrance helices to prevent protein translocation by the bacterial type I export apparatus. *J. Mol. Biol.* **327:**309–315.
86. **Farrell, D. J., and D. Felmingham.** 2004. Activities of telithromycin against 13,874 *Streptococcus pneumoniae* isolates collected between 1999 and 2003. *Antimicrob. Agents Chemother.* **48:**1882–1884.
87. **Farrell, D. J., I. Morrissey, S. Bakker, L. Morris, S. Buckridge, and D. Felmingham.** 2004. Molecular epidemiology of multiresistant *Streptococcus pneumoniae* with both *erm*(B)- and *mef*(A)-mediated macrolide resistance. *J. Clin. Microbiol.* **42:**764–768.
88. **Federici, L., D. Du, F. Walas, H. Matsumura, J. Fernandez-Recio, K. S. McKeegan, M. I. Borges-Walmsley, B. F. Luisi, and A. R. Walmsley.** 2005. The crystal structure of the outer membrane protein VceC from the bacterial pathogen *Vibrio*

cholerae at 1.8 A resolution. *J. Biol. Chem.* **280**:15307–15314.

89. **Fernandes, P., B. S. Ferreira, and J. M. Cabral.** 2003. Solvent tolerance in bacteria: role of efflux pumps and cross-resistance with antibiotics. *Int. J. Antimicrob. Agents* **22**: 211–216.
90. **Ferrandiz, M. J., J. Oteo, B. Aracil, J. L. Gomez-Garces, and A. G. De La Campa.** 1999. Drug efflux and *parC* mutations are involved in fluoroquinolone resistance in viridans group streptococci. *Antimicrob. Agents Chemother.* **43**: 2520–2523.
91. **Fleishman, S. J., S. E. Harrington, A. Enosh, D. Halperin, C. G. Tate, and N. Ben Tal.** 2006. Quasi-symmetry in the Cryo-EM structure of EmrE provides the key to modeling its transmembrane domain. *J. Mol. Biol.* **364**:54–67.
92. **Fraise, A. P.** 2002. Biocide abuse and antimicrobial resistance — a cause for concern? *J. Antimicrob. Chemother.* **49**:11–12.
93. **Fraise, A. P.** 2002. Susceptibility of antibiotic-resistant cocci to biocides. *Symp. Ser. Soc. Appl. Microbiol.* **2002**:158S-162S.
94. **Fraise, A. P.** 2002. Susceptibility of antibiotic-resistant cocci to biocides. *J. Appl. Microbiol.* **92**(Suppl.):158S-162S.
95. **Gay, K., W. Baughman, Y. Miller, D. Jackson, C. G. Whitney, A. Schuchat, M. M. Farley, F. Tenover, and D. S. Stephens.** 2000. The emergence of *Streptococcus pneumoniae* resistant to macrolide antimicrobial agents: a 6-year population-based assessment. *J. Infect. Dis.* **182**:1417–1424.
96. **Gerken, H., and R. Misra.** 2004. Genetic evidence for functional interactions between TolC and AcrA proteins of a major antibiotic efflux pump of *Escherichia coli*. *Mol. Microbiol.* **54**:620–631.
97. **Gibbons, S., M. Oluwatuyi, and G. W. Kaatz.** 2003. A novel inhibitor of multidrug efflux pumps in *Staphylococcus aureus*. *J. Antimicrob. Chemother.* **51**:13–17.
98. **Gilbert, P., A. J. McBain, and S. F. Bloomfield.** 2002. Biocide abuse and antimicrobial resistance: being clear about the issues. *J. Antimicrob. Chemother.* **50**:137–139.
99. **Giovanetti, E., A. Brenciani, R. Burioni, and P. E. Varaldo.** 2002. A novel efflux system in inducibly erythromycin-resistant strains of *Streptococcus pyogenes*. *Antimicrob. Agents Chemother.* **46**:3750–3755.
100. **Giraud, E., G. Blanc, A. Bouju-Albert, F. X. Weill, and C. Donnay-Moreno.** 2004. Mechanisms of quinolone resistance and clonal relationship among *Aeromonas salmonicida* strains isolated from reared fish with furunculosis. *J. Med. Microbiol.* **53**:895–901.
101. **Graham, L. L., R. Harris, W. Villiger, and T. J. Beveridge.** 1991. Freeze-substitution of gram-negative eubacteria: general cell morphology and envelope profiles. *J. Bacteriol.* **173**:1623–1633.
102. **Grkovic, S., M. H. Brown, and R. A. Skurray.** 2002. Regulation of bacterial drug export systems. *Microbiol. Mol. Biol. Rev.* **66**:671–701.
103. **Grkovic, S., K. M. Hardie, M. H. Brown, and R. A. Skurray.** 2003. Interactions of the QacR multidrug-binding protein with structurally diverse ligands: implications for the evolution of the binding pocket. *Biochemistry* **42**:15226–15236.
104. **Grohs, P., S. Houssaye, A. Aubert, L. Gutmann, and E. Varon.** 2003. In vitro activities of garenoxacin (BMS-284756) against *Streptococcus pneumoniae*, viridans group streptococci, and *Enterococcus faecalis* compared to those of six other quinolones. *Antimicrob. Agents Chemother.* **47**:3542–3547.
105. **Guan, L., and T. Nakae.** 2001. Identification of essential charged residues in transmembrane segments of the multidrug transporter MexB of *Pseudomonas aeruginosa*. *J. Bacteriol.* **183**:1734–1739.
106. **Guerin, F., E. Varon, A. B. Hoi, L. Gutmann, and I. Podglajen.** 2000. Fluoroquinolone resistance associated with target mutations and active efflux in oropharyngeal colonizing isolates of viridans group streptococci. *Antimicrob. Agents Chemother.* **44**:2197–2200.
107. **Gupta, A., K. Matsui, J. F. Lo, and S. Silver.** 1999. Molecular basis for resistance to silver cations in *Salmonella*. *Nat. Med.* **5**:183–188.
108. **Gutman, N., S. Steiner-Mordoch, and S. Schuldiner.** 2003. An amino acid cluster around the essential Glu-14 is part of the substrate- and proton-binding domain of EmrE, a multidrug transporter from *Escherichia coli*. *J. Biol. Chem.* **278**:16082–16087.
109. **Hagman, K. E., C. E. Lucas, J. T. Balthazar, L. Snyder, M. Nilles, R. C. Judd, and W. M. Shafer.** 1997. The MtrD protein of *Neisseria gonorrhoeae* is a member of the resistance/nodulation/division protein family constituting part of an efflux system. *Microbiology* **143**(Pt. 7):2117–2125.
110. **Hartman, A. B., I. I. Essiet, D. W. Isenbarger, and L. E. Lindler.** 2003. Epidemiology of tetracycline resistance determinants in *Shigella* spp. and enteroinvasive *Escherichia coli*: characterization and dissemination of *tet*(A)-1. *J. Clin. Microbiol.* **41**:1023–1032.
111. **He, G. X., T. Kuroda, T. Mima, Y. Morita, T. Mizushima, and T. Tsuchiya.** 2004. An H(+)-coupled multidrug efflux pump, PmpM, a member of the MATE family of transporters, from *Pseudomonas aeruginosa*. *J. Bacteriol.* **186**:262–265.
112. **Hearn, E. M., M. R. Gray, and J. M. Foght.** 2006. Mutations in the central cavity and periplasmic domain affect efflux activity of the resistance-nodulation-division pump EmhB from *Pseudomonas fluorescens* cLP6a. *J. Bacteriol.* **188**:115–123.
113. **Heymann, J. A., T. Hirai, D. Shi, and S. Subramaniam.** 2003. Projection structure of the bacterial oxalate transporter OxlT at 3.4A resolution. *J. Struct. Biol.* **144**:320–326.
114. **Heymann, J. A., R. Sarker, T. Hirai, D. Shi, J. L. Milne, P. C. Maloney, and S. Subramaniam.** 2001. Projection structure and molecular architecture of OxlT, a bacterial membrane transporter. *EMBO J.* **20**:4408–4413.
115. **Higgins, M. K., E. Bokma, E. Koronakis, C. Hughes, and V. Koronakis.** 2004. Structure of the periplasmic component of a bacterial drug efflux pump. *Proc. Natl. Acad. Sci. USA* **101**:9994–9999.
116. **Higgins, M. K., J. Eswaran, P. Edwards, G. F. Schertler, C. Hughes, and V. Koronakis.** 2004. Structure of the ligand-blocked periplasmic entrance of the bacterial multidrug efflux protein TolC. *J. Mol. Biol.* **342**:697–702.

117. **Higgins, P. G., H. Wisplinghoff, D. Stefanik, and H. Seifert.** 2004. Selection of topoisomerase mutations and overexpression of *ade*B mRNA transcripts during an outbreak of *Acinetobacter baumannii*. *J. Antimicrob. Chemother.* **54:** 821–823.

118. **Hirai, T., J. A. Heymann, D. Shi, R. Sarker, P. C. Maloney, and S. Subramaniam.** 2002. Three-dimensional structure of a bacterial oxalate transporter. *Nat. Struct. Biol.* **9:**597–600.

119. **Hirai, T., and S. Subramaniam.** 2004. Structure and transport mechanism of the bacterial oxalate transporter OxlT. *Biophys. J.* **87:**3600–3607.

120. **Hirakata, Y., R. Srikumar, K. Poole, N. Gotoh, T. Suematsu, S. Kohno, S. Kamihira, R. E. Hancock, and D. P. Speert.** 2002. Multidrug efflux systems play an important role in the invasiveness of *Pseudomonas aeruginosa*. *J. Exp. Med.* **196:**109–118.

121. **Hirata, T., A. Saito, K. Nishino, N. Tamura, and A. Yamaguchi.** 2004. Effects of efflux transporter genes on susceptibility of *Escherichia coli* to tigecycline (GAR-936). *Antimicrob. Agents Chemother.* **48:**2179–2184.

122. **Hirata, T., R. Wakatabe, J. Nielsen, Y. Someya, E. Fujihira, T. Kimura, and A. Yamaguchi.** 1997. A novel compound, 1,1-dimethyl-5(1-hydroxypropyl)-4,6,7-trimethylindan, is an effective inhibitor of the *tet*(K) gene-encoded metal-tetracycline/H+ antiporter of *Staphylococcus aureus*. *FEBS Lett.* **412:**337–340.

123. **Hoban, D. J., A. K. Wierzbowski, K. Nichol, and G. G. Zhanel.** 2001. Macrolide-resistant *Streptococcus pneumoniae* in Canada during 1998–1999: prevalence of *mef*(A) and *erm*(B) and susceptibilities to ketolides. *Antimicrob. Agents Chemother.* **45:**2147–2150.

124. **Hocquet, D., X. Bertrand, T. Kohler, D. Talon, and P. Plesiat.** 2003. Genetic and phenotypic variations of a resistant *Pseudomonas aeruginosa* epidemic clone. *Antimicrob. Agents Chemother.* **47:**1887–1894.

125. **Hocquet, D., C. Vogne, F. El Garch, A. Vejux, N. Gotoh, A. Lee, O. Lomovskaya, and P. Plesiat.** 2003. MexXY-OprM efflux pump is necessary for adaptive resistance of *Pseudomonas aeruginosa* to aminoglycosides. *Antimicrob. Agents Chemother.* **47:**1371–1375.

126. **Huang, Y., M. J. Lemieux, J. Song, M. Auer, and D. N. Wang.** 2003. Structure and mechanism of the glycerol-3-phosphate transporter from *Escherichia coli*. *Science* **301:**616–620.

127. **Huda, M. N., J. Chen, Y. Morita, T. Kuroda, T. Mizushima, and T. Tsuchiya.** 2003. Gene cloning and characterization of VcrM, a Na+-coupled multidrug efflux pump, from *Vibrio cholerae* non-O1. *Microbiol. Immunol.* **47:**419–427.

128. **Inoue, M., S. Kohno, M. Kaku, K. Yamaguchi, J. Igari, and K. Yamanaka.** 2005. PROTEKT 1999–2000: a multicentre study of the antimicrobial susceptibility of respiratory tract pathogens in Japan. *Int. J. Infect. Dis.* **9:**27–36.

129. **Ip, H., K. Stratton, H. Zgurskaya, and J. Liu.** 2003. pH-induced conformational changes of AcrA, the membrane fusion protein of *Escherichia coli* multidrug efflux system. *J. Biol. Chem.* **278:**50474–50482.

130. **Islam, S., S. Jalal, and B. Wretlind.** 2004. Expression of the MexXY efflux pump in amikacin-resistant isolates of *Pseudomonas aeruginosa*. *Clin. Microbiol. Infect.* **10:**877–883.

131. **Jalal, S., O. Ciofu, N. Hoiby, N. Gotoh, and B. Wretlind.** 2000. Molecular mechanisms of fluoroquinolone resistance in *Pseudomonas aeruginosa* isolates from cystic fibrosis patients. *Antimicrob. Agents Chemother.* **44:**710–712.

132. **Johnson, A. P., C. L. Sheppard, S. J. Harnett, A. Birtles, T. G. Harrison, N. P. Brenwald, M. J. Gill, R. A. Walker, D. M. Livermore, and R. C. George.** 2003. Emergence of a fluoroquinolone-resistant strain of *Streptococcus pneumoniae* in England. *J. Antimicrob. Chemother.* **52:**953–960.

133. **Johnson, J. M., and G. M. Church.** 1999. Alignment and structure prediction of divergent protein families: periplasmic and outer membrane proteins of bacterial efflux pumps. *J. Mol. Biol.* **287:**695–715.

134. **Johnston, N. J., J. C. De Azavedo, J. D. Kellner, and D. E. Low.** 1998. Prevalence and characterization of the mechanisms of macrolide, lincosamide, and streptogramin resistance in isolates of *Streptococcus pneumoniae*. *Antimicrob. Agents Chemother.* **42:**2425–2426.

135. **Kaatz, G. W.** 2002. Inhibition of bacterial efflux pumps: a new strategy to combat increasing antimicrobial agent resistance. *Expert Opin. Emerg. Drugs* **7:**223–233.

136. **Kaatz, G. W.** 2005. Bacterial efflux pump inhibition. *Curr. Opin. Investig. Drugs* **6:**191–198.

137. **Kaatz, G. W., F. McAleese, and S. M. Seo.** 2005. Multidrug resistance in *Staphylococcus aureus* due to overexpression of a novel multidrug and toxin extrusion (MATE) transport protein. *Antimicrob. Agents Chemother.* **49:**1857–1864.

138. **Kaatz, G. W., V. V. Moudgal, and S. M. Seo.** 2002. Identification and characterization of a novel efflux-related multidrug resistance phenotype in *Staphylococcus aureus*. *J. Antimicrob. Chemother.* **50:**833–838.

139. **Kaatz, G. W., V. V. Moudgal, S. M. Seo, J. B. Hansen, and J. E. Kristiansen.** 2003. Phenylpiperidine selective serotonin reuptake inhibitors interfere with multidrug efflux pump activity in *Staphylococcus aureus*. *Int. J. Antimicrob. Agents* **22:**254–261.

140. **Kern, W. V., M. Oethinger, A. S. Jellen-Ritter, and S. B. Levy.** 2000. Non-target gene mutations in the development of fluoroquinolone resistance in *Escherichia coli*. *Antimicrob. Agents Chemother.* **44:**814–820.

141. **Kobayashi, N., K. Nishino, T. Hirata, and A. Yamaguchi.** 2003. Membrane topology of ABC-type macrolide antibiotic exporter MacB in *Escherichia coli*. *FEBS Lett.* **546:**241–246.

142. **Kobayashi, N., K. Nishino, and A. Yamaguchi.** 2001. Novel macrolide-specific ABC-type efflux transporter in *Escherichia coli*. *J. Bacteriol.* **183:**5639–5644.

143. **Köhler, T., S. F. Epp, L. K. Curty, and J. C. Pechere.** 1999. Characterization of MexT, the regulator of the MexE-MexF-OprN multidrug efflux system of *Pseudomonas aeruginosa*. *J. Bacteriol.* **181:**6300–6305.

144. **Köhler, T., C. Van Delden, L. K. Curty, M. M. Hamzehpour, and J. C. Pechere.** 2001. Overexpression of the MexEF-OprN multidrug efflux system affects cell-to-cell signaling in *Pseudomonas aeruginosa*. *J. Bacteriol.* **183:**5213–5222.

145. **Koronakis, V., A. Sharff, E. Koronakis, B. Luisi, and C. Hughes.** 2000. Crystal structure of the bacterial membrane protein TolC central to multidrug efflux and protein export. *Nature* **405:**914–919.

146. **Kozlov, R. S., T. M. Bogdanovitch, P. C. Appelbaum, L. Ednie, L. S. Stratchounski, M. R. Jacobs, and B. Bozdogan.** 2002. Antistreptococcal activity of telithromycin compared with seven other drugs in relation to macrolide resistance mechanisms in Russia. *Antimicrob. Agents Chemother.* **46:**2963–2968.

147. **Kumar, A., and H. P. Schweizer.** 2005. Bacterial resistance to antibiotics: active efflux and reduced uptake. *Adv. Drug Deliv. Rev.* **57:**1486-1513.

148. **Kutschke, A., and B. L. de Jonge.** 2005. Compound efflux in *Helicobacter pylori*. *Antimicrob. Agents Chemother.* **49:** 3009–3010.

149. **Lambert, O., H. Benabdelhak, M. Chami, L. Jouan, E. Nouaille, A. Ducruix, and A. Brisson.** 2005. Trimeric structure of OprN and OprM efflux proteins from *Pseudomonas aeruginosa*, by 2D electron crystallography. *J. Struct. Biol.* **150:**50–57.

150. **Le Themas, I., G. Couetdic, O. Clermont, N. Brahimi, P. Plesiat, and E. Bingen.** 2001. In vivo selection of a target/efflux double mutant of *Pseudomonas aeruginosa* by ciprofloxacin therapy. *J. Antimicrob. Chemother.* **48:**553–555.

151. **Levy, S. B.** 1992. Active efflux mechanisms for antimicrobial resistance. *Antimicrob. Agents Chemother.* **36:**695–703.

152. **Levy, S. B.** 2001. Antibacterial household products: cause for concern. *Emerg. Infect. Dis.* **7:**512–515.

153. **Levy, S. B.** 2002. Active efflux, a common mechanism for biocide and antibiotic resistance. *J. Appl. Microbiol.* **92**(Suppl.):65S–71S.

154. **Lewis, K.** 2001. In search of natural substrates and inhibitors of MDR pumps. *J. Mol. Microbiol. Biotechnol.* **3:**247–254.

155. **Lewis, K., and O. Lomovskaya.** 2002. Drug efflux, p. 61–90. *In* K. Lewis, A. A. Salyers, H. W. Taber, and R. G. Wax (ed.), *Bacterial Resistance to Antimicrobials.* Marcel Dekker, Inc., New York, N.Y.

156. **Li, X. Z., and H. Nikaido.** 2004. Efflux-mediated drug resistance in bacteria. *Drugs* **64:**159–204.

157. **Li, X. Z., H. Nikaido, and K. Poole.** 1995. Role of *mexA-mexB-oprM* in antibiotic efflux in *Pseudomonas aeruginosa*. *Antimicrob. Agents Chemother.* **39:**1948–1953.

158. **Li, X. Z., H. Nikaido, and K. E. Williams.** 1997. Silver-resistant mutants of *Escherichia coli* display active efflux of Ag^+ and are deficient in porins. *J. Bacteriol.* **179:**6127–6132.

159. **Li, X. Z., and K. Poole.** 2001. Mutational analysis of the OprM outer membrane component of the MexA-MexB-OprM multidrug efflux system of *Pseudomonas aeruginosa*. *J. Bacteriol.* **183:**12–27.

160. **Liu, R., and F. J. Sharom.** 1996. Site-directed fluorescence labeling of P-glycoprotein on cysteine residues in the nucleotide binding domains. *Biochemistry* **35:**11865–11873.

161. **Livermore, D. M.** 2003. Bacterial resistance: origins, epidemiology, and impact. *Clin. Infect. Dis.* **36:**S11–S23.

162. **Livermore, D. M.** 2003. The threat from the pink corner. *Ann. Med.* **35:**226–234.

163. **Livermore, D. M.** 2004. The need for new antibiotics. *Clin. Microbiol. Infect.* **10**(Suppl. 4):1–9.

164. **Livermore, D. M.** 2005. Tigecycline: what is it, and where should it be used? *J. Antimicrob. Chemother.* **56:**611–614.

165. **Llanes, C., D. Hocquet, C. Vogne, D. Benali-Baitich, C. Neuwirth, and P. Plesiat.** 2004. Clinical strains of *Pseudomonas aeruginosa* overproducing MexAB-OprM and MexXY efflux pumps simultaneously. *Antimicrob. Agents Chemother.* **48:**1797–1802.

166. **Locher, K. P., A. T. Lee, and D. C. Rees.** 2002. The *E. coli* BtuCD structure: a framework for ABC transporter architecture and mechanism. *Science* **296:**1091–1098.

167. **Lomovskaya, O., and K. A. Bostian.** 2006. Practical applications and feasibility of efflux pump inhibitors in the clinic—a vision for applied use. *Biochem. Pharmacol.* **71:**910–918.

168. **Lomovskaya, O., M. S. Warren, A. Lee, J. Galazzo, R. Fronko, M. Lee, J. Blais, D. Cho, S. Chamberland, T. Renau, R. Leger, S. Hecker, W. Watkins, K. Hoshino, H. Ishida, and V. J. Lee.** 2001. Identification and characterization of inhibitors of multidrug resistance efflux pumps in *Pseudomonas aeruginosa*: novel agents for combination therapy. *Antimicrob. Agents Chemother.* **45:**105–116.

169. **Lomovskaya, O., and W. Watkins.** 2001. Inhibition of efflux pumps as a novel approach to combat drug resistance in bacteria. *J. Mol. Microbiol. Biotechnol.* **3:**225–236.

170. **Lonks, J. R., J. Garau, and A. A. Medeiros.** 2002. Implications of antimicrobial resistance in the empirical treatment of community-acquired respiratory tract infections: the case of macrolides. *J. Antimicrob. Chemother.* **50**(Suppl. S2): 87–92.

171. **Lynch, A. S.** 2005. Efflux systems in bacterial pathogens: an opportunity for therapeutic intervention? An industry view. *Biochem. Pharmacol.* **71:**949–956.

172. **Lynch, J. P., III, and F. J. Martinez.** 2002. Clinical relevance of macrolide-resistant Streptococcus pneumoniae for community-acquired pneumonia. *Clin. Infect. Dis.* **34**(Suppl 1): S27–S46.

173. **Ma, C., and G. Chang.** 2004. Structure of the multidrug resistance efflux transporter EmrE from *Escherichia coli*. *Proc. Natl. Acad. Sci. USA* **101:**2852–2857.

174. **Mao, W., M. S. Warren, D. S. Black, T. Satou, T. Murata, T. Nishino, N. Gotoh, and O. Lomovskaya.** 2002. On the mechanism of substrate specificity by resistance nodulation division (RND)-type multidrug resistance pumps: the large periplasmic loops of MexD from *Pseudomonas aeruginosa* are involved in substrate recognition. *Mol. Microbiol.* **46:** 889–901.

175. **Marchandin, H., H. Jean-Pierre, E. Jumas-Bilak, L. Isson, B. Drouillard, H. Darbas, and C. Carriere.** 2001. Distribution of macrolide resistance genes *erm*(B) and *mef*(A) among 160 penicillin-intermediate clinical isolates of *Streptococcus pneumoniae* isolated in southern France. *Pathol. Biol.* (Paris) **49:**522–527.

176. **Markham, P. N.** 1999. Inhibition of the emergence of ciprofloxacin resistance in *Streptococcus pneumoniae* by the multidrug efflux inhibitor reserpine. *Antimicrob. Agents Chemother.* **43:**988–989.

177. **Markham, P. N., and A. A. Neyfakh.** 1996. Inhibition of the multidrug transporter NorA prevents emergence of norfloxacin resistance in *Staphylococcus aureus*. *Antimicrob. Agents Chemother.* **40:**2673–2674.

178. **Markham, P. N., and A. A. Neyfakh.** 2001. Efflux-mediated drug resistance in Gram-positive bacteria. *Curr. Opin. Microbiol.* **4:**509–514.

179. **Markham, P. N., E. Westhaus, K. Klyachko, M. E. Johnson, and A. A. Neyfakh.** 1999. Multiple novel inhibitors of the NorA multidrug transporter of *Staphylococcus aureus*. *Antimicrob. Agents Chemother.* **43:**2404–2408.

180. **Marquez, B.** 2005. Bacterial efflux systems and efflux pumps inhibitors. *Biochimie* **87:**1137–1147.

181. **Maseda, H., I. Sawada, K. Saito, H. Uchiyama, T. Nakae, and N. Nomura.** 2004. Enhancement of the *mexAB-oprM* efflux pump expression by a quorum-sensing autoinducer and its cancellation by a regulator, MexT, of the *mex*EF-*opr*N efflux pump operon in *Pseudomonas aeruginosa*. *Antimicrob. Agents Chemother.* **48:**1320–1328.

182. **Mason, E. O., Jr., E. R. Wald, J. S. Bradley, W. J. Barson, and S. L. Kaplan.** 2003. Macrolide resistance among middle ear isolates of *Streptococcus pneumoniae* observed at eight United States pediatric centers: prevalence of M and MLSB phenotypes. *Pediatr. Infect. Dis. J.* **22:**623–627.

183. **McAleese, F., P. Petersen, A. Ruzin, P. M. Dunman, E. Murphy, S. J. Projan, and P. A. Bradford.** 2005. A novel MATE family efflux pump contributes to the reduced susceptibility of laboratory-derived *Staphylococcus aureus* mutants to tigecycline. *Antimicrob. Agents Chemother.* **49:**1865–1871.

184. **McGee, L., K. P. Klugman, A. Wasas, T. Capper, and A. Brink.** 2001. Serotype 19f multiresistant pneumococcal clone harboring two erythromycin resistance determinants (*erm*(B) and *mef*(A)) in South Africa. *Antimicrob. Agents Chemother.* **45:**1595–1598.

185. **McMurry, L., R. E. Petrucci, Jr., and S. B. Levy.** 1980. Active efflux of tetracycline encoded by four genetically different tetracycline resistance determinants in *Escherichia coli*. *Proc. Natl. Acad. Sci. USA* **77:**3974–3977.

186. **Mechetner, E. B., B. Schott, B. S. Morse, W. D. Stein, T. Druley, K. A. Davis, T. Tsuruo, and I. B. Roninson.** 1997. P-glycoprotein function involves conformational transitions detectable by differential immunoreactivity. *Proc. Natl. Acad. Sci. USA* **94:**12908–12913.

187. **Mendes, C., M. E. Marin, F. Quinones, J. Sifuentes-Osornio, C. C. Siller, M. Castanheira, C. M. Zoccoli, H. Lopez, A. Sucari, F. Rossi, G. B. Angulo, A. J. Segura, C. Starling, I. Mimica, and D. Felmingham.** 2003. Antibacterial resistance of community-acquired respiratory tract pathogens recovered from patients in Latin America: results from the PROTEKT surveillance study (1999–2000). *Braz. J. Infect. Dis.* **7:**44–61.

188. **Mendez, C., and J. A. Salas.** 1998. ABC transporters in antibiotic-producing actinomycetes. *FEMS Microbiol. Lett.* **158:**1–8.

189. **Middlemiss, J. K., and K. Poole.** 2004. Differential impact of MexB mutations on substrate selectivity of the MexAB-OprM multidrug efflux pump of *Pseudomonas aeruginosa*. *J. Bacteriol.* **186:**1258–1269.

190. **Mikolosko, J., K. Bobyk, H. I. Zgurskaya, and P. Ghosh.** 2006. Conformational flexibility in the multidrug efflux system protein AcrA. *Structure* **14:**577–587.

191. **Mitchell, B. A., I. T. Paulsen, M. H. Brown, and R. A. Skurray.** 1999. Bioenergetics of the staphylococcal multidrug export protein QacA. Identification of distinct binding sites for monovalent and divalent cations. *J. Biol. Chem.* **274:**3541–3548.

192. **Miyamae, S., O. Ueda, F. Yoshimura, J. Hwang, Y. Tanaka, and H. Nikaido.** 2001. A MATE family multidrug efflux transporter pumps out fluoroquinolones in *Bacteroides thetaiotaomicron*. *Antimicrob. Agents Chemother.* **45:**3341–3346.

193. **Mokhonov, V. V., E. I. Mokhonova, H. Akama, and T. Nakae.** 2004. Role of the membrane fusion protein in the assembly of resistance-nodulation-cell division multidrug efflux pump in *Pseudomonas aeruginosa*. *Biochem. Biophys. Res. Commun.* **322:**483–489.

194. **Morita, Y., K. Kodama, S. Shiota, T. Mine, A. Kataoka, T. Mizushima, and T. Tsuchiya.** 1998. NorM, a putative multidrug efflux protein, of *Vibrio parahaemolyticus* and its homolog in *Escherichia coli*. *Antimicrob. Agents Chemother.* **42:**1778–1782.

195. **Mortimer, P. G., and L. J. Piddock.** 1991. A comparison of methods used for measuring the accumulation of quinolones by *Enterobacteriaceae*, *Pseudomonas aeruginosa* and *Staphylococcus aureus*. *J. Antimicrob. Chemother.* **28:**639–653.

196. **Mullin, S., N. Mani, and T. H. Grossman.** 2004. Inhibition of antibiotic efflux in bacteria by the novel multidrug resistance inhibitors biricodar (VX-710) and timcodar (VX-853). *Antimicrob. Agents Chemother.* **48:**4171–4176.

197. **Murakami, S., R. Nakashima, E. Yamashita, T. Matsumoto, and A. Yamaguchi.** 2006. Crystal structures of a multidrug transporter reveal a functionally rotating mechanism. *Nature* **443:**173–179.

198. **Murakami, S., R. Nakashima, E. Yamashita, and A. Yamaguchi.** 2002. Crystal structure of bacterial multidrug efflux transporter AcrB. *Nature* **419:**587–593.

199. **Murakami, S., N. Tamura, A. Saito, T. Hirata, and A. Yamaguchi.** 2004. Extramembrane central pore of multidrug exporter AcrB in *Escherichia coli* plays an important role in drug transport. *J. Biol. Chem.* **279:**3743–3748.

200. **Muth, T. R., and S. Schuldiner.** 2000. A membrane-embedded glutamate is required for ligand binding to the multidrug transporter EmrE. *EMBO J.* **19:**234–240.

201. **Narita, S., S. Eda, E. Yoshihara, and T. Nakae.** 2003. Linkage of the efflux-pump expression level with substrate extrusion rate in the MexAB-OprM efflux pump of *Pseudomonas aeruginosa*. *Biochem. Biophys. Res. Commun.* **308:**922–926.

202. **Nehme, D., X. Z. Li, R. Elliot, and K. Poole.** 2004. Assembly of the MexAB-OprM multidrug efflux system of *Pseudomonas aeruginosa*: identification and characterization of mutations in *mex*A compromising MexA multimerization and interaction with MexB. *J. Bacteriol.* **186:**2973–2983.

203. **Nies, D. H.** 2003. Efflux-mediated heavy metal resistance in prokaryotes. *FEMS Microbiol. Rev.* **27:**313–339.

204. **Nikaido, H., M. Basina, V. Nguyen, and E. Y. Rosenberg.** 1998. Multidrug efflux pump AcrAB of *Salmonella typhimurium* excretes only those beta-lactam antibiotics containing lipophilic side chains. *J. Bacteriol.* **180:**4686–4692.

205. **Nikaido, H., and H. I. Zgurskaya.** 2001. AcrAB and related multidrug efflux pumps of *Escherichia coli*. *J. Mol. Microbiol. Biotechnol.* **3:**215–218.

206. **Ninio, S., D. Rotem, and S. Schuldiner.** 2001. Functional analysis of novel multidrug transporters from human pathogens. *J. Biol. Chem.* **276:**48250–48256.

207. **Ojo, K. K., C. Ulep, N. Van Kirk, H. Luis, M. Bernardo, J. Leitao, and M. C. Roberts.** 2004. The *mef*(A) gene predominates among seven macrolide resistance genes identified in gram-negative strains representing 13 genera, isolated from healthy Portuguese children. *Antimicrob. Agents Chemother.* **48:**3451–3456.

208. **Pages, J. M., M. Masi, and J. Barbe.** 2005. Inhibitors of efflux pumps in Gram-negative bacteria. *Trends Mol. Med.* **11:**382–389.

209. **Paulsen, I. T.** 2003. Multidrug efflux pumps and resistance: regulation and evolution. *Curr. Opin. Microbiol.* **6:**446–451.

210. **Paulsen, I. T., M. H. Brown, T. G. Littlejohn, B. A. Mitchell, and R. A. Skurray.** 1996. Multidrug resistance proteins QacA and QacB from *Staphylococcus aureus*: membrane topology and identification of residues involved in substrate specificity. *Proc. Natl. Acad. Sci. USA* **93:**3630–3635.

211. **Paulsen, I. T., R. A. Skurray, R. Tam, M. H. Saier, Jr., R. J. Turner, J. H. Weiner, E. B. Goldberg, and L. L. Grinius.** 1996. The SMR family: a novel family of multidrug efflux proteins involved with the efflux of lipophilic drugs. *Mol. Microbiol.* **19:**1167–1175.

212. **Pearson, J. P., C. Van Delden, and B. H. Iglewski.** 1999. Active efflux and diffusion are involved in transport of *Pseudomonas aeruginosa* cell-to-cell signals. *J. Bacteriol.* **181:**1203–1210.

213. **Perez-Trallero, E., C. Fernandez-Mazarrasa, C. Garcia-Rey, E. Bouza, L. Aguilar, J. Garcia-de-Lomas, and F. Baquero.** 2001. Antimicrobial susceptibilities of 1,684 *Streptococcus pneumoniae* and 2,039 *Streptococcus pyogenes* isolates and their ecological relationships: results of a 1-year (1998–1999) multicenter surveillance study in Spain. *Antimicrob. Agents Chemother.* **45:**3334–3340.

214. **Piddock, L. J.** 1999. Mechanisms of fluoroquinolone resistance: an update 1994–1998. *Drugs* **58**(Suppl. 2)**:**11–18.

215. **Piddock, L. J.** 2006. Clinically relevant chromosomally encoded multidrug resistance efflux pumps in bacteria. *Clin. Microbiol. Rev.* **19:**382–402.

216. **Piddock, L. J.** 2006. Multidrug-resistance efflux pumps — not just for resistance. *Nat. Rev. Microbiol.* **4:**629–636.

217. **Pihlajamaki, M., J. Jalava, P. Huovinen, and P. Kotilainen.** 2003. Antimicrobial resistance of invasive pneumococci in Finland in 1999–2000. *Antimicrob. Agents Chemother.* **47:**1832–1835.

218. **Poole, K.** 2000. Efflux-mediated resistance to fluoroquinolones in gram-negative bacteria. *Antimicrob. Agents Chemother.* **44:**2233–2241.

219. **Poole, K.** 2000. Efflux-mediated resistance to fluoroquinolones in gram-positive bacteria and the mycobacteria. *Antimicrob. Agents Chemother.* **44:**2595–2599.

220. **Poole, K.** 2001. Multidrug efflux pumps and antimicrobial resistance in *Pseudomonas aeruginosa* and related organisms. *J. Mol. Microbiol. Biotechnol.* **3:**255–264.

221. **Poole, K.** 2002. Mechanisms of bacterial biocide and antibiotic resistance. *J. Appl. Microbiol.* **92**(Suppl.)**:**55S–64S.

222. **Poole, K.** 2004. Efflux-mediated multiresistance in Gram-negative bacteria. *Clin. Microbiol. Infect.* **10:**12–26.

223. **Poole, K.** 2005. Efflux-mediated antimicrobial resistance. *J. Antimicrob. Chemother.* **56:**20–51.

224. **Poole, K., and O. Lomovskaya.** 2006. Can efflux inhibitors really counter resistance? *Drug Discov. Today* **3:**145–152.

225. **Pornillos, O., Y. J. Chen, A. P. Chen, and G. Chang.** 2005. X-ray structure of the EmrE multidrug transporter in complex with a substrate. *Science* **310:**1950–1953.

226. **Pos, K. M., and K. Diederichs.** 2002. Purification, crystallization and preliminary diffraction studies of AcrB, an inner-membrane multi-drug efflux protein. *Acta Crystallogr. D. Biol. Crystallogr.* **58:**1865–1867.

227. **Pos, K. M., A. Schiefner, M. A. Seeger, and K. Diederichs.** 2004. Crystallographic analysis of AcrB. *FEBS Lett.* **564:**333–339.

228. **Putman, M., H. W. van Veen, and W. N. Konings.** 2000. Molecular properties of bacterial multidrug transporters. *Microbiol. Mol. Biol. Rev.* **64:**672–693.

229. **Quale, J., S. Bratu, D. Landman, and R. Heddurshetti.** 2003. Molecular epidemiology and mechanisms of carbapenem resistance in *Acinetobacter baumannii* endemic in New York City. *Clin. Infect. Dis.* **37:**214–220.

230. **Quentin, Y., and G. Fichant.** 2000. ABCdb: an ABC transporter database. *J. Mol. Microbiol. Biotechnol.* **2:**501–504.

231. **Ramos, J. L., E. Duque, M. T. Gallegos, P. Godoy, M. I. Ramos-Gonzalez, A. Rojas, W. Teran, and A. Segura.** 2002. Mechanisms of solvent tolerance in gram-negative bacteria. *Annu. Rev. Microbiol.* **56:**743–768.

232. **Randall, L. P., and M. J. Woodward.** 2002. The multiple antibiotic resistance (*mar*) locus and its significance. *Res. Vet. Sci.* **72:**87–93.

233. **Reinert, R. R., C. Franken, M. van der Linden, R. Lutticken, M. Cil, and A. Al Lahham.** 2004. Molecular characterisation of macrolide resistance mechanisms of *Streptococcus pneumoniae* and *Streptococcus pyogenes* isolated in Germany, 2002–2003. *Int. J. Antimicrob. Agents* **24:**43–47.

234. **Reinert, R. R., R. Lutticken, S. Reinert, A. Al Lahham, and S. Lemmen.** 2004. Antimicrobial resistance of *Streptococcus pneumoniae* isolates of outpatients in Germany, 1999–2000. *Chemotherapy* **50:**184–189.

235. **Reinert, R. R., R. Lutticken, J. A. Sutcliffe, A. Tait-Kamradt, M. Y. Cil, H. M. Schorn, A. Bryskier, and A. Al Lahham.** 2004. Clonal relatedness of erythromycin-resistant *Streptococcus pyogenes* isolates in Germany. *Antimicrob. Agents Chemother.* **48:**1369–1373.

236. **Renau, T. E.** 2001. Efflux pump inhibitors to address bacterial and fungal resistance. *Drugs Future* **26:**1171–1178.

237. **Renau, T. E., R. Leger, L. Filonova, E. M. Flamme, M. Wang, R. Yen, D. Madsen, D. Griffith, S. Chamberland, M. N. Dudley, V. J. Lee, O. Lomovskaya, W. J. Watkins, T. Ohta, K. Nakayama, and Y. Ishida.** 2003. Conformationally-restricted analogues of efflux pump inhibitors that potentiate the activity of levofloxacin in *Pseudomonas aeruginosa*. *Bioorg. Med. Chem. Lett.* **13:**2755–2758.

238. **Renau, T. E., R. Leger, E. M. Flamme, J. Sangalang, M. W. She, R. Yen, C. L. Gannon, D. Griffith, S. Chamberland, O. Lomovskaya, S. J. Hecker, V. J. Lee, T. Ohta, and K. Nakayama.** 1999. Inhibitors of efflux pumps in *Pseudomonas aeruginosa* potentiate the activity of the fluoroquinolone antibacterial levofloxacin. *J. Med. Chem.* **42:**4928–4931.

239. **Renau, T. E., R. Leger, R. Yen, M. W. She, E. M. Flamme, J. Sangalang, C. L. Gannon, S. Chamberland, O. Lomovskaya, and V. J. Lee.** 2002. Peptidomimetics of efflux pump inhibitors potentiate the activity of levofloxacin in *Pseudomonas aeruginosa*. *Bioorg. Med. Chem. Lett.* **12:**763–766.

240. **Reuter, G., T. Janvilisri, H. Venter, S. Shahi, L. Balakrishnan, and H. W. van Veen.** 2003. The ATP binding cassette multidrug transporter LmrA and lipid transporter MsbA have overlapping substrate specificities. *J. Biol. Chem.* **278:**35193–35198.

241. **Reyes, C. L., and G. Chang.** 2005. Structure of the ABC transporter MsbA in complex with ADP.vanadate and lipopolysaccharide. *Science* **308:**1028–1031.

242. **Reyes, C. L., A. Ward, J. Yu, and G. Chang.** 2006. The structures of MsbA: insight into ABC transporter-mediated multidrug efflux. *FEBS Lett.* **580:**1042–1048.

243. **Reynolds, E., J. I. Ross, and J. H. Cove.** 2003. Msr(A) and related macrolide/streptogramin resistance determinants: incomplete transporters? *Int. J. Antimicrob. Agents* **22:** 228–236.

244. **Ribera, A., J. Ruiz, M. T. Jiminez de Anta, and J. Vila.** 2002. Effect of an efflux pump inhibitor on the MIC of nalidixic acid for *Acinetobacter baumannii* and *Stenotrophomonas maltophilia* clinical isolates. *J. Antimicrob. Chemother.* **49:**697–698.

245. **Rice, L. B.** 2006. Challenges in identifying new antimicrobial agents effective for treating infections with *Acinetobacter baumannii* and *Pseudomonas aeruginosa*. *Clin. Infect. Dis.* **43**(Suppl. 2):S100–S105.

246. **Roberts, M. C.** 2003. Tetracycline therapy: update. *Clin. Infect. Dis.* **36:**462–467.

247. **Roberts, M. C.** 2005. Update on acquired tetracycline resistance genes. *FEMS Microbiol. Lett.* **245:**195–203.

248. **Roberts, M. C., J. Sutcliffe, P. Courvalin, L. B. Jensen, J. Rood, and H. Seppala.** 1999. Nomenclature for macrolide and macrolide-lincosamide-streptogramin B resistance determinants. *Antimicrob. Agents Chemother.* **43:**2823–2830.

249. **Rosen, B. P.** 1999. The role of efflux in bacterial resistance to soft metals and metalloids. *Essays Biochem.* **34:**1–15.

250. **Rosenberg, M. F., G. Velarde, R. C. Ford, C. Martin, G. Berridge, I. D. Kerr, R. Callaghan, A. Schmidlin, C. Wooding, K. J. Linton, and C. F. Higgins.** 2001. Repacking of the transmembrane domains of P-glycoprotein during the transport ATPase cycle. *EMBO J.* **20:**5615–5625.

251. **Rotem, D., N. Sal-man, and S. Schuldiner.** 2001. *In vitro* monomer swapping in EmrE, a multidrug transporter from *Escherichia coli*, reveals that the oligomer is the functional unit. *J. Biol. Chem.* **276:**48243–48249.

252. **Rouquette-Loughlin, C., S. A. Dunham, M. Kuhn, J. T. Balthazar, and W. M. Shafer.** 2003. The NorM efflux pump of *Neisseria gonorrhoeae* and *Neisseria meningitidis* recognizes antimicrobial cationic compounds. *J. Bacteriol.* **185:**1101–1106.

253. **Ruan, Z. S., V. Anantharam, I. T. Crawford, S. V. Ambudkar, S. Y. Rhee, M. J. Allison, and P. C. Maloney.** 1992. Identification, purification, and reconstitution of OxlT, the oxalate: formate antiport protein of *Oxalobacter formigenes*. *J. Biol. Chem.* **267:**10537–10543.

254. **Rubin, R. A., and S. B. Levy.** 1990. Interdomain hybrid Tet proteins confer tetracycline resistance only when they are derived from closely related members of the *tet* gene family. *J. Bacteriol.* **172:**2303–2312.

255. **Russell, A. D.** 2001. Mechanisms of bacterial insusceptibility to biocides. *Am. J. Infect. Control* **29:**259–261.

256. **Russell, A. D.** 2002. Antibiotic and biocide resistance in bacteria: introduction. *J. Appl. Microbiol.* **92**(Suppl.):1S–3S.

257. **Russell, A. D.** 2004. Whither triclosan? *J. Antimicrob. Chemother.* **53:**693–695.

258. **Saidijam, M., G. Benedetti, Q. Ren, Z. Xu, C. J. Hoyle, S. L. Palmer, A. Ward, K. E. Bettaney, G. Szakonyi, J. Meuller, S. Morrison, M. K. Pos, P. Butaye, K. Walravens, K. Langton, R. B. Herbert, R. A. Skurray, I. T. Paulsen, J. O'Reilly, N. G. Rutherford, M. H. Brown, R. M. Bill, and P. J. Henderson.** 2006. Microbial drug efflux proteins of the major facilitator superfamily. *Curr. Drug Targets* **7:**793–811.

259. **Saier, M. H., Jr.** 2000. Families of transmembrane sugar transport proteins. *Mol. Microbiol.* **35:**699–710.

260. **Saier, M. H., Jr., J. T. Beatty, A. Goffeau, K. T. Harley, W. H. Heijne, S. C. Huang, D. L. Jack, P. S. Jahn, K. Lew, J. Liu, S. S. Pao, I. T. Paulsen, T. T. Tseng, and P. S. Virk.** 1999. The major facilitator superfamily. *J. Mol. Microbiol. Biotechnol.* **1:**257–279.

261. **Sanchez, P., E. Moreno, and J. L. Martinez.** 2005. The biocide triclosan selects *Stenotrophomonas maltophilia* mutants that overproduce the SmeDEF multidrug efflux pump. *Antimicrob. Agents Chemother.* **49:**781–782.

262. **Saurin, W., M. Hofnung, and E. Dassa.** 1999. Getting in or out: early segregation between importers and exporters in the evolution of ATP-binding cassette (ABC) transporters. *J. Mol. Evol.* **48:**22–41.

263. **Sawada, I., H. Maseda, T. Nakae, H. Uchiyama, and N. Nomura.** 2004. A quorum-sensing autoinducer enhances the *mex*AB-oprM efflux-pump expression without the MexR-mediated regulation in *Pseudomonas aeruginosa*. *Microbiol. Immunol.* **48:**435–439.

264. **Schmitz, F. J., A. C. Fluit, S. Brisse, J. Verhoef, K. Kohrer, and D. Milatovic.** 1999. Molecular epidemiology of quinolone resistance and comparative in vitro activities of new quinolones against European *Staphylococcus aureus* isolates. *FEMS Immunol. Med. Microbiol.* **26:**281–287.

265. **Schmitz, F. J., M. Perdikouli, A. Beeck, J. Verhoef, and A. C. Fluit.** 2001. Molecular surveillance of macrolide, tetracycline and quinolone resistance mechanisms in 1191 clinical European *Streptococcus pneumoniae* isolates. *Int. J. Antimicrob. Agents* **18:**433–436.

266. **Schmitz, F. J., R. Sadurski, A. Kray, M. Boos, R. Geisel, K. Kohrer, J. Verhoef, and A. C. Fluit.** 2000. Prevalence of macrolide-resistance genes in *Staphylococcus aureus* and *Enterococcus faecium* isolates from 24 European university hospitals. *J. Antimicrob. Chemother.* **45:**891–894.

267. **Schumacher, M. A., M. C. Miller, and R. G. Brennan.** 2004. Structural mechanism of the simultaneous binding of two drugs to a multidrug-binding protein. *EMBO J.* **23:**2923–2930.

268. **Schumacher, M. A., M. C. Miller, S. Grkovic, M. H. Brown, R. A. Skurray, and R. G. Brennan.** 2001. Structural

mechanisms of QacR induction and multidrug recognition. *Science* **294:**2158–2163.

269. **Schwarz, S., C. Kehrenberg, B. Doublet, and A. Cloeckaert.** 2004. Molecular basis of bacterial resistance to chloramphenicol and florfenicol. *FEMS Microbiol. Rev.* **28:**519–542.
270. **Schweizer, H. P.** 2003. Efflux as a mechanism of resistance to antimicrobials in *Pseudomonas aeruginosa* and related bacteria: unanswered questions. *Genet. Mol. Res.* **2:**48–62.
271. **Seeger, M. A., A. Schiefner, T. Eicher, F. Verrey, K. Diederichs, and K. M. Pos.** 2006. Structural asymmetry of AcrB trimer suggests a peristaltic pump mechanism. *Science* **313:**1295–1298.
272. **Sharff, A., C. Fanutti, J. Shi, C. Calladine, and B. Luisi.** 2001. The role of the TolC family in protein transport and multidrug efflux. From stereochemical certainty to mechanistic hypothesis. *Eur. J. Biochem.* **268:**5011–5026.
273. **Silver, S.** 2003. Bacterial silver resistance: molecular biology and uses and misuses of silver compounds. *FEMS Microbiol. Rev.* **27:**341–353.
274. **Sobel, M. L., G. A. McKay, and K. Poole.** 2003. Contribution of the MexXY multidrug transporter to aminoglycoside resistance in *Pseudomonas aeruginosa* clinical isolates. *Antimicrob. Agents Chemother.* **47:**3202–3207.
275. **Someya, Y., A. Yamaguchi, and T. Sawai.** 1995. A novel glycylcycline, 9–(*N*,*N*-dimethylglycylamido)-6-demethyl-6-deoxytetracycline, is neither transported nor recognized by the transposon Tn10-encoded metal-tetracycline/H+ antiporter. *Antimicrob. Agents Chemother.* **39:**247–249.
276. **Stegmeier, J. F., G. Polleichtner, N. Brandes, C. Hotz, and C. Andersen.** 2006. Importance of the adaptor (membrane fusion) protein hairpin domain for the functionality of multidrug efflux pumps. *Biochemistry* **45:**10303–10312.
277. **Stermitz, F. R., P. Lorenz, J. N. Tawara, L. A. Zenewicz, and K. Lewis.** 2000. Synergy in a medicinal plant: antimicrobial action of berberine potentiated by 5′-methoxyhydnocarpin, a multidrug pump inhibitor. *Proc. Natl. Acad. Sci. USA* **97:**1433–1437.
278. **Stermitz, F. R., L. N. Scriven, G. Tegos, and K. Lewis.** 2002. Two flavonols from *Artemisa annua* which potentiate the activity of berberine and norfloxacin against a resistant strain of *Staphylococcus aureus*. *Planta Med.* **68:**1140–1141.
279. **Sum, P. E., F. W. Sum, and S. J. Projan.** 1998. Recent developments in tetracycline antibiotics. *Curr. Pharm. Des.* **4:**119–132.
280. **Tamura, N., S. Konishi, S. Iwaki, T. Kimura-Someya, S. Nada, and A. Yamaguchi.** 2001. Complete cysteine-scanning mutagenesis and site-directed chemical modification of the Tn10-encoded metal-tetracycline/H+ antiporter. *J. Biol. Chem.* **276:**20330–20339.
281. **Tamura, N., S. Murakami, Y. Oyama, M. Ishiguro, and A. Yamaguchi.** 2005. Direct interaction of multidrug efflux transporter AcrB and outer membrane channel TolC detected via site-directed disulfide cross-linking. *Biochemistry* **44:**11115–11121.
282. **Tate, C. G., I. Ubarretxena-Belandia, and J. M. Baldwin.** 2003. Conformational changes in the multidrug transporter EmrE associated with substrate binding. *J. Mol. Biol.* **332:**229–242.
283. **Tavio, M. M., J. Vila, M. Perilli, L. T. Casanas, L. Macia, G. Amicosante, and M. T. Jimenez de Anta.** 2004. Enhanced active efflux, repression of porin synthesis and development of Mar phenotype by diazepam in two *Enterobacteria* strains. *J. Med. Microbiol.* **53:**1119–1122.
284. **Tegos, G., F. R. Stermitz, O. Lomovskaya, and K. Lewis.** 2002. Multidrug pump inhibitors uncover remarkable activity of plant antimicrobials. *Antimicrob. Agents Chemother.* **46:**3133–3141.
285. **Thanabalu, T., E. Koronakis, C. Hughes, and V. Koronakis.** 1998. Substrate-induced assembly of a contiguous channel for protein export from *E. coli*: reversible bridging of an inner-membrane translocase to an outer membrane exit pore. *EMBO J.* **17:**6487–6496.
286. **Tikhonova, E. B., Q. Wang, and H. I. Zgurskaya.** 2002. Chimeric analysis of the multicomponent multidrug efflux transporters from gram-negative bacteria. *J. Bacteriol.* **184:** 6499–6507.
287. **Tikhonova, E. B., and H. I. Zgurskaya.** 2004. AcrA, AcrB, and TolC of *Escherichia coli* form a stable intermembrane multidrug efflux complex. *J. Biol. Chem.* **279:**32116–32124.
288. **Touze, T., J. Eswaran, E. Bokma, E. Koronakis, C. Hughes, and V. Koronakis.** 2004. Interactions underlying assembly of the *Escherichia coli* AcrAB-TolC multidrug efflux system. *Mol. Microbiol.* **53:**697–706.
289. **Tseng, T. T., K. S. Gratwick, J. Kollman, D. Park, D. H. Nies, A. Goffeau, and M. H. Saier, Jr.** 1999. The RND permease superfamily: an ancient, ubiquitous and diverse family that includes human disease and development proteins. *J. Mol. Microbiol. Biotechnol.* **1:**107–125.
290. **Tuckman, M., P. J. Petersen, and S. J. Projan.** 2000. Mutations in the interdomain loop region of the *tet*A(A) tetracycline resistance gene increase efflux of minocycline and glycylcyclines. *Microb. Drug Resist.* **6:**277–282.
291. **Ubarretxena-Belandia, I., J. M. Baldwin, S. Schuldiner, and C. G. Tate.** 2003. Three-dimensional structure of the bacterial multidrug transporter EmrE shows it is an asymmetric homodimer. *EMBO J.* **22:**6175–6181.
292. **Ubarretxena-Belandia, I., and C. G. Tate.** 2004. New insights into the structure and oligomeric state of the bacterial multidrug transporter EmrE: an unusual asymmetric homo-dimer. *FEBS Lett.* **564:**234–238.
293. **Vaccaro, L., V. Koronakis, and M. S. Sansom.** 2006. Flexibility in a drug transport accessory protein: molecular dynamics simulations of MexA. *Biophys. J.* **91:**558–564.
294. **van Amsterdam, K., A. Bart, and A. van der Ende.** 2005. A *Helicobacter pylori* TolC efflux pump confers resistance to metronidazole. *Antimicrob. Agents Chemother.* **49:**1477–1482.
295. **Van Bambeke, F., E. Balzi, and P. M. Tulkens.** 2000. Antibiotic efflux pumps. *Biochem. Pharmacol.* **60:**457–470.
296. **van Veen, H. W., R. Callaghan, L. Soceneantu, A. Sardini, W. N. Konings, and C. F. Higgins.** 1998. A bacterial antibiotic-resistance gene that complements the human multidrug-resistance P-glycoprotein gene. *Nature* **391:**291–295.
297. **van Veen, H. W., K. Venema, H. Bolhuis, I. Oussenko, J. Kok, B. Poolman, A. J. Driessen, and W. N. Konings.** 1996. Multidrug resistance mediated by a bacterial homolog

of the human multidrug transporter MDR1. *Proc. Natl. Acad. Sci. USA* **93**:10668–10672.

298. **Vardy, E., I. T. Arkin, K. E. Gottschalk, H. R. Kaback, and S. Schuldiner.** 2004. Structural conservation in the major facilitator superfamily as revealed by comparative modeling. *Protein Sci.* **13**:1832–1840.

299. **Venter, H., R. A. Shilling, S. Velamakanni, L. Balakrishnan, and H. W. van Veen.** 2003. An ABC transporter with a secondary-active multidrug translocator domain. *Nature* **426**:866–870.

300. **Verdon, G., S. V. Albers, B. W. Dijkstra, A. J. Driessen, and A. M. Thunnissen.** 2003. Crystal structures of the ATPase subunit of the glucose ABC transporter from *Sulfolobus solfataricus*: nucleotide-free and nucleotide-bound conformations. *J. Mol. Biol.* **330**:343–358.

301. **Visalli, M. A., E. Murphy, S. J. Projan, and P. A. Bradford.** 2003. AcrAB multidrug efflux pump is associated with reduced levels of susceptibility to tigecycline (GAR-936) in *Proteus mirabilis*. *Antimicrob. Agents Chemother.* **47**:665–669.

302. **Viveiros, M., C. Leandro, and L. Amaral.** 2003. Mycobacterial efflux pumps and chemotherapeutic implications. *Int. J. Antimicrob. Agents* **22**:274–278.

303. **Vogne, C., J. R. Aires, C. Bailly, D. Hocquet, and P. Plesiat.** 2004. Role of the multidrug efflux system MexXY in the emergence of moderate resistance to aminoglycosides among *Pseudomonas aeruginosa* isolates from patients with cystic fibrosis. *Antimicrob. Agents Chemother.* **48**:1676–1680.

304. **Wang, G., R. Pincheira, and J. T. Zhang.** 1998. Dissection of drug-binding-induced conformational changes in P-glycoprotein. *Eur. J. Biochem.* **255**:383–390.

305. **Watkins, R. E., J. M. Maglich, L. B. Moore, G. B. Wisely, S. M. Noble, P. R. Davis-Searles, M. H. Lambert, S. A. Kliewer, and M. R. Redinbo.** 2003. 2.1 A crystal structure of human PXR in complex with the St. John's wort compound hyperforin. *Biochemistry* **42**:1430–1438.

306. **Webber, M. A., and L. J. Piddock.** 2003. The importance of efflux pumps in bacterial antibiotic resistance. *J. Antimicrob. Chemother.* **51**:9–11.

307. **Widdowson, M. A., A. Bosman, E. van Straten, M. Tinga, S. Chaves, L. van Eerden, and W. van Pelt.** 2003. Automated, laboratory-based system using the Internet for disease outbreak detection, the Netherlands. *Emerg. Infect. Dis.* **9**:1046–1052.

308. **Wiggins, P.** 2004. Efflux pumps: an answer to Gram-negative bacterial resistance? *Expert. Opin. Investig. Drugs* **13**:899–902.

309. **Wolter, D. J., E. Smith-Moland, R. V. Goering, N. D. Hanson, and P. D. Lister.** 2004. Multidrug resistance associated with *mex*XY expression in clinical isolates of *Pseudomonas aeruginosa* from a Texas hospital. *Diagn. Microbiol. Infect. Dis.* **50**:43–50.

310. **Wong, K. K., F. S. Brinkman, R. S. Benz, and R. E. Hancock.** 2001. Evaluation of a structural model of *Pseudomonas aeruginosa* outer membrane protein OprM, an efflux component involved in intrinsic antibiotic resistance. *J. Bacteriol.* **183**:367–374.

311. **Xu, X. J., X. Z. Su, Y. Morita, T. Kuroda, T. Mizushima, and T. Tsuchiya.** 2003. Molecular cloning and characterization of the HmrM multidrug efflux pump from *Haemophilus influenzae* Rd. *Microbiol. Immunol.* **47**:937–943.

312. **Yerushalmi, H., M. Lebendiker, and S. Schuldiner.** 1995. EmrE, an *Escherichia coli* 12-kDa multidrug transporter, exchanges toxic cations and H+ and is soluble in organic solvents. *J. Biol. Chem.* **270**:6856–6863.

313. **Yin, C. C., M. L. Aldema-Ramos, M. I. Borges-Walmsley, R. W. Taylor, A. R. Walmsley, S. B. Levy, and P. A. Bullough.** 2000. The quarternary molecular architecture of TetA, a secondary tetracycline transporter from *Escherichia coli*. *Mol. Microbiol.* **38**:482–492.

314. **Yoneyama, H., H. Maseda, T. A. Yamabayashi Ta, S. Izumi, and T. Nakae.** 2002. Secondary-site mutation restores the transport defect caused by the transmembrane domain mutation of the xenobiotic transporter MexB in *Pseudomonas aeruginosa*. *Biochem. Biophys. Res. Commun.* **292**:513–518.

315. **Yoshida, K., K. Nakayama, N. Kuru, S. Kobayashi, M. Ohtsuka, M. Takemura, K. Hoshino, H. Kanda, J. Z. Zhang, V. J. Lee, and W. J. Watkins.** 2006. MexAB-OprM specific efflux pump inhibitors in *Pseudomonas aeruginosa*. Part 5. Carbon-substituted analogues at the C-2 position. *Bioorg. Med. Chem.* **14**:1993–2004.

316. **Yu, E. W., J. R. Aires, G. McDermott, and H. Nikaido.** 2005. A periplasmic drug-binding site of the AcrB multidrug efflux pump: a crystallographic and site-directed mutagenesis study. *J. Bacteriol.* **187**:6804–6815.

317. **Yu, E. W., G. McDermott, H. I. Zgurskaya, H. Nikaido, and D. E. Koshland, Jr.** 2003. Structural basis of multiple drug-binding capacity of the AcrB multidrug efflux pump. *Science* **300**:976–980.

318. **Yu, J. L., L. Grinius, and D. C. Hooper.** 2002. NorA functions as a multidrug efflux protein in both cytoplasmic membrane vesicles and reconstituted proteoliposomes. *J. Bacteriol.* **184**:1370–1377.

319. **Zgurskaya, H. I., G. Krishnamoorthy, E. B. Tikhonova, S. Y. Lau, and K. L. Stratton.** 2003. Mechanism of antibiotic efflux in Gram-negative bacteria. *Front Biosci.* **8**:s862–s873.

320. **Zgurskaya, H. I., and H. Nikaido.** 1999. Bypassing the periplasm: reconstitution of the AcrAB multidrug efflux pump of *Escherichia coli*. *Proc. Natl. Acad. Sci. USA* **96**:7190–7195.

321. **Zgurskaya, H. I., and H. Nikaido.** 2000. Cross-linked complex between oligomeric periplasmic lipoprotein AcrA and the inner-membrane-associated multidrug efflux pump AcrB from *Escherichia coli*. *J. Bacteriol.* **182**:4264–4267.

322. **Zgurskaya, H. I., and H. Nikaido.** 2000. Multidrug resistance mechanisms: drug efflux across two membranes. *Mol. Microbiol.* **37**:219–225.

323. **Zhanel, G. G., K. Homenuik, K. Nichol, A. Noreddin, L. Vercaigne, J. Embil, A. Gin, J. A. Karlowsky, and D. J. Hoban.** 2004. The glycylcyclines: a comparative review with the tetracyclines. *Drugs* **64**:63–88.

324. **Zheleznova, E. E., P. N. Markham, A. A. Neyfakh, and R. G. Brennan.** 1999. Structural basis of multidrug recognition by BmrR, a transcription activator of a multidrug transporter. *Cell* **96**:353–362.

Dissemination of Antibiotic Resistance and Its Biological Cost

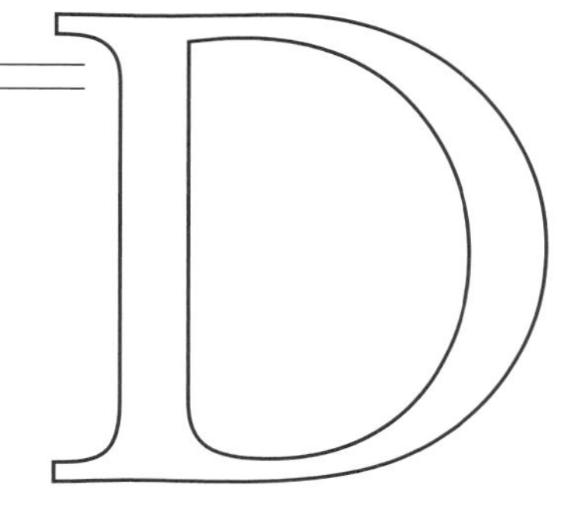

Enzyme-Mediated Resistance to Antibiotics: Mechanisms, Dissemination, and Prospects for Inhibition
Edited by Robert A. Bonomo and Marcelo E. Tolmasky

Marcelo E. Tolmasky

16

Overview of Dissemination Mechanisms of Genes Coding for Resistance to Antibiotics

Multiresistance to antibiotics is becoming common in a wide variety of pathogens (15, 33, 37, 40). Most resistance genes in pathogenic bacteria have been acquired by horizontal transfer, as opposed to evolving de novo (9). The emergence of antibiotic resistance determinants would not be as devastating if it were not for the inherent ability of bacteria to exchange genes at the cellular and molecular level. Hence, the study of the mechanisms by which resistance genes are moved between cells and exchanged between DNA molecules is as important as the research on the mechanisms mediating antibiotic resistance. Since the pioneering experiments that led to the discovery of gene exchange between cells by conjugation (18) and transduction (41) in the mid-1940s and early 1950s, new discoveries on the mechanisms and genetic elements involved in horizontal gene exchange have continued at a regular pace. The increment in antibiotic-resistant pathogenic bacteria led to an acceleration in the discovery and understanding of gene exchange mechanisms, new types of genetic elements, and their molecular bases. Thus, these advances can be considered a "by-product" that resulted from the multiple uses of antibiotics since the first utilization of penicillin for treatment of human infection in 1942 (19).

In 1952 Lederberg proposed the name "plasmids" for "any extrachromosomal hereditary determinant" (17). About 10 years later it was determined that *Shigella dysenteriae* strains isolated from diseased people in Japan were able to resist the action of four antibiotics. Surprisingly, *Escherichia coli* isolates from the same patients were resistant to the same antibiotics, (38, 39). This finding led to the discovery that plasmids played an important role in dissemination of multiple drug resistance among bacteria by conjugation. Plasmids harboring genetic determinants for antibiotic resistance were referred to as "R factors," and the initial experiments showing molecular evidence of their existence as independent DNA molecules were not reported until 1966 (11, 29). The tremendous impact of plasmids on dissemination of antibiotic resistance led to the isolation and the characterization of their biological properties, which in turn led to a plethora of discoveries in molecular genetics, transcending the plasmid biology or antibiotic resistance fields (6). A wide variety of DNA replication and regulation strategies were described, as well as a large number of ingenious stability mechanisms including partition systems, toxin-antitoxin, also known as "plasmid addiction" or "genetic addiction," and multimer resolution systems which usually occur through site-specific recombination (1, 12, 13, 35). Some plasmids, like RK2, encode their own site-specific resolution system including the recombinase and the target site (10). In many cases

Marcelo E. Tolmasky, Department of Biological Science, College of Natural Sciences and Mathematics, California State University–Fullerton, 800 N State College Blvd., Fullerton, CA 92831-3599.

these systems are derived from the cointegrate resolution system of replicative transposons. Other plasmids like ColE1 include a target site for the host-encoded Xer site-specific recombination system (32, 36). In these cases the target site has been modified and requires accessory proteins to form a nucleoprotein complex of specific topology, which ensures that recombination is preferentially intramolecular (7). A plasmid genome database containing all sequenced plasmids was recently established (23).

While plasmids participate in bacterial antibiotic resistance mainly by disseminating genes coding for drug resistance at the cellular level, other elements promote gene exchange at the molecular level. Transposable elements, which often include resistance genes, mediate their own transfer from replicon to replicon, allowing dissemination of the resistance genes among diverse microbial populations that may not support replication of the same plasmids (8). Prokaryotic transposons have usually been divided in classes according to their properties and characteristics: (i) insertion sequences and composite transposons such as Tn*10* or Tn5, which carry two copies of an insertion sequence at their ends (IS*10* and IS*50* in the cases of Tn*10* and Tn5, respectively); these transposons commonly include an antibiotic resistance gene and require only one protein for transposition; (ii) complex transposons with short inverted repeats at their ends; these elements usually transpose through a replicative mechanism requiring two proteins, a transposase and a resolvase (e.g., Tn3-like transposons); most of them harbor a variable number of antibiotic resistance genes; (iii) bacteriophages that utilize transposition as their way of life like Mu; and (iv) transposons that do not fit any of the mentioned classes.

Some transposable elements, "conjugative transposons," mainly found in gram positives, can excise from the host DNA molecule and transfer themselves to another cell and insert into a new DNA molecule (5, 30). More recently, other mobile genetic elements related to conjugative transposons were identified. However, since they are larger than typical gram-positive conjugative transposons and have other unique characteristics such as a mosaic structure and relation to pathogenicity islands, it was proposed that they be classified as "integrative and conjugative elements" (ICE) (4, 5, 14, 24). ICEs encode diverse excision, recombination, and conjugation systems, in addition to specific functions, including resistance to antibiotics. Their specificity of integration into a replicon present in the recipient cell varies widely; a large variety of sequences used as targets for integration of ICEs have been identified.

While transposons induce variability of the structure of plasmids and chromosomes by inserting themselves into these molecules, other processes induce variability of the structures of transposons. They can acquire resistance genes by different events that include but are not limited to random introduction of insertion sequences that generate composite transposons, transposition inside another transposable element to generate a larger mobile element with several resistance genes, complex combinations of transposable elements generating ICEs, or acquisition of integrons (20, 25, 26, 28). The latter elements were first described in 1989 as a consequence of comparative analyses of the nucleotide sequences of resistance genes and their environments in diverse mobile elements and replicons (34). They act as natural expression vectors and provide a highly efficient mechanism to capture resistance genes by site-specific recombination. Later the superintegrons, characterized by a large number of gene cassettes in their structure, were found in the *Vibrio cholerae* chromosome (22). Further research led to the recognition that superintegrons are an integral part of the genomes of γ-proteobacteria (27).

The combination of dissemination at the cellular level through conjugation, natural transformation, transduction, with dissemination at the molecular level, and mutagenesis permits genes coding for antibiotic resistance to reach virtually all bacterial cells, resulting in a virtual elimination of barriers between types of bacteria (19). However, while the acquisition of resistance genes provides an advantage to the bacterial cells when in the presence of antibiotics, it has been shown that their presence comes with an associated fitness cost when the cells are growing in the absence of antibiotic selective pressure (3). The mutation of chromosomal genes (usually essential) that modify target molecules or the acquisition of plasmids harboring the resistance genes creates burdens on the cell that result in a fitness cost that is typically reflected in a reduced growth rate in culture medium or animal hosts (3, 16). In some cases the fitness cost can be reduced by subsequent compensatory mutations (2). Recent cases in which there was little or no fitness costs for the mutations leading to resistance have also been reported (31), and in at least one case there was an enhanced fitness associated with acquisition of resistance (21).

The intense research on the elements mentioned above has generated a volume of knowledge that makes it impossible to be covered in one book section. In this section we include chapters on selected topics related to the dissemination of antibiotic resistance. In addition a chapter is devoted to discuss the biological cost of antibiotic resistance. The reader should refer to specialized publications for deep as well as extensive description of genetic elements and mechanisms not covered in this book (8, 12, 35).

References

1. **Actis, L. A., M. E. Tolmasky, and J. H. Crosa.** 1999. Bacterial plasmids: replication of extrachromosomal genetic elements encoding resistance to antimicrobial compounds. *Front Biosci* **4:**D43-D62.
2. **Andersson, D. I.** 2003. Persistence of antibiotic resistant bacteria. *Curr. Opin. Microbiol.* **6:**452–456.
3. **Andersson, D. I., and B. R. Levin.** 1999. The biological cost of antibiotic resistance. *Curr. Opin. Microbiol.* **2:**489–493.
4. **Burrus, V., and M. K. Waldor.** 2004. Formation of SXT tandem arrays and SXT-R391 hybrids. *J. Bacteriol.* **186:**2636–2645.
5. **Churchward, G.** 2002. Conjugative transposons and related mobile elements, p. 177–191. *In* N. Craig, R. Craigie, M. Gellert, and A. Lambowitz (ed.), *Mobile DNA II.* ASM Press, Washington, D.C.
6. **Cohen, S.** 1993. Bacterial plasmids: their extraordinary contribution to molecular genetics. *Gene* **135:**67–76.
7. **Colloms, S. D., R. McCulloch, K. Grant, L. Neilson, and D. J. Sherratt.** 1996. Xer-mediated site-specific recombination in vitro. *EMBO J.* **15:**1172–1181.
8. **Craig, N., R. Craigie, M. Gellert, and A. Lambowitz (ed.).** 2002. *Mobile DNA II*, 2nd ed. ASM Press, Washington, D.C.
9. **Davies, J.** 1997. Origins, acquisition and dissemination of antibiotic resistance determinants. *Ciba Found. Symp.* **207:**15–27.
10. **Easter, C. L., H. Schwab, and D. R. Helinski.** 1998. Role of the *parCBA* operon of the broad-host-range plasmid RK2 in stable plasmid maintenance. *J. Bacteriol.* **180:**6023–6030.
11. **Falkow, S., R. Citarella, J. Wohlhieter, and T. Watanabe.** 1966. The molecular nature of R factors. *J. Mol. Biol.* **17:**102–116.
12. **Funnell, B., and G. Phillips (ed.).** 2004. *Plasmid Biology.* ASM Press, Washington, D.C.
13. **Helinski, D., A. Toukdarian, and R. Novick.** 1996. Replication control and other stable maintenance mechanisms of plasmids, p. 2295–2324. *In* F. Neidhardt, R. Curtis III, J. Ingraham, E. Lin, K. Low, B. Magasanik, W. Rezikoff, M. Riley, M. Schaechter, and H. Umbarger (ed.), Escherichia coli *and* Salmonella: *Cellular and Molecular Biology*, 2nd ed., vol. 2. ASM Press, Washington, D.C.
14. **Hochhut, B., Y. Lotfi, D. Mazel, S. M. Faruque, R. Woodgate, and M. K. Waldor.** 2001. Molecular analysis of antibiotic resistance gene clusters in *Vibrio cholerae* O139 and O1 SXT constins. *Antimicrob. Agents Chemother.* **45:**2991–3000.
15. **Karchmer, A. W.** 2004. Increased antibiotic resistance in respiratory tract pathogens: PROTEKT US—an update. *Clin. Infect. Dis.* **39**(Suppl. 3)**:**S142–S150.
16. **Kim, C., J. Y. Cha, H. Yan, S. B. Vakulenko, and S. Mobashery.** 2006. Hydrolysis of ATP by aminoglycoside 3′-phosphotransferases: an unexpected cost to bacteria for harboring an antibiotic resistance enzyme. *J. Biol. Chem.* **281:**6964–6969.
17. **Lederberg, J.** 1952. Cell genetics and hereditary symbiosis. *Physiol. Rev.* **32:**403–430.
18. **Lederberg, J., and E. Tatum.** 1946. Gene recombination in *Escherichia coli. Nature* **158:**558.
19. **Levy, S.** 2002. *The Antibiotic Paradox. How the Misuse of Antibiotics Destroys Their Curative Powers*, 2nd ed. Perseus Publishing, Cambridge, Mass.
20. **Liebert, C. A., R. M. Hall, and A. O. Summers.** 1999. Transposon Tn21, flagship of the floating genome. *Microbiol. Mol. Biol. Rev.* **63:**507–522.
21. **Luo, N., S. Pereira, O. Sahin, J. Lin, S. Huang, L. Michel, and Q. Zhang.** 2005. Enhanced in vivo fitness of fluoroquinolone-resistant *Campylobacter jejuni* in the absence of antibiotic selection pressure. *Proc. Natl. Acad. Sci. USA* **102:**541–546.
22. **Mazel, D., B. Dychinco, V. Webb, and J. Davies.** 1998. A distinctive class of integron in the *Vibrio cholerae* genome. *Science* **280:**605–608.
23. **Mølbak, L., A. Tett, D. Ussery, K. Wall, S. Turner, M. Balley, and D. Field.** 2003. The plasmid genome database. *Microbiology* **149:**3043–3045.
24. **Pembroke, J., C. MacMahon, and B. McGrath.** 2002. The role of conjugative tranposons in *Enterobacteriaceae. Cell. Mol. Life Sci.* **59:**2055–2064.
25. **Recchia, G., and D. Sherratt.** 2002. Gene acquisition in bacteria by integron-mediated site-specific recombination, p. 162–176. *In* N. Craig, R. Craigie, M. Gellert, and A. Lambowitz (ed.), *Mobile DNA II.* ASM Press, Washington, D.C.
26. **Rice, L.** 2002. Association of different mobile elements to generate novel integrative elements. *Cell. Mol. Life Sci.* **59:**2023–2032.
27. **Rowe-Magnus, D., A. Guerot, P. Ploncard, B. Dychinco, and J. Davies.** 2001. The evolutionary history of chromosomal super-integrons provides an ancestry for multiresistant integrons. *Proc. Natl. Acad. Sci. USA* **98:**652–657.
28. **Rowe-Magnus, D., and D. Mazel.** 1999. Resistance gene capture. *Curr. Opin. Microbiol.* **2:**483–488.
29. **Rownd, R., N. Nakaya, and A. Nakamura.** 1966. Molecular nature of the drug-resistance fators of the *Enterobacteriaceae. J. Mol. Biol.* **17:**376–393.
30. **Salyers, A., and N. Shoemaker.** 1997. Conjugative transposons, p. 89–99. *In* K. Setlow (ed.), *Genetic Engineering*, vol. 19. Plenum Press, New York, N.Y.
31. **Sander, P., B. Springer, T. Prammananan, A. Sturmfels, M. Kappler, M. Pletschette, and E. C. Bottger.** 2002. Fitness cost of chromosomal drug resistance-conferring mutations. *Antimicrob. Agents. Chemother.* **46:**1204–1211.
32. **Sherratt, D., P. Dyson, M. Boocock, L. Brown, D. Summers, G. Stewart, and P. Chan.** 1984. Site-specific recombination in transposition and plasmid stability. *Cold Spring Harb. Symp. Quant. Biol.* **49:**227–233.
33. **Smolinski, M., M. Hamburg, and J. Lederberg (ed.).** 2003. *Microbial Threats to Health; Emergence, Detection, and Response.* The National Academies Press, Washington, D.C.
34. **Stokes, M., and R. Hall.** 1989. A novel family of potentially mobile DNA elements encoding site-specific gene-integration functions: integrons. *Mol. Microbiol.* **3:**1669–1683.

35. **Summers, D.** 1996. *The Biology of Plasmids*. Blackwell Science Ltd., Oxford, United Kingdom.
36. **Summers, D. K., and D. J. Sherratt.** 1984. Multimerization of high copy number plasmids causes instability: ColE1 encodes a determinant essential for plasmid monomerization and stability. *Cell* **36:**1097–1103.
37. **Tenover, F. C.** 2001. Development and spread of bacterial resistance to antimicrobial agents: an overview. *Clin. Infect. Dis.* **33**(Suppl. 3):S108–S115.
38. **Watanabe, T., and T. Fukasawa.** 1962. Episome-mediated transfer of drug resistance in *Enterobacteriaceae* IV. Interactions between resistance transfer factor and F-factor in *Escherichia coli* K-12. *J. Bacteriol.* **83:**727–735.
39. **Watanabe, T., and T. Fukasawa.** 1961. Episome-mediated transfer of drug resistance in *Enterobacteriaceae*. I. Transfer of resistance factors by conjugation. *J. Bacteriol.* **81:**669–678.
40. **Weber, J. T., and P. Courvalin.** 2005. An emptying quiver: antimicrobial drugs and resistance. *Emerg. Infect. Dis.* **11:**791–793.
41. **Zinder, N., and J. Lederberg.** 1952. Genetic exchange in *Salmonella*. *J. Bacteriol.* **64:**679–699.

Enzyme-Mediated Resistance to Antibiotics: Mechanisms, Dissemination, and Prospects for Inhibition
Edited by Robert A. Bonomo and Marcelo E. Tolmasky

Louis B. Rice

17 Conjugative Transposons

Conjugative transposons are integrative, nonreplicative mobile elements that encode all of the functions necessary for their own intercellular transfer. They have been implicated in the transmission of a variety of antimicrobial resistance determinants among clinically important bacteria. The widespread availability and affordability of molecular biologic techniques in the past decade have greatly enhanced our ability to characterize and compare conjugative elements from different bacterial species. As this knowledge evolves, it is becoming clearer that many diverse bacterial species use conjugative transposons to disseminate evolutionarily useful genetic determinants. Because of the ease with which they are identified, most efforts have focused on the characterization of conjugative transposons that confer resistance to antibiotics. However, conjugative transposons may also be vehicles for dissemination of bacteriocins, virulence determinants, and other evolutionarily advantageous genetic material.

The explosion of molecular data on bacterial species that has occurred over the past decade has led to some concerns regarding the nomenclature of conjugative transposons. The original conjugative transposon, Tn*916*, was characterized as integrative, nonreplicative, capable of encoding its own conjugal transfer, and relatively nonselective in its sites for integration. Elements that shared all of these characteristics except that they integrated in a site-specific manner were not designated conjugative transposons. It is now clear that site specificity is more or less species dependent. For example, Tn*916* integration is relatively nonspecific in enterococcal species (14) but highly site specific in *Clostridium difficile* (60), whereas integration of other elements that may be site specific in their native species may be far less selective in others. This breakdown of artificial distinctions has led some investigators to suggest reclassifying at least some of these conjugative mobile elements as conjugative self-transmissible, integrating elements (constins), (27) or ICEs (integrative conjugative elements) (7). It is not my intent to address the merits of these arguments in this chapter. The reader is referred to a recent reference for a thorough discussion of the subject (7). In this chapter, I use the term conjugative transposon to refer to all nonreplicating mobile elements that encode their own conjugation functions.

Growing sequence databases have also made it clear that many conjugative elements are the results of modular evolution. Integration and excision functions are often borrowed or evolved from those well characterized in bacteriophages. Transfer functions may be derived from those characterized on transferable plasmids. Some conjugative transposons, such as Tn*916*, clearly evolved long ago and have become well-honed tools of genetic exchange (20). Others, such as Tn*5397* from *C. difficile*, are more recently constructed elements that incorporate genes of very different lineages to take advantage of

Louis B. Rice, Louis Stokes Cleveland Department of Veterans Affairs Medical Center and Case Western Reserve University, Cleveland, OH 44106.

specific advantages offered by each (47). Selective pressure exerted by the presence of antimicrobial agents in the environment will no doubt continue to select for these creative combinations, so it would be wise to avoid being too strict in our definitions.

Transfer of conjugative transposons requires cell-to-cell contact and occurs at relatively low frequency. However, the host range of the various elements is, in some cases, quite diverse. Conjugative transposons are in many respects well suited for function in the complex and diverse environment of the mammalian gastrointestinal tract (52). It is therefore perhaps not surprising that many genera in which the conjugative transposon classes that I detail in this chapter are found (*Enterococcus*, *Bacteroides*, *Clostridia*, and *Vibrio*) are all colonizers of the human gastrointestinal tract. In this chapter, I review the current state of knowledge of the epidemiology, resistance profiles, and transposition mechanisms of the major types of characterized conjugative transposons found in human pathogenic bacteria. The specific transposons that I cover in some detail are listed in Table 17.1.

EPIDEMIOLOGY OF CONJUGATIVE TRANSPOSONS

Tn*916*-Like Conjugative Transposons

Tn*916* is an 18-kb conjugative transposon that confers resistance to tetracycline and minocycline and was first identified in a strain of *Enterococcus faecalis* (22). At around the same time, a tetracycline/minocycline, erythromycin, and kanamycin resistance conjugative transposon designated Tn*1545* was described in a strain of *Streptococcus pneumoniae* (17). Work by separate laboratories on these two elements soon revealed several similarities, including virtual identity of the genes that encoded integration and excision of the elements, as well as identity of the ends of the elements for at least 250 nucleotides on each end (8). Of the two elements, Tn*916* has been the most completely characterized and serves as the focus of our discussion. When differences between Tn*916* and its close relatives occur, they will be described.

Since the early discoveries of Tn*916* and Tn*1545*, Tn*916*-like transposons have been described in a wide variety of gram-positive and gram-negative species (44). This broad host range is explained in part by the fact that Tn*916* is relatively devoid of restriction sites. In fact, 70% of the unique restriction sites found in Tn*916* are present in the 2 kb surrounding the *tet*(M) gene, leading some investigators to conclude that the non-*tet*(M) regions of conjugative transposons are ancient structures honed for broad-host-range transfer (39). In this scenario, the presence of *tet*(M) reflects a relatively recent arrival of a resistance gene, perhaps prompted by increasing antimicrobial selective pressure. If that evolutionary progression is true, then one might anticipate finding other types of conjugative transposons with different resistance determinants present in the same relationship to the rest of the element, or perhaps some with other resistance determinants present along with the *tet*(M) genes. In fact, such a circumstance appears to be the case for Tn*5382*, a VanB glycopeptide resistance

Table 17.1 Conjugative transposons discussed in this chapter

Transposon	Species of origin	Resistance phenotype	Reference(s)
Streptococcal/enterococcal			
Tn*916*	*E. faecalis*	Tetracycline	15
Tn*1545*	*S. pneumoniae*	Tetracycline, erythromycin, kanamycin	17
Tn*5253*	*S. pneumoniae*	Tetracycline, chloramphenicol	2
Tn*5283/1549*	*E. faecium/faecalis*	Vancomycin	10, 21
Tn*5385*	*E. faecalis*	Penicillin, erythromycin, gentamicin, mercury, streptomycin, tetracycline	45
Bacteroides			
CTnDOT		Erythromycin, tetracycline	12
CTnERL		Tetracycline	65
CtnXBU4422		None	5
CTnGERM1		Erythromycin	61
Enterobacterial			
SXT	*Vibrio cholerae*	Florfenicol, trimethoprim	4
R391	*Proteus rettgeri*	Kanamycin, mercury	6
Clostridial			
Tn*5397*	*C. difficile*	Tetracycline	47

transposon with significant structural similarity to Tn*916* (10). In Tn*5382* (likely identical to the now fully sequenced Tn*1549*) (21), the VanB resistance operon is placed in approximately the same position relative to the integration and excision genes as *tet*(M) occupies in Tn*916* (21). The guanine + cytosine content of the open reading frames (ORFs) of Tn*5382* (ca. 50%) differs markedly from those of Tn*916* (ca. 36%), however, suggesting that their divergence occurred a very long time ago and their individual development occurred in different species.

The vast majority of described Tn*916*-like transposons encode *tet*(M) genes, conferring resistance to tetracycline and minocycline by a ribosomal protection mechanism (44). In fact, Tn*916*-like elements represent the major mechanisms by which *tet*(M) genes spread among gram-positive bacteria and beyond (44). The list of diverse bacterial species into which Tn*916*-like elements transfer in vitro is extensive. Similarly, the range of species within which Tn*916* or its remnants have been identified in clinical isolates is extensive, including gram-positive, gram-negative, anaerobic, and cell wall-deficient species (44). It is likely that the dissemination of Tn*916*-like elements has been facilitated by the widespread use of tetracyclines for the treatment of human and animal infections, as well as the inclusion of these antimicrobial agents in animal feed as a means to promote growth. Exposure to tetracycline promotes not only the selection of *tet*(M)-containing strains, but also the dissemination of Tn*916*-like determinants (11, 46, 54). Incubation of Tn*916*-containing enterococci with tetracycline results in increased expression of *tet*(M), with downstream effects on the genes encoding both excision and transfer of the element (see below).

It has also become increasingly obvious that Tn*916*-like elements are associated with the dissemination of genes other than *tet*(M). In some cases, as in Tn*1545* and Tn*5382*, non-*tet*(M) resistance genes are present within the transposon itself (10, 17). In other instances, single [usually *tet*(M)] resistance conjugative transposons insert within larger regions (or other transposons) yielding multiresistance. Tn*5253* is a large transferable element in *S. pneumoniae* that confers resistance to tetracycline/minocycline and chloramphenicol (2). Tetracycline-minocycline resistance is conferred by Tn*5251*, a Tn*916*-like element, embedded within Tn*5252*, a larger element that encodes chloramphenicol resistance. Both transposons encode their own transfer functions, although because of the structure of the element, chloramphenicol resistance is transferred with tetracycline resistance, whereas tetracycline resistance can transfer alone.

The complexity of these composite elements is exemplified by Tn*5385* (45), a multiresistance conjugative element identified in *E. faecalis* (Fig. 17.1). Tn*5385* consists of several different resistance-encoding mobile elements, including the Tn*916*-like transposon Tn*5381* and the gentamicin resistance-conferring transposon Tn*4001*, which can also combine with a downstream IS*256*, forming the erythromycin, gentamicin, and mercury resistance transposon designated Tn*5384*. Also included in Tn*5385* is a genetic region indistinguishable from the staphylococcal β-lactamase transposon Tn*552*. Interspersed between these transposons is a remnant of a replication region from a broad-host-range plasmid. Analysis of the Tn*5385* structure suggests that it is a composite element derived from staphylococci and enterococci, perhaps accumulated during the travels of the broad-host-range plasmid. Tn*5385* transfers between *E. faecalis* strains at a low frequency and most commonly inserts into the recipient genome in a site-specific fashion (45). The mechanism of transfer remains poorly characterized, as it is for all conjugative transposons. Non-*tet*(M) elements have also been found within composite elements. Most recently, Tn*5382* has been shown to transfer between *Enterococcus faecium* strains within a larger element that also confers resistance to ampicillin through expression of a low-affinity PBP5 (10).

Bacteroides Conjugative Transposons

Several conjugative transposons have been described in *Bacteroides* species. These transposons also frequently confer resistance to tetracycline by expressing *tet*(Q) or *tet*(X) genes (65). *Bacteroides* conjugative transposons have also been described that confer resistance to erythromycin through a variety of erythromycin ribosomal methylase (*erm*) genes. Surprisingly, these transposons have also been shown to carry a gene for an adenylating enzyme that confers resistance to streptomycin, even though *Bacteroides* species are naturally resistant to clinically achievable concentrations of this antibiotic. A recently published survey directly implicates conjugative transposons in the steady and significant rise in erythromycin and tetracycline resistance in *Bacteroides* species over the past three decades (53).

Unlike Tn*916* family transposons, *Bacteroides* conjugative transposons can mobilize coresident nonconjugative transposons. In many cases, these mobilized transposons do not themselves contribute to antimicrobial resistance (53). However, one mobilizable transposon (Tn*4555*) confers clinically significant levels of resistance to the β-lactam antibiotic cefoxitin (55). Insertion of the conjugative transposon or its mobilized counterpart into a nonconjugative or nonresistance plasmid can confer transferability onto previously nontransferable structures. *Bacteroides* conjugative transposons can be transferred in vitro into distantly related species, such as *Escherichia coli* (65). The finding of common resistance genes in *Bacteroides* and

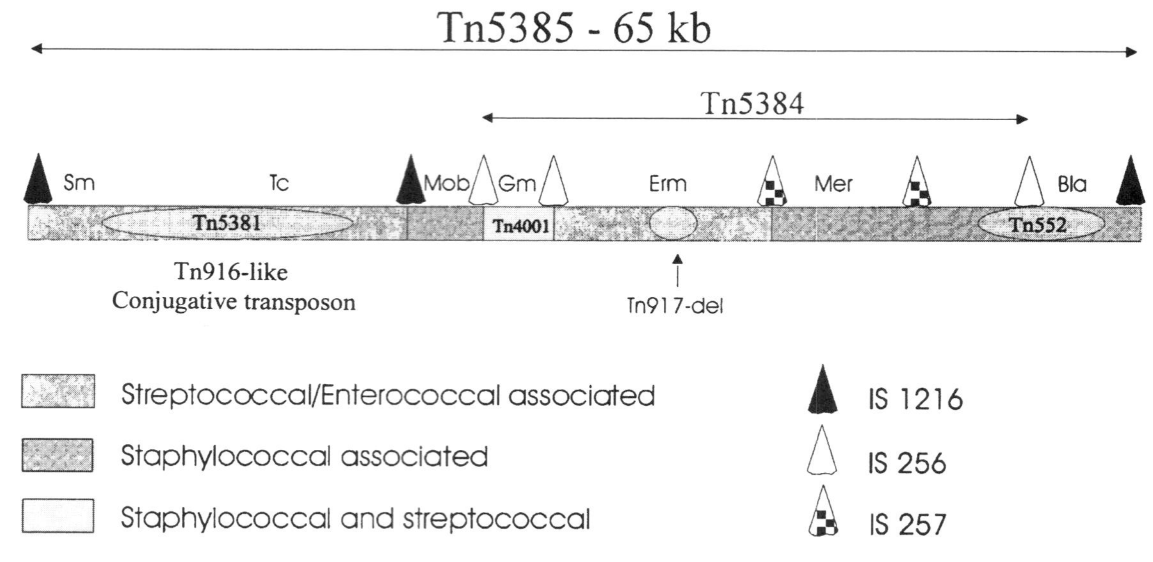

Figure 17.1 Graphic depiction of 65-kb composite transposon Tn*5385* from *E. faecalis*. Tn*5385* is a conjugative element that transfers between enterococci at low frequency and integrates in most cases into the recipient chromosome in a site-specific manner. Within Tn*5385* lies a typical Tn*916*-like conjugative transposon (Tn*5381*) that can transfer by itself or with the larger element. When transferring by itself, Tn*5381* is relatively nonselective in its insertion sites. Regions of different lineages are connected within Tn*5385* by different insertion elements (see the text). Reprinted with permission from reference 43.

non-*Bacteroides* anaerobic and even gram-positive species is suggestive of interspecies transfer in vivo but may also reflect the mining of common species by different pathogens of human importance.

Conjugative Transposons in *Enterobacteriaceae*

A general appreciation of the existence and importance of conjugative transposons in *Enterobacteriaceae* has been a relatively recent development, despite the fact that the first *Proteus rettgeri* strains containing these elements were described in 1972 (16). At the time of the first description, the integrative nature of these elements was not appreciated. They were assumed to be extrachromosomal, and since they were compatible with plasmids of all known incompatibility groups at the time, they were designated incompatibility group J (IncJ) plasmids. In the past few years, other similar elements have been reported. Because their behavior is somewhat different from classically described conjugative transposons, investigators have suggested that they be designated constins (28).

The extent to which conjugative transposons have been important vehicles for the transmission of antimicrobial resistance between *Enterobacteriaceae* is not known. The prominent role of transmissible plasmids among these species has been well documented, so it is not entirely clear whether the dearth of information regarding conjugative transposons reflects a lack of importance or merely a lack of attention. In any case, these elements have been described in relatively few genera to date (*Proteus/Providencia, Salmonella, and Vibrio*) (40), and they have incorporated within them relatively few resistance determinants, so their overall importance for the spread of antimicrobial resistance among *Enterobacteriaceae* is probably minor in comparison to transferable plasmids.

Clostridial Conjugative and Mobilizable Transposons

One conjugative transposon, Tn*5397*, has been described in *C. difficile*. This transposon is in many respects identical to Tn*916* [*tet*(M) plus homology over much of the rest of the transposon] (47). However, its integration and excision functions, and its ends, are entirely different from those of Tn*916*. It therefore appears to be a modular transposon that uses integration and excision functions similar to those of several clostridial transposons but the transfer functions of Tn*916*. Other transferable transposons in *Clostridium* are not formally conjugative transposons, because they do not encode their own transfer functions (1). These mobilizable transposons fall into two general families. The Tn*4451/4453* family transposons encode resistance to chloramphenicol and functions that allow excision, site-specific integration, and mobilization by other conjugative elements (37). The other major recently described element is Tn*5398*, an erythromycin resistance genetic region that is clearly mobile but does not encode any obvious mobilization or transposition functions (38). These functions are presumably supplied in *trans* by other mobile elements. The extent to which these mobile elements are present in many different clostridia and are responsible for dissemination of resistance is undetermined.

STRUCTURES AND TRANSPOSITION MECHANISMS OF CONJUGATIVE TRANSPOSONS

All conjugative transposons thus far described transpose by a conservative mechanism in which the first step is excision of the element, followed by circularization, transfer, and ultimately integration into a target replicon. This general mechanism distinguishes conjugative transposons from many other common types of mobile elements (for example, Tn*3*-family transposons and most insertion IS elements) that transpose in a replicative fashion, yielding copies of the transposon at both the initial and target integration sites.

Tn*916* Family Transposons

The Tn*916* family elements are the most thoroughly studied and characterized of the conjugative transposons. The complete nucleotide sequence of Tn*916* was first published by Flannagan et al. in 1994 (20). It is 18,032 nucleotides in length and consists of 24 ORFs (Fig. 17.2 [top]). The ORFs that are responsible for excision and integration of the element (*int-Tn* and *xis-Tn*) lie at the left end of the transposon, separated from the genes for transfer by the *tet*(M) gene. The ends of Tn*916* form a 26-bp imperfect inverted repeat, and integration of the element does not generate a duplication of the target sequence. The origin of transfer is located toward the right end of the element as shown in Fig. 17.2 (top) (30). Several ORFs to the right of *tet*(M) in Fig. 17.2 (top) are noteworthy (20, 50). *orf18* resembles restriction protection protein ArdA, potentially providing extra protection against restriction digestion upon entry into a new cell. *orf14* resembles *iap* of *Listeria monocytogenes*, an autolytic protein, suggesting that it could be involved in establishing a connection between two bacteria during the mating event. *orf23* resembles MbeA, a member of the MobA family of proteins that has been associated with relaxosome complexes containing nicked ColE1 DNA. It could therefore be involved in nicking the Tn*916* circular intermediate in preparation for transfer. At the present time these interpretations of gene function remain speculative.

The first step in conjugative transposition of Tn*916*-type transposons is excision from the original replicon (usually the chromosome). Excision begins with the creation of staggered nicks 5 or 6 bp beyond the ends of the element (50). The ends are then brought together, resulting in a nonreplicative circular intermediate. The recognition of this circular intermediate first occurred when Tn*916* was cloned into high copy number vectors in *E. coli* (under these circumstances, the excision frequency is very high) (51). Subsequent studies confirmed the presence of circular forms of Tn*916*-like elements in enterococci as well (46). Early studies with *E. coli* suggested that the overlapping staggered nicks of the circular form (referred to as the joint) consisted of a heteroduplex of mismatched DNA strands (representing the flanking sequences of the prior insertion) (9). Subsequent studies with enterococci suggested that the heteroduplex was repaired, resulting in a joint region consisting of a sequence from either of the two previous flanking regions (33).

Excision of Tn*916* is catalyzed by the Int protein encoded by the left end of the element (42). Tn*916* Int is a member of the λ-Int family of site-specific recombinases. λ-Int family recombinases mediate DNA cleavage/ligation reactions through transient covalent phosphotyrosine intermediates. Two sequential pairs of strand exchanges first generate and then resolve Holiday junction intermediates. Int exhibits heterobivalent DNA binding. The C-terminal portion of the Int protein binds to the ends of Tn*916* and performs the recombination reactions that allow integration and excision (31, 57). The N-terminal portion binds to direct repeats designated DR2 at each end of the transposon. The result of Int binding is that the ends of the transposon are brought into close proximity, allowing strand exchange to occur (for a model of how this would look, please see reference 50).

Considerable analysis of the N-terminal portion of Int and the thermodynamics of its DNA binding has been performed in the past two years (23, 35, 36). The minimal arm-type DNA binding domain of λ-Int has recently been defined, and its structure has been solved, revealing that it is a member of the growing family of three-stranded β-sheet DNA-binding proteins (66). The comparable region of Tn*916*-Int covers 74 residues. DNA binding is facilitated by hydrophobic interactions between the target DNA strand and the three-β-sheet portion of the protein. Protein-DNA binding is accompanied by a conformational shift in both the protein and the target DNA (35).

The binding of two Int molecules to each end of the transposon (the C termini to the ends of the element and the N termini to the DR2 direct repeats) facilitates an approximation of the ends and a bending of the DNA molecule. Strand exchange then occurs, creating the temporary heteroduplex at the joint of the excised element. It is presumed that an initial heteroduplex is also created at

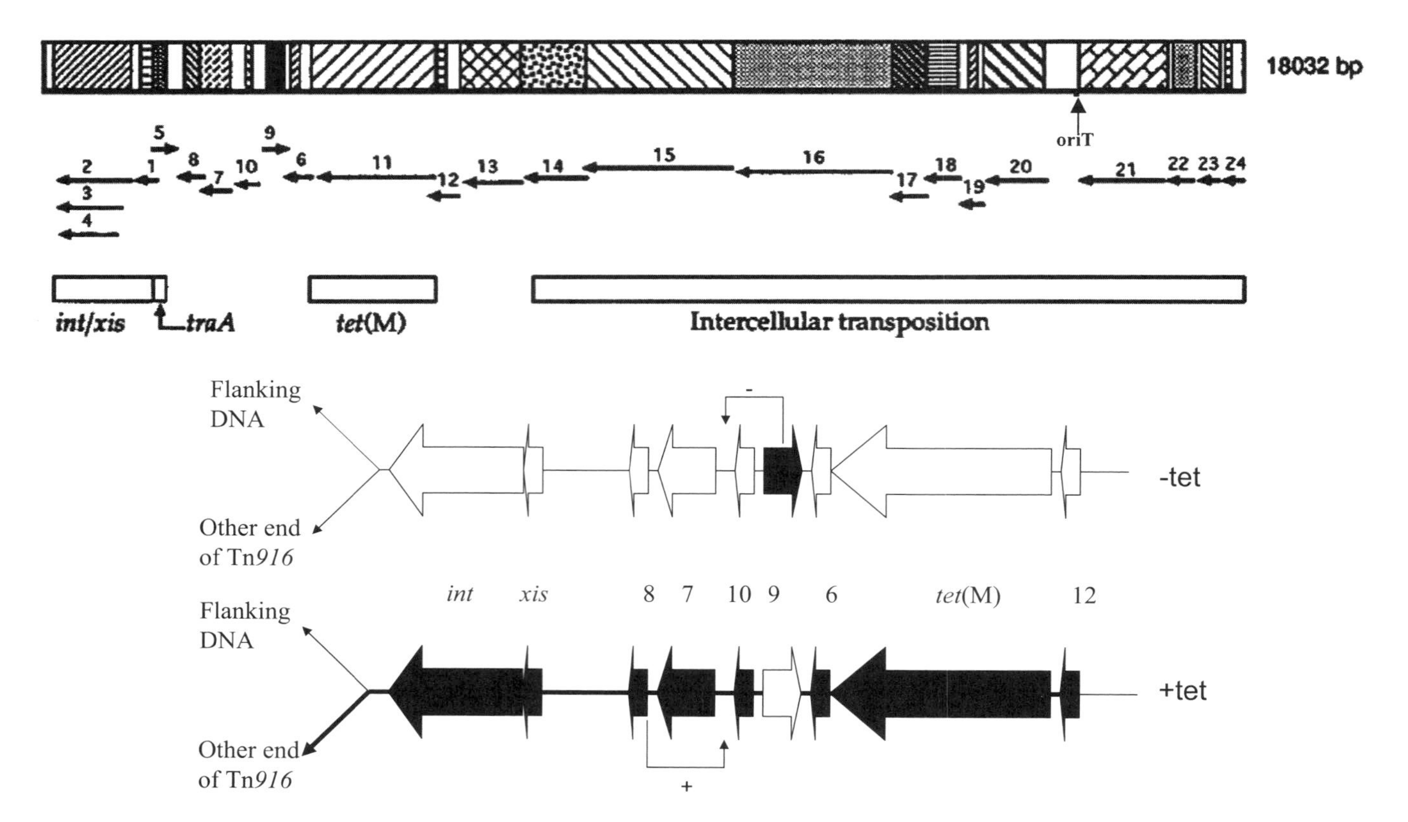

Figure 17.2 (Top) Map of Tn*916*. Identity of the specific ORFs and their directions of transcription are identified below the map, as are the functions of the regions, where that is known. The origin of transfer is indicated (*oriT*) at the right-hand end of the transposon. Reprinted with permission from reference 30. (Bottom) Graphic depiction of expression of ORFs in the left end of Tn*916* with or without exposure to tetracycline. Black arrows represent genes that are expressed under the identified conditions. In the absence of tetracycline, expression of *orf9* negatively regulates transcription of *orf7* and *orf8*. Transcription from the *tet* promoter (upstream of *orf12*) is also terminated within *orf12*. In the presence of tetracycline, transcription from the *tet* promoter moves through the palindromes within *orf12* and into *tet*(M). Transcription then continues through *tet*(M), negatively regulating expression of the gene products of *orf9*, which ultimately leads to transcription of *orf7* and *orf8*, which positively regulate their own transcription. This results in increased expression of *int* and *xis*, stimulating excision and circularization. Once circularized, transcription can continue through the newly formed joint and into the transfer genes found in the right-hand end of the transposon. Adapted from data published in reference 11.

the excision site, with eventual resolution through replication of the chromosome. Integration proceeds by reversing the above process.

Tn*916* encodes an additional protein, designated Xis, that in some species is absolutely required for excision. Xis promotes excision under most circumstances but has been shown to inhibit excision when present at high concentrations. Xis binding sites for the two ends of Tn*916* have been defined. Interestingly, Xis binding to the left end of Tn*916* occurs with a higher affinity than does binding to the right end (25). Xis binding to the left end of Tn*916* facilitates excision, whereas binding to the right end inhibits excision. The Xis binding site on the left end of Tn*916*, as defined by DNase protection assays, is located between the N-terminal and C-terminal Int binding sites (the "arm" and "core" binding sites in λ lexicon) (49). It has been proposed that Xis of Tn*916* facilitates bending of the DNA molecule in a manner that promotes the activity of Int. Xis could also interact with Int directly in a manner that promotes excision. Such a function would be similar to that of λ-Int and λ-Xis. Recent work has identified the specific residues in the N terminus of λ-Int that are responsible for Int and Xis binding during λ excision. The residue critical for Int-Xis binding is also critical for binding of Int protomers bound at adjacent arm-type sites (62). On the right end of Tn*916*, the Xis binding site is adjacent to the Int N-terminal binding site, suggesting that binding of Xis to this site could inhibit Int binding (25). Taken together, these structural details are consistent with a scenario in which Xis promotes excision when present in low or moderate concentrations (by binding to the left end of Tn*916* preferentially) but is inhibitory when present in high concentrations (by binding to the right end of the transposon).

Transcriptional regulation of Int and Xis expression is complex. It is tied to regulation of *tet*(M) and has consequences for expression of the conjugation genes present in the right portion of Tn*916*. Several ORFs in the left end of Tn*916* (*orf7*, *orf8*, *orf9*, and *orf12*) impact transcription of other genes within the region (Fig. 17.2[bottom]) (11). The gene products of *orf7* and *8* promote transcription from their own promoter (P_{orf7}), but transcription from this promoter is inhibited by the gene product of *orf9*. Transcription of *tet*(M) is directed from the P_{tet} promoter that resides upstream of *orf12*. Under noninducing (no tetracycline in the media) conditions, transcripts originating at P_{tet} terminate prior to reaching *tet*(M), likely at palindromic sequences present within *orf12*. Exposure to tetracycline results in increased transcription from P_{tet} and transcription through to the *tet*(M) gene, presumably through a transcriptional attenuation mechanism. Transcription then continues through *tet*(M), resulting in a decrease in *orf9* activity, perhaps through an antisense mechanism. Reductions in expression of *orf9* activity then result in increased transcription from P_{orf7} and through *xis* and *int*. Finally, this transcription should stimulate excision and subsequent formation of a circular intermediate. Formation of the circle then allows transcription through the joint region and into the transfer genes present in the right end of Tn*916* (11). As a result, growth of Tn*916*-containing bacteria in tetracycline is expected to promote transfer of the element to recipient cells, a finding that has been demonstrated repeatedly (18, 46, 54). Excision of Tn*916* catalyzed by an overexpression of Int and Xis from an inducible promoter in *trans* does not increase the frequency of conjugative transposition, however, suggesting that the regulation of transfer requires more than just an abundance of the excised form of the transposon (34).

Since it is postulated that Tn*916* transfers to recipient cells as a single strand, there should be a nick site that allows that single strand to be created. Based on analogy to transfer of *E. coli* plasmids, Scott and Churchward predicted in a 1995 review that the putative *ard* gene should transfer early, since its putative protection against restriction digestion would be necessary early in the recipient cell (50). On the other hand, the gene related to *mobA* (*orf23*) should transfer relatively late, to allow the continued presence of conjugation gene function in the donor for the duration of the transfer event. They therefore predicted that the origin of transfer would be present somewhere between *orf18* and *orf23*. The precise origin of transfer was subsequently identified between ORFs 20 and 21, confirming their logic and providing supportive circumstantial evidence for the putative functions of these genes (30). Finally, it has been shown that Int also binds to the origin of transfer of Tn*916* (24). The functional importance of this binding is not clear, but it may provide an additional measure of control over the timing or frequency of transfer by preventing transcription through the *oriT* into the transfer genes within the right end of Tn*916*.

Integration of Tn*916*-like elements into the recipient *E. faecalis* chromosome is relatively random, but there are sequences that serve as "hot spots" for Tn*916* integrations (29). Excision frequencies are also affected by the sequences flanking the inserted transposon (29). In general, these sequences are A-T rich and bear a resemblance to the ends of Tn*916*. This proclivity for AT-rich sequences has limited the utility of Tn*916* as an insertional mutagenic agent, since it tends to integrate into AT-rich promoter regions in gram-positive bacteria. Only a single site for Tn*916* integration exists in *C. difficile* (*attTn916*) (60), suggesting that host factors have importance for the selectivity of integration.

The presence of Tn*916* within a chromosome does not preclude entry of another copy of the element, and in fact excision of one Tn*916* copy has been shown to stimulate excision and transfer of a coresident Tn*916* (19). In vitro, it is common to observe transfer of multiple copies of Tn*916* to recipient cells (46). Despite this in vitro observation, the discovery of multiple copies of Tn*916* within clinical isolates is a rare event. This rarity may simply reflect a failure to look for multiple copies in a systematic fashion, or it may reflect efficiency on the part of clinical strains.

Bacteroides Conjugative Transposons

Several conjugative transposons have been described in *Bacteroides* species. All of them are similar to Tn*916* in that they transpose by a conservative mechanism whose first step is excision of the element from the donor replicon. They are also similar in that exposure to tetracycline induces excision and transfer of the elements. In fact, transfer of the tetracycline resistance transposons is virtually undetectable in the absence of tetracycline (65). They differ from Tn*916*-like elements in their size (ranging from 40 to 150 kb) and in their site selectivity for integration (generally about 8 sites per chromosome) (5).

The single "family" of *Bacteroides* conjugative transposons described to date is the CTnDOT family

(Figure 17.3 [top]) (65). This family is characterized by the presence of right and left ends that differ from each other but are conserved within the family. They also exhibit extensive sequence identity in their transfer, excision, and mobilization genes. CTnDOT is a 65-kb element that encodes resistance to tetracycline [*tet*(Q)] and erythromycin (*ermF*) (65). Tetracycline-induced resistance and transfer are encoded within a central region of CTnDOT that contains the *tet*(Q) gene and three additional genes designated *rteA*, *rteB*, and *rteC*. *ermF* lies within a 13-kb region near one end of the transposon. A second *Bacteroides* conjugative transposon, designated CTnERL, is virtually identical to CTnDOT except that it lacks the 13-kb *ermF* region (65). Since this region is not present in CTnERL, it is not involved in the integration/excision or transfer functions of CTnDOT. A final member of the family is the cryptic CTnXBU4422, which lacks both the *ermF* and *tet*(Q) regions (65).

Integration of CTnDOT is catalyzed by the products of a gene designated *int*, which is present near one end of the element (12). Int has homology to the λ family of site-specific recombinases, which includes the Tn*916* Int protein. Int is sufficient for integration of CTnDOT. It is required, but not sufficient, for excision, however. Excision requires the presence of *exc*, which lies within a 13-kb region separated from *int* by the *ermF* insertion in CTnDOT (13). Downstream of *int* in CTnDOT lies an ORF predicted to encode a small basic protein that suggested that it was similar to the Tn*916 xis*, but experimental data indicate that this ORF (designated *orf2*) does not play a role in excision (13). Homology searches performed on the deduced amino acid sequence of the Exc protein indicate a significant degree of homology with previously described topoisomerase genes. This would be the first example of a topoisomerase gene implicated in the excision of a mobile element.

The *rteA*, *rteB*, and *rteC* genes are also involved in excision, presumably through a regulatory role (13). *rteA* and *rteB* are required for expression of *rteC* (56). Expression of *rteC* increases transcription of *exc*, thereby increasing excision of CTnDOT. Located within the *exc* region are both positive and negative regulators of transfer functions (64). The positive regulators are cotranscribed with *exc*, whereas the negative regulator is not. Expression of the positive regulators and *exc* is enhanced by exposure to tetracycline, whereas the negative regulator is expressed in the absence of tetracycline. The end result is that regulation of excision and transfer is linked and affected by tetracycline, but by a strategy considerably different from that employed by Tn*916* (64).

CTnDOT also differs from Tn*916* in that the number of insertion sites is limited (less than 10) (5). Insertion of CTnDOT always occurs 5-bp 5′ to a 10-bp sequence that is identical to a sequence within the end of the transposon (12). The nature of sequences flanking CTnDOT insertions does not appear to influence the excision frequency, as seen with Tn*916*. However, the formation of the circular intermediate does appear to proceed through the creation of 4- or 5-base staggered cuts within the flanking sequences (12). The details of CTnDOT binding and recombinase activity have not been well characterized. However, CTnDOT Int is in the same tyrosine recombinase family as Tn*916* Int, so it is reasonable to presume that many of the details are similar.

Four *Bacteroides* conjugative transposons that are not formally members of the CTnDOT family have been described (65). CTn12256 has an intact copy of CTnDOT integrated within a larger (ca. 150-kb) conjugative transposable element. Although the integrated CTnDOT appears to be intact, transfer of the larger element is not regulated by exposure to tetracycline. CTn7853 is a 70-kb conjugative transposon whose only similarity to CTnDOT is the presence of a *tet*(Q) and truncated *rteA* gene within it. It also confers erythromycin resistance via the *ermG* gene. CTnGERM1 is a conjugative transposon of unknown size that confers erythromycin/clindamycin resistance through *ermG* (Fig. 17.3 [bottom]), whereas CTnBst is an erythromycin/clindamycin resistance conjugative transposon that confers resistance via the *ermB* gene. These elements all transfer constitutively at roughly the rate of CTnDOT transfer after tetracycline induction (65). Moreover, they transfer to a broader range of anaerobic gram-negative bacilli than does CTnDOT, suggesting that they may ultimately play a more important role in resistance dissemination. Little is known of the details of excision and transfer for these transposons.

IncJ-Family Elements

Molecular characterizations of conjugative transposons from *Enterobacteriaceae* are rare. In recent years, however, considerably more structural information has become available for some members of the IncJ mobile elements. The entire structures of two such elements, SXT from *Vibrio cholerae* and R391 from *P. rettgeri*, have been determined (3, 4, 6). Both excise from their chromosomal location through the action of an *int* gene that bears resemblance to the phage-encoded tyrosine recombinases. They form a nonreplicating circular intermediate prior to transfer and integrate in a site-specific manner into the recipient chromosome. A comparison of these structures is instructive.

SXT and R391 are 99.5 and 89 kb, respectively (3). The size differences largely reflect different modular components, with 65 kb of each element representing a "backbone" in which there is greater than 95% identity of

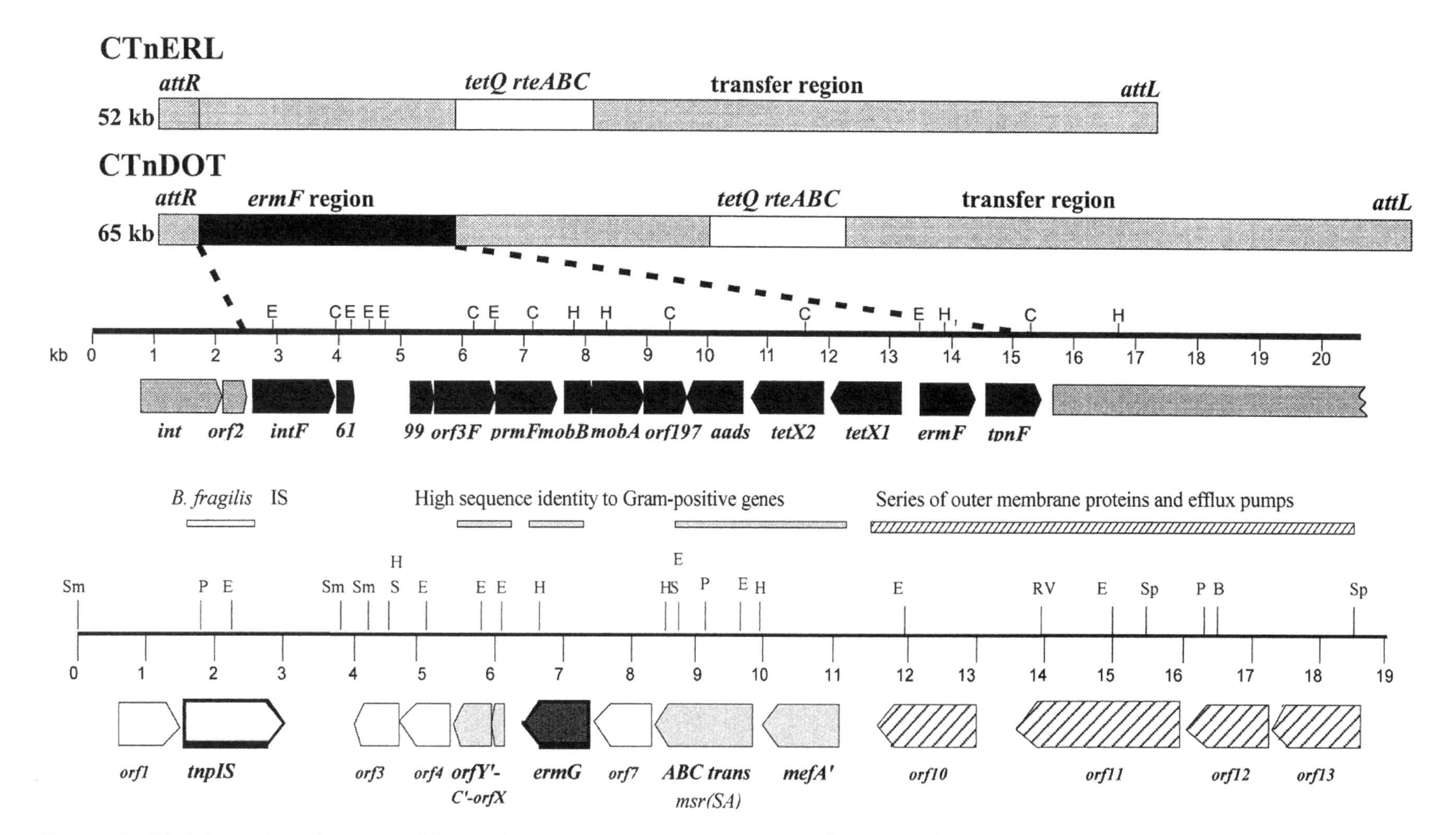

Figure 17.3 (Top) Comparison of structures of *Bacteroides* conjugative transposons CTnDOT and CTnERL. The transposons are identical except for the insertion of the *ermF* region in CTnDOT. The individual ORFs that make up the *ermF* region of CTnDOT are detailed. Reprinted with permission from reference 63. (Bottom) Structure of CTnGERM1, a non-CTnDOT-family *Bacteroides* conjugative transposon. Reprinted with permission from reference 61.

nucleotides. Three modules exist within the two elements. The important components of the integration module are the attachment sites and the *int* gene. The conjugation module is identical in both elements and encodes several genes related to conjugation genes found in the F plasmid and other previously described conjugative elements (Fig. 17.4). The regulation module contains activators SetC and SetD, related to master regulators of flagellar transcription and SetR, a negative regulator related to repressor C1 from bacteriophage λ. Two additional conserved modules of unknown function are present in both elements. Within these conserved modules are integrated different regions with specific characteristics. SXT contains a large insert within the *rumA* gene that encodes several different antimicrobial resistance determinants, including those for florfenicol, streptomycin, sulfamethoxazole, and trimethoprim (27). These are the only resistance genes encoded by SXT. R391 contains a region inserted between *traG* and s079 that encodes a mercury resistance operon and a separate insertion of a kanamycin resistance gene located within a cluster of inserted genes between s026 and s027 (3). There also appear to be three hot spots in which different insertions are found in the two elements.

The presence of the SXT-encoded Int gene is required for SXT excision and integration (28). The SXT Int is similar to the lambda Int and to other phage-encoded tyrosine recombinases. Integration of SXT occurs into a specific site within the 5′ end of the *prfC* gene, a nonessential gene encoding RF3, a protein involved in termination of translation (28). Because SXT encodes a novel 5′ coding sequence for *prfC*, as well as a functional promoter, *prfC* function remains intact when SXT integrates. Excision restores the wild-type *prfC* gene. Recent work indicates that R391 encodes a virtually identical integrase and inserts into the same site as SXT (26). The details of Int interactions with the ends of SXT have not been characterized.

When present in the same cell, SXT and R391 are often found in tandem inserted into their typical chromosomal site (26). This tandem insertion is unstable in the absence of selection for both elements. When R391 was inserted into a recA strain in which the related R997 IncJ element was present, a circular form of R391 was identified. The converse experiment yielded circular forms of R997 (41). While suggestive that these extrachromosomal forms represent replicative elements, the ability to isolate circular forms may also reflect relative differences in expression of integrase enzymes (40). R997 exhibits a higher transfer frequency than the other elements and therefore may have greater expressions of *int*, since excision is the first step in transfer. A more active integrase would lead to more frequent excision and therefore to a greater likelihood of observing circular forms under standard isolation conditions. The failure to identify genetic regions within either SXT or R391 that appear to encode replication functions lends credence to this alternative explanation (4, 6).

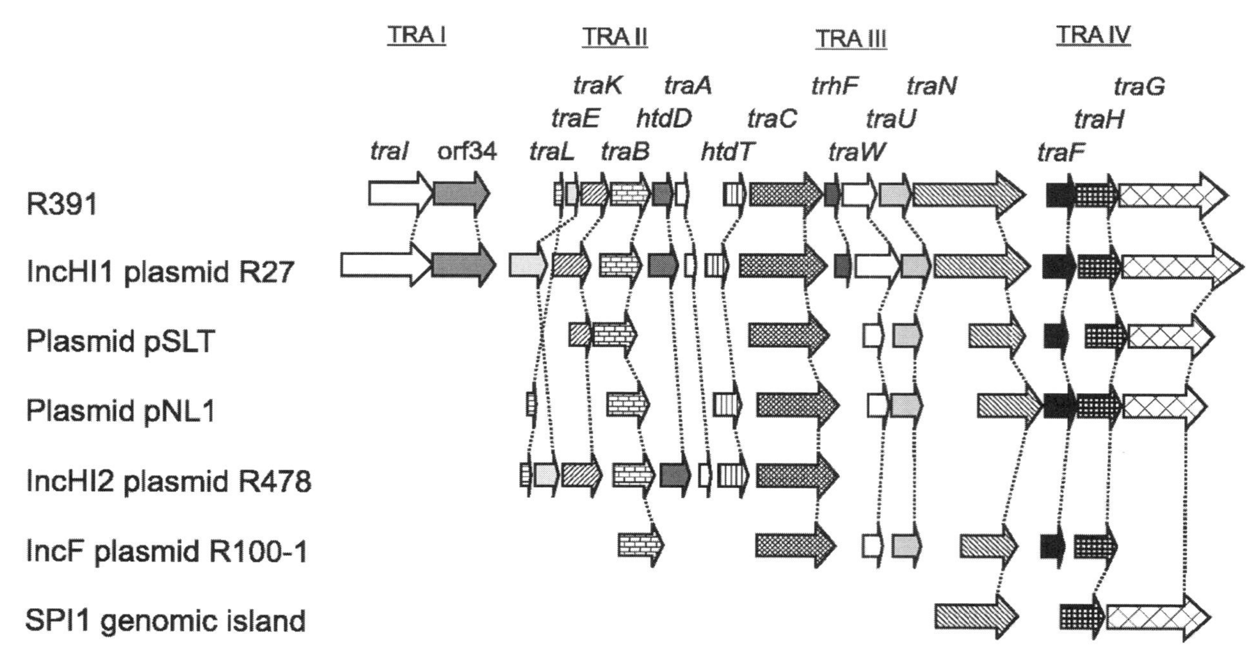

Figure 17.4 Similar regions found within CTnR391 conjugation region (identical to corresponding region of SXT) and other transferable elements from gram-negative bacteria. Reprinted with permission from reference 6.

Clostridium difficile Conjugative Transposon Tn*5397*

Tn*5397* represents the sole example of a conjugative transposon described in *Clostridium* (48). The structure of this element is of considerable interest, since the central portions of the transposon are very similar to Tn*916*, yet the integration/excision genes have been replaced by a protein of the large class V serine recombinase family (Fig. 17.5) (58, 59). This recombinase, designated TndX, is similar to the TnpX recombinases that catalyze integration and excision of chloramphenicol resistance clostridial mobilizable (but nonconjugative) transposons Tn*4451*, Tn*4453a*, and Tn*4453b* (59). The clostridial mobilizable transposons are transferable because they possess an origin of transfer recognized by several classes of conjugative plasmids (1). As such, their excised circular form can be mobilized by coresident transferable plasmids. Moreover, their integration into nonconjugative plasmids can create mobilizable plasmids as well.

tndX is required and sufficient for integration and excision of the Tn*5397* (58). The specifics of this enzyme's action have not been studied in detail. Like Tn*4451* and the clostridial mobilizable transposons, chromosomal insertions of Tn*5397* are always flanked by a GA dinucleotide (1). A similar dinucleotide characterizes the target sites for integration and the joint region of the circular intermediate. It is therefore likely that some conclusions about TndX activity can be drawn from work performed on Tn*4451* TnpX. However, it should be noted that TndX and Tn*4451* TnpX are only 37% identical and 61% similar in amino acid sequence (59). In comparison, the deduced amino acid sequences of Tn*4451* and Tn*4453a* are 87.7% identical. Moreover, while the TnpX enzyme can promote excision of Tn*4453a* in *trans*, the TndX enzyme supplied in *trans* cannot (59). Finally, there are no similarities between the ends of Tn*5397* and the ends of the mobilizable transposons, nor are their similarities in sequences of their target sites (which resemble their transposon ends). Nevertheless, comparisons are instructive.

The Tn*4451* TnpX protein has an N-terminal region similar to those of members of the resolvase/invertase family of site-specific recombinases, a group that includes CcrA and CcrB from the staphylococcal methicillin resistance chromosomal cassette (SCCmec) (32). The TnpX N-terminal region has been shown to be essential for excision of Tn*4451*. DNA cleavage by the large resolvase/invertase family of enzymes results from formation of a covalent phosphoserine linkage between the enzyme and the 5′ end of the cleaved DNA strand. All of the conserved amino acid residues found to be essential for TnpX activity are present in TndX (58). Tn*5397* represents the only described conjugative transposon to date that uses a serine rather than a tyrosine recombinase (37).

Expression of Tn*4453a* TnpX from a multicopy plasmid increases the frequency of Tn*4451* and Tn*4453a* excision to form a circular intermediate (32). Analysis of *tnpX* from Tn*4451* and Tn*4453a* indicates that transcription of these genes occurs at low levels when it is integrated into another replicon. Transcription of these genes increases significantly, however, when the transposon is excised as a nonreplicating circular intermediate. It is presumed that a

Tn*5397*

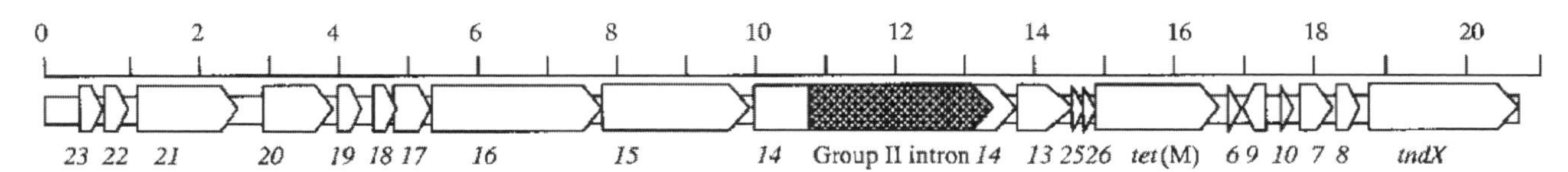

Tn*916*

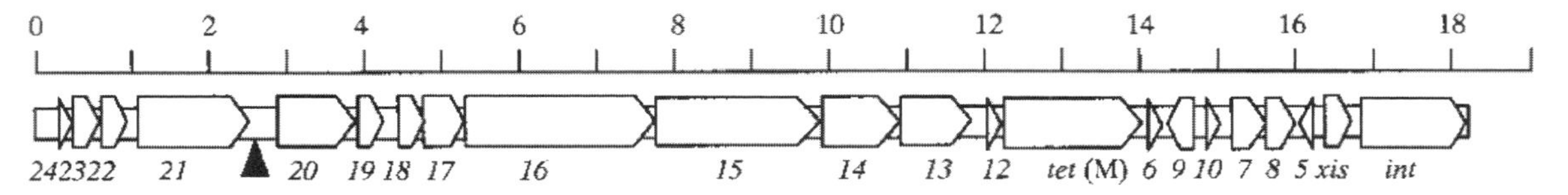

Figure 17.5 Comparison of clostridial conjugative transposons Tn*5397* and Tn*916*. Differences include the ends (where *orf5*, *orf25*, *xis*, and *int*) are missing from Tn*5397*. The ends of the transposons also differ. Tn*5397* also contains a group II intron within *orf14*, a serine recombinase (*tndX*) gene in place of *int* and *xis*, and two additional ORFs (*orf25 and orf26*) upstream of the *tet*(M) gene. Reprinted with permission from reference 47.

potent –35 region is present in the opposite end of the transposon, directing this transcription (32). As such, increased expression of TnpX when the element is in its circular form would favor integration. Integration would then favor survival of the element. The lack of a functional promoter within the transposon itself would indicate that transcription of *tnpX* while integrated would be dependent on the ability of the flanking sequences providing an adequate –35 region the appropriate distance from the transcriptional start, or the existence of a functional promoter somewhere upstream. As such, excision of Tn*4451* and Tn*4453a* would vary depending on the site of integration. There are data to indicate that overexpression of TnpX is deleterious to cells, so minimizing expression during stable times may be an advantage, whereas increased expression during circularization would favor return to the integrated form (32). Since the ends of Tn*5397* are different from those of Tn*4451*/*4453a*, it is not clear whether these regulatory mechanisms are operative with this transposon.

PROSPECTS FOR INHIBITION

When considering the advisability of embarking on major efforts to develop chemical inhibitors of conjugative transposition, there are two questions that must be addressed. The first is, Can chemical inhibitors be developed that will effectively interrupt the biochemical processes underlying conjugative transposition? The remarkable success of therapeutic agents developed for interrupting the transmission of human immunodeficiency virus (HIV) over the past 20 years suggests that the prospects for developing such inhibitors are quite good. As more of the specific details of HIV attachment, invasion, and replication become known, additional therapeutic strategies continue to be developed. That many of the conjugative transposon classes use variants of the λ-type bacteriophage integrase and excision processes for movement offers further hope that relatively broad-based inhibitors can be developed for clinical use. Recent advances in understanding the structural details of λ and related integrase interactions with DNA targets and cooperative proteins have the potential to stimulate intelligent, structure-based design of such chemical inhibitors. We should, however, be circumspect about the long-term efficacy of single agents, since the HIV experience suggests the potential for rapid mutational evolution to resistance. Moreover, it is possible that effective inhibitors for Tn*916* integrase will be ineffective at inhibiting other types of conjugative transposons. The remarkable plasticity and transmissibility of the bacterial genome that we have come to appreciate in the past few years would suggest that it may be a very short time before gene shuffling occurs to promote continued gene transfer through different vehicles.

The second and more difficult question is, would interruption of gene transmission through conjugative transposons, even if successful, have a significant impact on the overall prevalence of resistant bacteria causing human infections? Interrupting transmission of resistance would not impact the expression of resistance in bacteria that already possess a resistance determinant, so it would be unlikely to impact the treatment of individual infectious diseases. Moreover, the optimal time of therapy to interrupt transfer would not necessarily be clear. It is certainly arguable, perhaps even likely, that most resistance exchange occurs in places like the gastrointestinal tract, where numerous species coexist under a variety of external selective pressures. These selective pressures may be obvious to us (clinical treatment with antibiotics for a specific infection) or much less obvious (ingestion of antimicrobial agents from the environment). Moreover, it is likely that gene exchange is occurring constantly, with or without selective pressure, and that the primary role antimicrobial agents play in the process is to select out the recipients of these favorable genetic exchanges. It should be kept in mind that these exchanges can also occur in the gastrointestinal tracts of nonhuman mammals and then be transferred to humans through the food chain.

If the exchange of resistance determinants can occur at any time in a variety of settings, the only way to prevent the transfer is to treat universally and constantly. Such a strategy would be a prescription for the rapid emergence of resistance to any agent used. I therefore remain somewhat skeptical of the long-term feasibility of creating chemical agents whose function will be to inhibit genetic exchange between bacteria. The optimal strategy for preventing the transfer and emergence of resistance is, and always has been, reducing the selective pressure that favors survival of resistant transconjugants. This can only be accomplished through the rational and parsimonious use of antimicrobial agents.

References

1. **Adams, V., D. Lyras, K. A. Farrow, and J. I. Rood.** 2002. The clostridial mobilisable transposons. *Cell. Mol. Life Sci.* **59:**2033–2043.
2. **Ayoubi, P., A. O. Kilic, and M. N. Vijayakumar.** 1991. Tn*5253*, the pneumococcal omega (*cat tet*) BM 6001 element, is a composite structure of two conjugative transposons, Tn*5251* and Tn*5252*. *J. of Bacteriol.* **173:**1617–1622.
3. **Beaber, J. W., V. Burrus, B. Hochhut, and M. K. Waldor.** 2002. Comparison of SXT and R391, two conjugative integrating elements: definition of a genetic backbone for the mobilization of resistance determinants. *Cell. Mol. Life Sci.* **59:**2065–2070.
4. **Beaber, J. W., B. Hochhut, and M. K. Waldor.** 2002. Genomic and functional analyses of SXT, an integrating antibiotic

resistance gene transfer element derived from *Vibrio cholerae*. *J. Bacteriol.* **184:**4259–4269.

5. **Bedzyk, L. A., N. B. Shoemaker, K. E. Young, and A. A. Salyers.** 1992. Insertion and excision of *Bacteroides* conjugative chromosomal elements. *J. Bacteriol.* **174:**166–72.
6. **Boltner, D., C. MacMahon, J. T. Pembroke, P. Strike, and A. M. Osborn.** 2002. R391: a conjugative integrating mosaic comprised of phage, plasmid, and transposon elements. *J. Bacteriol.* **184:**5158–5169.
7. **Burrus, V., G. Pavlovic, B. Decaris, and G. Guedon.** 2002. Conjugative transposons: the tip of the iceberg. *Mol. Microbiol.* **46:**601–610.
8. **Caillaud, F., and P. Courvalin.** 1987. Nucleotide sequence of the ends of conjugative shuttle transposon Tn*1545*. *Mol. Gen. Genet.* **209:**110–115.
9. **Caparon, M. G., and J. R. Scott.** 1989. Excision and insertion of the conjugative transposon Tn*916* involves a novel recombination mechanism. *Cell* **59:**1027–1034.
10. **Carias, L. L., S. D. Rudin, C. J. Donskey, and L. B. Rice.** 1998. Genetic linkage and cotransfer of a novel, *vanB*-containing transposon (Tn*5382*) and a low-affinity penicillin-binding protein 5 gene in a clinical vancomycin-resistant *Enterococcus faecium* isolate. *J. Bacteriol.* **180:**4426–4434.
11. **Celli, J., and P. Trieu-Cuot.** 1998. Circularization of Tn916 is required for expression of the transposon-encoded transfer functions: characterization of long tetracycline-inducible transcripts reading through the attachment site. *Mol. Microbiol.* **28:**103–117.
12. **Cheng, Q., B. J. Paszkiet, N. B. Shoemaker, J. F. Gardner, and A. A. Salyers.** 2000. Integration and excision of a *Bacteroides* conjugative transposon, CTnDOT. *J. Bacteriol.* **182:**4035–4043.
13. **Cheng, Q., Y. Sutanto, N. B. Shoemaker, J. F. Gardner, and A. A. Salyers.** 2001. Identification of genes required for excision of CTnDOT, a *Bacteroides* conjugative transposon. *Mol. Microbiol.* **41:**625–632.
14. **Clewell, D. B., E. Senghas, J. M. Jones, S. E. Flannagan, M. Yamamoto, and C. Gawron-Burke.** 1986. Transposition in *Streptococcus*: structural and genetic properties of the conjugative transposon Tn*916*. *Soc. Gen. Microbiol.* **43:** 43–58.
15. **Clewell, D. B., S. E. Flannagan, L. O. Zitzow, Y. A. Su, P. He, E. Senghas, and K. E. Weaver.** 1991. Properties of conjugative transposon Tn*916*, p. 39–44. *In* G. M. Dunny, P. Patrick, and L. L. Cleary (ed.), *Genetics and Molecular Biology of Streptococci, Lactococci, and Enterococci*. American Society for Microbiology, Washington, D.C.
16. **Coetzee, J. N., N. Datta, and R. W. Hedges.** 1972. R factors from *Proteus rettgeri*. *J. Gen. Microbiol.* **72:**543–552.
17. **Courvalin, P., and C. Carlier.** 1987. Tn*1545*: a conjugative shuttle transposon. *Mol. Gen. Genet.* **206:**259–264.
18. **Doucet-Populaire, F., P. Trieu-Cuot, I. Dosbaa, A. Andremont, and P. Courvalin.** 1991. Inducible transfer of conjugative transposon Tn*1545* from *Enterococcus faecalis* to *Listeria monocytogenes* in the digestive tracts of gnotobiotic mice. *Antimicrob. Agents Chemother.* **35:**185–187.
19. **Flannagan, S. E., and D. B. Clewell.** 1991. Conjugative transfer of Tn*916* in *Enterococcus faecalis*: *trans* activation of homologous transposons. *J. Bacteriol.* **173:**7136–7141.
20. **Flannagan, S. E., L. A. Zitzow, Y. A. Su, and D. B. Clewell.** 1994. Nucleotide sequence of the 18-kb conjugative transposon Tn*916* from *Enterococcus faecalis*. *Plasmid* **32:** 350–354.
21. **Garnier, F., S. Taourit, P. Glaser, P. Courvalin, and M. Galimand.** 2000. Characterization of transposon Tn1549, conferring VanB-type resistance in Enterococcus spp. *Microbiology* **146:**1481–1489.
22. **Gawron-Burke, C., and D. B. Clewell.** 1982. A transposon in *Streptococcus faecalis* with fertility properties. *Nature* **300:**281–284.
23. **Gorfe, A. A., and I. Jelesarov.** 2003. Energetics of sequence-specific protein-DNA association: computational analysis of integrase Tn916 binding to its target DNA. *Biochemistry* **42:**11568–11576.
24. **Hinerfeld, D., and G. Churchward.** 2001. Specific binding of integrase to the origin of transfer (*oriT*) of the conjugative transposon Tn*916*. *J. Bacteriol.* **183:**2947–2951.
25. **Hinerfeld, D., and G. Churchward.** 2001. Xis protein of the conjugative transposon Tn916 plays dual opposing roles in transposon excision. *Mol. Microbiol.* **41:**1459–1467.
26. **Hochhut, B., J. W. Beaber, R. Woodgate, and M. K. Waldor.** 2001. Formation of chromosomal tandem arrays of the SXT element and R391, two conjugative chromosomally integrating elements that share an attachment site. *J. Bacteriol.* **183:**1124–1132.
27. **Hochhut, B., Y. Lotfi, D. Mazel, S. M. Faruque, R. Woodgate, and M. K. Waldor.** 2001. Molecular analysis of antibiotic resistance gene clusters in *Vibrio cholerae* O139 and O1 SXT constins. *Antimicrob. Agents. Chemother.* **45:**2991–3000.
28. **Hochhut, B., and M. K. Waldor.** 1999. Site-specific integration of the conjugal Vibrio cholerae SXT element into prfC. *Mol. Microbiol.* **32:**99–110.
29. **Jaworski, D. D., and D. B. Clewell.** 1994. Evidence that coupling sequences play a frequency-determining role in conjugative transposition of Tn*916* in *Enterococcus faecalis*. *J. Bacteriol.* **176:**3328–3335.
30. **Jaworski, D. D., and D. B. Clewell.** 1995. A functional origin of transfer (*oriT*) on the conjugative transposon Tn*916*. *J. Bacteriol.* **177:**6644–6651.
31. **Lu, F., and G. Churchward.** 1994. Conjugative transposition: Tn*916* integrase contains two independent DNA binding domains that recognize different DNA sequences. *EMBO J.* **13:**1541–1548.
32. **Lyras, D., and J. I. Rood.** 2000. Transposition of Tn4451 and Tn4453 involves a circular intermediate that forms a promoter for the large resolvase, TnpX. *Mol. Microbiol.* **38:** 588–601.
33. **Manganelli, R., S. Ricci, and G. Pozzi.** 1997. The joint of Tn*916* circular intermediates is a homoduplex in *Enterococcus faecalis*. *Plasmid* **38:**71–78.
34. **Marra, D., B. Pethel, G. G. Churchward, and J. R. Scott.** 1999. The frequency of conjugative transposition of Tn*916* is not determined by the frequency of excision. *J. Bacteriol.* **181:**5414–5418.
35. **Milev, S., A. A. Gorfe, A. Karshikoff, R. T. Clubb, H. R. Bosshard, and I. Jelesarov.** 2003. Energetics of sequence-

specific protein-DNA association: binding of integrase Tn916 to its target DNA. *Biochemistry* **42:**3481–3491.

36. **Milev, S., A. A. Gorfe, A. Karshikoff, R. T. Clubb, H. R. Bosshard, and I. Jelesarov.** 2003. Energetics of sequence-specific protein-DNA association: conformational stability of the DNA binding domain of integrase Tn916 and its cognate DNA duplex. *Biochemistry* **42:**3492–3502.
37. **Mullany, P., A. P. Roberts, and H. Wang.** 2002. Mechanism of integration and excision in conjugative transposons. *Cell. Mol. Life Sci.* **59:**2017–2022.
38. **Mullany, P., M. Wilks, and S. Tabaqchali.** 1995. Transfer of macrolide-lincosamide-streptogramin B (MLS) resistance in *Clostridium difficile* is linked to a gene homologous with toxin A and is mediated by a conjugative transposon, Tn*5398*. *J. Antimicrob. Chemother.* **35:**305–315.
39. **Oggioni, M. R., C. G. Dowson, J. M. Smith, R. Provvedi, and G. Pozzi.** 1996. The tetracycline resistance gene *tet*(M) exhibits mosaic structure. *Plasmid* **35:**156–163.
40. **Pembroke, J. T., C. MacMahon, and B. McGrath.** 2002. The role of conjugative transposons in the Enterobacteriaceae. *Cell. Mol. Life Sci.* **59:**2055–2064.
41. **Pembroke, J. T., and D. B. Murphy.** 2000. Isolation and analysis of a circular form of the IncJ conjugative transposon-like elements, R391 and R997: implications for IncJ incompatibility. *FEMS Microbiol. Lett.* **187:**133–138.
42. **Poyart-Salmeron, C., P. Trieu-Cuot, C. Carlier, and P. Courvalin.** 1989. Molecular characterization of the two proteins involved in the excision of the conjugative transposon Tn*1545*: homologies with other site specific recombinases. *EMBO J.* **8:**2425–2433.
43. **Rice, L. B.** 2000. Bacterial monopolists: the bundling and dissemination of antimicrobial resistance genes in gram-positive bacteria. *Clin. Infect. Dis.* **31:**762–769.
44. **Rice, L. B.** 1998. Tn*916*-family conjugative transposons and dissemination of antimicrobial resistance determinants. *Antimicrob. Agents Chemother.* **42:**1871–1877.
45. **Rice, L. B., and L. L. Carias.** 1998. Transfer of Tn*5385*, a composite, multiresistance element from *Enterococcus faecalis*. *J. Bacteriol.* **180:**714–721.
46. **Rice, L. B., S. H. Marshall, and L. L. Carias.** 1992. Tn*5381*, a conjugative transposon identifiable as a circular form in *Enterococcus faecalis*. *J. Bacteriol.* **174:**7308–7315.
47. **Roberts, A. P., P. A. Johanesen, D. Lyras, P. Mullany, and J. I. Rood.** 2001. Comparison of Tn5397 from Clostridium difficile, Tn916 from Enterococcus faecalis and the CW459tet(M) element from Clostridium perfringens shows that they have similar conjugation regions but different insertion and excision modules. *Microbiology* **147:**1243–1251.
48. **Roberts, A. P., J. Pratten, M. Wilson, and P. Mullany.** 1999. Transfer of a conjugative transposon, Tn5397 in a model oral biofilm. *FEMS Microbiol. Lett.* **177:**63–66.
49. **Rudy, C. K., J. R. Scott, and G. Churchward.** 1997. DNA binding by the *Xis* protein of the conjugative transposon Tn*916*. *J. Bacteriol.* **179:**2567–2572.
50. **Scott, J. R., and G. G. Churchward.** 1995. Conjugative transposition. *Annu. Rev. Microbiol.* **49:**367–397.
51. **Scott, J. R., P. A. Kirchman, and M. G. Caparon.** 1988. An intermediate in the transposition of the conjugative transposon Tn*916*. *Proc. Natl. Acad. Sci. USA* **85:**4809–4813.
52. **Scott, K. P.** 2002. The role of conjugative transposons in spreading antibiotic resistance between bacteria that inhabit the gastrointestinal tract. *Cell. Mol. Life Sci.* **59:**2071–2082.
53. **Shoemaker, N. B., H. Vlamakis, K. Hayes, and A. A. Salyers.** 2001. Evidence for extensive resistance gene transfer among *Bacteroides* spp. and among *Bacteroides* and other genera in the human colon. *Appl. Environ. Microbiol.* **67:**561–568.
54. **Showsh, S. A., and R. E. Andrews.** 1992. Tetracycline enhances Tn*916*-mediated conjugal transfer. *Plasmid* **28:** 213–224.
55. **Smith, C. J., and A. C. Parker.** 1993. Identification of a circular intermediate in the transfer and transposition of Tn*4555*, a mobilizable transposon from *Bacteroides* spp. *J. Bacteriol.* **175:**2682–2691.
56. **Stevens, A. M., N. B. Shoemaker, L.-H. Li, and A. A. Salyers.** 1993. Tetracycline regulation of genes on *Bacteroides* conjugative transposons. *J. Bacteriol.* **175:**6134–6141.
57. **Taylor, K. L., and G. Churchward.** 1997. Specific DNA cleavage mediated by the integrase of conjugative transposon Tn*916*. *J. Bacteriol.* **179:**1117–1125.
58. **Wang, H., and P. Mullany.** 2000. The large resolvase TndX is required and sufficient for integration and excision of derivatives of the novel conjugative transposon Tn*5397*. *J. Bacteriol.* **182:**6577–6583.
59. **Wang, H., A. P. Roberts, D. Lyras, J. I. Rood, M. Wilks, and P. Mullany.** 2000. Characterization of the ends and target sites of the novel conjugative transposon Tn*5397* from *Clostridium difficile*: excision and circularization is mediated by the large resolvase, TndX. *J. Bacteriol.* **182:**3775–3783.
60. **Wang, H., A. P. Roberts, and P. Mullany.** 2000. DNA sequence of the insertional hot spot of Tn916 in the Clostridium difficile genome and discovery of a Tn916-like element in an environmental isolate integrated in the same hot spot. *FEMS Microbiol. Lett.* **192:**15–20.
61. **Wang, Y., G. R. Wang, A. Shelby, N. B. Shoemaker, and A. A. Salyers.** 2003. A newly discovered *Bacteroides* conjugative transposon, CTnGERM1, contains genes also found in gram-positive bacteria. *Appl. Environ. Microbiol.* **69:** 4595–4603.
62. **Warren, D., M. D. Sam, K. Manley, D. Sarkar, S. Y. Lee, M. Abbani, J. M. Wojciak, R. T. Clubb, and A. Landy.** 2003. Identification of the lambda integrase surface that interacts with Xis reveals a residue that is also critical for Int dimer formation. *Proc. Natl. Acad. Sci. USA* **100:**8176–8181.
63. **Whittle, G., B. D. Hund, N. B. Shoemaker, and A. A. Salyers.** 2001. Characterization of the 13-kilobase *ermF* region of the *Bacteroides* conjugative transposon CTnDOT. *Appl. Environ. Microbiol.* **67:**3488–3495.
64. **Whittle, G., N. B. Shoemaker, and A. A. Salyers.** 2002. Characterization of genes involved in modulation of conjugal transfer of the *Bacteroides* conjugative transposon CTnDOT. *J. Bacteriol.* **184:**3839–3847.
65. **Whittle, G., N. B. Shoemaker, and A. A. Salyers.** 2002. The role of *Bacteroides* conjugative transposons in the dissemination of antibiotic resistance genes. *Cell. Mol. Life Sci.* **59:**2044–2054.
66. **Wojciak, J. M., D. Sarkar, A. Landy, and R. T. Clubb.** 2002. Arm-site binding by lambda-integrase: solution structure and functional characterization of its amino-terminal domain. *Proc. Natl. Acad. Sci. USA* **99:**3434–3439.

Enzyme-Mediated Resistance to Antibiotics: Mechanisms, Dissemination, and Prospects for Inhibition
Edited by Robert A. Bonomo and Marcelo E. Tolmasky

Virginia L. Waters

18

The Dissemination of Antibiotic Resistance by Bacterial Conjugation

BACTERIAL GENETICS IN THE HISTORY OF ANTIBIOTIC RESISTANCE

Antibiotic resistance history in bacterial infection generally begins with penicillin resistance. With large-scale production ongoing by the early 1950s, penicillin resistance soon emerged. It was first found in enteric organisms. In 1955, Bill Kirby documented the clinical appearance of penicillin resistance in *Staphylococcus aureus*. Ten years later, in 1965, penicillin resistance appeared in *Neisseria gonorrhoeae*. In 1972, penicillin resistance appeared in *Haemophilus influenzae*; in 1980, it appeared in *Streptococcus pneumoniae*; and in 1983, it appeared in *Streptococcus faecalis*. Penicillin resistance in *Streptococcus pyogenes* has been expected for some time, but, interestingly, it is yet to appear.

To illustrate the emergence of resistance another way, events could be written in terms of the organism rather than the antibiotic. For example, for the organism *Neisseria gonorrhoeae*, penicillin resistance was first noted in 1965. Then, in 1985, high-level tetracycline resistance appeared, and, in 1993, ciprofloxacin resistance appeared.

The subject of this chapter fits well into this kind of context: a more complex history began with the simultaneous dissemination of multiple drug resistance (MDR) determinants. In the 1950s, clinical strains that were able to simultaneously transfer several resistance genes to other bacteria were isolated. Gene dissemination occurred within species and across species. The rapid acquisition of MDR determinants could not be explained by mutation alone. Early on, in an epidemic caused by a strain of *Shigella flexneri*, the organism suddenly acquired resistance to tetracycline, chloramphenicol, streptomycin, and sulfanilamide. These same resistance determinants were also found in isolates of *Escherichia coli*. Bacterial conjugation was suspected as the mechanism of gene transfer, and it was ultimately attributed to transferable plasmids encoding all these resistance determinants. The phenomenon of conjugative gene transfer is not limited to enteric organisms. While multiply resistant strains were first seen in Japan and Mexico, conjugative MDR transfer is common all over the world.

During the 1958 Asian flu pandemic, about half of the *Staphylococcus aureus* isolates were penicillin resistant. Five years later, the isolation rate had increased to 95%. Such rapid appearance was attributed to the existence of β-lactamase-encoding plasmids in *S. aureus* before penicillin had been used clinically for this organism. The hypothesis was that penicillin-producing soil organisms had selected for resistance in bacteria long before there was any clinical selection. To test the hypothesis,

Virginia L. Waters, School of Medicine, University of California San Diego, La Jolla, CA 92093-0640.

antibiotic-resistant enteric organisms were isolated from indigenous human populations. Human populations were first identified that had received no form of standard medical treatment. One of the more adventurous studies of such populations involved the analysis of stool samples coaxed from the natives of the remote highlands of Borneo (18). It was found that individuals carried *E. coli* encoding resistance to penicillin, sulfadiazine, tetracycline, and streptomycin. All of these resistance markers were subsequently transferred from the sampled bacterial isolates to drug-sensitive strains in laboratory experiments. It was concluded that the genetics had been up and running in bacteria for centuries, before antibiotics were used in treatment. It is not surprising that soil organisms have been the primary resource for antibiotic purification. This resource is now essentially exhausted.

Analysis of individual examples of multiple resistance gives insight into the history of resistance and also into the ways that large MDR plasmids are built from smaller DNA segments. Recent analyses of transferable plasmids from clinical isolates have shown that these plasmids can collect a diversity of DNA segments. Diversity in antibiotic resistance determinants is shown in terms of DNA sequences, molecular mechanisms, and bacterial species of origin. Continued strong selective pressure has resulted in a greater creativity and complexity within this more recent assortment of transferable plasmids. This is wonderful for those of us who study these things. However, the clinical consequences of the world-traveling, multiresistant organisms demand that we become more creative in research and development of new strategies against MDR dissemination. These observations provide more reason to pursue the subject from every angle.

One or two mechanisms of genetic exchange usually predominate in a bacterial strain in a given milieu. The efficiencies of such mechanisms vary with the organism, population dynamics, and the environment. Contrasting the rates of emergence of penicillin resistance in two different organisms, *S. aureus* and *N. gonorrhoeae*, can illustrate some of the rate-limiting factors. The proportion of penicillin-resistant *S. aureus* isolates rapidly went from 50 to 95% in 5 years. Meanwhile, for *N. gonorrhoeae*, resistance gradually went from MICs of 0.3 to 3.0 μg/ml. The gradual rise over a 10-year period reflected several factors, among them the different levels of penicillin used to treat the infection. It was recognized that six different β-lactamases had come from *E. coli* by transfer to *N. gonorrhoeae*, demonstrating genetic exchange across those species via a number of events. Also, the organism *H. influenzae* was found eventually to have an *E. coli* transposon-encoded enzyme, indicative of low-frequency genetic events. The histories of these different organisms demonstrate what has been learned in the laboratory over the years, that genetic exchange is in fact more readily accomplished in organisms such as *E. coli* and *S. aureus* than in organisms such as *H. influenzae* and *N. gonorrhoeae*. The latter set of organisms are considered less genetically flexible and tend to have smaller-sized genomes. Finally, such historical accounts demonstrate the fact that these slower-evolving organisms nonetheless do become resistant.

S. pneumoniae has a unique drug resistance story. In 1960, penicillin resistance was low in Papua New Guinea. Twenty years later there was high-level resistance, with no oral treatment in some areas. In 1983 it was found that an enterococcal β-lactamase had been transferred from *S. aureus* to *S. pneumoniae*. After years of strong selection by an increased incidence and treatment of otitis media and meningitis caused by *S. pneumoniae* (16), vaccine development became imperative. But why had penicillin resistance emerged more slowly in *S. pneumoniae* than in *S. aureus* and gram-negative organisms? In the pneumococcus there is a gene encoding the transpeptidase enzyme in the last step in cell wall synthesis, the peptidoglycan cross-linking step. Mutations within the gene led to the same resistance mechanism all over the world. This had occurred by acquisition of a nonpneumococcal mutant gene, and the DNA was traced back to natural oral flora. The resulting gene is a mosaic of several DNA subsequences from a variety of different sources. There are also many hybrids, indicative of multiple events and explaining the derivation of different MICs. Among the first resistant strains to emerge, there were only 10 or 12 clones, with a few clones dominating. Nowadays, penicillin resistance in the pneumococcus is multiclonal and commonplace.

Why has *Streptococcus pyogenes* remained penicillin sensitive? Is sensitivity due to the fact that the organism is not naturally transformable? Natural transformation may be one factor. *S. pyogenes* appears to be the only clinical member of the *Streptococcus* genus which is not naturally transformable. A unique feature of this organism is its antiphagocytic M protein layer on the cell wall surface. The M protein might be a barrier to natural transformation and to efficient gene acquisition.

Another factor is bacterial conjugation. *S. pyogenes* is conjugative, but at a very low frequency. Transfer of the macrolide resistance gene *ermA* from donor *S. pyogenes* to recipient *S. pyogenes* and other bacteria has revealed that transfer frequencies are about a million-fold less than those of *E. coli* (29). The M protein may keep *S. pyogenes* bacterial cells apart just as it keeps phagocytes at a distance. However, low-frequency conjugation is adequate for the eventual emergence of resistance.

Conjugative transposition of the macrolide efflux gene *mefA* has been assessed for *S. pyogenes* and *S. pneumoniae*. It was found that conjugative transfer of this element from *S. pyogenes* to *S. pneumoniae* occurs at a rate of 10^{-6} transconjugant per donor cell, while the efficiency from *S. pneumoniae* to *S. pyogenes* occurs at a rate of 10^{-3} transconjugant per donor (78). Therefore, the M protein may have its impact on the donor in conjugation. As noted, low-frequency conjugation cannot adequately explain long-term penicillin sensitivity, because one such low-frequency event will eventually happen. Thereafter, the resistance gene provides cell survival in populations of organisms experiencing selective pressure

S. pyogenes is unique in several ways, including its unusual abilities to resist the host. It has been recognized for some time that the organism resists phagocytosis. More recently it has been found that it induces neutrophil necrosis, and its ability to do so is greater than that of the other organisms tested (43). We have no reason to believe that this defense mechanism would somehow relate to penicillin resistance, but it would be interesting to test the possibility.

In addition, *S. pyogenes* may have other, undefined properties. There are two major strategies for penicillin resistance in bacteria: β-lactamase gene transfer and the alteration of penicillin-binding protein activity in cell wall peptidoglycan cross-linking. It is possible that there is an anomaly in *S. pyogenes* wall building. An inability to resist the effect of penicillin could be due to aspects of wall building that are only recently being explored (25). An undefined peculiarity in *S. pyogenes* cell wall synthesis might make it intolerant of mutations to penicillin resistance. Because both *ermA* and *mefA* have been conjugated to and from *S. pyogenes*, the key is more probably a cell wall factor rather than inadequate gene transfer.

Macrolide use in *S. pneumoniae* is another interesting case in point. Selective pressure can lead to collateral damage caused by antibiotics aimed at another organism carried by the same person. This occurred in an instance of rapid selection for macrolide resistance in *S. pneumoniae*. Originally, newly treated populations could be adequately treated with one dose. In 1997, Leach and her collaborators documented how trachoma treatment with azithromycin affected the rate of resistant pneumococci culture soon after the trachoma treatment (51). In areas where trachoma is highly endemic, that is, where about 70% of the population is infected, many individuals typically carry *S. pneumoniae*. Of these, 2% were originally macrolide resistant. After 2 weeks of trachoma treatment, the carrier rate for macrolide-resistant pneumococci rose to 50%, with an increase in the carrier rate of the more highly resistant strains. This study dramatically illustrates the potential for rapid in vivo selection of resistance.

In this study it was also found that one chronically ill child was carrying a strain with very high level resistance. Interestingly, the strain was not spread to the others in the community, over the short term or over the long term (52). The drift to higher resistance in the strain had apparently occurred within the one individual. The strain was not spread by person-to-person transmission and remained a clonal population. Such well-documented observations indicate that host factors impact the development of antibiotic resistance in bacteria.

Epidemiological studies have shown that resistant bacteria are generally disseminated throughout a host community in three ways: by person-to-person dissemination, by nosocomial dissemination, and by transmission through food and veterinary sources. Bacterial species may use different mechanisms to disseminate resistance genes within each of the corresponding microbial communities. Similarly, different infecting organisms may share the same dominant genetic mechanism in clinical environments more than in other environments. The same MDR plasmid may be carried by two or three bacterial strains (or species) isolated from patients in the same hospital unit. This is indicative of conjugative transfer and perhaps broad-host-range conjugation. The many bacteria known for this kind of gene dissemination include *Clostridium*, *E. coli*, *Klebsiella*, *Pseudomonas*, *Salmonella*, *Shigella*, and *Vibrio*. The dominant mechanism for these bacterial species has indeed been found to be conjugation. For *S. pneumoniae*, the dominant mechanism is natural transformation; for *H. influenzae*, it is transposon and plasmid conjugation; for *Salmonella enterica* serovar Typhi, it is plasmid conjugation and mutation; and for *Mycobacterium tuberculosis*, it appears to be mediated by mutation alone (Table 18.1). MDR in *Enterobacter* species appears to be mediated mostly by mutation and efflux systems. Penicillin resistance is most often plasmid mediated in *S. aureus*, and methicillin resistance in *S. aureus* is typically transposon mediated. The stories of resistance emergence reflect the dominant genetic mechanism(s) as well as the organism, the patient, and the antibiotic.

The probability of simultaneously generating resistance to three drugs by mutation alone is quite low, the product of each of the mutation rates. The mutation rate per generation for one resistance gene could be, hypothetically, one mutation per million cells per generation. For three concurrent mutations, the rate would then be $(10^{-6})^3$, i.e., 10^{-18}. *M. tuberculosis* cells divide every 12 h under laboratory culture conditions and more slowly in vivo. With such slow growth, drug resistance by mutation would be

Table 18.1 Genetic mechanisms in bacteria[a]

Mutation	Natural transformation	Conjugative transfer (plasmid and conjugative transposon)
All bacteria	*Acinetobacter*	*Enterobacteriaceae*
M. tuberculosis	*Bacillus*	*Enterobacter*
	E. coli	*E. coli*
	Enterococcus	*Klebsiella*
	Haemophilus	*Proteus*
	Helicobacter	*Salmonella*
	Neisseria	*Shigella*
	Pseudomonas	*Serratia*
	Staphylococcus	*Acinetobacter*
	Streptococcus	*Bacteroides*
		Campylobacter
		Clostridium
		Enterococcus
		Haemophilus
		Helicobacter
		Listeria
		Mycoplasma
		Neisseria
		Pseudomonas
		Staphylococcus
		Streptococcus
		Vibrio
		Yersinia

[a] Bacterial genetic mechanisms are grouped as mutation, natural transformation, and conjugative transfer. Pathogenic bacteria demonstrating antibiotic resistance gene acquisition are listed according to dominant genetic mechanism(s). Gene acquisition in *M. tuberculosis* by transduction and transformation has been observed under laboratory conditions, but clinical acquisition of resistance has been attributed to mutation.

expected to emerge slowly. In fact, MDR tuberculosis (TB) is a recent phenomenon. For triple-resistant TB, selective pressure must have been high, whether for three simultaneous events or for three successive events. An atypical form of conjugative transfer has been documented for the more rapid-growing nonpathogenic organism *Mycobacterium smegmatis* (96), but this kind of conjugative transfer has not been found in *M. tuberculosis*.

The unlikely emergence of triple-drug resistance in TB was attributed entirely to mutation. It is possible that *M. tuberculosis* has some sort of undetected genetic mechanism, but mutation alone is most consistent with the data. Those who understand these dynamics in MDR emergence were dismayed to learn of the predicament, that is, drug resistance that could have been prevented, and TB strains for which there was no treatment. Over the years, it had become widely known that TB treatment in high-risk environments is often not prudent. In the developing world, isoniazide (INH) has been available over the counter for years and allowed to be sold as a cold remedy. More recently, certain drug-resistant TB strains have been labeled "XDR" to indicate extensive drug resistance. First seen in human immunodeficiency virus-infected populations in Africa, these strains carry resistance to first-line drugs and second-line drugs, that is, INH, rifampicin, fluoroquinolones, aminoglycosides, and capreomycin. The situation elicited a call for urgent international intervention.

Examination of vancomycin-resistant enterococci indicated that gene dissemination is mediated by conjugation (92). Genome analysis of strain V583 confirmed the role of conjugation, apparently by plasmids and by conjugative transposons. It also revealed that mobile elements and foreign DNA comprise a full quarter of the strain's genome. Analysis of mobile elements indicated that several virulence determinants had also been transferred into the strain. For this one strain, three plasmids and several conjugative and composite transposons were found. The gene for vancomycin resistance in this strain is thought to be originally from the conjugative transposon CTn*1549* (68).

Vancomycin-resistant enterococci comprise a significant and rising proportion of isolates, and the story is noteworthy. The enterococcal cell wall is sequentially constructed by the addition of D-Ala D-Ala to the murein layer. Vancomycin binds to the D-Ala D-Ala group, changes it to D-Ala D-lactate and then to D-Ala L-lactate. Without the continued presence of vancomycin, it reverts back to sensitivity. So why then is it such a resistance problem? The answer is related to high-level ongoing selective pressure. Vancomycin has been highly effective in veterinary and human populations. Wide use of broad-spectrum cephalosporins has resulted in resistance in enterococcal and normal flora strains, narrowing treatment options. Finally, the rise of methicillin-resistant *S. aureus* has increased the clinical usefulness of vancomycin and has thereby also provided another source of selective pressure for vancomycin resistance.

This overview has described some of the relevant history of antibiotic resistance in infectious disease. The historical vignettes reflect diverse bacterial species, diverse gene origins, and diverse mechanistic variations, and they reveal important lessons for us. One lesson is that bacterial resistance follows sooner or later after the development of an antibiotic. It is world news when a pathogen becomes resistant to the drug of choice that has been the mainstay of treatment for years. Another lesson is that resistance often involves MDR. Another lesson is that antibiotic resistance emerges most rapidly in those organisms most proficient in genetic change and exchange. What would be the kind of genetic proficiency that leads to MDR? This chapter defines gene dissemination mechanisms and shows how and why conjugative mechanisms are the most proficient in MDR transfer. Ultimately, however, resistance also slowly emerges in the "genetically impaired" bacteria. Exceptions are rare for the rule that for every useful antibiotic, there will be bacterial resistance.

DEVELOPING NEW APPROACHES IN TARGETING ANTIBIOTIC RESISTANCE

Can the trend towards greater resistance be reversed? The question has driven the search for new antimicrobial agents. A search for new agents, however, is not a new strategy. Each new drug is targeted to some cellular property, to stop growth or viability. To combat the spread of antibiotic resistance among bacterial strains, in addition to treating the infection, I have suggested that bacterial conjugation itself be targeted (103). Such an antidissemination agent could be given as an adjunct to treatment of the ongoing infection. The goal would be the prevention of resistance gene dissemination in the affected individual and in the surrounding community. This type of approach is being used, in a sense, with β-lactamase inhibitors that are often given in conjunction with traditional antibiotics.

Bacterial conjugation, conjugative transposition, and natural transformation are the most important genetic exchange mechanisms for resistance dissemination in clinical bacteria. It would be interesting and potentially fruitful to find more shared properties among these mechanisms. This could lead to targeting more than one mechanism of gene dissemination with the same agent. Although resistance emergence usually occurs by means of the dominant mechanism of the organism (Table 18.1), mechanistic aspects may be shared by these mechanisms. Addressing horizontal transfer by any approach could not replace the continued development of antibiotics that target cell viability. Rather, an additional strategy would be aimed at gene exchange. Such intervention in genetic exchange, if successful, could reverse the trend.

Targeting transfer mechanisms could help control virulence factor dissemination as well as antibiotic resistance dissemination. In vivo transfer of resistance genes by bacterial conjugation has been experimentally documented in the human colon and on skin (2, 60). Clinical gene transfer has been inferred for sites such as in blood, sputum, and wounds. The isolation of bacterial strains from a pediatric patient with chronic sepsis was followed over a 2-year period, as treatment progressed through a series of antibiotics. First, a *Salmonella* strain was isolated, followed by an *E. coli* strain, and still later a *Klebsiella* strain was isolated from this patient. It was hypothesized for this patient that in vivo transfer of virulence genes encoded on a conjugative plasmid was accomplished via successive events for the successive isolates (99). Why this hypothesis? Antibiotic pressure and the antibacterial pressure of serum had selected for antibiotic resistance and virulence, over and above the survival of any of the three bacterial species. It is reasonable to believe that cotransfer of genes was coselected, since all of these organisms were blood isolates. Genome sequencing has also supported the idea that there is such coselection of virulence and resistance genes in vivo.

The antibiotic paradox is that the antibiotics used to treat an infection simultaneously select for resistance in the bacteria causing the infection (53). The antibiotic dilemma is that the direct way to reduce resistance is to use these drugs less often, whereas treatment may be the best option for the individual. The immediate solutions are reducing antibiotic use and discovering new antibiotics. Using less does not have to jeopardize health care and includes reducing the promiscuous use of broad-spectrum drugs. Primary care often includes treatment of sinusitis, acute bronchitis, urinary tract infections, and upper respiratory infections. Although such illnesses are often viral, antibiotics are nonetheless quickly prescribed. In September 2003, the Centers for Disease Control and Prevention reminded health care providers of an earlier survey that found that 40% of outpatient antibiotic prescriptions were for viral infections.

Antibiotic resistance has been answered with the development of new derivatives such as methicillin for penicillin-resistant organisms. This in turn provided the selecting force for the corresponding resistance, and another antibiotic derivative became necessary. This cycle of resistance and new antibiotic development has gone on now for over 50 years, and the problem continues to escalate worldwide (Fig. 18.1). The increasing incidence and rapid spread of resistance have vindicated those who saw them coming.

Novel antibiotics, such as antimicrobial peptides, can be used for certain types of infections. Antimicrobial peptides cannot be taken orally, but there are studies designed to explore ways to protect these drugs from digestive enzymes. Similarly, lytic enzymes isolated from phages could be used for intravenous delivery but not taken orally (56). Another approach is the mining of the ocean floor for antibiotic-producing organisms. It is encouraging that newer drugs are working more consistently. This may be the case because many of the newer antibiotics are not derived from soil organisms. The availability of antibiotics from soil organisms, a ready resource for many years, has also brought a number of the corresponding preexistent resistance genes.

The problem has also involved the fact that no new class of antibiotics has been discovered since the 1970s. Some have predicted an end of the antibiotic era.

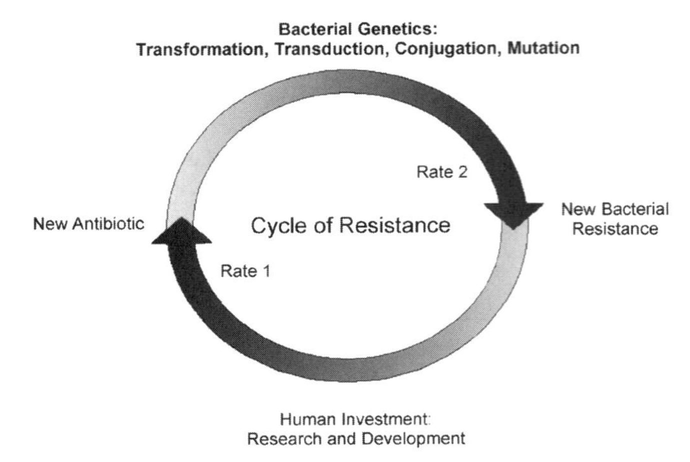

Figure 18.1 The cycle of antibiotic development and resistance. The cycle illustrates the inevitability of bacterial resistance. Rate 1 indicates the time required for the development of a new antibiotic, and Rate 2 is the rate at which clinical bacteria develop resistance to the new antibiotic.

Innovative types of vaccines are being developed, but there are organisms for which vaccine development cannot be a feasible approach. This applies to new pathogens for which a vaccine cannot be anticipated and developed and to the disease-causing but normally friendly flora and opportunistic organisms affecting the compromised host. The cycle of antibiotic resistance will continue for these kinds of infectious disease. As might be expected, studies indicate that mass prophylactic treatment, in order to prevent infection from spreading throughout a defined population, is counterproductive and probably irresponsible (75).

The cycle of resistance illustrates the dynamic of emerging resistance followed by new drug development, again followed by another resistance, a new drug development, another new resistance, and so on (Fig. 18.1). It also illustrates that we have not adequately addressed the inevitability of antibiotic resistance. The emergence of resistance and new antibiotic development occur in measurable time periods. The rates involved are the rates of resistance emergence and the rate of antibiotic development. These rates are critical, because they determine who wins. The rate for the organism is a function of the mutation rate, efficiency of gene acquisition, and selective pressure. The rate of drug development is a function of the quality and quantity of research. New methods are being used to screen literally millions of substances for suitable antibiotic activity. As the rate of drug development decreases for treatment of diseases caused by certain organisms, a subset will become untreatable. This is the case in some instances in the developing world. In a sense, then, we are seeing the beginning of the end of the antibiotic era. Innovative approaches are therefore needed to address the cycle itself, rather than the emergence of each new resistance.

BACTERIAL GENETICS: DNA MUTATION, TRANSFORMATION, TRANSDUCTION, AND CONJUGATION

Bacterial genetics courses have traditionally taught that there are three mechanisms of genetic acquisition and exchange in bacteria, that is, transformation, transduction, and conjugation. These mechanisms give bacteria the genetic diversity needed to survive changing environments. Bacterial genetic diversity is somewhat analogous to eukaryotic genetic diversity as gained by allelic variation, the consequence of chromosome pairing variations in offspring. However, bacteria are asexual, necessitating different means of new gene acquisition. Bacteria have fluid population dynamics, and the population itself can be viewed as an organism. Large numbers of microbial cells and species compete in an environment, undergoing genetic change and exchange. Each bacterial genome represents a bank of alleles for the population at large to tap into as needed. Genes are indeed disseminated in bacteria in nature by transformation, transduction, and conjugation. Since bacterial species vary in efficiency in gene acquisition, the blend of dominant species within a population has a fluid dynamic as well.

DNA mutation

DNA mutation, a change in DNA sequence, brings about genetic variation and can be considered the first event in the emergence of antibiotic resistance. Mutation is a mechanism for genetic change but not genetic exchange. Microbial populations adapt more slowly when mutation is the only force for change. To most quickly capitalize on newly mutated DNA, a mutant gene is transferred throughout a population. The recipient population then tests a mutant gene product for enhanced fitness. Greater fitness brings growth advantage in an environment. Without gene transfer, mutation can effect major changes in a cell population by outgrowing the nonmutant cells of the population, but it takes more time than gene transfer events. Without the dissemination of mutant alleles, overall change is slower per organism and slower in clinical impact. Mutation is therefore critical for genetic flexibility, but genetic exchange is critical for providing rapid adaptation. In light of these observations, it is notable that mutation can nonetheless serve as the only force for genetic change in certain organisms. It is therefore reasonable to treat mutation alone as a comparable genetic force, along with transformation, transduction, and conjugation.

The rapid development of resistance is more a function of an organism's genetic prowess than a function of variations in mutation rate. There are bacterial species that have higher mutation rates than others (8). However, high mutation rates are considered costly to an organism, and there is controversy about the potential advantage of high mutability. Inducible, temporary increases in mutation rate could provide a way to reduce the cost of deleterious mutations, but this could also reduce the benefit potential. It has been suggested that bacterial resistance can result from induced increase in mutation rates, but it has not been established as a significant force (57). It is conceivable that mutation rates are higher in natural environments than under laboratory conditions and that mutation rate is a greater worldwide force than is commonly thought (7). However, the clinical history of antibiotic resistance and known mutation rates support the contention that mutation is necessary but rarely sufficient in the clinical emergence of drug resistance. The less frequent the acquisition of genes, the longer it takes for a resistance phenotype to appear clinically.

Transformation

Transformation is the uptake of external free DNA by the bacterial cell. Transformation does not require cell-cell contact and is sensitive to DNase in the milieu. Transformation has been optimized in the laboratory for many bacterial species, but there is a growing list of organisms that are transformed in nature. These bacteria are called naturally transformable or competent for transformation, and the phenomenon was first discovered in *Streptococcus pneumoniae*. The capacity for natural transformation appears to be genus specific or species specific. In addition to *Staphylococcus*, *Streptococcus pneumoniae*, and other oral streptococci, naturally transformable organisms include *H. influenzae*, the *Neisseria* species, *Helicobacter pylori*, *Acinetobacter baumannii*, *Pseudomonas aeruginosa*, and the nonpathogenic organism *Bacillus subtilis* (Table 18.1). Genes homologous to competence genes have been identified in *E. coli*, and experiments have demonstrated low-level natural transformation in *E. coli* (94). Other naturally transformable pathogens will probably be identified, with efficiencies more like those of *E. coli* than of *S. pneumoniae*. Interestingly, *Streptococcus pneumoniae* organisms have been shown to lyse surrounding *S. pneumoniae* cells that have lost competence proficiency (87).

What allows bacteria to be naturally transformable? It appears that no one has yet cloned competence genes and engendered transformability in a cell background lacking competence. The components for natural transformation may require a compatible membrane structure that is specific to a given bacterial cell wall. There is a diversity of competence genes among the streptococci, and such genes are distinct from those for *Helicobacter* and *Neisseria*. Some competence genes of gram-negative organisms are homologous to type IV secretion system genes. In *H. pylori*, one of the *com* gene products is thought to be at the inner side of the outer membrane, with another at the outer side of the outer membrane and another at the outer side of the inner membrane (40). There is now a model for the *H. pylori* competence membrane structure, and it may be found to resemble aspects of the conjugative membrane bridge structure. A remarkable observation that is relevant to this discussion is that natural transformation takes up DNA in single-stranded form. Such DNA is more likely to survive in the newly transformed cell. Other factors may also enhance natural transformation efficiency. DNA may be less subject to degradation within transformable cells than in nontransformable bacteria. It is thought that there is less ongoing mismatch DNA repair in *H. pylori* than in enteric bacteria. Incoming DNA molecules would then be expected to have a higher survival rate in this organism.

Transduction

Transduction is the "accidental" packaging of bacterial DNA with phage DNA by an infecting bacteriophage. The bacterial DNA is cotransferred along with the phage DNA as the phage packages DNA and infects the next cell. Phage transduction does not require cell-cell contact but does require a cell receptor to which the phage attaches. Transduction has been used as a mechanism of gene transfer in laboratory experiments and continues to be a laboratory option in developing genetics in bacteria (22). Apparently, however, phage transduction has not been shown to be an in vivo mechanism of antibiotic resistance gene dissemination. This does not mean that it has not happened. Importantly, three resistance genes in a multiresistant epidemic strain of *Salmonella enterica* serovar Typhimurium have been cotransduced by a P22-like phage in laboratory experiments (80). This type of phenomenon, one that requires a cotransducing phage, may have happened in the clinical world, but the clinical significance remains inferential. While transduction of antibiotic resistance has not been documented regarding either animal or human infection, it is presumed to be an active mechanism in antibiotic gene dissemination because it is well known in dissemination of other genes.

Transduction in a general sense has been inferred from phage-like flanking DNA sequences in the genomes of pathogens. Conjugative resistance dissemination in patient populations has also been inferred from sequence data. Conjugation, but not transduction, has been performed in in vivo experiments that corroborate the apparent clinical drug resistance transfer events (2, 60). Indeed, conjugative transfer of resistance genes may be the actual mechanism of transfer of such genes with flanking phage-like sequences. Furthermore, it has been noted that plasmids, as well as phage DNA, integrate by site-specific recombination into chromosomal tRNA sequences (20). Therefore, the role of phage in clinical environments may not be what is inferred from sequence data.

It remains noteworthy that virtually all of the well-known bacterial toxins associated with specific conditions appear to be encoded by mobile genetic elements, including phage. Examples are the toxins for diphtheria, anthrax, tetanus, botulism, cholera, toxic shock, scarlet fever, and exfoliating dermatitis (62). None of the mobile elements for these toxins have been shown to also carry antibiotic resistance genes. However, other genes have been presumed to be phage transferred, based on nearby sequences encoding phage integrases and helicases, but such sequences have been found in integrated conjugative transposons as well (6, 9). It has been shown in vitro that phages (with helper phages) can mediate transfer of *S. aureus* superantigen genes (62). Thirty years ago it was suggested that phages had mediated macrolide resistance transfer in *S. pyogenes*, but since then the evidence for this resistance transfer has been attributed exclusively to conjugation and conjugative transposition (4, 29, 78). The DNA that can undergo conjugative transfer is not limited in size as is the DNA that is packaged by a phage particle. This alone may be the important factor in rapid resistance gene transfer, especially in multiple resistance gene dissemination. Regardless, the potential role of transduction needs to be clarified.

Transduction may occur more readily in environments unlike the mammalian host. Conjugative transfer has been shown to occur within and upon the human host, while phages are typically recovered from sewage plants. For RK2-specific phages (phages that infect bacteria carrying the plasmid RK2), phage sensitivity and conjugation are inversely regulated. Spontaneous bacteriophage resistance occurs in a predictable small proportion of RK2-carrying bacteria growing on solid media at 37°C. If isolated by phage resistance, this phage-resistant subpopulation shows a concomitant decrease in conjugation frequency (45). This is more interesting in light of the fact that incoming phage DNA and exiting conjugative DNA appear to share the same bacterial membrane apparatus (34, 100). At least for RK2 plasmid-carrying bacterial cells, phage transduction appears to be more suited to nonclinical environments. Transduction may rarely facilitate antibiotic resistance gene transfer in clinical bacteria, but conclusive experiments have yet to settle the issue.

Conjugation

Conjugation is a mechanism for cell-cell transfer of DNA by passage through a specialized connecting structure called the membrane bridge. All essential conjugative processes are encoded by donor bacterium DNA. Rolling-circle replication in the donor cell produces single-stranded DNA that is pumped through the bridge into the recipient cell by an energy-driven process. The transferred DNA is made double stranded in the recipient cell. Unlike DNA transformation, DNA conjugation is unaffected by the presence of DNase in the milieu.

The uptake of bacteria into phagocytic vacuoles of cultured mammalian cells can lead to stable expression of bacterium-encoded DNA by the phagocytizing cell (32, 85). This is not conjugative transfer. It is DNA acquisition by bacterial uptake by mammalian cells. Bacterium-to-mammalian cell conjugation is DNase-resistant and unaffected by the presence of high levels of cytochalasin D, an inhibitor of phagocytosis (104). In the presence of cytochalasin D, recipient mammalian cells are unable to divide. These cells nonetheless serve as recipients for conjugation and resume cell division after the inhibitor is removed. From these and other observations, conjugation is considered a donor-specific, recipient-independent process; that is, any cell can serve as recipient. Conjugation is now more precisely defined and distinguished from transformation and transduction.

Conjugation as a mechanism of cell-cell DNA transfer has been studied as encoded by extrachromosomal plasmid DNA molecules. Such conjugative plasmids carry and disseminate antibiotic resistance genes to diverse bacteria, but nonplasmid conjugative transfer may be more ubiquitous in drug resistance dissemination than originally thought. Conjugative transfer functions can be encoded by chromosomal or extrachromosomal DNA. DNA molecules that can be mobilized for transfer do so by virtue of an *oriT* sequence from which the transfer process begins. Self-transmissible conjugative plasmids are large plasmids that often mobilize smaller plasmids out of the same cell. The larger helper plasmid supplies the gene products for the mating bridge and for DNA processing. The smaller plasmids require only the *cis*-acting *oriT* sequences that are recognized by enzymes which mediate relaxosome formation. The relaxosome, the relaxed protein-DNA complex, is essential for transfer by mobilization.

Plasmids are categorized according to replication-specific incompatibility (Inc) groups. Plasmids with the same or very similar replication regions belong to the same Inc group and are said to be incompatible. Incompatibility is attributed to cross-control of replication and ultimate dominance of one plasmid over the other. Incompatibility groupings are used to identify plasmid-encoded conjugation systems. This system of classification for conjugation systems is imperfect because it is based on similarities in plasmid replication, copy number control, and partitioning. These are properties related to vegetative replication, not properties of conjugation or conjugative replication. In fact, related conjugation systems may be encoded by plasmids of different incompatibility groups. Incompatibility indirectly impacts conjugation because maintenance of a newly acquired plasmid requires compatibility with any established, preexisting plasmid. This would impact potential recipients already carrying large plasmids. Plasmid incompatibility is not a limiting factor when selective pressure for acquired genes maintains those genes on other, nonplasmid DNA molecules.

The most studied conjugation systems are the RK2/RP4 (IncP) and F (IncF1) plasmids systems. The plasmid F was named for fertility factor and RK2 was named for resistance in *Klebsiella*. RK2 was found in a series of isolates from hospital burn unit patients in Birmingham, England. This plasmid was found in several *Pseudomonas aeruginosa* strains in the same hospital unit, with the result that RK2 was and is still also called RP1 and RP4. The 100-kb F plasmid is the prototype for narrow-host-range conjugative plasmids, and the 60-kb RK2 plasmid is the prototype for broad-host-range conjugative plasmids (Fig.18.2). RK2 is broad host range with respect to conjugation and also broad host range with respect to replication; it transfers DNA to diverse species and can replicate in most of these bacteria. F and RK2 were the first conjugative plasmids to be entirely sequenced (27, 64).

The IncF-like plasmids include the IncF series (IncFI, IncFII, IncFIII, and IncFIV) and the IncC, IncD, IncH, IncJ, and IncS plasmids. The IncP-like plasmids include IncP, IncM, IncQ, IncU, IncW, and IncT tumor-inducing (Ti) plasmids. A third group, similar to the IncP group but with a different pilus structure, could be called the IncT-like plasmids and include IncX, IncN, and IncT plasmids. A fourth group is the IncI group, with IncI, IncB, and IncK. This group is also similar to the IncP group, with its own pilus distinctions. What plasmid incompatibility groups are the most important, clinically? The question could be addressed in terms of antibiotic resistance plasmids and in terms of virulence plasmids. Neither has been well defined. Antibiotic resistance genes are rarely found in IncFI plasmids but commonly found in IncFII plasmids. Most virulence plasmids of common enteric organisms are IncFI plasmids. Antibiotic resistance plasmids have been identified in the IncFII, IncH, IncJ, IncM, IncN, IncP, and IncW groups. There are probably many hybrid plasmids among the large plasmids in clinically significant organisms. Analysis of the IncHI1 antibiotic resistance plasmid R27, of *Salmonella enterica* serovar Typhi, has shown that it is a chimera of F (IncF1) and RK2 (IncP) plasmids (49). It is not unusual for clinical strains to carry large plasmids encoding virulence factors and antibiotic resistance.

Plasmid conjugation is normally a replicative event. The transferring plasmid is maintained in the donor and replicated into the recipient. DNA exits a donor cell in conjugation; DNA also enters the recipient cell in conjugation, as well as in transformation. DNA enters into a bacterial cell via *com* gene transformation systems. In conjugation, the donor cell builds the membrane bridge. Is there a polarity of structure, based on orientation in the membrane, or, to put it another way, do the membrane structures provide two-way traffic? For conjugation, the matter was unclear until Dave Figurski's group engineered a system using an RK2 plasmid lacking *oriT* that could mobilize other plasmids but was incapable of self-transfer (84). They demonstrated that transfer proceeds in a one-way fashion from donor to recipient through a membrane bridge. Figurski's group therefore demonstrated that there is indeed membrane bridge polarity. In a phenomenon called retrotransfer, after a recipient cell receives a conjugative plasmid (becoming a potential donor), the cell builds a membrane bridge and transfers the plasmid "back" to the original donor. Retrotransfer occurs only with a membrane bridge of proper polarity. (Retrotransfer events are normally rare because of the phenomenon called "surface exclusion.") DNA entry is also expected to traverse a polar *com* membrane structure into cells that are competent for natural transformation.

What is the significance of the "strandedness" of transferred DNA? DNA is transferred in single-stranded form in conjugation and in natural transformation. Conjugative transfer occurs across divergent bacterial species, and single-stranded DNA is advantageous because it is not recognized as foreign DNA by recipient restriction/modification systems. The single-to-double strand replication and subsequent modification, such as methylation, is accomplished in the recipient cell. The new DNA molecule is therefore not regarded as foreign DNA in the recipient cell.

In natural transformation, the surface-attached double-stranded DNA is unwound and imported in single-stranded form (15). DNA transformation in the laboratory is accomplished by methods for double-stranded and preferably supercoiled DNA. The three major mechanisms for antibiotic resistance gene dissemination are conjugation, conjugative transposition, and natural transformation. All three of these mechanisms mediate exclusively single-stranded DNA transfer. These observations are consistent with the view that restriction/modification systems do impact DNA exchange in the natural and clinical worlds. They also strengthen the view that conjugation and

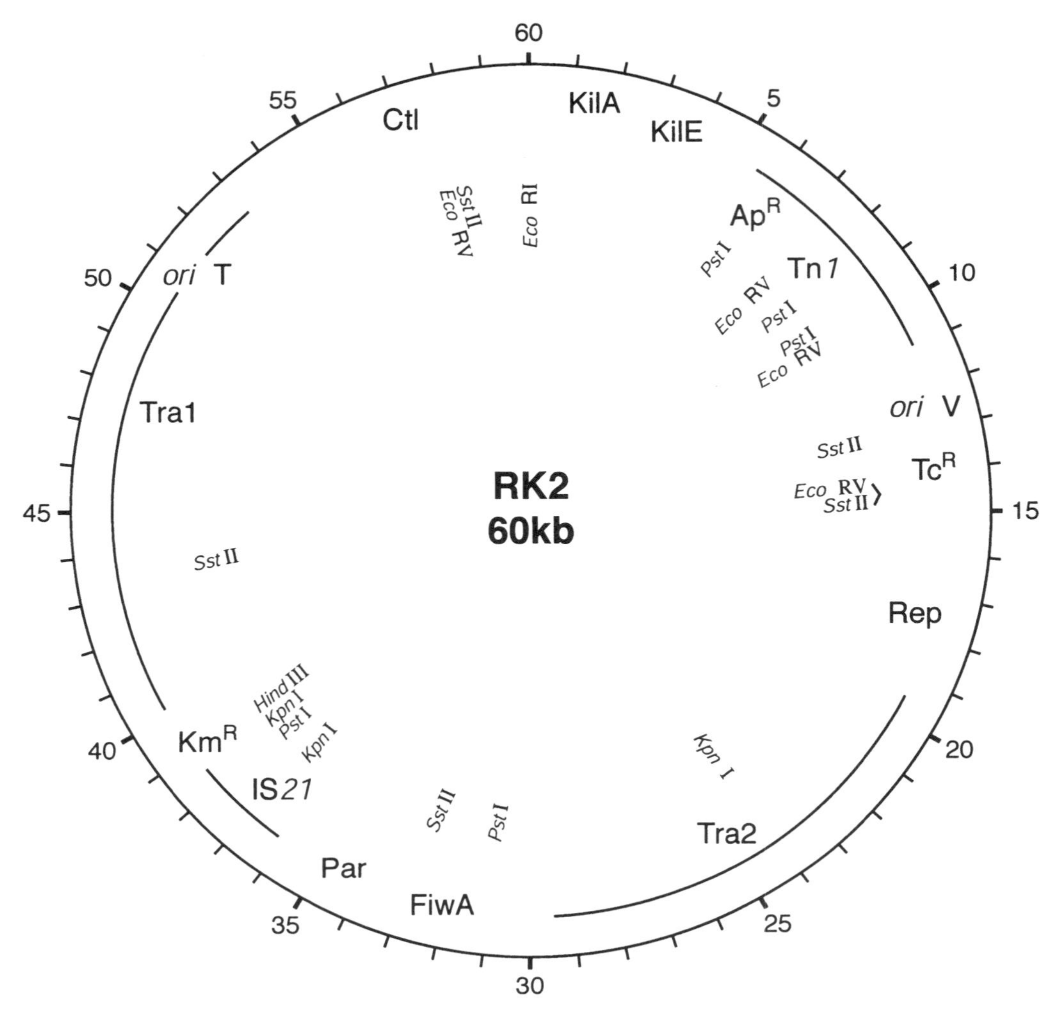

Figure 18.2 Genetic map of the RK2 plasmid. Conjugative transfer genes are clustered in two regions, Tra1 and Tra2. The transfer origin, *oriT*, is a 250-bp noncoding region within Tra1. The genes for DNA processing and coupling are in Tra1, and the genes for the type IV secretion system are in Tra2. Also shown are the genes encoding resistance to tetracycline (Tc^R), kanamycin (Km^R), and ampicillin (Ap^R), the origin of vegetative replication (*oriV*), and selected restriction sites.

conjugative transposition are well suited for MDR dissemination.

Some bacterial strains are superior conjugative donors, promoting high transfer efficiency, and some strains appear to be more efficient recipients. Little is known about the factors affecting efficiency for donor or recipient, including cell-cell interaction, relaxosome interaction with the membrane bridge, and DNA survival in the recipient. Interestingly, efficient plasmid conjugative transfer does not require ongoing donor cell viability. *E. coli* maxicell donors appear to be as efficient as viable donors in plasmid transfer (36; V. Waters, unpublished data). DNA transfer has been assessed in bacterium-to-yeast mating experiments using donor *E. coli* carrying a shuttle vector plasmid as mobilized by one of three different plasmid helper systems. Results demonstrated that only one of the plasmid systems, the RK2 IncP system, successfully transferred DNA to the yeast cells (5). These results support the accepted view that RK2 conjugation is highly efficient and broad host range.

Bacterial conjugation may be as ubiquitous as the microbial world. It occurs between free-living cells, between cells within organisms, within the digestive vacuoles of parasites (79), within the insect midgut (38, 39), and within phagocytic vacuoles of leukocytes (24). There are conjugative bacteria in saltwater and freshwater environments. It is noteworthy that conjugation is not comparable to eukaryotic RNA cell-to-cell transfer. In *Caenorhabditis elegans*, in which double-stranded RNA transfer has been studied, one important distinction is that the transfer proceeds in a two-way fashion (90).

BACTERIAL CONJUGATION AND MDR

While the important gene transfer mechanisms in clinical bacteria are transformation and conjugation, the most important MDR transfer mechanism is clearly conjugation. Multiple genes encode MDR, so a mechanism with potential for high-molecular-weight DNA transfer is not surprising. DNA transferred by transformation or by phage transduction is size limited. There is another kind of MDR that does not involve horizontal gene transfer, that is, the drug efflux systems. Most efflux systems are specific for the bacterial species. However, an efflux system was recently found to be encoded by an MDR plasmid, an IncP broad-host-range conjugative plasmid. This finding indicates that there probably are efflux systems serving different bacterial species. The plasmid is an interesting variation on the theme of MDR dissemination (89).

This chapter reviews aspects of bacterial conjugation in terms of basic science and clinical relevance. As a phenomenon of bacterial genetics, conjugation has recently regained interest. This is due to rapid genome sequencing, by which conjugative element genes are routinely discovered. It may not be appreciated, however, how or why conjugation is more efficient than other mechanisms. It can be most easily illustrated by comparing conjugation with mutation as a force for drug resistance emergence. Conjugation transfers several genes as efficiently as it transfers one gene. Efficiency is measured by transfer frequencies, the number of transfer events in a given time period per bacterial cell. For RK2-mediated *E. coli*-to-*E. coli* 1-h mating experiments, using a donor-to-recipient ratio of one to four, transfer frequencies are typically between 0.1 and 1.0 per recipient cell per generation. This means that between 10 and 100% of the recipients acquire genes. Conjugation rate is not significantly affected by the size of the DNA transferred in 1 h. Mutation rates can vary from 10^{-5} to 10^{-11} per gene per cell per generation. For MDR acquisition, the rate for simultaneous mutations is the product of each rate, as follows.

	1 gene	3 genes
rate of gene acquisition by conjugation:	10^{-1}	10^{-1}
rate of gene acquisition by mutation:	10^{-6}	10^{-18}

This dramatic difference illustrates the capacity of conjugative gene acquisition and its potential for rapid response to antibiotic pressure. Furthermore, conjugation efficiently transfers DNA in the megabase range. It was by interrupted conjugation experiments that the entire *E. coli* chromosome was originally mapped. Conjugative transfer of large DNA molecules increases homologous recombination and gene targeting frequency. Also, RecA-mediated homologous recombination pathways are found in diverse bacteria. Broad-host-range conjugative transfer therefore provides better DNA expression and survival rate in the recipient than DNA delivery by other mechanisms. These two features are summarized as follows: (i) conjugation transfers single-stranded DNA, the preferred substrate for recombination; (ii) conjugation transfers DNA of unlimited size, for potentially large regions of DNA homology.

Better targeting of DNA to its native site by homologous recombination enhances the survival of the DNA arriving in the recipient. By targeting to homologous regions, the organism is spared the hazard of insertional inactivation of essential genes by random integration. Insertional inactivation has been a troublesome outcome in gene therapy approaches, as in DNA delivery by viral vectors. Conjugative transfer of DNA is more likely to be targeted to regions of extensive homology and thereby avoid suicidal transfer events. If the gene provides greater fitness, the organism outgrows the other, less fit bacteria (10).

BACTERIAL CONJUGATION: DNA PROCESSING, TYPE IV SECRETION AND COUPLING

Studies of RK2 and F have led to the following model for the events of bacterial conjugation.

1. Donor and recipient cell contact, cell-cell stabilization and membrane bridge formation
2. Relaxosome formation by DNA nicking and protein attachment at the *oriT*
3. Relaxosome coupling to the membrane bridge in the donor cell
4. Rolling-circle replication and single-stranded DNA transfer from donor to recipient
5. DNA recircularization at the membrane bridge and release of DNA into the recipient cell
6. Complementary strand synthesis and vegetative replication in the recipient
7. Conversion of recipient cell to donor cell and surface exclusion of another conjugative event

The events of conjugation are grouped here into seven general steps. The first two steps could be reversed in sequence, and the overall chronological order is uncertain for this rapid process. For the F plasmid, cell contact followed by pilus retraction would be an initial event. Studies with RK2 are consistent with a model in which pili are not retracted. Bacterial conjugation is a cell-to-cell DNA transfer mechanism encoded by DNA in the donor bacterium. With F plasmid, the transfer genes are carried in one contiguous genetic region comprising about one-third of the entire plasmid. In RK2, conjugative functions are encoded in two genetic regions, Tra1 and Tra2, separated by about 10 kb on the RK2 map (Fig. 18.2). Tra1 encodes functions involved in "DNA transfer replication" (Dtr), and Tra2 encodes what has been called mating pair formation (Mpf). The Dtr region encodes the events of relaxosome formation, transfer replication, and the actual DNA transfer. The generalized term recently given to the Mpf region is type IV secretion systems (T4SS) based on sequence homologies with other DNA and protein transfer systems.

Erich Lanka and his group have provided an in-depth analysis of the RK2 sequence, genetic regions and encoded functions of RK2. For a better description of conjugative gene function, three processes, rather than two, were named. To DNA processing and T4SS, they added "coupling." Coupling proteins hook up the relaxosome to the membrane bridge. In RK2, coupling genes are carried in the DNA processing region, but the coupling process can also be considered part of T4SS. The conjugative processes for the donor cell are therefore as follows.

1. DNA processing (DNA relaxosome formation and DNA transfer by rolling-circle replication)
2. Type IV secretion (mating pair stabilization or membrane bridge formation)
3. Coupling (relaxosome DNA interaction with the membrane bridge)

The previously discussed seven steps provide a more detailed list of conjugative processes. These three processes present a more conceptual and functional understanding. Neither grouping gives a chronological order of events. However, the relaxosome is probably preformed, before a mating signal begins a cascade of events resulting in DNA transfer. Coupling proteins are thought to form part of the pore structure of the membrane bridge and attach the relaxosome to the bridge. A cell-cell contact signal may be transmitted down the bridge structure to the inner membrane, where transfer is then triggered. The single-stranded nick may be the response to the cell contact signal, followed by DNA unwinding and replicative displacement of the nicked strand. The displaced DNA strand is transferred into the recipient cell (Fig. 18.3).

DNA Processing

DNA processing includes relaxosome formation, rolling-circle replication, and DNA transfer. Three proteins interact at the transfer origin in relaxosome formation. The relaxosome is a plasmid molecule whose structure

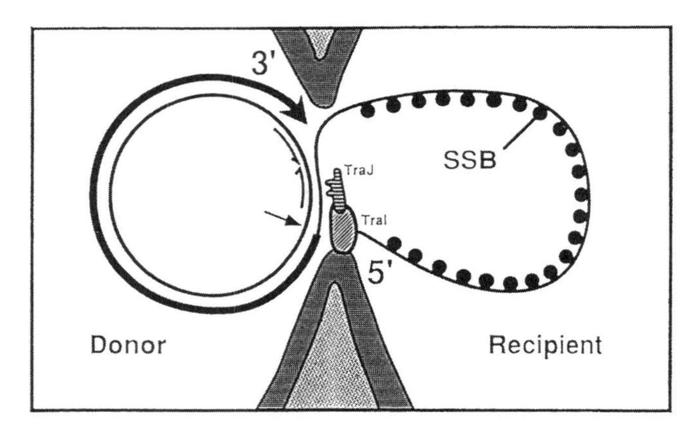

Figure 18.3 Conjugative transfer in cartoon form to illustrate cell-to-cell DNA transfer in concert with rolling-circle replication. Single-stranded DNA cleavage at *nic* precedes rolling-circle replication. Relaxation complex proteins TraI and TraJ are shown interacting with the inverted repeat and nick region, and the arrow indicates the nick site. Single-stranded binding (SSB) proteins are shown bound to the incoming single-stranded DNA.

is relaxed rather than supercoiled. The relaxed structure is maintained by covalent and noncovalent protein interaction. The TraI relaxase nicks DNA at the nick site *nic* (Fig. 18.3), covalently attaching to the 5′ end. It remains attached for one replicative cycle, when it mediates ligation of the 5′ and 3′ DNA ends. Rolling-circle replication begins and ends at the *nic* site and provides the transferred strand. The relaxase enzyme is considered a relaxase, a ligase, and a strand transferase because it mediates DNA nicking, strand exchange, and recircularization. The relaxosome is also called a relaxation complex, a DNA and protein complex which shows a characteristic density in CsCl density gradient centrifugation.

The nick site of RK2 is within the 250-bp origin of transfer (*oriT*). More precisely, the nick site (G′C) is part of a 6- to 10-bp region called the "nick region":... CCTATCCT-G′C... (Fig. 18.4). The nick region shares DNA homology with nick regions of related conjugative plasmids such as the *Agrobacterium tumefaciens* plasmid pTiC58 (100, 102). Homologous nick regions of such conjugative plasmids and of phages and certain small plasmids have been grouped together and presumably have a common ancestry (Fig. 18.5). The nick region, spanning about one helical turn, is important because it is the sequence recognized by relaxosome proteins in order to accomplish precise nicking, transfer, and ligation.

The 250-bp *oriT* of RK2 has all the *cis*-acting DNA structures required for full transfer efficiency by mobilization by an RK2 helper plasmid, including sequence and tertiary structure recognized by the relaxase and coupling

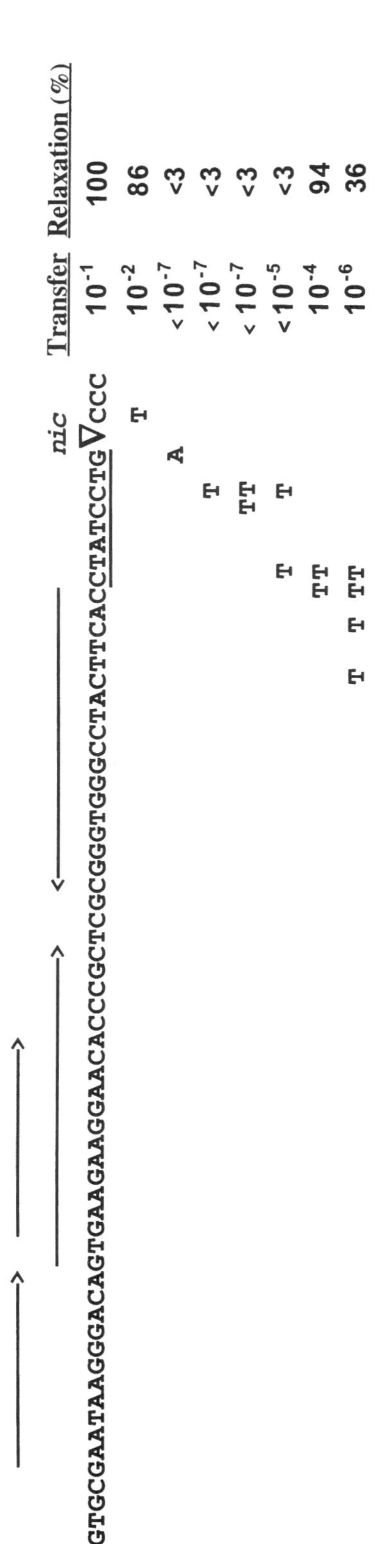

Figure 18.4 DNA transfer and DNA relaxation of RK2 *oriT*-containing shuttle vectors with mutations in the nick region and surrounding DNA. G-C–to–A-T transitions were generated by hydroxylamine mutagenesis. The wild-type strain is shown with high transfer proficiency and 100% relaxosome formation, and mutant donor base pair change and values are shown below for comparison. *E. coli*–to–*E. coli* conjugative transfer in a 1-h experiment was measured as the relative number of recipient cells that received DNA (100).

proteins. In RK2, one of the three relaxosome-forming proteins, TraK, serves as a histone-like protein (65). Another, TraJ, binds to the 38-bp imperfect inverted repeat near the nick site. In RK2, the distal arm of the inverted repeat overlaps two directly repeated sequences (Fig. 18.4). These direct repeats and the asymmetry of the inverted repeat orient TraJ to the arm near the nick site. TraJ simultaneously interacts with the inverted repeat and the relaxase, TraI, to position it for accurate nicking.

An interesting aspect of relaxation is the role of the 38-bp inverted repeat. When the DNA is double stranded, this inverted repeat forms an asymmetric cruciform structure, and when the DNA is single stranded, the inverted repeat forms a hairpin structure. It has been established for the IncN plasmid R388 that the analogous IncN *oriT* inverted repeat forms a cruciform structure (33). With the cointeraction of TraJ, the TraI relaxase/ligase recognizes the single-stranded hairpin structure for ligation of the newly synthesized DNA (63). TraJ also binds double-stranded DNA, and this complex is recognized by the TraI to facilitate accurate nicking. Thus, the DNA structure, whether double-stranded or single-stranded, directs the appropriate activity of TraI.

In RK2, the nick site is 9 bp from the base of the 38-bp inverted DNA repeat (Fig. 18.4). These sequence structures, the inverted and direct repeats, the nick site, and the DNA connecting them (the signature nick region) are located in the approximate center of the 250-bp *oriT*. Point mutations within the nick region reduce nicking and transfer efficiency. Point mutations near the nick region may slightly affect nicking efficiency and subsequent transfer, and point mutations beyond the nick region and repeats do not affect nicking (Fig. 18.4). DNA relaxation and conjugation experiments with such mutants have shown a direct correlation between nicking and transfer (100). Without nicking and relaxosome formation, conjugative DNA transfer becomes undetectable in an otherwise highly efficient conjugation system (102). These studies support the idea that conjugation is a precisely regulated process. Without strict control, DNA could conceivably "leak" out of the cell through the membrane bridge. Until it is known to what extent the donor is prepared for transfer before a mating signal, the precise order of the earliest conjugative events will be uncertain. These and other studies suggest that the relaxosome and the mating bridge are preexistent and reside at the cell membrane bridge at the time of cell-cell contact.

Conjugation may then begin with single-stranded DNA nicking to produce the 5′ and 3′ DNA ends. The 3′ end serves in priming DNA extension, and the growing DNA strand displaces the 5′ end DNA from the circular molecule as it rolls around the template strand (Fig. 18.3). The

IncP Conjugative *(oriT)* Nick Region

RK2	c t t c a C C T **A** T **C** C **T G** ▼C c c g g c
R751	c t t c a C A C **A** T **C** C **T G** ▼C c c g c c
pTiC58	c a c a a T A T **A** T **C** C **T G** ▼C c c a c c
pTFFC	c a a c g G T C **A** T **C** C **T G** ▼T a t t g c
R64	a a t t g C A C **A** T **C** C **T G** ▼T c c c g t

RC Replication *(oriV)* Nick Region

pC194	t c t t t C T T **A** T **C** T **T G** ▼A t a a t a
pUB110	t c t t t C T T **A** T **C** T **T G** ▼A t a c a t

Bacteriophage Nick Region

ΦX174	g c t c c C C C **A** A **C** T **T G** ▼A t a t t a

Figure 18.5 DNA nick regions, flanking sequences, and nick sites (...G ▼...) of IncP *oriT* and related sequences. Shown are sequences of gram-negative bacterial plasmids, including plant tumor-inducing plasmids, gram-positive plasmids, and phage ϕX174 DNA. All of the molecules represented by these sequences have rolling-circle replication originating at the nick site. Consensus nucleotides are in upper case, and four invariant nucleotides are in boldface.

displaced, transferring DNA becomes protein bound during this process. In RK2, there are potentially three DNA-binding proteins: single-stranded binding (SSB) protein, RecA protein, and the RK2 plasmid-encoded primase. RecA and SSB proteins bind cooperatively but can also bind competitively, because RecA is capable of displacing SSB (23). In RK2 conjugation, the RK2-encoded primase molecule is transferred to the recipient cell, where it primes single-stranded to double-stranded DNA synthesis (48, 74). There are in fact two distinct RK2-encoded primase molecules, synthesized from start sites of overlapping genes of the Tra1 region. With *E. coli* recipients, neither plasmid-encoded primase is needed since the *E. coli* chromosome-encoded DnaG can serve to prime RK2 DNA synthesis. Without DnaG, in other bacterial recipient species, one of the RK2-encoded primase molecules may be needed. This illustrates how RK2 can adapt to different recipients and how its transfer can be appreciated as broad host range.

The mating signal for *Agrobacterium tumefaciens* T-DNA transfer to plant cells is part of a two-component regulatory system. The DNA transfer of some conjugative plasmids of gram-positive bacteria is pheromone induced. The cell-cell contact event may be the mating signal in RK2 and other plasmid systems, and perhaps cell contact serves in addition to pheromones or other signals. Before the signal, relaxosomes are thought to be poised for transfer on the inner side of the membrane bridge. Rolling-circle transfer replication originating at the *oriT* and vegetative replication originating at the *oriV* both take place at the inner cell membrane (58). The detection of inner membrane-attached *oriT* DNA is consistent with the idea that plasmid molecules are at membrane bridge sites (Waters, unpublished). Evidence for this has also been provided by cell permeability data. Donor cells with membrane bridges but without relaxosomes are more permeable to ATP and lypophilic compounds than donor cells with both relaxosomes and membrane bridges (17). This suggests a membrane pore structure that is permeable when lacking a relaxosome to serve as a gate or barrier. A relaxosome then would have a critical role as a gate for each membrane bridge. This is a fascinating and fitting hypothesis. To further investigate, one might be able to assess the dynamic flux of plasmids in time and space, to discern plasmids poised for conjugative transfer or for plasmid partitioning as bacterial cells divide.

T4SS

T4SS are a family of transporters for the transfer of protein and DNA into and out of the bacterial cell. This family shares genetic homology and functional analogy, and plasmid conjugation is thought to predate the other systems (14). T4SS are grouped as families, and their subclasses are accumulating (21). In 1994 Steve Winans' group found homology between the genes for conjugative plasmid transfer and for whooping cough toxin secretion (72). Homologies had been noted previously among plasmid conjugation systems, those of the IncP, IncW, and IncT plasmids. After Winans' finding, homologies were also found between conjugation proteins and proteins of secretion systems such as CagA of *H. pylori*. The small genome of *H. pylori* is now known to encode four distinct T4SS.

The prototype T4SS is the plant tumor-inducing (Ti) plasmid conjugation system of *A. tumefaciens* (97, 108). These large plasmids carry two types of conjugation systems, one for bacterium-to-bacterium DNA transfer and the other for bacterium-to-plant DNA transfer. The latter transfers T-DNA into wounded plant leaf tissue that ultimately induces tumor formation and provides a carbon source for bacterial growth. T4SS contribute to the virulence of many varieties of pathogenic bacteria, including *A. tumefaciens*, *Bartonella* species, *Bordetella pertussis*, *Brucella* species, *Campylobacter jejuni*, *Coxiella burnetii*, *E. coli*, *H. pylori*, *Legionella pneumophila*, *N. gonorrhoeae*, *Rickettsia prowazekii*, and *Wolbachia* species (19, 76, 98). Conjugative plasmid T4SS are grouped by homologies. There are IncF-like, IncP-like, and IncI-like systems (50). In light of more recent experiments, secretion systems could be also be classified according to their mediated events, as follows:

1. Bacterium-to-bacterium conjugation, with cell-cell contact and transfer of DNA
2. Bacterium-to-eukaryotic cell conjugation, with cell-cell contact and transfer of DNA
3. Bacterium-to-eukaryotic cell effector secretion, with cell-cell contact and transfer of protein
4. Bacterium-to-extracellular milieu transfer of protein or DNA
5. Extracellular milieu-to-bacterium uptake of DNA

For gene transfer involving bacterial pathogens, examples of some of these events include DNA transfer from bacterium to bacterium, such as the conjugation among enteric bacteria; transfer from bacterium to extracellular milieu, such as the conjugative-like DNA export in *N. gonorrhoeae*); and from extracellular milieu to bacterium, such as the natural transformation of *Streptococcus pneumoniae*. An example of bacterium-to-eukaryotic cell transfer is given by *H. pylori* CagA protein injection into the adjacent gastric epithelial cells. Transfer occurs from bacterium to extracellular milieu to eukaryotic cell with the export and uptake of pertussis toxin.

These events have different levels of complexity and may not be readily regarded as comparable. In studying T4SS family members, we do find important parallels and potential ways to target subprocesses. For clinical usefulness and for basic science, the comparative approach can add information regarding conserved mechanisms, active-site amino acids, similarities and differences in DNA versus protein transfer, and molecular phylogenetic trees. Over the years, researchers of plasmid conjugation who have demonstrated the power of the comparative approach include Fernando de la Cruz, Dave Figurski, and Erich Lanka (55, 71, 73).

There are good reasons to keep the term "type IV secretion system" for conjugative export. Mating pair stabilization by pilus interaction and membrane bridge pore construction are two distinct functional processes encoded by the same set of genes. This has presented a conundrum. Bacterial pili, considered essential for mating pair formation, are long appendages that extend from the surface of the cells and mediate attachment to other cells or to surfaces. However, it has been found that not all of these subprocesses are absolutely necessary for *E. coli*-to-*E. coli* conjugation.

In defining essential RK2 conjugation subprocesses, it has been argued that pilus-mediated mating-pair stabilization should be reconsidered because mutant studies suggest that it is not required for transfer. Conjugation requires cell-cell contact, but this contact need not be mediated by visible pili. Mutant donor cells incapable of constructing extended pili have basal membrane bridge structures that can function as phage receptors. Cells carrying the mutant gene (*trbK*) are proficient in conjugation and in phage adsorption and uptake but have no visible pili (31, 35, 65). These studies are consistent with what is seen in photomicrographs of *E. coli* mating pairs (Fig. 18.6).

The type of bacterial pilus of a conjugation system can impact donor efficiency, surface exclusion, recipient host range, phage sensitivity, and optimal conjugation conditions. F plasmid conjugation is optimal in liquid media, while RK2 conjugation is optimal on solid media. The F pilus retraction model, with thin flexible pili, does not apply to the thick brittle pili of the IncP plasmids. There are additional finer distinctions among similar pili types, but the two general types are the thicker, more brittle pili and the thinner, more flexible pili (28, 65). IncP pilus interaction may be a first step in mating-pair formation in low cell density. In liquid, pilus interaction could facilitate mating pair formation between rapidly moving donor and recipient cells. In high cell density, extended pili are unnecessary (and perhaps counterproductive) in stabilizing cell contact. The F pilus also serves to recognize an enteric recipient cell surface, to perhaps "reel in" a prospective recipient. Such cell contact would seem to work best in a liquid milieu, which is the environment for higher-efficiency transfer for the F plasmid. For RK2 high-efficiency conjugation, bacterial cells are placed in high density on a membrane filter on an agar plate. Filter mate conditions for IncP plasmids might be considered comparable in cell density to biofilms found in nature. Notably, the RP4 plasmid was first found in the prototypic biofilm-forming organism *P. aeruginosa*.

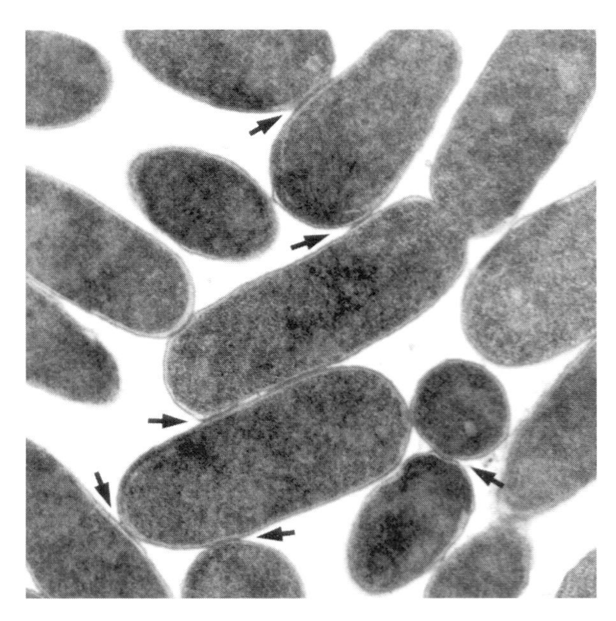

Figure 18.6 Electron micrograph of conjugating cells as mediated by RK2 on a solid surface. *E. coli*–to–*E. coli* filter mating is visualized by cryofixation. Arrows indicate junctions between donor and recipient cells, showing the points of contact and fusion of outer membranes. Fusion events require the presence of the RK2 plasmid in the donor cells (73).

Coupling

Coupling proteins attach a DNA/protein relaxosome to the cytosolic side of the membrane bridge. RK2 coupling proteins TraF and TraG are essential and interact with DNA and protein (100). TraG, the prototype for all TraG-like proteins, interacts with TraI relaxase and binds to DNA without sequence specificity (81). TraI "nickase" interacts with *oriT*-bound TraJ, nicks at the *nic* site, and covalently attaches to the 5′ DNA end. By simultaneous interaction with TraI, DNA, and the bacterial membrane, TraG attaches the relaxosome to the inner membrane. TraG proteins of some conjugation systems are interchangeable. TraG-like proteins are found in all conjugation systems, while TraF-like coupling proteins are not found in all conjugation systems. TraF, but not TraG, is required for RK2-specific phage infection (101). Sequence homology to T7 phage DNA polymerase, a processive enzyme, suggests that TraF facilitates DNA translocation for phage DNA injection into the cell and for conjugative DNA transfer out of the cell (37, 101). In RK2, TraF and TraG are cotranscribed and cotranslated and may functionally interact in a 1:1 molecular ratio at the membrane bridge. Homologies with other proteins also suggest a role for TraG in DNA translocation (30). Tn*PhoA* fusion experiments suggest that TraG resides entirely at the inner membrane and that TraF is further out, in contact with both inner and outer membrane preparations (Waters, unpublished). TraG proteins have inner-membrane-spanning and cytoplasmic domains, consistent with this coupling activity.

Coupling function may be explained by plasmid mobilization, the phenomenon in which non-self-transmissible plasmids are mobilized by self-transmissible "helper" plasmids. The native helper plasmid encodes the conjugative membrane bridge, coupling proteins and DNA processing functions. The mobilized plasmid encodes an *oriT*, and either or both plasmids encode cognate DNA relaxosome functions. As a helper plasmid, RK2 can mobilize any $oriT_{RK2}$ -carrying plasmid, or a relaxosome sufficiently related to the RK2 relaxosome. Conjugative transfer of such a relaxosome mediates transfer via coupling proteins that can sufficiently recognize the relaxosome in order to attach it to the membrane bridge.

In mobilizing IncQ plasmid RSF1010 by the IncP plasmid RK2, TraF and TraG are required in addition to the membrane bridge components (34, 81). RSF1010 encodes cognate relaxosome-forming proteins, and the RK2 helper plasmid supplies IncP coupling proteins and the membrane bridge. RSF1010 mobilization illustrates the role of coupling proteins and how they are sometimes interchangeable. TraF and TraG are essential for attaching either homologous or heterologous relaxosomes to the membrane bridge. Coupling proteins keep the relaxosome at the base of the bridge until the mating signal initiates transfer. The greater the affinity of the coupling proteins for the relaxosome and the bridge, the greater the transfer efficiency (55). Higher transfer efficiencies were found in experiments in which the RK2 TraG protein was used to replace the native TraG-like protein of the IncW group plasmid R388 (11). The RK2 TraG coupling protein provided better transfer efficiency for R388 DNA than did the R388 native plasmid coupling protein.

In nature, antibiotic resistance genes are found encoded on self-transmissible large plasmids and smaller mobilizable plasmids. Small plasmids without the capacity for self-transfer can be shuttled among pathogenic strains and thereby disseminate antibiotic resistance. One example of this is *Klebsiella pneumoniae* EK105, a causative agent of sepsis in neonates (91). This organism carries gentamicin resistance encoded on a large conjugative plasmid that mobilizes a small plasmid encoding resistance determinants for ampicillin and aminoglycosides.

DNA REPLICATION OF CONJUGATIVE PLASMIDS

There are three kinds of DNA replication to consider in regard to conjugative plasmids: (i) vegetative replication, originating at *oriV*; (ii) transfer replication, originating at *oriT*; and (iii) single- to double-stranded replication, in the recipient, originating at multiple sites.

Vegetative and transfer replication, the first two types, occur in the donor cell. Vegetative replication maintains the plasmid in the cell often by means of a bidirectional theta style of replication (Fig. 18.7). As noted, conjugative transfer replication is by rolling-circle replication. Vegetative replication and single-stranded-to-double-stranded DNA replication, the third type, occur in the recipient cell after transfer. The latter is a repair-like replication, originating at several sites on the newly acquired DNA. For the 60-kb RK2 plasmid, *oriT* and *oriV* are separated by about 20 kb of DNA (Fig. 18.2). In RK2, there is only one vegetative replication region, within which is the vegetative origin, *oriV*. Large conjugative plasmids may have two or three vegetative replication regions. One replication region may dominate in certain environments, and some replication regions are nonfunctional.

Theta replicating DNA can resemble the Greek letter θ in electron photomicrographs and is visualized in two dimensions as a bidirectional growing loop within the plasmid DNA circle. The center of the loop enlarges as the two replication forks separate. Replication terminates where the forks meet, completing the duplication of the entire molecule.

The rolling-circle replicative machinery rolls around a circular DNA "plus" strand. Single-stranded cleavage of the "minus" strand supplies the 3′ end that is then extended around the circle, with the plus strand serving as a template (Fig. 18.7). Rolling-circle replication is a simple, continuous, and specialized form of replication. In addition to replication for conjugative transfer and for vegetative replication of certain small plasmids, rolling-circle replication is used by phages. The prototypic systems are ΦX174 for bacteriophages and pT181 for the small plasmids (61). In fact, the nick region for initiation for ΦX174 replication is in the RK2 family of nick regions (Fig. 18.4). For phages, rolling-circle replication produces many genome copies for packaging progeny phages. For the small plasmids, it is the mechanism for vegetative replication. Minus strands are produced and then used as templates for complementary plus strands. Late mechanistic differences reflect the outcomes for phage or plasmid propagation. Infecting phages produce large numbers of progeny phage particles, while plasmids such as pT181 must regulate replication within plasmid copy number upper and lower limits. In both systems, the final product in the host cell is a double-stranded DNA genome.

In conjugative rolling-circle replication, nicking and ligating are coordinated for each unit length single-stranded molecule. Due to the fact that rolling-circle replication was initially studied in the phage λ, production of DNA multimers has been considered a hallmark of rolling-circle replication. In late-stage λ replication, a long multimeric DNA product appears as a tail extending out from the circular template. It has therefore been called sigma (σ) replication to describe the growing "tail" observed microscopically. In λ production, a head-full packaging process converts long multimeric DNA into unit lengths of 50 kb. The DNA is made double stranded, and each unit length is cut and packaged within a protein phage head. For rolling-circle replication in the small plasmids, each newly synthesized

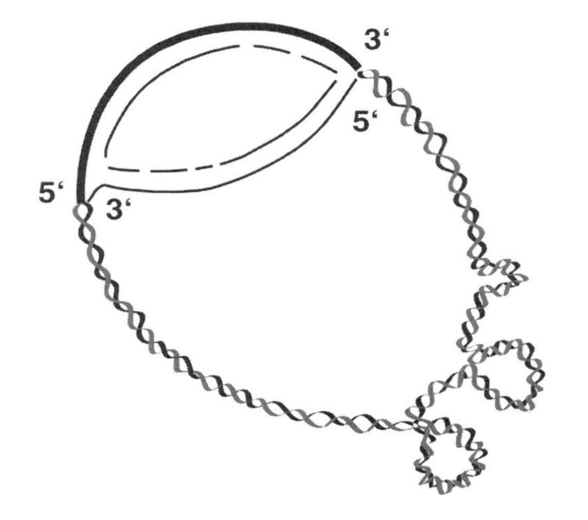

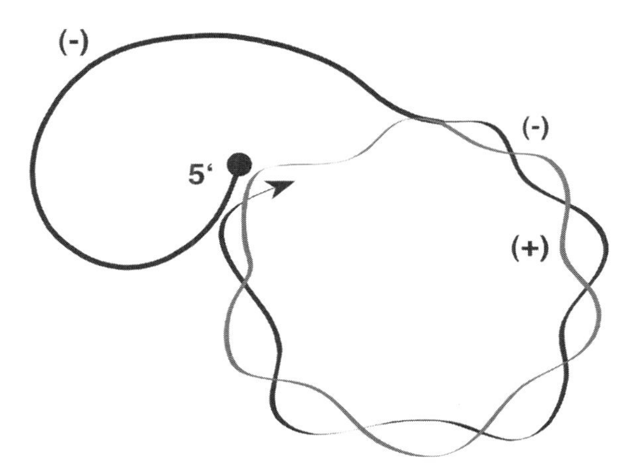

Figure 18.7 Theta (left) and rolling-circle (right) DNA replication in cartoon form. In theta bidirectional replication, growing forks enlarge the loop, resulting in a structure that resembles a Greek letter θ. Theta replication is a common form for plasmid vegetative replication. In rolling-circle replication, the 3′ (–) end is indicated by the arrow. The 3′ end uses the (+) strand DNA as the template as it leads continuous DNA synthesis around the template.

unit length is recircularized without a multimeric intermediate.

Different replication regions and mechanisms can become a source of confusion. Some plasmids can be mobilized by more than one mechanism. For example, mobilization can involve both site-specific recombination and cointegrate formation with the larger self-transmissible plasmid (102). One mobilizable plasmid, pC221, has two different rolling-circle replication systems, a vegetative replication region with a *rep* gene and an *oriV* origin and a transfer replication region with *mobA/mobB* genes and an *oriT* origin (102). Such plasmids illustrate how cointegrate formation could have created the large conjugative plasmids that have multiple replication regions.

Using the IncQ plasmid R1162, Parker and Meyer and Parker et al. have provided good evidence that replication originating at the *oriT* is indeed rolling-circle replication (66, 67). By manipulating the position and number of *oriT* sequences in plasmid constructs, they first demonstrated that multimeric forms are not produced during conjugation. The first *oriT* sequence supplied the initiating nick, and a downstream *oriT* served as a termination site. Then, they determined that the nicked strand was indeed the strand that transferred to the recipient. They also found that conjugative replication and transfer occur in a system in which the plasmid is incapable of vegetative replication. The capacity for vegetative replication in either the donor or the recipient cell was found to be unnecessary for bacterial conjugation. This experimentally accomplished vegetative-replication-free conjugation affirms models for bacterial conjugation. It also shows that plasmid conjugation and conjugative transposition are logical variations on the conjugation theme.

The plasmid RK2 is broad host range with respect to conjugative transfer and is broad host range with respect to vegetative replication. The RK2 conjugative host range is in fact "broader" than its replicative host range, that is, it can transfer into recipients in which it cannot vegetatively replicate. Such transfer events are shown using a shuttle vector plasmid constructed to include a recipient-specific replication region. The added vegetative replication capacity ensures that the shuttle vector will replicate upon arrival in the recipient cell. Suicide vectors, plasmids lacking the capacity to replicate in an intended recipient, can transfer and be detected by selecting marker genes encoded by the suicide plasmid. Marker genes gain stability by insertion into the genome.

Experiments using shuttle vector constructs suggest that bacterial conjugation has unlimited host range, that is, it can deliver DNA to any biological cell. Conjugation experiments with RK2 and similar broad-host-range donor systems have been used with a wide variety of recipient cells, including diverse gram-negative and gram-positive bacteria, mycobacteria, different species of yeast, plant tissues, and cultured mammalian cells. In bacterium-to-mammalian-cell experiments, it was found that mammalian recipient-specific replication was not necessary for successful transfer of the shuttle vector (104). The transferred DNA is stabilized by integration into mammalian cell DNA. Bacterium-to-mammalian-cell conjugation has generated interest in the possibility that these events occur in nature. Natural prokaryote-to-eukaryote events could explain the discovery in the human genome of a number of bacterial genes not found in the genomes of *C. elegans* or *Drosophila melanogaster* (41). These and other observations support the notion that conjugation to eukaryotic cells other than plant cells has occurred in nature.

More recently, immunofluorescence microscopy has provided visual evidence for single-stranded DNA transfer. Conjugation was performed using DNA methylation-minus recipient cells and DNA methylation-plus donors. After transfer, the DNA was hemimethylated and specifically visualized with green fluorescent protein. True recipients were counted, and it was found that 75% of all recipients had received the incoming DNA (44). These results, taken together with *oriT* nick region data, provide additional good evidence for single-stranded DNA transfer.

After transfer, vegetative DNA replication maintains the plasmid in the recipient over the long term. For RK2, DNA plasmid replication in the recipient is primed by a protein primase that binds DNA and is transferred with the DNA. In the *A. tumefaciens*-to-plant cell system, there are two single-stranded DNA binding proteins that have nuclear localization sequences to localize the DNA to plant cell nuclei (97). It is thought that the transferred single-stranded DNA is single stranded as it is delivered into the plant cell and that it remains single stranded as it is transported through the nuclear membrane. In bacterial conjugation of mobilized suicide vectors, transferred DNA is efficiently recombined into the recipient cell genome. With plasmid conjugation, the transferred DNA is likewise thought to remain single stranded until it undergoes recombination.

Reviewing plasmid replication raises the question of whether or not conjugation-specific replicative events could be targeted to diminish conjugative frequency. Little research is being done to explore such targets, including DNA-processing proteins operating at the *oriT* or the activities of plasmid-encoded DNA primase. Targeting conjugation might be accomplished by the pursuit of replication-related events occurring at the donor cell membrane bridge.

CONJUGATIVE TRANSPOSONS

Conjugative transposons are mobile DNA elements with conjugation ability and with transposition ability. These elements are critically important in drug resistance dissemination, due to the fact that they provide conjugative DNA transfer among bacteria. It is primarily conjugation that disseminates MDR and much of the single-drug resistance among bacteria. Consistent with this view, other conjugative elements could be grouped with bacterial conjugation. Subgroups of bacterial conjugation would then be plasmid-mediated conjugation and conjugative transposon-mediated conjugation. Until major conceptual differences are found among bacterial conjugation systems, it remains reasonable to group all conjugative elements together. Conjugative transposon subgroups may have distinct mechanisms for DNA insertion or excision, but such differences would not move the subgroups outside the conjugative transposon family.

Conjugative transposons are transposable conjugative elements. Like transposons, they are incapable of independent vegetative replication. Like conjugative plasmids, they are proficient in conjugative DNA transfer from a donor to a recipient cell. After transfer, the DNA element integrates into recipient cell DNA. For conjugative plasmids and conjugative transposons there are the same two functional categories: the self-transmissible and mobilizable elements. Self-transmissible conjugative transposons can be large, ranging from 20 to over 100 kb, as large as self-transmissible conjugative plasmids and larger than most transposons. Conjugative transposition was first observed in *Streptococcus* and *Bacteroides* species. There were two original designations: from the elements found in *Streptococcus* came the term conjugative transposon (CTn), and from the elements found in *Bacteroides* came the term nonreplicating bacteroides unit (NBU). Other conjugative transposons have been identified and have other designations (6, 9), but "CTn" has become the generally accepted designation.

Conjugative transposons excise from one DNA molecule, circularize, conjugatively transfer into a recipient cell, and integrate into recipient cell DNA. Without autonomous replication of the circularized molecule in the cell, conjugative transposons were initially difficult to isolate and characterize. The first conjugative transposon, Tn*916*, was discovered in 1981 in a strain of *Streptococcus faecalis* (26). At first it was thought that all conjugative transposons encoded tetracycline resistance and that their transfer was induced by tetracycline. More recently, conjugative transposons have been found in *Proteus*, *Providencia*, *Pseudomonas*, *Salmonella*, *Vibrio*, and other species (9, 70). Conjugative transposons have been found to encode virulence determinants and resistance to kanamycin, chloramphenicol, and erythromycin, and finally, there are conjugative transposons for which excision and transfer are regulated by factors other than tetracycline. Some conjugative transposons have been found to be integrated into tRNA-encoding sequences, suggesting an ancestral role for these elements in the evolution of pathogenic islands.

Conjugative transposons are conjugative, while conventional transposons are not. Conventional transposons transpose exclusively among DNA molecules within the host cell. In the process of excision, some conventional transposons duplicate DNA at the donor site, while others do not. For example, Tn*3* DNA is replicated as it excises. Transposons that do not duplicate, such as Tn*5*, are able to excise and transfer without preserving the transposon sequence of the donor site. These elements are called nonreplicative transposons, and the duplicated transposons are called replicative transposons. This terminology was adopted for conventional transposons and does not apply to the inability of conjugative transposons to replicate as the circular intermediate.

Conjugative plasmids, such as F and RK2, and conjugative transposons have been found integrated into bacterial chromosomes. Bacteria carrying integrated F plasmids are called Hfr (for high frequency) strains, referring to the high-frequency transfer of the donor cell chromosome into a recipient cell. Transfer originates at the *oriT* of the integrated element. Do conjugative transposons ever transfer the entire chromosome? Presumably not, because they excise and circularize before transfer.

Some have argued that conjugative transposons should be grouped with conventional transposons. The term "conjugative transposon" (CTn) is more descriptive and preferable for several reasons, including the following.

1. Transposons and conjugative transposons create different terminal ends upon insertion.
2. Transposons mediate only intracellular transfer, while conjugative transposons mediate both intracellular and intercellular transfer.
3. Transposons transpose as double-stranded DNA, while conjugative transposons excise as double-stranded DNA and then conjugatively transfer as single-stranded DNA.
4. Transposons require few gene products to mediate transposition, while conjugative transposons require a number of gene products for conjugation functions.
5. Transposons and conjugative transposons do not share homologous transfer genes while some conjugative transposons and conjugative plasmids do share homologous transfer genes, suggesting a common ancestry.
6. Transposons may duplicate DNA sequences at the insertion site, whereas conjugative transposons do not.

7. Transposons are low-molecular-weight elements, while conjugative transposons and conjugative plasmids are large molecues.

Conjugative transposons represent an unsurprising variation on the theme of conjugative resistance dissemination. Differences in conjugative transposons, as in transposons and conjugative plasmids, can serve to identify subgroups, such as differences in the excision process. A conjugative transposon was recently discovered that has no proteins related to any other elements, indicating that it represents a new and unique family (1).

At the insertion target site, many conventional transposons create a staggered cleavage as they insert. Then, the overhanging sequences are filled in, to create a short duplication at each flanking end of the inserted transposon. Conjugative transposons do not duplicate the target sequence. Instead, the remarkable discovery was made that their termini contain mismatched base pairs, arising from the fact that the circular intermediate is not replicated after excision and is transient enough to avoid mismatch repair. These heteroduplexes were identified by the isolation of circular intermediates in enough quantity to sequence. Analysis of terminal sequences of the integrated forms and comparing the sequences of the circular intermediates led to a greater understanding of conjugation transposition (12, 82). Since the cell-to-cell DNA transfer events of conjugative transposons appear to be accomplished via the same process as that used by plasmid-mediated conjugation, there are no reasons as yet to reject the plasmid conjugation model as a model for conjugative transposition (69). The steps for conjugative transposition would reflect this similar process, with of course the additional steps for excision and integration (Fig. 18.8).

Conjugative transposons integrate preferentially in AT-rich (106) regions of DNA molecules. For CTn*916* and related elements, excision has been likened to that of the lysogenic λ phage. Excision and circularization produce a double-stranded circular DNA molecule that undergoes relaxosome formation as in plasmid conjugation (Fig. 18.8). There are more complex excision events in some elements. In NBU1, the *oriT* is near an excision terminus and there is an unusual excision process. Such a strategy may prevent premature nicking at the nearby *oriT* during excision (83).

Some conjugative transposons integrate at specific sites in DNA molecules, while others integrate randomly. Indeed, a factor that gives specificity to an otherwise randomly integrating element has been discovered (93). Conjugative transposons have filled a niche in antibiotic resistance transfer that is unsuitable for plasmid-mediated conjugation. Thus far, conjugative transposons have been found in a limited number of bacterial species, while conjugative plasmids have been found in many. The broad host range IncP conjugative plasmids have been shown to mobilize conjugative transposons (54, 59, 86), suggesting that these events occur in nature. *P. aeruginosa* can carry both types of elements.

There could be selective advantages for a transmissible element lacking the capacity for vegetative replication. Conjugative transposons are committed to transfer and rapid integration into an established genome. Without independent replication, there is no plasmid incompatibility. Rapid integration into a stable, preexistent genome provides automatic stability. And rapid events, per se, are advantageous for bacteria under strong selective pressure.

CONJUGATION MODEL BUILDING

There seems to be every possible molecular variation on the theme of conjugation, conjugative transposition, and transposition among clinical bacteria. There are many conjugative plasmids that encode nonconjugative transposons, and there are nonconjugative transposons integrated into conjugative transposons. Self-transmissible and mobilizable forms of both plasmids and conjugative transposons have been found, and each of the two types of elements can mobilize the other. Some conjugative transposons undergo integration and excision using prophage-type mechanisms, while others use a resolvase enzyme (95). Conjugative transposons have also demonstrated high frequencies of recombination, with DNA within the element and beyond the element (69). A unifying model can be developed because the general mechanism for classical bacterial conjugation appears to be conserved in conjugative transposons. The broad-host-range systems might be considered to be the most generalized.

Most conjugation microbiologists do not adhere to a model for bacterial conjugation in which the DNA passes within and through long pilus appendages in transferring from the donor to the recipient cell. In textbooks currently in use, however, there are electron micrographs showing pilus-connected conjugating bacteria, with implications in the text that the connecting appendage is the conduit for transferring DNA. This pilus-as-conduit model has been controversial for decades, with a series of inconclusive experimental results. On the other hand, it has not been definitively ruled out. In light of what is known, what is a reasonable model? Experiments with donors without extended pili can transfer DNA, as discussed above. For RK2, the conjugative pilus conundrum is the fact that pilus-forming genes are necessary for conjugative transfer

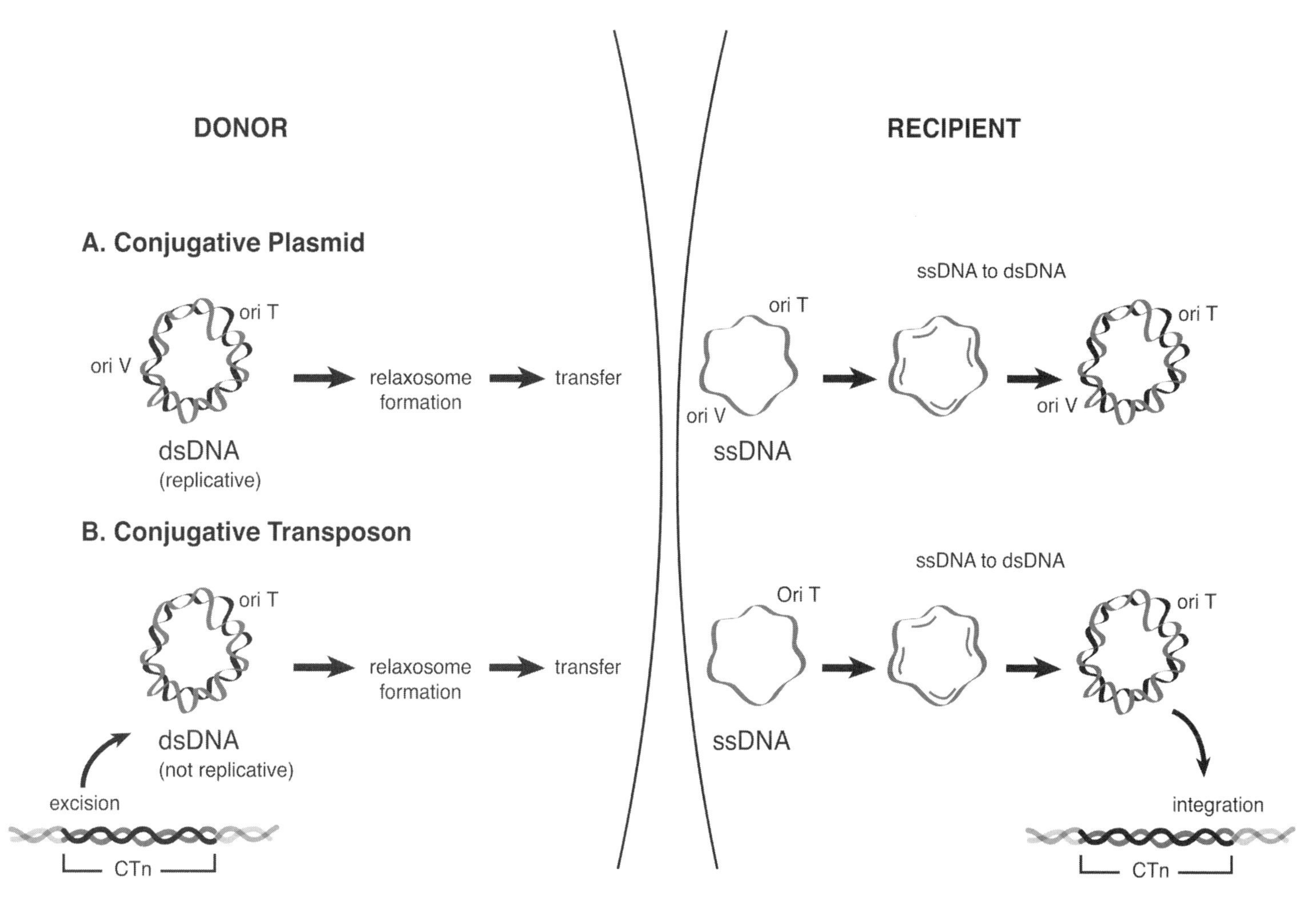

Figure 18.8 DNA transfer by plasmid conjugation and conjugative transposons. The basic steps are shown for plasmid conjugation (A), starting with a double-stranded plasmid undergoing vegetative replication in the donor cell, followed by relaxosome formation, cell-cell transfer of single-stranded DNA molecule, single-to-double-stranded DNA synthesis, and vegetative replication in the recipient. Conjugative transposition (B) has these steps plus excision and insertion. The conjugative transposon excises in the host cell as a double-stranded molecule incapable of vegetative replication. This is shown by a lack of *oriV* sequence. DNA transferred to a recipient cell integrates into host cell DNA. Not shown is an alternative event in which the CTn transposes into another donor cell DNA site without first transferring to another cell.

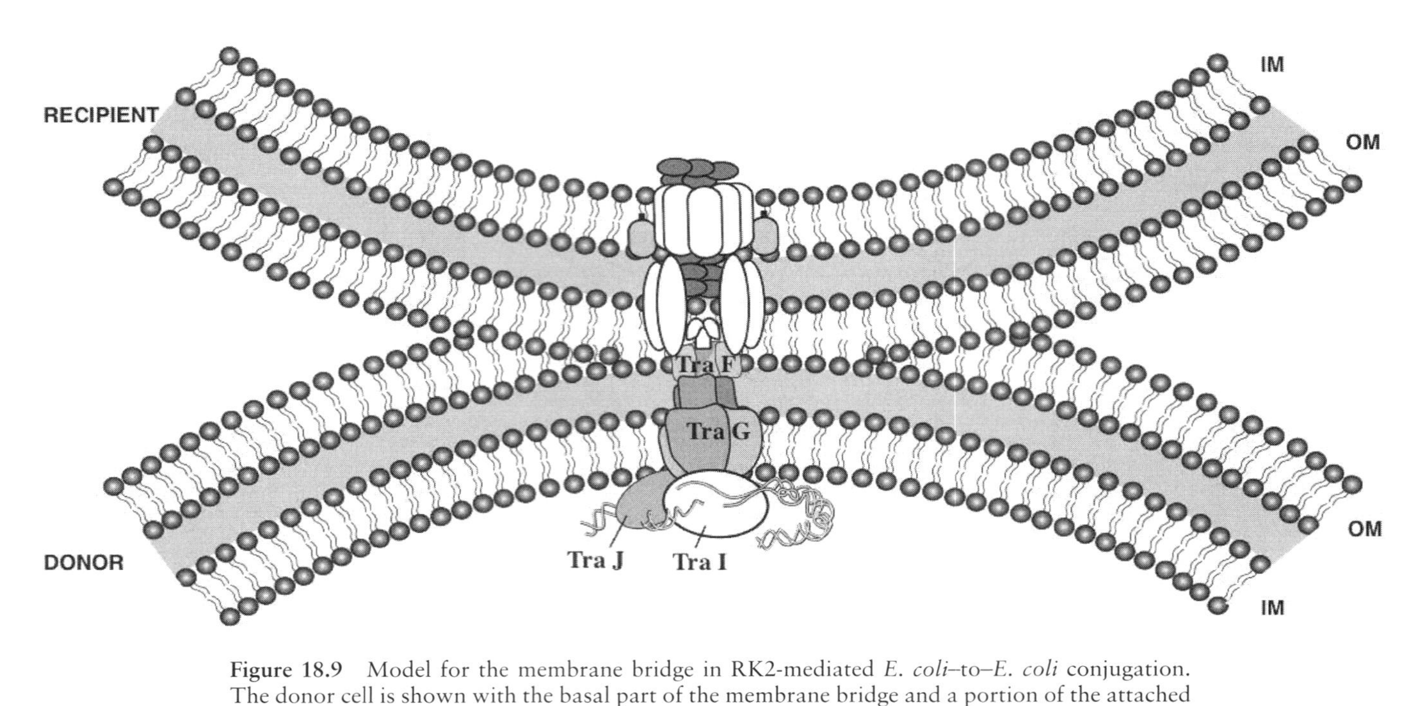

Figure 18.9 Model for the membrane bridge in RK2-mediated *E. coli*–to–*E. coli* conjugation. The donor cell is shown with the basal part of the membrane bridge and a portion of the attached relaxosome at the inner membrane (IM). TraJ interacts with *oriT* DNA to align the TraI relaxase enzyme with the nick region. TraG and TraF interact with the relaxosome and inner membrane. TraF is further out, shown at the merged outer membranes (OM) of the donor and recipient cells. Residual pilin subunits are shown at the top of the bridge, at the inner membranes of the recipient cell. Nicking by TraI at the nick site, within the underwound portion of DNA, is thought to initiate the DNA transfer event.

but that visible pili themselves are not necessary. In F plasmid conjugation, it was found that three mutant donor strains lacking the pilus-forming genes could mediate conjugative transfer (3). Two of these mutants were found to be defective in pilus extension. From the F and RK2 data, a model can be put forward that addresses the pilus conundrum. It is proposed that the DNA does not go through an extended pilus but rather traverses a basal membrane structure that is formed early in pilus construction or that remains after the pilus is broken off or retracted. In other words, the conjugative membrane bridge is this proposed basal structure upon which the pilus is built or somehow removed. This model must be reconciled with at least a momentary pilus presence.

Another more basic question regarding the membrane bridge should be addressed first. Is the conjugative membrane bridge a discrete unit embedded in the cell wall or perhaps a structural continuum along the cell surface wherever the donor and recipient contact? This became a question with the observed fusion of outer membranes of mating *E. coli* (77). For gram-negative bacterial conjugation, the mating bridge has been proposed to bridge the double membranes of both the donor and recipient. This comprises two sets of double membranes, for a total of four membranes. In micrographs of mating *E. coli*, the outer membranes appear to be fused together into one outer membrane of greater density (Fig. 18.6). Outer membrane fusion is not observed in donors lacking the RK2 plasmid (77). The proposed membrane bridge model includes outer membrane fusion seen in electron micrographs (Fig. 18.9).

A mating bridge continuum is probably not the case, however. Mobilization and phage infection experiments have demonstrated a competition for a limited number of surface sites. For RK2-specific phage, "mating bridge structures" serve as phage receptors that are limited in number. From phage adherence studies, it has been estimated that there are about 20 phage receptor sites per *E. coli* host cell (46). The copy number of plasmid RK2 is 7 to 15 molecules per chromosome, depending on temperature. There are normally one or two chromosomes per cell. In cells carrying the F plasmid, there are about three pili per cell and one plasmid per chromosome (3). These numbers are consistent with the idea that relaxosomes occupy and "gate" the membrane bridge sites. These observations are not consistent with a mating bridge continuum all along regions of outer membrane fusion. There are unanswered questions. For example, what specifically causes the apparent outer membrane fusion in conjugating bacteria? Is the RK2-dependent fusion present in all conditions?

Detection of certain DNA molecular forms indirectly addressed this question of the membrane bridge surface sites. Employing fluorescence in situ hybridization to locate supercoiled and relaxed circular DNA molecules, supercoiled RK2 plasmid DNA was localized to mid-cell and quarter-cell positions in three different bacterial species. Relaxed chromosomal DNA was localized at the cell poles (42). These results support other data suggesting that supercoiled plasmid DNA resides at the inner membrane, as relaxosomes poised for transfer, or as vegetatively replicating plasmids poised for partitioning, or in the cytoplasm doing neither. It would be of interest to use this type of method to visualize relaxosome DNA in cultures of donors, mating cells, and phage-infected cells, to more precisely determine DNA locations.

Another study provided for the first time a way to visualize the T4SS apparatus on a bacterial surface. With immunofluorescence microscopy, *H. pylori* cell surface export structures were seen in conditions for protein transfer to epithelial cells. The T4SS apparatus appeared to be a short filamentous structure protruding through the cell envelope, elevated slightly above the bacterial surface (88). This is what one might expect, and genetic and functional similarities suggest that conjugative bridge structures resemble this *H. pylori* structure. Although the conjugative bridge is a channel for DNA/protein transfer rather than for protein transfer, it is known that a primase protein that binds to IncI1 plasmid DNA and a protein that binds to T-DNA in conjugation to plant cells can be transferred as "DNA-less" proteins (13, 105). Conjugation has been investigated for decades, but this more recent pursuit of the many T4SS might more quickly bring a greater understanding of the transfer membrane structure.

The well-studied plasmid systems now reveal a conserved basic mechanism for conjugation, and models can accommodate different types of pili, such as the retractable and nonretractable. F-like plasmids encode thin, flexible, retractable pili, while IncP-like plasmids encode thick, brittle, nonretractable pili. Brittle pili have been seen in electron micrographs as broken-off bundles located in areas apart from the cells (65). One model invokes a basal pilus structure, perhaps the result of breaking off. The trauma of cell contact could break off pili and produce residual basal structures that could bridge cells. The concept of a generalized T4SS for transport of protein and DNA suggests a membrane structure dedicated to export of DNA/protein complexes and of pilin protein subunits during pili construction. DNA could be pumped out in a way that mimics pilus construction, which is the sequential addition of pilin subunits from the bottom up. The relaxosome may block subunits from queuing up, and/or the transferring DNA may in fact push out any residual subunits. The two processes use the same machinery, and after the mating signal, relaxosomes could outcompete for the bridge to block access of pilin subunits.

The membrane bridge structure is also the receptor and channel for plasmid-specific phage. PRD1 RK2-specific phage injects linear DNA with attached 5′-bound protein. Phages have been visualized attached to pili and also to the cell surface, suggesting that these phages attach to pilin subunits wherever they are found. The model does not address IncP-specific phage DNA entry, whether single- or double-stranded, and how the pore structure accommodates DNA traffic direction change. One candidate in RK2 for expediting traffic reversal might be TraF, the more external coupling protein that is required for phage infection. In RK2, it appears that phage resistance is up-regulated when conjugation is down-regulated, but the membrane bridge has not been examined for such changes.

Experiments using mutants and immunofluorescence microscopy could address these kinds of questions about the membrane bridge regarding its complex activities in conjugation, phage adsorption, and pilus production (47, 101, 107). Is there a donor cell inner-outer membrane fusion structure, as well as outer-outer membrane fusion of donor and recipient? How do DNA and protein traverse the bridge? How does the membrane bridge accommodate different recipients, such as gram-negative and gram-positive bacteria and yeast and mammalian cells? Of interest for this model building is the only paradigm known to occur in nature for prokaryote-to-eukaryote conjugation, the transfer of *A. tumefaciens* DNA/protein to plant cells. The sequence of events in the bacterium and recipient plant cell is being worked out in detail, and some parallels may be seen using other eukaryotic recipients. Comparing differences and similarities will contribute to model building and the development of strategies to target conjugative dissemination of antibiotic resistance.

CONCLUDING REMARKS

It is reasonable to suggest that bacterial conjugation is the greatest mover of genes in the microbial world and, in the clinical world, that these genes are often antibiotic resistance genes. This overview of bacterial conjugation and other gene acquisition mechanisms shows how and why this could be true. Bacteria adapted to antibiotics in the soil years before clinical bacteria adapted to antibiotics in hospital patients. Awareness of the conjugative process as a tool of genetics in nature and in the laboratory was critical for the microbial geneticists who first hypothesized that bacterial conjugation mediated MDR dissemination. It was a timely convergence of medicine and molecular biology. As provided by years of ongoing antibiotic selective pressure, the resulting clinical science has shown the prowess of bacterial conjugation as no experiment could be designed to do.

More recently the story is being told by means of bacterial genome sequencing and analysis. In bacterial genomes carrying multiple resistance determinants, the origin of genes can be traced back to a native source. The totality of mobile genetic elements in MDR pathogens can constitute 25% of the genome. Mobile elements are genetic regions that are mobilized by transfer among DNA molecules or genomes within the cell or between cells. Cell-cell bacterial gene transfer always involves conjugation. Sequence analysis can also reveal the pathways taken by acquired genes as inferred from patchworks of mobile elements, indicating again that the primary mechanism for dissemination of MDR is bacterial conjugation.

The clinical histories of resistant organisms illustrate that the most genetically efficient organisms become resistant first. These examples demonstrate the inevitability of resistance. The high efficiencies of bacterial conjugation and conjugative transposition provide rapid multiple gene transfer and gene stability in those bacteria. A logical next step is going beyond targeting bacterial viability to the targeting of subprocesses of gene dissemination. Relaxosomes, replication, secretion, the membrane bridge, and transfer regulation do in fact bear directly on the conjugative process and dissemination of antibiotic resistance among pathogens. The subprocesses of bacterial conjugation and conjugative transposition reveal a precise, multistep cascade of events that is clearly distinguished from activities such as naked DNA uptake. Conjugation, a complex mechanism for both single-drug and multidrug resistance dissemination, is employed in some form by most clinical strains.

Apart from the dissemination of antibiotic resistance, the capacity for rapid gene transfer also gives pathogenic bacteria access to the microbial gene pool, for adaptation to other kinds of selective pressure. This means that virulence factors can be efficiently acquired and coselected wherever resistance markers are readily disseminated. Sequences of the more genetically endowed organisms have shown a mosaic nature of genome sequence and reveal that certain high-risk and high-impact genes are intermingled among genomes. The earlier teaching was that bacterial plasmids were preeminent among mobile and dispensable carriers of high impact genes. These genes might save the day in one environment but also carry a cost, such as slower growth, in the next environment. Plasmids do provide this mobility and are dispensable, but it is clear that nonplasmid conjugative elements are also mobile, dispensable, and ubiquitous. Such elements can also acquire surrounding genetic regions.

Addressing the conjugative transfer of antibiotic resistance genes will also address the conjugative transfer of virulence factors. Both factors impact the nosocomial emergence of pathogens that are highly resistant and virulent. DNA processing, still the most understood aspect of conjugation, is one potential target aimed at reducing gene transfer. Targeting events of other processes, such as type IV secretion, may also yield strategies that are specific for conjugation. The cycle of antibiotic resistance and pathogen genome sequencing are showcasing the prominent role of bacterial conjugation in gene dissemination. Understanding more of the conjugative process, including the aspects showing no immediate application, will ultimately bring novel and long-term strategies to target conjugative DNA transfer of antibiotic resistance genes. These kinds of studies will also benefit gene transfer as a science to be applied for other purposes.

POSTSCRIPT ON SEXUAL EUPHEMISMS

Can asexual organisms participate in sexual activity? Can asexual organisms be promiscuous? For bacterial conjugation, the answers to such questions have been yes, but also no, depending on how strictly the word "sex" is used. What exactly is sex? Could it loosely be construed as horizontal genetic exchange? Conjugative DNA transfer was at first envisioned as a sex-like phenomenon, evidenced in the naming of the fertility factor F. The donor cell was originally called the male cell, and the recipient cell was originally called the female cell. DNA is in fact donated to the recipient cell without donor DNA loss, because it is replicated as it is transferred. Genetic diversity is accomplished in the prokaryotic world largely by bacterial conjugation. Offspring arise that are genetically distinct from the parental male and female cells. All this amused attendees of meetings dedicated to the prokaryotic world.

Fickle societal correctness then entered the discussion and first demanded that the male and female bacterial cells be renamed the donor and recipient cells. This seemed to be a stricter interpretation of sex. Soon thereafter, however, it was declared that broad-host-range conjugative plasmids were in fact promiscuous, for a seemingly looser interpretation of sex. Regardless, the terminology has continued to elevate public interest in bacterial conjugation and inspired the creation of clever manuscript titles and cartoons (Fig. 18.10). However, the loose terminology has also resulted in some unexpected confusion. A popular science writer recently wrote that bacteria could now have sex with hamsters. A photo of a hamster was shown. In an

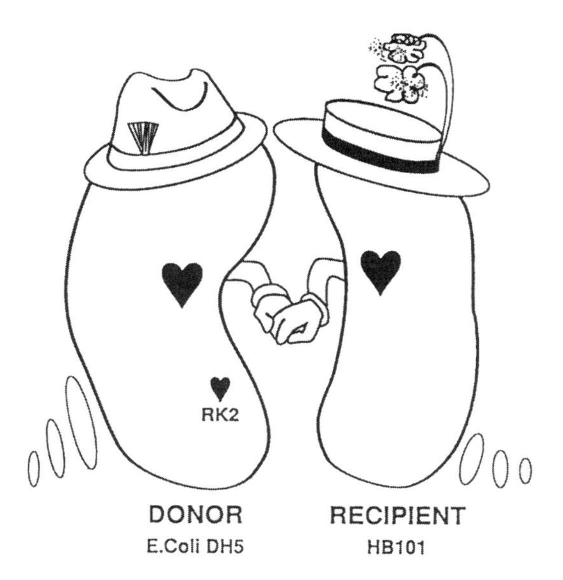

Figure 18.10 Mr. and Mrs. *E. coli* bacterial cells in cartoon form. Both cells have chromosomes as indicated by the large hearts. The male cell, Mr. *E. coli* DH5, carries the RK2 plasmid, ready for transfer to Mrs. *E. coli* HB101, the female cell.

interview about bacterial conjugation to mammalian tissue culture cells, another popular science writer seemed to be very disappointed to learn that the terms had been nothing more than euphemisms.

I gratefully acknowledge Marcelo Tolmasky, Moselio Schaechter, Joshua Fierer, Don Guiney, Monica Bray, Daniel Yee, Jennifer Holt, Patty Hasegawa, the Sam and Rose Stein Institute for Research on Aging, the National Cancer Institute, and the National Institute of Allergy and Infectious Diseases.

References

1. **Adams, V., D. Lyras, K. A. Farrow, and J. I. Rood.** 2002. The clostridial mobilizable transposons. *Cell. Mol. Life Sci.* **59:**2033–2043.
2. **Anderson, E. S.** 1975. Viability of, and transfer of a plasmid from, *E. coli* K12 in the human intestine. *Nature* **255:** 502–504.
3. **Anthony, K. G., W. A. Klimke, J. Manchak, and L. S. Frost.** 1999. Comparison of proteins involved in pilus synthesis and mating pair stabilization from the related plasmids F and R100-1: insights into the mechanism of conjugation. *J. Bacteriol.* **181:**5149–5159.
4. **Banks, D., S. B. Beres, and J. M. Musser.** 2002. The fundamental contribution of phage to GAS evolution, genome diversification, and strain emergence. *Trends Microbiol.* **10:**515–521.
5. **Bates, S., A. Cashmore, and B. M. Wilkins.** IncP plasmids are unusually effective in mediating conjugation of *Escherichia coli* and *Saccharomyces cerevisiae*: involvement of the Tra2 mating system. *J. Bacteriol.* **180:**6538–6543.
6. **Beaber, J. W., B. Hochhut, and M. K. Waldor.** 2002. Genomic and functional analyses of SXT, an integrating antibiotic gene transfer element from *Vibrio cholerae*. *J. Bacteriol.* **184:**4259–4269.
7. **Beck, M.A., Q. Shi, V.C. Morris, and O.A. Levander.** 1995. Rapid genomic evolution of a non-virulent Coxsackie B3 in selenium-deficient mice results in selection of identical virulent isolates. *Nat. Med.* **1:**433–436.
8. **Bjorkholm, B., M. Sjolund, P. G. Falk, O. Berg, L. Engstrand, and D. I. Andersson.** 2001. Mutation frequency and biological cost of antibiotic resistance in *Helicobacter pylori*. *Proc. Natl. Acad. Sci. USA* **98:**14607–14612.
9. **Boltner, D., C. MacMahon, J. T. Pembroke, P. Strike and A. M. Osborn.** 2002. R391: conjugative integrating mosaic comprised of phage, plasmid, and transposon elements. *J. Bacteriol.* **184:**5158–5169.
10. **Bushman, F. D.** 2003. Targeting survival: integration site selection by retroviruses and LTR retro transposons. *Cell* **115:**135–138.
11. **Cabezon, E., J. I. Sastre, and F. de la Cruz.** 1997. Genetic evidence of a coupling role for the TraG protein family in bacterial conjugation. *Mol. Gen. Genet.* **254:**400–406.
12. **Caparon, M. G., and J. R. Scott.** 1989. Excision and insertion of the conjugative transposon Tn*916* involves a novel recombination mechanism. *Cell* **59:**1027–1034.
13. **Christie, P. J., J. E. Ward, S. C. Winans, and E. W. Nester. 1988.** The *Agrobacterium tumefaciens virE2* gene product is a single-stranded-DNA-binding protein that associates with T-DNA. *J. Bacteriol.* **170:**2659–2667.
14. **Christie, P. J.** 2001. Type IV secretion: intercellular transfer of macromolecules by systems ancestrally related to conjugation machines. *Mol. Microbiol.* **40:**294–305.
15. **Claverys, J.-P., and B. Martin.** 2003. Bacterial 'competence' genes: signature of active transformation, or only remnants? *Trends Microbiol.* **11:**161–165.
16. **Craig, A. S., P. C. Erwin, W. Schaffner, J. A. Elliott, W. L. Moore, X. T. Ussery, L. Patterson, A. D. Dake, S. G. Hannah, and J. C. Butler.** 1999. Carriage of multi-drug resistant *Streptococcus pneumoniae* and the impact of chemophrophylaxis during an outbreak of meningitis at a day care center. *Clin. Infect. Dis.* **29:**1257–1264.
17. **Daugelavicius, R., J. Bamford, A. M. Grahn, E. Lanka, and D. H. Bamford.** 1997. The IncP plasmid-encoded cell envelope-associated DNA transfer complex increases cell permeability. *J. Bacteriol.* **179:**5195–5202.
18. **Davis, C. E., and J. Anandan.** 1970. The evolution of an R factor: a study of a "preantibiotic" community in Borneo. *N. Engl. J. Med.* **282:**117–122.
19. **DelVecchio, V.G., V. Kapatral, R. Redkar, G. Patra, C. Mujer, T. Los, N. Ivanova, I. Anderson, A. Bhattacharya, A. Lykidis, R. Reznik, K. Jablonski, N. Larsen, M. D'Souza, A. Bernal, M., Mazur, E. Goltsman, E. Selkov, P. Elzer, S. Hagius, D. O'Callaghan, J.-J. Letesson, R. Haselkorn, N. Kyropides, and R. Overbeek.** 2002. The genome sequence of the facultative intracellular pathogen *Brucella melitensis*. *Proc. Natl. Acad. Sci. USA.* **99:**443 448.
20. **Dimopoulou, I. D., J. E. Russell, Z. Mohd-Zain, R. Herbert, and D. W. Crook.** 2002. Site-specific recombination with the chromosomal tRNA (Leu) gene by the large conjugative *Haemophilus* resistance plasmid. *Antimicrob. Agents Chemother.* **46:**1602–1603.

21. **Ding, Z., K. Atmakuri, and P.J. Christie.** The ins and outs of bacterial type IV secretions substrates. *Trends Microbiol.* **11:**527–535.
22. **Egger, C. H., B. J. Kimmel, J. L. Bono, A. F. Elias, P. Rosa, and D. S. Samuels.** 2001. Transduction by ΦBB-1, a bacteriophage of *Borrelia burgdorferi. J. Bacteriol.* **183:** 4771–4778.
23. **Eggler, A. L., S. L. Lusetti, and M. M. Cox.** 2003. The C terminus of the *E. coli* RecA protein modulates the DNA binding competition with single-stranded DNA-binding protein. *J. Biol. Chem.* **278:**16389–16396.
24. **Ferguson, G. C., J. A. Heinemann, and M. A. Kennedy.** 2002. Gene transfer between *Salmonella enterica* serovar Typhimurium inside epithelial cells. *J. Bacteriol.* **184:**2235–2242.
25. **Fiser, A., S. R. Filipe, and A. Tomasz.** 2003. Cell wall branches, penicillin resistance and the secrets of the MurM protein. *Trends Microbiol.* **11:**547–553.
26. **Franke, A., and D. B. Clewel.** 1981. Evidence for a chromosome-borne resistance transposon (Tn*916*) in *Streptococcus faecalis* that is capable of "conjugal" transfer in the absence of a conjugative plasmid. *J. Bacteriol.* **145:**494–502.
27. **Frost, L., K. Ippen-Ihler, and R. Skurray.** 1994. Analysis of the sequence and gene products of the transfer and gene products of the transfer region of the F sex factor. *Microbiol. Rev.* **58:**162–210.
28. **Fullner, K. J., J. C. Lara, and E. W. Nester.** 1996. Pilus assembly by *Agrobacterium* T-DNA transfer genes. *Science* **273:**1107–1109.
29. **Giovanetti, E., G. Magi, A. Brenciani, C. Spinaci, R. Lupidi, B. Facinelli, and P. E. Varaoldo.** 2002. Conjugative transfer of the *erm(A)* gene from erythromycin-resistant *Streptococcus pyogenes* to macrolide-susceptible *S. pyogenes, Enterococcus faecalis* and *Listeria innocua. J. Antimicrob. Chemother.* **50:**249–252.
30. **Gomis-Ruth, F.X., G. Moncalian, R. Perez-Luque, A. Gonzalez, E. Cabezon, F. de la Cruz, and M. Coll.** 2001. The bacterial conjugation protein TrwB resembles ring helicases and F_1 ATPase. *Nature* **409:**637–641.
31. **Grahn, A. M., J. Haase, E. Lanka, and D. H. Bamford.** 1997. Assembly of a functional phage PRD1 receptor depends on 11 genes of the IncP plasmid mating pair formation complex. *J. Bacteriol.* **179:**4733–4740.
32. **Grillot-Courvalin, C., S. Goussard, F. Huet, D. M. Ojcius, and Courvalin, P.** 1998. Functional gene transfer from intracellular bacteria to mammalian cells. *Nat. Biotech.* **16:** 862–866.
33. **Guasch, A., M. Lucas, G. Moncalian, M. Cabezas, R. Perez-Luque, F. X. Gomis-Ruth, F. de la Cruz, and M. Coll.** 2003. Recognition and processing of the origin of transfer DNA by conjugative relaxase TrwC. *Nat. Struct. Biol.* **12:** 1002–1010.
34. **Haase, J., R. Lurz, A. M. Grahn, D. Bamford, and E. Lanka.** 1995. Bacterial conjugation mediated by plasmid RP4: RSF1010 mobilization, donor-specific phage propagation, and pilus production require the same Tra2 core components of the proposed DNA transport complex. *J. Bacteriol.* **177:**4779–4791.
35. **Haase, J., M. Kalkum, and E. Lanka.** 1996. TrbK, a small cytoplasmic membrane lipoprotein, function in entry exclusion of the IncPα plasmid RP4. *J. Bacteriol.* **178:** 6720–6729.
36. **Heinemann, J. A., and R. G. Ankenbauer.** 1993. Retrotransfer of IncP plasmid R751 from *Escherichia coli* maxicells: evidence for the genetic sufficiency of self-transferable plasmids for bacterial conjugation. *Mol. Microbiol.* **10:** 57–62.
37. **Himawan, J. S., and C. C. Richardson.** 1996. Amino acid residues critical for the interaction between bacteriophage T7 DNA polymerase and *Escherichia coli* thioredoxin. *J. Biol. Sci.* **271:**19999–20008.
38. **Hinnebusch, B. J., M.-L. Rosso, T. G. Schwan, and E. Carniel.** 2002. High-frequency conjugative transfer of antibiotic resistance genes to *Yersinia pestis* in the flea midgut. *Mol. Microbiol.* **2:**349–354.
39. **Hoffmann, A., T. Thimm, M. Droge, E. R. B. Moore, J. C. Munch, and C. C Tebbe.** 1998. Intergeneric transfer of conjugative and mobilizable plasmids harbored by *Escherichia coli* in the gut of the soil microarthropod *Folsomia candida* (*Collembola*). *Appl. Environ. Microbiol.* **64:**2652–2659.
40. **Hofreuter, K., A. Karnholz, and R. Haas.** 2003. Topology and membrane interaction of *Helicobacter pylori* ComB proteins involved in natural transformation competence. *Int. J. Med. Microbiol.* **293:**153–165.
41. **International Human Genome Sequencing Consortium.** 2001. Initial sequencing and analysis of the human genome. *Nature* **409:**860–921.
42. **Kahng, L.S., and L. Shapiro.** 2003. Polar localization of replicon origins in the multipartite genomes of *Agrobacterium tumefaciens* and *Sinorhizobium meliloti. J. Bacteriol.* **185:**3384–3391.
43. **Kobayashi, S. D., K. R. Braughton, A. R. Whitney, J. M. Voyish, T. G. Schwan, J. M. Musser, and F. R. DeLeo.** 2003. Bacterial pathogens modulate an apoptosis differentiation program in human neutrophils. *Proc. Natl. Acad. Sci. USA* **100:**10948–10953.
44. **Kohiyama, M., S. Hiraga, I. Matic, and M. Radman.** 2003. Bacterial sex: playing voyeurs 50 years later. *Science* **301:** 802–803.
45. **Kornstein, L. B., V. L. Waters, and R. C. Cooper.** 1992. A natural mutant of plasmid RP4 that confers phage resistance and reduced conjugative transfer. *FEMS Microbiol. Lett.* **91:**97–100
46. **Kotilainen, M. M., A. M. Grahn, J. K. H. Bamford, and D. H. Bamford.** 1993. Binding of an *Escherichia coli* double-stranded DNA virus PRD1 to a receptor coded by an IncP-type plasmid. *J. Bacteriol.* **175:**3089–3095.
47. **Krall, L., U. Wiedemann, G. Unsin, S. Weiss, N. Komke, and C. Baron.** 2002. Detergent extraction identifies different VirB protein subassemblies of the type IV secretion machinery in the membranes of *Agrobacterium tumefaciens. Proc. Natl. Acad. Sci. USA* **99:**11405–11410.
48. **Lanka, E., and P. Barth.** 1981. Plasmid RP4 specifies a deoxyribonucleic acid primase involved in its conjugal transfer and maintenance. *J. Bacteriol.* **148:**769–775.
49. **Lawley, T. D., M. W. Gilmour, J. E. Gunton, D. M. Tracz, and D. Taylor.** 2003. Functional and mutational analysis of conjugative transfer region 2 (Tra2) from the IncHI1 plasmid R27. *J. Bacteriol.* **185:**581–591.

50. **Lawley, T. D., W. A. Klimke, M. J. Gubbins, and L. S. Frost.** 2003. F factor conjugation is a true type IV secretion system. *FEMS Microbiol. Lett.* **224:**1–15.

51. **Leach, A. J., T. M. Shelby-James, M. Mayo, M. Gratten, A. C. Laming, B. J. Currie, and J. D. Mathews.** 1997. A prospective study of the impact of community-based azithromycin treatment of trachoma on carriage and resistance of *Streptococcus pneumoniae. Clin. Infect. Dis.* **24:**356–362.

52. **Leach, A. J., P. S. Morris, H. Smith-Vaughan, and J. D. Mathews.** 2001. In vivo penicillin MIC drift to extremely high resistance in serotype 14 *Streptococcus pneumoniae* persistently colonizing in nasopharynx of an infection with chronic suppurative lung disease: a case study. *Antimicrob. Agents Chemother.* **46:**3648–3649.

53. **Levy, S.** 1992. *Antibiotic Paradox: How Miracle Drugs Are Destroying the Miracle.* Harper Collins, New York, N.Y.

54. **Li, L.-Y., N. B. Shoemaker, G. R. Wang, S. Cole, M. K. Hashimoto, J. Wang, and A. A. Salyers.** 1995. The mobilization regions of two integrated *Bacteroides* elements, NBU1 and NBU2, have only single mobilization protein and may be on a cassette. *J. Bacteriol.* **177:**3940–3945.

55. **Llosa, M., S. Zunzunegui, and F. de la Cruz.** 2003. Conjugative coupling proteins interact with cognate and heterologous VirB10-like proteins while exhibiting specificity for cognate relaxosomes. *Proc. Natl. Acad. Sci. USA* **100:** 10465–10470.

56. **Loeffler, J. M., S. Djurkovic, and V. A. Fishetti.** 2003. Phage lystic enzyme Cpl-1 as a novel antimicrobial for pneumococcal bacteremia. *Infect. Immun.* **71:**6199–6204.

57. **Massey, R. C., and A. Buckling.** 2002. Environmental regulation of mutation rates at specific sites. *Trends Microbiol.* **10:**580–584.

58. **Michaels, K., J. Mei, and W. Firshein.** 1994. TrfA-dependent inner-membrane-associated plasmid. RK2 DNA synthesis in *E. coli* maxicells. *Plasmid* **32:**19–31.

59. **Murphy, C.G., and M.H. Malamy.** 1995. Requirements for strand- and site-specific cleavage within the *oriT* region of Tn*4399*, a mobilizing transposon from *Bacteroides fragilis. J. Bacteriol.* **177:**3158–3165.

60. **Naidoo, J.** 1984. Interspecific co-transfer of antibiotic resistance plasmids in staphylococci in vivo. *J. Hyg.* **93:**59–66.

61. **Novick, R. P.** 1995. Contrasting lifestyles of rolling circle phages and plasmids. *Trends Biol. Sci.* **23:**434–438.

62. **Novick, R. P.** 2003. Mobile genetic elements and bacterial toxinoses: the superantigen-encoding pathogenicity islands of *Staphylococcus aureus. Plasmid* **49:**93–105.

63. **Pansegrau, W., W. Schroder, and E. Lanka.** 1993. Relaxase (TraI) of IncP plasmid RP4 catalyzes a site-specific cleaving-joining reaction of single-stranded DNA. *Proc. Natl. Acad. Sci. USA* **90:**2925–2929

64. **Pansegrau, W., W. Lanka, P. T. Barth, D. H. Figurski, D. G. Guiney, D. Haas, D. Helinski, H. Schwab, V. Stanisch, and C. M. Thomas.** 1994. Complete nucleotide sequence of Birmingham IncPα plasmids. *J. Mol. Biol.* **239:**623–663.

65. **Pansegrau, W. and E. Lanka.** 1996. Enzymology of DNA transfer by conjugative mechanisms. *Prog. Nucleic Acids Res.* **54:**197–251.

66. **Parker, C., and R. J. Meyer.** 2002. Selection of plasmid molecules for conjugative transfer and replacement strand synthesis in the donor. *Mol. Microbiol.* **46:**761–768.

67. **Parker, C., X. Zhang, D. Henderson, E. Becker, and R. Meyer.** 2002. Conjugative DNA synthesis: R1162 and the question of rolling circle replication. *Plasmid* **48:**186–192.

68. **Paulsen, I. T., L. Banerjei, G. S. A. Myers, K. E. Nelson, R. Seshadri, T. D. Read, D. E. Fouts, J. A. Eisen, S. R. Gill, J. F. Heidelberg, H. Tettelin, R. J. Dodson, L. Umayam, L. Brinkac, M. Beanan, S. Daugherty, R. T. DeBoy, S. Durkin, J. Kolanay, R. Madupu, W. Nelson, J. Vamathevan, B. Tran, J. Upton, T. Hansen, J. Shetty, H. Khouri, T. Utterback, D. Radune, K. A. Detchum, B. A. Daugherty, and C. M. Fraser.** 2003. Role of mobile DNA in the evolution of vancomycin-resistant *Enterococcus faecalis. Science* **299:** 2071–2074.

69. **Pembroke, J. T., and D. B. Murphy.** 2000. Isolation and analysis of a circular form of the IncJ conjugative transposon-like elements R391 and R997: implications for IncJ incompatibility. *FEMS Microbiol. Lett.* **187:**133–138.

70. **Pembroke, J.T., C. MacMahon, and B. McGrath.** 2002. The role of conjugative transposons in the *Enterobacteriaceae. Cell. Mol. Life Sci.* **59:**2055–2064.

71. **Planet, P. J., S. C. Kachlany, R. DeSalle, R., and D. H. Figurski.** 2001. Phylogeny of genes for secretion of NTPases: identification of the widespread *tadA* subfamily and development of diagnostic key for gene classification. *Proc. Natl. Acad. Sci. USA* **98:**2503–2508.

72. **Pohlman, R. F., H. D. Genetti, and S. C. Winans.** 1994. Common ancestry between IncN conjugal transfer genes and macromolecular export systems of plant and animal pathogens. *Mol. Microbiol.* **14:**655–668.

73. **Rabel, C., A. M. Grahn, R. Lurz, and E. Lanka.** 2003. The VirB4 family of proposed traffic nucleoside triphosphatases: common motifs in plasmid RP4 TrbE are essential for conjugation and phage adsorption. *J. Bacteriol.* **185:**1045–1058.

74. **Rees, C. E. D. and B. Wilkins.** 1990. Protein transfer into the recipient cell during bacterial conjugation: studies with F and RP4. *Mol. Microbiol.* **4:**1199–1206.

75. **Rekart, M.L, D. M. Patrick, B. Chakraborty, J. L. Maginley, H. G. Jones, C. D. Bajdik, B. Pourbohiol, and R. C. Brunham.** 2003. Targeted mass treatment for syphilis with oral azithromycin. *Lancet* **361:**313–314.

76. **Ridenour, D. A., S. L. G. Cirillo, S. Feng, M. Samrakandi, and J. D. Cirillo.** 2001. Identification of a gene that affects the efficiency of host cell infection by *Legionella pneumophila* in a temperature dependent fashion. *Infect. Immun.* **71:**6256–6263.

77. **Samuels, A. L., E. Lanka, and J. E. Davis.** 2000. Conjugative junctions in RP4-mediated mating of *Escherichia coli. J. Bacteriol.* **182:**2709–2715.

78. **Santagati, M., F. Iannelli, C. Cascone, F. Campanile, M. Oggioni, S. Stefani, and G. Pozzi.** 2003. The novel conjugative transposon Tn*1207.3* carries the macrolide efflux gene *mef(A)* in *Streptococcus pyogenes. Microb. Drug Resist.* **9:**243–247.

79. **Schlimme, W., M. Marchiani, K. Hanselmann, and B. Jenni.** 1997. Gene transfer between bacteria within digestive vacuoles of protozoa. *FEMS Microb. Ecol.* **23:**239–247.

80. **Schmieger, H., and P. Schicklmaier.** 1999. Transduction of multiple drug resistance of *Salmonella enterica* serovar Typhimurium DT104. *FEMS Microbiol. Lett.* **170:**251–256.

81. **Schroder, G., S. Krause, E. Zechner, B. Traxler, H.-J. Yeo, R. Lurz, G. Waksman, and E. Lanka.** 2002. TraG-like proteins of DNA transfer systems and the *Helicobacter pylori* type IV secretion system: inner membrane gate for exported substrates? *J. Bacteriol.* **184:**2767–2779.

82. **Scott, J. R., F. Bringel, D. Marra, G. Van Alstine, and C. Rudy.** 1994. Conjugative transposons of Tn*916*: preferred targets and evidience for conjugative transfer of a single strand and for a double-stranded circular intermediate. *Mol. Microbiol.* **11:**1099–1108.

83. **Shoemaker, N., G.-R. Wang, and A. A. Salyers.** 2000. Multiple gene products and sequences required for excision of the mobilizable integrated *Bacteroides* element NBU1. *J. Bacteriol.* **182:**928–936.

84. **Sia, E. A., D. M. Kuehner, and D. H. Figurski.** 1996. Mechanism of retrotransfer in conjugation: prior transfer of the conjugative plasmid is required. *J. Bacteriol.* **178:**1457–1464.

85. **Sizemore, D., A. Branstrom, and J. Sadoff.** 1995. Attenuated *Shigella* as a DNA delivery vehicle for DNA-mediated immunization. *Science* **270:**299–302.

86. **Smith, C. J., A. C. Parker, and M. Bacic.** 2001. Analysis of a *Bacteroides* conjugative transposon using a novel "targeted capture" model system. *Plasmid* **46:**47–56.

87. **Steinmoen, H., A. Teigen, and L. S. Havarstein.** 2003. Competence-induced cells of *Streptococcus pneumoniae* lyse competence-deficient cells of the same strain during cocultivation. *J. Bacteriol.* **185:**7176–7183.

88. **Tanaka, J., T. Suzuki, H. Mimuro, and C. Sasakawa.** 2003. Structural definition on the surface of *Helicobacter pylori* type IV secretion apparatus. *Cell. Microbiol.* **5:**395–404.

89. **Tauch, A., A. Schluter, N. Bischoff, A. Goesmann, F. Meyer, and A. Puhler.** 2003. The 79,370-bp conjugative plasmid pB4 consists of an IncP1ββ backbone loaded with a chromate resistance transposon, the *strA-strB* streptomycin resistance gene pair, the oxacillinase gene bla_{NPS1}, and a tripartaite antibiotic efflux system of the resistance-nodulation-division family. *Mol. Genet. Genomics* **268:** 570–584.

90. **Timmons, L., D. L. Court, and A. Fire.** 2001. Ingestion of bacterially expressed dsRNAs can produce specific and potent genetic interference in *C. elegans*. *Gene* **263:**103–112

91. **Tolmasky, M. E., M. Roberts, M. Woloj, and J. H. Crosa.** 1986. Molecular cloning of amikacin resistance determinants from a *Klebsiella pneumoniae* plasmid. *Antimicrob. Agents Chemother.* **30:**315–320.

92. **Tomita, H., C. Pierson, S. K. Lim, D. B. Clewel, and Y. Ike.** 2002. Possible connection between a widely disseminated conjugative gentamicin resistance (pMG1-like) plasmid and the emergence of vancomycin resistance in *Enterococcus faecium*. *J. Clin. Microbiol.* **40:**3326–3333.

93. **Tribble, G. D., A. C. Parker, and C. J. Smith.** 1999. Transposition of the *Bacteroides* mobilization transposon Tn4555: role of a novel targeting gene. *Mol. Microbiol.* **34:**385–394.

94. **Tsen, S. D., S. S. Fang, , M. J. Chen, J. Y. Chien, C. C. Lee, and D. H. Tsen.** 2002. Natural transformation in *E. coli*. *J. Biomed. Sci.* **9:**246–252.

95. **Wang, H., A. P. Roberts, D. Luras, J. I. Rood, M. Wilks, and P. Mullany.** 2000. Characterization of the ends and target sites of the novel conjugative transposon Tn*5397* from *Clostridium difficile*: excision and circularization is mediated by a large resolvase, TndX. *J. Bacteriol.* **182:**3775–3783.

96. **Wang, J., L. M. Parsons, and K. M. Derbyshire.** 2003. Unconventional conjugal DNA transfer in mycobacteria. *Nat. Genet.* **34:**80–83.

97. **Ward, D.V., and P. C. Zambryski.** 2001. The six functions of *Agrobacterium* VirE2. *Proc. Natl. Acad. Sci. USA* **98:** 385–386.

98. **Ward, D. V., O. Draper, J. R. Zupan, and P. Zambryski.** 2002. Peptide linkage mapping of the *Agrobacterium tumefaciens* vir-encoded type IV secretion system reveals protein subassemblies. *Proc. Natl. Acad. Sci. USA* **99:**11493–11500.

99. **Waters, V., and J. Crosa.** 1988. Divergence of the aerobactin systems encoded by plasmids pColV-K30 in *E. coli* and pSMN1 in *Aerobacter aerogenes*. *J. Bacteriol.* **170:** 5153–5160

100. **Waters, V., K. Hirata, W. Pansegrau, E. Lanka, and D. Guiney.** 1991. Sequence identity in the nick region of IncP plasmid transfer origins and T-DNA border of *Agrobacterium* Ti plasmids. *Proc. Natl. Acad. Sci. USA* **88:** 1456–1460.

101. **Waters, V., B. Strack, W. Pansegrau, E. Lanka, and D. Guiney.** 1992. Mutational analysis of essential IncP plasmid transfer genes *traF* and *traG*, and involvement of *traF* in phage sensitivity. *J. Bacteriol.* **174:**6666–6673.

102. **Waters, V., and D. Guiney.** 1993. Processes at the nick region link conjugation, T-DNA transfer, and rolling circle replication. *Mol. Microbiol.* **9:**1123–1130.

103. **Waters, V. L.** 1999. Conjugative transfer in the dissemination of β-lactam and aminoglycoside resistance. *Front. Biosci.* **4:**D433–D456.

104. **Waters, V. L.** 2001. Conjugation between bacterial and mammalian cells. *Nat. Genet.* **29:** 375–376.

105. **Wilkins, B. M., and A. T. Thomas.** 2000. DNA-independent transport of plasmid primase protein between bacteria by the I1 conjugation system. *Mol. Microbiol.* **38:**650–657.

106. **Wojciak, J. M., K. M. Connolly, and R. T. Clubb.** 1999. NMR structure of the Tn*916* integrase DNA complex. *Nat. Struct. Biol.* **6:**366–373.

107. **Yeo, H.-J., Q. Yuan, M. R. Beck, C. Baron, and G. Waksman.** 2003. Structural and functional characterization of the VirB5 protein from the type IV secretion system encoded by the conjugative plasmid pKM101. *Proc. Natl. Acad. Sci. USA* **100:**15947–15952.

108. **Zupan, J. R., D. Ward, and P. Zambryski.** 1998. Assembly of the VirB transport complex for DNA transfer from *Agrobacterium tumefaciens* to plant cells. *Curr. Opin. Microbiol.* **1:**649–655.

Enzyme-Mediated Resistance to Antibiotics: Mechanisms, Dissemination, and Prospects for Inhibition
Edited by Robert A. Bonomo and Marcelo E. Tolmasky

Juan C. Alonso, Dolors Balsa, Izhack Cherny, Susanne K. Christensen, Manuel Espinosa, Djordje Francuski, Ehud Gazit, Kenn Gerdes, Ed Hitchin, M. Teresa Martín, Concepción Nieto, Karin Overweg, Teresa Pellicer, Wolfram Saenger, Heinz Welfle, Karin Welfle, and Jerry Wells

19

Bacterial Toxin-Antitoxin Systems as Targets for the Development of Novel Antibiotics

PROTEIC TA SYSTEMS LOCATED ON PLASMIDS

In the past, toxin-antitoxin (TA) systems had been found only on low-copy-number plasmids and were shown to play a role in the postsegregational killing of bacterial cells that have lost the plasmid. Typically, bacterial plasmids are not essential for survival of the host cell except in specific environments where they might confer a selective advantage as in the case of antibiotic resistance or the virulence plasmids of pathogens, and several mechanisms that prevent plasmid loss during cell division have been reported. Plasmid TA coding genes are part of a very efficient plasmid maintenance mechanism that inhibits the proliferation of daughter cells that have not inherited a plasmid copy (42). While the T gene invariably encodes a protein, the A gene either is transcribed to produce an antisense RNA that inhibits translation of the T gene mRNA or encodes a protein that forms a stable TA complex with the toxin, thereby inhibiting its activity (33). This review focuses only on the proteic TA systems, but antisense RNA-regulated TA systems have been extensively studied and reviewed by Jensen and Gerdes (42). If a daughter cell fails to inherit a plasmid after cell division, then the differential stability of the A relative to the T will result in postsegregational killing or growth inhibition of the plasmid-free cell. This is a consequence of the presence of the toxin and antitoxin in the cytosol of all daughter cells during cell division and of the higher rate of turnover of the A by cellular proteases such as Lon and ClpP.

The well-characterized TA systems found on plasmids are listed in Table 19.1 and include CcdAB of plasmid F, Phd-Doc of prophage P1, and PemIK of plasmid R100 from *Escherichia coli*, as well as the ε-ζ, the Axe-Txe, and the RelBE2 and YefM-YoeB systems from gram-positive *Streptococcus*, *Enterococcus*, and *Streptococcus pneumoniae*, respectively (10, 27, 36, 71, 72). The plasmid TA systems

Juan C. Alonso and M. Teresa Martín, Department of Microbial Biotechnology, Centro Nacional de Biotecnología, CSIC, Darwin 3, 28049 Madrid, Spain. **Dolors Balsa and Teresa Pellicer,** Department of Pharmacological Biochemistry, Laboratorios SALVAT S.A., Barcelona, Spain. **Izhack Cherny and Ehud Gazit,** Department of Molecular Microbiology and Biotechnology, George S. Wise Faculty of Life Sciences, Tel-Aviv University, Tel-Aviv 69978, Israel. **Susanne K. Christensen and Kenn Gerdes,** Department of Biochemistry and Molecular Biology, South Denmark University, Odense M, Denmark. **Manuel Espinosa and Concepción Nieto,** Department of Protein Science, Centro de Investigaciones Biológicas, CSIC, Ramiro de Maeztu, 9, 28040-Madrid, Spain. **Djordje Francuski and Wolfram Saenger,** Institute for Chemistry and Biochemistry/Crystallography, Freie Universität Berlin, Takustr. 6, D-14195 Berlin, Germany. **Ed Hitchin and Karin Overweg,** Institute of Food Research, Norwich Research Park, Norwich, United Kingdom. **Heinz Welfle and Karin Welfle,** Max Delbrück Center for Molecular Medicine Berlin-Buch, Robert-Roessle-Str. 10, D-13125 Berlin, Germany. **Jerry Wells,** Swammerdam Institute for Life Sciences, University of Amsterdam, 1018 WV, Amsterdam, The Netherlands.

Table 19.1 TA systems identified on plasmids and chromosomes

Name of TA system (plasmid or chromosome); species	Other nomenclatures cited in literature	Antitoxin (length [aa])	Toxin (length [aa])	Cellular target of toxin	Protease degrading antitoxin
*ccd*AB[a] (plasmid F); *E. coli*	H/G letAB	CcdA (72)	CcdB (101)	DNA gyrase	Lon
phd-doc[b] (plasmid P1); *E. coli*		PhD (73)	Doc (126)	Unknown	ClpPX
kis-kid[a,c] (plasmid R1); *E. coli*	PemIK (plasmid R100)	Kis (84)	Kid (110)	mRNA	Unknown
parDE[b] (plasmid RK2/RP4); *E. coli*		ParD (83)	ParE (103)	DNA gyrase	Unknown
higAB (plasmid Rts1); *E. coli*		HigA (104)	HigB (92)	Unknown	Unknown
relBE[b] (plasmid R307); *E. coli*		RelB (83)	RelE (95)	mRNA	Unknown
mvpAT (pMYSH6000); *Shigella flexneri*		MvpA (75)	MvpT (133)	Unknown	Unknown
pasABC[b] (pTF-FC2); *Thiobacillus ferrooxidans*		PasA (74)	PasB (90)	Unknown	Unknown
axe-txe[b] (pRUM); *Enterococcus faecium*		Axe (89)	Txe (85)	Unknown	Unknown
εζ (pSM19035); *S. pyogenes*		ε (90)	ζ (287)	Unknown	LonA
relBE (chromosome[b,d]); *E. coli*		RelB (79)	RelE (95)	mRNA in ribosomal A site	Lon
$relBE_{III}$ (chromosome[b,d]); *E. coli*	YefM-YoeB	$RelB_{III}$ (83)	$RelE_{III}$ (84)	mRNA	Unknown
chpAIK (chromosome[c,d]); *E. coli*	MazEF, $RelBE_{IV}$	ChpAI (82)	ChpAK (111)	mRNA	ClpAP
ChpBIK (chromosome[c,d]); *E. coli*	$RelBE_{V}$	ChpBI (83)	ChpBK (116)	mRNA	Unknown

[a]Structure conservation.
[b]Sequence conservation with the chromosomally located RelBE system.
[c]Sequence conservation with the ChpAIK system.
[d]The chromosomal TA systems have been identified in gram-positive and gram-negative bacteria and in *Archaea*.

share common functional and organizational characteristics. The genes encoding the antitoxins (72 to 84 amino acids long) precede the genes encoding the toxins (90 to 130 amino acids long) except in the case of the *higA-higB* system of plasmid Rts1 (96). The *pasABC* and ωεζ TA systems are unique in that they consist of three components and not two. The ε-ζ PasA-PasB proteins form a TA complex as in the other TA systems; PasC is also required to effectively neutralize the effects of PasB (92), but the role of ω is poorly characterized.

Expression of several TA systems is autoregulated at the level of transcription, as in the *ccdAB* system, where the TA complex binds to operator sites in the *ccd* promoter (95). In the *kis-kid*, *pasABC*, and *phd-doc* systems, the antitoxin can serve as a repressor on its own, but full repression requires a combination of both toxin and antitoxin (54, 85). In contrast, the *parDE* system of *E. coli* plasmid RK2 is regulated solely by the ParD antitoxin (84). In the case of the TA encoded by streptococcal plasmids of the *inc18* incompatibility group there is a third component (protein ω) of the ωεζ operon that regulates TA expression, plasmid copy number, and accurate segregation (23, 25, 98, 99).

The first and most extensively studied system is that of CcdB, which has been shown to target the GyrA subunit of DNA gyrase, a tetramer formed by the association of two GyrA and two GyrB subunits (6, 20, 60). Structural and genetic studies of mutants resistant to killing by CcdB suggest that the CcdB toxin forms a symmetric dimer that binds into the head of the GyrA interface, thus inhibiting DNA gyrase activity during DNA replication (57). Other evidence suggests that GyrA is also the target of ParE toxin (38, 43). The mechanism of other plasmid-located TA toxins is not known, although the crystal structure and site-directed mutagenesis of the ζ toxin from *Streptococcus pyogenes* pSM19035 showed that free ζ acts as a phosphotransferase by using ATP/GTP (64).

Despite the functional parallels of the different plasmid TAs, there is only a low level of similarity in amino acid sequence between the toxin and antitoxin proteins of the different systems. Interestingly, DNA gyrase has been shown to be the target of both ParE and CcdB toxins even though they do not share even weak homology or similarity in amino acid sequence (84). Weak protein similarity exists between the antitoxins ParD of pRK2 and PasA of pTF-FC2 and between the antitoxins CcdA and Kid/PemI. The crystal structure of the Kid toxin showed that it resembles the CcdB dimer despite the lack of protein similarity (38), and biochemical and structural studies have shown that the Kid toxin acts as a specific endoribonuclease (47), as in the case of toxins RelE (15), MazF (70, 104), YoeB (46), and HigB (18).

CHROMOSOMALLY LOCATED PROTEIC TA SYSTEMS

Intriguingly, TA systems are also encoded in the genomes of a variety of both eubacteria and archaea (37, 80), suggesting that vertical transmission of a common ancestor of the TA systems that appeared early in evolution could account for the ubiquity of the TA systems. It is likely, therefore, that the TA cassettes present on chromosomes could have become the source of genes for the plasmid TA systems (19, 32, 83).

E. coli has several well-characterized TA systems located in the chromosome, the *relBE* locus being the most extensively studied (14–16) (reviewed in reference 32). Another studied TA system is *yefM–yoeB* (11, 12), which has homologs in a plasmid from the gram-positive (G+) bacterium *Enterococcus faecium* (36) and in the chromosome of the G+ pathogen *S. pneumoniae* (71). In addition to those, other studied chromosomal TA loci have been detected in the *E. coli* K-12 chromosome, namely *dinJ–yafQ* (31, 80), *mazEF* (or *chpA-chpBK*, named for chromosomal homologue of plasmid-encoded genes [46, 62]). Cytotoxins MazF and ChpB-K (MazF-2) both belong to a toxin family distinct from the RelE family (31) and are frequently found in bacteria (80). In the case of the *relBE* loci, they are very common both among *Bacteria* and *Archaea* (35, 80), a fact that has allowed the determination of the crystal structure of the archaeal equivalent of the RelBE complex, termed aRelBE (94). Homologies are found between plasmid Kis-Kid/PemIK TA and the chromosomal ChpAIK/MazEF and ChpBI-ChpBK systems. For example, Kis-Kid/PemIK shares 34% sequence identity and 69% sequence similarity to the MazEF/ChpAIK system found in the *E. coli* chromosome. The *relBE* and *chp* loci are present on a large number of prokaryotic chromosomes, often in multiple copies. For example, the *E. coli* chromosome has three *relBE* loci and two *chp* loci and the chromosomes of the archaea *Methanococcus jannaschii* and *Archaeoglobus fulgidus* each have four *relBE* loci. In the case of the G+ bacterium *S. pneumoniae*, two *relBE* loci have been found, although only one of them, RelBE2*Spn*, was shown to be functional (72). The RelE and ChpAK (MazF) of *E. coli* have both been shown to inhibit protein translation (81). Furthermore, a previous report indicated that Kid toxin, a homolog of ChpAK TA, targets a factor in replication (87), a fact that now is understood since Kid has RNase activity (70) and would cleave the RNA primer of plasmid ColE1, the replicon assayed for Kid activity.

RelE and ChpK homologs were divided into two separate protein families, because database searches and pairwise sequence alignments did not reveal any obvious sequence similarity between the two groups of proteins (30, 69). On the other hand, using the Hidden Markov Model algorithm we have been able to find significant similarity between YoeB, encoded by the *E. coli yefM-yoeB* locus (11, 36), and RelE, which makes it likely that both toxins belong to the RelE family of proteins. The finding that YoeB proteins are related to RelE leads to a large expansion of the RelE family (30, 36, 69). The recent cloning and characterization of the pneumococcal *yefM-yoeB* TA locus and the solution of the structure of the YefM-YoeB complex (45) allowed the molecular modeling of YefM*Spn* and YoeB*Spn* proteins. The results indicated that the toxins of the family were homologous, whereas the antitoxins seem to be specifically designed for each toxin counterpart (71).

In general, it seems that three chromosome-encoded TA systems (*relBE*, *yefM-yoeB*, and *dinJ-yafQ*) are homologous to the plasmid-encoded *relBE* system and therefore are members of this family (30, 35, 80). However, it has been discussed that the *yefM-yoeB* and *relBE* TA systems may belong to different families as their toxins and antitoxins differ substantially either functionally or structurally (11). In this sense it is also worth noting that the *E. coli* YefM antitoxin could not complement its pneumococcal counterpart as an efficient antidote of the YoeB*Spn* toxin (71).

There are also similarities in the organization of the genes and functional regions within the TA proteins, suggesting a common origin for these systems. Like the plasmid TA systems, the chromosomal *relBE* and *mazEF/chpAIK* operons are negatively autoregulated by the combined action of the T and the A and the A is sensitive to cellular proteases in its rapid turnover. Translation of the *mazEF/chpAIK* and *relBE* operons is inhibited during conditions of nutritional starvation, leading to diminished levels of the MazE and RelB antitoxins and growth inhibition by the toxins MazE/ChpAI and RelE. Originally it was proposed that overproduction of the bacterial toxins leads to cell death, helping to ensure the survival of the population through the sacrifice of starved cells (the so-called "altruistic" response [27]). However, recent evidence has shown that expression of MazF/ChpAK, RelE, or ζ induces a bacteriostatic condition that can be fully reversed by ectopic production of the cognate A within a certain time interval (<180 min of halted cell growth [9, 72, 82]). This indicates a potential role for these TA systems in temporarily inhibiting specific cell processes under severe stress (i.e., nutritional starvation) until more favorable conditions return (30, 32, 73). This is consistent with the observation that the toxins target macromolecular synthesis pathways, mainly affecting translation by mRNA cleavage or an uncharacterized target as the ζ toxin; these effects, if left unchecked, might perpetrate cell death of a fraction of the cell population (6, 31, 35, 70).

MECHANISTIC STUDIES ON THE TOXINS OF TA SYSTEMS

The chromosomal RelEB system originally identified in *E. coli* is currently the best characterized and understood of the chromosomal TA systems. Nutritional starvation induces degradation of the A (RelB) in a Lon protease-dependent manner, resulting in accumulation of the T (RelE) that inhibits protein translation and results in growth arrest of the cell (82). As the *relBE* promoter is negatively autoregulated by the TA complex, transcription of the TA genes is stimulated under conditions, such as nutrient starvation, that contribute to increased proteolytic turnover of the RelB protein. However, under favorable growth conditions there is accumulation of both proteins, resulting in formation of the TA complex and repression of *relBE* transcription.

RelE has been shown to bind in the A site of the ribosome and to induce a ribonucleotidyl activity that results in cleavage of mRNAs bound to the ribosome both in vitro and in vivo (82). Cleavage occurred between the second and third bases of both sense and stop codons and was highly dependent on the actual codon present in the ribosomal A site. Thus, the stop codons UAG and UAA were cleaved 1,000- and 100-fold more specifically than the UGA stop codon in terms of relative K_{cat}/K_m values. The catalytic nature of this activity in vitro (82) is consistent with the very high toxicity of the protein in vivo (16, 81). Addition of release factor I (RF I) inhibited RelE-mediated mRNA cleavage in vitro (82). Since RF I binds firmly to the ribosomal A site (103), this observation indicates that RelE must have access to the A site in order to induce mRNA cleavage. The mechanism of mRNA cleavage is of considerable interest because it was not previously shown that translation can be inhibited by mRNA cleavage. Cleavage of mRNA at the ribosomal A site leads to stalling of the ribosomes at the damaged mRNA. In bacteria, such complexes are rescued by a regulatory RNA called tmRNA. The ribosomes trapped in an inactive state at the 3′-end of the RelE-cleaved mRNA can then be rescued by a regulatory tmRNA, and the translated tmRNA-tagged proteins are degraded by intracellular proteases to recycle amino acids into the available pool (49, 50, 82). Consistent with this hypothesis, RelE (and ChpAK) toxin was counteracted by the simultaneous overproduction of tmRNA (15). Overproduction of ChpAK, ChpBK, and YoeB toxins also induces mRNA cleavage, indicating that they function through a mechanism similar to that of RelE (17). However, the RNA cleavage patterns induced by the four toxins were all different and in some cases the requirement for the ribosomal A site is not obvious, suggesting that the toxins play a direct role in the cleavage reaction.

In the case of the *ccd* locus of the *E. coli* plasmid F, it encodes CcdB toxin and CcdA antitoxin (77). CcdB toxin is an efficient inhibitor of DNA gyrase in vivo (6) and in vitro (7). Thus, overexpression of CcdB leads to arrest of replication and inhibition of cell growth, whereas in the case of the ωεζ locus of the streptococcal plasmids of the *inc18* group (namely, pSM19035, pIP501, pAMβ1, pRE25, etc.), they encode the toxin ζ and the antitoxin ε (9, 64). The primary target of the toxin ζ is not DNA, RNA, or protein synthesis and remains to be characterized (56, 64).

BIOPHYSICAL ANALYSIS OF TA SYSTEMS

The proteins of three TA systems from gram-negative bacteria, namely, CcdA-CcdB (21, 22, 97), Phd-Doc (28, 29, 53, 59), ParD-ParE (44, 74, 75, 84), YefM-YoeB (11, 12, 71), and one system from a plasmid from a G+ bacterium (ω-ε-ζ) (9, 24, 56, 67, 105, 106), have been studied in vitro with respect to their properties in solution (as isolated proteins and complexes) and binding to DNA. We summarize the knowledge accumulated on these proteins in the next paragraphs.

CcdA-CcdB

Toxin CcdB and antitoxin CcdA, the two proteins of the plasmid addiction system *ccd* of the F plasmid, have been analyzed by guanidinium chloride and urea unfolding at different temperatures and pH values. Thermal unfolding was monitored by changes in circular dichroism (CD) and by differential scanning calorimetry (DSC), and the stabilizing effect of DNA was also investigated (22). This study revealed that both proteins denature in a two-state equilibrium (native dimer versus unfolded monomer), and CcdA has a significantly lower thermodynamic stability than CcdB. The concentration dependence of the denaturation transition temperature for both proteins was calculated, as it is possibly physiologically significant because the expression of the *ccd* addiction proteins is autoregulated and their concentration cannot exceed a certain limit. Whereas DNA containing the *ccd* operator sequences increased the thermal stability of CcdA, there was no effect of DNA on the thermal stability of CcdB. In solution, alternate interactions between CcdA and CcdB are possible, leading to various complexes with different stoichiometries (21). CcdA has two binding sites for CcdB and vice versa, permitting soluble hexamer formation but also causing precipitation. The presence of CcdB was shown to enhance the affinity and the specificity of CcdA-DNA binding and results in a stable CcdA:CcdB-DNA complex with a CcdA:CcdB ratio of one (21).

Phd-Doc

Phd binds directly to Doc, enabling the purification of the proteins as a complex. Gel filtration and analytical ultracentrifugation revealed a trimeric complex (P_2D) with one molecule of Doc and two molecules of Phd. CD experiments showed that changes in secondary structure accompany the formation of the heterotrimeric complex (28). Studies of Phd and Doc molecules labeled with fluorescent energy donor and acceptor groups gave an equilibrium dissociation constant of about 0.8 μM and a very short, subsecond half-life of complex dissociation (28). The secondary structure and thermal stability of Phd, the effect of operator DNA binding on the structure and stability of Phd, and the stoichiometry, affinity, and cooperativity of Phd binding to operator subsites and intact operator DNA were investigated (29). Phd folds as a monomer at low temperatures or in the presence of 2 M trimethylamine *N*-oxide but exists predominantly in an unfolded conformation at 37°C. Phd is a substrate of the ClpXP serine protease of *E. coli* (53), and it seems probable that the low thermodynamic stability of Phd facilitates its in vivo degradation. The native state of Phd is stabilized by operator binding. Two Phd monomers bind to each operator subsite, and four monomers bind to the intact operator. Experimental evidence suggests that two Phd monomers bind cooperatively to each operator subsite, but there seems not to be significant dimer-dimer cooperativity (29).

ParD-ParE

Complex formation between ParD and ParE was studied by using a ParE protein derivative, designated ParE′ which has, on the N terminus, three additional amino acids and a methionine in place of the wild-type leucine residue. Covalent cross-linking with glutaraldehyde produced band patterns which were consistent with the presence of dimeric forms of ParD and ParE′ in solution when either ParD alone or ParE′ alone was incubated; when the two proteins were mixed and cross-linked, bands appeared, which suggested the formation of heterodimers, trimers, and tetramers (84). Light-scattering studies and gel filtration chromatography showed the existence of a stable dimer of ParD in solution (74). The CD spectra of ParD showed a typical α-helical pattern with minima at 222 and 208 nm and a strong maximum at 195 nm; calculation of the secondary structure yielded 33 to 36% α-helix content (74). Analysis of secondary structure based on the chemical-shift indices, sequential nuclear Overhauser enhancements, and $^3J_{H\alpha NH}$ scalar coupling data showed that the N-terminal domain of ParD consists of a short β-ribbon followed by three α-helices (75), suggesting the presence of the ribbon-helix-helix fold as present in the Arc/MetJ superfamily. Further nuclear magnetic resonance (NMR) data showed that ParD is divided into two separate domains, a well-ordered N-terminal domain and a very flexible C-terminal domain. Thermal stability was analyzed by monitoring the CD at 201 and 222 nm and revealed a remarkably high thermostability with a melting temperature (T_m) of 60.7°C; the complex of ParD with its cognate DNA site has an even higher thermostability with a T_m of 71°C (74). DSC measurements of ParD indicate a single two-state transition of ParD upon heating with a transition temperature T_m of 63.5°C and an unfolding enthalpy ΔH of 25.8 kcal/mol of the monomer (74). The ratio of calorimetric heat change, ΔH, over van't Hoff heat change, ΔH_v, is approximately 0.45, indicative of a single coupled two-state transition of two monomers. DSC measurements of ParD:DNA complexes yielded transition temperatures of 73 and 76°C at ParD:DNA ratios of 2:1 and 4:1, respectively, suggesting the binding of cognate DNA by tetramers of ParD (74).

YefM-YoeB

Two chromosomally encoded TA systems from *E. coli* and from *S. pneumoniae* have been characterized. In the case of the YefM*Eco* antitoxin, it has been shown that the protein is natively unfolded (11). Purification of YoeB and YefM proteins showed that they coeluted as single peaks in chromatographic columns, indicating the formation of a YoeB-YefM complex. Studies of fluorescence anisotropy of purified YefM-YoeB proteins showed a 2:1 stoichiometry of the complex, thus demonstrating the generation of a physical complex between both proteins. CD spectroscopy showed that, similar to the Doc toxin (see above), YoeB is a well-folded protein, a state that was confirmed by thermal denaturation experiments (12).

In the case of the pneumococcal TA system, CD analyses showed that protein YefM*Spn* was also partially unfolded, but to a lesser extent than the *E. coli* antitoxin (71). Thermal stability of YefM*Spn* and of the (YefM-YoeB)*Spn* complex showed that the proteins underwent unfolding with temperature increases, but the unfolding rate of the antitoxin was faster than that of the complex, with a thermal stability of the antitoxin of ~45°C and that of the complex of ~70°C. In addition, the antitoxin YefM*Spn* was relatively resistant to heat (about 85°C), suggesting that the antitoxin lacks a significant hydrophobic core, which, in turn, could be relevant to maintain it proteolitically unstable (71).

ω-ε-ζ

In gram-positive bacteria so far only one proteic TA system has been biophysically characterized in detail (9, 25, 64, 67, 98, 99). This system is encoded by the ω-ε-ζ operon of the low-copy-number, broad-host-range plasmid pSM19035 from *S. pyogenes* (105) and differs in several important properties from those of gram-negative bacteria. It is composed of three proteins, namely the transcriptional regulator ω, the antitoxin ε, and the toxin ζ. Neither the A nor the TA complexes are involved in the regulation of the *ω-ε-ζ* operon, which is exclusively achieved by ω. The ω protein occurs as a homodimer in solution (ω_2) (67, 98, 99). Protein ω_2 consists of a short β-ribbon followed by two α-helices, a typical ribbon-helix-helix fold as present in the Arc-MetJ superfamily (78). Protein ω_2 unfolds thermally with half-transition T_ms between ~43 and ~78°C depending on the ionic strength of the buffer and binds its operator binding sites with high affinity (67). A Raman spectroscopic analysis of ω_2-operator DNA complexes revealed a sequence-specific induced fit of both interacting macromolecules with binding of ω_2 to the major groove (24, 25). Recombinant ε and ζ proteins coelute in chromatographic purification steps and form a stable complex that, according to analytical ultracentrifugation and gel filtration, exists as an $\varepsilon_2\zeta_2$ heterotetramer in solution. Unfolding studies monitoring CD and fluorescence changes show that ζ has a significantly lower thermodynamic stability than ε both in free state and in the complex (9).

The functioning of postsegregational growth inhibition systems is based on different in vivo decay rates of the two proteins involved: in a Lon protease-dependent manner the A is degraded, and the T, which is much more stable, accumulates (9, 13). An interesting aspect of the physicochemical studies on TA systems concerns the possible correlation between in vivo lifetime and thermodynamic stability. Unfolded or thermodynamically unstable proteins are candidates for fast degradation in vivo; thus, one might expect to find a correlation between the in vivo stability and thermodynamic stability of antitoxin and toxin proteins. Consistent with this idea was the finding that the antitoxins CcdA and Phd have a lower thermodynamic stability than the toxins CcdB and Doc. However, this generalization is obviously not justified in view of the high thermostability of antitoxin ParD and the significantly higher thermodynamic stability of antitoxin ε in comparison to toxin ζ. Nevertheless, the lack of hydrophobic core in the antitoxins (leading to a higher thermal stability) could account for a higher proteolytic instability of the antitoxins. It is clear that more physicochemical studies are needed to get a clear understanding of the correlation between thermal stability and proteolytic cleavage.

THREE-DIMENSIONAL STRUCTURES OF TA SYSTEMS

Although many different TA systems have been characterized biochemically (see reviews in references 19, 27, 30, and 32), knowledge on their three-dimensional structures is still scarce. The presently published structures of TA systems are (i) the homodimeric toxins (without antitoxins) $CcdB_2$ (57) and the antitoxin CcdA in complex with DNA (58); (ii) the crystal structure of toxin Kid_2 (38) and the secondary structure elements of both Kid and Kis in solution, determined by NMR spectroscopy (48); (iii) the heterohexameric $MazF_2$-$MazE_2$-$MazF_2$ formed by antitoxin MazE and toxin MazF (45, 46); (iv) the heterocomplex of the archaeal RelBE TA system (94); (v) the YefM-YoeB complex from the *E. coli* chromosome (45); (vi) the heterotetrameric $\varepsilon_2\zeta_2$ formed by antitoxin ε and toxin ζ (64) and the regulator of the operon protein ω in complex with DNA (98); and (vii) FitAB from *Neisseria gonorrhoeae* bound to DNA (63). The functions of most of these toxins are known based on genetic, biochemical, and structural studies. CcdB interacts with the *E. coli* DNA-gyrase subunit A, thereby inhibiting DNA supercoiling (6, 57, 66); RelE, Kid, MazF, and YoeB are RNases that cleave mRNA (15, 45, 70, 104), and ζ is a putative phosphotransferase as deduced from the crystal structure (64).

Structures of CcdB and Kid

The amino acid sequences of the 101- and 110-amino-acid-long toxins CcdB and Kid, respectively, are only 11% identical (38). The three-dimensional structures of the respective homodimers have been determined by X-ray crystallography, and the structure of Kid_2 in solution has been studied additionally by NMR spectroscopy (38). The topographies of $CcdB_2$ and Kid_2 are very similar as both polypeptide chains fold into seven β-strands, of which strands β1, β2, β3, β6, and β7 form an antiparallel, twisted β-sheet (strand sequence, β7-β1-β2-β3-β6) (Fig. 19.1). This central β-sheet is decorated with two very short and one long C-terminal α-helices and with a small β-sheet (β4 and β5) that extrudes between β3 and β6. The arrangement of the short α-helices is different in $CcdB_2$ and Kid_2 (see the legend to Fig. 19.1).

In the crystals (and also in solution) CcdB and Kid occur as structurally similar homodimers (38, 57) (Fig. 19.1). They are each stabilized by hydrogen bonds between strands β6 and β6′ from two monomers to form an

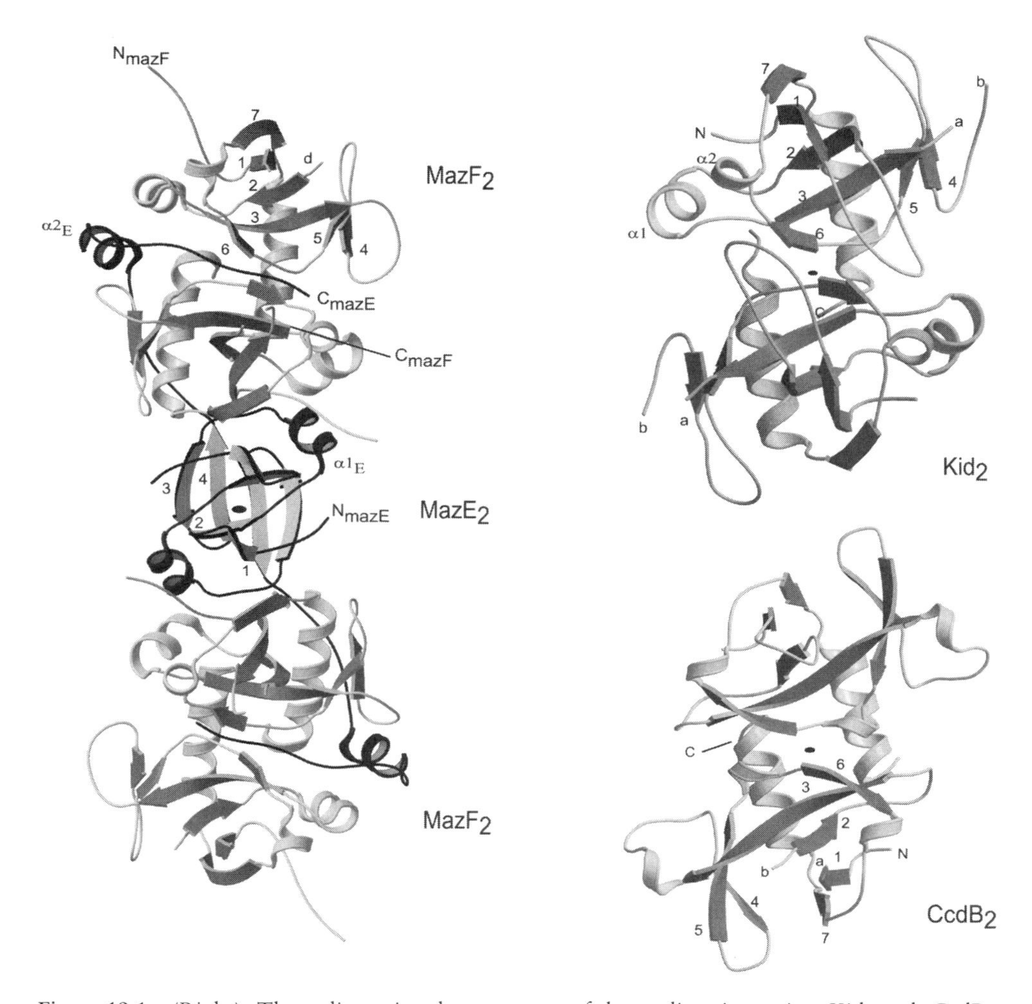

Figure 19.1 (Right) Three-dimensional structures of homodimeric toxins Kid_2 and $CcdB_2$; α-helices are shown as spirals and β-strands are shown as arrows (7, 20; PDB codes 1VUB [9] and 1M1F [5]). The noncrystallographic twofold axes relating the monomers are vertical to the paper plane and indicated by ellipses, β-strands are numbered, N and C termini are labeled, and a and b mark the positions where loops are disordered and not seen in the electron density. (Left) The TA complex $MazF_2$-$MazE_2$-$MazF_2$ (PDB code 1UB4 [8]) has twofold crystallographic symmetry indicated by the ellipse in the center of $MazE_2$. The orientation of the two $MazF_2$ is similar to that of $CcdB_2$ and Kid_2 in order to illustrate their structural homology. β-Strands of MazF and MazE are numbered, and α-helices of MazE are labeled $\alpha1_E$ and $\alpha2_E$; α-helices and loops of $MazE_2$ are drawn darker than for $MazF_2$. The $MazE_2$ C termini (one is labeled C_{mazE}) bind to the two $MazF_2$ homodimers. The loop between β-strands β1 and β2 in MazF is disordered (labeled d) and not seen in the electron density. The sequence of α-helices and β-strands is β1-β2-α1-β3-β5-β6-α2-β7-α3 in Kid and MazF, and in CcdB α1 is shifted and located between β5 and β6. For MazE the sequence is β1-β2-α1-β3-β4-α2.

extended, 10-stranded twisted β-sheet with noncrystallographic twofold rotation symmetry, the symmetry axis being oriented normal to the β-sheet. When viewed perpendicular to the twofold axis, the $CcdB_2$ and Kid_2 homodimers have a flat surface on one side formed by the two C-terminal α-helices and a convex surface on the other side (towards the viewer in Fig. 19.1). The surface electrostatic potentials of $CcdB_2$ and Kid_2 are very different (38).

Although $CcdB_2$ and Kid_2 have similar crystal structures, their targets are different; $CcdB_2$ interacts with the GyrA, and Kid_2 is an endoribonuclease. The crystal structures of

$CcdB_2$ and Kid_2 provided explanations for the results of several mutation studies. Thus, the three C-terminal amino acids on the flat surface of the $CcdB_2$ homodimer are important for interaction with the target GyrA. By contrast, the convex surface of the Kid_2 homodimer interacts not only with the target mRNA but also with the antitoxin Kis, suggesting that Kis is an antagonist for the toxic action of Kid_2 (38, 47, 48). The crystal structure of Kid_2 in solution has been solved and showed that both monomers of Kid are needed to form one mRNA binding site (47). As Kid_2 is symmetric, two mRNA binding sites are present on a dimer. However, several studies showed that only one molecule of mRNA is bound at each time. In this way, the C-terminal tail of the antitoxin Kis will partially occupy one of the RNA-binding sites, thereby disrupting the second RNA-binding pocket and inactivating the toxin (47). On the convex surface of both toxins, a number of conserved residues cluster near a highly flexible loop (Ser47 to Arg53 in Kid and Ser38 to Arg48 in CcdB) that is disordered in Kid_2 (indicated by a and b in Fig. 19.1) and implicated in binding of antitoxin CcdA to toxin $CcdB_2$; however, the antitoxins of CcdB and Kid do not cross-react with the toxins (86).

Structure of TA Complex $MazF_2$-$MazE_2$-$MazF_2$

MazF (111 amino acids) is a homodimer ($MazF_2$) that adopts a structure similar to that of Kid_2, with one major difference in the conserved loop between strands β1 and β2 (amino acids 17 to 27) that is ordered in Kid_2 and $CcdB_2$ but disordered in $MazE_2$-bound $MazF_2$ (46). The antitoxin MazE (82 amino acids) is composed of four antiparallel β-strands and two short α-helices (see the legend to Fig. 19.1) forming a globular $MazE_2$ core located right on a crystallographic twofold rotation axis (C2). The long C termini (amino acids 47 to 82) extend away from the C2 axis, and each wraps around and binds one $MazF_2$. On the basis of sequence similarities between Kid, CcdB, and MazF and Kis, CcdA, and MazE, it has been proposed that their TA complexes are of the form T_2-A_2-T_2 found for MazE/MazF (46).

These TA complexes bind the promoter DNA of their own operator, recognize the respective operator sequences, and regulate transcription of the associated genes (27, 30, 32, 105). The structure of $MazF_2$-$MazE_2$-$MazF_2$ suggests that Lys7 and Arg8, which are conserved in Kis and (partially) in CcdA, serve as primary DNA anchor for the T_2-A_2-T_2 complexes. Subsequent, stronger binding to DNA is, however, associated with structural rearrangement of $MazF_2$-$MazE_2$-$MazF_2$ so that multiple copies of T_2-A_2 cover the 47-bp operator motif containing two overlapping palindromes (61), in the form T_2-A_2-T_2-A_2-T_2 (46).

Structure of TA Complex $\varepsilon_2\zeta_2$

The X-ray structure of the nontoxic $\varepsilon_2\zeta_2$ complex (64) shows that ε forms a three-helix bundle, whereas ζ is composed of a six-stranded β-sheet in which the five N-terminal β-strands are parallel and the C-terminal β-strand is antiparallel, with strand sequence β2-β3-β1-β4-β5-β6 (Fig. 19.2, top panel). The $\varepsilon_2\zeta_2$ complex is dumbbell shaped with twofold, noncrystallographic symmetry. The twofold axis intersects with the handle that is formed by the ε dimer, and the two ζ subunits are bound distal to the twofold axis. The central β-sheet of ζ is decorated on both sides with nine α-helices, and at its C terminus, two additional solvent-exposed α-helices, K and L, are appended.

A search in the Protein Data Bank showed that ζ and phosphotransferases or kinases feature comparable topographies (64). The scoring factors Z (9.2 to 5.4) are low but suggestive of structural similarities, whereas the amino acid sequence identities of only 4 to 15% are not. The most similar of these proteins to ζ is chloramphenicol phosphotransferase (Cmp) from *Streptomyces venezuelae* (41). If the structures of ζ and Cmp are superimposed, the enzymatically important amino acids of Cmp correspond to the following amino acids of ζ: Lys46 and Thr47 in the Walker A motif ^{40}GQPGSGKT^{47} of ζ; Arg158 and Arg171 in ζ would bind to nucleoside triphosphate (NTP); Asp67 is found in the same position as Asp37 in Cmp that transfers γ-phosphate from NTP to the substrate chloramphenicol (Fig. 19.3). These amino acids were mutated in ζ and led to variants that were nontoxic, suggesting that ζ indeed acts as phosphotransferase (64). A comparison of the structures of Cmp and ζ suggested the binding site for NTP, and a narrow cleft in ζ could be the binding site for a small-molecule substrate to which the γ-phosphate of NTP is transferred by Asp67, which is located between these two binding sites. In $\varepsilon_2\zeta_2$, the N-terminal α-helix of ε is inserted into the binding site of NTP and interferes with NTP binding to the Walker A motif. This explains on a structural basis why the $\varepsilon_2\zeta_2$ complex is nontoxic.

A BLAST search indicated that several genes encoding proteins in *Enterococcus*, *Streptococcus*, and *Lactococcus* are closely related to ε and ζ, suggesting that structure and functionality of the ε/ζ system occur commonly in gram-positive bacteria (64).

In all TA systems derived from plasmids or chromosomes of gram-negative bacteria, T and A are small proteins of similar size (27, 30, 32). Although the amino acid sequences and functions of the different systems are dissimilar, one might postulate on the basis of the $CcdB_2$, Kid_2, and $MazF_2$ structures that all these TA systems feature a comparable topography and are derived from a common ancestor. ε/ζ is the only well-known TA system

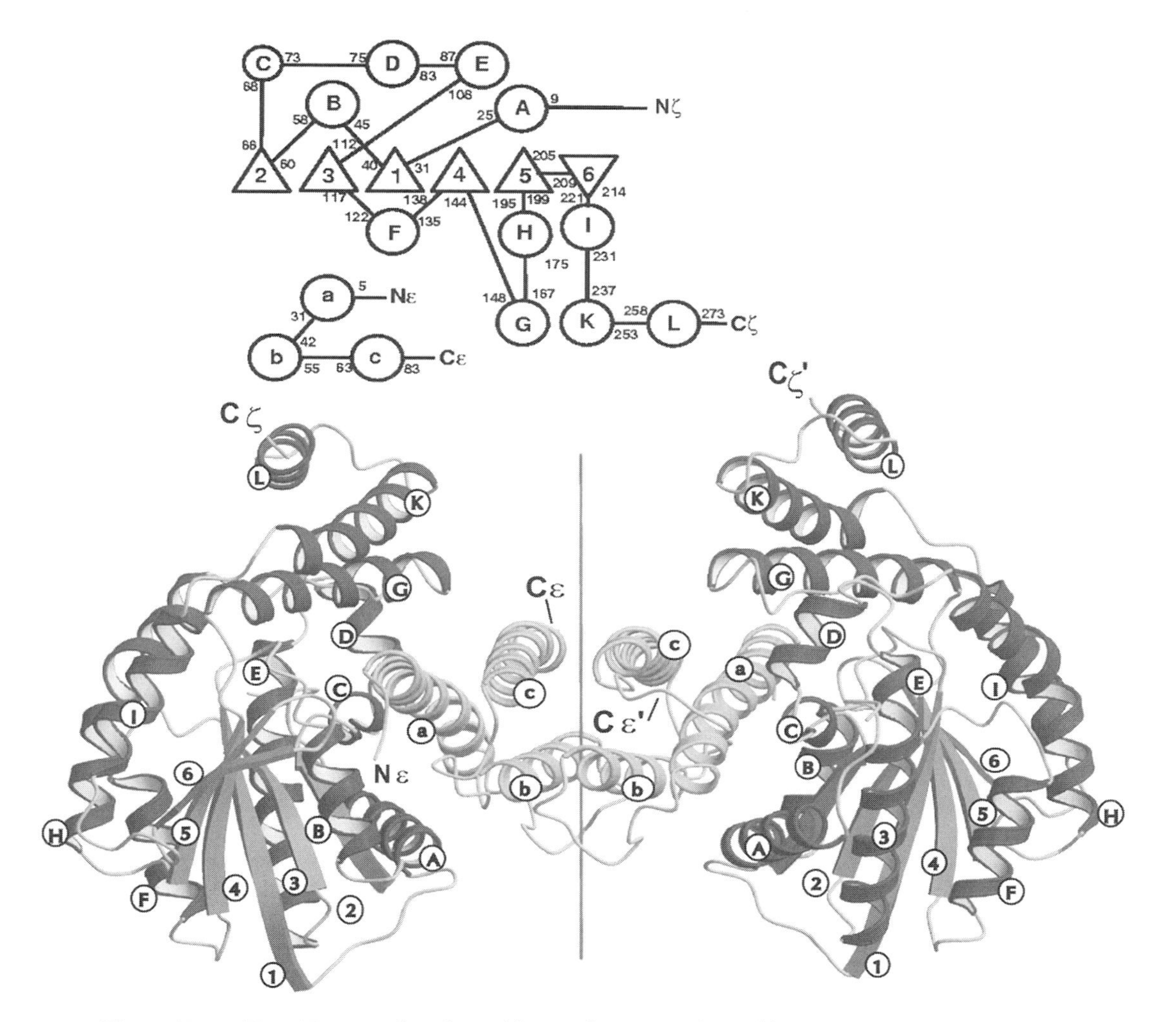

Figure 19.2 (Top) Topography of ε and ζ. α-Helices are indicated by circles and labeled a to c in ε and A to L in ζ. The polarity of β-strands (large numbers) is given by up and down pointing triangles, and small numbers mark positions in amino acid sequences. (Bottom) Three-dimensional dumbbell-shaped structure of complex $\varepsilon_2\zeta_2$ (PDB code 1gvn). α-Helices and β-strands are labeled as in the top panel. The noncrystallographic twofold axis relating εζ dimers in the heterotetrameric $\varepsilon_2\zeta_2$ complex is indicated by a vertical line; termini are labeled N, C, and C′.

from a plasmid from a G+ host (10, 105). Antitoxin ε with 90 amino acids is much shorter than toxin ζ with 287 amino acids. As the three-dimensional structure of ζ is very different from those of CcdB, Kid, and MazF except for the general architecture of a β-sheet (with parallel and antiparallel β-strands, respectively) decorated with α-helices, it appears that the TA systems from gram-negative and -positive bacteria are not derived from a common ancestor. Recent evidence shows that ζ has a cellular target different from those of CcdB, Kid, and MazF (56), but all of the toxins derived from TA systems have been shown to inhibit cell growth and/or result in cell killing following overexpression in their bacterial hosts.

TA SYSTEMS AS ANTIBIOTIC TARGETS

Pathogenic bacteria are subjected to an enormous selective pressure because of the indiscriminate overuse and misuse of broad-spectrum antibiotics. As a response to such pressure, bacteria develop resistance to one or more antibiotics, and these genetic traits can be transferred among different bacterial species by means of horizontal DNA

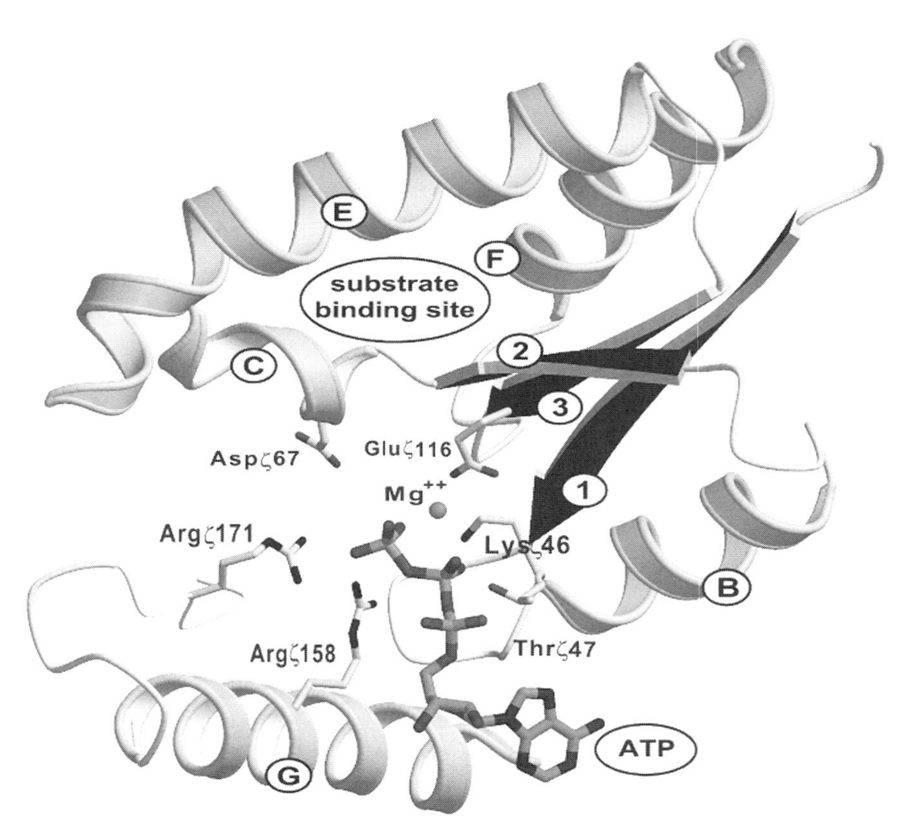

Figure 19.3 Putative active site of ζ (64). Functional amino acids are drawn and labeled, ATP and the essential Mg^{2+} are modeled according to the structure of Cmp, and the binding site for the yet unknown substrate is indicated by an ellipse.

transfer. Most antibiotics were originally isolated from organisms found in natural environments (e.g., the soil bacteria of the genus *Streptomyces*), and although many of these compounds have been chemically modified for clinical use, the genetic factors giving rise to resistance against these compounds were already present in microorganisms found in the environment. The introduction of DNA into microbial genomes, referred to as horizontal DNA transfer, can occur in several ways, i.e., by transformation, by bacteriophages (transduction), or via conjugative transposons. In addition, the multiple-antibiotic-resistance genes have been shown to be clustered into mobile DNA elements, called integrons, which are responsible for the recruiting of multiple smaller mobile gene cassettes that carry antibiotic resistance genes (52). Surveys of microbial genomes have revealed that up to 20% of the genome of some bacteria constitutes horizontally transferred DNA and the retention of this DNA over evolutionary time contributes to species diversification.

Some G+ pathogenic bacteria, like *Staphylococcus aureus* and *S. pneumoniae*, are considered to be the most important bacterial cause of human mortality and morbidity throughout the world (91). The recent appearance of vancomycin-resistant strains of pneumococcus poses an even greater threat to human life in the form of untreatable disease. Pneumococcus is naturally transformable with exogenous DNA, and horizontal DNA transfer among closely related streptococci is evident from the mosaic structure of certain virulence factors, such as the choline-binding proteins and penicillin-binding proteins (76). Therapy of pneumococcal disease is hampered by the increasing prevalence of antibiotic-resistant strains and suboptimal clinical efficacy of the available vaccines (5, 79). To tackle these problems, a number of avenues have been followed, vaccination being perhaps the most common strategy although in many cases the available vaccines are of limited use, e.g., due to their restricted serotype specificity (100). Rotation schemes for the use of antibiotics have also been proposed as a measure to prevent spread of resistance, the idea being that alleviation of the selective pressure for a particular antibiotic will lead to elimination of antibiotic-resistant bacteria from the environment (88). Synthesis of new chemical entities to search for new antibiotic compounds is still a very active field of research, especially in cases where macromolecular structures of drug targets are available to perform drug docking experiments. The use of naturally occurring peptide antibiotics or peptide-based inhibitors as design templates for the synthesis of compounds with similar physicochemical properties is also an approach used in pharmaceutical research (39). Other strategies include the targeting of key virulence factors, bacteriophage therapy, and the use of genomic approaches to identify genes that are essential for microbial survival or virulence. Now that more than 200 microbial genome sequences have been completed or are under way, it is possible to search for genes conserved among bacterial species that cause similar diseases or survive in similar sites within the body or environment; these represent potential targets for antimicrobials or anti-infective and vaccines. Genes such as the chromosomal TA systems that are involved in responses to environmental stresses are also potential targets for the development of drugs that would interfere with the microbes' ability to adapt to environmental stresses. The TA systems discussed above have been found in many species of eubacteria and archaea, and often more than one TA system resides in a genome (e.g., *S. pneumoniae* has two RelBE-like systems, although one of them is inactive [72]). Drugs that could interfere with the formation of TA complexes in bacteria would inhibit cell replication through the release of free toxin and increased expression of TA genes by preventing negative autoregulation of transcription via the TA complex.

Although drug-induced release of free toxin may not kill bacteria immediately, they would most likely be eliminated by the host's defense systems within a few days. Clearly, structural studies on individual components of TA systems as well as their complexes may provide insights into the nature of the molecular interactions and assist in the rational design of inhibitors of TA binding. Mechanistic and structural studies on the toxins themselves may also enable the design of small compounds that mimic the activity of the toxin in the active site of the cellular targets.

STRATEGIES FOR THE IDENTIFICATION OF TA INHIBITORS

Protein-protein interactions play a central role in almost every physiological event in any living cell. The nature of these interactions is diverse, but it always involves the specific binding of two or more proteins in a defined cellular compartment. TA systems in prokaryotes are an example of a protein-protein interaction whereby the toxin activity is prevented by its binding to the antidote antitoxin. The recognition of the importance of protein-protein interactions within the cell has led to their investigation as targets for novel inhibitors. Here, the approaches that can be used for screening of inhibitors of protein-protein interactions are highlighted by recent research on the TA systems.

The identification of small-molecule inhibitors represents one of the major challenges in the high-throughput screening (HTS) laboratory. There are three critical rate-limiting steps in the drug discovery process: (i) the availability of a high-purity compound library, (ii) the production of high-quality X-ray crystals of the protein partners, and (iii) the development of a rapid and reliable biological assay of the binding of both proteins that is amenable to high-throughput technology. It is often beneficial if the library has been designed with structures relevant for screening specifically against important target classes. The assay of a large compound library, by itself, has limited value for creating high-quality leads. Instead, it is proposed to use designed Lead Generation Libraries, so that any leads generated require less optimization and will result in candidates with a greater likelihood of clinical success (51).

The first steps in the design process will be taken at the computational level. In the beginning, compounds which have undesirable functional groups have to be filtered out. Then, the next step is to search for "drug-like" qualities, such as attractive Lipinski's rule-of-five; adsorption, distribution, metabolism, and elimination (ADME) or ADME-toxicity; and solubility properties, after which the most drug-like compounds could be chosen. Subsequently, pharmacophore analysis will be performed on the selected compounds. At this point, researchers have a group of in silico compounds that are drug-like and have desirable pharmacophore groups (68). The issue of chemical diversity is, however, a subject of ongoing debate and investigation in the scientific community, which has led to tremendous advances in the field (1). Nevertheless, at this time, there is still no uniformly accepted interpretation of "chemical similarity." One way to address the issues of chemical diversity is to accommodate these concerns by combining both empirical and mathematical approaches, such as the Tanimoto's coefficient (see, for instance, http://www.daylight.com/dayhtml/doc/theory/theory.finger.html). Diversity analysis is the final in silico stage of the selection process after which the most diverse subsets of the collection can be selected. After biological testing, quantitative structure-activity relationship studies could be performed.

The virtual screening can be carried out for the Advanced Libraries targeted to disrupt protein-protein interaction (93). For this, it is needed to produce the highest quality X-ray crystal structures of these systems, as in the case of the epsilon-zeta TA system discussed above (64). The compounds selected in the screen will then be synthesized and tested in vitro for initial antimicrobial activity. Then, it would be possible to obtain high-value compounds, based on the results of molecular docking, the structural comparison with known leads, and hits found in the biological assays.

Another major factor in the drug discovery process is the establishment of suitable screens to detect specific inhibition of the target protein-protein interaction. Currently, there are multiple experimental approaches to study protein interactions. These techniques range from the classical biochemical methods (purification, cross-linking, coprecipitation, and fractionation) to the latest resonance methods (fluorescent resonance energy transfer [FRET], bioluminescence resonance energy transfer [BRET], or surface plasmon resonance) (89). The study of protein-protein interactions was, until the introduction of genetic methods, restricted to tedious, time-consuming classical biochemical techniques, involving the isolation of proteins from natural sources, often with the loss of activity and/or associated cofactors. The introduction of the yeast-two hybrid system (YTH) and coimmunoprecipitation allowed researchers to study proteins inside a living cell. Thanks to these sensitive techniques, thousands of binding partners have been described. However, for several reasons these techniques have proved not to be robust enough to be considered the first choice to study the interaction between two proteins in an HTS laboratory: first, the high number of false positives (which lead to irrelevant biological interactions) and false negatives (predicted interactions are not always detected); secondly, the fact that interactions that depend upon cell-type-specific posttranslational modifications that do not occur in yeast cannot be assayed by this method; and thirdly, cofactors that cannot pass the nuclear membrane or whose activity is independent of transcription would never be found in a YTH system. Although coimmunoprecipitation has been shown to be highly suitable for localizing proteins within cellular compartments, the technique is laborious and time-consuming and requires a large number of cultured cells, making the process difficult to automate for HTS. In addition, coimmunoprecipitation and the YTH systems require confirmation of the physiological relevance of the found interactions through other classical biochemical methods. For all these reasons, these techniques are not applicable to an HTS format and are virtually confined to basic scientific research (4).

A group of techniques, commonly referred to as resonance energy transfer, provide an exciting new approach to study protein-protein interactions. We focus here on two resonance energy transfer techniques, namely, FRET and, especially, BRET, since they have demonstrated to be highly useful for studying interactions between two proteins that have been shown to form complexes. These techniques, which are around 50 years old, have recently flourished thanks to new labeling techniques, to the synthesis of new fluorescent dyes, and to the increase in sensitivity of optical instrumentation (90). These technologies are based on the distance-dependent transfer of energy from a donor molecule to an acceptor moiety with overlapping emission spectra and can be carried out both in vitro (with purified proteins) and in vivo (with intact bacterial or eukaryotic cells). These methods are very useful for large-scale screening because the assays can be miniaturized, allowing HTS formats and real-time kinetic studies to be performed (8).

Regarding FRET, when two fluorophores (the donor and the acceptor) with overlapping emission/absorption spectra are within ~50 Å of one another (from 10 to 100 Å) and in the correct orientation, the donor is able to transfer its excited-state energy to the acceptor. Therefore, if appropriate fluorophores are linked to the protein pair, the proximity of these proteins in the formed complex can be measured by determining whether fluorescence resonance energy is transferred from the donor to the acceptor. The incident light (at a discrete wavelength) excites the donor, which, if in close proximity to its partner pair, transfers the energy of its excited state to the acceptor that emits light of a distinct wavelength. The instrumentation registers the change in the intensity of both wavelengths and calculates the interaction under different physiological conditions. However, FRET presents some disadvantages, mainly due to the excitation of the donor by monochromatic light that directly excites the acceptor as a consequence of the overlapping spectra of both moieties. Photobleaching of the donor fluorophore by illumination of the sample can result in loss of resonance with time and can also induce autofluorescence of cellular components, which results in a high background noise. The same problem can be encountered when the assay is carried out in the presence of molecular entities whose potential activity is to be assayed in this system. A large number of compounds with very different origins, sources, and chemical structures have proven to emit light when excited with light of the same wavelength as that used with the fluorescent proteins in the assay. The resulting high background renders FRET too insensitive for quantitative purposes. For all these reasons, FRET is currently being used more often to visualize the dynamics of protein interactions with high spatial resolution (55).

In 1999, a variant of FRET, called BRET, which may overcome the problems associated with fluorescence excitation, was developed. In BRET, the donor fluorophore of the FRET pair is replaced by a luciferase, in which bioluminescence from the luciferase in the presence of a substrate excites the acceptor fluorophore through the same resonance energy transfer mechanism as FRET (102). BRET has one singular advantage over FRET: it does not require the use of excitation illumination, since the excitement is achieved by the enzymatic transformation of a luminogenic substrate by *Renilla* luciferase (101). An additional advantage of BRET is that the relative levels of expression of the donor and the acceptor can be independently quantified by measuring the luminescence of the donor and the fluorescence of the acceptor. Moreover, in identical immunoassays, BRET has proven to be 10-fold more sensitive than FRET in HTS (2). Furthermore, the instrumentation needed for BRET is simpler than that for FRET (no external light source is needed), though it requires a higher degree of sensitivity.

BRET is currently being used as the method of choice to unambiguously assess interactions between therapeutically relevant proteins. For instance, recent works have shown that BRET is a reliable quantitative and highly sensitive method to (i) monitor the activation state of the insulin receptor in vivo (40), (ii) assess dynamic hormone receptor interactions (26), (iii) observe the dimerization of the β_2-adrenergic receptor (65) and of a principal human immunodeficiency virus coreceptor (3), and (iv) study the interaction between TIF1 transcriptional regulators with ZNF74 nuclear matrix protein (34). Furthermore, BRET is also a powerful tool in the research pipeline of many pharmaceutical companies that seek protein-protein inhibitors through HTS in a broad range of therapeutic fields (e.g., glucocorticoid receptors homodimerization and EGF receptor/EGF binding). Prokaryotic TA systems are of considerable interest as drug targets, because the possible compounds that would prevent TA interactions may have the potential to be used as antimicrobial drugs with a new action mechanism and a lower incidence of acquired resistance than classical antibiotics. In fact, BRET assays have been developed and shown to be functional in the interaction of the pneumococcal RelBE*Spn* TA system (72).

SUMMARY

TA systems function as vitally important regulatory systems in bacteria and represent ideal targets for the development of novel antibiotic therapeutic agents. A broad mechanistic understanding of TA systems at physiological, biochemical, biophysical, and structural levels provides the scientific framework needed both for rational drug design and for elegant selection schemes using large pools of compounds. The BRET approach appears to be especially promising and is currently used by our team, BAS (Bacterial Apoptotic Systems), as part of the Quality of Life and Management of Living Resources in the Fifth EU Framework Programme (QKL3-CT-2001-00277).

These studies were supported by EU-project QKL3-CT-2001-00277, Deutsche Forschungsgemeinschaft (We 1745/5-1 and GK 80/2) (to H.W.), Deutsche Forschungsgemeinschaft (Sa 196/38-2) and Fonds der Chemischen Industrie (to W.S.), Spanish Ministerio de Ciencia y Tecnología (Acción Especial SAF2001-5040-E, to J.C.A. and M.E.). J.W. and K.O. are also grateful for financial support from the European Bacterial Proteomics (EBP) Network (EC contract QLK2-CT-2000-01536) and TCS-TARGETS (EC contract QLK2-CT-2000-00543). We thank S. K. Burley and K. Kamada for atomic coordinates of the MazEF complex prior to release by PDB.

References

1. **Agrafiotis, D. K., J. C. Myslik, and F. R. Salemme.** 1999. Advances in diversity profiling and combinatorial series design. *Mol. Divers.* **4:**1–22.
2. **Arai, R., H. Nakagawa, K. Tsumoto, W. Mahoney, I. Kumagai, H. Ueda, and T. Nagamune.** 2001. Demonstration of a homogeneous noncompetitive immunoassay based on bioluminescence resonance energy transfer. *Anal. Biochem.* **289:**77–81.
3. **Babcock, G. J., M. Farzan, and J. Sodroski.** 2003. Ligand-independent dimerization of CXCT4, a principal HIV co-receptor. *J. Biol. Chem.* **278:**3378–3385.
4. **Baler, R.** 2001. Clockless yeast and the gears of the clock: how do they mesh? *J. Biol. Rhythms* **16:**516–522.
5. **Bartlett, J. G., R. F. Breiman, L. Mandell, and T. M. File.** 1998. Community-acquired pneumonia in adults: guidelines for management. *Clin. Infect. Dis.* **26:**811–838.
6. **Bernard, P., and M. Couturier.** 1992. Cell killing by the F plasmid CcdB protein involves poisoning of DNA-topoisomerase II complexes. *J. Mol. Biol.* **226:**735–745.
7. **Bernard, P., K. E. Kezdy, L. van Melderen, L. Steyaert, L. Wyns, M. L. Pato, P. N. Higgins, and M. Couturier.** 1993. The F plasmid CcdB protein induces efficient ATP-dependent DNA cleavage by gyrase. *J. Mol. Biol.* **234:** 534–541.
8. **Boute, N., R. Jockers, and T. Issad.** 2002. The use of resonance energy transfer in high-throughput screening: BRET versus FRET. *Trends Pharmacol. Sci.* **23:**351–354.
9. **Camacho, A. G., R. Misselwitz, J. Behlke, S. Ayora, K. Welfle, A. Meinhart, B. Lara, W. Saenger, H. Welfle, and J. C. Alonso.** 2002. In vitro and in vivo stability of the epsilon2zeta2 protein complex of the broad host-range *Streptococcus pyogenes* pSM19035 addiction system. *Biol. Chem.* **383:**1701–1713.
10. **Ceglowski, P., A. Boitsov, S. Chai, and J. C. Alonso.** 1993. Analysis of the stabilization system of pSM19035-derived plasmid pBT233 in *Bacillus subtilis*. *Gene* **136:**1–12.
11. **Cherny, I., and E. Gazit.** 2004. The YefM antitoxin defines a family of natively unfolded proteins: implications as a novel antibacterial target. *J. Biol. Chem.* **279:**8252–8261.
12. **Cherny, I., L. Rockah, and E. Gazit.** 2005. The YoeB toxin is a folded protein that forms a physical complex with the unfolded YefM antitoxin: implications for a structural-based differential stability of toxin-antitoxin systems. *J. Biol. Chem.* **280:**30063–30072.

13. **Christensen, S. K., G. Maenhauf-Michel, N. Mine, S. Gothesman, K. Gerdes, and L. Van Melderen.** 2004. Overproduction of the Lon protease triggers inhibition of translation in *Escherichia coli*: involvement of the *yefM-yoeB* toxin-antitoxin system. *Mol. Microbiol.* **51:**1705–1717.

14. **Christensen, S. K., and K. Gerdes.** 2004. Delayed-relaxed response explained by hyperactivation of RelE. *Mol. Microbiol.* **53:**587–597.

15. **Christensen, S. K., and K. Gerdes.** 2003. RelE toxins from bacteria and Archaea cleave mRNAs on translating ribosomes, which are rescued by tmRNA. *Mol. Microbiol.* **48:**1389–1400.

16. **Christensen, S. K., M. Mikkelsen, K. Pedersen, and K. Gerdes.** 2001. RelE, a global inhibitor of translation, is activated during nutritional stress. *Proc. Natl. Acad. Sci. USA* **98:**14328–14333.

17. **Christensen, S. K., K. Pedersen, F. G. Hansen, and K. Gerdes.** 2003. Toxin-antitoxin loci as stress-response-elements: ChpAK/MazF and ChpBK cleave translated RNAs and are counteracted by tmRNA. *J. Mol. Biol.* **332:**809–819.

18. **Christensen-Dalsgaard, M., and K. Gerdes.** 2006. Two *higBA* loci in the *Vibrio cholerae* superintegron encode mRNA cleaving enzymes and can stabilize plasmids. *Mol. Microbiol.* **62:**397–411.

19. **Condon, C.** 2006. Shutdown decay of mRNA. *Mol. Microbiol.* **61:**573–583.

20. **Couturier, M., M. Bahassiel, and L. van Melderen.** 1998. Bacterial death by gyrase poisoning. *Trends Microbiol.* **6:**269–275.

21. **Dao-Thi, M.-H., D. Charlier, R. Loris, D. Maes, J. Messens, L. Wyns, and J. Backmann.** 2002. Intricate interactions within the *ccd* plasmid addiction system. *J. Biol. Chem.* **277:**3733–3742.

22. **Dao-Thi, M.-H., J. Messens, L. Wyns, and J. Backmann.** 2000. The thermodynamic stability of the proteins of the *ccd* plasmid addiction system. *J. Mol. Biol.* **299:**1373–1386.

23. **de la Hoz, A. B., F. Pratto, R. Misselwitz, C. Speck, W. Weihofen, K. Welfle, W. Saenger, H. Welfle, and J. C. Alonso.** 2004. Recognition of DNA by omega protein from the broad-host range *Streptococcus pyogenes* plasmid pSM19035: analysis of binding to operator DNA with one to four heptad repeats. *Nucleic Acids Res.* **32:**3136–3147.

24. **Dostál, L., R. Misselwitz, S. Laettig, J. C. Alonso, and H. Welfle.** 2003. Raman spectroscopy of regulatory protein Omega from *Streptococcus pyogenes* plasmid pSM19035 and complexes with operator DNA. *Spectroscopy* **17:**435–445.

25. **Dostál, L., F. Pratto, J. C. Alonso, and H. Welfle.** Binding of regulatory protein Omega from *Streptococcus pyogenes* plasmid pSM19035 to direct and inverted 7-base pair repeats of operator DNA. *J. Raman Spectrosc.* in press.

26. **Eidne, K. A., K. M. Kroeger, and A. V. Hanyaloglu.** 2002. Application of novel resonance energy transfer techniques to study dynamic hormone receptor interactions in living cells. *Trends Endocrinol. Metabol.* **13:**415–421.

27. **Engelberg-Kulka, H., and G. Glaser.** 1999. Addiction modules and programmed cell death and antideath in bacterial cultures. *Annu. Rev. Microbiol.* **53:**43–70.

28. **Gazit, E., and R. T. Sauer.** 1999. The Doc toxin and the Phd antidote proteins of the bacteriophage P1 plasmid addiction system form a heterotrimeric complex. *J. Biol. Chem.* **274:**16813–16818.

29. **Gazit, E., and R. T. Sauer.** 1999. Stability and DNA binding of the Phd protein of the phage P1 plasmid addiction system. *J. Biol. Chem.* **274:**2652–2657.

30. **Gerdes, K.** 2000. Toxin-antitoxin modules may regulate synthesis of macromolecules during nutritional stress. *J. Bacteriol.* **182:**561–572.

31. **Gerdes, K., S. Ayora, I. Canosa, P. Ceglowski, R. Díaz-Orejas, T. Franch, A. P. Gultyaev, R. Bugge Jensen, I. Kobayashi, C. Macpherson, D. Summers, C. M. Thomas, and U. Zielenkiewicz.** 2000. Bacterial plasmids and gene spread, p. 49–85. *In* C. M. Thomas (ed.), *The Horizontal Gene Pool.* Harwood Academic Publishers, Amsterdam, The Netherlands.

32. **Gerdes, K., K. S. Christensen, and A. Lobner-Olensen.** 2005. Prokaryotic toxin-antitoxin stress response loci. *Nat. Rev. Microbiol.* **3:**371–382.

33. **Gerdes, K., A. P. Gultyaev, T. Franch, K. Pedersen, and N. D. Mikkelsen.** 1997. Antisense RNA-regulated programmed cell death. *Annu. Rev. Genet.* **31:**1–31.

34. **Germain-Desprez, D., M. Bazinet, M. Bouvier, and M. Aubry.** 2003. Oligomerization of transcriptional intermediary factor 1 regulators and interaction with ZNF74 nuclear matrix protein revealed by bioluminescence resonance energy transfer in living cells. *J. Biol. Chem.* **278:**22367–22373.

35. **Gotfredsen, M., and K. Gerdes.** 1998. The *Escherichia coli relBE* genes belong to a new toxin-antitoxin gene family. *Mol. Microbiol.* **29:**1065–1076.

36. **Grady, R., and F. Hayes.** 2003. Axe-Txe, a broad-spectrum proteic toxin-antitoxin system by a multidrug-resistant, clinical isolate of *Enterococcus faecium. Mol. Microbiol.* **47:**1419–1432.

37. **Gronlund, H., and K. Gerdes.** 1999. Toxin-antitoxin systems homologous with relBE of Escherichia coli plasmid P307 are ubiquitous in prokaryotes. *J. Mol. Biol.* **285:**1401–1415.

38. **Hargreaves, D., S. Santos-Sierra, R. Giraldo, R. Sabariegos-Jareño, G. de la Cueva-Méndez, R. Boelens, R. Díaz-Orejas, and J. B. Rafferty.** 2002. Structural and functional analysis of the Kid toxin protein from *E. coli* plasmid R1. *Structure* **10:**1425–1433.

39. **Huttner, K. M., and C. L. Bevins.** 1999. Antimicrobial peptides as mediators of epithelial host defense. *Pediatr. Res.* **45:**785–794.

40. **Issad, T., N. Boute, and K. Pernet.** 2002. A homologous assay to monitor the activity of the insulin receptor using bioluminescence resonance energy transfer. *Biochem. Pharmacol.* **64:**813–817.

41. **Izard, T., and J. Ellis.** 2000. The crystal structures of chloramphenicol phosphotransferase reveal a novel inactivation mechanism. *EMBO J.* **11:**2690–2700.

42. **Jensen, R. B., and K. Gerdes.** 1995. Programmed cell death in bacteria: proteic plasmid stabilization systems. *Mol. Microbiol.* **17:**205–210.

43. **Jiang, Y., J. Pogliano, D. R. Helinski, and I. Konieczny.** 2002. ParE toxin encoded by the broad-host-range plasmid RK2 is an inhibitor of *Escherichia coli* gyrase. *Mol. Microbiol.* **44:**971–979.

44. **Johnson, E. P., A. R. Ström, and D. R. Helinski.** 1996. Plasmid RK2 toxin protein ParE: purification and interaction with the ParD antitoxin protein. *J. Bacteriol.* **178:**1420–1429.

45. **Kamada, K., and F. Hanaoka.** 2005. Conformational change in the catalytic site of the ribonuclease YoeB toxin by YefM antitoxin. *Mol. Cell* **19:**497–509.

46. **Kamada, K., F. Hanaoka, and S. K. Burley.** 2003. Crystal structure of the MazE/MazF complex: molecular bases of antidote-toxin recognition. *Mol. Cell* **11:**875–884.

47. **Kamphuis, M. B., A. M. J. J. Bonvin, M. C. Monti, M. Lemonnier, A. Muñoz-Gómez, R. H. H. van den Heuvel, R. Díaz-Orejas, and R. Boelens.** 2006. Model for RNA binding and the catalytic site of the RNase Kid of the bacterial *parD* toxin-antitoxin system. *J. Mol. Biol.* **357:**115–126.

48. **Kamphuis, M. B., M. C. Monti, R. H. H. van den Heuvel, S. Santos-Sierra, G. E. Folkers, M. Lemonnier, R. Díaz-Orejas, A. J. R. Heck, and R. Boelens.** Interactions between the toxin Kid of the bacterial *parD* system and the antitoxins Kis and MazE. *Proteins,* in press.

49. **Karzai, A. W., E. D. Roche, and R. T. Sauer.** 2000. The SsrA-SmpB system for protein tagging, directed degradation and ribosome rescue. *Nat. Struct. Biol.* **7:**449–455.

50. **Karzai, A. W., M. M. Susskind, and R. T. Sauer.** 1999. SmpB, a unique RNA-binding protein essential for peptide-tagging activity of SsrA (tmRNA). *EMBO J.* **18:**3793–3799.

51. **Kirkpatrick, D. L., S. Watson, and S. Ulhaq.** 1999. Structure-cased drug design: combinatorial chemistry and molecular modeling. *Comb. Chem. High Throughput Screen.* **2:**211–221.

52. **Lawrence, J. G.** 1999. Selfish operons: the evolutionary impact of gene clustering in prokaryotes and eukaryotes. *Curr. Opin. Genet. Dev.* **9:**642–648.

53. **Lehnherr, H., and M. B. Yarmolinsky.** 1995. Addiction protein Phd of plasmid prophage P1 is a substrate of the ClpXP serine protease of Escherichia coli. *Proc. Natl. Acad. Sci. USA* **92:**3274–3277.

54. **Lemonnier, M., S. Santos-Sierra, C. Pardo-Abarrio, and R. Díaz-Orejas.** 2004. Identification of residues of the Kid toxin involved in autoregulation of the *parD* system. *J. Bacteriol.* **186:**240–243.

55. **Li, H., E. Ng, S. Lee, M. Kotaka, S. Tsui, C. Lee, K. Fung, and M. Waye.** 2001. Protein-protein interaction of FHL3 with FHL2 and visualization of their interaction by green fluorescent proteins (GFP) two-fusion fluorescence resonance energy transfer (FRET). *J. Cell. Biochem.* **80:**293–303.

56. **Lioy, V. S., M. T. Martin, A. G. Camacho, R. Lurz, H. Antelmann, M. Hecker, E. Hitchin, Y. Ridge, J. M. Wells, and J. C. Alonso.** 2006. pSM19035-encoded {zeta} toxin induces stasis followed by death in a subpopulation of cells. *Microbiology* **152:**2365–2379.

57. **Loris, R., M.-H. Dao Thi, L. Bahassi, L. Van Melderen, F. Poortmans, R. Liddington, M. Couturier, and L. Wyns.** 1999. Crystal structure of CcdB, a topoisomerase poison from *E. coli. J. Mol. Biol.* **285:**1667–1677.

58. **Madl, T., L. van Melderen, N. Mine, M. Respondek, M. Oberer, W. Keller, L. Khatai, and K. Zangger.** 2006. Structural basis for nucleic acid and toxin recognition of the bacterial antitoxin CcdA. *J. Mol. Biol.* **364:**170–185.

59. **Magnuson, R., H. Lehnherr, G. Mukhopadhyay, and M. B. Yarmolinsky.** 1996. Autoregulation of the plasmid addiction operon of bacteriophage P1. *J. Biol. Chem.* **271:** 18705–18710.

60. **Maki, S., S. Takiguchi, T. Miki, and T. Horiuchi.** 1992. Modulation of DNA supercoiling activity of *Escherichia coli* DNA gyrase by F plasmid proteins. *J. Biol. Chem.* **267:**12244–12251.

61. **Marianovsky, I., E. Aizenman, H. Engelberg-Kulka, and G. Glaser.** 2001. The regulation of the *Escherichia coli mazEF* promoter involves an unusual palindrome. *J. Biol. Chem.* **276:**5975–5984.

62. **Masuda, Y., K. Miyakawa, Y. Nishimura, and E. Ohtsubo.** 1993. *chpA* and *chpB*, *Escherichia coli* chromosomal homologs of the *pem* locus responsible for stable maintenance of plasmid R100. *J. Bacteriol.* **175:**6850–6856.

63. **Mattison, K., J. S. Wilbur, M. So, and R. G. Brennan.** 2006. Structure of FitAB from *Neisseria gonorrhoeae* bound to DNA reveals a tetramer of toxin-antitoxin heterodimers containing Pin Domains and ribbon-helix-helix motifs. *J. Biol. Chem* **281:**37942–37951.

64. **Meinhart, A., J. C. Alonso, N. Strater, and W. Saenger.** 2003. Crystal structure of the plasmid maintenance system epsilon/zeta: functional mechanism of toxin zeta and inactivation by epsilon 2 zeta 2 complex formation. *Proc. Natl. Acad. Sci. USA* **100:**1661–1666.

65. **Mercier, J. F., A. Salahpour, S. Angers, A. Breit, and M. Bouvier.** 2002. Quantitative assessment of beta 1- and beta 2-adrenergic receptor homo- and heterodimerization by bioluminescence resonance energy transfer. *J. Biol. Chem.* **277:**44925–44931.

66. **Miki, T., J. A. Park, K. Nagao, N. Murayama, and T. Horiuchi.** 1992. Control of segregation of chromosomal DNA by sex factor F in *Escherichia coli.* Mutants of DNA gyrase subunit A suppress *letD* (*ccdB*) product growth inhibition. *J. Mol. Biol.* **225:**39–52.

67. **Misselwitz, R., A. B. de la Hoz, S. Ayora, K. Welfle, J. Behlke, K. Murayama, W. Saenger, J. C. Alonso, and H. Welfle.** 2001. Stability and DNA-binding properties of the omega regulator protein from the broad-host range *Streptococcus pyogenes* plasmid pSM19035. *FEBS Lett.* **505:**436–440.

68. **Mitchell, T., and G. A. Showell.** 2001. Design strategies for building drug-like chemical libraries. *Curr. Opin. Drug Discov. Devel.* **4:**314–318.

69. **Mitenhuber, G.** 1999. Occurrence of *mazEF*-like antitoxin-toxin systems in bacteria. *J. Mol. Microbiol. Biotechnol.* **1:**295–302.

70. **Muñoz-Gomez, A. J., S. Santos-Sierra, A. Berzal-Herranz, M. Lemonnier, and R. Diaz-Orejas.** 2004. Insights into the specificity of RNA cleavage by the *Escherichia coli* MazF toxin. *FEBS Lett.* **567:**316–320.

71. **Nieto, C., I. Cherny, S. K. Khoo, M. García de Lacoba, W. T. Chan, C. C. Yeo, E. Gazit, and M. Espinosa.** 27 October 2006. The *yefM-yoeB* toxin-antitoxin systems of *Escherichia coli* and *Streptococcus pneumoniae*: functional and structural correlation. *J. Bacteriol.* doi:10.1128/JB. 01130–06.

72. **Nieto, C., T. Pellicer, D. Balsa, S. K. Christensen, K. Gerdes, and M. Espinosa.** 2006. The chromosomal *relBE2* toxin-antitoxin locus of *Streptococcus pneumoniae*: characterization and use of a bioluminescence resonance energy transfer assay to detect toxin-antitoxin interaction. *Mol. Microbiol.* **59:**1280–1296.

73. **Nyström, T.** 1999. Starvation, cessation of growth and bacterial aging. *Curr. Opin. Microbiol.* **2:**214–219.

74. **Oberer, M., H. Lindner, O. Glatter, C. Kratky, and W. Keller.** 1999. Thermodynamic properties and DNA binding of the ParD protein from the broad host-range plasmid RK2/RP4 killing system. *Biol. Chem.* **380:**1413–1420.

75. **Oberer, M., K. Zangger, S. Prytulla, and W. Keller.** 2002. The anti-toxin ParD of plasmid RK2 consists of two structurally distinct moieties and belongs to the ribbon-helix-helix family of DNA binding proteins. *Biochem. J.* **361:**41–47.

76. **Obregón, V., P. García, E. García, A. Fenoll, R. López, and J. L. García.** 2002. Molecular peculiarities of the *lytA* gene isolated from clinical pneumococcal strains that are bile insoluble. *J. Clin. Microbiol.* **40:**2545–2554.

77. **Ogura, T., and S. Hiraga.** 1983. Mini-F plasmid genes that couple host cell division to plasmid proliferation. *Proc. Natl. Acad. Sci. USA* **80:**4784–4788.

78. **Pabo, C. O., and R. T. Sauer.** 1992. Transcription factors: structural families and principles of DNA recognition. *Annu. Rev. Biochem.* **61:**1053–1095.

79. **Pallares, R., P. F. Viladrich, J. Linares, C. Cabellos, and F. Gudiol.** 2000. Impact of antibiotic resistance on chemotherapy for pneumococcal infections, p. 157–168. *In* A. Tomasz (ed.), Streptococcus pneumoniae. Mary Ann Liebert Inc. Publishers, New York, N.Y.

80. **Pandey, D. P., and K. Gerdes.** 2005. Toxin-antitoxin loci are highly abundant in free-living but lost from host-associated prokaryotes. *Nucleic Acids Res.* **33:**966–976.

81. **Pedersen, K., S. K. Christensen, and K. Gerdes.** 2002. Rapid induction and reversal of a bacteriostatic condition by controlled expression of toxins and antitoxins. *Mol. Microbiol.* **45:**501–510.

82. **Pedersen, K., A. V. Zavialov, M. Y. Pavlov, J. Elf, K. Gerdes, and M. Ehrenberg.** 2003. The bacterial toxin RelE displays codon-specific cleavage of mRNAs in the ribosomal A site. *Cell* **112:**131–140.

83. **Rawlings, D. E.** 1999. Proteic toxin-antitoxin, bacterial plasmid addiction systems and their evolution with special reference to the *pass* system of pTF-FC2. *FEMS Microbiol. Lett.* **176:**269–277.

84. **Roberts, R. C., C. Spangler, and D. R. Helinski.** 1993. Characteristics and significance of DNA binding activity of plasmid stabilization protein ParD from the broad-host-range plasmid RK2. *J. Biol. Chem.* **268:**27109–27117.

85. **Ruiz-Echevarría, M. J., A. Berzal-Herranz, K. Gerdes, and R. Díaz-Orejas.** 1991. The *kis* and *kid* genes of the *parD* maintenance system of plasmid R1 form an operon that is autoregulated at the level of transcription by the co-ordinated action of the Kis and Kid proteins. *Mol. Microbiol.* **5:**2685–2693.

86. **Ruiz-Echevarría, M. J., G. de Torrontegui, G. Giménez-Gallego, and R. Diaz-Orejas.** 1991. Structural and functional comparison between the stability systems *parD* of plasmid R1 and *ccd* of plasmid F. *Mol. Gen. Genet.* **225:** 355–362.

87. **Ruiz-Echevarría, M. J., G. Giménez-Gallego, R. Sabariegos-Jareño, and R. Díaz-Orejas.** 1995. Kid, a small protein of the *parD* stability system of plasmid R1 is an inhibitor of DNA replication acting at the initiation of DNA synthesis. *J. Mol. Biol.* **247:**568–577.

88. **Salyers, A., and C. F. Amabile-Cuevas.** 1997. Why are antibiotic resistance genes so resistant to elimination? *Antimicrob. Agents Chemother.* **41:**2321–2325.

89. **Sambrook, J., and D. W. Russell.** 2001. Protein interaction technologies, chapter 18. *In Molecular Cloning: a Laboratory Manual*, 3rd ed. Cold Spring Harbor Laboratory Press, Cold Spring Harbor, N.Y.

90. **Selvin, P. R.** 2000. The renaissance of fluorescence resonance energy transfer. *Nat. Struct. Biol.* **7:**730–734.

91. **Shann, F.** 1990. Modern vaccines. Pneumococcus and influenza. *Lancet* **335:**898–901.

92. **Smith, A. S., and D. Rawlings.** 1997. The poison-antidote stability system of the broad-host-range *Thiobacillus ferrooxidans* plasmid pTF-FC2. *Mol. Microbiol.* **26:**961–970.

93. **Stahura, F. L., L. Xue, J. W. Godden, and J. Bajorath.** 2002. Methods for compound selection focussed on hits and application in drug discovery. *J. Mol. Graph. Model.* **20:** 439–446.

94. **Takagi, H., Y. Kakuta, T. Okada, M. Yao, I. Tanaka, and M. Kimura.** 2005. Crystal structure of archaeal toxin-antitoxin RelE-RelB complex with implications for toxin activity and antitoxin effects. *Nat. Struct. Mol. Biol.* **12:**327–331.

95. **Tam, J. E., and B. C. Kline.** 1989. The F plasmid *ccd* autorepressor is a complex of CcdA and CcdB proteins. *Mol. Gen. Genet.* **219:**26–32.

96. **Tian, Q. B., M. Ohnishi, A. Tabuchi, and Y. Terawaki.** 1996. A new plasmid-encoded proteic killer gene system: cloning, sequencing, and analyzing *hig* locus of plasmid Rts1. *Biochem. Biophys. Res. Commun.* **220:**280–284.

97. **Van Melderen, L., M.-H. Dao Thi, P. Lecchi, S. Gottesman, M. Couturier, and M. R. Maurizi.** 1996. ATP-dependent degradation of CcdA by Lon protease. Effects of secondary structure and heterologous subunit interactions. *J. Biol. Chem.* **271:**27730–27738.

98. **Weihofen, W. A., A. Cicek, F. Pratto, J. C. Alonso, and W. Saenger.** 2006. Structures of omega repressors bound to direct and inverted DNA repeats explain modulation of transcription. *Nucleic Acids Res* **34:**1450–1458.

99. **Welfle, K., F. Pratto, R. Misselwitz, J. Behlke, J. C. Alonso, and H. Welfle.** 2005. Role of the N-terminal region and of beta-sheet residue Thr29 on the activity of the Omega2

global regulator from the broad-host range *Streptococcus pyogenes* plasmid pSM19035. *Biol. Chem.* **386:**881–894.

100. **Whitney, C. G.** 2000. Vaccination against pneumococcal disease: current questions and future opportunities. *BioMed Central* **1:**7.

101. **Xu, Y., A. Kanauchi, A. G. von Arnim, D. W. Piston, and C. J. Johnson.** 2003. Bioluminescence resonance energy transfer (BRET): a new technique for monitoring protein-protein interactions in living cells. *Methods Enzymol.* **360:**289–301.

102. **Xu, Y., D. W. Piston, and C. J. Johnson.** 1999. A bioluminescence energy transfer (BRET) system: application to interacting circadian clock proteins. *Proc. Natl. Acad. Sci. USA* **96:**151–156.

103. **Zavialov, A. V., L. Mora, R. H. Buckingham, and M. Ehrenberg.** 2002. Release of peptide promoted by the GGQ motif of class 1 release factors regulates the GTPase activity of RF3. *Mol. Cell* **10:**789–798.

104. **Zhang, Y., J. Zhang, K. P. Hoeflich, M. Ikura, and M. Inouye.** 2003. MazF cleaves cellular mRNAs specifically at ACA to block protein synthesis in *Escherichia coli*. *Mol. Cell* **12:**913–923.

105. **Zielenkiewicz, U., and P. Ceglowski.** 2001. Mechanisms of plasmid stable maintenance with special focus on plasmid addiction systems. *Acta Biochim. Polon.* **48:**1003–1023.

106. **Zielenkiewicz, U., and P. Ceglowski.** 2005. The toxin-antitoxin system of the streptococcal plasmid pSM19035. *J. Bacteriol.* **187:**6094–6105.

Enzyme-Mediated Resistance to Antibiotics: Mechanisms, Dissemination, and Prospects for Inhibition
Edited by Robert A. Bonomo and Marcelo E. Tolmasky

Robert A. Bonomo
Andrea M. Hujer
Kristine M. Hujer

20 Integrons and Superintegrons

OVERVIEW

Integrons were first described in 1989 by Stokes and Hall (19). Two groups of integrons are now recognized: resistance integrons (RIs) and superintegrons (SIs). Generally speaking, RIs carry gene cassettes that confer resistance against antibiotics and disinfectants. RIs can be located either in the bacterial chromosome or on mobile genetic elements (i.e., transposons and/or conjugative plasmids). Although RIs usually contain three to five antibiotic resistance genes, integrons containing up to 10 resistance gene cassettes have been found in multidrug resistant (MDR) clinical isolates (11). Transposons and/or conjugative plasmids facilitate the intra- and interspecies transmission of resistance genes that have accumulated in RIs. As evidenced by differences in codon usage among gene cassettes within the same RI, antibiotic resistance determinants can be of diverse origin (15). The likely ancestors of RIs are SIs.

SIs are large integrons located in the bacterial chromosome that contain gene cassettes endowing a variety of metabolic or homeostatic functions. Many SIs have been found in *Vibrio* spp. and *Pseudomonas* spp., both pathogens and nonpathogens.

Numerous studies have appeared that describe the importance of RIs and SIs in antibiotic resistance, microbial physiology, and environmental adaptation in phylogenetically diverse gram-negative bacteria. Due to their ability to acquire new genes and modify bacterial phenotypes, integrons have a clear role in the evolution of the bacterial genomes. It may be stated that the evolutionary success of an organism is determined by the resistance gene cassettes (RIs) and its SI organization.

The purpose of this chapter is to (i) describe the genetic organization of an integron; (ii) summarize the different classes of RIs and highlight their importance in antibiotic resistance; (iii) describe the organization of an SI; and (iv) highlight the structure of the key component of the integron, the integrase, and its binding to an *attC* site.

GENETIC ORGANIZATION OF INTEGRONS

It is easiest to think of integrons as natural cloning and expression systems that incorporate single or multiple open reading frames (ORFs). Considered in this respect, integrons (RIs especially) are the "root of antibiotic resistance gene acquisition" (15). Integrons contain a 5′ conserved integrase gene, *IntI* (integron integrase is a site-specific recombinase of the tyrosine recombinase family), mobile antibiotic resistance genes (called gene cassettes), an integration site for the gene cassette, *attI* (att = attachment), and a promoter (Fig. 20.1).

Robert A. Bonomo, Andrea M. Hujer, and Kristine M. Hujer, Research Service, Louis Stokes Cleveland VAMC, Cleveland, OH 44106.

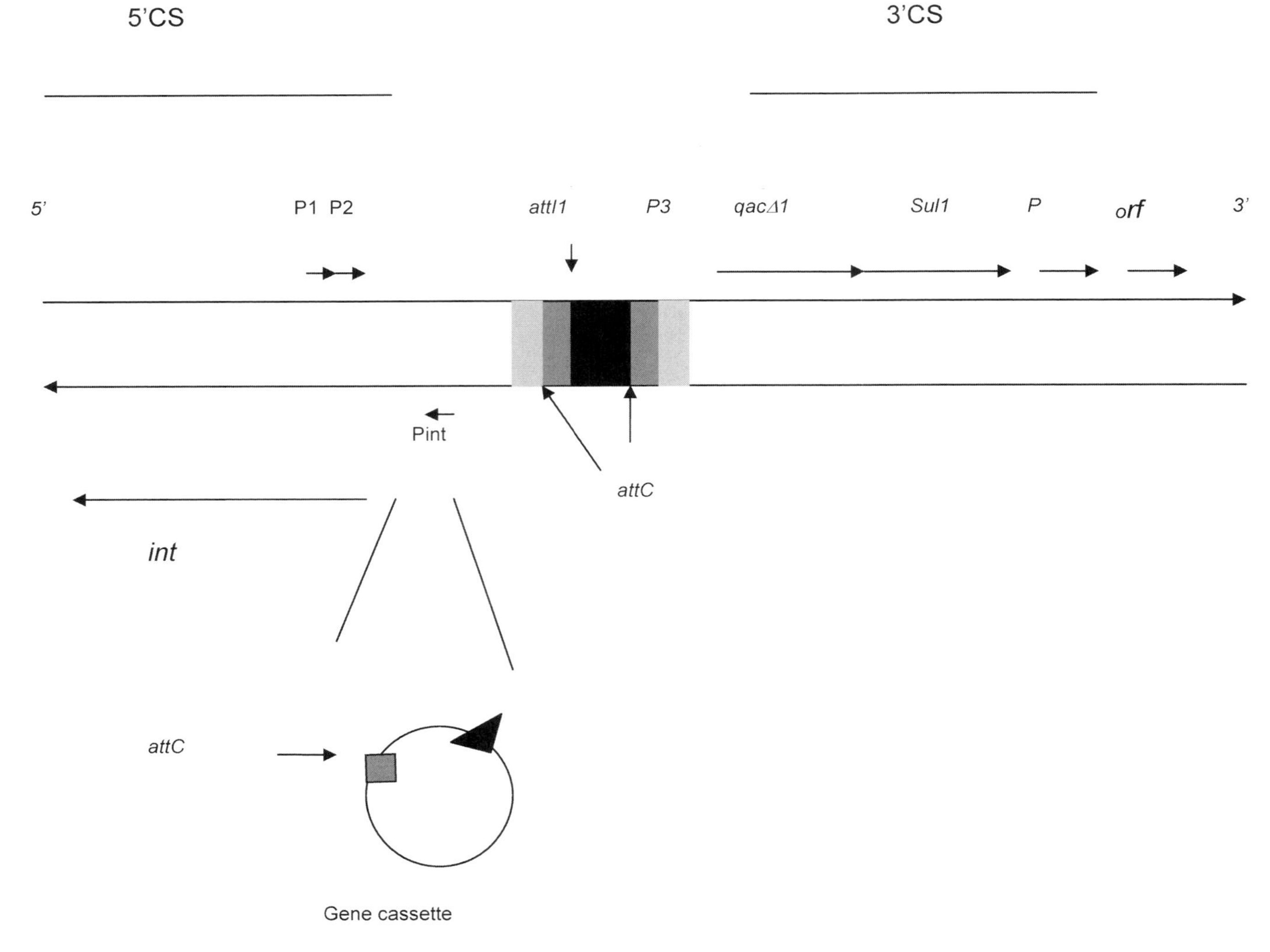

Figure 20.1 Organization of a class 1 integron.

The integrase, the defining element of the integron, excises and integrates genes from and into the integron. The integrase catalyzes recombination between the *attI* site and the *attC* site (or 59-bp element). The *attC* site is normally found associated with a single ORF, and the ORF-*attC* structure is termed a gene cassette (15).

As stated above, integron integrases (*intI*) belong to the tyrosine recombinase family. To review, there are two types of recombinases: tyrosine recombinases (integrases) and serine recombinases (resolvases or invertases). Tyrosine recombinases rearrange DNA in diverse processes such as viral integration, chromosome partitioning, selective gene activation, and conjugative transposition reactions. The integrase of integrons and the integrase protein from bacteriophage λ are among the most studied tyrosine recombinases. In each case, tyrosine (or serine) covalently attacks the phosphodiester backbone of DNA and forms a covalent bond with the phosphate group. The energy to rejoin DNA is conserved within this covalent protein-DNA bond. Tyrosine recombinases break and rejoin two strands at a time and generate a Holliday junction. A Holliday junction is a four-way branched structure that is the central intermediate in homologous recombination (Fig. 20.2). It is formed as a result of a reciprocal exchange of DNA between two nearly identical DNA molecules (8).

Integrases require short matching DNA sequences between the donor and genomic DNA. Unlike the other studied tyrosine recombinases, integron integrases mediate

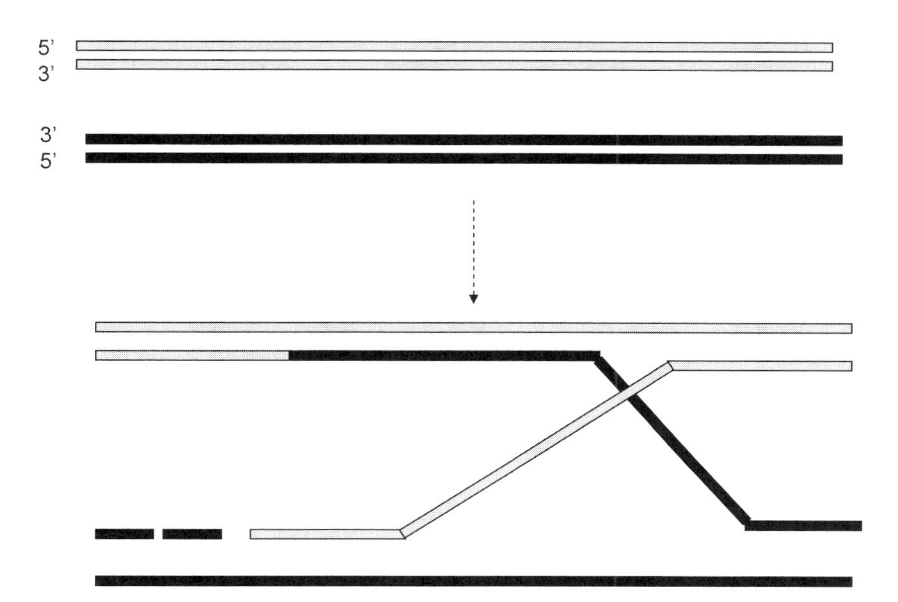

Figure 20.2 Strand exchange produces a Holliday junction. A single-stranded (ss) nick in the donor DNA is followed by DNA synthesis to cause strand displacement. The ss DNA pairs with the recipient DNA to form a heteroduplex. The unpaired recipient DNA forms a "D loop." Nuclease removes the unpaired "D loop DNA." Strand exchange (X over) then produces a Holliday junction. This is followed by isomerization (adapted from www.sci.sdsu.edu/.../Rec-HollidayJunction1.gif).

the exchange of DNA between two architecturally distinct sites even though homology predicts only one DNA binding domain (8).

The *attC* site is an imperfect inverted repeat located at the 3′ end of the gene. These *attC* sites are a diverse family of sequences that function as recognition sites for the site-specific integrase. The *attC* sites vary in size from 60 to 141 bp (hence, the term *attC* site is preferred to 59-bp element). The *attC* site has a core sequence (GTTRRRY) and an inverse core sequence (RYYYAAC). These two sites are called R′ (RYYYAAC) and R″ (GTTRRRY), where R is a purine and Y is a pyrimidine. The point of recombination is between the G site and the T site. Gene cassettes carry different *attC* sites.

Figure 20.1 illustrates the general organization of an integron (class 1). As the reader can see, we have illustrated the 5′ conserved segment (5′CS), which contains the integrase gene (*intI*), and the 3′ conserved segment (3′CS) (contains *qacE*Δ*1* and *sul1*); *attI*, integration site; *qacE*Δ*1*, a functional deletion derivative of the *qacE* gene, which specifies resistance to antiseptics and disinfectants; the *sul1* gene, which confers sulfonamide resistance; *orf*, an ORF of unknown function; P3, the promoter of the *qacE*Δ*1* and *sul1* genes; and *attC*, the sequence on the gene cassette recognized by the integrase. The *sul1* gene is not part of resistance gene cassettes of the integron; it is always contained within the 3′CS. Cassettes are always integrated in the same orientation and are cotranscribed from one or two common promoters (P1 or P2) located in the 5′CS.

In general, gene cassettes consist of a coding sequence, which usually lacks a promoter, followed by an *attC* site (3). Gene cassettes can exist either free in a circularized form or integrated at the *attI* site. As is addressed below, a single site-specific recombination event involving the integron-associated *attI* site and a cassette-associated *attC* leads to insertion of a free circular cassette into a recipient integron.

An increasing number of new gene cassettes are being described. A particularly noteworthy example is the gene cassettes found in In53 (12). In53 (see Figure 20.3) was found in *Escherichia coli* MG-1, an MDR strain, isolated from a South Asian patient who had been hospitalized in France (14). *E. coli* MG-1 was resistant to tetracycline, netilmicin, tobramycin, amikacin, gentamicin, ethidium bromide, sulfonamide, streptomycin, spectinomycin, rifampin, and chloramphenicol. When analyzed, In53 was found to be part of a composite transposon, Tn*2000*, that was inserted on a large self-transferable plasmid (pNLT-1, 160 kb). In addition, In53 contains a fusion between two previously known gene cassettes (*bla*$_{OXA-10}$,

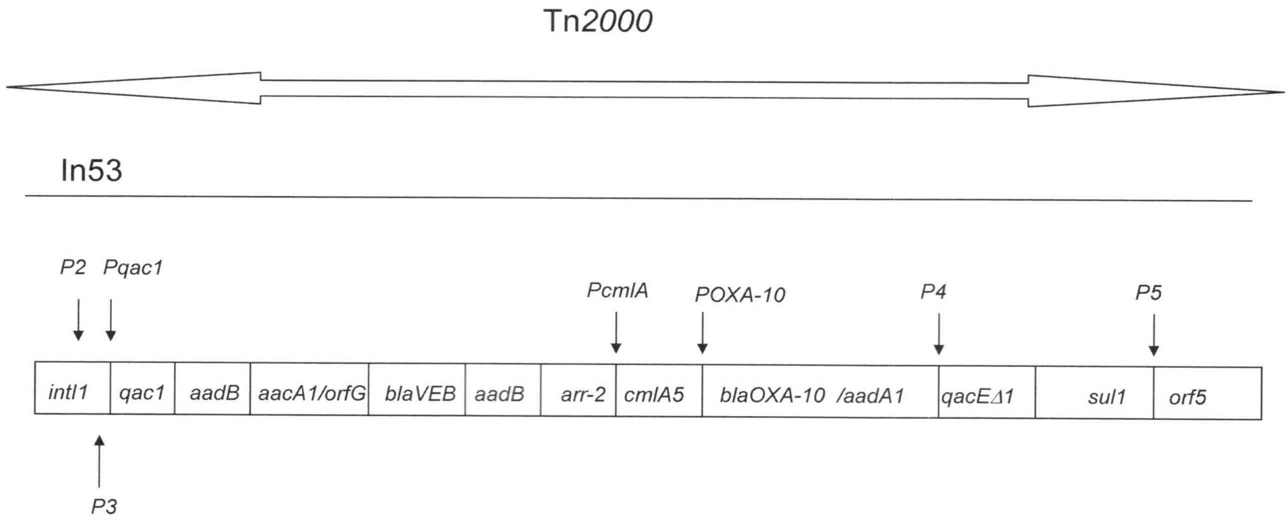

Figure 20.3 Representation of In53, a class 1 integron from *Escherichia coli*. Listed are the 10 different antibiotic resistance genes: *qac1*, a member of the small multidrug resistance family of proteins that serves as an exporter protein that mediates resistance to intercalating dyes and quaternary ammonium compounds; *aadB*, a 2″-aminoglycoside nucleotidyltransferase that confers resistance to kanamycin, tobramycin, and gentamicin; *aacA1/orfG*, a novel 6′-*N*-acetyltransferase; bla_{VEB-1}, an ESBL; *aadB*, see above; *arr-2*, an enzyme that inactivates rifampin by ribosylation; *cmlA5*, a novel nonenzymatic chloramphenicol resistance gene of the *cmlA* family; bla_{OXA-10}/*aadA1*, a fused gene cassette of an ESBL and an aminoglycoside adenyltransferase as a single cassette. The IS*26*-related inverted right and left repeats are not shown (adapted from Naas et al. [12]).

a gene encoding an extended-spectrum beta-lactamase [ESBL], with *aadA1*, a gene encoding an aminoglycoside-modifying enzyme) and another containing the gene encoding an ESBL, bla_{VEB-1}. This formation of novel and functional resistance enzymes that result from gene fusions is an extraordinary method of resistance gene evolution. It is unclear why many cassettes contain genes encoding resistance to spectinomycin and streptomycin, aminoglycoside antibiotics that are no longer in use.

Although the factors defining this are unknown, the order of gene cassettes can also change within an integron. The integrase can excise the gene cassette, and each cassette can then form a circular intermediate. The subsequent reintegration of the gene cassettes at an alternate or preferred *attI* site can lead to cassette rearrangement. The position of a cassette in the integron relative to its promoter may influence the level of antibiotic resistance observed. The *attC* sites form imperfect inverted repeats (or stem loop structures) that attenuate downstream expression of cassettes. Translation attenuation and translation initiation regions have also been described (3).

The origin of integrons and their cassettes remains unknown. It must be kept in mind that integrons themselves are not mobile (3). When RIs are part of transposons and plasmids, they are sometimes referred to as mobile integrons. The transposons and plasmids make these elements mobile.

CLASSES OF INTEGRONS

To date, five distinct integron classes have been found associated with cassettes that contain antibiotic resistance genes. Three main classes of integrons (classes 1, 2, and 3) have been described in gram-negative bacteria. These "classes" of integrons are implicated in multidrug resistance and are defined based on the divergence of their integrase sequences (*IntI1*, *IntI2*, and *IntI3*, respectively). The similarity between the three integrases (40 to 58% genetic identity) suggests that their evolutionary divergence extended beyond the introduction of antibiotics into clinical medicine (10, 13).

Class 1 Integrons

Class 1 integrons are the most commonly found (and clinically important) integrons. Class 1 integrons code for an integrase, IntI1 (337 amino acids in length), and are generally borne on elements derived from Tn*5090*, such as that found in the central part of Tn*21*. As these genetic elements carry multiple resistance determinants and can be readily mobilized, their impact on antibiotic resistance cannot be overstated. Successive recombination events in class 1 integrons allow the accumulation of gene cassettes in a variety of orders. As part of a conjugative plasmid or a transposon, class 1 integrons can spread rapidly. A common feature of class 1 integrons is the presence of sulfamethoxazole resistance determinants (see Fig. 20.3 also). This may reflect the worldwide use of this antibiotic combination.

The class 1 integrase (see below) recognizes three different types of recombination sites (*attI1*, *attC*, and secondary sites) (5). The class 1 integron integrase *IntI1* has been shown to recombine several types of *attC* sites. It has a stronger affinity for a single-stranded rather than a double-stranded recombination substrate (4); *IntI1* recombination is also RecA-independent and does not seem to need any accessory protein (9).

Class 1 integrons have also been associated with a variety of insertion sequence (IS) elements. IS elements are small (<2.5-kb), generally phenotypically silent segments of DNA that have a simple organization and are capable of insertion at multiple sites in a target DNA molecule. IS*6100* has commonly been found at the 3′ end of integrons.

Class 1 integrons have been reported in many gram-negative bacteria including *Pseudomonas aeruginosa*, *Acinetobacter baumannii*, *Achromobacter xylosoxydans*, *Aeromonas salmonicida*, *Acinetobacter* spp., *Alcaligenes* spp., *Burkholderia cepacia*, *Campylobacter* spp., *Citrobacter* spp., *Enterobacter* spp., *E. coli*, *Klebsiella pneumoniae*, *Salmonella* spp., *Serratia* spp., *Shigella* spp., and *Vibrio* spp. They have also been found in other bacteria such as *Corynebacterium glutamicum* and *Mycobacterium fortuitum*, and a gene cassette has been discovered in *Enterococcus faecalis* (3).

Class 1 Integrons and Resistance: Beta-Lactamase Genes

Among gram-negative bacteria, class 1 integrons that have come to clinical attention have been discovered in isolates resistant to extended-spectrum cephalosporins and carbapenems. Investigations have uncovered the presence of ESBLs and metallo-beta-lactamase (MBL) genes. Among the class 1 integron-associated *bla* genes are bla_{IMP}, bla_{VEB}, bla_{GES}, bla_{CTX-M}, bla_{VIM}, bla_{OXA}, bla_{PER}, bla_{DHA}, bla_{SIM}, bla_{BEL}, bla_{GIM}, bla_{IBC}, bla_{CMY}, bla_{CARB-4}, and bla_{AER-1}. Regarding the MBLs, nearly all genes encoding IMP- and VIM-type as well as GIM-1 are found as gene cassettes in class 1 integrons (22). It is notable that bla_{VIM} is also found in conjunction with genes encoding aminoglycoside-modifying enzymes such as *aacA4* that confer resistance to kanamycin, neomycin, amikacin, and streptomycin.

Class 1 Integrons and Resistance: Genes Encoding Aminoglycoside-Modifying Enzymes

Resistance to aminoglycosides can be mediated by aminoglycoside-modifying enzymes (AMEs) whose genes are encoded in class 1 integrons. It is notable that most RIs contain the *aadA* resistance determinant. This gene encodes an aminoglycoside-modifying enzyme that confers streptomycin and spectinomycin resistance. Among class 1 integron-associated aminoglycoside-modifying enzyme genes are *aacA4*, *aacA7*, *aacA29a*, *aacA29b*, *aac(6′)-Ib*, *aadA1*, *aadA2*, *aadA5*, *aadA8*, *aadB*, ant(2″), *ant(2″)-I*, *ant(3″)-1a*, and *aphA15*.

Class 1 Integrons and Resistance: Other Antibiotic Resistance Genes Found in RIs

In addition to *bla* and genes encoding aminoglycoside-modifying enzymes, other antibiotic resistance genes have been discovered in class 1 RIs. These include determinants that code for chloramphenicol resistance (*cmlA4* and *cmlA5*, which encode efflux pumps), rifampin resistance (*arr*-2, ADP ribosylation), plasmid-mediated quinolone resistance (*qnr*), and fosfomycin, erythromycin, and lincomycin resistance.

orf513, Common Regions, and Class 1 Integrons: ISCR Elements

Recently attention has been focused on *orf513* and its association with class 1 integrons ("complex *sul1*-type integrons"). *orf513* is a gene encoding a transposase belonging to the IS*91* family, which is involved in the recruitment of novel resistance genes. *orf513* genes are linked to sequences termed "common regions" (CRs), which are often found beyond, but close to, the 3′ conserved sequences of class 1 integrons. Common regions were first discovered and reported in the early 1990s (20). A comparative analysis of these CRs has revealed that they are related to each other and resemble an atypical class of ISs, designated IS*91*-like elements (see below). IS*91* and IS*91*-like elements lack terminal inverted repeats and are thought to transpose by a rolling-circle transposition mechanism. In a class 1 integron, this element can confer resistance to chloramphenicol, trimethoprim, aminoglycosides, and tetracycline and may carry a range of beta-lactamase genes as well as the *qnrA* (plasmid-mediated quinolone resistance) gene (20). Elements such as "*orf513*" are now renamed "insertion sequence CRs" (ISCRs). The genetic context surrounding ISCRs indicates that they obtain 5′ sequences via a process called "unchecked transposition." ISCRs are increasingly linked to MBLs in *P. aeruginosa* and trimethoprim-sulfamethoxazole resistance in *Stenotrophomonas maltophilia*. This mechanism provides antibiotic resistance genes with a highly mobile genetic vehicle that could greatly exceed the effects of previously reported mobile genetic mechanisms.

Class 2 integrons are part of the Tn7 family of nonreplicative transposons. Tn7 (Fig. 20.4) is a remarkably widespread transposon which commonly accounts for trimethoprim resistance in gram-negative bacteria. Hansson et al. sequenced the Tn7 integrase gene, *intI2* (6). The predicted peptide sequence (IntI2*) is 325 amino acids long and is 46% identical to IntI1. They discovered that the class 2 integrase gene contains a defective codon, *ochre179*, which yields a nonfunctional truncated protein. Converting this defective codon into a glutamic acid restores full integrase activity. At the time of this writing, the class 2 integrase genes thus far sequenced have not been functional due to an internal stop codon. Although gene cassettes found in class 2 integrons are identical to those in class 1 integrons, the functional IntI2 was not able to excise gene cassettes from class 1 integrons. In contrast, IntI2 is able to both excise and insert gene cassettes into class 2 integrons. Often class 1 intregrons are found in the presence of class 2 integrons. The functional dynamics between class 1 and class 2 integrons still remains to be fully explained.

Tn7 includes the gene cassettes *dfrA1-sat2-aadA1-orfX* in its variable region. Class 2 integrons have been found in *Acinetobacter*, *Shigella*, and *Salmonella*. A limited number of antibiotic resistance gene cassettes have been described in class 2 integrons.

Class 3 integrons have been described in *Serratia marcescens*, *P. aeruginosa*, *Alcaligenes xylosoxidans*, *Pseudomonas putida*, and *K. pneumoniae*. Although not as widespread, the class 3 integrases (as defined by *intI3*) are similar to class 1 integrases. The class 3 integrases can recognize different *attC* sites and integrate the cassettes into the *attI3* site. A notable example follows: bla_{IMP} was first described in a class 3 integron but has been isolated subsequently from a class 1 integron (2, 7).

SIs

The integron originally designated class 4 is now named *Vibrio cholerae* SI. The term "superintegron" was coined by Mazel et al. (10) after studying a very large integron with an array of gene cassettes found in the chromosome of *V. cholerae*. It was observed that the *V. cholerae* genome possessed repeated sequences (*V. cholerae* repeats [VCRs]) in clusters that have a genetic organization similar to that

Figure 20.4 Map of the integron region of Tn7.

of integron-gene cassette structures. The repeat sequences of SIs are usually named after the species they are found in (e.g., VCRs, *V. cholerae* repeats). These VCRs repeated multiple times (up to 100 copies/cell) and were located in a single restriction fragment comprising 10% of the *V. cholerae* genome. DNA analyses found 90% sequence identity between the VCR sequences and the *attC* site associated with a carbenicillinase, CARB-4, isolated from *P. aeruginosa*. Using nucleic acid hybridization techniques to isolate a gene for an integron integrase, Mazel et al. identified the *intI4* gene, a gene with 45 to 50% identity to the previously described three known integrases (this integrase is now referred to as VchIntIA). By careful genetic mapping studies, Mazel et al. showed that VCR islands are integron-like structures that were formed using an integrase-mediated mechanism (10). Functional analysis of the SI cassettes of *V. cholerae* strain N16961 revealed a variety of adaptive functions, including metabolic activities, virulence traits, and potential antibiotic resistance determinants (16, 17).

SIs have also been described in other *Vibrio* spp., xanthomonads, and pseudomonads (especially environmental organisms) (8). To date, more than 20 SIs have been analyzed and just five *Vibrio* SIs contain a transferable genetic reservoir that is equivalent in size to a small genome (16, 21).

All these SIs share similar features: they contain multiple gene cassettes (many times greater than 20), and there is extensive homology between the *attC* sites of their endogenous cassettes. Two examples are presented: (i) in *Vibrio vulnificus* there are more than 200 gene cassettes that are part of the SI; and (ii) the SI of *V. cholerae* serotype O1 biotype El Tor strain N16961 contains at least 215 ORFs and occupies up to 3% of the genome.

Recognizing that there is remarkable similarity in the sequence of the *attC* sites (e.g., in one case, 149 of the 179 gene cassettes were more than 90% identical over their entire length of 122 to 124 nucleotides) (8), it is postulated that the high level of species-specific sequence identity seen in the *attC* sites in each SI suggests that gene cassettes are assembled through the physical association of an SI *attC* site with an incoming DNA fragment.

Not all SIs share these properties. In a manner characteristic of RIs, the SI of *Shewanella* spp. contains only three cassettes with structurally different *attC* sites. As can be seen, SIs are truly distinct from RIs, since they differ in size, placement of promoters, replicon location, and the nature of the genes found within cassettes (e.g., transport molecules, pathogenesis, and restriction endonuclease activity) (21).

SIs are also present among pseudomonads. Vaisvila et al. described the functions of an SI, In55044, found in *Pseudomonas alcaligenes* ATCC 55044. The SI is 18 kb in size, with a total of 33 repeats called PARs (for *P. alcaligenes* repeat) (21).

SIs are located exclusively in the bacterial chromosome. It is believed that the gene cassettes in RIs probably originated from SIs. Rowe-Magnus et al. have shown that gene cassette reservoirs of SIs are a source of the gene cassettes identified within RI and that RIs randomly recruit gene cassettes directly from SIs (15–18).

STRUCTURE OF THE INTEGRASE

The paradigm for integron recombination is undoubtedly the class 1 integrase, IntI1. In a series of elegant studies, Bouvier et al. studied the mechanism of insertion by using a recombination assay that delivered either the top or bottom strand of different integrase recombination substrates (1). These studies revealed that recombination of a single-stranded *attC* site with an *attI* site was 1,000-fold higher for one strand than for the other. On the contrary, following conjugative transfer of either *attI* strand, recombination with *attC* is highly unfavorable. These results supported a novel integron cassette insertion model in which the single bottom *attC* strand adopts a folded structure, generating a double-stranded recombination site. Thus, recombination would insert a single-stranded cassette, which must be subsequently processed.

MacDonald et al. working at the Institute Pasteur in Paris, France, recently reported the crystal structure of *IntI* from *Vibrio cholerae* (VchIntIA) bound to a complex substrate (*attC*) that was derived from the bottom strand of a (VCR) sequence (8) (Color Plate 11). The structure (2.8-Å resolution) showed that the site of recombination along the DNA backbone is determined by the position of two preexisting extrahelical DNA bases that serve as molecular markers to position the integrase along the DNA. In other words, the integrase (a protein made up of four identical subunits) actually recognizes extrahelical bases — ones that do not pair up with the other strand — inside the *attC*. While two of the enzyme's molecules are involved in the recognition of these bases, the other two cleave at a specific site on the DNA backbone (Color Plates 12 and 13). These two functions lead to the excision of the single-stranded gene, which can then be "captured" by another integron. The remainder of the protein-DNA interface is composed almost entirely of nonspecific protein-to-DNA phosphate interactions. The mechanism shows how sequence-degenerate single-stranded genetic material is recognized and exchanged between bacteria mediated by integrase.

CONCLUSION

In this brief chapter we have demonstrated that integrons are the genetic elements responsible for the capture and spread of antibiotic resistance determinants among diverse gram-negative clinical isolates. It is clear that the widespread occurrence of the MDR phenotype in gram-negative bacteria is due to the dissemination of integrons in the last 60 years. The discovery of RIs and SIs has also demonstrated the importance of integrons in adaptive bacterial evolution and genomics. These recombination systems for gene transfer have successfully adapted to each new environmental and antibiotic challenge in ecology and human and veterinary medicine (18). As such, the combination of conjugative plasmids, transposons, and phages and numerous integron gene cassettes creates an extremely vast and complex gene pool in bacteria.

The Veterans Affairs Merit Review Program and the National Institutes of Health (RO1 AI063517-01) supported these studies. Special thanks to Louis B. Rice for careful review of the manuscript.

References

1. **Bouvier, M., G. Demarre, and D. Mazel.** 2005. Integron cassette insertion: a recombination process involving a folded single strand substrate. *EMBO J.* **24:**4356–4367.
2. **Collis, C. M., M. J. Kim, S. R. Partridge, H. W. Stokes, and R. M. Hall.** 2002. Characterization of the class 3 integron and the site-specific recombination system it determines. *J. Bacteriol.* **184:**3017–3026.
3. **Fluit, A. C., and F. J. Schmitz.** 2004. Resistance integrons and super-integrons. *Clin. Microbiol. Infect.* **10:**272–288.
4. **Francia, M. V., J. C. Zabala, F. de la Cruz, and J. M. Garcia Lobo.** 1999. The IntI1 integron integrase preferentially binds single-stranded DNA of the *attC* site. *J. Bacteriol.* **181:**6844–6849.
5. **Hall, R. M., and C. M. Collis.** 1995. Mobile gene cassettes and integrons: capture and spread of genes by site-specific recombination. *Mol. Microbiol.* **15:**593–600.
6. **Hansson, K., L. Sundstrom, A. Pelletier, and P. H. Roy.** 2002. IntI2 integron integrase in Tn7. *J. Bacteriol.* **184:**1712–1721.
7. **Laraki, N., M. Galleni, I. Thamm, M. L. Riccio, G. Amicosante, J. M. Frere, and G. M. Rossolini.** 1999. Structure of In31, a bla_{IMP}-containing *Pseudomonas aeruginosa* integron phyletically related to In5, which carries an unusual array of gene cassettes. *Antimicrob. Agents Chemother.* **43:**890–901.
8. **MacDonald, D., G. Demarre, M. Bouvier, D. Mazel, and D. N. Gopaul.** 2006. Structural basis for broad DNA-specificity in integron recombination. *Nature* **440:**1157–1162.
9. **Martinez, E., and F. de la Cruz.** 1990. Genetic elements involved in Tn21 site-specific integration, a novel mechanism for the dissemination of antibiotic resistance genes. *EMBO J.* **9:**1275–1281.
10. **Mazel, D., B. Dychinco, V. A. Webb, and J. Davies.** 1998. A distinctive class of integron in the *Vibrio cholerae* genome. *Science* **280:**605–608.
11. **Naas, T., F. Benaoudia, L. Lebrun, and P. Nordmann.** 2001. Molecular identification of TEM-1 beta-lactamase in a Pasteurella multocida isolate of human origin. *Eur. J. Clin. Microbiol. Infect. Dis.* **20:**210–213.

12. **Naas, T., Y. Mikami, T. Imai, L. Poirel, and P. Nordmann.** 2001. Characterization of In53, a class 1 plasmid- and composite transposon-located integron of *Escherichia coli* which carries an unusual array of gene cassettes. *J. Bacteriol.* **183:**235–249.
13. **Ochman, H., and A. C. Wilson.** 1987. Evolution in bacteria: evidence for a universal substitution rate in cellular genomes. *J. Mol. Evol.* **26:**74–86.
14. **Poirel, L., T. Naas, M. Guibert, E. B. Chaibi, R. Labia, and P. Nordmann.** 1999. Molecular and biochemical characterization of VEB-1, a novel class A extended-spectrum beta-lactamase encoded by an *Escherichia coli* integron gene. *Antimicrob. Agents Chemother.* **43:**573–581.
15. **Rowe-Magnus, D. A., A. M. Guerout, and D. Mazel.** 2002. Bacterial resistance evolution by recruitment of super-integron gene cassettes. *Mol. Microbiol.* **43:**1657–1669.
16. **Rowe-Magnus, D. A., A. M. Guerout, and D. Mazel.** 1999. Super-integrons. *Res. Microbiol.* **150:**641–651.
17. **Rowe-Magnus, D. A., and D. Mazel.** 1999. Resistance gene capture. *Curr. Opin. Microbiol.* **2:**483–488.
18. **Rowe-Magnus, D. A., and D. Mazel.** 2002. The role of integrons in antibiotic resistance gene capture. *Int. J. Med. Microbiol.* **292:**115–125.
19. **Stokes, H. W., and R. M. Hall.** 1989. A novel family of potentially mobile DNA elements encoding site-specific gene-integration functions: integrons. *Mol. Microbiol.* **3:** 1669–1683.
20. **Toleman, M. A., P. M. Bennett, and T. R. Walsh.** 2006. ISCR elements: novel gene-capturing systems of the 21st century? *Microbiol. Mol. Biol. Rev.* **70:**296–316.
21. **Vaisvila, R., R. D. Morgan, J. Posfai, and E. A. Raleigh.** 2001. Discovery and distribution of super-integrons among pseudomonads. *Mol. Microbiol.* **42:**587–601.
22. **Walsh, T. R., M. A. Toleman, L. Poirel, and P. Nordmann.** 2005. Metallo-beta-lactamases: the quiet before the storm? *Clin. Microbiol. Rev.* **18:**306–325.

Enzyme-Mediated Resistance to Antibiotics: Mechanisms, Dissemination, and Prospects for Inhibition
Edited by Robert A. Bonomo and Marcelo E. Tolmasky

Dan I. Andersson, Sophie Maisnier Patin, Annika I. Nilsson, and Elisabeth Kugelberg

21

The Biological Cost of Antibiotic Resistance

The use of antibiotics for the last 60 years has been a combination of a medical success story and a large-scale evolutionary experiment in which many bacterial species have adapted to antibiotics by developing resistance (35). Since resistance surveillance was introduced, we have observed a gradual increase in the frequency of resistance for virtually every combination of bacterium and antibiotic. As a result, today in most countries antibiotic resistance is a significant and costly health problem for both hospital- and community-acquired infections (43). Whether we can slow down, reverse, or prevent antibiotic resistance development is unclear because many of the parameters (i.e., microbial and host factors) that affect the emergence, spread, and stability of resistance are outside our control. On the other hand, certain measures, i.e., infection control and regulation of the volume and choice of the antibiotics consumed and treatment regimens (dosage, time, etc.), are in principle controllable and could conceivably be altered to minimize resistance development without compromising treatment efficacy.

For any resistance mechanism the bacterium is altered not only in its ability to withstand the drug but also potentially in its growth, survival, and transmission inside and outside a host. That is, the resistance may confer a biological (fitness) cost. This chapter is focused on the impact of antibiotic resistance caused by chromosomal mutations on bacterial fitness. Generally, the conclusions regarding costs of chromosomal resistances also apply to plasmid-encoded resistances, and the reader is referred to recent reviews for a more specific discussion of costs associated with plasmids (8, 17). For many bacteria and resistances there has been considerable interest in determining the mechanism and magnitude of the biological cost. The main reason for the interest in this parameter is that it has a major influence on the dynamics of resistance development. As shown by theoretical modeling, the fitness cost affects both the rate by which resistance ascends in a population and the steady-state frequency of resistance at a given antibiotic pressure as well as the rate at which resistant bacteria will disappear if antibiotic use is reduced (1–4, 8, 12, 31–33, 39). Thus, the larger the cost, the slower resistance will develop and the faster it will disappear if antibiotic use is reduced. The question of reversibility is of special interest since it has been proposed that if a resistance mutation or plasmid reduces fitness, the resistant strains will be outcompeted by the susceptible strains if antibiotic use is discontinued. As a result, the prevalence of resistant bacteria will decline in the population. However,

Dan I. Andersson and Annika I. Nilsson, Dept. of Medical Biochemistry and Microbiology, Uppsala University, S-751 23 Uppsala, Sweden. **Sophie Maisnier Patin and Elisabeth Kugelberg,** Dept. of Bacteriology, Swedish Institute for Infectious Disease Control, S-171 82 Stockholm, and Microbiology, Tumour and Cell Biology Center, Karolinska Institute, S-171 77 Stockholm, Sweden.

as discussed below, reversibility of resistance in community settings is unlikely to occur and our resistance problems are probably here to stay.

MEASURING THE BIOLOGICAL COSTS OF ANTIBIOTIC RESISTANCE

The biological fitness of an antibiotic-resistant bacterium is determined by several factors, including the relative rates at which resistant and sensitive bacteria (i) grow and die inside and outside a host, (ii) are transmitted between hosts and between the host and the environment, and (iii) are cleared from infected hosts. To accurately estimate the overall fitness of resistant bacteria, one needs to experimentally quantify all these parameters. In principle, at least three approaches can be used to estimate the costs associated with resistance. One approach uses a combination of mathematical modeling and epidemiological data, and the fitness cost of resistance can be indirectly estimated from determining which fitness cost best predicts the observed decrease in the frequency of resistant bacteria in response to reduced drug consumption. Even though this approach gives only an indirect estimate of fitness costs, it is useful in that these studies help to identify and quantitatively evaluate the roles of different factors contributing to the appearance and maintenance of resistance as well as to predict and evaluate the results of intervention strategies. Epidemiological studies that have examined changes in both the frequency of resistance and the volume of antibiotic use are rare. One example is the study of penicillin-resistant *Streptococcus pneumoniae* in Iceland, in which epidemiological data and mathematical modeling were used to infer the costs of resistance (4). To account for the observed dynamic it had to be assumed that the resistant mutant had an increased transmission between hosts and decreased growth in the host compared to susceptible strains. Since there were no independent experimental measurements of the cost parameters in this study, it is unclear how well this approach predicted the fitness burden of penicillin resistance.

Another approach is to use epidemiological data to prospectively follow the rate at which a patient infected with a resistant or susceptible bacterial strain transmits it to other people. Such experiments would allow one to measure the basic reproductive number, which is the most relevant parameter to use when predicting the relative rate of spread of the resistant and susceptible bacteria. These types of measurements are generally applicable only to strains with unusual resistances or causing serious infections since they require a close surveillance of transmission and methods to easily identify the primary as well as secondary cases. For example, this approach has been used for resistant and susceptible *Mycobacterium tuberculosis* strains in several studies, but at present the results from such fitness estimations are complex and contradictory (14).

A third, and the most common, approach is to estimate the costs of resistance by experimental studies. Here the rates of growth, survival, and competitive performance of susceptible and resistant bacteria (isolated from patients or in vitro) are determined. By pairwise competition experiments, where mixtures of otherwise isogenic susceptible and resistant strains are grown in chemostats, batch cultures, laboratory animals, or humans, changes in their relative frequencies can be followed. Typically, one uses a neutral genetic tag (e.g., a transposon) or the resistance itself as the selective marker, and the changes in relative frequencies of susceptible and resistant bacteria can easily be followed by plating on selective medium. Depending on the specific experimental setup, these competition experiments can be used to estimate several components of the competitive fitness of susceptible and resistant bacteria: for example, their lag periods, rates of exponential growth, survival in stationary phase, adherence to surfaces, and death rates in the presence or absence of host defenses. When properly repeated, these competition experiments can ideally allow detection of differences in fitness as low as 1%. Since fitness differences less than 1% are difficult to detect, it remains unclear if so-called "no cost" mutations actually exist, as has been claimed (see below). One central question when estimating fitness costs by experimental studies is to what extent these measures correlate to fitness in real life, i.e., to healthy humans or clinical situations. We do not have a good answer to this question, but by generalizing from our present knowledge it is likely that if a cost is seen in vitro there is also a cost in vivo. However, the real problem concerns resistances, which appear to confer no cost under favorable conditions in vitro since it is still possible that they are costly under more stressful conditions in vivo. An illustration of this point is high-level fusidic acid resistance due to *fusA* mutations in *Salmonella enterica* serovar Typhimurium. Here no cost is seen during growth in laboratory medium, but during growth in mice this resistance drastically decreases fitness (10). At present, only very few comparisons of the cost of a specific resistance in different assay systems have been done. A recent analysis of costs of fusidic acid and ciprofloxacin resistance in coagulase-negative staphylococci showed that survival on human skin and growth rates in laboratory medium correlated well, supporting the use of in vitro assays, at least for these resistances and bacteria (26). Nevertheless, additional comparisons of different assay systems are needed to validate the use of in vitro experimental measurements to estimate costs. Finally, a more serious limitation is the fact that these studies measure only growth rates.

Controlled experiments to determine transmission and clearance rates are even more rare. Since the overall fitness of a resistant bacterium is strongly influenced by any impact of resistance on transmission, it is, in spite of the technical difficulties, important to try to develop relevant transmission models for cost measurements.

WHAT ARE THE COSTS?

A number of theoretical studies of both hospital- and community-acquired infections show that the volume of antibiotic use is the major selective pressure driving changes in the frequency of resistance (1–4, 30–33, 39). However, at a given antibiotic pressure the cost of resistance is the main bacterial factor that determines how rapidly and to which frequency resistance will develop (31–33). Thus, determinations of fitness costs are important to allow a rational prediction and description of the rate and extent of resistance development at a given volume of antibiotic use. During recent years a number of studies have been performed to estimate the costs of resistance, especially of resistances conferred by chromosomal mutations that cause a modification of the target molecules (summarized in Table 21.1). A main conclusion from these studies is that most resistances confer a cost, typically observed as a reduced growth rate in culture medium or in experimental animals. The extent of the costs varies considerably depending on the specific resistance mutation, the affected gene, the bacterium examined, and the assay system used. At present, it is not possible to predict the effect of these mutations on fitness from the site and type of mutation.

Even though most resistances are associated with a fitness cost, some resistance mutations appear to be cost-free. For example, certain *rpoB* mutations (conferring rifampicin resistance) in *M. tuberculosis* (6), *Staphylococcus aureus* (57), and *Escherichia coli* (49); *rpsL* mutations (conferring streptomycin resistance) in *M. tuberculosis* (13, 51), *E. coli* (52, 53), and *S. enterica* serovar Typhimurium (9–11), *katG* mutations (conferring isoniazid resistance) in *M. tuberculosis* (47) as well as *gyrA* and *parC* mutations (conferring fluoroquinolone resistance) in *S. pneumoniae* (24) confer no measurable cost in the in vitro assay systems used. However, as mentioned before, these assays will typically only detect fitness differences >1%. That is, smaller but potentially still very important costs might not be detected. In addition, these resistances might confer costs in other, more stressful environments. An interesting case in this context is the phenotypic switching of gentamicin resistance observed in S. *aureus*. Here an inducible and rapidly reversible resistance mechanism circumvents the permanent fitness cost associated with the resistance (42).

Table 21.1 Examples of cases in which biological costs of chromosomal resistances have been estimated

Bacterium	Resistance	Mutation	Cost	Assay system	Reference(s)
S. enterica serovar Typhimurium	Streptomycin	*rpsL*	Yes/no	Mice, in vitro	9–11, 41
	Rifampicin	*rpoB*	Yes	Mice, in vitro	9
	Nalidixic acid	*gyrA*	Yes	Mice, in vitro	9
	Fusidic acid	*fusA*	Yes/no	Mice, in vitro	10, 28
E. coli	Streptomycin	*rpsL*	Yes/no	In vitro	52, 53
	Fluoroquinolones	*gyrA, parC*	Yes	In vitro	5
	Rifampicin	*rpoB*	Yes/no	In vitro	49
M. tuberculosis/bovis	Isoniazid	*katG*	Yes	Mice, guinea pigs, in vitro	27, 36, 47, 58
	Streptomycin	*rpsL, rRNA*	Yes/no	In vitro	13, 51
	Rifampicin	*rpoB*	Yes/no	In vitro	6
S. aureus	Fusidic acid	*fusA*, non-*fusA*	Yes/no	Rats, in vitro	44
	Gentamicin				42
	Rifampicin	*rpoB*	Yes/no	In vitro	57
Helicobacter pylori	Clarithromycin	23S rRNA	Yes	Mice	7
Pseudomonas aeruginosa	Fluoroquinolones	*nalB*	Yes	In vitro, *Caenorhabditis elegans*	50
Pseudomonas fluorescens/putida	Rifampicin	Unknown	Yes	Soil	15
Listeria monocytogenes	Class IIa bacteriocin	Unknown	Yes	In vitro	16
S. pneumoniae	Fluoroquinolones	*gyrA, parC*	No	In vitro	24
	Penicillin	*pbp*	Yes	In vitro	22
Neisseria meningitidis	Sulfonamide	*sul2*	Yes	In vitro	19

One hitherto neglected aspect of resistance is how it affects bacterial virulence and disease pathology (e.g., inflammatory responses, histopathology, and course of infection). When both growth rates and virulence have been measured, resistant mutants often show both decreased fitness and virulence. For example, studies of isoniazid-resistant *katG* mutants of *Mycobacterium bovis* and *M. tuberculosis* show that both the growth rate and the virulence as measured either by lethality or histopathology are reduced (36, 58). However, there are exceptions to this finding and there is no reason a priori to expect that growth rate and virulence should always be positively correlated. Also, disease pathology might be different in the resistant mutant and in the wild type. An example is cephalosporin-resistant gram-negative bacteria, such as *Citrobacter freundii* and *Enterobacter cloacae*, which can express a chromosomal β-lactamase that is inducible by β-lactams. These organisms are susceptible to broad-spectrum cephalosporins, but mutations arise at a high frequency, leading to constitutive β-lactamase production and resistance to broad-spectrum cephalosporins. These mutations are located in the *ampD* gene, encoding a cytosolic anhydro-muramyl-peptide amidase required for peptidoglycan recycling (37). Mutants in the *ampD* gene cannot remove the peptide from recycling anhydro-muramyl peptides and as a result accumulate anhydro-*N*-acetyl-muramyl-tripeptide and show constitutive high-level cefotaxim resistance (46). Anhydro-muramyl-peptides are involved in the pathogenicity of *Bordetella pertussis* (25), and their accumulation in an *ampD* mutant results in induction of nitric oxide in both epithelial and phagocytic cells infected with *E. cloacae* (21). This could conceivably change the inflammatory response.

Since chromosomal mutations causing resistance often alter target molecules that are responsible for essential cellular activities, it is not surprising that they often confer a reduced fitness. Three such systems have been examined in detail for the mechanisms that cause a fitness decrease. Mutations in the *rpoB* gene encoding the β-subunit of RNA polymerase are usually responsible for rifampicin resistance in *E. coli*. These mutations reduce the rate of transcription, and as a result the growth rate is reduced (49). Similarly, certain amino acid substitutions in the *fusA* gene encoding elongation factor G (EF-G) can cause fusidic acid resistance in *S. enterica* (28) and *S. aureus* (44). Most of these mutations affect the GDP-GTP exchange rate on EF-G, resulting in a slowed translation elongation and growth rate (10, 28). In addition, these mutations alter the intracellular level of the transcriptional regulator molecule ppGpp, which might cause additional pleiotrophic fitness effects (40). Finally, certain mutational alterations in ribosomal protein S12 (encoded by the *rpsL* gene) causing streptomycin resistance reduce translational efficiency. So-called nonrestrictive mutants have normal ribosomal proofreading and translational fidelity and grow at a rate similar to that of the wild type both in vitro and in experimental animals. In contrast, restrictive mutants show an increased rate of ribosomal proofreading which results in a higher translational accuracy, a slowed translation elongation rate, and a decreased growth rate both in vitro and in animal experiments (9–11, 52).

COMPENSATION OF THE BIOLOGICAL COSTS OF ANTIBIOTIC RESISTANCE

A deleterious resistance mutation has several possible fates. It may exist at some frequency in the population, go extinct, revert back to the susceptible state, or be compensated by additional second-site suppressor mutations. Compensation is of special interest since antibiotic-resistant bacteria may genetically adapt to the costs by acquiring mutations that restore fitness and as a result stabilize the resistance in the population. The process of compensation is usually studied experimentally by monitoring the growth of the resistant strain (in the absence or presence of antibiotic) in chemostats, by serial transfer in laboratory cultures, in cultured cell lines, or in experimental animals. By this procedure, from initially slow-growing bacteria mutants that show a faster growth rate or increased growth yield are selected. After different time points samples are withdrawn to detect fast-growing compensated mutants. This initial screen for fast growers can often be done by visual inspection of colony size on plates. Subsequently, the exact degree of compensation can be measured by pairwise competitions of the compensated mutant with the ancestral resistant mutant or, alternatively, a reference strain that is genetically tagged with a resistance/metabolic marker. By using these types of procedures, compensated mutants have been isolated in several systems (see Table 21.2 for a summary). It is always possible to detect compensation to some extent, but the degree of restoration of fitness can vary considerably between systems studied. Often it is possible to eliminate the costs completely and restore fitness to the parental wild-type level. A detailed study of compensation of the costs of a specific *rpsL* mutation in *S. enterica* showed that among the compensated mutants the relative fitness varied between 0.86 and 1.0 (whereas the wild type had a relative fitness of 1.0 and that of the *rpsL* mutant was 0.74) (41). Similar results have been obtained also for other resistances (9, 10, 49). Compensated mutants typically maintain their high-level resistance; i.e., the compensatory mutation has no effect on the level of resistance caused by the ancestral resistance mutation. However, there are also exceptions such that the compensated mutants show a

Table 21.2 Examples of cases in which genetic compensation of fitness cost caused by chromosomal mutations has been demonstrated or inferred

Bacterium	Resistance mutation (resistance)	Compensatory mutation (resistance in compensated mutant)	Compensatory evolution in:	Reference(s)
S. enterica serovar Typhimurium	*rpsL* (streptomycin)	Intragenic, *rpsL* (maintained)	Mice	9–11
	rpsL (streptomycin)	Extragenic, *rpsD/E, rplS* (maintained)	Laboratory medium	9–11, 41
	gyrA (nalidixic acid)	Intragenic, *gyrA* (maintained)	Mice	9
	rpoB (rifampicin)	Intragenic, *rpoB* (maintained)	Mice	9
	fusA (fusidic acid)	True reversion, *fusA* (lost)	Mice, laboratory medium	10, 28
	fusA (fusidic acid)	Intragenic, *fusA* (often maintained)	Laboratory medium	10, 28
S. aureus	*fusA* (fusidic acid)	Extragenic, ?[a] (maintained)	Rats	44
	fusA (fusidic acid)	Intragenic, *fusA* (often maintained)	Laboratory medium, humans	44
E. coli	*rpsL* (streptomycin)	Extragenic, *rpsD/E* (maintained)	Laboratory medium	52, 53
	rpoB (rifampicin)	Intragenic, *rpoB* (maintained)	Laboratory medium	49
M. tuberculosis	*katG* (isoniazid)	Extragenic, *ahpC* (maintained)	Humans	55
	Several mutations	Unknown (maintained)	Humans	23
H. pylori	23S rRNA (clarithromycin)	Unknown (maintained)	Humans	7
N. meningitidis	*sul2* (sulfonamide)	Intragenic, *sul2* (maintained)	Humans	19

[a]?, mutation unknown.

reduced resistance, e.g., fusidic acid resistance in serovar Typhimurium (10). Several tests can be performed to show that the mutations are specific compensatory mutations and not generally fitness-increasing mutations that appeared during serial passage. First, one may serially passage the susceptible parent strain under similar conditions. Thus, if the putative compensatory mutations are not found in the wild-type background, it indicates that the identified mutations were not generally increasing fitness but rather specifically suppressing the resistance mutation. Secondly, one may genetically move the compensatory mutations to a susceptible genetic background where they should not increase fitness if they are specific compensatory mutations. On the contrary, under such circumstances it is often observed that the compensatory mutations by themselves decrease fitness (41).

The number of potential compensatory mutations for a given resistance mutation can vary considerably. To date only two studies have been performed in which the compensatory targets were thought to be nearly saturated. In one case low-fitness, fusidic acid-resistant *fusA* mutants in *S. enterica* were compensated, and about 20 different amino acid substitutions within the EF-G protein were identified that could increase fitness (8, 28). Similarly, compensation of a specific *rpsL* mutation in *S. enterica* yielded 35 different mutations located in *rpsD* (16), *rpsE* (13), *rpsL* (2), or *rplS* (4) (41). From the multiple-recovery frequency of mutations the total number of possible mutations can be calculated. If one makes the rough approximation that all mutations occur at the same rate, the distribution of the number of occurrences is expected to be Poissonian. Thus, the number of possible mutations (m) can be estimated from $k = m(1 - e^{-n/m})$, where k is the observed number of different types of mutants and n is the number of independent tries. For the *rpsL* case mentioned above, this calculation implies that the total number of compensatory mutations possible is about 46 (41). Generalizing from the above two cases, it is predicted that in most systems the target size for compensation is >20-fold larger than that for reversion.

The rate of adaptation can be determined by evolving several independent lineages of the resistant, slow-growing mutant by serial transfer. From the rate of fixation of the compensated mutants, their relative fitness, and the population sizes, one may then calculate the mutation rate for compensatory mutations. The fixation process is divided into two phases, establishment and growth. The length of the establishment phase can be treated in a quasideterministic manner and is determined by the mutation rate and the growth advantage of the compensated mutant. The second phase, growth, is treated deterministically, and here the time required to grow from a presence at a fraction of ca. $1/N_0$ to 0.5 is given approximately by $T_{50} = (1/s)\ln(0.5N_0)/\ln(2)$, where N_0 is the size of the sample (bottleneck), s is the selection coefficient for the mutant, and T_{50} is the average growth time in generations. Using this model, we calculated the mutation rate to compensation for a specific *rpsL* mutant to be approximately 10^{-7}/cell/generation (41).

This relatively high mutation rate is a consequence of the large mutational target combined with a possible mutagenic SOS induction during serial passage.

Which mutants are selected during serial passage depends on several factors: the mutation rates for the different mutant types, the fitness of the different mutants, and the bottlenecks associated with the serial transfer. It is predicted that the size of the bottleneck will influence which mutants are fixed. Thus, if the population is large (no bottlenecks) it is expected that the fitter mutants (revertants) will be fixed, whereas if bottlenecks are introduced, the most common mutants (compensated) will be fixed. This notion has been examined theoretically, by simulation (34), and by actual experiment (41). As shown in these studies, with a large bottleneck the fixed mutants have a higher average fitness than those found with a small bottleneck. Thus, compensated mutants are much more commonly found than revertants during serial passage since the combination of a large mutation target for compensation compared to reversion and the presence of bottlenecks will bias mutant selection towards compensated mutants. This finding is probably also relevant for the clinical situation since a general characteristic of many pathogens and commensals is the occurrence of bottlenecks during both transmission and growth.

In most cases, it is unknown how the compensatory mutations restore fitness to the resistant mutants, but for a few resistances the compensatory mechanisms have been determined. In principle, compensatory mutations can restore fitness by reducing the need for the affected function, by substituting the impaired function with an alternative, or by directly restoring the efficiency of the function itself. For resistance mutations, the last two have been observed. For example, in isoniazid-resistant *M. tuberculosis*, *katG* mutants with decreased fitness can be compensated by overproduction of another enzyme that may substitute for the defective catalase (55). But the most common compensation mechanism is restoration of the function itself, either by intragenic or extragenic mutations. For example, the cost of streptomycin resistance in *rpsL* mutations in *E. coli* and *S. enterica* can be compensated for by extragenic *rpsD*, *rpsE*, and *rplS* mutations or intragenic *rpsL* mutations that restore the efficiency and rate of translation to wild-type or near-wild-type levels (11, 41, 53). Similarly, both intragenic and extragenic compensatory mutations can restore the rate of transcription to wild-type levels in rifampicin-resistant *rpoB* mutants of *E. coli* (49). At present, in none of the cases where the function itself is restored can it be explained in structural terms how the compensatory mutations act to compensate. For *rpsL* and *fusA* mutants in *S. enterica* attempts have been made to structurally rationalize how the compensatory mutations restore fitness, but until the X-ray crystallographic structures of the resistant and compensated proteins have been determined, we are still a far way from precise explanations at the molecular level.

The types of fitness-restoring mutations that are selected may be influenced by the environmental conditions under serial passage. This has been found for a fusidic acid-resistant *fusA* mutant and a streptomycin-resistant *rpsL* mutant in *S. enterica*, where the compensated mutants obtained in mice were different from those obtained in laboratory medium (10). For the *fusA* mutant, in mice only revertants were found whereas in laboratory medium intragenic compensatory mutations were more common. For the *rpsL* mutant, only one specific intragenic compensatory mutation was found in mice whereas in laboratory medium mostly extragenic compensatory mutations were found. In principle, two explanations can account for the different mutation spectra. Either the relative fitness of the compensated mutants is dependent on the environment, or the environmental conditions differentially affect the rates for formation of different mutations. An example of the former is compensation of the *fusA* mutant. The rare reversion mutation was predominantly selected in mice because it had a fully restored fitness, whereas the intragenically compensated mutants showed only partial compensation in mice. On the other hand, in laboratory medium the fitness of the reversion and that of compensated mutants were similar and therefore the more commonly occurring compensatory mutants were preferentially selected. In contrast, for the *rpsL* mutant there was no difference in the fitness of the compensated mutants found after evolution in laboratory medium and the fitness of the intragenically compensated mutant found after evolution in mice. These results suggest that the different spectra of compensatory mutations found in vitro and in vivo were caused by an environmentally dependent generation of the mutations. The environmental dependence of evolution to reduce costs suggests that in order to identify the most clinically relevant pathways for fitness restoration compensatory evolution should be done under many different culture conditions, in particular in experimental animals and in humans.

CLINICAL EVIDENCE FOR COMPENSATION

A crucial question is to what extent experimental data on compensatory evolution are of relevance for the clinical situation; i.e., does compensation occur in bacteria growing in patients and does compensation cause stabilization of low-fitness resistant bacterial strains? This question is experimentally not as easy to address as compensation in vitro, and essentially two approaches have been used to examine if resistant bacteria found in patients have been compensated. One approach is generally applicable but requires as a prerequisite an extensive genetic analysis of compensation in vitro. First, one identifies all the compensatory mutations for a specific resistance mutation in vitro, and then the resistant clinical isolates are examined for the presence of mutations at the same sites as those identified in vitro as being compensatory. If such mutations are found, it is a strong indication that compensation has occurred in the clinical isolate. The second approach can be used when it is possible to isolate from a patient an antibiotic-susceptible pretreatment strain and a resistant posttreatment strain that are clonally related; i.e., the resistant mutant was likely to be selected from the susceptible population during antibiotic treatment. One can then compare the fitness difference in the clinical strain pair with the fitness difference in a defined in vitro pair containing the same resistance mutation and in which the possibility of compensation has been minimized. If compensatory evolution has occurred in the resistant clinical isolate, the fitness difference in that pair should be smaller than for the defined in vitro-isolated pair.

By applying these approaches, compensation in bacteria isolated from patients has been suggested to occur in a few cases. The first example involves isoniazid-resistant *M. tuberculosis* strains that are resistant due to loss-of-function mutations in the *katG* gene (55). These mutations cause a loss of catalase activity, which results in slower growth in experimental animals and decreased virulence as measured by competition experiments, 50% lethal dose tests, and histopathology (36, 58). Many (55), but not all (27), clinical isolates with the *katG* null mutations also contain a promoter-up mutation in the *ahpC* gene which causes an increase in the level of alkyl hydroxyperoxidase reductase (AhpC). Thus, it is conceivable that overproduction of AhpC due to the promoter-up mutations can compensate for the lack of catalase. A second example of compensation in bacterial isolates from patients is fusidic acid resistance in *S. aureus* (44). Here an extensive in vitro analysis of which mutations can compensate for *fusA* mutations allowed us to examine fusidic acid-resistant blood isolates for the presence of similar or identical compensatory mutations. In three of five clinical isolates examined there was such evidence for compensatory mutations. Finally, compensation has been observed in clarithromycin-resistant *Helicobacter pylori* strains isolated from a patient that developed resistance during treatment (7). From these patients clonally related pairs of susceptible pretreatment and resistant posttreatment strains could be isolated. The fitness difference in the clinical strain pair was then compared with the fitness difference in a defined in vitro pair by competition experiments in mice. In this analysis, it appeared as if the cost of clarithromycin resistance had been compensated for after only 3 months of growth in a human stomach in two of the three resistant mutants examined. A similar approach was also used by Gillespie et al. to study how multidrug-resistant *M. tuberculosis* isolates changed their fitness as they were passaged through humans during a hospital outbreak (23).

IMPLICATIONS FOR RESISTANCE DEVELOPMENT

If resistance involves a fitness cost, it is predicted that the resistant bacteria will be outcompeted by susceptible strains if antibiotic use is reduced and as a result the frequency of resistant bacteria will decline. The expected rate (months to years before substantial changes occur) and extent of the decrease are, however, predicted to be moderate in community settings (1, 31, 33, 38, 39). In contrast, in hospital settings the driving force for a reduced frequency of resistance in response to reduced use of antibiotics is different (39). Here, the main driving force is a "dilution" effect caused by the continuous inflow of patients that are uninfected or infected with mainly susceptible bacteria, which tends to replace the more resistant population of bacteria resident in the hospital. With such dilution it is predicted and clinically observed (see references 39 and references therein) that changes in the frequency of resistance in response to reduced antibiotic use in a hospital are more rapid (substantial changes may occur in only weeks to months) and extensive than at the community level. At the community level, few studies have been performed to examine if reversibility occurs, and the interpretation of these interventions as successful has been overly optimistic. In Iceland, the frequency of penicillin-resistant pneumococci increased rapidly in the late 1980s due to the import of a specific resistant clone. Because of this, in the early 1990s measures were taken to reduce antibiotic consumption in children. After this intervention resistance peaked in 1993 at 19.8% and then declined and leveled off at 15% in 1996 (4). Similarly, in Finland after a nationwide reduction in the volume of use of macrolide antibiotics, there was a decline from 16.5% in 1992 to 8.6% in 1996 in the frequency of erythromycin resistance among group A streptococci (54). There are several things to note in these studies. First, in these cases considerable reductions in antibiotic use resulted in at best moderate decreases in the frequency of resistance. Secondly, the decline appeared to level off at a frequency substantially higher than before the increase occurred. Third, in these studies the apparent correlation between reduced antibiotic use and decreased frequency of resistance could have resulted from other factors. For example, in the latter case clonal shifts might have occurred, whereby a highly transmissible susceptible clone increased in frequency coincidentally with the reduction in antibiotic use (29). Also, in both of the above studies there were no independent experimental measurements of the fitness costs of these resistances to determine if the rate and extent of the decline in the frequency of resistance were correlated to the potential fitness cost of the resistance. Finally, as shown in this chapter, there are several processes that act against cost-driven reversibility of antibiotic resistance. Thus, one would a priori predict that such a process is rarely of importance at the community level. As discussed above, compensatory evolution to reduce the costs of resistance might be an important stabilizing force, even though it remains to be shown that it is a significant stabilizer in clinical situations. A second factor contributing to a reduced reversibility is the presence of cost-free mutations. Even though it remains uncertain if truly cost-free mutations exist, it is clear that the smaller the cost, the more likely the mutation is to be maintained in the population. Thus, for resistances where both high-cost and no- or low-cost mutations exist (e.g., *rpsL* mutations in *M. tuberculosis*), low-cost mutations are, as predicted, more frequent in clinical settings than high-cost mutations (13, 51). Finally, a third factor stabilizing resistance is the genetic linkage of different resistance markers. Thus, a nonselected resistance marker might by its genetic linkage to a selected marker increase in frequency in the population. This phenomenon is of particular importance for different genetic elements such as plasmids, transposons, and integrons but might also apply to a specific multiresistant clone. An interesting illustration of the effect of genetic linkage on the frequency of resistance was provided by Enne et al. (18). They showed that after a radical 75% reduction in the use of sulfonamide in the United Kingdom from 1991 to 1999 the frequency of sulfonamide-resistant *E. coli*, instead of decreasing as predicted, actually increased slightly (from 40 to 46%). This is in contrast to the experience in Sweden during the 1970s and 1980s, in which the resistance declined after the reduction in sulfonamide use (20). The most likely explanation for this difference is that the sulfonamide resistance gene present in isolates from the United Kingdom is genetically linked on large plasmids to other resistance genes that were continuously selected during this time period. In contrast, the classical sulfonamide resistance gene (probably predominant in Sweden) was commonly encoded by a small single resistance plasmid (48).

In the cases discussed above, the persistence of resistant bacteria was examined at the population level. In contrast, very few studies of the long-term (months to years) persistence of resistant bacteria in individuals have been performed. One notable exception is a study where it was determined how a widely used regimen for *H. pylori* eradication affects resistance development and persistence of enterococci present in the intestinal microflora (56). Here, it was shown that in three of five examined patients the resistant enterococci appearing after treatment persisted in the intestinal microflora for 1 to 3 years after treatment.

A final implication emerging from studies of fitness costs and their genetic compensation concerns the development of new antibiotics. At present, the key parameter from a resistance development point of view that is considered by drug developers is the rate of appearance of the initial resistance mutation (or plasmid). Even though these rates do influence the rate of resistance development, their importance might be overestimated. More significant might be any possible counterselection against the resistant variants and whether compensatory mutations can reduce these costs. In other words, an ideal antibiotic from the resistance point of view would be a drug for which, apart from resistance mutations being rare, the fitness costs of resistance is high and the rate and effect of fitness-restoring compensatory mutations are low. An illustration of the importance of costs was recently given by an analysis of fosfomycin resistance in *E. coli* (45). Fosfomycin resistance develops rapidly under experimental conditions, but in spite of the high mutation rate, resistance in clinical isolates is rare. The likely explanation for this is that fosfomycin resistance mutations are associated with a reduced fitness both in the absence and in the presence of fosfomycin. Thus, the resistant mutants appear at a high rate, but their reduced fitness prevents them from establishing in the bladder.

This work was supported by the Swedish Research Council, Swedish Institute for Infectious Disease Control, Swedish Strategic Research Foundation, EU 5th framework programme, AFA Research Fund, and Leo Pharmaceuticals. We thank Lars G. Burman and Cecilia Dahlberg for comments.

References

1. **Austin, D. J., and R. M. Anderson.** 1999. Studies of antibiotic resistance within the patient, hospitals and the community using simple mathematical models. *Philos Trans. R. Soc. Lond. B* **354:**721–738.
2. **Austin, D. J., and R. M. Anderson.** 1999. Transmission dynamics of epidemic methicillin-resistant *Staphylococcus aureus* and vancomycin-resistant enterococci in England and Wales. *J. Infect. Dis.* **179:**883–891.
3. **Austin, D. J., M. J. Bonten, R. A. Weinstein, S. Slaughter, and R. M. Anderson.** 1999. Vancomycin-resistant enterococci in intensive-care hospital settings: transmission dynamics, persistence, and the impact of infection control programs. *Proc. Natl. Acad. Sci. USA* **96:**6908–6913.
4. **Austin, D. J., K. G. Kristinsson, and R. M. Anderson.** 1999. The relationship between the volume of antimicrobial consumption in human communities and the frequency of resistance. *Proc. Natl. Acad. Sci. USA* **96:**1152–1156.

5. **Bagel, S., V. Hullen, B. Wiedemann, and P. Heisig.** 1999. Impact of *gyrA* and *parC* mutations on quinolone resistance, doubling time, and supercoiling degree of *Escherichia coli. Antimicrob. Agents. Chemother.* **43:**868–875.

6. **Billington, O. J., T. D. McHugh, and S. H. Gillespie.** 1999. Physiological cost of rifampin resistance induced in vitro in *Mycobacterium tuberculosis. Antimicrob. Agents Chemother.* **43:**1866–1869.

7. **Björkholm, B., M. Sjolund, P. G. Falk, O. G. Berg, L. Engstrand, and D. I. Andersson.** 2001. Mutation frequency and biological cost of antibiotic resistance in *Helicobacter pylori. Proc. Natl. Acad. Sci. USA* **98:**14607–14612.

8. **Björkman, J., and D. I. Andersson.** 2000. The cost of antibiotic resistance from a bacterial perspective. *Drug. Resist. Updat.* **3:**237–245.

9. **Björkman, J., D. Hughes, and D. I. Andersson.** 1998. Virulence of antibiotic-resistant *Salmonella typhimurium. Proc. Natl. Acad. Sci. USA* **95:**3949–3953.

10. **Björkman, J., I. Nagaev, O. G. Berg, D. Hughes, and D. I. Andersson.** 2000. Effects of environment on compensatory mutations to ameliorate costs of antibiotic resistance. *Science* **287:**1479–1482.

11. **Björkman, J., P. Samuelsson, D. I. Andersson, and D. Hughes.** 1999. Novel ribosomal mutations affecting translational accuracy, antibiotic resistance and virulence of *Salmonella typhimurium. Mol. Microbiol.* **31:**53–58.

12. **Blower, S. M., and J. L. Gerberding.** 1998. Understanding, predicting and controlling the emergence of drug-resistant tuberculosis: a theoretical framework. *J. Mol. Med.* **76:** 624–636.

13. **Bottger, E. C., B. Springer, M. Pletschette, and P. Sander.** 1998. Fitness of antibiotic-resistant microorganisms and compensatory mutations. *Nat. Med.* **4:**1343–1344.

14. **Cohen, T., B. Sommers, and M. Murray.** 2003. The effect of drug resistance on the fitness of *Mycobacterium tuberculosis. Lancet Infect. Dis.* **3:**13–21.

15. **Compeau, G., B. J. Al-Achi, E. Platsouka, and S. B. Levy.** 1988. Survival of rifampin-resistant mutants of *Pseudomonas fluorescens* and *Pseudomonas putida* in soil systems. *Appl. Environ. Microbiol.* **54:**2432–2438.

16. **Dykes, G. A., and J. W. Hastings.** 1998. Fitness costs associated with class IIa bacteriocin resistance in *Listeria monocytogenes* B73. *Lett. Appl. Microbiol.* **26:**5–8.

17. **Elena, S. F.** 2001. Evolutionary consequences and costs of plasmid-borne resistance to antibiotics, p. 163–180. *In* D. Hughes and D. I. Andersson (ed.), *Antibiotic Development and Resistance.* Taylor and Francis, London, United Kingdom.

18. **Enne, V. I., D. M. Livermore, P. Stephens, and L. M. Hall.** 2001. Persistence of sulphonamide resistance in *Escherichia coli* in the UK despite national prescribing restriction. *Lancet* **357:**1325–1328.

19. **Fermer, C., and G. Swedberg.** 1997. Adaptation to sulfonamide resistance in *Neisseria meningitidis* may have required compensatory changes to retain enzyme function: kinetic analysis of dihydropteroate synthases from *N. meningitidis* expressed in a knockout mutant of *Escherichia coli. J. Bacteriol.* **179:**831–837.

20. **Ferry, S., and L. G. Burman.** 1987. Urinary tract infection in primary health care in northern Sweden. *Scand. J. Prim. Health Care* **5:**233–240.

21. **Folkesson, A.** 2002. *On Extrinsic and Intrinsic Organizational Themes in Gram-Negative Bacteria and Their Role in Evolution and Virulence of the Bacterial Genus* Salmonella *spp.* Doctoral thesis, Karolinska Institutet, Stockholm, Sweden.

22. **Garcia-Bustos, J., and A. Tomasz.** 1990. A biological price of antibiotic resistance: major changes in the peptidoglycan structure of penicillin-resistant pneumococci. *Proc. Natl. Acad. Sci. USA* **87:**5415–5419.

23. **Gillespie, S. H., O. J. Billington, A. Breathnach, and T. D. McHugh.** 2002. Multiple drug-resistant *Mycobacterium tuberculosis*: evidence for changing fitness following passage through human hosts. *Microb. Drug. Resist.* **8:**273–279.

24. **Gillespie, S. H., L. L. Voelker, and A. Dickens.** 2002. Evolutionary barriers to quinolone resistance in *Streptococcus pneumoniae. Microb. Drug. Resist.* **8:**79–84.

25. **Goldman, W. E., and B. T. Cookson.** 1988. Structure and functions of the *Bordetella* tracheal cytotoxin. *Tokai. J. Exp. Clin. Med.* **13** (Suppl.):187–191.

26. **Gustafsson, I., O. Cars, and D. I. Andersson.** 2003. Fitness of antibiotic resistant *Staphylococcus epidermidis* assessed by competition on the skin of human volunteers. *J. Antimicrob. Chemother.* **52:**258–263.

27. **Heym, B., E. Stavropoulos, N. Honore, P. Domenech, B. Saint-Joanis, T. M. Wilson, D. M. Collins, M. J. Colston, and S. T. Cole.** 1997. Effects of overexpression of the alkyl hydroperoxide reductase AhpC on the virulence and isoniazid resistance of *Mycobacterium tuberculosis. Infect. Immun.* **65:**1395–1401.

28. **Johanson, U., A. Aevarsson, A. Liljas, and D. Hughes.** 1996. The dynamic structure of EF-G studied by fusidic acid resistance and internal revertants. *J. Mol. Biol.* **258:**420–432.

29. **Kataja, J., P. Huovinen, A. Muotiala, J. Vuopio-Varkila, A. Efstratiou, G. Hallas, H. Seppala, et al.** 1998. Clonal spread of group A *Streptococcus* with the new type of erythromycin resistance. *J. Infect. Dis.* **177:**786–789.

30. **Kristinsson, K. G.** 1997. Effect of antimicrobial use and other risk factors on antimicrobial resistance in pneumococci. *Microb. Drug. Resist.* **3:**117–123.

31. **Levin, B. R.** 2001. Minimizing potential resistance: a population dynamics view. *Clin. Infect. Dis.* **33**(Suppl. 3): S161–S169.

32. **Levin, B. R.** 2002. Models for the spread of resistant pathogens. *Neth. J. Med.* **60:**58–64, 64–66.

33. **Levin, B. R., M. Lipsitch, V. Perrot, S. Schrag, R. Antia, L. Simonsen, N. M. Walker, and F. M. Stewart.** 1997. The population genetics of antibiotic resistance. *Clin. Infect. Dis.* **24**(Suppl. 1):S9–S16.

34. **Levin, B. R., V. Perrot, and N. Walker.** 2000. Compensatory mutations, antibiotic resistance and the population genetics of adaptive evolution in bacteria. *Genetics* **154:**985–997.

35. **Levy, S. B.** 1992. *The Antibiotic Paradox: How Miracle Drugs Are Destroying the Miracle.* Plenum Press, New York, N.Y.

36. **Li, Z., C. Kelley, F. Collins, D. Rouse, and S. Morris.** 1998. Expression of *katG* in *Mycobacterium tuberculosis* is

associated with its growth and persistence in mice and guinea pigs. *J. Infect. Dis.* **177:**1030–1035.

37. **Lindberg, F., S. Lindquist, and S. Normark.** 1987. Inactivation of the *ampD* gene causes semiconstitutive overproduction of the inducible *Citrobacter freundii* beta-lactamase. *J. Bacteriol.* **169:**1923–1928.
38. **Lipsitch, M.** 2001. The rise and fall of antimicrobial resistance. *Trends Microbiol.* **9:**438–444.
39. **Lipsitch, M., C. T. Bergstrom, and B. R. Levin.** 2000. The epidemiology of antibiotic resistance in hospitals: paradoxes and prescriptions. *Proc. Natl. Acad. Sci. USA* **97:**1938–1943.
40. **MacVanin, M., U. Johanson, M. Ehrenberg, and D. Hughes.** 2000. Fusidic acid-resistant EF-G perturbs the accumulation of ppGpp. *Mol. Microbiol.* **37:**98–107.
41. **Maisnier-Patin, S., O. G. Berg, L. Liljas, and D. I. Andersson.** 2002. Compensatory adaptation to the deleterious effect of antibiotic resistance in *Salmonella typhimurium. Mol. Microbiol.* **46:**355–366.
42. **Massey, R. C., A. Buckling, and S. J. Peacock.** 2001. Phenotypic switching of antibiotic resistance circumvents permanent costs in *Staphylococcus aureus. Curr. Biol.* **11:** 1810–1814.
43. **McCormick, J. B.** 1998. Epidemiology of emerging/re-emerging antimicrobial-resistant bacterial pathogens. *Curr. Opin. Microbiol.* **1:**125–129.
44. **Nagaev, I., J. Björkman, D. I. Andersson, and D. Hughes.** 2001. Biological cost and compensatory evolution in fusidic acid-resistant *Staphylococcus aureus. Mol. Microbiol.* **40:** 433–439.
45. **Nilsson, A., O. G. Berg, O. Aspewall, G. Kahlmeter, and D. I. Andersson.** 2003. Biological costs and mechanisms of fosfomycin resistance in *Escherichia coli. Antimicrob. Agents Chemother.* **47:**2850–2858.
46. **Normark, S.** 1995. beta-Lactamase induction in gram-negative bacteria is intimately linked to peptidoglycan recycling. *Microb. Drug. Resist.* **1:**111–114.
47. **Pym, A. S., B. Saint-Joanis, and S. T. Cole.** 2002. Effect of *katG* mutations on the virulence of *Mycobacterium tuberculosis* and the implication for transmission in humans. *Infect. Immun.* **70:**4955–4960.
48. **Radstrom, P., G. Swedberg, and O. Skold.** 1991. Genetic analyses of sulfonamide resistance and its dissemination in gram-negative bacteria illustrate new aspects of R plasmid evolution. *Antimicrob. Agents Chemother.* **35:**1840–1848.
49. **Reynolds, M. G.** 2000. Compensatory evolution in rifampin-resistant *Escherichia coli. Genetics* **156:**1471–1481.
50. **Sanchez, P., J. F. Linares, B. Ruiz-Diez, E. Campanario, A. Navas, F. Baquero, and J. L. Martinez.** 2002. Fitness of in vitro selected *Pseudomonas aeruginosa nalB* and *nfxB* multidrug resistant mutants. *J. Antimicrob. Chemother.* **50:** 657–664.
51. **Sander, P., B. Springer, T. Prammananan, A. Sturmfels, M. Kappler, M. Pletschette, and E. C. Bottger.** 2002. Fitness cost of chromosomal drug resistance-conferring mutations. *Antimicrob. Agents. Chemother.* **46:**1204–1211.
52. **Schrag, S. J., and V. Perrot.** 1996. Reducing antibiotic resistance. *Nature* **381:**120–121.
53. **Schrag, S. J., V. Perrot, and B. R. Levin.** 1997. Adaptation to the fitness costs of antibiotic resistance in *Escherichia coli. Proc. R. Soc. Lond. B* **264:**1287–1291.
54. **Seppala, H., T. Klaukka, J. Vuopio-Varkila, A. Muotiala, H. Helenius, K. Lager, P. Huovinen, et al.** 1997. The effect of changes in the consumption of macrolide antibiotics on erythromycin resistance in group A streptococci in Finland. *N. Engl. J. Med.* **337:**441–446.
55. **Sherman, D. R., K. Mdluli, M. J. Hickey, T. M. Arain, S. L. Morris, C. E. Barry III, and C. K. Stover.** 1996. Compensatory *ahpC* gene expression in isoniazid-resistant *Mycobacterium tuberculosis. Science* **272:**1641–1643.
56. **Sjölund, M., K. Wreiber, D. I. Andersson, M. J. Blaser, and L. Engstrand.** 2003. Long-term persistence of resistant Enterococcus species after antibiotic to eradicate *Helicobacter pylori. Ann. Intern. Med.* **139:**483–487.
57. **Wichelhaus, T. A., B. Boddinghaus, S. Besier, V. Schafer, V. Brade, and A. Ludwig.** 2002. Biological cost of rifampin resistance from the perspective of *Staphylococcus aureus. Antimicrob. Agents Chemother.* **46:**3381–3385.
58. **Wilson, T. M., G. W. de Lisle, and D. M. Collins.** 1995. Effect of *inhA* and *katG* on isoniazid resistance and virulence of *Mycobacterium bovis. Mol. Microbiol.* **15:**1009–1015.

Index

A

AACs (aminoglycoside *N*-acetyltransferases)
  action of, 12–13, 39
  distribution of, 12
  genes of, 39–42
  inhibition of, 16
  location of, 43
  modification of, 40–42
  naming of, 36
  structures of, 12–13, 27–30, 39–40
  subclasses of, 40–41
ABC transporters, 236, 237, 247–248
Acceptors, in enzyme kinetics, 205–206
Acinetobacter
  β-lactamases of
    clinical relevance of, 133–134, 190
    detection of, 167
    epidemiology of, 164–167
    geographic distribution of, 72–74
  integrons of, 336
  penicillin-binding proteins of, 84, 86
Acinetobacter baumannii
  β-lactamases of
    epidemiology of, 164–166
    integron structure of, 124–125
    substrate specificity of, 190
  efflux system of, 250, 252–253
  resistance of, transformation in, 291
Acinetobacter calcoaceticus, penicillin-binding proteins of, 84, 86
Acinetobacter haemolyticus, AAC enzyme of, 42
AcrAB, in efflux system, 242–247
ACT-1 enzyme, 146
Actinobacillus pleuropneumoniae, *erm* genes of, 57
Actinomycetes, aminoglycoside resistance in, 11
Active drug efflux, *see* Efflux systems
Acylation
  of β-lactamases, 176, 228
  of penicillin-binding proteins, 199–202
Aeromonas, β-lactamases of, 115, 132–133
Aeromonas hydrophila, β-lactamases of
  epidemiology of, 166
  genes of, 116, 118
  substrate specificity of, 185, 187
Aeromonas jandaei, β-lactamases of, 164, 165
Aeromonas sobria, β-lactamases of, 166
Agrobacterium tumefaciens, conjugation in, 296–299, 302, 307
Alcaligenes xylosoxidans, integrons of, 336
  Amikacin
  AAC enzyme of, 40–42
  ANT enzyme of, 26
  in enzyme inhibition, 16
  resistance to, 46
    enzymatic, 11
    epidemiology of, 16
    molecular mechanisms in, 11
Aminoglycoside(s), *see also specific aminoglycosides*
  antibiotic effects of, 8
  biochemical properties of, 8
  clinical uses of, 16–17, 21, 35
  in combination therapy, 4–5
  discovery of, 7, 35–36
  list of, 3
  mechanism of action of, 4–5, 9–10, 36
  overview of, 3–5
  pharmacodynamics of, 35
  resistance to
    enzymatic, *see* Aminoglycoside-modifying enzymes; *specific enzymes*
    epidemiology of, 15–16
    mechanisms of, 10–11, 23–24, 36
    nonenzymatic, 14–15
    overview of, 23–24
  semisynthetic, 36
  spectrum of activity of, 3–4
  structures of, 7–8, 21–23
  toxicity of, 9
Aminoglycoside adenylyltransferases, *see* ANTs (aminoglycoside nucleotidyltransferases)
Aminoglycoside-modifying enzymes
  AAC group, *see* AACs (aminoglycoside *N*-acetyltransferases)
  ANT group, *see* ANTs (aminoglycoside nucleotidyltransferases)
  APH group, *see* APHs (aminoglycoside phosphotransferases)
  classification of, 11
  detection of, 16
  genes of, 43–46
  inhibition of, 16, 46–47
  integrons and, 335
  location of, 43
  overview of, 36
  structures of, 24–30
  subcellular location of, 43
  terminology of, 36
m-Aminophenylboronic acid, β-lactamase complex with, 154
Amoxicillin, β-lactamases of, 184, 188
AmpC β-lactamases
  boronic acid inhibition of, 154–156
  discovery of, 145
  extended-spectrum, 149–151
  noncovalent inhibitors of, 156–157
  structures of

AmpC β-lactamases (*Continued*)
active sites, 149–153
drug design based on, 153–159
three-dimensional, 146–149
variants of, 146
ampD gene, mutations of, biologic costs of, 342
Ampicillin, β-lactamases of, 179–182, 184, 188, 189
Anacardic acids, for resistant staphylococci, 92
Anaerobic organisms, *erm* genes of, 57
Antibiotic paradox, 289
Antimicrobial peptides, as novel agents, 289
Antisense oligonucleotides, in aminoglycoside-modifying enzyme inhibition, 47
Antiseptics, efflux systems for, 251
Antitoxins, *see* Toxin-antitoxin systems
ANTs (aminoglycoside nucleotidyltransferases)
action of, 11–12, 38–39
genes of, 11, 38–39
inhibition of, 16
modification of, 38–39
structures of, 11–12, 26–27, 38–39
APHs (aminoglycoside phosphotransferases)
action of, 13–14, 37–38
as bifunctional enzyme, 114
distribution of, 11
genes of, 37–38
inhibition of, 16
modifications of, 37–38
structures of, 13–14, 24–26, 37–38
Apramycin
AAC enzyme of, 40
structure of, 8
Arbekacin, AAC enzyme of, 40
Archaeoglobus fulgidus, RelBE2 system of, 315
ARE (antibiotic resistance) proteins, 248
att genes, in integrons, 331–335, 337
Axe-Txe system, 313
Azithromycin, *see* MLSKO antibiotic group
Aztreonam, β-lactamases of, 177–182, 184, 188, 189

B
Bacillus anthracis, β-lactamases of, 133
Bacillus cereus, β-lactamases of, 67, 115, 116, 208
Bacillus subtilis
efflux system of, 247, 249, 250
resistance of, transformation in, 291
Bacterial conjugation, *see also* Conjugative transposons
coupling in, 300
description of, 292–295
DNA processing in, 296–298
DNA replication in, 300–302
as drug target, 289–290
efficiency of, 294
environments for, 294–295
mechanisms of, 292–293
model for, 295–300, 304, 306–307
in multiple drug resistance, 295
vs. mutation, 295
sexual euphemisms and, 308–309
T4SS transporter family in, 298–299
Bacteriophages, 268, 291–292
Bacteroides
conjugative transposons of, 272–274, 277–278, 303
erm genes of, 57–58
penicillin-binding proteins of, 87
Bacteroides fragilis
β-lactamases of, 115
clinical relevance of, 132
genes of, 118
penicillin-binding proteins of, 87
Baicalin, for resistant staphylococci, 92
BBL numbering system, for β-lactamases, 167–168
Bearberry extract, for resistant staphylococci, 92
Benzo[b]thiophene-2-boronic acid, β-lactamase complex with, 154
Benzylpencillin
β-lactamases of, 179–182, 184, 188, 189
penicillin-binding protein susceptibility to, 84
β-Lactam(s)
acylation by, 199–202
hydrolysis of, 228
new, 93
β-Lactamase(s)
activity of, 202–205
biologic costs and, 342
in carbapenem resistance, 70
chromosomally encoded, 70, 71
class A, *see* Class A β-lactamases
class B, *see* Class B β-lactamases
class C, *see* Class C β-lactamases
class D, *see* Class D β-lactamases
classification of, 67–68, 163
clavulanic acid interaction with, 207–208
conformational change of, 208
families of, 69
geographic distribution of, 70, 72–74
kinetics of, *see under* Kinetics
measurement of, 68–69
mechanism of action of, 101
new phenotypes of, 69–70
suicide substrates of, 207
TEM, *see* TEM β-lactamases
β-Lactamase inhibitory proteins (BLIPs), 227–234
inhibitors of, 228–229
minimization of, 232–234
structures of, 229, 230
TEM β-lactamase interactions with, 230–233
types of, 227
BiaCore approach, 220
Biapenem, 107
Biocides, efflux systems for, 251
Biofilms, 251–252
Biological costs of resistance, 268, 339–348
compensation of, 342–345
examples of, 341–342
factors determining, 340
implications of, 345–346
measurement of, 340–341
Bioluminescence resonance energy transfer (BRET), 324–325
bla genes, of β-lactamases, 121–122, 136
inhibitors of, 232
in integrons, 335
BLIPs, *see* β-Lactamase inhibitory proteins (BLIPs)
Bordetella pertussis, resistance of, biologic costs of, 342
Boronate transition-state analogs, 109–111
Boronic acids, β-lactamase complex with, 154–156
BRET (bioluminescence resonance energy transfer), 324–325
Brittle pili, in conjugation, 307
BRL 42715 (penem), 108, 152
Burkholderia cepacia, aminoglycoside resistance to, 11
Burkholderia pseudomallei
β-lactamases of, 164–166
efflux system of, 252
Butirosin, AAC enzyme of, 40

C
Caenorhabditis elegans, conjugative transfer in, 294–295
Campylobacter coli, penicillin-binding proteins of, 86
Campylobacter fetus, penicillin-binding proteins of, 86
Campylobacter jejuni
β-lactamases of, 166
penicillin-binding proteins of, 86
Carbapenemases, 101
detection of, 134–136
epidemiology of, 164, 166–167
Carbapenems, 106–108
active against methicillin-resistant *Staphylococcus aureus*, 91, 93
resistance to, 167
Carbenicillin, β-lactamases of, 179, 181–182, 184, 188, 189
Cashew product, for resistant staphylococci, 92
Catalytic constant (k_{cat}), 196, 197, 202–203
CAU-1 enzyme, 127
Caulobacter vibrioides, β-lactamases of, genes of, 116
CBBLs, *see* Class B β-lactamases
CcdA-CcdB system
biophysical analysis of, 316–317
description of, 313–314
mechanisms of, 316
structures of, 314, 318–320
targets of, 319–320
Cefazolin, β-lactamases of, 188
Cefepime, β-lactamases of, 184, 188, 189
Cefotaxime, β-lactamases of, 177, 179, 180, 184, 188, 189
Cefotoxitin, resistance to, conjugative transposons in, 273
Cefoxitin, β-lactamases of, 179, 181–182
Cefpirome, β-lactamases of, 184, 189
Cefsulodine, β-lactamases of, 188
Ceftazidime, β-lactamases of, 179, 184, 188, 189
Ceftobiprole, penicillin-binding protein susceptibility to, 84
Ceftriaxone
β-lactamases of, 177, 179, 189
penicillin-binding protein susceptibility to, 84
Cefuroxime, β-lactamases of, 179, 180, 188, 189
Cephaloridine, β-lactamases of, 179–182, 184
kinetic parameters of, 184, 188, 189
substrate specificity of, 190
Cephalosporin(s)
active against methicillin-resistant *Staphylococcus aureus*, 90, 93

β-lactamase inhibitors based on, 111–112
β-lactamases and, 178
leaving group expulsion from, 206–207
Cephalosporinases, 101
action of, 184
discovery of, 69
new, 70
Cephalothin
β-lactamases of, 179, 180
AmpC, 149–150
kinetic parameters of, 184, 188, 189
penicillin-binding protein susceptibility to, 84
CfiA/CcrA enzyme, *Bacteroides fragilis*, 132
Chelating agents, in β-lactamase inactivation, 129
Chemical libraries, for drug development, 217–218, 323
Chlamydia, penicillin-binding proteins of, 87
Chlamydia suis, efflux systems of, 238
Chloramphenicol
efflux systems for, 250
resistance to, 335
conjugative transposons in, 272, 274, 281–282
Chloramphenicol phosphotransferase, 320
3-[(4-Chloroanilino)sulfonyl]thiophene-2-carboxylic acid, β-lactamase complex with, 156
Choline kinases, vs. aminoglycoside kinases, 26
ChpAIK (MazEF) system, 315, 320
Chromosomes, toxin-antitoxin systems of, *see* Toxin-antitoxin systems
Chryseobacterium gleum, β-lactamases of, 118
Chryseobacterium indologenes, β-lactamases of, 133
Chryseobacterium meningosepticum, β-lactamases of, 118, 133
Ciprofloxacin, resistance to, AAC enzyme in, 42
Citrobacter, β-lactamases of, 146
Citrobacter freundii
β-lactamases of
AmpC, 146, 147, 149, 151
geographic distribution of, 72
structures of, 171
resistance of, biologic costs of, 342
Clarithromycin, *see* MLSKO antibiotic group
Class A β-lactamases, 101–114
vs. class D, 145–146
inhibitors of
carbapenems, 106–108
cephalosporin-based, 111–112
clavulanic acid, 103
history of, 101–103
methylidene penem, 108
monobactams, 108–109
penicillanic acid derivatives, 103–106
transition-state analogs, 109–111
structures of, 101
subclasses of, 101
Class B β-lactamases, 115–144
clinical relevance of, 132–134
detection of, 134–136
discovery of, 115
distribution of, 118, 120–122
diversity of, 116
evolution of, 123
functions of, 123, 127–129
genes of, 116, 118, 120–122, 124–126
inactivation of, 129
kinetic parameters of, 127–129
kinetics of, 209–210
mechanisms of action of, 130–132
resident, 116–118
structures of, 116, 119, 130–132
subclasses of, 116, 117
kinetic parameters of, 128
structures of, 119, 130–132
in superfamily, 123
Class C β-lactamases, 145–161
boronic acid inhibition of, 154–156
vs. class A enzymes, 145–146
complexes with, 153–160
mechanisms of action of, 145–146, 149–153
noncovalent inhibitors of, 156–157
penem action on, 151–153
structures of, 145–161
active sites, 149–153
drug design based on, 153–159
extended-spectrum, 149–151
three-dimensional, 146–149
Class D β-lactamases, 163–194
catalytic mechanisms of, 176–178
classification of, 163–164, 167–170
clinical relevance of, 190
conserved elements of, 178
detection of, 167
epidemiology of, 163–164, 166–167
genetics of, 166–167
group I, 174–176, 178–184
group II, 176, 184–186
group III, 174, 185–186
group IV, 185–187
group V, 186, 187
group VI, 187, 190
kinetics of, 208–209
new, 70
phylogeny of, 167–168, 170
structures of, 167–168
quaternary, 174
tertiary, 171–174
three-dimensional, 168, 171–176
substrate specificity of, 178–190
group I, 178–184
group II, 184–186
group III, 185–186
group IV, 185–187
group V, 186, 187
group VI, 186, 187, 190
Clavulanic acid
action of, 103, 228
β-lactamase interactions with, 145, 207–208
β-lactamases of, 181–182
structure of, 103
Clindamycin, *see* MLSKO antibiotic group
Clostridium
conjugative transposons of, 272
resistance of, mechanisms of, 287
Clostridium difficile, conjugative transposons of, 271, 272, 274, 277, 281–282
Clostridium perfringens, penicillin-binding proteins of, 86
Clover-leaf test, for β-lactamases, 167
Cloxacillin, β-lactamases of, 178–182, 184, 188, 189
CmlA, in efflux system, 250
CMY-2 enzyme, 146
ColE1 plasmid, target site in, 268
Common regions, in integrons, 335–336
Complex transposons, 268
Composite transposons, 268
Conjugative self-transmissible, integrating elements, 271
Conjugative transposons, 271–284
vs. conventional transposons, 303–304
definition of, 268, 271
description of, 303–304
discovery of, 271
as drug targets, 289–290
epidemiology of, 272–275
evolution of, 271–272
functions of, 303–304
host range of, 272
inhibitors for, 282
mechanisms of, 275–282
mobilizable elements of, 273, 303
self-transmissible elements of, 303
structures of, 275–282
Constins, 271
Coptis chinensis extract, in sortase inhibition, 222
Corilagin, for resistant staphylococci, 92
Corynebacterium striatum, APH enzyme of, 37
Costs, *see* Biological costs
Coupling proteins, in bacterial conjugation, 296, 300
CphA/Imi enzyme, *Aeromonas*, 132–133
CTn7853 transposon, 278
CTn12256 transposon, 278
CTnBst transposon, 278
CTnDOT transposon family, 277–279
CTnERL transposon family, 278, 279
CTnGERM1 transposon, 278, 279
CTns, *see* Conjugative transposons
CTnXBU4422 transposon family, 278
Cycle of resistance and antibiotic development, 289–290
Cytochalasin D, 292

D

Deacylation, of β-lactamases, 176–178, 228
2-Deamino-2-nitro neamine, 16
2-Deoxystreptamine moiety, in aminoglycosides, 7–8, 21–22
Dibekacin
AAC enzyme of, 13, 40–41
in enzyme inhibition, 16
resistance to, 11, 46
3′-Dideoxykanamycin A, resistance to, 46
3′,4′-Dideoxykanamycin B, *see* Dibekacin
Diffusion assay, for β-lactamases, 135
DinJ-YafQ system, 315
Dirithromycin, *see* MLSKO antibiotic group
Disk diffusion assay, for *erm* genes, 59
Dissemination of resistance, *see also* Bacterial conjugation; Conjugative transposons
mechanisms for, 287–288
overview of, 267–270
DNA
conjugative transfer of, *see* Bacterial conjugation; Conjugative transposons
in Holliday junctions, 332
mutation of, in resistance dissemination, 290–291
in plasmids, *see* Plasmid(s)
replication of, 300–302
strandedness of, in conjugative transfer, 293

DNA gyrase, CcdB effects on, 314
DNA hybridization assays, for *erm* genes, 59
DNA microarray analysis, for *erm* genes, 60
Dog rose products, for resistant staphylococci, 92
Drug development, 217–225
approaches to, 217
class C β-lactamase complexes and, 153–159
cycle of, 289–290
difficulties with, 217–218, 289–290
targets for
conjugative transposons, 282
dissemination mechanisms, 289–290
essential pathways, 219
protein-protein interactions, 219–220
screening of, 218–219
toxin-antitoxin systems, 321–325
virulence, 220–223
Drug efflux systems, *see* Efflux systems

E

Efflux systems, 235–264
ABC transporters, 236, 237, 247–248
in antiseptics, 251
bacterial conjugation in, 295
in biocides, 251
biofilm formation and, 251–252
classification of, 235–236
clinical relevance of, 249–253
conjugative transposition in, 286–287
detection of, 238–239
epidemiology of, 236, 238
examples of, 237
MATE family, 236, 237, 247
MFS family, 235, 237, 239–241, 249–250
multidrug, 248–249
overcoming, 252–253
regulation of, 236
RND family, 236, 237, 241–247
SMR family, 236, 237, 241
EmrE, in efflux system, 241
Enterobacter, resistance of, 287
Enterobacter aerogenes
β-lactamases of, 72–73
efflux system of, 253
Enterobacter cloacae
β-lactamases of, 69
AmpC, 146, 147, 149, 150, 152
genes of, 122
geographic distribution of, 72–74
integron structure of, 126
resistance of, biologic costs of, 342
Enterobacteriaceae
AAC enzymes of, 42
APH enzymes of, 38
β-lactamases of, 69, 119, 120, 122
AmpC, 146
clinical relevance of, 133–134, 190
detection of, 134–136
epidemiology of, 164
conjugative transposons of, 272, 274, 278, 280–281
efflux system of, 250
Enterococcus
APH enzyme of, 37
erm genes of, 57
penicillin-binding proteins of, 82
resistance of, 55, 288
Enterococcus faecalis
AAC enzyme of, 41
conjugative transposons of, 272–273
drug targets in, 218
efflux system of, 253
erm genes of, 58
penicillin-binding proteins of, 81, 83, 87
Enterococcus faecium
AAC enzyme of, 27–28
Axe-Txe system of, *see* Axe-Txe system
penicillin-binding proteins of, 87–89
resistance to, conjugative transposons in, 273
YefM-YoeB system of, 313, 315, 317–318
Enterococcus hirae, penicillin-binding proteins of, 87
Enterococcus raffinosus, penicillin-binding proteins of, 87
Epi(gallo)catechin gallates, for resistant staphylococci, 92
5-Epi-gentamicin, in enzyme inhibition, 16
5-Epi-sisomicin, in enzyme inhibition, 16
ε-ζ system, *see* ωω-ε-ζ system
erm genes
detection of, 59–60
epidemiology of, 238
geographic distribution of, 55–57
in mobile elements, 57–58
Ertapenem, 107–108
Erythromycin, *see also* MLSKO antibiotic group
efflux systems for, 238
resistance to
conjugative transposons in, 272, 273, 278
control of, 345
Escherichia coli
APH enzyme of, 37
β-lactamases of
AmpC, 145–147, 149
detection of, 167
epidemiology of, 164, 165
geographic distribution of, 72–73
integron structure of, 125
measurement of, 68
substrate specificity of, 184, 185
TEM, 101
CcdA-CcdB system of, *see* CcdA-CcdB system
conjugation in, 295, 298, 306
conjugative transposons of, 275
DinJ-YafQ system of, 315
drug targets in, 218
efflux system of, 239, 241, 242–248, 251
HigA-HigB system of, 314
integrons of, 333
MazEF system of, 315, 320
ParD-ParE system of, 314, 317
PemIK system of, 314, 317
penicillin-binding proteins of, 81, 83, 84, 88
proteins of, interactions of, 220
RelBE2 system of, 313, 315, 316
resistance of
biologic costs of, 341–344
conjugative transposons in, 273
control of, 346
history of, 285–286
in macrolides, 55
mechanisms of, 286, 287
plasmids and, 267
transformation in, 291
YefM-YoeB system of, 313, 315, 317–318
2′-*N*-Ethylnetilmicin, AAC enzyme of, 40
6′-*N*-Ethylnetilmicin, AAC enzyme of, 40
Excision, in conjugation, 276–277
Extended-spectrum β-lactamases, 149–151

F

Fitness costs, *see* Biological costs
Flavinoids, for resistant staphylococci, 92
Florfenicol, resistance to, conjugative transposons in, 272
Florfenicol-chloramphenicol resistance marker (FloR), 240, 250
Fluorescence resonance energy transfer (FRET) assay, 222, 324–325
Fluoroquinolones, resistance to, biologic costs of, 341, 344
Fosfomycin, resistance to, 346
FRET (fluorescence resonance energy transfer) assay, 222, 324–325
Fritillaria verticillata extract, in sortase inhibition, 222
FtsZ protein, ZipA protein interactions with, 220
Fusidic acid, resistance to, 341–343
Fusobacterium nucleatum, β-lactamases of, 177

G

Genes
as drug targets, 218–219
exchange of, *see* Bacterial conjugation; Conjugative transposons
"Genetic addiction," 267; *see also* Toxin-antitoxin systems
Gentamicin
AAC enzyme of, 40–41
ANT enzyme of, 26
mechanism of action of, 9–10
resistance to
biologic costs of, 341
conjugative transposons in, 273
enzymatic, 12
molecular mechanisms in, 11
structure of, 8, 10, 22
Gentamicin B
APH enzyme of, 13
in enzyme inhibition, 16
Geraniol, for resistant staphylococci, 92
GIM enzymes, 121, 134
GlpT (glycerol-3-phosphate antiporter), 240
Glycerol-3-phosphate antiporter (GlpT), 240
Glycylboronic acids, β-lactamase complex with, 155
Glycylcyclines, efficacy of, 252

H

Haemophilus, penicillin-binding proteins of, 86
Haemophilus influenzae
β-lactamases of, 68
efflux system of, 249, 250, 253
penicillin-binding proteins of, 89
resistance of
history of, 285
mechanisms of, 286, 287
transformation in, 291
Hanes-Woolf equation, 196
Helicobacter pylori
efflux system of, 251
penicillin-binding proteins of, 86, 88, 89

resistance of
biologic costs of, 343, 345, 346
transformation in, 291
type 2 secretion system of, 299, 307
Henri-Michaelis equation, 196–197
HEPN (higher eukaryotes and prokaryotes nucleotide-binding) domain of proteins, 27
HigA-HigB system, 314
Higher eukaryotes and prokaryotes nucleotide-binding (HEPN) domain of proteins, 27
Hodge (clover-leaf) test, for β-lactamases, 167
Holliday junctions, 332
Hydrolysis, of β-lactams, 197
α-Hydroxyisopropylpenicillanate, 173
7-Hydroxytropolone, in enzyme inhibition, 16
Hygromycin
APH enzyme of, 38
resistance to, in ribosome defects, 15

I

ICEs (integrative and conjugative elements), 268, 271
Imipenem, 107–108
β-lactamases of, 179–182
AmpC, 149
kinetic parameters of, 184, 188, 189
penicillin-binding protein susceptibility to, 84
IMP enzymes
clinical relevance of, 133–134
distribution of, 118, 120
kinetic parameters of, 127, 129
Inc groups, of plasmids, 278, 280–281, 292–293
Insertion sequence common regions, 335–336
Insertion sequence element, 335
Int proteins, in conjugation, 275–277
Integrases
action of, 332–333
class 1, 334–336
class 2, 336
class 3, 336
structures of, 337
Integrative and conjugative elements, 268, 271
Integrons, 331–338; *see also* Integrases
in aminoglycoside-modifying enzymes, 44
aminoglycoside-modifying enzymes and, 335
of β-lactamases, 124–126, 164–166, 335
class 2, 336
class 3, 336
classes of, 334–337
discovery of, 268
genetics of, 331–334
insertion sequence common regions and, 335–336
overview of, 331
very large (superintegrons), 331, 336–337
Isepamicin, in enzyme inhibition, 16
Isoelectric focusing method, for β-lactamase detection, 69
Isoniazid, resistance to, biologic costs of, 341, 343

K

Kanamycin
AAC enzyme of, 40–42
ANT enzyme of, 26
APH enzyme of, 13–14, 24, 37
in enzyme inhibition, 16
resistance to
conjugative transposons in, 272
enzymatic, 11, 12
molecular mechanisms in, 11
structure of, 8
Kanamycin A, AAC enzyme of, 13
Kanamycin nucleotidyltransferase, action of, 11–12
k_{cat} (catalytic constant), 195, 202–203
Ketolides, *see* MLSKO antibiotic group
Kid toxin, *see* Kis-Kid system
Kinases, aminoglycoside, *see* APHs (aminoglycoside phosphotransferases)
Kinetics
of β-lactamases
activity of, 202–205
branched pathways in, 205–209
class B, 127–129, 209–210
class D, 184, 188, 189, 208–209
clavulanic acid and, 207–208
in conformational change, 208
linear models of, 197–199
order of reactions and, 210–211
principles of, 195–197, 210–211
suicide substrates in, 207
zinc and, 210
of penicillin-binding proteins, 88
acylation in, 199–202
branched pathways in, 205–209
k_2/k_3 ratio in, 203–205
linear models of, 197–199
order of reactions and, 210–211
principles of, 195–197, 210–211
Kis-Kid system
description of, 314–315
structures of, 315, 318–320
targets of, 319–320
Klebsiella, resistance of, 287
Klebsiella oxytoca, β-lactamases of, 72
Klebsiella pneumoniae
aminoglycoside-modifying enzymes of, 44
β-lactamases of
epidemiology of, 164, 165
genes of, 122
geographic distribution of, 72–74
integron structure of, 124
measurement of, 68
substrate specificity of, 184
conjugation in, 300
integrons of, 336
K_m (Michaelis constant), 196, 197, 202–203

L

L1 enzyme, *Stenotrophomonas maltophilia*, 132
Lactamases, beta-, *see* β-Lactamase(s)
Lactococcus lactis, efflux system of, 247
Lactose permease (LacY), in efflux system, 240
LCR-1 β-lactamase, 176, 187
Legionella gormanii, β-lactamases of, 164, 165
Legionella pneumophila, β-lactamases of, 164, 166
Levofloxacin, efflux inhibition in, 252–253
Libraries, for drug development, 217–218, 323
Licoricidin, for resistant staphylococci, 92
Lincomycin, *see* MLSKO antibiotic group
Lincosamide, *see* MLSKO antibiotic group
Linezolid, *see* MLSKO antibiotic group
Lipinsky rules, 217–218, 323
Listeria monocytogenes
conjugative transposons of, 275
penicillin-binding proteins of, 86
resistance of, biologic costs of, 341
sortase of, 220
Lividomycin
AAC enzyme of, 40
APH enzyme of, 13
LmrA, in efflux system, 247–248
LPXT-Gase, 223
Lys-Xxx-Gly element, of β-lactamases, 178

M

M protein, in antibiotic susceptibility, 286–287
MacB, in efflux system, 248
Macrolides, *see also* MLSKO antibiotic group; *specific macrolides*
efflux systems for, 238, 249–250; *see also mef* genes
overview of, 5
resistance to, 287
Major facilitator superfamily (MFS), 235, 237, 239–241, 249–250
Mar, in efflux system, 251
MATE (multidrug and toxic compound extrusion) family efflux system, 236, 237, 247
Mating (membrane) bridge, in DNA transfer, 292, 296, 306–307
Maveyraud classification, of β-lactamases, class D, 168
MazEF (ChpAIK) system, 315, 320
MbeA protein, in conjugation, 275
MBLs (metallo-β-lactamases), *see* Class B - lactamases
MDR/ABCB1, in efflux system, 247
Mecillinam, penicillin-binding protein susceptibility to, 84
mef genes
clinical relevance of, 249
conjugative transposition of, 286–287
detection of, 239
epidemiology of, 238
in MLSKO antibiotic resistance, 55
mel gene, in MLSKO antibiotic resistance, 55
Membrane bridge, in DNA transfer, 292, 296, 306–307
Mep proteins, in efflux system, 252
Mercury, resistance to, conjugative transposons in, 272–273
Meropenem, 107–108
β-lactamases of, 179, 180, 189
penicillin-binding protein susceptibility to, 84
Metallo-β-lactamases, *see* Class B β-lactamases
Methanococcus jannaschii, RelBE2 system of, 315
Methicillin
β-lactamases of, 181–182, 184, 188
for penicillin-resistant organisms, 289
Methylidene penems, 108, 151–153, 157, 159
Mex proteins, in efflux system, 242–247, 251–252
MFS (major facilitator superfamily) efflux system, 235, 237, 239–241, 249–250
Michaelis constant (K_m), 196, 197, 202–203
Microarray analysis, for *erm* genes, 60
Microbial surface components recognizing adhesive matrix molecules (MSCRAMMs), 220–222

Microbiological assays, for β-lactamases, 68
Microdilution test, for β-lactamases, 135
Minimum inhibitory concentration, measurement of, in efflux detection, 239
Minocycline, resistance to, conjugative transposons in, 272–273
MLSKO antibiotic group, 53–63
  clinical uses of, 53–54
  efficacy of, 252
  mechanisms of action of, 54–55
  resistance to
    biologic costs of, 341, 343
    detection of, 59–60
    evolutionary aspects of, 58
    geographic distribution of, 55–57
    mechanisms of, 55
    mobile elements in, 57–58
    molecular mechanisms of, 58–59
Mobile elements, in MLSKO resistance, 57–58
Monobactams, 108–109
Moraxella catarrhalis, efflux system of, 249
MreA, in efflux system, 249–250
MsbA, in efflux system, 247–248
MSCRAMMs (microbial surface components recognizing adhesive matrix molecules), 220–222
msr genes and proteins
  in efflux system, 250
  in MLSKO antibiotic resistance, 55
Multidrug and toxic compound extrusion (MATE) family efflux system, 236, 237, 247
Multidrug binding sites, 248–249
Multiple drug resistance, 250–251
  bacterial conjugation in, 295
  history of, 285–289
  integrons and, 333–334
Mutation
  vs. conjugation, 295
  DNA, in resistance dissemination, 290–291
Mycobacterium
  penicillin-binding proteins of, 87
  rRNA methylase of, 55
Mycobacterium bovis, resistance of, 342
Mycobacterium leprae, AAC enzyme of, 40
Mycobacterium smegmatis
  AAC enzyme of, 40
  resistance of, mechanisms of, 288
Mycobacterium tuberculosis
  AAC enzyme of, 28–29, 39–40
  efflux system of, 251
  resistance of
    biologic costs of, 340–345
    mechanisms of, 287–288

N

Nalidixic acid, resistance to, biologic costs of, 341, 343
Natural transformation, 291
  as drug target, 289–290
  Streptococcus pyogenes, 286
NBU (nonreplicating bacteroides unit), 303
Neisseria
  erm genes of, 57
  resistance of, transformation in, 291
Neisseria cinerea, penicillin-binding proteins of, 84
Neisseria denitrificans, penicillin-binding proteins of, 84
Neisseria flavescens, penicillin-binding proteins of, 84
Neisseria gonorrhoeae
  conjugation in, 299
  erm genes of, 58
  penicillin-binding proteins of, 84
  resistance of
    history of, 285
    mechanisms of, 286
Neisseria lactamica, penicillin-binding proteins of, 84
Neisseria meningitidis
  erm genes of, 58
  penicillin-binding proteins of, 84, 89
  resistance of
    biologic costs of, 341, 343
    history of, 285
Neomycin
  AAC enzyme of, 40
  APH enzyme of, 13–14, 24
  resistance to, enzymatic, 12
  structure of, 8
Netilmicin
  AAC enzyme of, 40–42
  in enzyme inhibition, 16
  resistance to, 46
Nick sites, of plasmids, in bacterial conjugation, 296–297, 301–302
Nitrocefin
  in acylation studies, 199–200
  in assay for β-lactamase detection, 68–69
  β-lactamases of, 188
  hydrolysis of, 209–210
NMC-A (non metallo-carbapenemase A), 108
Non metallo-carbapenemase A (NMC-A), 108
Nonreplicating bacteroides unit (NBU), 303
Nonretractable pili, in conjugation, 307
NorA, in efflux system, 250, 253
NorM, in efflux system, 247
Nucleophiles, transition-state analogs action against, 109–111
Nucleotide-binding domain, in ABC transporters, 247–248

O

ω-ε-ζ system
  biophysical analysis of, 318
  description of, 313–314
  mechanisms of, 316
  structures of, 320–321
Open reading frames
  in *Bacteroides* conjugative transposons, 278
  in integrons, 331, 335
  of Tn*916* transposon, 275–277
Opr proteins, in efflux system, 243–247, 252
OXA enzymes, *see* Class D β-lactamases
Oxacillin, β-lactamases of, 178–182, 184, 188, 189
Oxacillinases, *see* Class D β-lactamases
Oxalate transporter (OxlT), 240
Oxalobacter formigenes, efflux system of, 240
Oxazolidinones, *see* MLSKO antibiotic group

P

Pandoraeae pnomenusa, β-lactamases of, 166
ParD-ParE system, 314, 317
Paromomycin
  AAC enzyme of, 40
  APH enzyme of, 13–14
  clinical use of, 9
  mechanism of action of, 9–10
  resistance to, in ribosome defects, 15
  structure of, 8, 10, 21–22
PasA-PasB system, 314
PBPs, *see* Penicillin-binding proteins
PemIK system, 313, 315
Penam sulfones, 103–106
Penicillanic acid derivatives, 103–106
Penicillin, resistance to, 340
  biologic costs of, 341
  conjugative transposons in, 272
  history of, 285–286
  mechanisms of, 286
Penicillin-binding proteins, 81–99
  actions of, 81
  acylation of, 199–202
  antibiotic susceptibility of, 84
  clinical relevance of, 82, 84–87
  detection of, 87
  epidemiology of, 82, 84–87
  inhibition of, new products for, 90–92
  kinetics of, *see under* Kinetics
  molecular mechanisms of
    kinetic, 87–88
    resistant mutants and, 88, 89, 91–92
  nomenclature of, 81–82
  non-penicillin-binding domains of, 81
  in resistance, 82, 83
  structures of, 88, 89
Pernot classification, of β-lactamases, 168
Phd-Doc system, 313–314, 317
Phenothiazines, for resistant staphylococci, 92
Phosphonate transition-state analogs, 109–111
Pili, in bacterial conjugation, 299, 304–307
Piperacillin
  β-lactamases of, 179, 180, 184, 188, 189
  penicillin-binding protein susceptibility to, 84
Plasmid(s)
  in conjugative transfer, *see* Bacterial conjugation
  discovery of, 267
  replication-specific incompatibility of, 292–293
  toxin-antitoxin systems of, *see* Toxin-antitoxin systems
  types of, 292–293
"Plasmid addiction," 267; *see also* Toxin-antitoxin systems
Plasmid F
  in bacterial conjugation, 295–296, 306–307
  CcdA-CcdB system of, *see* CcdA-CcdB system
Plasmid pC221, replication in, 302
Plasmid R27, in conjugative transfer, 293
Plasmid R1162, replication in, 302
Plasmid R 100, PemIK system of, 313, 315
Plasmid RK2, in conjugative transfer, 292–302, 304, 306–307
Plasmid RP4, in conjugative transfer, 293
Plasmid Rts1, HigA-HigB system of, 314
Pneumococci, *see Streptococcus pneumoniae*
Polymerase chain reaction
  for β-lactamases, 136
  for efflux detection, 238
  for *erm* genes, 59
  for resistance enzyme detection, 16
Pre-steady state, in enzyme kinetics, 196–197
prfC gene, in conjugation, 280

Primases, in bacterial conjugation, 298
Prophage P1, Phd-Doc system of, 313–314, 317
Protein(s)
  interactions between, 219–220, 229–230
  synthesis of, inhibition of, MLSKO antibiotics in, 54–55
Proteus mirabilis
  β-lactamases of, 72–73, 164
  penicillin-binding proteins of, 86
Proteus rettgeri
  conjugative transposons of, 272, 274
  Inc-J family elements of, 278, 280–281
Proteus vulgaris, APH enzyme of, 37
Providencia
  AAC enzyme of, 40
  conjugative transposons of, 274
Providencia rettgeri, APH enzyme of, 37
Providencia stuartii, β-lactamases of, 72
Pseudomonas, resistance of, 287
Pseudomonas aeruginosa
  AAC enzyme of, 40
  aminoglycoside resistance in, 11, 13
  β-lactamases of, 69–70, 119–122
    AmpC, 145–146
    clinical relevance of, 133, 190
    detection of, 136, 167
    epidemiology of, 164–166
    geographic distribution of, 72–74
    integron structure of, 124–126
    substrate specificity of, 180, 183–185, 190
  conjugation in, 299
  conjugative transposons of, 304
  drug targets in, 218
  efflux system of, 244–247, 250–253
  integrons of, 335, 336
  penicillin-binding proteins of, 83, 84
  resistance of
    biologic costs of, 341, 343
    transformation in, 291
Pseudomonas alcaligenes, superintegron of, 337–338
Pseudomonas fluorescens
  AAC enzyme of, 41
  efflux system of, 249
  resistance of, biologic costs of, 341
Pseudomonas putida
  β-lactamases of, 72–74, 124–125
  integrons of, 336
  resistance of, biologic costs of, 341
Psychrobacter immobilis, β-lactamases of, 146, 149

Q

qac genes, in integrons, 333
Qac proteins, in efflux system, 249, 253

R

"R factors," 267
R391 mobile element, 278, 280–281
R997 mobile element, 280
Ralstonia pickettii, β-lactamases of, 164–166, 190
RecA protein, in bacterial conjugation, 298
Recombinases, types of, 332
Relaxases, in bacterial conjugation, 296–297
Relaxosomes, 296–298, 300
RelBE2 system, 313, 315, 316
Release factor 1, in toxin-antitoxin action, 316
Reporter constructs, in drug screening, 218
Reporter substrate method, 201–202
Reserpine
  in efflux detection, 239
  for resistant staphylococci, 92
Resistance integrons
  Class 1, 334–336
  genetics of, 331–334
  vs. superintegrons, 331
Resistance nodulation cell division (RND) family, 236, 237, 241–247
Resolvase, 44
Resonance energy transfer techniques, 222, 324–325
Retractable pili, in conjugation, 307
Retrotransfer, in conjugation, 293
Ribosomes
  defects of, aminoglycoside resistance in, 14–15
  function of, aminoglycoside effects on, 9–10
Ribostamycin
  AAC enzyme of, 30, 40
  APH enzyme of, 13
Rifampicin, resistance to, 341–343
RmtA protein, in aminoglycoside resistance, 11
RNA, bacterial, aminoglycoside binding to, 9–10
RND (resistance nodulation cell division) family efflux system, 236, 237, 241–247
  inner membrane transporter of, 241–242
  outer membrane factor of, 243–244
  periplasmic adaptor protein of, 242–243
  tripartite, 244–247
Rolling-circle replication, in bacterial conjugation, 296, 301–302
Roxithromycin, *see* MLSKO antibiotic group
rpoB gene, mutations of, biologic costs of, 342
rps genes
  in aminoglycoside resistance, 14–15
  mutations of, biologic costs of, 342–343
rRNA methylases
  antibiotics targeting, 54–55
  genes of
    alterations of, 55
    evolution of, 58
    geographic distribution of, 55–57
    laboratory detection of, 59–60
    mobile elements in, 57–58
    molecular mechanisms of, 58–59
Rte proteins, in conjugation, 278
Rts1 plasmid, HigA-HigB system of, 314
rumA gene, in conjugation, 280

S

Salmonella
  β-lactamases of, 146
  conjugative transposons of, 274
  resistance of, mechanisms of, 287
Salmonella enterica
  AAC enzyme of, 30, 40
  APH enzyme of, 37
  β-lactamases of
    epidemiology of, 165–166
    geographic distribution of, 72
    substrate specificity of, 185
Salmonella enterica serovar Typhi
  plasmid R27 of, 293
  resistance of, mechanisms of, 287
Salmonella enterica serovar Typhimurium
  efflux system of, 247–248, 253
  resistance of
  biologic costs of, 340–344
    transformation in, 291
Serine recombinases, 332
Serratia marcescens
  AAC enzyme of, 28, 41
  ANT enzyme of, 12, 39
  β-lactamases of
    clinical relevance of, 133–134
    geographic distribution of, 72–74
    integron structure of, 124, 126
Ser-Xxx-Lys element, of β-lactamases, 178
Ser-Xxx-Val element, of β-lactamases, 178
Set proteins, in conjugation, 280
Shewanella
  β-lactamases of, 165–166
  superintegron of, 337
Shewanella oneidensis, β-lactamases of, 166
Shigella
  β-lactamases of, 146
  resistance of, mechanisms of, 287
Shigella dysenteriae
  efflux systems of, 238
  resistance of, plasmids and, 267
Shigella flexneri
  efflux systems of, 238
  resistance of, 285
SHV β-lactamases, 101, 232
Sichuan pepper product, for resistant staphylococci, 92
Sigma replication, DNA, 301–302
Silver ions, efflux systems for, 251
Single- to double-stranded replication, DNA, 301–302
Single-stranded binding protein, in bacterial conjugation, 298
Sisomicin
  AAC enzyme of, 13, 40–41
  in enzyme inhibition, 16
Small multidrug resistance (SMR) family efflux system, 236, 237, 241
SME-1 β-lactamase, BLIP inhibition of, 232
SMR (small multidrug resistance) family efflux system, 236, 237, 241
Sortase, 220–222
Sox, in efflux system, 251
Spectinomycin
  APH enzyme of, 38
  resistance to, 14–15, 335
  structure of, 21
Spectrophotometry, for β-lactamases, 135–136
SPM enzymes, 120–121, 134
Staphylococcus
  erm genes of, 57
  penicillin-binding proteins of, 82
  resistance in, inhibition of, 92, 93
Staphylococcus aureus
  ANT enzyme of, 26–27, 39
  β-lactamases of, 68, 147
  efflux system of, 247, 249, 250, 252, 253
  methicillin-resistant
    carbapenems for, 91
    cephalosporins for, 90
  penicillin-binding proteins of, 83, 84, 86
    kinetics of, 87–88
    structure of, 88, 89
  resistance of, 322
    biologic costs of, 341, 342, 345
    history of, 285
    mechanisms of, 286, 287
    in MLSKO antibiotics, 55
    transduction in, 292
  sortase of, 220–222

Staphylococcus lentus, efflux system of, 250
Steady-state conditions, in enzyme kinetics, 195–197, 202–203
Stenotrophomonas maltophilia
  β-lactamases of, 132
  efflux system of, 250–253
  integrons of, 335–336
Streptococcus
  APH enzyme of, 37
  conjugative transposons of, 303
Streptococcus faecalis
  conjugative transposons of, 303
  resistance of, history of, 285
Streptococcus gordonii, sortase of, 220
Streptococcus mitis, penicillin-binding proteins of, 87
Streptococcus pneumoniae
  conjugation in, 299
  conjugative transposition in, 272–273, 286–287
  drug targets in, 218
  efflux system of, 238, 249, 250, 253
  erm genes of, 57, 58–59
  penicillin-binding proteins of, 83, 87
    kinetics of, 87–88
    mutants of, 88, 89, 91, 93
    structure of, 88, 89, 91
  RelBE system of, 313, 315, 316
  resistance of, 322–323, 340
    biologic costs of, 341
    history of, 285, 286
    in macrolides, 55
    mechanisms of, 287
    transformation in, 291
  YefM-YoeB system of, 313, 315, 317–318
Streptococcus pyogenes
  conjugative transposition in, 286–287
  efflux system of, 238
  erm genes of, 58–59
  ω-ε-ζ system of, *see* ω-ε-ζ system
  resistance of
    history of, 285–286
    to macrolides, 54
    transduction in, 292
  sortase of, 222
Streptogramin, *see* MLSKO antibiotic group
Streptomyces
  in macrolide production, 55
  methylases of, 58
Streptomyces clavuligerus, β-lactamase inhibitors of, 228–229
Streptomyces exfoliatus, β-lactamase inhibitors of, 229
Streptomyces griseus
  AAC enzyme of, 40
  APH enzyme of, 37
Streptomycin
  discovery of, 35–36
  resistance to, 335
    biologic costs of, 341, 343
    conjugative transposons in, 272, 273
    in ribosome defects, 14–15
  structure of, 8, 21
STX mobile element, 278, 280–281
sul genes, in integrons, 333
Sulbactam, 103–104
  action of, 228
  β-lactamases of, 181–182
Sulfolobus solfataricus, efflux system of, 248
Sulfonamide, resistance to
  biologic costs of, 341, 343
  control of, 346
Superintegrons, 268
SYN-1012 (methylidene penem), 108

T

Tazobactam, 103–106, 181–182
Tea extracts, for resistant staphylococci, 92
Telithromycin, *see* MLSKO antibiotic group
Tellimagrandin I, for resistant staphylococci, 92
TEM β-lactamases, 101
  BLIP interactions with, 230–233
  functions of, 227–228
  inhibitor resistant (IRTs), 103
  structures of, 227–228
tet genes
  clinical relevance of, 249
  epidemiology of, 236, 238
  in MF superfamily, 239–240
  mutations of, 252
  transposons and, 272–273
    Bacteroides, 278
    Tn*916* family, 275–277
Tetracycline
  efflux systems for, 236, 238
    clinical relevance of, 249
    MF superfamily, 239–240
  resistance to, conjugative transposons in, 272–273, 278
Theta replicating DNA, 301
Thienamycin, 106
Ticarcillin, β-lactamases of, 184, 188
Ticecycline, efficacy of, 252
tmRNA, in toxin-antitoxin action, 316
Tn*7* transposon, 336
Tn*552* transposon, 273
Tn*916* transposon and family, 272–273
  vs. CTnDOT, 278
  discovery of, 271, 303
  inhibitors of, 282
  mechanisms of, 275–277
  structures of, 275–276
Tn*1545* transposon, 272–273
Tn*4001* transposon, 273
Tn*4451* transposon, 281–282
Tn*4551* transposon, 274
Tn*4553* transposon, 274
Tn*4555* transposon, 273
Tn*5251* transposon, 273
Tn*5381* transposon, 273
Tn*5382* transposon, 272–273
Tn*5384* transposon, 273
Tn*5385* transposon, 273
Tn*5397* transposon, 271
Tn*5398* transposon, 274–275
Tn*4453a* transposon, 281–282
Tn*4453b* transposon, 281
TndX recombinase, in conjugation, 281–282
tnp genes, 44
TnpX recombinase, in conjugation, 281–282
Tobramycin
  AAC enzyme of, 13, 40–42
  ANT enzyme of, 26
  clinical use of, 9
  in enzyme inhibition, 16
  resistance to
    enzymatic, 11
    molecular mechanisms in, 11
    in ribosome defects, 15
  structure of, 21–22
TolC, in efflux system, 243–247, 253
Toxin(s), encoded by mobile genetic elements, 292
Toxin-antitoxin systems, 313–329
  biophysical analysis of, 316–318
  description of, 313–316
  inhibitors of
    identification of, 323–325
    potential targets for, 321–323
  mechanisms of, 316
  regulation of, 314
  structures of, 314–315, 318–321
Tra proteins, in bacterial conjugation, 297, 300
Transduction, in resistance dissemination, 291–292
Transfer replication, DNA, 300–301
Transformation, natural, 291
    as drug target, 289–290
    Streptococcus pyogenes, 286
Transient-state conditions, in enzyme kinetics, 196–198
Transition-state analogs, 109–111
Transmembrane domain, in ABC transporters, 247–248
Transposase, 44
Transposons
  in aminoglycoside-modifying enzymes, 44–46
  conjugative, *see* Conjugative transposons
  nonreplicative, 303
  types of, 268
Treponema denticola, *erm* genes of, 57
Triclosan, efflux systems for, 251
Trimethoprim, resistance to, conjugative transposons in, 272
Tuberculosis, *see Mycobacterium tuberculosis*
Type IV secretion systems, 296, 298–299, 307
Tyrosine recombinases, 332; *see also* Integrases

V

Vaccines, 290
Vancomycin, resistance to
  conjugative transposons in, 272–273
  mechanisms of, 288
VceC, in efflux system, 244
Vegetative replication, DNA, 300–302
VgaA, in efflux system, 248
Vibrio
  conjugative transposons of, 274
  penicillin-binding proteins of, 86
    resistance of, mechanisms of, 287
Vibrio cholerae
  APH enzyme of, 37
  β-lactamases of, geographic distribution of, 72
  conjugative transposons of, 272
  efflux system of, 247
  Inc-J family elements of, 278, 280–281
  integrases of, 337
  superintegron of, 268, 336–337
Vibrio parahaemolyticus, efflux system of, 247
Vibrio vulnificus, integrons of, 336

VIM enzymes, 122
clinical relevance of, 133, 134
distribution of, 120
kinetic parameters of, 127, 129
Virginiamycin, 54
Virulence
efflux systems and, 251–252
inhibitors of, 220–223

X

Xis proteins, in conjugation, 276–277
X-nan Huangqin product, for resistant staphylococci, 92

Y

YdhE, in efflux system, 247
YefM-YoeB system, 313, 315, 317–318

Z

Zinc center, of β-lactamases, 130–132, 210
ZipA protein, FtsZ protein interactions with, 220
Zymomonas mobilis, penicillin-binding proteins of, 84